GEOTECHNICAL SPECIAL PUBLICATIONS

PREFACE

The 1990 Specialty Conference on Design and Performance of Earth Retaining Structures was conceived to provide a major update on earth support technologies since the 1970 Specialty Conference on Lateral Stresses and Earth Retaining Structures. The goal of the conference and the resulting proceedings is to share the rapid and profound changes that design and construction practices for earth retaining structures have undergone worldwide in the past 20 years. The conference is organized to establish the current state-of-the-art and to forecast technological developments which are likely to carry the practice into the next century.

The conference was proposed by the Earth Retaining Structures Committee of the ASCE Geotechnical Engineering Division. A conference steering committee was established in early 1986 to plan for the conference, and a formal proposal was submitted to ASCE in September 1987. The members of the committee, who are listed below, developed and executed all planning for the conference:

D. R. McMahon, Goldberg-Zoino Associates of New York, P.C., Chairman
J. DiMaggio, Federal Highway Administration
J. P. Gould, Mueser Rutledge Consulting Engineers
L. A. Hansen, Sergent, Hauskins & Beckwith
P. C. Lambe, North Carolina State University
A. J. Nicholson, Nicholson Construction of America
T. D. O'Rourke, Cornell University
W. L. Schroeder, Oregon State University

A general call for abstracts of papers was issued in spring 1988, stressing the intent to survey the state of practice since the 1970 conference and to provide a review of current applications. The steering committee had established historical perspectives, wall selection, contracting practices, waterfront structures, gravity walls, mechanically stabilized systems, supported excavation, cast-in-place walls, soil nailing, tied-back support and seismic design as major topics to be addressed. Internationally known experts were invited to present abstracts in each of these areas.

A total of 134 abstracts were submitted for consideration. These abstracts were reviewed by the steering committee and 34 were accepted. Abstracts from 16 invited authors were also reviewed and accepted, with some being asked to change the scope of their papers to meet more fully the overall intent of the conference. The selected authors were requested to prepare papers based on their abstracts.

The completed papers were submitted for review and evaluation. These papers were reviewed employing the same technical standards as the ASCE Geotechnical Engineering Journal. All papers had at least two technical reviews, coordinated by the proceedings editors. The proceedings editors also reviewed the invited papers, which are indicated by an asterisk in the table of contents. Uniform editorial review forms and standards also were applied to ensure that the proceedings were of as high a quality as can be expected of camera-ready papers. The persons listed below, in addtion to the steering committee members, acted as referees of selected papers during the review process.

Loren Anderson	Robert M. Hoyt
Richard Barksdale	Edmund Johnson
James S. Barron	Ilan Juran
Rudy Bonaparte	Robert W. Meyers

Thomas Boni
Hubert Deaton, III
Victor Elias
Richard J. Finno
Alan G. Hobelman
Robert Holtz

Jean-Yves Perez
Larry Rayburn
C. K. Satyapriya
R. Steele
George Tamaro
Ed Ulrich

All of the papers published in these proceedings are available for discussion, just as any paper published in ASCE Journals. The due date for any discussions is the same as that for a June 1990 journal paper. All of these proceedings papers also are eligible for ASCE awards.

Philip C. Lambe, M. ASCE
Lawrence A. Hansen, M. ASCE
Proceedings Editors

CONTENTS

E. Gravity Walls

F. Mechanically Stabilized Systems

G. Supported Excavation

H. Cast-in-Place Walls

I.   Soil Nailing

J.   Tied-Back Support

K.   Seismic Design

FIFTY YEARS OF LATERAL EARTH SUPPORT
by
Ralph B. Peck[1], Hon. M. ASCE

Like most students of my generation, I was introduced to the subject of lateral earth support by learning the derivation of Coulomb's formula in about my junior year. The derivation was straightforward enough; without explanation or apology, the surface of sliding was taken as a plane, the properties of the soil were characterized by the angle of repose, and the point of application was said to be at one-third the height of the retaining structure. The objective of the exercise was apparently to determine the loading that, once obtained, would permit the more serious business of learning how to carry out the structural design of a retaining wall.

Two years later, in a course in harbor engineering during my first year of graduate work, my professor called attention to the series of articles in Engineering News-Record in which Terzaghi described the results of the MIT experiments showing the relationships among the magnitude of the earth pressure, its point of application, and the manner and extent of movement of the retaining structure. These articles seemed to introduce complications into what had heretofore been the simple concepts of active and passive earth pressure. I must confess that neither my professor nor his students attached much significance to the articles.

However, three years later, I found myself at Harvard where Arthur Casagrande enlightened us on the importance of the findings and made it clear that for any real retaining wall with a factor of safety greater than unity the earth pressure must be greater than the active

---

1 - Prof. of Foundation Engineering, Emeritus, University of Illinois at Urbana-Champaign; Consulting Engineer, Albuquerque, N.M.

value obtained by Coulomb's theory. He also directed us
to the famous 1881 paper in which Sir Benjamin Baker, a
practitioner of the greatest experience, cited example
after example of retaining structures that were in fact
perfectly stable, but that could not have escaped
failure if the pressures against them had been as great
as the Coulomb value. It is fair to say that we
students were confused by this seeming contradiction.
Were real earth pressures greater or smaller than the
Coulomb values?

Professor Casagrande also made sure that we were
familiar with Terzaghi's paper of 1936, with the
unequivocal title "A Fundamental Fallacy in Earth
Pressure Computations". The paper dealt with the
assumptions in Rankine's theory. Its first conclusion
was as follows: "The fundamental assumptions of
Rankine's earth pressure theory are incompatible with
the known relation between stress and strain in soils,
including sand. Therefore, the use of this theory
should be discontinued." This statement was quickly
absorbed into the new gospel of soil mechanics as
perceived by us students.

In the fall of 1938, Terzaghi returned to the United
States from Vienna and was given the title of Visiting
Lecturer at Harvard. We students were tremendously
impressed with the good fortune that we would have
first-hand contact with the man who was already known as
the father of soil mechanics. Quite naturally we all
turned out to his first set of lectures. Imagine our
surprise to find them devoted entirely to the Rankine
states of stress, to slip lines, to the use of the pole
for determining the states of stress when the ground
surface was other than horizontal, and other intricacies
of Rankine's theory. Our confusion was now almost
complete. Why was the great man wasting his time and
ours expounding a theory that he had already said should
no longer be used? We did not realize that Terzaghi was
working on the manuscript for Theoretical Soil
Mechanics, had drafted the chapter on plastic
equilibrium and the Rankine states of stress, and had
simply pulled this chapter off the shelf for his
lectures. So we could hardly be blamed for concluding
that the subject of lateral earth pressure was indeed
confusing and controversial.

Our confusion was but the latest episode in a long
history of confusion. Controversy had existed for many
decades over two principal questions about earth
pressure: its magnitude and its distribution. The
question of magnitude had been addressed by Sir Benjamin

Baker. He took for granted that the point of application was one-third the height of the retaining structure from its base, and he calculated the earth pressure very simply by the equivalent of Coulomb's formula. He found, as I have already indicated, that a good many major structures would not be standing if the pressure had been as great as that predicted by Coulomb's theory. We understand today that the discrepancy was the result, not of a serious defect in the theory, but of a lack of understanding concerning soil properties. We would agree today that if we correctly evaluate the effective friction angle, the effective cohesion, and the pore pressures involved, Coulomb's theory will give a reasonable answer. But in Benjamin Baker's time, and indeed for half a century longer, all information concerning the resistance of soil was lumped into the term "angle of repose", and values of the angle of repose were taken unquestioningly from various tables based almost exclusively on opinion.

Confusion and disagreement concerning the distribution of earth pressure were highlighted in the discussions to a paper by J. C. Meem in 1908. Meem was not particularly interested in retaining walls. His field was the construction of subways, as chief engineer for the contractor on many projects in Brooklyn and elsewhere in the New York City area. He knew from experience that the struts in the upper parts of his braced cuts were loaded more heavily than those in the lower parts, and he concluded that the center of pressure was more nearly at the upper third point than at the lower third point. His paper began with a theory, unfortunately incorrect, explaining why this should be so, and went on to present examples that demonstrated clearly the correctness of his conclusions. Needless to say, the discussions to his paper were uncomplimentary. At best the discussers recognized his experience, but dismissed his findings because they disagreed with accepted theory. One of the discussers, J. F. O'Rourke, who was also a New York contractor, cautioned the readers not to be misled by Meem's belief that strut loads decreased near the bottom and recommended that they should stick to putting the big timbers where theory said they belonged. Meem's frustration appears in his closing discussion, "Now, just a word to those who say that the assumptions made in this paper are contrary to all accepted theories. It appears to the writer to be beyond the range of successful contradiction, if the accepted theories are absolutely true, that there could have been no deep tunnels driven in soft ground by ordinary methods, and also that the sheeting at the bottom of deep trenches in sand would have been impractical, as the sand would have burst through at the bottom before the sheeting could have been set in place; and further, that all theory is

the result of practical experience, experiment, and
research tabulated and formulated . . . In conclusion,
it is well, as Mr. O'Rourke suggests, to 'stick to the
big timber', when in doubt. But it is the function of
the engineer to find where these big timbers properly
belong and to eliminate them elsewhere. It is the
earnest wish of the writer, particularly in these days
of forest conservation, that the Society shall not let
this matter rest until the responsibility on these big
timbers is properly placed and is defined so clearly
that 'those who run may read'."

Today, we understand that the distribution of the
earth pressure is related to the deformation conditions,
and we appreciate that the significance of these
conditions was indeed the outstanding contribution of
Terzaghi in his cigar box tests at Robert College and
his larger-scale tests at MIT. Terzaghi's attempt to
translate these findings into an arching concept appears
in a rather crude form in the Harvard Conference
Proceedings of 1936, where he demonstrated that various
deformation conditions could lead to a center of
pressure anywhere between the lower and upper third
points. The definitive discussion of the subject was,
however, Terzaghi's paper on the General Wedge Theory in
1941. Interestingly, this paper grew out of
measurements of strut loads in open cuts in sand for the
Berlin subway. The measurements fully confirmed Meem's
qualitative observations, and the theory that eluded
Meem took its place in practice.

However, the full significance of the deformations,
important as they were in explaining the position of the
center of pressure, did not receive the practical
attention it deserved until it was gradually realized
that the deformations experienced by the retaining
structure were part of general ground movements
extending far from the wall itself and leading to
settlements of the adjacent ground surface and sometimes
to damage of adjacent structures.

The need for tight bracing obtained by systematic
wedging was recognized and publicized in the late 1930's
in the New York area. It was closely related to the
"pretest" method of underpinning advocated by Lazarus
White and was emphasized by White and Prentis in their
1940 book on cofferdams. Not yet realized, however,
was the extent of the movements that took place,
particularly in clay soils, below as well as alongside
excavations as a result of removal of the weight and
lateral support of the earth. Perhaps the first
systematic evidence of such inevitable movements was
obtained during the excavation of the S-3 open cut on
the Chicago subway. The existence of a small tunnel
below excavation level beneath the cut permitted, in

those days before the invention of a practical inclinometer, installation of measuring devices that disclosed inward movement of the sheet piles well below excavation level. Moreover, elevations measured in the tunnel disclosed that the tunnel rose as the excavation progressed. It soon became apparent that some rise of the soil beneath the bottom of a cut and some lateral movement leading to nearby settlement could not be prevented by even the most conscientious prestressing of the bracing.

In the early 1950's plans were being developed for an underground railway in the deep soft clays of downtown Oslo. Construction of braced excavations in the city had led to several failures and near failures involving large movements. The investigation of these occurrences became one of the principal endeavors of the newly formed Norwegian Geotechnical Institute. The role of base failure was recognized and developed, and was described in the classical paper by Bjerrum and Eide in 1956. The significance of the base stability factor in determining the critical depth of excavation was established, as was the inevitability of large movements of the clay surrounding the excavation as the critical depth was approached. In the following decade, NGI produced a remarkable series of case histories under the guidance of Elmo DiBiagio that clearly showed the interrelations among movements of the clay, strength of the clay, depth of excavation, strut loads, and bending moments in the sheeting. Design of the bracing systems still remained semi-empirical, however, and when I prepared my state-of-the-art paper for the 1969 Mexico City ICSMFE, I could only summarize the semi-empirical procedures. Equal attention was given, however, to means for estimating strut loads and for predicting the settlements associated with the excavations.

Thus, by the time of the first Cornell Conference the next year, the distribution of earth pressures against retaining structures, whether permanent or temporary, was reasonably well understood. The magnitude of the pressures could be estimated, largely on a semi-empirical basis, in a fairly satisfactory way. Movements adjacent to excavations where base failure was a remote possibility could be estimated on an empirical basis. The role of the stability number or of the factor of safety against base failure was appreciated, but quantitative estimates of movements for high stability numbers or low factors of safety against base failure were most uncertain.

Design involved determination of earth pressures, checking against the possibility of base failure, and estimating movements on the basis of such precedents as might have been available. These three aspects of design remained almost independent of each other. Nevertheless, at the Conference there were signs that the integration of these aspects had begun. In two of the state-of-the-art papers, one on braced excavations by T. W. Lambe, and one on methods of estimating lateral loads and deformations by Morgenstern and Eisenstein, finite-element analyses were shown to hold promise.

The twenty years between the two Cornell Conferences have seen major programs of subway construction in the downtown areas of some of our largest cities as well as deep excavations for commercial buildings. In designing these facilities, the emphasis has shifted from the strength of the support systems to the movements of the adjacent ground and the potential for damage to nearby structures and utilities. Geotechnical Design Reports that contain one or two lateral pressure diagrams are likely to devote many pages to predicted settlements, their implications, and means for reducing them. These means may include restrictions on excavation depth below supports, close vertical spacing of supports, and increased stiffness of the walls. To a considerable extent the predictions are still based on empirical data, but in critical situations, particularly where deep clay deposits are involved, the empirical procedures are often inadequate because the data base is too limited.

One of the great achievements of the last 20 years has been the refinement of the finite-element method to the extent that the soil properties, including their strain dependence, can be taken into account realistically, the construction steps can be modeled including excavation and insertion of supports, and the stiffness of the support system can be introduced. The development of the analyses has gone hand in hand with comprehensive field observations of the behavior of prototype excavations and with improved in-situ methods for evaluating the appropriate soil properties. Parametric studies have extended the range of variables beyond the specific observational data. It is now possible, largely as a result of the work of Wayne Clough and his associates, to select the bracing system and excavation sequence at a given site to ensure that the settlement or lateral movement will not exceed predetermined limits. The economy of various possible arrangements to achieve this end can be investigated readily. For many problems a satisfactory solution can be reached by use of summary diagrams relating movement, wall stiffness, support spacing, soil stiffness, and factor of safety against base failure. More detailed studies can be made with an interactive computer program.

Thus, the last 20 years have culminated in a new milestone in the long history of lateral earth support - a workable practical theory, well based in empiricism, for designing retaining systems to limit adjacent movements as well as to indicate the loads to be carried by the supports. We can now deal with lateral earth support, not just lateral earth pressure.

I have chosen in this review of one lifetime of lateral earth support to dwell on one aspect, the development of basic understanding. Two other aspects are equally striking. One is the uncluttering of the space enclosed by the support system, from excavations 50 years ago virtually filled with timber bracing, to more open systems of steel supports, to tiebacks with virtually no encroachment on the excavated space, and to the use of stiff diaphragm-type walls requiring few lateral supports and sometimes serving as the final structural wall. The other is an expanding variety of retaining structures that utilize the strength of the backfill itself: crib walls, reinforced-earth walls, tieback walls and soil nailing. I leave these subjects to others; they are of no less interest than the one I have chosen.

EARTH RETAINING STRUCTURES - DEVELOPMENTS THROUGH 1970
James P. Gould[1], M.ASCE

ABSTRACT: This paper chiefly concerns developments in the 25 years from the end of the second war through 1970, a period of acceleration in techniques for retaining structures stimulated by rebuilding war-damaged cities of Western Europe. Innovations from Europe appeared only gradually in the United States, then applications proliferated and tended to giantism. Within this span of time were introduced cast-in-place concrete walls, tieback anchors in soil and rock, and the all-important concept of earth reinforcement. This paper follows the path of these developments in the United States. Some subjects have been traced to their earlier manifestation to give a broader picture of evolution.

INTRODUCTION

Under pressure of post-war reconstruction and burgeoning free markets in Europe and East Asia, the period up to 1970 was one of unprecedented progress in foundation construction. The United States contributed important concepts in soil mechanics because of the interest generated in America by Terzaghi and Casagrande in the post-war period. However, the main impetus for change came from urban reconstruction in Europe abetted by their fashion of engineering innovation by the constructor. For example, the 1970 Cornell conference did not consider reinforced or stabilized earth. Slurry trench construction, tieback anchors and finite element analysis were treated peripherally.

Because of the iron curtain in American practice between designers and constructors, design and academic developments almost never lead, and in fact are slow to follow progress in the field. Innovators emerge from the category of engineer-constructor, the classic American example being John A. Roebling. Traditionally, in American foundation construction, these innovators were represented by William Sooy Smith, Lazarus White and Daniel Moran. After the war, the same pattern of development was continued by engineer-constructors like Ben C. Gerwick, Jr. and Richard Robbins in their respective specialties.

---

1 - Partner, Mueser Rutledge Consulting Engineers, 708 Third Avenue, New York, NY 10017

CONCEPTS IN RETAINING STRUCTURE ANALYSIS

Terzaghi's classic ASCE papers on cellular cofferdams (20), anchored bulkheads (22) and coefficients of subgrade reaction (23) set the stage for improved design in their subject areas. In the 1940's he modified classic earth pressure theories by the recognition that pressures were to a great extent controlled by the mode and magnitude of movement of the retaining structure. The conferences at Harvard in 1936 and in 1939 at Purdue heavily emphasized earth pressure analytical procedures. The great revelation of definite design recommendations for lateral pressures on braced excavations were set forth by Terzaghi and Peck in their 1948 standard text (21). Here were introduced the key concepts of an enveloping apparent pressure diagram and certain multiplying factors on the computed "active" resultant.

Terzaghi's 1945 cellular cofferdam paper included a series of innovative rules which formulated an overall design scheme. In the discussion that followed (15), D.P. Krynine made the penetrating observation that on the axial plane of the cofferdam, with shear fully mobilized, the lateral pressure coefficient is no longer that of a minor principal stress, but can be readily computed from the Mohr's circle at failure. By the date of the Brussels 1958 conference on earth pressures, Jaky's expression for at-rest pressure (1-sin ϕ) was accepted for granular soils and normally consolidated clays. The Brooker and Ireland elucidation (3) demonstrated the influence of overconsolidation on residual lateral pressures in clays.

Terzaghi's anchored bulkhead paper of 1954 brought to the designer's attention Peter Rowe's evidence for reducing nominal positive moment in a flexible wall by the ability of retained soil to shift lateral pressures away from the point of maximum moment. This concept raised a debate between Tschbotarioff and Terzaghi over the appropriate rule for taking into account this reduced positive moment. The concept has been reflected in criteria which moderate the applied pressures or allow stress increases for positive moments computed by conventional methods on flexible retaining walls. Emphasis is placed on the capacity of the anchor or brace and the need to ensure stability of the embedded portion of the wall.

Braced Excavation Walls. Terzaghi's 1955 systematizing of the Winkler modulus concept (23) led on to the analysis of pile-soil interaction by workers at the University of Texas. This was adapted by Haliburton (11) into a finite difference procedure for retaining structures acted on by soil represented by Winkler spring coefficients and braced with yielding supports of varying stiffness. With the use of computers, this became the "BMCOL" solution for analysis of multi-braced retaining walls. Only recently has this been rivalled by finite element (FEM) analyses of wall-ground interaction. In the late 1960's appeared several FEM applications, including the Clough-Duncan analysis (5) of the flexure of U-frame locks and Morgenstern-Eisenstein study (17) of pressures from retained soil at the 1970 Cornell conference. The rivalry between these alternatives continues. Generally, the FEM can predict plausible overall movements but the loads and moments are insufficiently conservative for the structural designer because the analysis does not delineate a pattern of abrupt curvature in the wall envisioned by the designer. It is analogous to the case of a mat designed on Winkler springs compared to the long radius curvature created by overall dishing settlement. If the designer accepted that simple, smooth flexure diagram, he might simply omit all top tension reinforcing in the mat or outside steel in the wall.

The last major step in concepts to 1970 was provided by R.B. Peck's 1969 Mexico conference summary of shoring wall performance and associated ground movements (19). This laid the basis for the next two decades of instrumentation, observations and the prediction and control of shoring wall movement by monitoring procedures which at best exemplified Peck's definition of the "observational method."

CAST-IN-PLACE BASEMENT WALLS

American Beginnings. From the 1880's, the American construction industry was creating high-rise buildings, many including heavy exterior bearing walls and basements two to three levels deep. New York City and Chicago were the focus of these early tall buildings. In Manhattan, groundwater was high and sandy strata above bedrock were considered unstable. This set the stage for a special form of foundation construction which combined a deep perimeter bearing wall carried to rock or hardpan with the permanent basement wall in the form of continuous box caissons. These were usually constructed in a row of abutting boxes, sunk by digging with the use of air pressure to control unstable water-bearing sands without deep dewatering. Figure 1 illustrates a typical box caisson wall, in this case the basement/foundation wall of the New York Stock Exchange. Here, rock bearing was as deep as 60 feet, 43 feet below the water table. In fact, these box caissons served the same function as modern slurry wall concrete construction, adapted in its special manner to American urban construction of the time. The heyday of this method was up to the early 1930's before depression and war interrupted tall building construction for almost two decades.

When building construction renewed after the second war, efficient deep dewatering largely supplanted use of air pressure for New York foundations in difficult ground. Reference (13) gives the following instructive quotation:

"The wellpoint method of predrainage dates back to 1838, when Robert Stephenson used it in building a tunnel in England. In this country, it was used as early as 1889 in the construction of the Number 2 Ridgewood Pumping Station in Brooklyn, NY. However, it was not until 1927 that general recognition was given its inherent advantages through the experience gained in building the foundations for the Harriman Building in New York City. Prior to this date, it was believed that foundation piers could be safely sunk through the water bearing sands of lower Manhattan only by the pneumatic process. Now, by

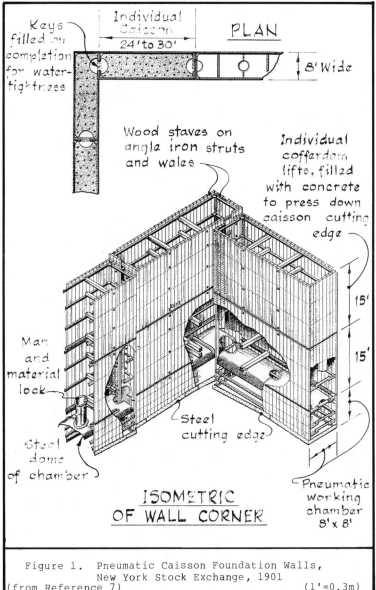

Figure 1. Pneumatic Caisson Foundation Walls,
New York Stock Exchange, 1901
(from Reference 7)   (1'=0.3m)

predraining the sand, foundations were placed safely, quickly and cheaply by means of either of open pier wells or full-lot excavation."

Double Wall Cofferdam Construction. High-rise construction did not thrive again on a grand scale until the early 1950's. An interesting example of the transition between the pre-war style of deep foundation walls and the future direction was in the construction of the 85 foot deep, five basement levels, of the Chase Manhattan Bank Central Office Building in southern Manhattan, 1956 to 1958 (14). The general cross-section is illustrated on Figure 2. After excavating to remove existing foundations, two parallel lines of steel sheeting, 13 to 16 feet apart, were driven entirely surrounding the perimeter of the site, tied into hardpan and boxed out at 55 foot intervals with cross walls.

The original contract package called then for excavating the remaining 25 feet through hardpan and coarse glacial till under air pressure in the old-fashioned style. When the four bids were received, it became apparent that labor cost of work under air had become unacceptable and another scheme was needed for safe advance through till to rock. After it was disclosed by a test pit excavation that there was a drawdown of about 15 feet of the piezometric levels in the lowermost till, W.H. Mueser agreed that the remainder of the double wall cofferdam could be excavated with the aid of chemical grouting in the till with deep well dewatering in the pervious lower section of the stratum. This case is also instructive as an early example of "top down" basement construction, utilizing the permanent steel floor beams as cross lot bracing with design preload jacked into them to restrict the perimeter wall movement.

Advent of Slurry Trench Concrete Walls. With improved dewatering techniques, construction of deep basements in difficult sites might have continued by this style of parallel wall cofferdams. However, at this juncture innovations emerged from the mid-1950's Europe's post-war building boom and had begun to infiltrate American practice in a form of slurry wall concrete construction. Use of stabilizing mud for deep borings was a product of the American petroleum industry and the earliest adaptions in Europe in the 1920's were for construction of cutoff walls of overlapping cylindrical caissons stabilized by bentonite mud. Veder's early paper in the 1953 Third ISSMFE Conference (24) may have been the first time this cutoff technique was noted by the American soil mechanics community.

The slurry trench structural wall construction that attracted the earliest attention was for the Milan subway in water-logged, coarse grained alluvium. That example generated a number of spectacular early uses from the mid-1950's. One of these was the Liverpool Seaforth dock whose wall consisted of a series of cusps, each with a tail wall and post-tensioned ties to sandstone to counter overturning. By 1957, the technique had been exported to Canada through French and Italian connections and was being considered in the United States.

American Slurry Walls. Our first structural slurry wall,

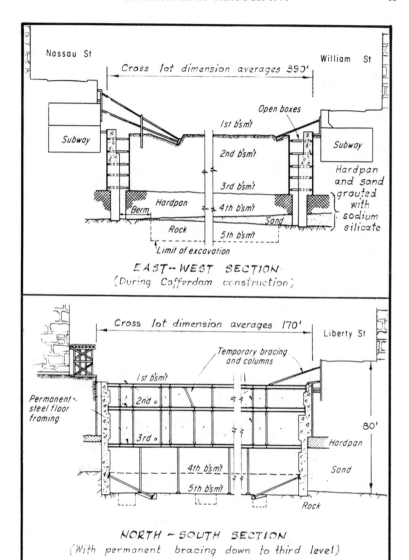

Figure 2. Foundations of Chase Manhattan Bank,
           Central Office Building, Manhattan, 1958
(from Reference 14)   (1'=0.3m)

utilizing unreinforced panels was constructed by Spencer, Rodio and
Soletanche, Inc., 1962 to 1963, for a 23-foot diameter shaft at
Hudson Avenue on the East River in Brooklyn. This was carried
through 80 feet of glacial outwash, much of it a torrential
deposit, to bedrock. The second use was by the same contractor for
deep basement construction in an apartment building in Boston from
1963 to 1964 (18). Still the best known structural slurry concrete
wall was that for the New York Port Authority World Trade Center
made by ICOS of Canada in 1967 to 1968. It incorporated
essentially the modern form of deep basement construction with
temporary ties carried down at a high angle to bedrock and
permanent bracing by the basement floor system.

Developing the SPTC Wall. Interestingly, in the early 1960's,
an American development started quite independently by Ben C.
Gerwick, Jr., who provided the following remembrance of his
experience (10):

"In 1962, I had the opportunity to inspect the construction of
the Milan Metro, where reinforced slurry walls were being
constructed. I knew of several planned deep foundation
projects for which the slurry wall would be appropriate, but
only if three improvements could be made:
1. Better joints: they must be essentially watertight.
2. Greater safety during construction: we would have to work
   immediately adjacent to existing, old, fragile, buildings.
3. Greater assurance as to moment and shear capacity at
   depths: could the steel be properly placed with small
   tolerance?
On my return, I turned the problem over to the then Chief
Engineer of Ben. C. Gerwick, Inc., Mr. S. Clifford Doughty,
who developed a procedure using wide flange beams as the
moment resisting member, which met the criteria. This system
was then applied by us on a number of major buildings in San
Francisco: the Bank of California, Pacific Insurance, Pacific
Gas and Electric, and Security Pacific Bank, among others.
BART engineers followed the development and selected it for
use on a number of major stations in downtown San Francisco.
It's since been used rather widely when the special conditions
required similar criteria."

These initiatives led to a series of applications in the Bay
area of the "SPTC" (Soldier-Pile-Tremie-Concrete) for deep
foundation construction. The first was the Bank of California in
1964 to 1966, which combined top down construction for a three
level basement with an early form of Gerwick's patented procedure.
Twenty inch diameter pile holes were drilled on four foot centers;
a 24 wide flange was installed and the drilled hole backfilled with
concrete. Then two overlapping 20-inch holes were drilled between
each pile, the space cleaned with a special cleaning tool, and
concrete tremied within alternate panels between the soldier piles.
The sequence is illustrated in Figure 3.

Powell Street and Civic Center Stations of BART were then
constructed with similar SPTC walls in 1968 to 1970 (9). For these

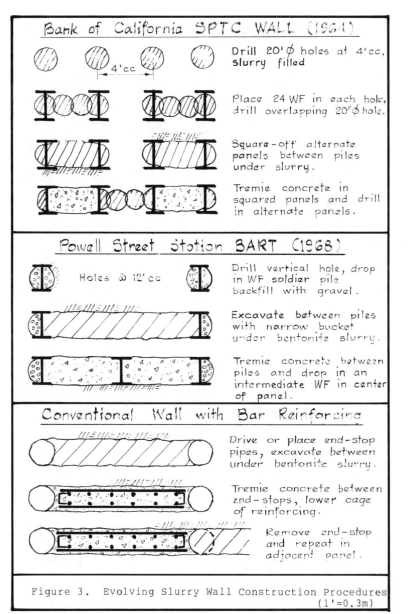

Figure 3.   Evolving Slurry Wall Construction Procedures
(1'=0.3m)

walls, piles were installed at about 12 foot centers, the intermediate panel excavated and then an additional pile placed in the center of that interval, Figure 3. Quality was not uniform for these walls, but they were given credit as a portion of the permanent interior structure. They were braced with permanent interior deck steel as cross lot struts, working from street grade downward to subgrade. These were the first example of what was to become a common American practice, slurry trench concrete as all or a portion of the final subway wall.

Continuous Pile Walls. An early variation of the slurry cutoff wall developed particularly in the southwestern United States using drilled-in-place concrete caissons in soil or soft rock in a contiguous tangent or secant layout to form an excavation support wall. It reached its highest manifestation in states where reasonably uniform flat-bedded compact alluvium or soft rock occupied the full height of the excavation. Various methods were used to stabilize holes during drilling, depending on groundwater conditions. As with any of the cast-in-place techniques, these walls had the advantage of eliminating a portion of the underpinning of neighboring structures that would otherwise be required with a flexible wall supporting the excavation.

By 1970, the basic cast-in-place procedures were well-established in the United States, to be followed by an enormous expansion in the variety of cast-in-place concrete walls built for temporary construction and used as all or part of the permanent foundation.

TIEBACK ANCHORAGE

In the same period, with similar European genesis, tiebacks developed through techniques of anchorage in rock, then in soil and from temporary to permanent installations. Andre Coyne, working with the first high capacity rock ties at Cheurfas Dam in Algeria, is credited with developing the technology of rock anchorage. With the introduction of the Freyssinet jack, permanent anchorage in rock for walls and floors of structures developed in and around the Alpine massif by the mid-1950's (12).

Shoring of Excavations. Through the building boom of 1920 and the early 1930's, the usual interior bracing of excavations in a crowded urban location was by structural timbers, relatively light weight and readily cut and fit to need. After the war when building construction revived, excavation support was by steel sections, cross-lot or raking. The great revolution in the post-war period was the application of the pre-stressed anchor to the support of deep excavation.

Taking note of the European experience, Spencer, White & Prentis employed tiebacks in rock for shoring on four building excavations in 1960 and 1961, the first in Milwaukee and two in New York City, using elementary grouting procedures, simply tremieing grout in the bonded length of uniform diameter.

Their New York Telephone Building was the first anchorage job for a building excavation involving the author's firm. This was celebrated by an ENR editorial in their June 8, 1961 issue as follows:

> "The cluttered obstacle course for excavating equipment formed
> by inclined braces shoring the sides of excavations presents a
> sad site in this age of technological marvels. ...With its
> disappearance, builders will get much more production out of
> their equipment without the irritations and hazards associated
> with internal (infernal) bracing ... Not alone the construc-
> tors, but sidewalk superintendents, with their improved view
> of excavation operations made possible by open building sites
> should be eternally grateful."

Shortly thereafter, use of tiebacks in shoring started in Los
Angeles under the initiative of LeRoy Crandall (16). They first
employed expanding metal anchors and screw-type anchors in 1963.
These low-capacity anchors were followed by underreamed double-cone
anchors with high strength rods which, in turn, were outmoded by
longer anchors relying on direct friction or bond. The typical
alluvial soil and soft rock strata of the central Los Angeles basin
with groundwater below the base of most excavations offered a
highly favorable setting for the use of anchored shoring. A
compelling advantage of the system was the ability to maintain the
integrity of adjacent structures without the need for costly
underpinning.

In the early 1960's in North America and Europe, a variety of
techniques were being used for installation of soil tiebacks
without high pressure grouting or regrouting. Hollow stem augers
were employed in soil creating relatively large diameter holes.
Recently, R.J. Barley (1) noted the following:

> "An anchor construction and design manual of the late 1960's
> opened its rock anchors section with the firm pronouncement:
> 'Anchors installed in wet or weak rocks such as marls,
> mudstones, shales and sandstone, and all anchors requiring
> loads of greater than 50 tons working capacity, should be
> constructed with underreamed enlargements. The enlargements,
> filled with grout (NB: merely poured in place), form a
> positive lock even in soft, weak or wet strata'."

By the mid-1960's, pretesting to prove the tie capacity was
widely employed. At the end of this decade, there were still
substantial reservations about the suitable anchor in various
overburden soils and the prospect of enhancing capacity by high
pressure grouting, fracturing, or regrouting had not yet been
adopted in American practice. There was concern over creep of the
anchor bond stress in plastic soils. W.S. Booth (2), describing a
notable early use of soil tiebacks in 1966 at the House of
Representatives underground garage, a job that led to protracted
litigation, cautioned that anchors in soil were not suitable for:

> "1. Soils of high plasticity, since anchor creep is a strong
>   possibility under such conditions.
> 2. Granular soils with no coherence, because there is diffi-

culty in guaranteeing the shape of the bell in such soil.
3. Areas where much water is present, as it will prevent
   drilling and belling. It is almost impossible to place
   concrete properly under such conditions.
4. Soils containing obstructions, such as large boulders.
   Frequent obstructions will increase the cost of tieback
   installation prohibitively."

In the mid-1960's, drilling equipment was being developed in
Europe with percussion and torque capacities, accommodating
driven or drilled casing which permitted pressure grouting and
increased anchorage capacity. Soletanche added regrouting and
post-grouting techniques. At dates near 1970 Peter Nicholson
adapted rotary drilling for advancing casing while Harry Schnabel
introduced the Bauer anchor with a driven casing for pressure
grouting. The stage was set for a proliferation of installation
methods and enhanced bond which greatly extended anchorage
applications in the 1970s.
   Driven Piles as Anchors. For shored excavations on land, piles
as anchors had a phase in the late 1960's. Included was Gerwick's
scheme for driven batter piles for the Wells Fargo Bank building in
San Francisco (6). They were installed at an orientation much like
that of a drilled tieback and joined by connecting wales. Bond was
enhanced by grouting injection through pre-placed grout pipes after
the piles were driven. A somewhat similar scheme was used by
Stuart Beall in the soldier pile cofferdam for the Washington Mall
Vehicular Tunnel in 1968. The scheme was simply to make a bench
cut at top of bank and below this to drive heavy soldier piles to a
very resistant base stratum, supporting them with one driven slant
anchor pile guided and driven on a beveled surface at the top of
plumb pile. On the same project, another contractor used a similar
scheme but welding the batter pile at the side of the soldier pile.
The inevitable torsion in this connection caused ruptures that were
not encountered in Beall's straightforward detail.

REINFORCED AND STABILIZED SOILS AS RETAINING STRUCTURES

Certainly the most dramatic innovation in our period was the
"reinforced earth" concept. From earliest times cribbing of
various types have been utilized with coarse backfill in a
container on the face of a cut slope to provide resistance to
sloughing or shallow slope failure. However, these bins appended
to the face merely provide a drained gravity section on the surface
of what had to be an essentially stable cut. It prevented erosion
or surface sloughing, promoted drainage and avoided an infinite
slope failure. However, cribbing does not greatly alter the path
of a potential failure surface in the underlying strata. The
reinforced earth technique and the subsequent array of nailing and
stabilizing methods differ in that they engage a sufficiently deep
section by the tensile capacity of the strips, sheets or nails to
create a new and presumably stable free body at the rear of the
zone in which the ties or dowels are placed.
   This is one innovation that has had a clear point of origin:

Henri Vidal in 1957 on a beach at Ibiza, pondering the stabilizing effect of pine needles in a pile of beach sand. After years of experimenting and probing the significance of the stabilizing effect, the first installation was completed in 1965 at Pragneres in the French Pyrenees. This initial application utilized face members of curved thin steel channels, connected to two-inch wide steel strips, four millimeters thick.

Recognition of the economic value of the method built a rapid record in our period of review, 500 projects worldwide, none in the United States. Vidal presented (25) his ideas with a theoretical explanation that was not entirely convincing at the annual Highway Research Board meeting in 1968. Not until Ken Lee became involved in investigations of the mechanics of the procedure were design concepts evolved which reassured American engineers.

The first U.S. application was by Caltrans on Highway 39 in the Angeles National Forest (8). This use was occasioned by a slide that occurred in the winter of 1969 which extended above the highway for a length of 800 feet. After considering alternatives, Caltrans, with FHWA concurrence and encouragement, selected reinforced earth as a remedy. This was a rather dramatic wall, 50 feet in height, built on 200 feet of supposedly stabilized slide debris.

While the method was widely accepted abroad, it was adapted only gradually in the U.S. Over 1000 projects were completed worldwide by 1975 (8). Up to that same time, the few U.S. applications chiefly involved remediation of slide problems where the retaining structure was placed in ground in which movements might be continuing. The first application for a relatively rigid bridge abutment creating a retained fill was in 1974. Applications then expanded rapidly to a vast array of earth retention.

Soil Nailing. Happily for this review, the only other innovation in stabilization which needs to be considered is the original form of soil nailing. The earliest example was (typically) generated by engineers for the French contractor Bouygues, in a temporary stabilization of a cut at the Versailles Chantier Station of the SNCF in 1972. This involved pinning a cut 72 feet high in cemented sand with about 25,000 nails up to 20 feet in length. Its success was followed by that contractor's development of a system of installing nails of angle section by a vibro-percussion technique. By the present date, perhaps half of the temporary support of excavations in France and Germany is accomplished with soil nailing. The vast array of other materials for stabilization was yet to appear upon the scene after 1970.

SUMMARY

This review concerned developments from the end of the second war to 1970. In that period, North America was introduced to new forms of cast-in-place below-grade walls, tiebacks in soil and rock both temporary and permanent, and the all-important concept of earth reinforcement which lead onto a wide variety of earth stabilization. Design procedures evolved by 1970 which set the direction for the next 20 years. The concept of the observational method applied to deep excavations, with progress in instrumentation, established an integrated approach to projects where performance is predicted in advance, measured in the field with controls available to guide the work toward the desired end product. Most of the procedures now being utilized for earth retaining structures were well developed by 1970 or had appeared in full scale examples.

REFERENCES

1. Barley, T., "Ten Thousand Anchorages in Rock,: <u>Ground Engineering</u>, October 1988, pp. 24 to 35.

2. Booth, W.S., "Tiebacks in Soil for Unobstructed Deep Excavation," <u>Civil Engineering</u>, September 1966, pp. 46-49.

3. Brooker, E.W. and H.O. Ireland, "Earth Pressures at Rest Related to Stress History," <u>Canadian Geotechnical Journal</u>, Vol. II, No. 1, February 1965.

4. Bruce, D.A. and R.A. Jewell, "Soil Nailing: Application and Practice - Part 2," <u>Ground Engineering</u>, January 1987, pp. 21-33 and 38.

5. Clough, G.W. and J.M. Duncan, "Finite Element Analysis of Port Allen and Old River Locks," Report to Waterways Experiment Station, Corps of Engineers, Draft of September 1969.

6. Engineering News Record, "Batter Piles Brace Sheeting Around Building Excavation," March 11, 1965, pp. 26-27.

7. Engineering Record, "Pneumatic Caisson Foundations of the New York Stock Exchange Building," Vol. 44, No. 13, September 28, 1901, pp. 289-292.

8. Gedney, D.S. and D.P. McKittrick, "Reinforced Earth: A New Alternative for Earth-Retention Structures," <u>Civil Engineering</u>, October 1975, pp. 58-61.

9. Gerwick, B.C.Jr., "The Use of Slurry-Trench (Diaphragm Wall) Techniques in Deep Foundation Construction," Conference Preprint 496, ASCE Structural Engineering Conference, Seattle, May 1967.

10. Gerwick, B.C.Jr., Personal Communication, March 15, 1989.

11. Halliburton, T.A., "Numerical Analysis of Flexible Retaining Structures," <u>Journal of the SM and FD</u>, ASCE, Vol. 94, No. SM6, 1968, pp. 1233-1251.

12. Hanna, T.H., <u>Foundations in Tension-Ground Anchors</u>, 1st Ed. McGraw-Hill Book Co., New York, 1982, pp. 361-363.

13. Jacoby, H.S. and R.P. Davis, <u>Foundations of Bridges and</u>

Buildings, 3rd Ed. McGraw-Hill Book Co., New York, 1941, pg. 382.

14. Johnston, R.C. and N.W. Koziakin, "Foundation Design and Methods Cut Skyscraper Cost," Engineering News Record, July 24, 1958, pp. 34-37.

15. Krynine, D. P., Discussion of "Stability and Stiffness of Cellular Cofferdams," Transactions ASCE, Vol. 110, 1945, pp 1175 - 1178.

16. Maljian, P. A. and J. L. VanBeveren, "Tied-Back Deep Excavations in Los Angeles Area," Journal of the Constr. Div. ASCE, Vol. 100, No. CO3, Sept 1974, pp 337-356.

17. Morgenstern, N. R. and Z. Eisenstein, "Methods of Estimating Lateral Loads and Deformations," Proceedings, ASCE Specialty Conference on Lateral Stresses in the Ground and Design of Retaining Structures, Cornell, 1970, pp. 51 to 102.

18. O'Neill, P.A., Personal Communication, August 8, 1989.

19. Peck, R.B., "Deep Excavations and Tunneling in Soft Ground," State-of-the-Art Report, Proceedings 7th International Conference, ISSMFE, Vol. III, Mexico 1969, pp. 225-281.

20. Terzaghi, K., "Stability and Stiffness of Cellular Cofferdams," Transactions ASCE, Vol. 110, 1945, pp. 1083-1202.

21. Terzaghi, K. and R.B. Peck, Soil Mechanics in Engineering Practice, 1st Ed., John Wiley and Sons, Inc., New York, 1948.

22. Terzaghi, K, "Anchored Bulkheads," Transactions ASCE, Vol. 119, 1954, pp. 1243-1342.

23. Terzaghi, K., "Evaluation of Coefficients of Subgrade Reaction," Geotechnique, No. 5, 1955, pp. 297-326.

24. Veder, C., "Procedure for the Construction of Impermeable Diaphragms at Great Depths by Way of Thixotropic Seals," Proceedings 3rd International Conference, ISSMFE, Zurich, 1953.

25. Vidal, H., "The Principle of Reinforced Earth," Highway Research Record, Number 282, 1969.

OVERVIEW OF EARTH RETENTION SYSTEMS: 1970-1990

by T.D. O'Rourke[1], M., ASCE and C.J.F.P. Jones[2]

ABSTRACT: There has been significant change in earth retention systems relative to those used twenty years ago. This paper explores the developments in methods and materials associated with these changes. Different procedures for soil support are reviewed, and a classification scheme is recommended as a basis for future description and reference. Special attention is devoted to the use of polymers. Design concepts are examined in light of strain compatibility between reinforcement and soil, ultimate strength, and serviceability criteria. Performance of mechanically stabilized walls is reviewed, and future directions in earth retention systems are projected.

INTRODUCTION

In 1970 engineering practice for retaining structures emphasized earth pressures and their application in choosing an appropriate design and support system. The principles of interaction between wall movement and the distribution of earth pressure were understood, and there was growing interest in new construction technologies. Since 1970 there has been dramatic growth in the methods and products for retaining soil. Now our concepts of earth pressure are as much molded by the diversity of retention systems as the structures themselves are influenced by our concepts of earth pressure.

Two trends in particular have emerged in the past twenty years. First, there has been increasing use of reinforcing elements, either by incremental burial to create reinforced soils or by systematic in-situ installation to produce soil nailing. Second, there has been increasing use of polymeric products to reinforce and control drainage in soils. Reinforced soils and soil nailing have changed the ways we construct built-up and in-situ walls, respectively, by providing economically attractive alternatives to conventional methods. Rapid developments in polymer manufacturing have supplied a wide array of geosynthetics. The use of these products in reinforced soil has encouraged a multitude of different earth retention schemes which seem to be limited only by the imagination of the engineer.

As pointed out by Gould (15), earth support systems in 1970 benefited from the construction boom of the late 1950s and 1960s.

---

1 Professor, School of Civil and Environmental Engineering, Cornell University, Ithaca, NY
2 Professor, Geotechnical Engineering, University of Newcastle-Upon-Tyne, Newcastle Upon Tyne, U.K.

New construction methods were developed at that time, largely through adaptations and improvements in specialized excavation and drilling equipment. This trend has continued, with the most notable advances occurring in reinforced soils and soil nailing. As new products and processes have been developed, the choices among specialists and special products have become more diverse and difficult to resolve on the basis of familiarity with conventional systems. Specification writing, selection of bidders, and the apportionment of liabilities often must be judged relative to products and procedures for which there is sketchy information and a lack of direct knowledge.

This paper begins by proposing a classification scheme for earth retention systems and using the scheme to identify important trends and concepts of design. A historical review then follows, with emphasis on braced and tied-back walls, in-situ wall construction, reinforced soils, and soil nailing. Materials for earth retention systems are examined with emphasis on creep and durability of polymers. Performance of mechanically stabilized walls is reviewed on the basis of centrifuge tests, field experiments, and actual job experience. Design concepts are reviewed by concentrating on ultimate strength and serviceability criteria. Finally, future directions in earth retention technologies are projected.

EARTH RETENTION SYSTEMS

It is helpful to organize earth retention systems into a general scheme. Such a process allows for classifying the systems according to basic mechanisms of support. It also permits a consolidated view in which interrelationships can be traced so that a coherent and logical perspective of the field emerges.

Table 1 presents a summary of current earth retention methods organized according to the two principal categories of externally and internally stabilized systems. An externally stabilized system uses an external structural wall, against which stabilizing forces are mobilized. An internally stabilized system involves reinforcements installed within and extending beyond the potential failure mass. In the last twenty years, all types of retention systems have enjoyed extensive development. Nevertheless, it has been in the area of internally stabilized systems that a fundamentally new concept has been introduced. Shear transfer to mobilize the tensile capacity of closely spaced reinforcing elements has freed retaining structures of the need for a structural wall, and has substituted instead a composite system of reinforcing elements and soil as the primary structural entity. A facing is required on an internally stabilized system, but its role is to prevent local raveling and deterioration, rather than to provide primary structural support.

Figure 1 shows that virtually all traditional walls may be regarded as externally stabilized systems. Gravity walls, in the form of a cantilever structure or gravity elements (e.g., bins, cribs and gabions), support the soil through weight and stiffness to resist sliding, overturning, and excessive shear and moments. Bracing, in the form of cross-lot struts and rakers, provides temporary support for in-situ walls. Tiebacks provide support through the pullout capacity of anchors established in stable soil outside of the zone of potential failure. Cellular cofferdams provide support primarily

Table 1. Classification Scheme for Earth Retention Systems

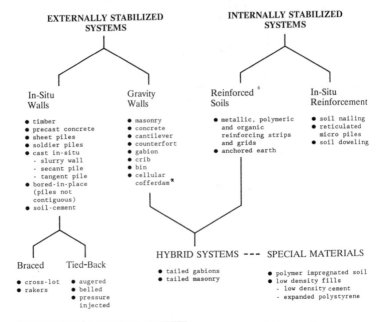

* cellular cofferdams involve consideration of internal cell stability as well as
gravity wall effects

through the gravity force mobilized by each cell, but they also
depend on the tensile and shear capacity of sheet pile interlocks to
sustain internal stability.

Internally stabilized walls are represented in the figure by reinforced soils with horizontally layered elements, such as metallic
strips or polymeric grids, and soil nailing, in which metallic bars
or dowels are installed during in-situ construction. A key aspect
of an internally stabilized system is its incremental form of construction. In effect, the soil mass is partitioned so that each par-
tition receives support from a locally inserted reinforcing element.
This process is just the opposite of what occurs in a conventional
backfilled wall where earth pressures are integrated to produce an
overall force resisted by the structure. The overall earth pressure
in reinforced soil, for example, is actually differentiated by the
multiple layers of reinforcement. In soil nailing, multiple levels
of bars or dowels interconnect the soil mass so that each potential
failure surface is crossed by enough reinforcing elements to maintain
stability.

Hybrid systems combine elements of both internally and externally
supported soil. They include tailed gabions, as illustrated in

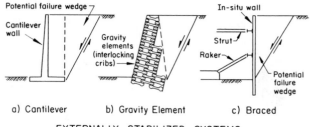

a) Cantilever   b) Gravity Element   c) Braced

EXTERNALLY STABILIZED SYSTEMS

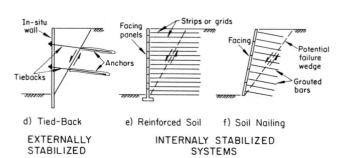

d) Tied-Back   e) Reinforced Soil   f) Soil Nailing

EXTERNALLY   INTERNALY STABILIZED
STABILIZED   SYSTEMS

Figure 1.   Examples of Externally and Internally Stabilized Earth
            Retention Systems

Figure 2.   In this system, geogrids are substituted for the wire
baskets conventionally employed in gabion structures, and geogrid
tails are extended behind the gabion elements for supplemental
tensile reinforcement of the soil.   Structures as high as 8 m have
been built using this technique in the U.K.   One advantage of tailed
gabions is increased stability; the reinforcing tail effectively
eliminates a progressive failure mechanism in which gabion settlement
may lead to gradual shear distortion and lateral deformation of the
wall (36).

     Another example of combining conventional retaining wall construc-
tion with the concept of reinforced soil is the use of concrete
blocks and geogrids.   In Norway, for instance, precast concrete
blocks have been used for the construction of low height gravity
retaining walls for some years.   The blocks are 500 mm long x 250 mm
high x 600 mm deep and require a small crane to lift them into place.
Once placed, they are stable and self-supporting.   As illustrated in
Figure 3, versatility of the block wall can be improved by the provision of polymeric grid reinforcement, fixed either to the block or
incorporated between blocks.   The result is a hybrid structure, part
gravity wall and part reinforced soil.   Walls of 5.5 m in height have
been constructed.

     Given the variety of reinforcing components and the possibilities
of combining elements of externally and internally stabilized

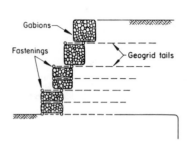

Figure 2. Tailed Gabion
with Geogrids

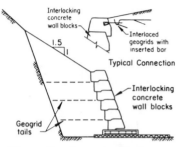

Figure 3. Concrete Block
Wall with Geogrid Tails

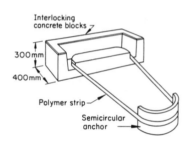

Figure 4. Wall System with
Concrete Blocks, Polymer Strips,
and Anchors

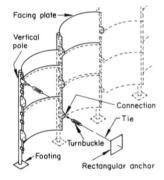

Figure 5. Wall System with
Facing Plates and Rectangular
Anchors

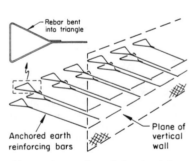

Figure 6. Anchored Earth with
Triangular Rebar Reinforcement

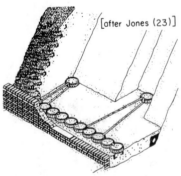

Figure 7. Earth Retention with
Waste Tires and Geotextiles

systems, there has been no shortage in the types of reinforcing schemes proposed and applied in the field. Anchored earth systems, for example, have been developed which involve aspects of reinforced soil and soil anchoring. Figures 4 and 5 illustrate applications originating from Austrian and Japanese practitioners. Both applications utilize multiple layers of closely spaced reinforcement and thus are similar in construction to the methods employed for reinforced soils. The Austrian application involves strips connecting concrete wall blocks and semicircular anchors, while the Japanese application exploits the local passive resistance of small rectangular anchor plates. A scheme developed in Britain (24,34), which is illustrated in Figure 6, employs reinforcing steel bent into triangular anchors. Pullout resistance is mobilized by friction along the straight portion of the steel and by passive pressure mobilized at the triangular anchor. Anchored earth concepts have been extended even to waste automobile tires, as illustrated in Figure 7 (23). The tires may be tied together with metal bars or polymer strips.

As shown in Table 1, the development of earth retention systems can be viewed as an evolutionary process, in which methods for supporting soil have involved progressively more alterations and insertions of reinforcing elements. One of the goals has been to transform soil into an engineered medium, which is less soil-like because of enhanced mechanical properties.

Recent innovations include polymer impregnation of soils formed by mixing soil with a small continuous filament of polymer. The reinforcing element described by Leflaivre, et al. (30) is a polyester filament of 0.1 mm diameter and a tensile strength of 10N. The placement requires the simultaneous projection of sand, water, and filament. Approximately 0.1 to 0.2 % of the composite material is filament, producing a total length of reinforcement of 200,000 mm per m^3. Cohesive strengths of 100 to 200 kPa have been reported for the material.

Similar mechanical behavior has been developed by the inclusion of small 60 mm x 40 mm polymeric grids into sand. As reported by Mercer, et al. (32), the inclusion of 0.2% by weight of grids can increase the shear strength of sandy soil by 25 - 60%.

In some cases, particularly when very soft compressible soils underlie highway walls and abutments, it may be advantageous to replace the soil entirely by a light weight material. Flaate (13) describes the use of expanded polystyrene (EPS) at several highway retaining structures in Norway. Low density cements and soil-cement products have been used for similar applications.

HISTORICAL PERSPECTIVE

Four areas in particular experienced significant changes from 1970 to 1990 and are selected for discussion as representing some of the most important trends in methods and materials. These areas include braced and tied-back excavations, construction of in-situ walls, reinforced soil, and soil nailing.

Braced and Tied-Back Excavation. The decade from 1970 to 1980 was a watershed for rapid transit construction. In the U.S., major underground rapid transit projects were executed or planned for Washington, D.C., Atlanta, Baltimore, Boston, New York City, Buffalo, and Los Angeles. In Western Europe, over twenty different metros either were initiated or expanded as part of a general program to improve urban transportation (35). Many high-rise buildings with deep basements were constructed, with notable deep basement excavations in cities such as Seattle, San Francisco, Chicago, Los Angeles, and Houston.

As Peck (41) indicates, the excavation landscape was transformed gradually from multiple levels of steel cross-lot bracing and rakers to one in which tiebacks resulted in substantially more room and maneuverability. Currently, tied-back excavations account for approximately 85% of in-situ wall construction in the U.S., with 5 to 10% supported by soil nailing and the remainder supported by cross-lot braces and rakers.

One excavation, which exemplifies the complexity and scale achieved with tied-back walls, was the deep cut for the Columbia Seafirst Center basement in Seattle. Figure 8 shows an aerial view of the site at a time in 1983 when the excavation had reached invert level at a maximum depth of 37 m. Excavation was performed in interbedded stiff clays, sands, and till above the water table. The wall was composed of soldier piles on 4-m centers, combined in different locations with timber and cast-in-place concrete lagging. Earth anchored tiebacks, with 6 to 9 m long, 310 mm diameter anchors, were installed on 1 to 2.4 m vertical spacings. The edge of the excavation was adjacent to the five-story Columbia House. An intricately criss-crossed system of tiebacks was used to stabilize the wall at this location and to underpin the foundations of the building.

The support system and excavation sequence were simulated with finite element analyses to evaluate potential ground movements and optimize construction for displacement control (16). Figure 9 shows a profile view of the excavation in which the finite element predictions and measured soil displacements are compared. The actual ground movements were surprisingly small, being less than those predicted by finite element simulations. The predictions, nonetheless, helped to establish a rationale for evaluating the magnitude and distribution of potential deformation and for planning remedial and protective measures.

In recent years increasing emphasis has been placed on selecting support systems and construction procedures with the primary goal of controlling ground movements (11,37). Predictive tools for doing this have been devised principally as a result of a) field observations of wall and adjacent displacements in a variety of soil conditions, often by means of sophisticated instrumentation, and b) numerical modeling, most often by the finite element method, of the excavation and support process and resulting ground movements. The semi-empirical methods so developed (10) provide a rational and systematic means of estimating soil movements based on the stiffness of the support system, factor of safety against basal heave, and an understanding of the different activities which make up the construction process.

As experience in urban settings has been acquired, attention has been focused on the performance of structures adjacent to deep excavations. Traditionally, civil engineers have been concerned about structural response to differential settlement. Horizontal soil

Figure 8.   Aerial View of Tied-Back Excavation for Columbia Seafirst
Center (courtesy of Shannon & Wilson, Inc.)

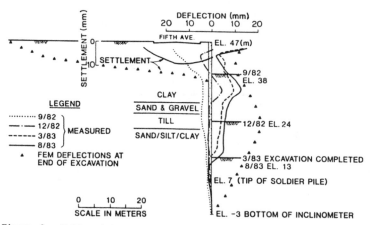

Figure 9.   Wall and Street Movements at Columbia Seafirst Center
[after Grant (16)]

movements, however, are also a major factor when excavations are
involved. Data from deep excavations, tunneling, and mining subsidence have been synthesized to evaluate how various structures
respond to combined horizontal and vertical ground movements.

Guidelines now exist for estimating building damage on the basis of angular distortion and horizontal strain (5,38).

Construction of In-Situ Walls. There have been improvements and adaptations in construction equipment. Figure 10 illustrates five different methods for installing cast in-situ walls and cutoffs which are representative of advances in the last two decades. Figure 10a shows two views of a clamshell rig for concrete slurry wall construction which was developed in France in the 1960s and applied throughout North America in the 1970s. A device, known as the Hydrofraise (40), was developed in France in the 1970s for the excavation of in-situ panels through soil and rock. The basic components of the Hydrofraise are illustrated in Figure 10b. The unit is operated from a crawler crane. It consists of a heavy metal frame at the base of which are two counter-rotating drums with tungsten-carbide teeth. A special pump, mounted centrally within the unit, removes spoil and cuttings suspended in a bentonite slurry. It is reported that the device can excavate rock with compressive strengths of 100 MPa and advance panels to 130 m depths with tight vertical tolerances (40).

Another method of in-situ wall construction is by soil-cement mixing, which is illustrated in Figure 10c. This method, which was developed in Japan, uses three or two-axis hollow stem augers which churn and mix the soil as water and cement are introduced through the base of the augers. Soil-cement mixtures can be adjusted to obtain various compressive strengths, usually less than 5 to 10 MPa. Structural walls are built by installing H-piles in the soil-cement columns before substantial hardening of the mixtures.

Figure 10d shows three-dimensional and profile views of the Benoto rig which is used to install secant pile walls. The rig, which was developed in Belgium, rotates a casing and special end tube, equipped with cutting teeth, into the ground as soil simultaneously is removed from inside by a mechanical grab. Reinforcing steel, if required, then is placed and concrete skipped or tremied into the pile as the casing and tube are withdrawn. This type of equipment was used for concrete diaphragm wall construction on the London, Tyne-Tees, and Hong Kong metros.

Among the more recent methods introduced to U.S. practice is jet grouting, illustrated in Figure 10e. The method, which was developed in Japan, involves drilling a vertical hole and then grouting by means of ultra high pressure water jets to cut, replace, and mix in-situ soil with a cementing agent. Jet grouting can be performed by single, double, and triple fluid systems, and readers are referred elsewhere (27) for a comprehensive discussion of the procedures. Soil-cement columns formed by jet grouting can be used as ground water cutoffs, stabilizing units at the base of an excavation, and underpinning elements to protect adjacent structures from excavation-induced ground movements.

The methods and equipment illustrated in Figure 10 show that the construction plant associated with in-situ wall installation can be quite sophisticated, requiring substantial skill and expertise for proper operation. This increased sophistication in construction plant parallels similar developments in drilling and grouting equipment. Variations in drilling devices have proliferated in an attempt to cope with different soil and groundwater conditions. Bruce (7), for example, has described at least seven different

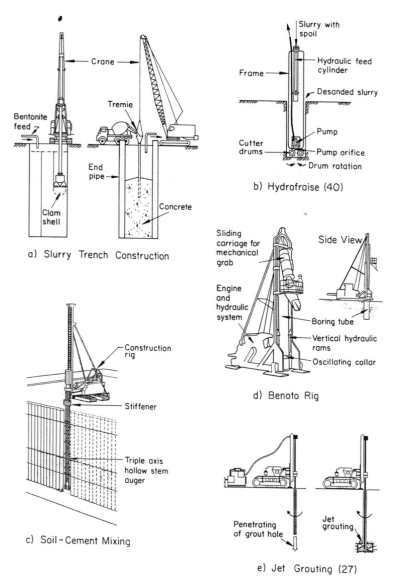

Figure 10.   Concrete Diaphragm Wall and Cutoff Construction Methods

classes of overburden drilling, with modifications and special adaptations in each class.

There is a paradox in modern construction in that increasing mechanization places more, not less, emphasis on the labor force. The labor force, however, must be skilled. Reliance on increasingly complex forms of construction and on the skills to operate equipment have made specialty contractors an indispensable part of the building process and has fostered greater emphasis on design-construct options.

The equipment illustrated in Figure 10 has played a key role in building stiff in-situ walls. The inherent bending stiffness of these walls has proved to be a valuable asset in controlling adjacent ground movements. In San Francisco, for example, the use of concrete diaphragm walls for deep basement and rapid transit excavations resulted in a 50 to 75% reduction in lateral wall movement compared with sheet pile walls braced similarly in approximately the same soil conditions (9).

A critically important factor affecting wall deformation is the stiffness and timely installation of bracing or tieback supports. Moreover, there are limits to the amounts of ground movement reduction which can be achieved with an in-situ wall. The semi-empirical methods developed in the past twenty years can provide effective guidance for these concerns. Predictive methods, for example, show that the ability to reduce wall movements approaches a limiting value as the system stiffness of the supports increases (10). Accordingly, it will not be productive to increase wall stiffness beyond certain values, and the measures for safeguarding adjacent structures beyond these bounds must be supplemented by other protective schemes.

Finally, the equipment illustrated in Figure 10 shows the strong influence which technologies developed abroad have had on U.S. practice. The past twenty years indicate, without doubt, that in-situ wall construction is a global enterprise and that future directions in U.S. practice should be viewed as being driven in a major way by overseas innovations.

Reinforced Soil. In 1970, reinforced soil was an established procedure for building retaining structures in Europe. It was introduced to North America in 1972 when it was used to repair and stabilize a landslide adjacent to a highway in California (14). These early forms of reinforced soil were built exclusively with metallic strips. In 1973, the first polymeric strips were introduced during the construction of a highway wall in Yorkshire, U.K. (22). The reinforcing elements were composed of continuous glass fibers embedded lengthwise in a protective coating of resin. In the mid-1970s experiments were conducted in several countries with woven and nonwoven geotextiles to build reinforced soils (18).

The first use of polymer grids can be traced to Japanese practitioners, who used the method in the 1960s to reinforce weak subsoils at embankments and runways (23). These early grids were composed of non-oriented polymers, and thus were relatively fragile, lacking the tensile strength required for reinforced soil walls. The first geogrids for reinforced soil walls were composed of cross-linked steel reinforcing bars. In 1974, this technique was used to build a 6 m high wall for a highway in California (14). In the 1970s advances in polymer manufacturing led to substantial improvements in stiffness and strength.   By stretching some polymer resins, high degrees of molecular orientation and tensile strength were achieved. In the late 1970s a manufacturing process was developed which allowed for the production of geogrids with oriented polymers in orthogonal directions.   These geogrids were first employed in 1979 for the construction of a reinforced soil wall at a railroad station in Yorkshire, U.K. In the early 1980s the geogrid manufacturing process was exported to the U.S. The first geogrid application in the U.S. occurred in 1982 for a slope stabilization project in Texas.   The first geogrid wall was built in Oregon in 1984 to stabilize a highway embankment.   Geogrid applications in North America have grown exponentially since 1982.

Reinforced soils are gravity structures built incrementally from soil, tensile reinforcement, and a facing.   A key aspect of design is the accommodation of settlement without damage to the facing. Three basic configurations, illustrated in Figure 11, have been developed to accommodate movement. Except for some special cases, all reinforced soil structures constructed above ground conform to one or the other of these forms of construction or a combination of them.

The 'concertina' construction, originally proposed by Vidal (45), is shown in Figure 11a. Differential settlement within the soil mass is accommodated by the front or face of the structure closing in a manner similar to a set of bellows or a concertina. Some of the largest reinforced soil structures have been built using this technique. It is the form of construction most frequently used with geotextiles.

In the telescope construction, which also was proposed by Vidal (46), the deformations within the soil mass are accommodated by the facing panels closing and moving forward an amount equivalent to the internal deformations. As illustrated in Figure 11b, this is made possible by supporting the facing panels during construction on wedges so that discrete horizontal gaps are left between them. Failure to provide a large enough gap can result in crushing and spalling of the facing.

In the sliding method of construction, which was developed by Jones (22), differential settlement and compaction within the fill can be accommodated by permitting the reinforcing elements to slide vertically relative to the facing.   Slidable attachments can be provided by grooves, slots, vertical poles, lugs, or bolts. Facings made up of discrete elements, as with the telescope method, may be used, as can full height facings displaying a range of architectural finishes. Figure 11c shows how the sliding method of construction is adapted to a continuous rigid facing.   Reinforcing strips can slide downwards on bolted connections which link strip and facing.

The use of reinforced soils has had a profound influence on the speed of construction. Construction rates can be as high as 40 to 200 m^2 per day per worker, with speed usually determined by the rate of placing and compacting the fill. Reinforced soils also allow for the use of nonconventional construction sequences.   For example, bridge abutments can be constructed with reinforced soil in advance of approach embankments and even before the final structural facing.

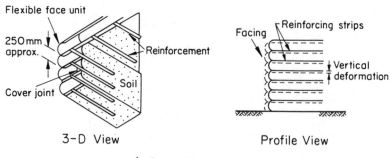

a) Concertina Method

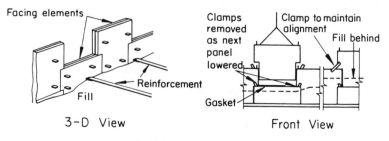

b) Telescope Method

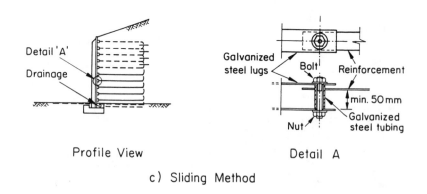

c) Sliding Method

Figure 11.　Different Methods of Reinforced Soil Construction

Soil Nailing. The development of soil nailing has been treated in considerable detail by Bruce and Jewell (8) and Mitchell and Villet (33), and reference to their work is encouraged for those seeking a more comprehensive overview of the subject. Soil nailing is a process in which excavation walls and slopes are stabilized in-situ by installing relatively short, fully bonded steel bars, according to a regular and relatively closely spaced pattern.

Early uses of soil nailing were adapted from procedures followed in the New Austrian Tunneling Method (NATM), and its application is often cited as an outgrowth of NATM (8). In 1972, soil nailing was first employed to stabilize a railway cut in heavily cemented sand above the water table near Versailles, France (19). The construction eventually resulted in an 18 to 22 m high reinforced soil slope, inclined at 70°. Benches, typically 100 m long, were excavated from the top down at regular height intervals of 1.4 m. The soil nails varied in length from 4 to 6 m. Each nail consisted of two 10-mm-diameter steel bars grouted in 100 mm diameter drilled holes. Facing along the slope was provided by a 50 to 80 mm thick mat of reinforced shotcrete.

The first U.S. application of soil nailing occurred in 1976 to provide temporary excavation support for basement construction at a hospital in Portland, Oregon (42). The soils at site were cohesive, dense silty sands above the water table. Excavation proceeded from the top down at height intervals of 1.5 m to a maximum depth of 13.7 m. The soil nails consisted of 25 to 38 mm diameter steel bars grouted in augered holes, approximately 7 to 8.5 m long. The face of the excavation was covered with mesh-reinforced shotcrete, roughly 25 to 50 mm thick.

Applications of soil nailing were carried out independently in France, West Germany, and the U.S. in the 1970s. Toward the end of the decade, there was considerable interchange among practitioners in the different countries, with consolidation and refinements in the general principles of practice and theory.

Soil nailing is an accepted practice now, and it is easy to forget that the method actually represents a radical departure from procedures for in-situ wall construction before 1970. Conventional wisdom used to require that in-situ excavations in soil be built with a structural wall which was established first, and then followed by the excavation and bracing process. Difficulties in establishing in-situ walls before excavation are often encountered in residual soils and weathered rock, and these materials represent one of the geologic environments in which soil nailing has been used to great advantage.

Soil nailing continues to challenge conventional thinking. For example, it provides ample evidence that in-situ walls can be built in some types of sand and cohesive soils from the top down without pre-existing embedment. On many jobs the depth of wall embedment is a major source of differing opinions, with detailed calculations offered by the design engineer to prove the necessity of an embedment which the contractor claims is not justified by experience. Clearly, wall embedment can play a significant role in promoting stability and controlling ground movement for excavations in soft clays. Equally clear is that pre-existing embedment is not necessarily required in certain stiff cohesive soils and cohesive sands above the water

table. Reconciliation is still needed between the practical virtues
of building from the top down with "nails" and the traditional
concerns for walls to achieve bottom stability by penetrating the
base of excavation.

MATERIALS

It has been claimed that the 1980s were a time of revolutionary,
not evolutionary, change in materials (29). Optical fibers and
superconducting oxides were innovations which encouraged great
changes in the telecommunications and electronic industries. Equally
important opportunities were provided for the civil construction and
environmental industries in the form of composites of all possible
combinations of metals, polymers, cementitious materials, and soils.
Traditionally, problems of soil-structure interaction have
involved structural materials such as steel, concrete, and timber.
Now, the structural materials are equally likely to be polymers,
especially for landfills, embankment stabilization, and earth
retention systems. Conventional designs always regarded the soil as
the most variable and uncertain entity in the soil-structure
partnership. Polymers also behave in complex and uncertain ways,
displaying time-dependent, irreversible, nonlinear, and
environmentally sensitive stress-strain properties. Moreover, there
are many different classes of polymers. Structures now may be
composed of polyethylene, polyvinyl chloride, epoxy, polyurethane,
polyester, and polypropylene, to name a few. The diversity of
materials, variations within a given class, effects of processing,
and the many different ways of modifying properties by additives,
fillers, and reinforcements have added a multitude of new dimensions
to soil-structure interaction. The increasing diversity and complexity of polymers has provided an ability to customize materials for
a given environment or application, essentially starting with a need
and developing a material with an optimal combination of properties
to meet the need.
Manufacturing processes have evolved to a stage where strong and
durable reinforcing elements can be produced en masse. The most
familiar products in earth retention systems are high density
polyethylene (HDPE) and polypropylene grids. Composite polymers are
also available. Composite reinforcing strips have been used in the
U.K. since the 1970s. The strips are composed of polyester bands
encased in polyolefine. The polyester bands provide tensile
strength, and the casing provides an environmentally resistant
exterior with frictional properties necessary for suitable pullout
capacity. These composite polymer strips are now manufactured in a
grid pattern, similar to their HDPE and polypropylene counterparts.
Polymers vary substantially with respect to surface hardness and
texture, both of which are important factors affecting frictional
characteristics. Different products within a given polymer class can
show substantial variations. For example, unplasticized polyvinyl
chloride (PVC) in underground piping will develop less than one-half
the interface frictional resistance of plasticized PVC used in
geomembranes, given similar sand density and normal stress.
Recent experimental investigations at Cornell University (39)
provide insight and help clarify some aspects of interface frictional

behavior between granular soil and polymers. Figure 12 shows the relationship between the interface friction angle, δ, and the Shore D Hardness, H_D, of the polymer. The interface friction angle is expressed as a fraction of the direct shear angle of soil friction, δ/ϕ'_{ds}. The results summarized in Figure 12 are based on over four hundred direct shear tests involving several different types of sands with densities ranging from loose to very dense at normal stress levels between 3.5 and 69 kPa. The Shore D Hardness test (1) is performed simply with a hand-held indentation device, and thus allows for widespread and expedient use in practice.

Two types of behavior were observed during testing, as illustrated in Figure 12. On relatively hard surfaces with $H_D \gtrsim 58$, a skidding mechanism controlled the frictional resistance. On relatively soft surfaces, a rolling mechanism controlled resistance. Rolling tended to reduce interlocking and interference among sand grains adjacent to the polymer such that very little dilatancy could develop. No dilatancy was observed when skidding occurred along relatively hard surfaces.

If the interface between soil and polymer is a plane of zero extension and no dilatancy is assumed, then the maximum value that δ can achieve is shown by plastic strain theory (e.g.,21) to be

$$\tan \delta = \sin \phi'_{cv} \qquad\qquad\qquad (1)$$

in which ϕ'_{cv} is the critical state angle of soil shear resistance, representing the residual or steady state shear resistance at constant volumetric strain. This simple condition sets a limiting value for relatively soft polymer surfaces. For relatively hard polymers, values of δ are considerably less than ϕ'_{cv}.

Figure 12 helps illustrate how even simple and basic properties, such as interface friction, can vary depending on special polymer characteristics, which in this case is the surface hardness. The figure also shows that δ/ϕ'_{ds} can be as small as 0.3 to 0.4 for certain unplasticized PVC and hard epoxies. These values are roughly one-half to two-thirds as much as those typically assumed for steel. Moreover, the test results summarized in Figure 12 apply only for smooth surfaces. Asperities, ribbed textures, and roughening can increase the interface shear resistance significantly by promoting dilatancy.

Polymers are viscoelastic materials, and hence their behavior is influenced substantially by time-dependent effects. Time-dependency is well recognized among geotechnical engineers. Nevertheless, the combination of time-dependent soil and structure involves a relatively complex interaction, and design for long-term effects must take account of special strength and viscoelastic characteristics.

An important consideration is that, for some loading conditions, irreversible damage can accelerate in polymers long after force has been applied and an apparently steady creep has been established. Figure 13, which has been adapted from tests on HDPE grids (31), illustrates this condition in which strains and strain rates increase significantly above a certain threshold. This threshold can be taken as a performance limit. For some HDPE grids the performance limit

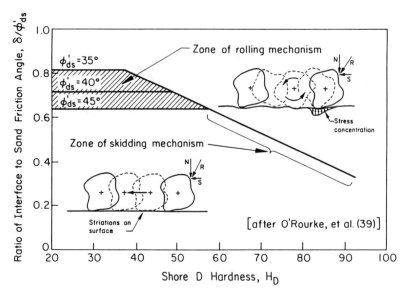

Figure 12. General Relationship Between the Ratio of Interface to Soil Friction Angle, δ/ϕ'_{ds}, and Shore D Hardness of a Smooth Polymer Surface, H_D.

strain is 10%, which corresponds to about 40% of the short-term breaking load.

Another important consideration for polymers is that stress-strain curves obtained under specific environmental and short-term loading conditions are insufficient for design purposes because duration of loading, temperature, and other environmental factors change behavior in a critical way. Accordingly, isochronous stress-strain curves are defined, as shown in Figure 14, to account for time-dependent effects at constant temperature and physico-chemical environments.

In selecting or accepting polymer products for a given job, questions inevitably arise about durability. With so many new products being introduced, it is usually difficult to make clear judgments about field performance. Decisions about durability must be based on composition, stress-strain-time characteristics, temperature dependency, resistance to ultraviolet light, and biological and chemical attack, as well as abrasion resistance, workability, and handling stresses. Such concerns pose challenges for specification writers and place substantial emphasis on the work of professional society committees in developing test standards and guidelines.

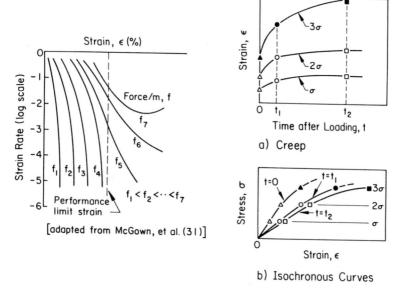

Figure 13. Strain vs. Log Strain
to Determine Perform-
ance Limit Strain

Figure 14. Creep Behavior and
Isochronous Stress-
Strain Curves

PERFORMANCE

Just as field observations and instrumented test sites were
crucial for our understanding of braced and tied-back excavations,
the same also is essential for advancing the state-of-the-art and
practice for reinforced soils and soil nailing. In the past twenty
years, our knowledge and design methods for reinforced soils have
been driven primarily through field performance observations,
augmented by centrifuge and analytical modeling.

The integrity of reinforcing members is critical to internally
stabilized structures and much attention has focused on the
durability of the reinforcing materials. Some reinforced soil
failures have occurred, principally among the metallic reinforced
structures which suffer from corrosion. In particular, pitting
corrosion is seen to be the most damaging form of attack, which can
lead to rapid loss of tensile strength (3,28). Reports of failures
involving synthetic reinforcement are few and relate to early forms
of polymeric materials (46). Although creep dominates much debate
on the use of polymers, the potential for creep failure is limited
by considerations of serviceability. As discussed previously,
strains needed to induce ductile creep rupture in HDPE are greater

than 10% which is substantially in excess of any serviceability
limit.

Failure involving the internal stability of reinforced soil is
different from that assumed in conventional retaining structure
design.   Observations of full-scale structures (e.g.,2,25) and
centrifuge tests (4) show that, when failure is caused by breakage
of reinforcements, the locations of maximum tensile forces coincide
with a potential failure surface different in shape and closer to the
facing than a conventional Coulomb failure plane. Moreover, field
measurements of systems with relatively stiff metallic reinforcements
have shown soil stresses near the top of the wall consistent with
earth pressure at rest, or K_o, conditions (33). Design procedures
have been developed which account directly for K_o earth pressures
near the top of wall (23,33), while others have been formulated
through limiting equilibrium analyses using a log spiral failure
surface (26). Guilloux and Jailloux (17) have described the failure
of a 6 m high wall in which plain steel strips were corroded by salt
water infiltrating the structure. In this case, the failure surface
closely resembled a log spiral.

Recent centrifuge tests provide an example of experimental work
which can lead to a change in perception. In a study performed as
a part of a full scale trial funded by the Federal Highway Administration, the robustness of polymeric reinforced structures was shown
to be greater than many designers had originally assumed (20).   It
was shown that the use of extensible polymeric reinforcement, with
its potential for creep, leads to load shedding and a uniform
increase in reinforcement stress. As a result, reinforcing elements
in this form of construction tend to approach failure simultaneously.

In 1986, two reinforced soil walls in Tennessee failed soon after
construction and a number of others have shown signs of serious
distress, Figure 15.   A particular feature of the Tennessee
structures is their size; some are over 18 m in height and support
large rock fill embankments. Failure was sudden and unexpected and
resulted in rupture of the connections between the reinforcement and
facing.   Differential settlements between the fill and the facing
were measured as exceeding 450 mm.

The cause of settlement appears to be related to the nature of the
fill, with consolidation, loss of fines, and fill degradation contributing to vertical movement.   Three-dimensional strains resulting
from both the behavior and geometry of the fill played a role in the
collapse mechanism.

As a matter of convenience, plane strain conditions are assumed
for the analysis of reinforced soil structures, and in the majority
of cases the assumption of two-dimensional analysis is adequate.
However, failure of the structures in Tennessee indicates that there
are conditions where three-dimensional geometry influences structural
behavior.

Figure 16 illustrates the case of a reinforced soil wall
constructed across a ravine in a mountainous region, which supports
a sloping embankment. Three-dimensional finite element analyses were
performed to determine the pattern of fill movement and corresponding
deformation of the wall. Although the details of the modeling are
outside the scope of this paper, it nonetheless is instructive to
examine the simplified isometric view provided in the figure. Vector

Figure 15.   Side View of Reinforced Soil Wall Failure, Tennessee

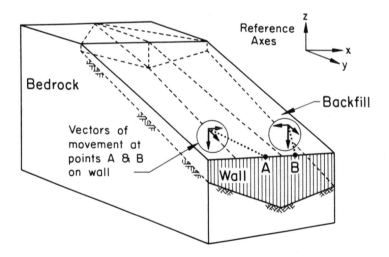

Figure 16.   Isometric View of a Reinforced Soil Wall with In-Plane Movements Conveyed to the Wall

displacements, determined by analysis, are shown for two points on the crest of the wall, either side of the tallest section of the structure. It can be seen that the movements in the ZX plane (i.e., in plane with the face of the wall) are both towards the center of the structure. In some conditions, such as large walls which are curved or articulated in plan or which cross deep and steep sided ravines, these in-plane movements are significant and must be considered in design.

DESIGN CONCEPTS

Given the observed performance of reinforced soils and the special characteristics of polymeric products, it is important to identify key design concepts which are likely to affect how materials are selected and earth pressures chosen. A prime consideration is the strength of the reinforcement. Reinforcing elements may lose strength either due to loss of section (e.g., by corrosion of metallic reinforcement) or through other forms of degradation such as physical or chemical aging of synthetics. In addition, synthetic reinforcements creep, and it is often necessary to consider limiting strain criteria.

The stiffness of the reinforcement has a fundamental influence on the behavior and performance of reinforced soil structures. Axially stiff reinforcement will strain little before taking up load. Stress in the reinforcement can accumulate rapidly and may occur at lower strains than those required to mobilize peak soil strength. On the other hand, extensible reinforcement requires greater deformation before it takes up the stresses imposed by the soil. This may lead to higher strains, and the peak shear strength of the soil may be approached or exceeded.

As discussed previously, field observations indicate that the upper level of fill in a mechanically stabilized structure can exhibit a coefficient of lateral earth pressure equivalent to the at rest pressure. The use of less stiff reinforcement may result in yielding where the strength of the soil is fully mobilized and stress conditions approximate the active case. Thus, the mobilized frictional resistance in the soil depends on the strain which develops in the soil, and the mobilized reinforcement forces depend on the strain in the reinforcement. The behavior of the soil and reinforcement are linked by strain compatibility.

Bransby and Milligan (6) have described the pattern of plastic soil strains associated with cantilever wall movements, as illustrated in Figure 17. The figure shows lines of zero linear strain, referred to as velocity characteristics. These lines are oriented at angles of 45° + Ψ/2 either side of the direction of maximum tensile strain, where Ψ is the angle of dilatancy of the soil.

A simple model for reinforced soil deformation is to assume that the wall rotates rigidly about its base, generating a displacement pattern similar to that in Figure 17. Horizontal tensile strains will accumulate, therefore, in proportion to the wall rotation. For these conditions, the horizontal strain, ϵ_H, in the reinforcement can be expressed as

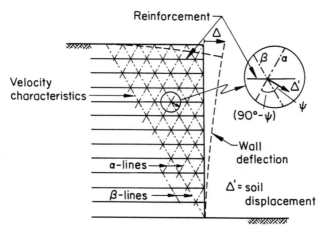

Figure 17.   Velocity Characteristic for Reinforced Soil Structure
with Cantilever Deformation of Facing

$$\epsilon_H = (\Delta/H) \tan (45° + \Psi/2) \qquad\qquad (2)$$

in which Δ/H is the dimensionless displacement, or rotation, of the
plane of the facing.

Although this model greatly simplifies the soil mass deformations,
it nevertheless provides a conceptual basis for linking horizontal
strain and wall movement when extensible reinforcements are used.
Rigid metallic reinforcements would be expected to result in a
slightly different pattern of facing deformation, with more restriction on movement in the upper part of the wall. The general methodology of relating plastic strains and wall deformation, however,
would still be valid.

Figure 18a shows the active earth force developed by a retaining
wall, supporting loose and dense sand, as a function of wall rotation
Δ/H. These patterns are consistent with experimental results for
wall rotation similar to that in Figure 17 (12,44). Because of
dilatancy, the dense sand mobilizes a peak angle of shear resistance,
corresponding to a minimum active force, after which the critical
state soil shear strength is approached as the magnitude of wall
movement increases. Figure 18b shows isochronous curves for a single
extensible reinforcement, which is representative of all reinforcements used in a particular reinforced soil structure.   The
isochronous curve at the design life of the structure is identified
by the time, t_d. Figure 18c shows the system compatibility diagram
in which the active earth force and isochronous curves for the entire
system of reinforcements are superimposed. The reinforcement system
is composed of elements, each with the same characteristics as shown
in Figure 18b. Accordingly, it is reasoned that the stress-strain-

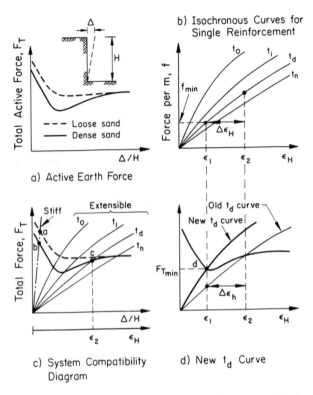

a) Active Earth Force

b) Isochronous Curves for Single Reinforcement

c) System Compatibility Diagram

d) New t_d Curve

Figure 18. System Compatibility of Strains and Forces for Reinforced Soil Structures

time response of multiple levels of reinforcement is approximately in the same form as that of a single reinforcement.

Equilibrium occurs when the curves representing the reinforcement tension intersect the curve representing the active earth force, which reflects the mobilized frictional resistance of the soil. For a system of stiff metallic reinforcements, this intersection may occur at low levels of strain and will be independent of time. This condition is illustrated in Figure 18c by a relatively high level of tension that is imposed at points a and b. In contrast, extensible reinforcements will creep so that equilibrium conditions change. A variety of forces might be sustained by an extensible system, depending on the design life. If relatively large creep deformation occurs, the active earth force that needs to be supported will approach a limit state near point c, defined by K_a and ϕ'_{cv}.

As shown in Figure 18b, a reduction in strain, $\Delta\epsilon_H$, can be achieved if the force, f, at time, t_d, is reduced. Such a reduction is possible if the number of reinforcement levels is increased, thereby diminishing the forces in individual members. Hence, the level of strain decreases while the tensile force in the system seeks a new equilibrium condition which is compatible with the reduced strain. In effect, this leads to a new t_d curve which is illustrated in Figure 18d. The case represented in the figure shows the effect of an average force reduction per meter of reinforcement which produces a strain at point d. This implies that serviceability limits based on maximum wall deformation can be adjusted by the number of reinforcement layers.

Limit states have been adopted for structural design in several European countries, and similar procedures in the form of load and resistance factor design are used in U.S. codes (43). In its simplest and most basic form, the limit state approach establishes a margin of confidence against a) structural collapse (ultimate limit state) and b) undesirable performance under service loads (serviceability limit state). Limit state definitions require that the designer set rational bounds on performance and for each limiting condition ascribe safety factors which are consistent with the uncertainties in choosing respective loads and resisting forces.

Figure 18c can be used to visualize ultimate and serviceability limit states for reinforced soils. Design for rigid reinforcement, associated with points a and b, would be described as an upper bound solution in which the reinforcement loads reflect K_o conditions. This is consistent with a serviceability limit state in which wall deformation is constrained. It also represents an ultimate limit state for rupture of the rigid reinforcing elements. Design of extensible reinforcement for residual soil strength (point c) would reflect an ultimate limit state for both rupture and pullout. The residual soil strength also leads to an ultimate limit state for the pullout of rigid reinforcement.

Figure 19 presents a flow diagram for a design approach consistent with limit state concepts and the use of current design models. For each type of reinforcement, the coefficient of earth pressure and angle of shear resistance are specified in accordance with a given limit state and mode of failure. The approach covers the two generic forms of soil reinforcement: extensible and inextensible. An extensible reinforcement now can be defined with greater precision as one where tensile deformation in the reinforcement is greater than the horizontal extension required to develop a minimum active earth force. Similarly, the strain in inextensible reinforcement is not sufficient to develop active conditions. To accommodate these different conditions, separate criteria are required for extensible and inextensible reinforcements used within the same analytical model.

The earth pressure conditions appropriate in granular soil for design with inextensible reinforcement are either related to a K_o state of stress, defined on the basis of a peak angle of soil shear strength, $\phi_p{}'$, or a K_a state of stress, defined on the basis of the critical void shear strength, $\phi_{cv}{}'$. Both the ultimate limit state for inextensible reinforcement rupture and the serviceability limit state are evaluated according to K_o and $\phi_p{}'$. The ultimate limit

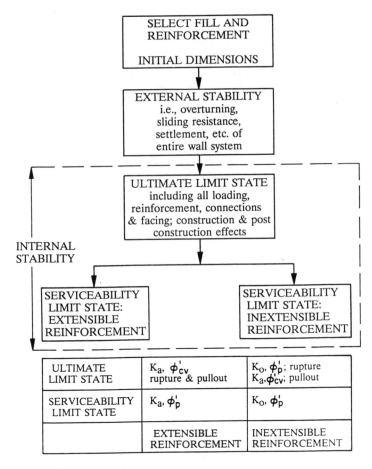

Figure 19. Design Approach for Reinforced Soil

state with respect to pullout failure is evaluated according to K_a and ϕ_{cv}'. In contrast, extensible reinforcement is expected to strain sufficiently such that its ultimate limit state for both rupture and pullout is defined on the basis of K_a and ϕ_{cv}'. The serviceability limit state for extensible reinforcement is referenced in the diagram to the minimum active earth pressure, defined according to K_a and ϕ'_p. Usually, this will correspond to relatively small and acceptable levels of wall deformation.

FUTURE DIRECTIONS

In 1970, it would have been difficult to foresee the widespread use of geosynthetics in structural reinforcing systems. Although geosynthetics were used at the time for improved subgrade and drainage performance, it would have been difficult to predict the important role that polymer grids and strips now play in earth retention systems. One would have been more comfortable in forecasting increased use and diversity in specialized drilling and in-situ wall construction methods. Likewise, in 1990 the same difficulty exists.

The most noteworthy future changes are likely to take place in areas which currently are experiencing rapid development. Opportunities, for example, exist in multidisciplinary activities between civil engineers and material scientists. New materials, such as conductive and self-reinforcing polymers, could allow for better systems of soil support. It may be possible to develop "smart" materials which change properties to suit requirements. Most polymers have relatively large coefficients of thermal expansion. If methods for integrating soil and heated polymeric elements were devised, subsequent thermal contraction would generate confining pressures, effectively prestressing the soil. Some materials, such as expanded polytetrafluoroethylene (PTFE), have a negative Poisson's ratio and thus expand when loaded in tension. Although PTFE lacks certain structural attributes, new materials currently are being developed with similar expansion characteristics and improved structural response. Polymer reinforcements, which expand under tension, would tend to generate higher pullout capacities in direct proportion to increased earth pressure loads.

Changes are likely to occur in methods of manufacturing. Knitting technologies are being refined which produce geotextile grids with strength and durability. Knitting procedures involve a less sophisticated plant than the polymer processing factories needed for HDPE and polypropylene grids. Accordingly, the knitting process is more mobile. It is predicted that new polymeric products could be manufactured and marketed as part of a franchise agreement. In this system the manufacturing plant would be set up near the end user market, and the local manufacturer (i.e., the franchise holder) would be supplied with the necessary technical information and application technology by the product developer. The system could be especially helpful in developing countries, where manufacturing would be adapted to local needs and allow for better scheduling and delivery services.

Waste management and disposal concerns are likely to encourage greater use of marginal soils and industrial refuse as backfill materials in earth retention systems. The use of backfills composed of minewaste, pulverized fuel ash, and even sawdust can provide economic benefits which may be matched with complimentary technical benefits. Pulverized fuel ash, for example, can develop a substantial amount of cohesive strength. Polymers used to reinforce this material would not be vulnerable to corrosion. Composite fills with waste polymers for reinforcement may also help in the beneficial reuse of discarded non-degradable municipal and industrial products.

Increased availability of computer systems and rapid data processing will improve the remote guidance, drilling, and mixing activities associated with in-situ wall construction, tieback installation, and grouting. Numerical analyses will become easier to perform, and design methods based on interactive and multi-component models will be widely used.

SUMMARY

This paper provides an overview of earth retention systems. A classification scheme is proposed in which externally and internally stabilized systems are identified as the primary forms of construction. It has been in the area of internally stabilized systems that a fundamentally new concept has been introduced. Shear transfer to mobilize the tensile capacity of closely spaced reinforcing elements has freed retaining structures of the need for a primary structural wall and has substituted instead a composite system of reinforcing elements and soil as the principal structural entity.

Developments in four areas have been reviewed, including braced and tied-back excavations, in-situ wall construction, reinforced soils, and soil nailing. The most rapid changes are occurring in reinforced soils, where the introduction of polymeric reinforcements and their combination with conventional walls as hybrid systems have encouraged a multitude of different earth retention schemes. Strain compatibility between soil and reinforcement is seen to be a critical aspect of design, and various limit states are proposed for design, based on strain compatibility concepts.

Future directions are projected for new polymers, manufacturing processes, increased utilization of marginal soil and waste products for retention systems, improvements in construction plant, and better management and analyses with computers. Earth retention systems are viewed as a global enterprise, with implication for U.S. practice as well as the best use of emerging technologies.

ACKNOWLEDGMENTS

The authors wish to thank several people for providing information and exchanging views on earth retention systems and current construction practices. Special thanks are extended to D.A. Bruce for his perceptive comments and assistance in gathering information on construction methods, and to E. Passaris for his contributions in three-dimensional finite element analyses. Others who contributed by way of discussion or information exchange include H. Deaton III, J.P. Gould, K. Lee, R.H. Murray, A.J. Nicholson, H. Parker, W.R. Sullivan, W.D. Trollinger, and D. Weatherby.

REFERENCES

1. ASTM, "Plastics (2): D1601-D3099," Annual Book of ASTM Standards, Vol. 08.02, Philadelphia, PA, 1988.
2. Baguelin, F., "Construction and Instrumentation of Reinforced Earth Walls in French Highway Administration," Symposium on Earth Reinforcement, ASCE, New York, NY, 1978, pp. 127-156.
3. Blight, G.E. and M.S.W. Drus, "Deterioration of a Wall Complex Constructed of Reinforced Earth," Geotechnique, Vol. 39, No. 1, 1989, pp. 47-53.

4. Bolton, M.D. and P.L.R. Pang, "Collapse Limit States of Reinforced Earth Retaining Walls," Geotechnique, Vol. 32, No. 4, 1982, pp. 349-367.
5. Boscardin, M.D. and E.J. Cording, "Building Response to Excavation-Induced Settlement," Journal of Geotechnical Engineering, ASCE, Vol. 115, No. 1, Jan. 1989, pp. 1-21.
6. Bransby, P.L. and G.W.E. Milligan, "Soil Deformations Near Cantilever Sheet Pile Walls," Geotechnique, Vol. 25, No. 2, 1975, pp. 175-195.
7. Bruce, D.A., "Methods of Overburden Drilling in Geotechnical Construction - A Generic Classification," Ground Engineering, Vol. 22, No. 7, Oct. 1989, pp. 25-32.
8. Bruce, D.A. and R.A. Jewell, "Soil Nailing: Application and Practice," Ground Engineering, Part 1: Vol. 19, No. 8, Nov. 1986, pp. 10-15; Part 2: Vol. 20, No. 1, Jan. 1987, pp. 21-33, 38.
9. Clough, G.W. and A.L. Buchignani, "Slurry Walls in the San Francisco Bay Area," Preprint 81-142, ASCE Spring Convention, New York, NY, May 1981, 14 p.
10. Clough, G.W. and T.D. O'Rourke, "Construction Induced Movements of Insitu Walls," Design and Performance of Earth Retaining Structures, ASCE, New York, NY, 1990. (found in these proceedings)
11. Clough, G.W., E.M. Smith, and B.P. Sweeney, "Movement Control of Excavation Support Systems by Iterative Design," Foundation Engineering: Current Principles and Practices, Vol. 2, ASCE, New York, NY, 1989, pp. 869-882.
12. Fang, Y-S. and I. Ishibashi, "Static Earth Pressures with Various Wall Movements," Journal of Geotechnical Engineering, ASCE, Vol. 112, No. 3, Mar. 1986, pp. 317-333.
13. Flaate, K., "The (Geo)Technique of Superlight Materials," The Art and Science of Geotechnical Engineering, Prentice Hall, Englewood Cliffs, NJ, 1989, pp. 193-205.
14. Forsyth, R.A., "Alternative Earth Reinforcements," Symposium on Earth Reinforcement, ASCE, New York, NY, 1978, pp. 358-370.
15. Gould, J.P., "Earth Retaining Structures - Developments Through 1970," Design and Performance of Earth Retaining Structures, ASCE, New York, NY, 1990. (found in these proceedings)
16. Grant, W.P., "Performance of Columbia Center Shoring Wall," Proceedings, 11th International Conference on Soil Mechanics and Foundation Engineering, Vol. 4, San Francisco, CA, 1985, pp. 2079-2082.
17. Guilloux, A. and J.M. Jailloux, "Comportement d'un Mur Experimental en Terre Armee vis-a-vis de la Corrosion," Proceedings, International Conference on Soil Reinforcement: Reinforced Earth and Other Techniques, Paris, France, 1979, pp. 503-512.
18. Holtz, R.D., "Special Applications," Symposium on Earth Reinforcement, ASCE, New York, NY, 1978, pp. 77-97.
19. Hovart, C. and R. Rami, "Elargissement de l'Emprise SNCF pour la Desserte de Saint-Quentin-en-Yvelines," Revue Travaux, Jan. 1975, pp. 44-49.

20. Jaber, M.B., "Behaviour of Reinforced Soil Walls in Centrifuge Model Tests," Ph.D. Thesis, University of California, Berkeley, CA, 1989.

21. Jewell, R.A. and C.P. Wroth, "Direct Shear Tests on Reinforced Sand," Geotechnique, Vol. 37, No. 1, 1987, pp. 53-68.

22. Jones, C.J.F.P., "The York Method of Reinforced Soil Construction," Symposium on Earth Reinforcement, ASCE, New York, NY, 1978, pp. 501-527.

23. Jones, C.J.F.P., Earth Reinforcement and Soil Structures, Revised Reprint, Butterworths, London, UK, 1988, 192 p.

24. Jones, C.J.F.P., R.T. Murray, J. Temporal, and R.J. Mair, "First Application of Anchored Earth," Proceedings, 11th International Conference on Soil Mechanics and Foundation Engineering, Vol. 3, San Francisco, CA, 1985, pp. 1709-1712.

25. Juran, I., F. Schlosser, N.T. Long, and G. Legeau, "Full Scale Experiment on a Reinforced Earth Bridge Abutment in Lille," Symposium on Earth Reinforcement, ASCE, New York, NY, 1978, pp. 556-584.

26. Juran, I. and F. Schlosser, "Theoretical Analysis of Failure in Reinforced Earth Structures," Symposium on Earth Reinforcement, ASCE, New York, NY, 1978, pp. 528-555.

27. Kauschinger, J.L. and J.P. Welsh, "Jet Grouting for Urban Construction," Proceedings, Seminar on Design, Construction, and Performance of Deep Excavations in Urban Areas, BSCE, Boston, MA, 1989, 60 p.

28. King, R.A., "Corrosion in Reinforced Earth," TRRL Supplemental Report 457, Transport and Road Research Laboratory, Crowthorne, UK, 1978, pp. 276-285.

29. Langer, E.I., "Advanced Materials in a Changing Materials World," ASTM Standarization News, Vol. 17, No. 8, Oct. 1989, pp. 32-34.

30. Leflaivre, E., M. Khay, and J.C. Blivet, "Un Nouveauu Material: Le Texsol," Bulletin de Liason du Laboratoire des Ponts et Chaussees, Paris, France, No. 125, 1983, pp. 105-114.

31. McGown, A., N. Paine, and D. DuBots, "Use of Geogrid Properties in Limit Equilibrium Analysis," Polymer Grid Reinforcement, Thomas Telford, London, UK, 1984, pp. 31-37.

32. Mercer, F.B., K.Z. Andrewes, A. McGown, and N. Hytiris, "A New Method of Soil Stabilisation," Polymer Grid Reinforcement, Thomas Telford, London, UK, 1984, pp. 244-249.

33. Mitchell, J.K. and W.C.B. Villet, "Reinforcement of Earth Slopes and Embankments," NCHRP Report 290, Transportation Research Board, Washington, DC, June 1987, 323 p.

34. Murray, R.T. and M.J. Irwin, "A Preliminary Study of TRRL Anchored Earth," TRRL Supplementary Report 674, Transport and Road Research Laboratory, Crowthorne, UK, 1981.

35. O'Rourke, T.D., "Systems and Practices for Rapid Transit Tunneling," Underground Space, Vol. 4, No. 1, Jul./Aug. 1979, pp. 33-44.

36. O'Rourke, T.D., "Lateral Stability of Compressible Walls," Geotechnique, Vol. 37, No. 2, 1987, pp. 145-149.

37. O'Rourke, T.D., "Predicting Displacement of Lateral Support Structures," _Proceedings_, Seminar on Design, Construction, and Performance of Deep Excavations in Urban Areas, BSCE, Boston, MA, 1989, 36 p.

38. O'Rourke, T.D., E.J. Cording, and M.D. Boscardin, "The Ground Movements Related to Braced Excavations and Their Influence on Adjacent Buildings," _Report DOT-TST 766T-23_, U.S. Department of Transportation, Washington, DC, Aug. 1976.

39. O'Rourke, T.D., S.J. Druschel, and A.N. Netravali, "Shear Strength Characteristics of Sand-Polymer Interfaces," _Journal of Geotechnical Engineering_, ASCE, Vol. 116, No. 3, Mar. 1990.

40. Parkinson, J.J. and C.M. Gilbert, "Design and Construction of Slurry Walls," _Proceedings_, Seminar on Design, Construction, and Performance of Deep Excavations in Urban Areas, BSCE, Boston, MA, 1989, 29 p.

41. Peck, R.B., "Fifty Years of Lateral Earth Support," _Design and Performance of Earth Retaining Structures_, ASCE, New York, NY, 1990. (found in these proceedings)

42. Shen, C.K., S. Bang, K.M. Romstad, L. Kulchin, and J.S. DeNatale, "Field Measurements of an Earth Support System," _Journal of the Geotechnical Engineering Division_, ASCE, Vol. 107, No. GT12, Dec. 1981, pp. 1625-1642.

43. Task Committee on Design, "Structural Plastics Design Manual," _ASCE Manuals and Reports on Engineering Practice No. 63_, ASCE, New York, NY, 1984.

44. Terzaghi, K., "Anchored Bulkheads," _Transactions_, ASCE, Vol. 119, 1954, pp. 1243-1280.

45. Vidal, H., "La Terre Armee," _Annales de L'Institute Technique du Batinent et Travaux Publics_, Vol. 19, July-Aug., 1966, pp. 223-224.

46. Vidal, H., "The Development and Future of Reinforced Earth," _Symposium on Earth Reinforcement_, ASCE, New York, NY, 1978, pp. 1-61.

SELECTION OF RETAINING STRUCTURES:
THE OWNER'S PERSPECTIVE
by: Richard S. Cheney[1]

ABSTRACT: An owner's selection of a particular type of earth retaining structure is motivated by various factors including, in common order of importance, aesthetics, cost, risk and durability. In the past 20 years, development of innovative technologies for earth retention have complicated the selection process and changed the priorities of many owners. The prudent owner must now establish a rational process for technical evaluation of innovative wall types. A suggested evaluation process is outlined for owners as well as owner consideration of certain limitations or advantages of innovative earth retaining systems.

INTRODUCTION

Prior to 1970, the predominant types of earth retaining walls for permanent structures were gravity and cantilever. Both wall types had decades of successful use in both cut and fill situations. Accepted design methodology existed which had been documented by comparing predicted to actual performance. In fact, the theoretical and physical aspects of such walls were so well known that standard design sheets were available for a wide variety of backslope, foreslope or wall face geometrics. Reliable standard specifications and experienced inspectors were available to insure proper construction of conventional walls, even by an inexperienced general contractor. Furthermore, the owners could easily and accurately predict the bid cost of such common retaining structures from previous years weighted average bid prices. Post-construction monitoring was only done in unusual situations where safety or litigation problems were expected. Owners were free to focus on the aesthetic qualities of different facing materials, secure in the knowledge that a reliable, safe earth support system would be installed.

Selection of a wall system in 1990 is considerably more complex than in 1970. An ever growing number of innovative alternates are now available to the conventional gravity and cantilever walls.

These alternate systems generally rely on either reinforcement of the earth mass to be retained or mobilization of the physical

1 - Highway Engineer, Bridge Division, Office of Engineering FHWA-HNG-31, 400 7th Street, S.W., Washington, D.C. 20590

properties of the ground. The introduction, and now widespread use, of these attractive, cost effective alternate retaining wall systems has renewed the need of the owner to insure the reliability and safety of the system.

As before, the owner's objectives for a completed earth retaining structure are a pleasing architectural facade, low cost, a reasonable safety factor against failure and low maintenance cost over a long life. However, while the owner's major consideration for conventional walls was exterior appearance, the major concern in selecting an alternate wall must be a comprehensive technical assessment of both the design basis and construction procedures. The consideration of alternate, often proprietary, wall types requires complete changes in the owner's philosophy of design, contracting, inspection, and performance monitoring for earth retaining structures. Motivating such changes in owner philosophy are documented cost savings of up to 40 percent achieved on projects where bidding of alternate wall designs is permitted (1). An example is the Interstate Route 75 project in Atlanta, Georgia where a 35 foot [10.7 m] high wall was required as part of a major highway widening. The Georgia Highway Department permitted contractors to bid either a permanent ground anchor wall design or a conventional cantilever wall design. The bid results, shown below for the 18,000 square foot [1672 sq. m] wall, indicate that only one bidder selected the conventional wall alternate.

Table 1 - Interstate Route 75 Alternate Wall Costs

Bidder Rank	Total Project Bid (Millions)	Alternate Wall Bids	
		Anchored	Conventional
1	$ 61.5	$ 725,000	
2	$ 63.3	$ 558,000	
3	$ 63.5		$1,135,000
4	$ 65.1	$ 550,000	
5	$ 65.3	$ 800,000	
6	$ 69.3	$ 735,000	
7	$ 74.6	$ 700,000	
8	$ 83.2	$ 857,000	

Engineer's estimate = $1,111,000

Owners must exercise extreme caution when evaluating new systems that have little performance history. A rigorous technical evaluation should be mandatory before accepting any new system. Even those new systems found theoretically acceptable should only be constructed with provisions for detailed monitoring of both construction and post construction performance. The monitoring of a particular system should continue through a sufficient number of projects to verify acceptable performance before approving system use. Adequate data on performance history permits the informed owner to make intelligent decisions on optimal wall type(s) for a particular project. Public agencies such as the Federal Highway Administration have instituted technology transfer efforts on selection procedures for alternate walls (2).

WALL SELECTION FACTORS

Individual wall types have specific features which impact selection by owners. The first step in wall type selection is a general knowledge of which wall types are best suited for either fill situations or cut situations. A partial listing follows:

Fill Retention	Cut Retention
Mechanically Stabilized	Permanent Ground Anchor
(extensible or inextensible)	(tieback)
Crib	Master Pile
Metal Bin	Nailed
Gabion	Reticulated Micro-Pile
	(Pali-Radice)
Gravity	Slurry
Cantilever	Cylinder Pile
Counterfort	Lagged Soldier Pile
Deadman Anchor	Sheet Pile

General reports describing these wall types include USDA Forest Service Retaining Wall Design Guide (12), Reinforced Soil Structures (13), and Reinforcement of Earth Slopes and Embankments (6).

Each of the above wall types may encompass several proprietary wall names. For example, inextensible mechanically stabilized includes among others the following trade names: Reinforced Earth, Retained Earth, Reinforced Soil Embankment, Mechanically Stabilized Embankment, and Georgia Stabilized Embankment. Although categorization of the wall type is not always possible from the trade name, a cursory review of the design basis for the wall will disclose the type.

Hybrid Wall Types. Occasionally hybrid wall types are proposed which embody the characteristics of both fill and cut wall types. These hybrid wall types require special attention by the owner. Systems designed for cut support characteristically require much smaller strains to mobilize the restraint mechanism than systems designed for fill support. Attachment of multiple systems with incompatible strain characteristics to a common wall face can result in overstress of the low strain elements and damaging differential movement of the wall face.

One example of this problem is a 34 foot [10.4 m] high by 200 foot [61 m] long hybrid retaining wall built in 1975 to preserve a pioneer cemetery near Hope, Idaho. The wall was designed with a flexible face of precast reinforced concrete segments (stretchers) interconnected with vertical steel rods. Lateral earth pressures were resisted by a hybrid system consisting of rock anchors in the lower 12 feet [3.7 m], a gravity bin wall in the middle 8 feet [2.4 m] and deadman anchors in the upper 14 feet [4.3 m] (See Figure 1). Following construction, cracks were noted in stretchers in the lower 12 feet [3.7 m] and the wall face above this level had bulged. Local resident's concern for the safety of

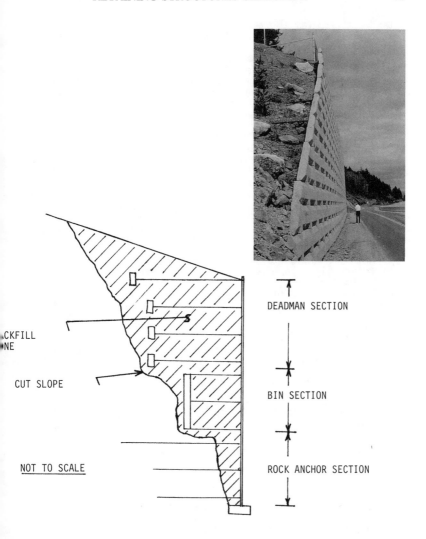

CKFILL
NE

CUT SLOPE

NOT TO SCALE

DEADMAN SECTION

BIN SECTION

ROCK ANCHOR SECTION

FIGURE 1 - HOPE, IDAHO HYBRID WALL

the wall prompted an analysis of wall conditions. The actual pressure distribution behind the wall was not measured. However, outward movement of the middle and upper wall section apparently reduced pressures in those sections to active values. These movements stabilized shortly after completion of wall construction, affecting both sections' appearance but not their stability. At rest pressures developed in the lower rock anchor section where no outward movement occurred. These high stresses, when transmitted to the face panels, caused break-out of the reinforcement in the stretchers. The $170,000 repair consisted of constructing a 12 foot [3.7m] high, 14 inch [0.4m] thick, permanent ground anchor wall against the lower section. The repair provided both structural stability and added width to the base which deemphasized the top overhang. Most hybrid walls combine different restraint mechanisms in terms of the extensibility of the restraining member which is necessary to mobilize the designed degree of restraint. Prestressed anchors require little deflection to mobilize resistance whereas extensible reinforcement elements such as geotextiles require substantial elongation to mobilize tensile resistance. Combining both elements in a hybrid wall would require consideration of differential movement of the wall facing panels.

Hybrid walls are often proposed in cut-fill situations. In such situations the potential for settlement of the wall panel members should be investigated. Internal fill settlement can produce non-design shear forces in rigid reinforcement members. Similarly, panel tilting will produce undesirable bending moments at rigid connections as well as changes to wall features. The characteristics of hybrid wall systems should be carefully studied on each project by owners. Particular attention should be given to non-uniform face distortions inherent in the system. Such distortions although not structurally damaging, as in the above example, should be considered in relation to the aesthetic appearance desired by the owner.

Aesthetics. A pleasing architectural facade is not only desired by owners but mandated by environmental impact statements on sensitive projects. Alternate retaining systems offer a wide variety of facing elements. Materials vary among steel, concrete and wood. Concrete facing can be poured, precast or pneumatically projected. Precast panels are available in a wide range of both shapes (from rectangular to cruciform) and face textures. Innovative wall types even permit plantings between individual face elements. In general, the owner can choose from a wide variety of aesthetic facades as long as the system design basis is not affected.

Aesthetic modifications to alternate wall types should be done with care. An owner can unintentionally complicate alternate wall construction by selection of non-conforming face elements. For example, mechanically stabilized systems are often selected because individual wall elements are able to articulate within the face to withstand differential wall settlements. Standard sizes of precast modular panels used in mechanically stabilized systems commonly range from 20 to 40 square feet [1.9 - 3.7sq.m]. On a few projects the architectural desirability of the repetitive joint pattern of these modular panels has been questioned. Full height panels exceeding 100 square feet [9.3 sq.m.] have been substituted to eliminate horizontal joints in the mechanically stabilized wall face. Unfortunately the owners involved with these projects failed to recognize that the smaller modular panels served important engineering functions as discussed below:

(a) When placed over compressible soils, modular panels which articulate reduce both the chance of panel cracking and the stress level in the reinforcement. However, small displacements between a rigid full height panel and the stabilized mass may induce stress levels in the reinforcement in excess of design values.

(b) The designed state of stress for mechanically stabilized walls is based on yielding of the modular panel system to produce a relatively uniform distribution of pressure behind the wall face. Full height panels may alter both the assumed state of stress in the stabilized soil mass and the internal failure mechanism which both depend on horizontal deformation of the face.

(c) Erection procedures for full height panels are difficult and uncertain to produce the designed alignment. Once the panel is set and backfilling begins, misalignment problems due to over or under compaction cannot be corrected.

Surveys of full height panels have found distorted panels to be concave, convex or out of plumb in either direction. Owners should carefully consider ramifications of substituting full height panels for modular panels on mechanically stabilized systems. Based on performance to date, full height panels should not be considered when post construction wall settlements are predicted to exceed 2 inches [51mm].

Cost. Retaining wall costs are frequently stated per square foot of wall face. The major elements in that cost are wall materials, erection, and, for fill retention systems, select backfill. The cost of wall materials and erection are relatively predictable for a particular wall system. However, backfill costs can vary dramatically depending on the backfill quality desired and its local availability. In general, the cost of backfill decreases as the allowable percent of backfill passing the No. 200 sieve increases. Whenever possible the owner's engineer should write realistic specifications to permit the use of local backfill

Determination of the actual cost of innovative walls is a two phase problem. The initial cost can be determined by bidding the innovative walls in competition with previously used wall systems until a sufficient data base of cost is developed. Initial costs can also be estimated from documented use of the innovative wall type. Such cost history is available from public agencies or from technical papers. Table 2 contains average costs in the Northwest United States (3).

Table 2 Average Wall Costs in Northwest U.S.
(Reprinted with permission of R.G. Chassie, 1988)

	Average Cost in Dollars per S.F. of Wall Face
Wall Type	
Reinforced Earth	$18 - $22
VSL Retained Earth	$18 - $22
Wire Wall	$13 - $20
Geotextile Walls	
w/o Permanent Facing	$ 6 - $12
w/ 3" (76mm)Shotcrete Face	$15 - $25
Tensar Walls	
w/o Permanent Facing	$15 - $25
w/ Permanent Facing	N/A
Metal Bin Wall (H=10'[3m] to 30'[9m])	$20 - $40
Gabion Wall (H=6'[1.8m] to 21'[6.4m])	$18 - $35
Doublewall	$20 - $25
Criblock	$12 - $25
Permanent Tieback Walls	$45 - $60
Soil Nailed Walls	
(U.S. Experience is Limited)	
Temporary	$17 - $25
Permanent (Shotcrete Face)	$25 - $30
Permanent (Architectural Facing)	$35 - $50
Cantilever Soldier Pile	
w/Wood Lagging (H=5'[1.5m]to[4.6m]15')	$12 - $18
Cantilever Sheet Pile Wall	
(H=5'[1.5m] to [4.6m]15')	$12 - $18
Concrete Cantilever	
(H=6'[1.8m] to 30'[9.1m])	$25 - $40

NOTES:
1. Average prices given above are total in-place costs including wall materials, erection, and select wall backfill. For MSE walls, the select backfill is the backfill contained within the reinforced volume. Since excavation quantities can vary greatly job-to-job, excavation cost is not included.
2. Unless otherwise noted, the prices given cover walls in the 10'[3m] to 40'[12m] height range.
3. Alaska prices are higher.
4. Not all systems listed above are marketed in all states.
5. N/A denotes no cost experience to date in Region 10 [Northwest U.S.].

Long term costs of alternate wall systems are difficult to assess when both monitoring and maintenance data are only available for a small percentage of the system's designed life. The owner must rely on the engineering staff to review the longevity of innovative designs. Estimation of the design life and a long term cost is more difficult for wall systems constructed with materials other than concrete, steel and wood. New wall materials such as plastics, fiberglass, and other geosynthetics require in-depth study of available performance history before approval. The engineers who evaluate these products should be familiar with the performance history of both the system under consideration and the conceptually similar systems. Current technical guidance can be obtained by contacting involved technical groups such as the AASHTO - AGC - ARTBA Joint Committee on New Materials[2] for the latest technical information and specifications. Owners are encouraged to track and publish information on the maintenance costs of innovative wall systems.

Durability. An owner must consider the design life of retaining structures in the selection process. Commonly, earth retaining structures are designed for a minimum life of 75 years, or if critical structures, 100 years. Conventional structures have performance history documented to beyond their design lives. Such historical documentation is not available for alternate retaining systems. The owner should understand that durability includes both resistance against physical damage during construction and corrosive deterioration.

The selection process should include an assessment of the durability of alternate wall components to withstand construction activities. The owner can obtain information on product durability from independent organizations. For instance, the Geosynthetic Research Institute has recently published a report (4) which contains both the results of constructibility tests on various extensible reinforcements and the recommended design safety factors to account for durability in construction. Often special provisions are required in the construction documents when extensible reinforcement materials are permitted. These provisions may restrict either construction procedures or materials which have been observed to damage the reinforcements. Approval of non-durable systems should be contingent on inclusion and enforcement of special provisions for protection of the reinforcement during construction.

Post construction durability of alternate wall systems has received substantial study. Numerous research efforts provide short term observations of many alternate systems (the oldest mechanically stabilized structure under study is about 25 years old). The results of such studies provide owners with conservative guidance on selection of durability criteria for various systems. For example, an extensive research study of epoxy coatings for steel reinforcement elements was completed in France by Terre Armee Internationale (5). The results indicated that if epoxy coating was selected for corrosion protection, a minimum epoxy coating thickness of 18 mils was necessary for the coating to survive transportation, installation and still provide an acceptable level of safety. Similar guidance for other alternate systems can be found in other publications (6,7,8,9).

2. AASHTO Subcommittee on New Material, Douglas Bernard, Secretary, FHWA-HHO-40, Washington, D.C. 20590.

Risk. An assessment of risk is necessary by the owner prior to wall type selection. Safety aspects in both design and construction including impact on adjacent facilities or utilities must receive consideration. Particular wall types are better suited to certain subsurface conditions than others. To assess the applicability of individual systems an adequate subsurface investigation must be completed at the project site. This investigation should provide the necessary factual data to accept or reject various wall types.

Most conventional wall design is completed in-house by the owner's engineering staff. However, alternate wall design is usually done by the supplier of the proprietary wall system and submitted to the owner for approval. To mitigate risk and to equate the design of different alternate wall systems the owner must establish formal guidelines for the design of alternate walls. Such guidelines must clearly separate the design risks to be handled by both the owner and the wall supplier. It is recommended that the owner be responsible for long term stability of the ground surrounding the retaining structure. Several spectacular failures involving alternate earth retaining systems have occurred in recent years. Subsequent analysis has determined latent subsurface conditions, not involved with internal wall stability, caused the failures.

In 1977, a failure occurred on a highway project near Coos Bay, Oregon involving a Reinforced Earth Wall. Reports of the failure (10, 11) indicated the wall top moved out 18 feet [5.5m] and dropped 12 feet [3.7m] while the wall base translated 23 feet [7.0m] (see Figure 2). Additional investigations disclosed that the failure surface was along a thin inclined seam of soft clay in the natural deposits below the wall base. This seam had not been found in the owner's subsurface investigation for the wall. Further study concluded that the failure had nothing to do with the alternate wall type selected. Failures of this type can cause an owner to eliminate a particular, or all, alternate wall systems from future consideration. To the credit of the Oregon owner, post failure investigations were made which clearly showed the wall system did not contribute to the failure.

An owner should be responsible for assigning the degree of risk to be used in design of the wall system. For example, the owner should establish design criteria such as minimum safety factors for overturning and sliding, design life, maximum allowable bearing pressure, depth of embedment, and earth pressure magnitude and distribution. The owner should predicate consideration of an alternate wall system on a supplier providing a design which conform to the stated criteria.

The owner's decision to permit supplier designed alternate wall systems should be made only when an in-depth technical review can be accomplished by the owner's staff. A comprehensive review requires experienced structural and geotechnical engineering staff versed in design and construction of alternate walls. The owner should establish a realistic timeframe for staff review. An in-depth review of alternate wall designs can take several

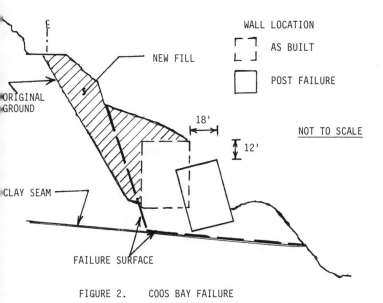

FIGURE 2.   COOS BAY FAILURE

days assuming all needed information is furnished by the supplier. Owners should never select alternate wall types without performing a comprehensive technical review of the design.

During construction, owners are frequently faced with proposed substitutions to previously selected wall types. Usually the substitute system offers attractive economic benefits to both owner and contractor. The key factor for the owner to assess is the risk of the substitute system versus the risk of the originally selected system. A comprehensive technical review must be done for the proposed system in addition to comparisons of risk elements between the selected system and the substitute system.

Many owners do not permit post bid substitutions due to the pressures of keeping the construction on schedule while allowing time to perform adequate review and make an intelligent decision. Owners can substantially alleviate these pressures by adopting formal procedures for general approval of wall systems. Early review and acceptance procedures permit a fair appraisal of system risk.

FORMAL REVIEW AND ACCEPTANCE PROCEDURES

The owner's selection of an earth retaining structure type must be based on a rational review and acceptance procedure. The acceptance of many alternate wall systems has been stymied for a number of reasons, including incorrect perceptions of how the system functions, rumors of unsatisfactory performance of a particular wall system, misapplication of wall type, poor specifications, lack of specification enforcement, inequitable bidding procedures, and inconsistent selection, review, and acceptance practices on the part of the owner. Although the actual causes of each particular problem are unique, the lack of formal, documented procedures which address the design and construction of earth retaining structures have repeatedly been an indirect cause.

To permit equitable assessment of alternate walls the owner must establish rational procedures to evaluate both the technical viability of the system and the system's constructibility/performance under varying project specific constraints and design criteria. The review process may have several stages depending on the owners familiarity with the particular system under study. General initial system approval guidelines are as follows:

(a) A supplier requests in writing to be placed on the list of approved alternate retaining walls.
(b) The owner considers approval of the system and the supplier, based on the following factors:
    o   The supplier's engineers ability to provide both rational documentation and timely responses to permit technical appraisal of the system.
    o   The supplier's plant capacity is sufficient to supply the necessary wall components in a timely fashion.

o   The system has a sound theoretical and practical basis
    for the owner's engineers to evaluate its claimed
    performance.
o   The constructibility and past performance of the proposed
    system is documented.

For this purpose, the supplier must submit a package which
satisfactorily addresses the following items:

(a) System theory and the year proposed.
(b) Where and how the theory was developed.
(c) Laboratory and field experiments which support the theory.
(d) Practical applications with descriptions and photos.
(e) Limitations and disadvantages of the system.
(f) List of users including names, addresses, and phone numbers.
(g) Details of wall elements, analysis of structural elements,
     design calculations, factors of safety, estimated life,
     corrosion design procedure for soil reinforcement elements,
     procedures for field and laboratory evaluation, including
     instrumentation and special requirements, if any.
(h) Sample material and construction control specifications
     showing material type, quality, certifications, field
     testing, acceptance and rejection criteria, and placement
     procedures.
(i) A well documented field construction manual describing in
     detail, and with illustrations where necessary, the step-by-
     step construction sequence. Copies of this manual should
     also be provided to the contractor and the project engineer
     at the beginning of wall construction.
(j) Typical unit costs, supported by data from actual
     projects.

This submission should be thoroughly reviewed by the owner's
geotechnical and structural engineers with regard to the design,
construction practicality, and anticipated performance of the
system. The owner's position on the submission, i.e., acceptance
or rejection, with technical comments should be provided by a
written notification to the applicant.

After all technical comments are resolved and initial approval
is recommended, the owner should be advised by the engineering
staff to place the new earth retaining structure on either an
experimental list or a general approval list. The experimental
designation usually attaches a few temporary limitations or
special requirements to the use of the system, such as:

(a) The owner requires performance monitoring of the system
     by his engineers.
(b) The retaining system will only be permitted for routine,
     low risk applications.
(c) The total number of installations will be limited.

Regarding performance monitoring, the owner may require the
contractor/supplier to provide instrumentation of the wall
elements or the surrounding ground. The extent of the
instrumentation, which would be determined by the owner's engineer
based on specific project needs, would likely be greater on the
initial projects and lesser as more installations occur. The
contractor/supplier would need to include the cost of the

instrumentation in the overall bid for the wall. This overall cost for the new system would be compared against the cost of other retaining structure types by the owner to determine if the new wall should be selected for a specific project. The owner's engineer would be responsible for data acquisition, evaluation and reporting the results to the owner.

The owner's philosophy in including the instrumentation cost in the new system cost relates to several items. First, the owner is taking additional risk in using a new system. Second, the owner will pay his engineer to evaluate the system. Third, the contractor/supplier benefits both from usage of the new system and the documented system performance data. Last, the motivation for the owner to even consider a new system is economy; if the system plus instrumentation cost is not less than other proven retaining systems, the new system probably is not worth considering.

The general approval list will contain wall systems that the owner deems to perform satisfactorily. This does not connote unrestricted use of every system on the list. Wall systems on the general approval list may have specific restrictions to their use such as maximum permitted wall height. Contractors/suppliers may ask to extend the use of a restricted system. In such cases the owner would require both the submission of supplemental documentation and the performance monitoring plan required for initial system approval.

All feasible, innovative, cost-effective alternates must be seriously considered by the owner to maximize the potential for cost savings.

A note of caution to small private owners who build walls infrequently and cannot undertake formal review and acceptance procedures. These owners must restrict their usage of innovative wall types to those that have a proven track record in their local geographical area. Design-construct contracts with contractor prequalification requirements should be employed by small private owners to insure reliable innovative wall performance. Maximum height of the innovative wall should be limited to that previously constructed by the contractor.

WALL SELECTION

The owner's planning for a specific project begins with a review of the approved wall lists and an assessment of which wall types best fit the site. Some factors which are routinely considered in selection are site risk potential, durability requirements, aesthetics and cost. The question of site design constraints and contracting procedures must be carefully evaluated. If several alternate wall designs are to be permitted, a framework of technical criteria must be established to insure all wall designs are made on a common basis and result in the desired end product for the site.

Finally, the wall selection will be made considering: cut or fill earthwork situation, size of wall area, average wall height, foundation conditions (i.e. would a deep or shallow foundation be appropriate for a cast-in-place concrete retaining wall), availability and cost of select backfill material, cost and availability of right-of-way needed, complicated horizontal and vertical alignment changes, need for temporary excavation support systems, maintenance of traffic during construction, and aesthetics.

CONCLUSION

The rapid emergence of alternates to conventional earth retaining systems have posed difficult problems for owners. While recognizing the need to adapt to the new technology, owners have encountered problems in differentiating between sound and unsound designs.

Owners retain the same objectives in regard to wall selection that were prevalent in the past: aesthetics, cost, safety and durability. Alternate retaining systems can be compatible with the owner's objectives once the strengths and limitations of individual systems are known. The key factor in the owner's acceptance of an alternative retaining system is personal, positive experience in using the system. However initial uses of any system must be controlled to assure safety. Included in this paper are the owner's perception of what data should be furnished to permit equitable alternate designs for dissimilar earth retaining structure types. Formal review and acceptance procedures are needed both to protect owners and to provide suppliers of soundly based innovative walls a fair chance to market their system.

REFERENCES

1. R. S. Cheney, "Permanent Ground Anchors Demonstration Project
    68 Final Report" FHWA-HHO-42, Washington, D.C. 20590,
    February 1990.

2. J. L. Walkinshaw, "Handout on Retaining Wall Alternates",
    FHWA- HST-09, San Francisco, CA 94105 April 1989.

3. R. G. Chassie, "Basic Primer on Retaining Wall Alternates",
    Alaska Department of Transportation, Retaining Wall
    Seminar p. 21, March 1988.

4. R. M. Koerner et al, "Installation Damage Assessment of
    Geosynthetics in Retaining Wall Applications", Geosynthetics
    Research Institute, November 1988.

5. J. M. Jailloux, "Organic Coatings for the Corrosion
    Protection of Reinforcing Strips (Synthesis)", Terre Armee
    Internationale, May 1984.

6. J.K. Mitchell, W.C.B. Villet, et al, "Reinforcement of Earth
    Slopes and Embankments", NCHRP Report 290, June 1987.

7. V. Elias et al, "Durability/Corrosion of Reinforced Soil
    Structures " FHWA/RD 89/186, July 1989

8. D. E. Weatherby, "Tiebacks", FHWA/RD/82/047, July 1982.

9. V. Elias, I. Juran, "Manual of Practice: Soil Nailing",
    FHWA/RD/89/198, October 1989.

10. D. P. McKittrick, "Soil Reinforcing and Stabilizing
    Techniques in Engineering Practice" Proceedings New South
    Wales University and Institute Technology Joint Symposium,
    Sydney Australia, October 1978.

11. R. G. Chassie, "Landslide Tests Reinforced Earth Wall",
    Highway Geology Symposium, Portland Oregon, August 1979.

12. D.D. Driscoll, "Retaining Wall Design Guide", U.S.D.A.
    Forest Service/Foundation Sciences Inc., Portland, Oregon
    December 1979.

13. B.R. Christopher, S.A. Gill, et al, "Reinforced Soil
    Soil Structures" FHWA RD 89/043 November 1989 (2 Volumes)

A Contractor's Perspective On Wall Selection And Performance Monitoring

By Harry Schnabel Jr.[1], M.ASCE

ABSTRACT: A discussion by an engineer, who is also a contractor, on modern walls. These new structures are rapidly replacing conventional concrete cantilever walls. They involve new concepts of structural analysis. All of them depend on soil/structure interaction. The differences between a mechanically stabilized wall, which is used to retain fill, and a tiedback wall, which is used to retain earth in-situ, are discussed. Also discussed is the structural function and analysis of key structural elements. The author also presents his claim that tiedback walls, which are dependent on good workmanship are best built by an experienced design/build contractor. The types of performance monitoring being used to understand the mechanisms involved and monitor wall performance is discussed. Then he directs his attention to thoughts on innovation.

INTRODUCTION

Twenty years ago the ASCE sponsored a conference at Cornell on earth retaining structures. No mention was made of reinforced earth and tiebacks were mentioned only incidentally in one slurry wall paper. As I drove here I passed a hundred reinforced earth walls. Surely they are one of the most significant developments in this field in the last twenty years. Also, earlier in our program many, many tiedback walls have been discussed. There is no need to discuss their acceptance as the preferred method of shoring for deep building excavations, which is the other most significant development. At that conference, the emphasis was on the lateral earth pressure, and especially computerized methods of computing it, yet little or no improvement has been made in the accuracy of our calculations. Also featured was a carefully instrumented subway excavation which several well known engineers analyzed. The design was unconstructable, which was not recognized, therefore, all of the computed values were incorrect, which the measurements confirmed. I say this only to emphasize that we did not recognize the significant innovations twenty years ago. That should make us humble. Nevertheless, I will predict a great many tiedback highway walls in the next twenty years if we successfully overcome a few problems.

In the next twenty years tiebacks should be used on most highway walls built in cuts. Mechanically stabilized walls, the generic for reinforced earth, will be used almost exclusively where the walls retain fill. All of the tiedback walls you have seen earlier

---

1 - Chairman Of The Board, Schnabel Foundation Company, Sterling, VA 22170

were temporary, that is, they were built to shore an excavation. When the building is built in that excavation, the wall is no longer needed. Probably the walls were not suitable for long term use. But, by addressing the short comings of the temporary wall, it can be made suitable for long term use. Already it is being done, but I predict it will be so advantageous it will soon become normal. When that happens, the extra cost of cantilevered walls will seldom be justified. Instead, the highway engineer will select a tieback or a reinforced earth wall, and the selection will be based on whether it's to be built in a cut or a fill. Many tieback highway walls are already being built, so let's first consider the present state of the art.

TYPICAL APPLICATIONS FOR HIGHWAY WALLS

Most of the tieback walls are being built as part of a depressed roadway, to retain a side hill cut or a landslide, or to modify an abutment. All of them have certain elements in common, such as soldier beams, lagging and a concrete facing, but the new, unique element is the tieback. We can now make tiebacks in most soils, make them last, and make them for reasonable capacities. All of that capability was first developed for the temporary walls, then extended to highway walls.

Figure 1. Depressed Highway.

The first typical wall is in Richmond, VA, and is for a grade separation. In this case one roadway was excavated so that it passes under a bridge which carries the other roadway. The excavation is typically only about twenty feet deep (6.01 m). Figure 1 shows both the poured concrete facing wall, and what the wall looks like without the concrete. The facing wall looks like any other poured in place wall. The soldier beams, lagging and tiebacks can all be seen before the concrete is poured, and look much like the shoring for a building excavation.

Figure 2. Landslide.

The purpose of the second typical wall was to correct a landslide. Driven soldier beams were used together with tiebacks and wales. Wood lagging was installed between the soldier beams as excavation progressed to the bottom of the wall. Then precast concrete panels were erected as shown in Figure 2. The space behind the panels was backfilled with gravel. All of the lateral pressure on the wall was resisted by tiebacks which were anchored beyond the slip surface.

Finally, Figure 3 shows an abutment being modified. The abutment is part of a bridge over the mainline of the Baltimore and Ohio Railroad. A rapid transit line was built alongside the railroad, and it was necessary to excavate the slope which went from one pier to an abutment, and build a new wall in front of the abutment. In this case, the soldier beams under the bridge were placed in hand dug pits, and the tiebacks were driven between the piles for the abutments, but the same soldier beams, lagging, tiebacks and facing elements are combined to build the wall.

These walls are not difficult to grasp in concept, but the selection of the various elements, and how they are combined, will govern their effectiveness and their cost.

Figure 3. Abutment Wall.

BASIC ELEMENTS OF TIED BACK WALLS

Tiedback walls have four main elements, the soldier beams, the lagging, the tiebacks and the facing. Soldier beams can be piles, caissons, etc. which are spaced along the wall. The earth between the soldier beams is retained by lagging. The tiebacks provide a force which resists the resultant force of the lateral earth pressure. Finally, the wall is completed by facing it with concrete. Tiebacks are the unique element which makes these walls possible. We are still learning better ways to make them, and ways to incorporate them with the rest of the wall.

Most of the walls are excavated to the bottom using the soldier beams, lagging and tiebacks before the concrete facing is added. Soldier beams, lagging and tiebacks permit many excavation bracing walls to be built in situ, and the permanent wall should be built in a similar way.

There is a great deal of literature on tiebacks and tiedback walls which can mislead an engineer who relies on it. Figure 4 is a section through the first wall shown, the one in Richmond, VA. These tiebacks are made in clay, and will hold as long as the wall lasts, yet it has been written that permanent tiebacks are not used in clay. All of the soldier beams were driven, yet it has been written that all of the soldier beams must be drilled to achieve the required tolerance. The wall moved less than a half inch, yet it has been written that tiedback walls move more than braced walls. One of the difficulties with design is that much of what is written is misleading, which often prevents the use of the appropriate element.

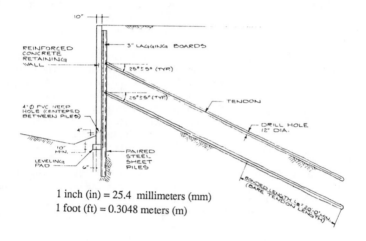

1 inch (in) = 25.4 millimeters (mm)
1 foot (ft) = 0.3048 meters (m)

Figure 4. Section Through Wall.

In recent years, the cost of tiebacks has declined. Most of the impetus for this cost reduction has come from the extensive use of tiebacks for building excavations. In the same way that computers have become more powerful and less costly, tiebacks have become stronger and less costly. Most of these cost reductions are the result of proprietary developments, so it is important to consider how we can use the proprietary products if we want the cost benefit. We engineers have been doing that for years. We recently celebrated the building of the Brooklyn Bridge by Roebling. As engineers we know the principle of suspension bridges, but converting that knowledge to a bridge over the East River required a Roebling. Earlier I referred to reinforced earth, which is a proprietary product. Most highway departments use it. So, the use of proprietary products is neither new nor unique. Tiedback walls often use proprietary methods, and the real challenge is to accept that fact and use them in the most effective way. Patents cover many of the items discussed in this paper, for instance.

In general, all of my comments about tiedback walls will apply to soil nailed walls. The tiedback wall develops its strength from an anchor outside of the critical failure wedge. In contrast, the soil nails increase the size of the failure wedge then a few nails into original soil stabilize it. In any case, either soil nailed or tiedback walls can be used for cut walls.

ADVANTAGES OF HIGHWAY WALLS

Tiedback walls are being used for highway cut walls and they have performed well. Usually they are considered as an alternative to a stem wall. Cost, time of construction, depth, required bypasses and ways of permitting proprietary systems all figure in the selection.

Tiedback walls cost less. The choice is usually between a conventional stem wall and the tiedback wall. The conventional wall is usually built by excavating, pouring a footing, pouring the stem, and backfilling. The footing must resist sliding and overturning of the wall. Often to do this, it must be on piles or drilled piers. The stem must be designed to resist the earth pressures by bending. It often must be quite thick at the bottom. Finally, earth is placed behind it so that as the earth is placed it deflects. A tiedback wall can be designed and built for substantially less money. A published report[3] of the Federal Highway Administration estimates the saving is about a third of the wall cost.

They also save construction time. They do this two ways. They can be built faster, and they often eliminate bypasses. After the soldier beams are in place, the lagging and the tiebacks are coordinated with the excavation. To construct the wall to this point requires about the same time as temporary sheeting, yet it is already performing its intended function. The wall facing can be added, at any convenient time, to complete the wall. The soldier beam and facing of a tiedback wall are only a few feet thick (less than 1.0 m), and can usually be placed in the space between the service road and the depressed roadway. As in the temporary case, they are built in situ to retain the soil outside of the excavation. This has allowed tiedback walls to eliminate bypasses.

Tiedback wall are also more efficient for deeper cuts. Since the tiebacks resist the earth pressure, the usual practice is to increase the tieback loads and the number of tiebacks as the depth increases. For conventional walls, which act as a cantilever, as the wall height increases the wall thickness increases dramatically, and the cost increment is greater than for a tiedback wall. One of the conventional walls for which we did the sheeting was a maximum of eight feet thick (2.44 m), counterforted, and thirty-five feet high (10.67 m). A tiedback wall would have been about a foot thick, (0.30 m) since it spans horizontally, and very little cost would have been added to the temporary sheeting.

Tiedback walls often involve some proprietary items. It is important to recognize this fact, and benefit from it. At the same time, it is important the proprietary systems be used in a way that does not exclude acceptable alternatives. Any patent is obviously proprietary, but all states use patented products. Usually this is done by allowing

---

3   Permant Ground Anchors (FHWA-DP-68IR) of November 1984

acceptable alternatives so that competition is not inhibited. But some items can be proprietary because special skills or equipment are needed which only one organization possesses. Allow them, but make them compete. Some of the ways this is done include soliciting designs to be included in the bid documents, allowing alternatives, or using performance specifications. All of these allow the use of proprietary items and allow the highway department to receive the cost benefit.

DESIGN AND CONSTRUCTION

Earth Pressure. A wall is built to retain the soil. Normal practice is to calculate the pressure on the wall from the earth. Then the wall elements are selected to resist those pressures. The earth pressure is one of the least certain parts of the calculation.

There are two schools of thought on the earth pressure, those who calculate it by Rankine or similar methods where the pressure increases with depth, and those who calculate it by Terzaghi and Peck or similar methods, where the pressure might be as great near the top of the wall as at its bottom. Obviously the walls would be different. The second method came into widespread use because walls designed the first way fell down. Yet today many walls are being designed each way. Often increased pressures are assumed which may compensate for the inaccurate distribution. At the 1970 conference Professor Lambe presented Figure 5[4]. The main thing to observe from Figure 5 is the wide spread of the calculated brace loads and that the preload is about equal to the measured brace loads. In other words, a load was applied by jacking the braces, and that brace load was not exceeded. I believe it is reasonable to infer that the loads which were required to resist the lateral earth pressure of the soil were less. We can of course continue to design safe structures by selecting lateral earth pressures which are very much in excess of the actual, and using that to calculate the resultant forces, as in this case. I believe a better course is to evaluate the lateral earth pressures more accurately and design accordingly. That evaluation must involve developing a theoretical understanding and then confirming that our calculations are correct. All methods cannot be equally correct. Professor Lambe observed "at the present time we cannot predict the strut loads".

Another reason the determination is difficult is that soil has most of the properties of a solid. A force can act at many locations on a solid and be effective. Thus, by measuring tieback or strut loads we may not learn the best location and magnitude of the force needed to retain the earth. We can only learn that the measured load, applied at that location, is large enough to retain the earth.

---

4   1970 Specialty Conference - Lateral Stresses in the Ground and Design of Earth Retaining Structures

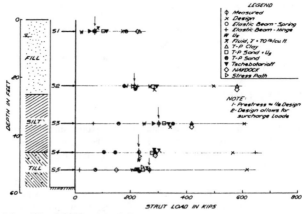

1 foot (ft) = 0.3048 meters (m)
1 KIP = 4.45 kiloNewtons (kn)

Figure 5. Strut Loads At Full Excavation - Section B.

Facing. The facing should be either cast in place concrete or precast concrete. It is designed to span horizontally between soldier beams, which are about eight to ten feet (2.44 to 3.05 m) apart. The thickness of the facing is usually calculated to be less than one foot (0.30 m). It is attached to the soldier beams. When the soldier beams are steel, welded studs are usually used. When the soldier beams are concrete, bars are generally used for this attachment. In general, the cast-in-place wall is poured against the soldier beam. Epoxy coated bars are often used for corrosion protection. Drainage is generally provided by placing prefabricated drainage panels. These are nailed to the lagging before the concrete is poured. A weep hole is provided through the wall to the prefabricated drainage panels, or a drain is connected to it.

I believe precast facing will be used more in the future. It can be made in a factory, so that better quality concrete can be assured. A wide range of architectural treatments is possible. Gravel can also be placed between the facing and soldier beams to provide drainage and resist soil freezing. We have used full height panels and small panels, textured and colored panels, and each time we have learned something to be incorporated in our next design.

Lagging. The lagging between soldier beams is not a part of the final wall. When the wall is completed, the facing is in front of the lagging. However, the facing is built after the wall is fully excavated. The soldier beams support the earth which arches between beams, but the lagging prevents the earth from falling out and destroying the arch. Shotcrete lagging is occasionally substituted for wood lagging. In either case if it retains the earth between the soldier beams until the facing is built it is no longer needed. Lagging and the earth behind it need constant attention. If the earth is displaced, it should be replaced. In any case, since the wall contractor is on the job until the facing is poured, he can replace the missing soil.

Soldier Beam The soldier beams are vertical members which are horizontally spaced along the wall. They may be driven, drilled or dug. H-piles or steel sheetpiling are often used as soldier beams. We have also drilled holes and placed channels or H-piles in the holes before backfilling with weak concrete. Other soldier beams are constructed using concrete and reinforcing steel placed in the drilled hole. In case of limited headroom, the holes may be dug by hand. In every case the soldier beam must be designed in bending. It must have at least the capacity to resist the calculated moment. They are also always designed to be in contact with the in-situ soil. In selecting the particular type of soldier beam it is very important to know what type of tieback will be used so that the selection can be compatible. Any method of installing the soldier beams will result in some deviation from the theoretical location, which must be recognized and dealt with in the design. Finally, the horizontal spacing of the soldier beams is determined by the soil and the method of lagging chosen.

Corrosion should not be a factor in the soldier beam design. As of this time I know of no wall failure because of soldier beam corrosion. The research that has been done on corrosion of other steel structures buried in earth would not indicate any special corrosion protection is necessary. All of us are familiar with buildings and bridges supported by H-piles which have no corrosion protection. In the completed wall the soldier pile is never exposed to air nor any kind of mechanism which would remove any protective rust which formed so it should not need any corrosion protection. The steel is also relatively mild so that stress corrosion is not a concern. All soldier beams have performed well, regardless of the type. A few of these have had some corrosion protection such as coal tar epoxy or galvanizing. Also, a few others have been furnished oversized to allow rusting. In a few cases the bars in drilled piers have been epoxy coated. These decisions to provide corrosion protection were probably influenced more by the numerous infrastructure failures because of corrosion and the absence of a long history for tiedback walls.

Tieback. The key element of the wall is the tieback. They should be close to horizontal, since they resist horizontal forces. They should be stressed against the soldier beams, and the design should consider the need for corrosion protection. There are many ways of making them, and every effort should be made to become familiar with the choices and know how they might affect the design of the other portions of the wall.

The tieback consists of a tendon stressed between the soldier beam and its anchor made in the soil behind the wall. Post-tensioning steel is used for all modern tendons. The stress is applied with hydraulic jacks acting on the soldier beam, and maintained with standard post-tensioning hardware. The tendon is usually bonded to the grout anchor. Between the soldier beam and the anchor, the tendon should not transfer load to the soil.

The wall is being built to resist lateral earth pressures. The facing and soldier beams concentrate those lateral pressures into approximately horizontal resultants which the tiebacks resist. In practice, most tiebacks are installed at some angle to the horizontal. For tiebacks where that angle is twenty or thirty degrees, a large horizontal component acts. The vertical component of the tieback exerts a thrust on the soldier beam, which is resisted by friction or soil below the subgrade. For both of these reasons, to increase the horizontal component and keep the vertical one low, it is advantageous to make the tieback as nearly perpendicular to the wall as feasible.

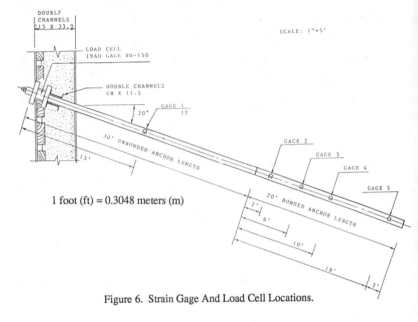

1 foot (ft) = 0.3048 meters (m)

Figure 6. Strain Gage And Load Cell Locations.

On one highway wall a load cell and five gauges were installed on a tieback, as shown in Figure 6. The load cell is of course at the head of the tieback. The first gauge was placed about in the middle of the unbonded length. The other four gauges were placed in the anchor length. All of the gauges and the load cell have been read over a long period of time, and the results are plotted in Figure 7. Note that the load cell and gauge number 1, which should agree, are in substantial agreement. Also note that the load read at gauge 2 is only about a third of that applied to the tieback, and that gauge 3 is less than a half of gauge 2. From these results, and many more similar ones, we have concluded that the rate at which the tieback load is transferred from the anchor to the soil is directly related to the movement of the anchor.

In Figure 8 we have rearranged data similar to Figure 7. The rate at which the anchor transfers the tendon load to the soil is assumed to be equal to the reduction in tendon load. The movement of any increment of anchor can be calculated if the end does not move, and we are assuming the tendon and anchor movement equal. By plotting one against the other Figure 8 was constructed. It shows two important design facts. First, it shows that the adhesion between the anchors and the earth is not a constant,

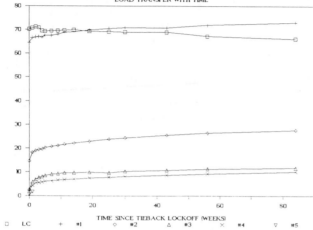

1 KIP = 4.45 kiloNewtons (kn)

Figure 7. Tieback Load Transfers.

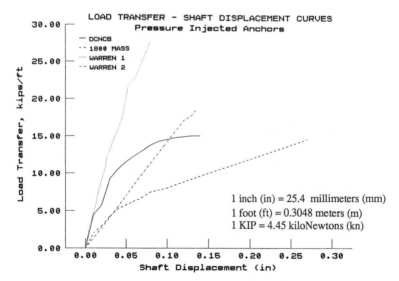

Figure 8. Measured Rate Of Transfer.

but is some function of the strain. It also shows there is a lot of adhesion. All of these anchors were three inches in diameter, so the contact area for the adhesion is only about 0.8 square feet per foot (0.07 m^2)of anchor. In other words, this grout to earth adhesion for all anchors exceeded 15,000 pounds per square foot (66.73 kN), and in one case exceeded 35,000 pounds per square foot (155.70 kN). I believe such high adhesion valves result from the use of high grout pressures which is discussed later.

When the stress in the tendon is reduced from the test load back to the alignment load, the front of the anchor does not return to its original position. This is because some strain of the tendon remains in the bonded length. Figure 9 shows the result of tests on one tendon. This residual strain results in some residual movement of the anchor which is observed in the performance test.

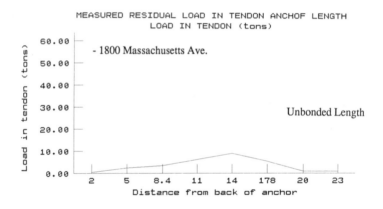

Figure 9. Measured Residual Loads.

The anchor is always made by grouting the tendon to the soil at the deep end of a hole made from the soldier beam. The simplest anchors have little or no pressure on the grout and have the lowest rate of transfer. Most anchors use grout pumped under pressure. The pressure applied to the grout is often as high as 600 psi (4,136.83 kN/m^2), but is dependent on the method of making the anchor. The pressure is also dependent on the type of soil in which the anchor is made. The best results are obtained and highest pressures are used in porous soils. The pressure creates a stress in the soil, and since water flows into the soil until the grout becomes a solid, that radial stress permanently acts on the anchor.

The other end of the tieback is referred to as the head. This end must be accessible for jacking against the soldier beam, and permanently locking the stress into the tendon. Since the tendon is post-tensioning steel, the manufacturer has plates, nuts and wedges available for this purpose.

The tendon between the head and anchor should be unbonded. This is the length that passes through soil which is being retained by the wall. For most walls, it must be at least long enough to assure the anchor is behind the critical failure plane. In the case of landslides, the anchor should be behind the slip surface. There is also a minimum unbonded length so that the wedges or nut are tightly seated. Referring again to Figure 7, the stress in this length is not exactly the jacking load. If the tendon is bent, it will transfer some small load. If the unbonded length is greased and sheathed, or if the sheath is crushed around the tendon, a slight load transfer may occur.

We test all tiebacks. The tendon strain is much more than the expected wall movement. Therefore, the strain is induced by stressing the tendon against the wall. Usually it is a simple matter to obtain a simple stress-strain curve. We then use that test as a means of verifying the anchor capacity, but in doing so also test not only the entire tieback but the ability of the structure to carry the load. Where creep is a concern, the testing can also evaluate that.

CORROSION PROTECTION

Corrosion of the tiebacks is a major design concern. The metallurgy of the tendon steel has been developed for long term dependable use. However, it is also designed to be used in concrete, which is an ideal protection. If corrosion were to occur it would probably be stress corrosion. In high strength steel of the type used for tendons, the corrosion is not on the surface of the steel, but occurs in deep cracks which reduce the cross section. Also, this generally occurs near the head, so special precautions are required there. For corrosion of the tendon to occur, a long electrolytic cell must develop. Corrosion protection is designed to prevent this by encasing the tendon in a plastic which is a non conductor, or interrupting the electrical current with an insulator.

I know of no tendon which was properly grouted and then failed in the anchor length because of corrosion. All the failures in service have been near the head, which justifies special attention to that portion of the tieback. Head protection must protect nuts, wedges and plates, and also allow for the tendon to be strained. The first I'll refer to as the anchorage protection, and the second the trumpet. Both must allow movements to occur as the tieback is locked off.

The tendon is stressed at the anchorage. Either strands or a bar passes through the plate against which they will be stressed. To provide access to tighten the wedges or nuts, a seat is placed over them on which a jack rests. No corrosion protection is possible on the tendon where the wedges or nut must grip it. So this end of the tendon is usually protected after the tieback is stressed and the excess jacking length removed. A cap is placed over the anchorage and filled with cement grout or the head is covered with concrete. The grout has excellent properties to prevent corrosion, and it will remain. I do not believe grease is a suitable protection.

Behind the plate against which the tendons are stressed they are also bare. To protect this length, a trumpet is usually welded to the plate. This trumpet bridges the bare strand, and allows grout to be pumped around the bare tendon. At the top, it is attached to the plate.   At the bottom, it overlaps the corrosion protection for the unbonded length and should be sealable.   Once again, our experience is that grout is more dependable corrosion protection than grease.

The tendon should be protected against corrosion in the unbonded length, and it should also be free to strain. Strand can be bought with factory applied grease coating and a sheath over the tendon in the unbonded length that is excellent protection. Bars usually have a plastic sleeve placed over this corrosion protection as a bond breaker. The corrosion protection can also be installed at the factory, and is usually a plastic sleeve filled with grout. We also install an epoxy directly on the bar and provide a Raychem sleeve in the unbonded length.

Of course some soils are more corrosive than others.   At this time tests should be run to determine the resistivity and chemical properties.   If favorable, the only corrosion protection needed is the grout in which the tieback is always encased.

In a short discussion like this, all design questions can not be addressed.   The development of tiebacks has recently made this type of wall possible, and as more developments take place it will become even more desirable.

PERFORMANCE

Tiedback walls have performed well. Wall movements are routinely checked with offset surveys. Occasionally movement is checked with a slope indicator. The load in the tieback can be checked with load cells, and compared to the design load. The axial stress in the soldier beam can be checked with strain gauges.

Temporary walls built in-situ deflect. This deflection occurs with little time lag as the excavation becomes deeper. Thus, the maximum deflection occurs shortly after the wall is excavated. It is usually less than .1% of the height; but may be as great as .25% of the height. A twenty foot (6.10 m) high wall which deflects .1% will move about a quarter of an inch (6.35 mm). Movement of that magnitude is detected and measured with an ordinary surveyors transit.

Slope indicators are installed on the soldier beams to show the deflections. Figure 10 shows the results of one set of readings. The movements shown are the total since the start of construction, and this particular set was taken months after the wall was completed. The maximum wall movement was about 0.2 inches (5.08 mm), .06% of the height, which for a 26 foot (7.93 m) high wall is good. Since the beam bends most where the net pressures on it are greatest, we can infer those are at the wall. We also know there is no net pressure where the beam is straight, so only the first few feet (approximately 1.0 m) of the soldier beam toe has any pressure.

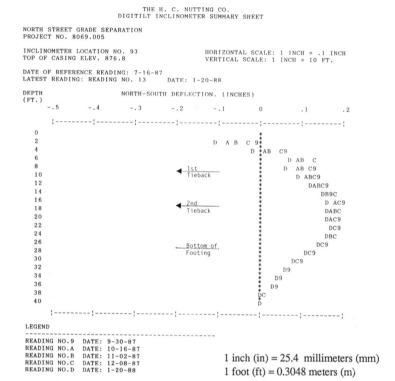

Figure 10. Inclinometer Readings.

We can also measure the tieback load. Load cells can be placed under the anchorage hardware. Earlier I stressed that there is no agreement on the earth pressures which act on a wall. Values calculated for the required tieback forces to make the wall stable will therefore vary. But we can measure the actual load and compare it with the calculated capacity. Figure 11 shows the measured load in a tieback at various depths of excavation, both plotted to the same time scale. The results show that the tieback load never increased beyond the lock-off load, nor did it increase as excavation proceeded. The maximum calculated design load was calculated for the full depth of excavation. The lock-off load was set at sixty percent of the design load. The tieback is installed and tested when the excavation is at or near the head elevation. Thus, the load at the lock-off greatly exceeds the load required for that stage of excavation. It also exceeded the load required at any subsequent stage of excavation.

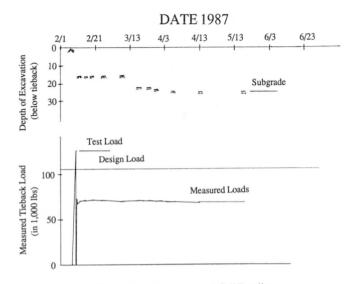

Figure 11. A Typical Load Cell Reading.

INNOVATION

In conclusion, I'd like to discuss innovation. Successful innovation results in better ways of doing things. Reinforced Earth and tiebacks are innovations. When successful innovation occurs, we see it on the jobs. What we don't see are the design discussions, the testing and modification of methods, the training, the new equipment, the development of standards, etc, that are all a part of the innovative process. The innovation is a result of these activities. I also believe innovation requires a champion, perhaps an organization, but usually an individual who will work to make it happen.

Having a champion is extremely important. He must not only see all the problems to be overcome, he must be able to visualize the innovation. But, he is only a daydreamer unless he can also accomplish and pay for the necessary work. For example, Dave McKittrick and Henri Vidal were the champions of the innovation we now know as mechanically stabilized embankments. In that case Vidal, who owned a key patent, was an obvious person to be the champion. There seems to be a champion for most innovations.

One of the advantages that most foreigners enjoy is that their governments encourage innovation by helping pay for it if they can be shown they will receive a benefit. They may benefit from cost savings, safety improvements, increased export sales or in a variety of ways.   If they expect a benefit, they help with the development of the innovation, generally by paying some of the costs. The champion must still see that the innovation is pursued.

Our contracting practices also discourage innovation.   Since most innovations involve design and construction, we divide the responsibility for successful innovation. In doing so, we certainly hamper and perhaps prevent the innovation.   Almost all of the steps in the innovation process are affected by both the design and the builder.   For example, a design may be improved by using a promising, but new method, for which men still must be trained. Who decides to use it, or not use it? If this method does not work, who shall pay to use another, more expensive method.   If the new method is proprietary, how can a designer specify it.   All of these, and similar considerations, make innovation difficult.

Our litigious society also discourages innovation.   It is best to follow accepted practice, if you are an engineer.   If you are a builder, try to build exactly what the designer specified.   Both policies discourage innovation.

I personally am encouraged by the small number of highway walls which have already being designed and built as tiedback walls.   As I said earlier, I believe this will be one of the next significant innovations in our field.   Our firm is also particularly able to combine the necessary engineering and construction skills, so I hope we will benefit from this innovation.   Most of all, I plan on being a participant in the innovation.

We now build better looking walls, where the key design element is the tieback, and build them for less money and with less inconvenience to the motorists. Yet, many states have had bad experiences.   In some cases, they have designed the work in the conventional way, and taken bids.   One wall required four years to build because the engineer's design was poor.   Another state has had such large cost overruns that the walls are not economical.   Their walls cost four times as much as those built by an adjoining state.   We certainly have not achieved the innovative breakthrough we seek when many states are reluctant to use tiedback walls for these and other reasons. Tiedback walls will be a significant innovation when they are normally used for highway walls in cut situations, and when the reinforced concrete stem wall is seldom used.

How should tiedback walls be built? I believe the design and the construction should be done by the same firm, and that firm should be in the business regularly. Because no innovation emerges complete from the mind of man, we must experiment with various facets until we are satisfied. To do that requires constant attention to both design and construction.   It often requires modifying the design based on observations.   It certainly will result in using more economical details.   But innovation is a process, as distinguished from an invention.   It will certainly involve doing things differently, but may also involve training men, developing equipment, feedback, and in general becoming proficient in those things necessary to build a safe, economical wall.   There are many ways of doing this.   Some states solicit designs for the walls from several firms and review them prior to seeking bids.   This allows the state to know well in advance what the bids will be limited to.   Also, the contractor has the same benefit. Others seek alternate bids.   In this case, a prudent bidder will want to review his proposal with the state, but much less time is allowed.

It is reasonable for the innovator to describe the innovation so the user can evaluate it. Very few of us want to use something without properly evaluating it, so I feel the burden should be on the innovator to explain the application. That may involve drawings, pictures or models. It is also essential to the process that the innovator listens to the user. He must fully consider and answer any concerns expressed. This means he must know not only how to design it, but how to build it.

Some of the engineering should be done by the owner or his representative, and some can't. For example, I quoted Professor Lambe as saying we do not know how to calculate the lateral earth pressures from the soil. Certainly that is a problem that could be addressed by civil engineering faculty and researchers. The entrepreneur however can do the engineering to integrate the best sub-systems, such as tieback methods, into the wall design.

Most of the walls which are being built are being designed by the same organization that builds them. We are one, but we have a number of competitors. All of us are able to integrate our peculiar skills into the design and construction. We benefit by prompt feedback, repeated applications which allow us to refine our techniques, and any proprietary methods we own.

SUMMARY

Tied back walls are now being built for a number of states. In general, any wall built to retain an excavated cut is a good candidate for this type of wall. They are usually less expensive than a conventional wall, at least as attractive, and often cause less inconvenience to motorists during the construction phase. Their construction requires new skills and specialized equipment. The designer must have an extensive knowledge of tiebacks and the other elements, and know the equipment and skills which are available. Often firms which build these walls combine the design and construction knowledge.

INNOVATIVE EARTH RETAINING STRUCTURES: SELECTION, DESIGN & PERFORMANCE

George A. Munfakh[1], M.ASCE

ABSTRACT: The factors affecting the selection of an earth retaining structure are discussed from the point of view of the design engineer. Technical, practical, economical and political selection factors are outlined. A selection matrix is proposed involving a qualitative evaluation and rating of the most preferred alternatives. Ten case histories are presented in the paper where innovative earth retaining structures are used for support of cuts or fills in surface, marine and underground construction. The selection, design and performance of the retaining structures are discussed with a particular emphasis on the selection process. Alternate solutions are often specified for a particular project. The economical merits of the innovative solutions are briefly discussed. General recommendations for selecting earth retaining structures are presented at the end of the paper.

INTRODUCTION

When a fill is required to be placed in a relatively limited space or at near-vertical slope, an earth-retaining structure is often called for. This structure is mostly required when adding more traffic lanes within the same right of way of a highway or when building waterfront facilities. Earth-retaining structures are also used for support of temporary or permanent excavations, particularly in urban areas.

To design an earth-retaining structure, one has to deal with three basic elements: the structure, the ground, and any external factor affecting their behavior (traffic load, ship impact, adjacent structure, etc.). In conventional civil engineering practice, the structure is assumed to support the

---

1 - Vice President and Head of Geotechnical Department, Parsons Brinckerhoff Quade & Douglas, Inc., One Penn Plaza, New York, N.Y. 10119

horizontal load from the ground (and groundwater) behind it and any superimposed surcharge. Recently however, a new concept has emerged where the soil supports itself or is incorporated into the structure and assumes a major structural or load-carrying function. The successful use of this concept and the many benefits derived from its application have led to the rapid rise of "earth walls" (Mitchell, 1988) in civil engineering construction.

The design and construction aspects of non-conventional earth retaining structures, used for support of cuts or fills, have been thoroughly studied and well documented in a number of recent publications (Goldberg et al., 1976; Mitchell and Villet, 1987; Munfakh, 1987). In this paper, the factors affecting the selection of a particular type of structure are discussed from the point of view of the design engineer. A number of case histories are presented in the paper to demonstrate the selection process and document the behavior of some recently built non-conventional structures.

RECENT DEVELOPMENTS IN EARTH-RETAINING STRUCTURES

The non-conventional earth retaining structures discussed in this paper are divided into two major categories: those supporting compacted earth and those supporting ground excavations.

The recent developments in the first category include: (a) embankment-type reinforced compacted earth walls (Reinforced Earth wall, VSL retained earth wall, geotextile wall, geogrid wall, welded wire wall, etc.) and (b) gravity earth walls (Doublewal, gabion wall, Evergreen wall, stresswall, etc.).

In the second category, the non-conventional solutions include: (a) in-situ ground reinforcement structures (soil nailing, element wall, reticulated micro pile system, jet grouted wall, etc.) and (b) pile-type structures (secant piles, H-Z systems, drilled-shaft walls, etc.).

Each earth-retaining system has its own advantages and limitations. To minimize construction cost, more than one system is often designed and alternate bids are allowed for the same project.

HOW TO SELECT AN EARTH-RETAINING STRUCTURE

With the variety of systems available on the market, building an earth-retaining structure should be a relatively easy, and not too expensive, task. With the rapid influx of new techniques, however, and the many restrictions imposed on construction, it is becoming more difficult for the design engineer to determine which of the available systems is the most suitable for the project. To do so, a thorough evaluation of many factors involving the design, construction, use and maintenance of each system is required. Following is a brief summary of the ten most important factors affecting selection:

The Ground. An earth-retaining structure is obviously influenced by the earth it is designed to retain, and the one it rests on. The influence of the earth is mostly important in "earth walls" where the retained earth itself has a major load-carrying function. In these walls, which usually involve some sort of reinforcement, the pull-out force in the reinforcement is resisted by (1) the friction along the soil-reinforcement interface and (2) the passive resistance along the transverse members of the reinforcement, if any (grid reinforcement). Therefore, these systems are mostly suitable for soils with high internal friction such as sands and gravels.

Strain compatibility is another element affecting the design of an earth wall. In systems using inextensible soil-reinforcing elements (Reinforced Earth, VSL, Tensar, etc.), the strains required to mobilize the full strength of the reinforcing elements are much smaller than those needed to mobilize the full strength of the soil. For extensible reinforcing materials (geotextiles), the required strains are much larger. Therefore, relatively large internal deformations usually occur in these systems and the soil properties are measured at large strains (residual strength). Obviously, these systems are less compatible with soils of relatively low residual strength.

When in-situ reinforcement is used to support excavations (soil nailing, element wall, etc.), the possible saturation and creep of the in-situ soil can have a large negative impact on the long-term performance of the system. Therefore, these systems are not suitable for clayey soils.

Gravity-type structures are less influenced by the type of soil than the systems involving soil reinforcement. For soils with large vertical and horizontal deformations, a very flexible system such as a gabion wall may be chosen in lieu of a more rigid system that attempt to resist such deformations. In jet-grouted walls, the specific soil type is not important since the concept of jet grouting is to break down the soil structure and replace it with a self supporting composite mass of soil and grout sometimes called soilcrete (Munfakh, et al., 1987).

Groundwater. Generally, the groundwater table behind an earth-retaining structure is lowered to: (1) reduce the hydrostatic pressure acting on the structure, (2) reduce the likelihood of corrosion of any metal reinforcing and facing elements used in the system, and (3) prevent saturation of the soil which may significantly increase displacements and cause instability during or after excavation, particularly in systems involving staged excavation and support such as soil nailing. To reduce the negative impact of groundwater, a free-draining system such as Reinforced Earth can be used. Alternatively, for relatively watertight structures (soil nailing, element wall, geotextile wall, etc.), an appropriate groundwater drainage system is provided.

Sometimes, it is desirable to keep the water table high to prevent settlement of adjacent structures or protect existing untreated timber pile foundations from fungus decay due to exposure to oxygen. In these cases, a relatively rigid watertight structure is used (slurry wall, tangent piles, jet-grouted wall, etc.). These structures usually are designed to support the full hydrostatic pressure.

Construction Considerations. Construction schedule, availability of material, site accessibility, equipment availability and labor considerations are important factors affecting the selection of an earth-retaining structure. Construction of a Doublewal or similar prefabricated modular system, for instance, is fast and simple. With a crane and a crew of four, a construction rate of about 200 square meters per day (2000 sq. ft.) may be accomplished. When the site is inaccessible to heavy equipment, such as in rough mountainous terrain, a system that can be installed with a minimum of equipment (geotextile wall) is preferred. On the other hand, labor-intensive systems usually are not cost-effective in areas with strong labor union requirements.

Material availability is an important factor affecting the selection of an earth-retaining structure. Where rock is abundant, a gabion wall may be suitable and economical. This may not be the case, however, where suitable rock backfill has to be hauled a long distance to the project site. Where suitable aggregate is not available, structures built of concrete usually are avoided.

A major construction-related advantage of earth walls, as compared to conventional cast-in-place concrete structures, is that no form work is required on the job, and construction can be accomplished regardless of temperature. Some walls, however, are influenced by the freeze-thaw behavior of the compacted soil.

Right-of-Way. In embankment-type reinforced compacted earth systems, a relatively large space is required behind the structure face as compared to that needed for construction of conventional walls (the length of the reinforcing elements is at least 0.7 times the wall height).

To support an excavation in a very tight space, a top-down staged excavation and support system such as soil nailing may be the most suitable. The feasibility of such a structure, however, is influenced by the presence of utilities and buried structures nearby and the additional cost of permanent underground easement for placement of the reinforcing elements.

Aesthetics. In addition to being functional and economical, the permanent earth-retaining structures, in most cases, have to be aesthetically pleasing. Different types, shapes and color facings are used in construction of earth walls. The types of facings range from built on-site continuous facings (shotcrete, welded wire mesh, cast-in-place concrete) to prefabricated concrete or steel panels. These panels usually are more attractive than the traditional cast-in-place concrete or soldier pile and lagging walls. In permanent drilled shaft retaining walls, an architectural wall facing usually is provided, but at an additional cost to the structure.

The aesthetic factor is extremely important when building a retaining structure in parks, forests and natural habitat. A number of attractive wall systems (Criblock, Evergreen, etc.) are usually considered for those areas because of their aesthetic, accoustic and anti-graffiti advantages. The Evergreen wall, for instance, consists of precast concrete units with open spaces at the face into which are planted shrubs, vines, etc. With adequate water supply for the foliage, the concrete facing will no longer be visible a few years after construction.

Environmental Concerns. Like most structures, the selection of an earth-retaining system is influenced by its potential environmental impact during and after construction. Excavation and disposal of contaminated material at the project site, and discharge of large quantities of water are of primary concern. Structures that encroach on wetlands or have negative ecological impacts usually are not welcome. To reduce noise and vibration impacts, the systems that use pile driving or heavy construction machinery may be rejected. On a recent excavation support project adjacent to the Johns Hopkins University hospital in Baltimore, Maryland, for instance, an extremely tight vibration control policy (peak particle velocity of 0.004 cm/sec) was adopted due to the presence of sensitive laboratory optical equipment nearby.

To reduce traffic noise in environmentally-sensitive areas, the gravity-type gabion and Evergreen walls appear to be the most suitable. The open nature of the first and the presence of foliage covering in the second are effective in absorbing the noise hitting their facings, making these walls accoustically superior as compared to other earth-retaining structures where the traffic noise is reflected on hard or smooth continuous surfaces.

<u>Durability and Maintenance</u>. An earth-retaining structure built of concrete has a higher durability against corrosion and weathering effects than a structure constructed of metal, or which uses metal or synthetics for reinforcement and/or facing. Corrosion of the reinforcement under the influence of chemicals or stray currents is one of the major problems facing walls that use metal reinforcement. This problem is usually handled by using galvanized steel and considering a sacrificial thickness when determining the reinforcement's cross-section. Epoxy-coated steel is also used, but at a cost-penalty to the system. A detailed discussion of the corrosion problem was provided by King (1977). Gabion walls have similar durability problems. When exposed to sea water or to corrosive chemical agents, PVC-coated gabions are recommended.

When geosynthetics are used for reinforcement, their long-term creep behavior and resistance to deterioration due to chemical attack and exposure to ultraviolet light are major concerns. Significant reduction in tensile strength and elongation to failure may result from exposing the geotextile to ultraviolet radiation. Wall facings are usually used to cover the reinforcement. Durability of concrete structures is influenced by the quality of aggregates and water used in the mix, and by the casting procedures.

The durability factor is extremely important when selecting a maintenance-free earth-retaining structure in highly corrosive surroundings, or when the structure is subjected to attack by non-conventional elements such as waves, chemicals or marine borers.

<u>Cost</u>. The total cost of an earth-retaining system has many components including the structure, right-of-way, temporary or permanent easement, overexcavation and disposal of unsuitable material, drainage, etc.

In general, the construction cost of earth walls in fill areas is less than 60% that of conventional cast-in-place concrete walls. According to Mitchell (1984), the non-conventional walls become more economical when the wall is more than 3m (10 ft) in height. McKibben

(1984) reported actual costs of $220 to $380 per square meter ($20 to $35 per square foot) of finished wall for projects throughout the freeway system in Atlanta, as compared to $650 per square meter ($60 per square foot) for cast-in-place concrete walls. In New York State, the basic cost of reinforced compacted earth walls is about $325 per square meter ($30 per square foot) as compared to $545 per square meter ($50 per square foot) for conventional concrete walls.

The cost of non-conventional earth-retaining structures used for excavation support is usually more than that of conventional systems (soldier pile and lagging, sheet piles, etc.), except for permanent cuts where soil nailing or similar structures may be more economical. In many cases, however, the non-conventional structures are selected for reasons other than cost. Due to the increasing competition among the various wall systems, substantial savings have been realized on recent projects when alternate bidding was allowed.

Politics. National policies, trade barriers and political influences sometimes affect the selection. The "Buy America" law in the U.S.A., for instance, prevents the use of foreign steel for permanent structures on federally-funded projects, even if the required steel sections are not available through U.S. industries. This precludes the use of some superior-types of earth-retaining structures, such as the Arbed H-Z wall or similar, in waterfront construction. In many countries where steel has to be imported, very little emphasis is placed on steel structures. Near steel industries, on the other hand, pressures are sometimes applied by politicians and labor unions for use of steel alternatives in any proposed construction.

Tradition. Like politics, tradition may dictate, or prevent, the use of a certain-type of structure irrespective of its technical rating. Although earth walls are very popular in certain states, for instance, there are rarely considered in others. Where drilled caissons are traditionally used as deep foundations, drilled shaft retaining walls are popular and generally economical since the local contractors are equipped for, and experienced with, that type of construction. Tradition plays a greater role in construction in underdeveloped countries.

THE SELECTION PROCESS

The first step in the selection process is to determine the feasibility of the considered structures. Once the feasible alternatives are identified, they can be rated with respect to some or all of the above

discussed factors. A selection matrix can be used for a
qualitative evaluation of these alternatives. Based on
each evaluation factor, a qualitative rating between
one and four can be given for each alternative with the
highest number representing the best rating. The
qualitative ratings are usually multiplied by weighting
factors reflecting the importance of the selection
factors -- usually, cost and durability are given
higher weights than the rest. The alternative(s) with
the highest score is then selected for final design and
detailed cost estimate.

CASE APPLICATIONS

    Following are 10 case studies of non-conventional
earth retaining structures designed by Parsons
Brinckerhoff Quade & Douglas, Inc. in the past 10
years. In all of those projects, conventional and non-
conventional alternatives were considered. Although the
design and construction aspects of these structures are
briefly discussed, a greater emphasis is placed on
their selection process which is the basic theme of
this paper.

CASE 1. AVIATION CORRIDOR HIGHWAY, TUCSON, ARIZONA

    The new Aviation Corridor Highway (SR 210) in
Tucson, Arizona has an extensive requirement for earth-
retaining structures, mostly in the downtown area where
the roadway is in a depressed section. Retaining walls
are also required in the fill areas, but to a much
lesser extent.
    Following a qualitative evaluation of a number of
alternatives, a preliminary design and cost study was
performed on the wall systems believed to be the most
appropriate for the soil conditions and construction
practice in the Tucson area.
    Two wall configurations were considered: the single
wall system and the double wall system (stepped-back
wall). The double wall system was recommended by the
project's landscape architect to mitigate the canyon's
effect of two high parallel walls on each side of the
roadway.
    The single walls studied included: (a) cast-in-place
retaining walls with spread footings, (b) cantilever
drilled shaft retaining walls (c) circular drilled
shafts with tie-back anchors and (4) nailed-soil walls
(Fig. 1). The double wall alternatives consisted of
combinations of the single wall systems (Fig. 2).
    The typical soil profile used in the design
consisted of cohesionless medium dense to dense sand,
characterized by an angle of internal friction of 36°

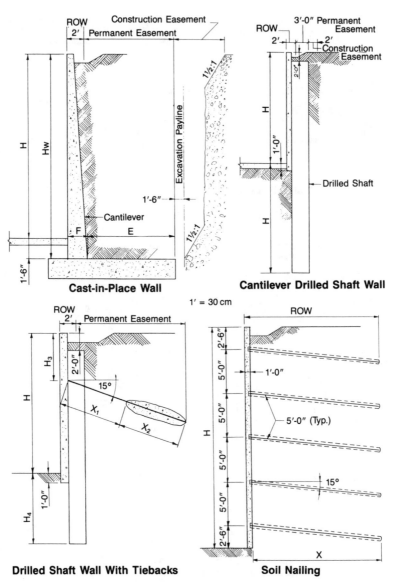

Fig. 1.   Single wall alternatives considered for
Aviation Corridor Highway.

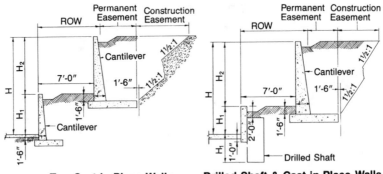

Two Cast-in-Place Walls   **Drilled Shaft & Cast-in-Place Walls**

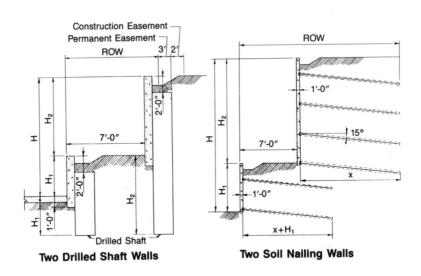

Two Drilled Shaft Walls   **Two Soil Nailing Walls**

$1' = 30$ cm

Fig. 2.   Double wall alternatives considered for Aviation Corridor Highway.

and a total unit weight of 2185 kg/m^3 (130 pcf) to a
depth below the base of the wall equal to the wall
height. This soil profile is found at the locations of
the majority of cut walls within the project limits.
"Non-typical" soil conditions were also analyzed at a
few wall locations where cohesive clayey soils were
present beneath the wall, or cemented sands behind it.
The cemented sands were treated as a cohesive material
with an undrained shear strength of 100 kPa (2000 psf).
The clay was assigned an undrained shear strength of 75
kPa (1500 psf).

The cast-in-place concrete walls were designed using
classical soil mechanics theory. The cantilever drilled
shaft walls were designed as soldier pile and lagging
but using drilled piers instead of the conventional
steal H-piles. Movements in the drilled shafts and wall
deflections were calculated by treating the drilled
shafts as laterally-loaded piles (using the program COM
624 developed by Reese and Sullivan at the University
of Texas). The anchored drilled shaft walls were
analyzed using the U.S. Army Engineers computer program
CSHTWAL and the FHWA handbook on permanent anchors
(Cheney, 1984). A reinforced drilled shaft, 600 mm (24
inch) in diameter, was sufficient for all cases.
However, for walls 9 m (30 ft) or higher, a 900 mm (36
inch) diameter was used to reduce the required
embedment and the wall deformation.

The nailed-soil walls were analyzed for internal and
external stability according to procedures presented in
Mitchell and Villet (1987). A typical spacing of 1.5 m
(5 ft) was considered for the 25 mm (1 inch) steel bar
nails. The nail length was governed by the external
stability of the composite soil-nail structure.

Double cantilevered drilled shaft walls were
selected for all areas where the cut depth is greater
than 3 m (10 ft). These structures were the least
costly among the double wall configurations dictated by
the aesthetic requirements of the landscape architect.
Their estimated construction costs were from $305 to
$338 per square meter ($28 to $31 per sq. ft.) for wall
heights of 4.5 m to 9 m (15 ft to 30 ft). Where the
depth of cut was 3 m (10 ft) or less, a single
cantilevered drilled shaft was selected.

Although nailed-soil walls appeared to be the most
economical for heights less than 3 m (10 ft), the tight
right-of-way limitations in the downtown area, and the
difficulty in obtaining permanent easements to the tips
of the nails, prohibited their use. For the few areas
where soil nailing was feasible, it was not believed to
be cost effective to require the contractor to mobilize
specialized equipment needed for this type of
construction. The equipment required for construction

of drilled shaft walls, on the other hand, was readily available at the site for construction of bridge foundations. Unfortunately, due to political reasons, the construction of the project was put on hold indefinitely.

Selection of the earth-retaining structures for this project was based on the following factors: <u>cost</u>, <u>aesthetics</u>, <u>right-of-way</u>, <u>construction requirements</u> and <u>tradition</u>.

CASE 2. H-3 TUNNEL ACCESS ROADS, HAWAII

Building of Interstate Route H-3 through the Koolau Mountain Ridge of the Hawaiian Island of Oahu involved construction of a major tunnel. To reach the remote tunnel portal location during construction, access roads having extensive retaining walls were required on both sides of the mountain. Since the retaining walls were to be constructed in mountainous terrain with difficult accessibility, alternatives requiring heavy machinery were ruled out, and the wall selection concentrated on all three most promising alternatives – – a Reinforced Earth wall, a gabion wall and a geotextile wall. To minimize cost, all three alternatives were designed for the North Halawa Valley side and the prospective bidders were requested, but not required, to bid on all three. All walls were required to have a minimum service life of 10 years and to be resistant to the moderately to highly acidic in-situ soils.

Fig. 3 presents typical cross-sections of the three alternatives. The gabion wall, which was designed as a gravity structure, consisted of steel wire baskets filled with good quality gravel or crushed rock resistant to degradation and chemical deterioration. This type of rock, however, was difficult to find in Oahu. Due to the temporary nature of the walls, cold rolled galvanized steel face panels were specified for the Reinforced Earth alternative instead of the conventional but more expensive precast concrete panels. For the geotextile wall alternative, woven or non-woven sheets of polypropylene were required to be placed between layers of compacted granular backfill. The fabric sheets were wrapped upward then overlapped for anchorage.

Design of the geotextile walls is described in detail by Castelli and Munfakh (1986). The fabric spacing was based on the ultimate tensile strength of the fabric assuming a safety factor of 1.5. Both internal and external stabilities were considered in determining the required fabric width. The external stability of the reinforced soil system governed the

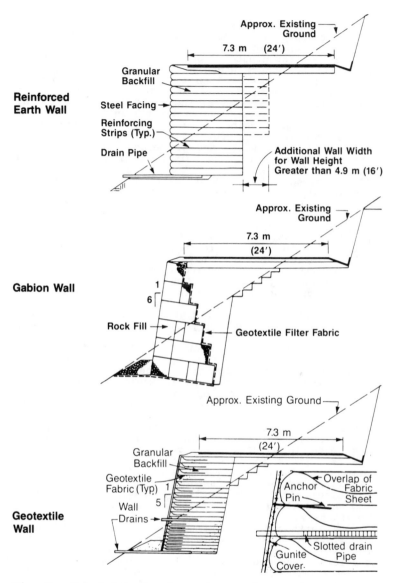

Fig. 3. Retaining structure alternatives for North Halawa Access Road.

design, requiring a minimum width to height ratio of
0.6. For depths less than 1.5 m (5 ft) below the top of
the wall, an overlap length of 1.4 m (4.5 ft) was
calculated. For greater depths, a minimum length of 0.9
m (3 ft) was used. When the required earth-retaining
structure was greater than 6 m (20 ft) in height, a
two-level stepped wall was used. The bench between the
two walls and the faces of the walls were protected by
a wire-mesh reinforced gunite layer (Fig. 4).

The average bidding price for the geotextile wall
was approximately 32 percent less than that for the
Reinforced Earth wall, and 42 percent less than that
for the gabion wall. The unit cost of the geotextile
wall from the overall low bid was $155 per square meter
($14 per square foot) of wall face (1983 price).
Immediately after the bids were received, the
construction was halted due to a legal action relating
to the environmental aspects of the project. The issue
was later resolved and the geotextile wall contract was
awarded in 1987 at a low bid price of 175 per square
meter ($16.10 per square foot) for both the North
Halawa Valley and the Haiku (Fig. 4) access roads.
Several wall sections have been constructed since and
are functioning properly. The same design was also
applied for protection of a landslide area nearby.

Selection of the earth retaining structure was based
on the following factors: cost, material availability,
right-of-way, construction requirements and durability.

CASE 3. NEWARK TRANSIT REHABILITATION, NEW JERSEY

To accommodate widening of the track beds at a
depressed section of the Newark Rapid Transit System,
steepening of the side slopes was required. To maintain
an adequate safety factor against deep-seated failures,
retaining walls 2 to 3m (7 to 10 ft) high were required
for partial support of the excavated slopes.

The subsurface profile along the slopes consisted of
2 to 6 m (10 to 20 ft) of loose fill, underlain by a
0.6 to 1 m (2 to 3 ft) layer of soft to firm silty
clay, over glacial till. Angles of internal friction of
29 and 38 degrees were assumed for the loose fill and
the glacial till, respectively. The silty clay layer
had an undrained shear strength of 21 to 37.5 kPa (420
to 750 psf).

The earth retaining structures considered included
cantilevered concrete or steel sheeting, a Reinforced
Earth wall, a geotextile wall and a gravity-type
Doublewal. The sheet pile wall required steel sections
of PS28 and PDA27, or concrete panels 180 mm (7 inch)
and 360 mm (14 inch) thick. One disadvantage of the
sheet pile wall was the presence of boulders in the

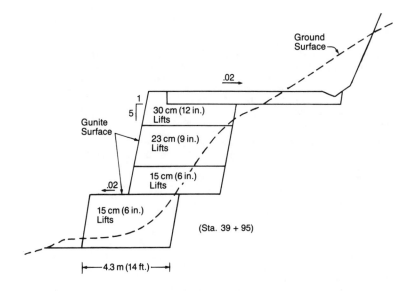

Fig. 4. Haiku Access Road typical wall cross-section.

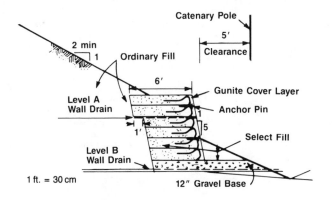

Fig. 5. Cross-section of geotextile wall for Newark Transit.

glacial till which could damage the concrete panels or split the interlocks of the steel sheeting.

The Reinforced Earth wall and the geotextile wall were analyzed for internal and external stability. A major concern of the Reinforced Earth wall was the possible accelerated corrosion of the metal strips due to stray currents from existing electric lines located above the transit right-of-way.

A cost comparison of the five alternatives showed the geotextile wall to be the least expensive. Its preliminary engineering cost was about 80 percent that of the Doublewal, 66 percent that of the Reinforced Earth wall, 65 percent that of the conventional steel sheet pile wall, and 55 percent that of the cantilevered concrete sheet pile wall.

Based on the above <u>cost</u> analysis and considering <u>durability</u> and <u>constructibility</u>, the geotextile wall shown in Fig. 5 was selected for the project.

CASE 4. MT. MCDONALD TUNNEL VENTILATION, CANADA

As part of the ventilation system for the 15-km railroad tunnel passing through Mt. McDonald (Rogers Pass) in British Colombia, Canada, a 300 m (1000 ft) deep ventilation shaft connects the tunnel to a building housing the ventilation equipment at the ground surface. To construct the ventilation building, a 7.8 m (25.5 ft) deep excavation was made at a remote and rugged hillside location.

The overburden soil at the site generally consisted of dense sand and gravel containing some cobbles and boulders. The angle of internal friction of the soil was estimated to vary between 35 and 40 degrees. No permanent groundwater table was present above the excavation level.

Several earth-retaining structure alternatives were considered. The presence of cobbles and boulders in the soil made the constructibility of a conventional soldier pile and lagging structure questionable. Due to difficult topography and potential access problems at the site, a top-down staged excavation and support method that requires a minimum right-of-way was desirable. Because of the severe cold climate, the use of cast-in-place concrete was avoided as much as possible. Based on these and other factors, and after pricing the most feasible alternatives, a precast concrete "element wall" was selected.

The design, construction and performance of the element wall were discussed in detail by Abramson and Hansmire (1988). This earth support system used prestressed rod reinforcements installed in the ground in a manner similar to that of construction of soil anchors. Each reinforcement had an independent facing in the form of a precast concrete panel (Fig. 6). The difference between this system and that of a tied-back wall is that, in the element wall, the soil and the reinforcement form a coherent material that act as the retaining wall and, therefore, the conventional structural elements (soldier piles, wales, lagging) are not necessary. As an added advantage, the excavation support system became part of the permanent building structure, thus reducing the space needed for construction of the permanent wall at the back of the building.

A long-term earth pressure diagram based on the average of the active and at-rest earth pressure envelopes was used to design the element wall. Water pressure was not considered in the design since the permanent groundwater table is below the excavation level.

The performance of the element wall was monitored during and, for a long period of time, after construction. The wall deflections, measured by inclinometers placed behind the wall facing, were between 0.02 and 0.07 percent of the wall height as compared to the 0.25 percent of the wall height expected from a conventional soldier pile and lagging system. This relatively low deflection can be attributed to the top-down level-by-level excavation and support method of construction. Shotcreting the wall face and placing the wall element immediately after excavation, and the high tie-back prestress force

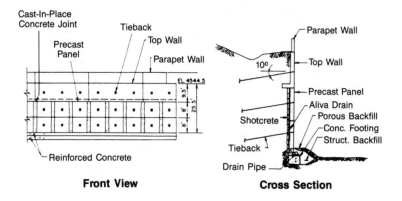

Fig. 6. Element wall for Mt. McDonald ventilation building.

apparently prevented decompression of the soil and reduced the associated horizontal deformation. Another important observation was the appearance of wet spots on the wall face during snow melt. Although horizontal drains were used, water continued to seep between concrete panels and along the tie-backs. For two snow-melt seasons after construction, water seepage was relatively low and was handled by sumping. In the long-run, however, water seepage may be eliminated by placing additional horizontal drains behind the wall.

Selection of the element wall was based on cost, right-of-way, climate and construction considerations.

CASE 5. STERLING MOUNTAIN TUNNEL PORTAL, NORTH CAROLINA

The Sterling Mountain Tunnel in North Carolina carries Interstate Highway I-40 through the Blue Ridge Mountains. In March, 1985, a large rock slide destroyed the tunnel portal. After removal of the debris, an emergency action was implemented to stabilize the side of the mountain and to reopen the tunnel to traffic.

A combination of permanent rock reinforcement and horizontal drainage was used for the large-scale slope protection. To provide protection from small rock falls, a 10.6 m (35 ft)-high retaining wall supporting an embankment was constructed parallel to the roadway with a 90 degree end wall abutting the rock face. The wall and its embankment were designed to serve as a buffer zone between the roadway and the rock slope, absorbing the kinetic energy of any falling rock and accumulating loose rock behind a fence erected above the wall.

To meet the requirements of the emergency situation, three types of earth retaining structures were analyzed: a Reinforced Earth (or similar) structure which incorporated the embankment in its design, a gravity-type module structure (Doublewal), and a concrete-faced tieback wall. All three were required to be constructed in a very limited time and to be aesthetically pleasing. The cost differential between the Reinforced Earth wall and the Doublewal was within the degree of precision of a preliminary cost estimate. The tie-back wall was 60% higher in construction cost, mainly because it required difficult drilling through very hard quartzite for installation of the soldier piles and tiebacks. To speed up construction, the material had to be procured in a separate advance contract and be stored at the site. The space required for material storage was greatest for the Doublewal and least for the tieback wall.

The tieback wall alternative was eliminated because of its relatively high cost. Because of the tight construction schedule and the limited space available for operation and storage of construction material, the Reinforced Earth wall was judged to be the most appropriate. Construction of the wall involved placement of steel reinforcing strips within layers of compacted granular material, supported by a wall facing of prefabricated concrete panels. The total construction time, including material procurement, was about two months for the 63 m (208 ft) long wall. The wall and its embankment are functioning properly.

Selection of this earth-retaining structure was based on the following factors: <u>construction schedule</u>, <u>right-of-way</u>, <u>aesthetics</u> and <u>cost</u>.

CASE 6. NEW JERSEY TURNPIKE WIDENING, NEW JERSEY

At Interchange 15W of Section D of the New Jersey Turnpike, a ramp was required to be added under an existing bridge to accommodate traffic volume increases. To allow construction of the new lane while maintaining traffic on the bridge deck above, a top-down construction method was selected for the earth retaining structure using soil nailing for support. Conventional earth retaining structures, such as sheet piles, soldier piles and lagging or reinforced concrete walls, were more expensive and required removal of the bridge deck and interruption of the bridge traffic. Maintaining that traffic was the highest priority placed on the selection process by the Turnpike Authority.

Fig. 7 illustrates a cross-section of the nailed-soil wall. Three rows of nails were required with vertical spacings of 0.75 to 1 m (2.5 to 3 ft) and a horizontal spacing of 1.5 m (5 ft). Each nail consists of a 3.5 m (15 ft) long, No. 8 steel reinforcing bar, installed in a 150 mm (6 inch) diameter drilled hole. The entire lengths of the nails were within the compacted granular fill material of the bridge abutment.

The internal stability of the system was checked for tensile rupture of the nail and adhesion failure of the soil-nail interface. The external stability of the coherent nailed-soil structure, however, governed its design. A conservative approach using Ko was followed for determining the horizontal pressure acting on the wall face in order to prevent any horizontal soil movement that may have a negative impact on the existing bridge piles behind the wall.

The earth retaining structure is constructed in three repetitive stages. Each involves excavation of a soil layer, installation of a row of nails and shotcreting of the exposed face. A cast-in-place

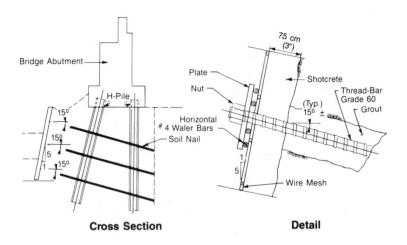

Cross Section   **Detail**

Fig. 7. Nailed-soil retaining structure for New Jersey
         Turnpike widening.

concrete facing provides the final wall finish. No
internal drainage was required since the permanent
groundwater table was below the base of the nailed-soil
mass.

Selection of the nailed-soil wall was based on the
following   parameters: right-of-way,   ease   of
construction and cost.

CASE 7. SHOT TOWER METRO STATION, BALTIMORE, MARYLAND

To construct the Shot Tower Station of the Northeast
Line of the Baltimore Metro, three types of retaining
structures are being used: a soldier pile tremie
concrete (SPTC) slurry wall, a steel soldier pile and
timber lagging wall, and a jet-grouted wall (Fig. 8).
These walls were necessitated by the ground conditions
at the project site, the presence of three high voltage
electric transmission lines across the site, and the
proximity of existing structures to the excavation.
Construction of the three types of retaining structures
at the Shot Tower Station is presently underway.

In general, the soil profile at the station consists of 4.5 to 7.5 m (15 to 25 ft) of fill, underlain by 6 to 10.5 m (20 to 35 ft) of Cretaceous soils, followed by residual material and rock. The station excavation is mainly in the Cretaceous soils which consist of dense to very dense sands and gravels with up to 49 percent fines (primarily homogeneous silt binder with varying percentages of clay). The design groundwater table is about 16.5 to 21 m (55 to 69 ft) above the bottom of the excavation.

Due to the high density of the soil (standard penetration resistances of more than 100 blows/ft were common), the use of sheet piling was not feasible. To use conventional soldier pile and lagging walls, the ground would have to be dewatered ahead of excavation. The effectiveness of dewatering, however, was questionable and its impact on nearby structures was of concern. Based on that, the SPTC slurry wall was selected for excavation support. This excavation support structure was designed also to be the permanent station wall.

The design of the SPTC wall is discussed in detail in a geotechnical design summary report prepared for the project (DKP, 1989). Lateral earth pressures were calculated for short-term, long-term and unbalanced loading conditions. The theoretical active earth pressures were converted to rectangular apparent earth pressure diagrams for the design of the temporary bracing system. Seepage analyses were performed to determine the design water pressure distribution behind the wall. Water seepage under the wall was a factor in reducing the full hydrostatic pressure applied to the wall during construction. The maximum horizontal and vertical ground movements behind the SPTC wall are expected to be less than 19 mm (0.75 inch).

Four 115 Kv electric transmission lines cross the station excavation (Fig. 8). Due to the difficulty in constructing the SPTC wall around and below these live lines (which could not be relocated), an alternative excavation support system using soldier piles and lagging was selected for that location. The soldier piles consist of steel H-piles placed in pre-drilled holes to ensure damage-free installation and proper alignment in the very dense soils, and to minimize noise and vibration during construction. To achieve contact between the back flange of each pile and the soil, the pre-drilled hole is filled with concrete after the pile is seated plumb in its center. The soldier pile and lagging system was designed for the temporary loading condition using a rectangular apparent earth pressure diagram appropriate for design of braced excavations. The final station wall at that

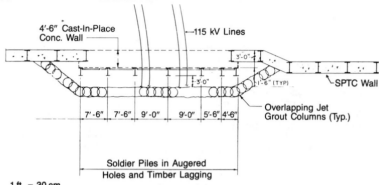

Fig. 8.   Shot Tower Station retaining structures.

location   consisted   of   cast-in-place   reinforced
concrete.

Alternative methods of preventing seepage into the
excavation   through   the   lagging   were   considered.
Chemical grouting was eliminated because of the high
fines content in the Cretaceous soils at that location.
Ground freezing was rejected due to the limited space
available adjacent to the excavation for placement of
freezing equipment and pipes. A seepage cut-off wall
comprised of overlapping jet-grouted columns, on the
other hand, was determined to be feasible for the
subsurface conditions at the site.

Jet grouting is a relatively new method of ground
improvement that is less dependent on the type of
ground than conventional grouting methods. It involves
applying   high   speed   water   jets   vertically   and
horizontally through a special drill bit to excavate
the in-situ soil, then mixing it with grout pumped at
high pressure through the nozzles of the drill bit as
it is withdrawn. By rotating the drill bit, a high-
strength jet-grouted column is created (Munfakh, et
al.,   1987).   Substantial   reductions   in   the
permeabilities of sandy soils (up to 0.0001 of their
original values) were reported due to jet grouting. The
specified jet-grouted columns were 0.75 m (2.5 ft) in
diameter, placed about 1 m (3 ft) behind the soldier
pile and lagging wall. A field test of the hydraulic

conductivity of the jet-grouted wall will be performed prior to the station excavation.

Although the jet grouted wall could have been designed to provide structural support (by placing a reinforcing element in the grouted column), thus eliminating the need for the soldier pile and lagging wall, a decision was made to use it as a cut-off wall only due to the sensitivity of the proposed excavation, and the lack of prior experience in this type of construction in the Baltimore area.

Selection of the retaining structures at this project was dictated by <u>ground and groundwater conditions</u>, <u>right-of-way limitations</u>, and the presence of <u>existing structures</u> at, and adjacent to, the excavation.

CASE 8. KISMAYO PORT, SOMALIA

The Port of Kismayo lies 45 km (28 miles) below the Equator in the East African country of Somalia. As part of a major reconstruction of the port, funded by the United States Agency for International Development (USAID), a bulkhead was constructed outboard of an existing deteriorated wharf, with granular backfill placed underwater behind it.

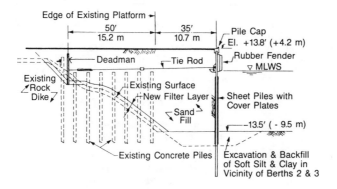

Fig. 9.   Cross-section of bulkhead for Kismayo Port.

The subsurface conditions at the site consisted generally of a surface layer of soft silt and clay about 1 m (3 ft) thick, underlain by approximately 20 m (67 ft) of loose to very dense silty sand followed by a hard coral surface. No suitable sources of aggregate material were present locally or at an economical hauling distance from the project site. The low quality aggregates used in the original construction were major contributors to the rapid deterioration of the concrete deck and pile caps.

Three types of bulkhead structure were judged to be capable of meeting the design criteria established for the project: (1) a cellular structure consisting of a series of large diameter steel sheet pile cells filled with granular backfill and connected by sheet pile arcs, (2) a gravity-type unreinforced concrete block wall and (3) an anchored steel sheet pile wall supported by an A-frame pile anchorage or a continuous sheet pile deadman.

The cellular structure was rejected due to its relatively high cost. The concrete block wall was the most desirable since it utilizes local material, requires unskilled labor, and requires minimum maintenance. This structure, however, was also rejected due to the unavailability of the large quantities of suitable aggregates required to cast the concrete blocks. The anchored steel sheet pile wall, on the other hand, required a minimum quantity of structural concrete, had a relatively low cost and provided a low maintenance structure. This scheme was selected for construction of the bulkhead (Fig. 9).

To build the new bulkhead, the existing wharf platform was demolished and the tops of the existing concrete piles were cut off. A steel sheet pile structure was then constructed 10.7 m (35 ft) outboard of the existing platform, supported by continuous steel sheet pile deadman located near the back of the platform. Behind the bulkhead, a sand backfill was placed underwater then compacted from the surface using deep vibratory compaction. The deep compaction was specified to minimize future settlement of the backfill and provide a stable foundation for a maintenance-free rigid concrete pavement above it.

The bulkhead design is discussed in detail by Castelli and Secker (1987). Briefly, the sheet pile structure was analyzed using the free earth support method and Rowe's moment reduction factor. Angles of internal friction of 30 and 35 degrees were used for the granular soils. A factor of safety of 1.5 was applied to the passive soil resistance on the outboard side of the bulkhead. A water level differential of 0.5 m (1.5 ft) was used in the design.

The required section modulus of the sheet pile exceeded available standard American sections. The use of a European master pile system of interlocking H and Z sections, though well suited for the bulkhead, was excluded by the "Buy America" requirement of the USAID funded project. A master pile system fabricated from American sections on the other hand, was uneconomical. The bulkhead structure finally selected utilized standard Z sections (PZ-40) reinforced with cover plates on the outer faces of both flanges. A multiple corrosion protection system using coal-tar epoxy coating and cathodic protection was provided.

The spacings, equipment and procedures used in the deep vibratory compaction of the hydraulic fill were selected on the basis of field trials performed at the beginning of construction. Terra probe, vibrowing and two types of vibroflots were tested, and compaction was investigated with or without the use of water jet and with or without the addition of backfill during compaction. The best results were accomplished using two vibroflots rigidly connected together and applying water jets during compaction (Munfakh, 1989). Relative densities of more than 80 percent were measured following compaction. Construction of the Kismayo Port was completed in early 1989. The new bulkhead is functioning properly.

In addition to cost, selection of the earth retaining structure for this project was influenced by durability, availability of material, maintenance requirements and political considerations.

CASE 9. CATON-SEAGIRT FACILITY, BALTIMORE MARYLAND

In conjunction with construction of the Fort McHenry Tunnel in Baltimore, Maryland, an offshore disposal and containment facility was built in the Baltimore Harbor to receive approximately 900,000 cubic meters (1.5 million cubic yards) of dredged material generated by construction of the immersed tube tunnel. Due to stability and environmental considerations, construction of a dike was not allowed and an earth retaining structure was needed to contain the hydraulically placed dredged material.

Borings drilled along the alignment of the retaining structure, where the harbor bottom was about 3.7 to 4.6 m (12 to 15 ft) below water, revealed very soft to medium stiff organic silts or silty clays varying in thickness from 7.6 to 12 m (25 to 43 ft), underlain by dense sands and hard Cretaceous clays.

The design criteria (loading, elevations, etc.) were set with the objective of turning the disposal facility into a container port operated by the Maryland Port

Administration. The containment structure, therefore, had to provide complete separation of the dredged material from the harbor water and to support the horizontal loads from the hydraulic fill and an additional surcharge load of 50 kPa (1000 psf).

The least costly retaining structure suitable for the severe loading criteria was an H-Z wall of the European ARBED system, which uses interlocking H-pile sections to provide the required high section modulus. The structure would be built at a batter to reduce the applied pressure and, in turn, the required section modulus. Due to political reasons, however, the use of foreign steel was precluded by the Maryland State Specifications and a similar system had to be fabricated using domestic steel H-pile and sheet pile sections (Fig. 10). Since the inclined master pile-sheet pile structure was unfamiliar to U.S. contractors and required building a special template for driving both the H and Z sections at a batter, and to meet the extremely tight construction schedule (the facility had to be completed prior to the start-up of tunnel dredging), an alternate containment structure consisting of 76 19m (62 ft) diameter steel sheet pile cells was built by the contractor.

Normal construction practice for cellular structures would require removal of all soft soils within the cell and replacement with granular backfill. The presence of a soft cohesive layer sandwiched between the hard bottom and the granular fill would reduce both the vertical and horizontal shear resistances. In addition, this layer would settle under the superimposed load resulting in downdrag forces on the steel sheet piles. Since the vertical stress applied on the outboard side of the cell is larger than that applied on the inboard side due to the unbalanced loading condition, settlement of the clay and the resulting downdrag force would be larger at the outboard side causing tilting of the cell. Therefore, removal of all cohesive soil from within the cell was desirable.

The environmental restrictions on construction in the Baltimore Harbor and the unavailability of a disposal site precluded predredging of the soft soils. Excavation of all clays from within the cells was not believed feasible due to the potential collapse of the cell. Therefore, a decision was made to leave some of the clays inside the cells. It was also decided to monitor the cellular structure and to implement remedial measures once the lateral movement of any cell reached 0.6 m (2 ft).

As the contractor filled the contained site with dredged material, several cells moved. Once the allowable limits were exceeded, different remedial

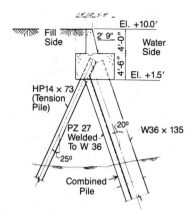

Cross-Section of Retaining Structure

1′ = 30 cm

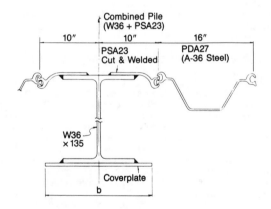

Detail of Master Pile-Sheet Pile System

Fig. 10. Master pile retaining structure proposed for Canton–Seagirt facility.

measures were undertaken to stabilize the structure. The contractor installed deep wells, excavated behind the cells, installed underdrains behind the cells and placed a berm in front of the cells. A few months later, the outboard lateral movement had reached 1.2 m (4 ft) and the settlement had reached 0.9 m (3 ft) in some cells. The remedial measures applied to that date, therefore, were only partially successful. As a result, it was decided to install stone columns inside the cells. The presence of stone columns within the clay layer would (1) increase its resistance to horizontal shear, (2) reduce the total settlement and resulting downdrag forces and (3) accelerate consolidation and strength gain in the cohesive soil.

Fig. 11 illustrates a layout of stone columns inside a cell. The column spacing varied from 1.7 to 3.1 m (5.7 to 10 ft) across the cell to accommodate the distribution of vertical stresses throughout the cell. A column diameter of 1.1 m (3.5 ft), a friction angle of 42° and a stress ratio of 3 were used in the design (Munfakh, 1984). The resistance of the clay layer to horizontal shear was increased by 67 percent and its anticipated additional settlement was reduced by approximately 300 percent due to reinforcement with stone columns. Movements of the cells stopped within 30 days after installation of the stone columns.

Selection of the cellular earth retaining structure was seriously influenced by the following factors: construction schedule, environmental restrictions, politics and tradition.

CASE 10. JOURDAN ROAD TERMINAL, NEW ORLEANS, LOUISIANA

Construction of Berth 4 of the Jourdan Road Terminal of the Port of New Orleans presented an innovative use of a combination of earth retaining structures. The generalized soil profile at the site consisted of approximately 18 m (60 ft) of soft clays and silty clays with interbeds of silt and fine sand, underlain by medium to very dense sands which act as the bearing layer. The undrained shear strength of the clay ranged from about 10 kPa (200 psf) at the top to approximately 22.5 kPa (450 psf) immediately above the sand stratum.

Several alternative wharf structures were considered for the project, many of which involved the use of earth retaining structures including a steel sheet pile bulkhead, a cellular structure, a concrete caisson wall, a slurry wall relieving bulkhead and others. A selection matrix was prepared for a qualitative evaluation of the feasible alternatives. Twelve selection factors were considered representing construction, operation, maintenance and relative cost.

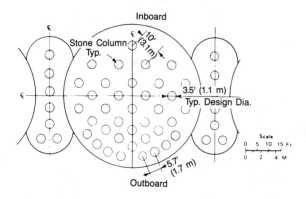

Fig. 11. Cellular structure reinforced with stone
columns used for the Canton-Seagirt facility.

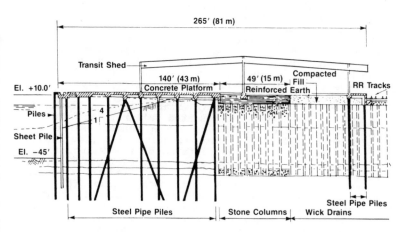

Fig. 12. Composite wharf structure for Jourdan Road
Terminal.

The factors related to operation and cost were given higher weights than the rest. As a result of the preliminary engineering study, an innovative, cost-effective scheme was selected using a combination of earth retaining structures and ground improvement methods.

Fig. 12 illustrates the selected wharf structure. Basically, it involves construction of a narrow pile-supported deck with a steel sheet pile wall connected to its outboard edge. The sheet pile wall allows dredging in front of the wharf while retaining the material behind it. To improve the overall stability of the system, the soft cohesive soil behind the deck is reinforced with stone columns extending from the ground surface to the top of the bearing stratum. On top of the stone columns, a Reinforced Earth wall is constructed. The wall and its reinforcing embankment have three functions: (1) to provide a semi-rigid mat above the stone column zone, thus forcing potential failure surfaces to pass beyond the mat and intersect the maximum number of stone columns, (2) to transfer the superimposed loads to the stone columns by arching over the in-situ soils and (3) to allow vertical filling behind the deck without a bearing capacity or shallow type failure. The combined use of in-situ reinforcement and earth retaining structures improved the overall stability of the wharf structure, eliminating the need for a wide pile-supported deck.

The outboard sheet pile wall was designed to support the horizontal loads from the soil behind it (the deck and operating loads were carried by the piles) using conventional methodology. The Reinforced Earth wall was designed to meet internal stability requirements. Except for a possible deep-seated failure, the external stability of the system was not a concern since the width to height (L/H) ratio was quite high (the length of the reinforcement was dictated by the width of the stone-column reinforced area). The overall stability of the stone column/Reinforced Earth system was analyzed using a composite shear strength value for the reinforced soil, determined from weighted average material properties. The stone columns had a 2.1 m (7 ft) spacing, an average diameter of 1.11 m (3.65 ft), a friction angle of 42 degrees and a stress ratio of 3.

The vertical and horizontal deformations of the wharf structure and adjoining soils were monitored during and after construction (Castelli, et al., 1983). No stability or settlement problems were experienced.

The presence of stone columns below the reinforced embankment reduced the anticipated settlements by about 70 percent. Although the horizontal deformations of the subsoil were relatively high -- a maximum value of 127 mm (5 inch) -- they did not cause instability. The use of this composite structure realized an estimated saving of $1.25 million per berth as compared to the cost of a conventional pile-supported platform.

Selection of the Jourdan Road Terminal Wharf structure was based on the following factors: cost, stability, constructibility, impact on an existing levee and operation & maintenance requirements.

SUMMARY

The use of innovative earth retaining structures for support of cuts or fills is gaining wide acceptance in surface, marine and underground construction. A notable innovation is the concept of earth walls where the soil is dependent upon itself for support, or is incorporated into the structure assuming a major load-carrying function.

Selection of the earth retaining structure is influenced by many factors involving technical, practical, economical and political considerations. Alternate bids are often allowed to secure a cost-effective solution.

The ten cases discussed in the paper demonstrated clear advantages of the innovative solutions as compared to conventional structures. Up to 45 percent reductions in construction cost were realized on these projects. Table 1 summarizes the factors that governed the selection of these innovative structures.

Table 1. Summary of Selection Factors

Project	Ground	Groundwater	Construction	Right-of-way	Aesthetics	Environment	Durability	Cost	Politics	Tradition
Aviation Corridor	■		■	■	■			■		■
H–3 Access Roads	■		■	■		■	■	■		
Newark Transit	■		■	■			■	■		
Mt. McDonald	■		■	■		■		■		
Sterling Mountain	■		■	■	■			■		
N.J. Turnpike	■		■	■				■		
Shot Tower Station	■	■	■	■		■		■		
Kismayo Port	■		■				■	■	■	
Canton–Seagirt	■		■			■		■	■	■
Jourdan Road	■		■				■	■		

Based on the lessons learned from these case studies and others, the following phylosophical approach to selecting earth retaining structures is recommended:

(a) Select a structure compatible with the soil it supports and the one it rests on -- Strain compatibility with the soil is particularly important when the soil is incorporated within the structure.

(b) Beware of groundwater -- It is the geotechnical engineer's enemy.

(c) Consider the contractor's capabilities, constraints and experience -- Remember, he is the one who will implement your design. Perform a constructibility review including assessment of construction schedule and staging.

(d) Consider politics and tradition when making your selection -- There is almost always someone trying to prevent you from accomplishing your planned construction for one political or traditional reason or another.

(e) Be sensitive to the environment -- One day, you or your fellow man may be living adjacent to that structure.

(f) Place a special emphasis on aesthetics -- People seem to always remember an attractive structure.

(g) Consider the long-term behavior and function of the structure -- A structure with the least construction cost does not necessarily mean an economical solution if it has to be replaced within a few years after construction, or if it requires continuous maintenance and rehabilitation.

(h) Provide the contractor with sufficient information and enough flexibility -- It avoids contingencies and future claims.

(i) Allow alternate bidding on your project -- It encourages competition and assures minimum cost.

(j) Monitor the behavior of your structure and the soil it supports. Document your instrumentation results and make them available to the profession -- They can assist in future innovative designs and state-of-the-art applications.

ACKNOWLEDGEMENT

The ten case histories presented in this paper were designed by Parsons Brinckerhoff Quade & Douglas, Inc. The cellular structure alternative of the Canton-Seagirt Facility was designed by Mueser Rutledge Engineers for the construction joint venture of Kiewit-Raymond-Tidewater. Special acknowledgement is made to Arizona DOT, Hawaii DOT, N.J. Transit, C.P. Rail, North

Carolina DOT, N.J. Turnpike, Maryland Transit Administration, USAID, Maryland DOT and the Port of New Orleans for their positive attitude toward the use of non-conventional, state-of-the-art, technology. For the Jourdan Road Terminal and the Sterling Mountain tunnel portal projects, Parsons Brinckerhoff Quade & Douglas, Inc. received engineering excellence awards from the New York Association of Consulting Engineers. The Fort McHenry Tunnel and its disposal facility won the Grand Conceptor Award from the U.S. Consulting Engineers Council.

REFERENCES

Abramson, L.W. and Hansmire, W.H. (1988), "Three Examples of Innovative Retaining Wall Construction," Third International Conference on Case Histories in Geotechnical Engineering, St. Louis, Missouri.

Castelli, R.J. and Munfakh, G.A. (1986), "Geotextile Walls in Mountainous Terrain," Third International Conference on Geotextiles, Vienna, Austria.

Castelli, R.J. and Secker, N. (1986), "Port of Kismayo (Somalia) Rehabilitation," ASCE Specialty Conference, Ports 86, Oakland, California.

Castelli, R.J., Sarkar, S.K. and Munfakh, G.A. (1983), "Ground Treatment in the Design and Construction of a Wharf Structure," Advances in Piling and Ground Treatment for Foundations, Thomas Telford, London.

Cheney, R.S. (1984), "Permanent Ground Anchors," Federal Highway Administration, Report FHWA-DP-68-1.

DKP (1989), "Baltimore Metro Northeast Line-Shot Tower Line and Station Structure, Phase I," Geotechnical Design Summary Report prepared for the Maryland DOT, Mass Transit Administration, Baltimore, Maryland.

Goldberg, D.T., Jaworski, W.E. and Gordon, M.D. (1976) "Lateral Support Systems and Underpinning" Report No. FHWA-RD-75-128, Federal Highway Administration, Washington, D.C.

King, R.A. (1977), "A Review of Soil Corrosiveness with Particular Reference to Reinforced Earth," Transport and Road Research Laboratory, SR 316, U.K.

McKibben, J.R. (1984), "Using Alternative Retaining Walls in Rebuilding the Freeway System in Atlanta," TR News Number 114, Transportation Research Board, Washington, D.C.

Mitchell, J.K. (1984), "Earth Walls," TR News Number 114, Transportation Research Board, Washington, D.C.

Mitchell, J.K. and Villet, W.C.B. (1987), "Reinforcement of Earth Slopes and Embankments," NCHRP Report 290, Transportation Research Board, Washington, D.C.

Munfakh, G.A. (1988), "Non-Conventional Retaining Walls - From Theory to Practice," Chicago Geotechnical-Structural Lecture Series on "Soil-Structure Interaction," Illinois Section, ASCE.

Munfakh, G.A. (1984), "Soil Reinforcement by Stone Columns - Varied Case Applications," International Conference on In-Situ soil and Rock Reinforcement, Paris.

Munfakh, G.A. (1989), "Soil Stabilization at Depth-Practical Applications," Seminar on "Foundations in Difficult Soils, State of the Practice," Met Section, ASCE.

Munfakh, G.A., Abramson, L.W., Barksdale, R.D. and Juran, I. (1987), "In-Situ Ground Reinforcement," State-of-the-Art Report, "Soil Improvement - A Ten Year Update," ASCE Geotechnical Special Publication No. 12.

SELECTING RETAINING WALL TYPE AND
SPECIFYING PROPRIETARY RETAINING WALLS
IN NYSDOT PRACTICE

Austars R. Schnore [1]

ABSTRACT: The main factors New York State Department of Transportation (NYSDOT) considers when selecting retaining walls for a
project prior to the detailed design phase are presented. The
procedures utilized to incorporate proprietary retaining wall
systems in NYSDOT projects are described. The fact that most of
these systems are not fully interchangeable with others has
impeded the development of a generic retaining wall specification.

INTRODUCTION

The development in recent years of a number of innovative non-
proprietary and proprietary retaining wall systems has made it
possible for the highway designer to choose from a wide variety of
retaining walls the one that fits best within the relevant economic, construction time and right-of-way constraints. General
engineering considerations that enter into the selection of a
retaining wall type will be presented here. However, it must be
emphasized that the designer should work with the geotechnical
engineer in order to properly take into account the effects of
soil conditions on the viability of various retaining wall types.

The use of newly developed retaining wall systems, not yet
proven in practice, must be controlled to protect travelers and
taxpayers. NYSDOT has established a procedure for approval of new
proprietary retaining wall systems that includes review by structures, materials and geotechnical groups. Economic considerations
dictate the need to promote maximum competition in the use of
proprietary retaining walls.

SELECTING A RETAINING WALL TYPE

Many factors have to be considered in selecting a retaining
wall type for a location. The most important ones are given in
Tables 1 and 2:

1 – Associate Soils Engineer, New York State Dept. of Transpor-
    tation, 1220 Washington Ave., Albany, NY 12232.

TABLE 1: ENGINEERING CONSIDERATIONS FOR SELECTION OF NON-PROPRIETARY PERMANENT RETAINING WALLS[1]

Engineering considerations / Wall type	Relative construction time	Maximum height[2]	Sensitivity to differential settlement	Approx. base width	Potential settlement of retained mass	Typical cost per sq.ft. of wall face[3]
Cast-in-place concrete	long	50± ft.	high	0.5 x wall height	medium	$60 to $100
Concrete cribbing	short	16 ft. (24 ft.[7])	low	0.5 x wall height	medium to high	$35 to $60
Sheet piling or H-pile and lagging--cantilever	short	15± ft.	low[5] (high[5])	< 3 ft.[7]	high	$30 to $60[6] $50 to $90[5,6]
Sheet piling or H-piles and lagging--tied back	medium	50+ ft.	high	< 5 ft.[7]	low	$40 to $ 80[6] $60 to $110[5,6]
Gabion wall	medium	25± ft.[4]	low	0.5 x wall height	high	$20 to $60
Geotextile wall	medium	20± ft.[4]	low (medium[8])	0.5 x wall height	medium to high	insufficient data

[1] Based on NYSDOT experience
[2] Greater heights require special study
[3] Does not include cost of excavation and temporary excavation support. Includes cost of backfill.
[4] Requires batter
[5] With cast-in-place concrete facing
[6] Does not include cost of pre-drilling for H-piles where this is required.
[7] Easement needed for tiebacks if they extend outside of right-of-way
[8] With shotcrete facing.

TABLE 2: ENGINEERING CONSIDERATIONS FOR SELECTION OF PROPRIETARY PERMANENT RETAINING WALLS[1]

Engineering considerations / Wall type	Relative construction time	Maximum height[2]	Sensitivity to differential settlement	Approx. base width	Potential settlement of retained mass	Typical cost per sq.ft. of wall face[3]
MSES (Reinforced Earth[R] or Retained Earth[TM]) (2, 3)	medium	50± ft.	medium[4]	0.75 × wall height[5]	low to medium	$30 to $45[6]
Doublewal[R] (4)	short	40± ft.	medium	0.5 × wall height	medium	insufficient data
Evergreen[R] (5)	short	30± ft.[7]	medium	0.5 × wall height	medium to high	insufficient data
Tensiter N (6)	medium	16 ft.	high	0.5 × wall height	medium	insufficient data

(1 ft. = 30.48 cm)

[1] Based on NYSDOT experience
[2] Greater heights require special study
[3] Does not include cost of excavation and temporary excavation support. Includes cost of backfill. 1989 prices
[4] Reinforced Earth Company criterion: diff. sett. must be less than 1 ft. per 100 ft. of wall (7)
[5] Greater for walls supporting bridge abutments
[6] For walls with a face area of at least 1500 sq.ft.
[7] Requires batter

Relative construction time. The present-day construction program consists almost entirely of the rehabilitation and/or upgrading of existing facilities. Construction time may be an important consideration in order to reduce the duration of inconvenience to the traveling public. Cast-in-place concrete walls require the longest construction time since forms have to be erected and the concrete poured and permitted to gain strength before backfill loads can be applied to the wall. In contrast, the concrete facing for H-pile and lagging and sheet pile walls can be constructed while the walls support their design lateral loads. For the most part NYSDOT takes into account the time required for construction of retaining walls in a qualitative manner only.

Maximum height. The maximum heights shown in the table are for walls retaining a horizontal ground surface.

Sensitivity to differential settlement. If retaining walls sensitive to differential settlement are to be used in an area of compressible soils, special treatment, such as deep foundations, foundation preloading or foundation densification, all of which involve an increase in expense and construction time, will be required.

Approximate base width. An important consideration on urban projects is the width of the work area during construction.

Potential settlement of the retained mass. In the case of cuts made adjacent to existing facilities the potential settlement of the earth mass retained by the wall and supporting these facilities has to be taken into account.

Cost. The cost per square foot of a given wall type can vary greatly. Some of the factors affecting it are:
- Height of the wall. Higher walls are designed to resist greater forces.
- Extent of the wall. The unit price of walls with a large total face area is generally lower than that of smaller walls.
- Location. Walls built in urban areas are generally more expensive. However, a wall built at a remote location may have high transportation costs.
- Contractor's bidding strategy. The contractor may decide to bid higher for items of work to be completed and paid for early in the contract.

Among other conditions that may affect the choice of a wall are:

Wall alignment. Some of the retaining wall types are better suited than others where the alignment of the wall includes tight horizontal curves. Cast-in-place concrete walls, sheet piling, H-pile and lagging, gabion, geotextile and Tensiter walls can be designed for almost any radius of curvature encountered in practice. Mechanically Stabilized Earth System (MSES) walls are limited to a 60 ft (18m) radius of curvature while NYSDOT specifies a minimum radius of 200 ft (60m) to 600 ft (180m) for cribbing. Doublewal[R] and Evergreen[R] walls require modifications to standard units in order to accommodate any significant horizontal curves.

Aesthetics. A valid reason for limiting the choice of retaining wall types is the desire to harmonize the wall$_{(R)}$with the surrounding area or with existing walls. The Evergreen$^{(R)}$ wall is preferred in certain situations because its face has provisions for the establishment of shrubs or other plants. However, without long-term care of the plantings, the wall loses much of its appeal.

Corrosion potential. In corrosive environments, steel should be protected by suitable methods. Where the long term effectiveness of coatings is in question, retaining walls that provide sufficient cover of concrete over reinforcing steel should be selected.

Future access to utilities. The possible need to gain access in the future to buried utilities may eliminate MSES and retaining walls supported by tiebacks from consideration.

SPECIFYING PROPRIETARY RETAINING WALLS

In accordance with FHWA regulations, the use of a proprietary retaining wall on a Federal Aid contract without the inclusion of proprietary or non-proprietary alternates is permitted under the following conditions(8):
1) Justification by the State highway agency that the proprietary product is essential for synchronization with existing highway facilities, or that no equally suitable alternate exists.
2) The product is used for research or experimental purposes.

Another avenue open to suppliers of proprietary retaining walls is through a Value Engineering proposal by the contractor. This approach has not met with success in cases of walls not previously accepted by NYSDOT because a thorough review of a new retaining wall system requires more time than is generally available during construction.

Most proprietary retaining walls constructed on NYSDOT projects are specified under the item Mechanically Stabilized Earth System (MSES). Currently, only two systems are considered to be sufficiently interchangeable to be used under this generic item: Reinforced Earth$^{(R)}$ and Retained Earth$^{(TM)}$. MSES is used only if a detailed cost comparison shows it to be more economical than a conventional concrete wall design, total face area is not less than 1500 ft^2 (139m^2), there are no utilities underneath the system, and it is used to retain an embankment instead of a cut.

The design of MSES has to consider the overall stability, the external stability and the internal stability of the system. NYSDOT evaluates the overall stability and designs any treatment that may be necessary to ensure it. NYSDOT provides the designers-suppliers of the approved systems with soil strength parameters and minimum acceptable safety factors for external stability calculations. The designer-supplier uses methods that have been reviewed and approved by NYSDOT to calculate internal stability.

The contract plans prepared by NYSDOT show the lateral and vertical limits of the MSES, appurtenances (such as traffic barrier and drainage) and the need for special features (joints that permit localized differential settlement, for example).

After the contract has been awarded, the contractor must notify the Deputy Chief Engineer (Structures) of the name and address of the designer-supplier chosen for the MSES. The MSES must be designed by a Professional Engineer licensed to practice in New York State. The designer-supplier submits to NSYDOT plan, elevation and section views of the MSES installation and shop drawings of the individual units. NYSDOT reviews the submittal to verify that the criteria agreed to with the designer-supplier have been met.

Before construction of the MSES is permitted to begin, the designer-supplier is required to supply the contractor and the engineer with two copies of his Installation Manual and provide a qualified technical representative to advise the contractor and the engineer concerning installation procedures.

MSES is paid for by square foot of face area computed from the plans between the top of the leveling footing and the top of the facing units and within the horizontal limits shown on the plans. The cost of the backfill, the leveling footing and the coping is included in the unit price for MSES. Excavation is paid for under a separate item.

CONCLUSIONS

NYSDOT has found that the selection of the most appropriate wall type can result in monetary savings, decreased time to completion of the project and a reduction in construction-caused disruption to traffic and the surrounding community. A geotechnical engineer should be involved in the selection process.

NYSDOT has a generic specification for retaining walls, which is limited to Reinforced Earth$^{(R)}$ and Retained Earth$^{(TM)}$ because, in general, one wall type cannot be substituted freely for another.

REFERENCES

1. GE Douglas, Design and Construction of Fabric-Reinforced Retaining Walls by New York State, TRR 872, TRB, 1982, 32–37.
2. Reinforced Earth$^{(R)}$ An Accepted Construction Technology, The Reinforced Earth Company, 1700 N. Moore St., Arlington, VA 22209.
3. Retained Earth$^{(TM)}$ Installation Manual, VSL Corporation, 101 Albright Way, Los Gatos, CA 95030.
4. Doublewal$^{(R)}$ Interlocking Precast Retaining Wall System, Doublewal Corporation, 59 East Main St., Plainville, CT 06062.
5. Evergreen – The Planted Wall – Economical and Ecological, Evergreen Systems, Inc., P.O. Box 345, Kings Park, NY 11754.
6. Retaining Wall Systems, TTW, Inc., 4912 Windom Rd., Bladensburg, MD 20710.
7. Reinforced Earth$^{(R)}$ Fifteen Years of Experience and Lessons for the Future, The Reinforced Earth Company, 1986.
8. Federal-Aid Highway Program Manual, 6-4-1-16, USDOT, FHWA, April 24, 1984.

A KNOWLEDGE BASE
FOR RETAINING WALL REHABILITATION DESIGN

Teresa M. Adams[1], A.M. ASCE,
Paul Christiano[2], M. ASCE and Chris Hendrickson[3], M. ASCE

ABSTRACT: Rehabilitating failing retaining walls is a common engineering problem involving consideration of numerous strategies ranging from crack patching to insertion of tiebacks and underpinning. While most designers are skillful at determining an optimal configuration of a particular design once the appropriate strategies have been identified, deciding which strategies are feasible and desirable in a given case is not easy since many different options usually are available. Selecting appropriate and technically feasible rehabilitation strategies requires considerable judgment and expertise. This paper describes the formal organization and model of the RETAIN knowledge base system (KBS) for retaining wall failure diagnosis and rehabilitation design synthesis. The kinds of knowledge contained in the model are: diagnostic knowledge used to guide field investigation and infer influencing failure mechanisms; design synthesis knowledge used to identify and preliminarily design rehabilitation strategies; and evaluation knowledge used to analyze design adequacy and to estimate costs.

INTRODUCTION

Retaining walls provide an excellent example of the problems and processes associated with preliminary engineering design and cost estimation. Retaining wall rehabilitation is difficult because many different retaining systems exist, failure modes are complex, many treatments are possible, and many project constraints and site conditions must be considered. In the last few decades, more alternatives to conventional cast-in-place concrete walls have been constructed due to a rising popularity of specialty proprietary systems such as earthwork reinforcement and prefabricated modular systems that can be constructed for less cost and construction time (Leary and Klinedinst 1984). Retaining wall designers must be knowledgeable of various wall types and alternative methods of repair and strengthening of structural elements and systems. Assessing the technical feasibility, cost and overall desirability of different treatment options requires considerable judgment and expertise, because the choice of treatment is directly related to the wall's physical condition and attendant failure mechanisms as well as

---

[1] Assistant Professor, Dept. of Civil and Environmental Engineering, University of Wisconsin-Madison, Madison, WI 53706, formerly Graduate Assistant, Dept. of Civil Engineering, Carnegie Mellon University

[2] Dean, Carnegie Institute of Technology, Carnegie Mellon University, Pittsburgh, PA 15213

[3] Professor, Dept. of Civil Engineering, Carnegie Mellon University, Pittsburgh, PA 15213

125

the project constraints and site conditions. Table 1 contains a check list of project constraints and site conditions that influence the rational selection, design and rehabilitation of retaining structures.

Table 1: Check List of Retaining Wall Rehabilitation Constraints

construction time	availability of material and equipment
soil characteristics	potential disruptions to traffic
cost	loading requirements
aesthetics	proximity to neighboring structures
equipment access	maintenance requirements
service life	anticipated settlement
space required	adaptability to field changes
reliability	environmental requirements

The hierarchy of retaining wall treatment options, shown in Figure 1, was organized according to the basic treatment options for existing infrastructure facilities (Harty and Peterson 1984; Schuster and Krizek 1978). The treatment options include do nothing, abandon sell or lease, reduce loading, maintenance and replacement. The treatment option selected by a design engineer is usually influenced by an agency's infrastructure management policy. The *do nothing* option is the result of a crisis maintenance policy to perform emergency repairs when and if the need arises. The *abandon, sell or lease* option is to abandon a wall site unless the land is attractive to private developers, in which case local governments may obtain additional revenues from lease or sale and through additional property taxes. The *reduce loading* option reduces wear-and-tear on walls, for example, by removing surcharge loads or by posting weight limits on walls supporting moving loads. *Maintenance* options serve to restore a wall to its design level of performance. Maintenance can be divided into three categories:

1. *Preventive or Routine Maintenance* of retaining walls increase the service life and aesthetics by reducing the rate of deterioration. Examples include cleaning existing wall drains, installing through-wall or surface drains to reduce hydrostatic pressure, and patching concrete spalls.

2. *Partial Reconstruction* involves replacing failed wall components or even a failed wall segment, but does not involve extensive synthesis and comparison of design alternatives. Typically the original wall design is modified to compensate for influencing failure mechanisms by resizing components. New construction materials may be used or existing materials may be reused. For example, a failing crib wall or drystone wall may be disassembled and then reconstructed using larger components.

3. *Rehabilitation* involves repair and upgrade to improve the overall and/or internal stability. Rehabilitation strategies such as underpinning, buttressing and installing tiebacks provide additional strength and capacity to compensate for wall deficiencies or failures.

The *replacement* option requires synthesis and comparisons of complete designs as potential alternatives for replacing an existing wall. For example, a failing wall may be removed and replaced with soldier piles and lagging, a precast modular system or the embankment may be

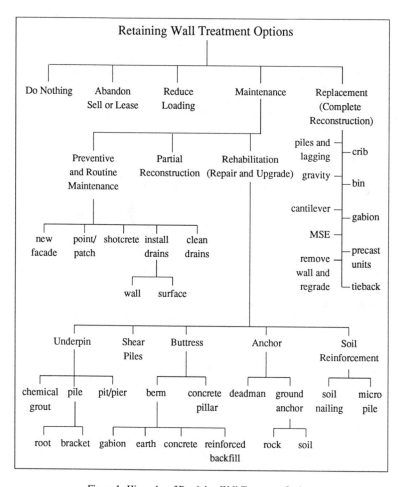

Figure 1: Hierarchy of Retaining Wall Treatment Options

regraded. The replacement and partial reconstruction treatment options differ in that the former requires the generation of alternative new wall designs while the later requires the generation of alternative modifications to the existing wall design.

In this paper, computer methods are illustrated for organizing and codifying retaining wall failure diagnosis and design synthesis of the rehabilitation-maintenance treatment options presented in Figure 1. The knowledge base for failure diagnosis is modeled as inference networks of assertions. The knowledge base for design synthesis is modeled as relational database tables and program functions. Examples of organized retaining wall failure diagnosis and rehabilitation design knowledge are given.

THE RETAIN SYSTEM

RETAIN is a knowledge-based retaining wall diagnosis and rehabilitation strategy design system. The system is designed to be used by an engineer called upon to evaluate a retaining wall failure, to survey the condition of an existing wall, or to design alternative rehabilitation strategies. The system architecture is shown in Figure 2. RETAIN consists of a database management system (DBMS) and a series of program modules, Site Identification, Failure Diagnosis, Design Synthesis, Cost Estimation and Evaluate/Consistency. The numerous programs which comprise the RETAIN modules are controlled by a menu system and interact with the user and the system database. Communication between modules is possible through shared information in the database.

- SITE IDENTIFICATION. Problem specific information including geometric models and properties of walls, soils, surcharge loads, and site constraints are collected and stored in the database. Problem specific data is used for wall stability analysis, strategy feasibility checking, strategy design, and selection of construction technology.

- FAILURE DIAGNOSIS. All influencing surcharge loads are assembled and the retaining wall is analyzed for overall stability. Field observations and the results of stability analysis comprise the input evidence used to infer a set of influencing failure modes for a given retaining wall problem.

- DESIGN SYNTHESIS. A set of possible rehabilitation strategies are synthesized from all known strategies based on the influencing failure modes found by the diagnosis module. The set of strategies is pruned by applying feasibility constraints then preliminary strategy designs are generated. Complete rehabilitation solutions are formed by searching and combining strategy designs. Solutions comprising incompatible strategies are eliminated.

- COST ESTIMATION. The cost estimation module, RETCOST (Ciarico et al 1989), estimates the cost of each preliminary strategy design, then aggregate strategy costs to obtain design solution costs.

- EVALUATE/CONSISTENCY. The rehabilitation solutions are ranked according to how well they fulfill project objectives and constraints including those listed in Table 1. Currently, this module of the RETAIN system has not been implemented.

The system architecture emulates a *rehabilitation design process*. The necessary tasks in the conventional design process are synthesis, analysis and evaluation. The rehabilitation design process includes an additional first step, failure diagnosis. The RETAIN modules are

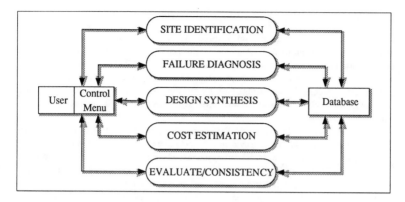

Figure 2: RETAIN System Architecture

intended to be executed consecutively in the top down order shown in Figure 2, however the architecture allows looping and backtracking between modules in compliance with an actual implementation of the rehabilitation process. Then, for example one may revise soil properties, wall parameters or even field observation information in order to gain a more accurate failure diagnosis. This flexibility is available within the modules as well as between modules.

Upon entering the RETAIN system, the main control menu is used to invoke the module menus. Screen forms and report generators on the module menus allow the user to review, modify and print the diagnosis, design synthesis and cost estimation results. A typical session with the RETAIN system may proceed as follows.

- The user invokes the site identification module menu and uses a form-filling user interface to enter problem specific data including the wall, soil, loads and construction constraints which are stored in the database. The user returns to the main control menu.

- The user invokes the failure diagnosis module menu and selects the option to diagnose a wall failure. The system queries the user for a wall location called the *design station.* Subsequently, all stability analysis, field observations, failure diagnosis and design synthesis are focused at the design station. The diagnosis program queries the user about the physical condition of the wall and infers the influencing failure modes at the design station. After failure diagnosis, the design station, field observations and influencing failures modes are stored in the database. The user returns to the main control menu.

- The user invokes the design synthesis module menu and selects to synthesize preliminary designs. The design synthesis program generates rehabilitation designs based on the influencing failure modes at the design station. Results of design synthesis are stored in the system database. The user returns to the main control menu.

- The user invokes the cost estimation module menu, chooses the desired unit cost data, city and overtime indices, and selects to cost estimate the possible rehabilitation designs. The user quits the RETAIN program by returning to the main control menu and exiting.

A KNOWLEDGE BASE FOR FAILURE DIAGNOSIS

Diagnosing a failing retaining wall involves evaluating the wall's stability and failure characteristics to determine the influencing mechanisms. An engineer performing a field survey must know what evidence to look for such as cracking, sliding, settlement and overturning. Evidence may indicate the symptoms or causes of wall failure and may be fuzzy or imprecise. A complete knowledge base for retaining wall failure diagnosis contains all possible failure modes of all possible wall types, as well as the possible evidence associated with each failure mode.

The RETAIN knowledge base for failure diagnosis contains sets of predefined retaining wall types, failure modes and failure evidence. Associated with each wall type is a subset of the failure modes and associated with each failure mode is a body of diagnostic knowledge relating various pieces of failure evidence. The relationships between wall types and failure modes and between failure modes and diagnostic knowledge are encoded in an inference network with connections between the wall types and failure modes and between failure modes and input evidence.

The levels of organization in the RETAIN diagnosis knowledge base are shown in Figure 3. At the lowest level, failure evidence is organized into two categories: symptoms and causes. The failure symptoms are all the possible wall distresses which motivate the need to rehabilitate. The symptoms of a particular wall failure are collected by field inspection. Failure causes may be computed by stability analysis, site characterization or gathered by field inspection. Pieces of evidence are collected and organized into models for diagnosing individual failure modes. In Figure 3, failure mode k has possible symptoms j and n and causes i, j and n. A single piece of evidence may be shared by several failure mode models. A portion of the failure mode model for longitudinal settlement is described later in this section.

A group of failure mode models is collected and organized into a model for diagnosing the failures modes, such as longitudinal settlement and soil shear failure, of an individual wall type. In Figure 3, failure modes i, j and k are the possible failure modes of wall type i. Failure modes may be unique for a particular wall type such as differential lateral shear in drystone walls (Cooper 1986); other failure modes such as sliding, overturning, and bearing capacity failure are manifested in many wall types. The failure mode models in the RETAIN diagnostic knowledge base may be common to several wall types.

At the highest level of organization, the RETAIN diagnostic knowledge base is a collection of wall diagnosis models for walls such as drystone, concrete gravity and concrete cantilever, joined by the top node as in Figure 3. When applied to a particular wall failure, the inference network is traversed using an inexact inferencing scheme to determine the certainty measure of each failure mode based on the certainty measures of failure evidence. Certainty measures are defined on a scale of 5 to -5 ranging from certainly true to certainly false, respectively. In Figure 3, the inference network above the failure mode level is necessary to organize and control the diagnosis process. Retaining wall diagnostic knowledge is encoded in the network below the failure mode level.

Diagnostic inferencing follows a backward chaining, depth first, left to right sequence. The inference engine starts at the top node and checks the left branch of the left most wall diagnosis model to determine whether the input wall type matches the wall type associated with the diagnosis model. The wall type nodes, *wall type i* and *wall type n* in Figure 3, are used for this comparison. If wall types do not match, the inference engine returns to the top node and

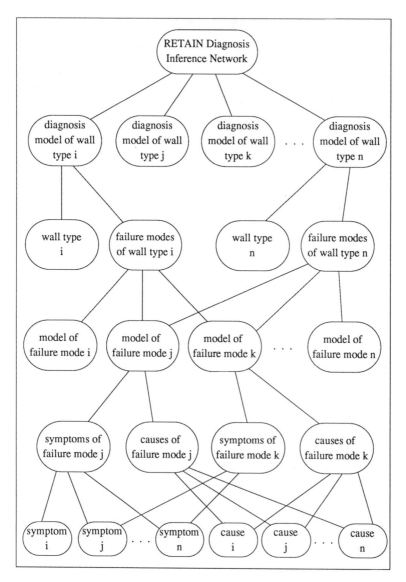

Figure 3: Organization of Diagnosis Knowledge Base

successively checks each wall diagnosis model until the appropriate model is found. When the correct wall diagnosis model is found, the right branch of the model is traversed. The certainty measure of each failure mode associated with the input wall type is determined in sequence by the inference engine, starting with the leftmost failure mode. For example, if the input wall type is n, the certainty measures of its possible failure modes j, k and n are to be determined. The certainty measure of each failure mode is computed from the combined certainty measure of its symptoms and the combined certainty measure of its causes. The certainty measure of evidence is queried from the user or the system database as evidence is needed during the diagnosis process. To reduce the possibility of conflicting evidence, the inference engine does not make redundant queries. For example, although the certainty of symptom n is required to infer both of failure modes j and k, the inference engine queries once while computing the certainty of failure mode j then uses the symptom n certainty measure again to compute the certainty measure of failure mode k. After the failure mode models of the input wall type have been traversed, the inference engine returns to the top node and failure diagnosis is complete.

The diagnostic inference networks contain three types of nodes for combining certainty measures: AND, OR and RULE (Duda et al 1976; Adams et al 1988). When several pieces of evidence must exist, a logical conjunction (AND) is used. When only one of several pieces of evidence must exist, the logical disjunction (OR) is used. To allow for some "partial credit" for unknown evidence, RULE nodes are used. RULE nodes combine certainty measures based on Bayesian decision theory, assuming the evidence is conditionally independent. The presence or absence of evidence combined in a RULE node confirms or refutes the possibility of a failure mode.

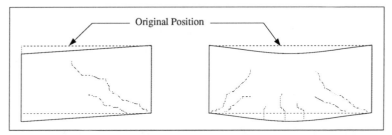

Figure 4: Inclination of Cracks to Estimate Direction of Differential Settlement

Consider a case of diagonal cracking along the length of a wall, as shown in Figure 4. A cast-in-place concrete wall may experience this type of cracking failure as a result of nonuniform longitudinal settlement. A partial inference network for diagnosing differential longitudinal settlement is shown in Figure 5. The inclination of the diagonal cracking pattern on the wall face indicates this type of failure and provides clues to the direction of differential settlement (Tschebotarioff 1973). Figure 4 shows cracking patterns caused by wall end settlement and by greater settlement in the central part of the wall. Differential settlement may be the result of several causes (Terzaghi and Peck 1948) including:

- Thawing Failure. If the settlement failure occurs as temperatures rise above freezing, the failure may be due to thaw below the wall weakening and washing out foundation soil. This failure occurs only in certain climates in the Northern Hemisphere.

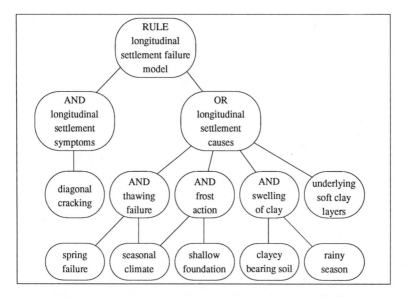

Figure 5: Inference Network for Diagnosing Longitudinal Settlement Failure

- Frost Action. The base of the wall foundation should be below the depth of frost penetration and the depth to which soil is broken up by seasonal volume changes. If the wall footing has been founded above the suggested frost depth of the bearing soil, repeated frost action may weakened the bearing stratum, thus causing settlement.

- Swelling of Clay. Failure may be due to the swelling and lifting power of dry clays during a rainy season.

- Underlying Soft Clay Layers. The differential settlement may be due to natural compression of underlying soft clay layers.

The network in Figure 5 shows that the primary symptom of longitudinal settlement is diagonal cracking. The alternative possible causes of longitudinal settlement failure are thawing, frost action, swelling of clay or underlying soft clay layers. Field evidence is combined to determine whether one or more of the failure causes is present. For example, if a wall is founded on clayey bearing soil and failure occurs during a rainy season, then a possible cause of failure is swelling of clay.

The certainty that longitudinal settlement has occurred is represented by the certainty measure of the top node in Figure 5 and is computed by backward chaining network traversal. First the left branch is traversed to compute the certainty measure of failure symptoms. If the presence of diagonal cracking is sufficiently certain, then the right branch of the network is traversed to determine the certainty of failure causes. The combined certainty of symptoms and causes indicates how well the input evidence matches the longitudinal settlement failure model.

A KNOWLEDGE BASE FOR REHABILITATION DESIGN SYNTHESIS

The synthesis of retaining wall rehabilitation solutions is controlled by wall type, failure modes and site constraints. In addition, an engineer must be familiar with strategy design procedures and understand which strategies may be combined and how multiple strategies influence one another and overall stability. Retaining wall problems may be diagnosed into failure modes such as sliding and overturning and complete rehabilitation solutions may be formed by combining strategies such as ground anchors and underpinning. Then relationships between failure modes and rehabilitation strategies may be represented in tabular form and stored in relational database tables, reducing the synthesis process largely to database manipulation. An example is illustrated in Figure 6. The directional arrows indicate the flow of information and the result of data manipulation in the five phases of the synthesis process.

1. SELECT FAILURE SET. A set of pertinent failure modes is selected from the results of failure diagnosis. The failure set contains those failure modes with certainty measures greater than some minimum value. In Figure 6, the minimum certainty is 3 (corresponding to a verbal description of "certain") and the selected failure set includes sliding and bearing capacity failures.

2. SELECT STRATEGY SET. A set of possible strategies for rehabilitating the failure modes in the failure set is selected from all known strategies in the knowledge base, Table A. This two-column database table represents the many-to-many relationship between failure modes and rehabilitation strategies. From Table A, the strategy set for rehabilitating sliding and bearing capacity failure modes are gabion berm buttress, soil berm buttress, bracket pile underpin, root pile underpin, pit or pier underpin, ground anchor and shear piles.

3. PRUNE STRATEGY SET. Strategies which cannot be designed or constructed because of soil, geometrical or environmental constraints are pruned from the strategy set. Suppose for example, the strategies soil berm buttress, root pile underpin and pit or pier underpin are deleted from the strategy set in Figure 6 because they violate one or more constraints. Pruned strategies are no long consider as candidates for rehabilitation.

4. STRATEGY DESIGN GENERATE-AND-TEST. For each remaining strategy in the strategy set, preliminary designs are generated and the existing wall combined with the strategy design is analyzed for stability. Preliminary designs are generated and tested until a design is found which satisfies stability criteria. If no stable design is found, the strategy is removed from the strategy set.

5. SOLUTION FORMATION. Alternative complete rehabilitation solutions are formed by searching and combining strategy designs. Strategy combinations which cannot be contained in a multiple strategy solution are listed in Table B of Figure 6. Solution number 3 which combines shear piles with bracket pile underpin is incompatible and listed in Table B. The reason these strategies are incompatible is that they both require the space below the wall toe. Solutions comprising incompatible strategies are eliminated.

Knowledge for strategy selection, strategy design and solution formation is stored in Tables A and B. The tables contains three types of heuristic knowledge: candidate rehabilitation strategies for a given set of influencing failure modes, failure modes which are influenced by

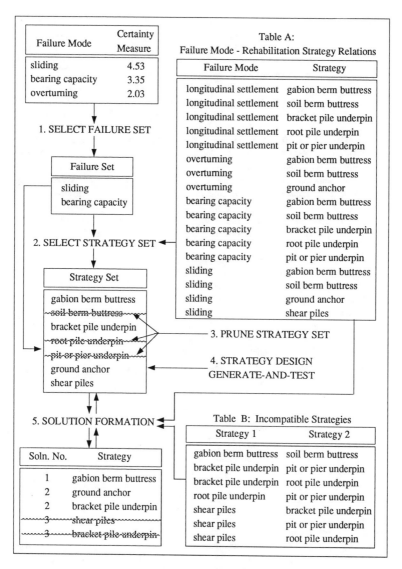

Figure 6: Example Rehabilitation Design Synthesis

the application of a strategy, and strategy combinations which form complete and compatible solutions when several failure modes are influencing.

Table A contains knowledge of how the application of a rehabilitation strategy may influence the overall stability of a retaining wall. Alternative strategy designs are generated then applied to the existing wall and the system is tested for stability against all the failure modes which the strategy may influence even if all the modes are not present. This ensures that increasing stability against one failure mode does not weaken resistance against another. Thus for stability testing, all of the failure modes which may be influenced by each strategy must be predefined. For example, a gabion berm buttress is a possible strategy for rehabilitating sliding and bearing capacity failures. However, because the gabion berm may also be used to rehabilitate overturning and longitudinal settlement, each berm design is tested for overturning and longitudinal settlement stability as well as sliding and bearing capacity stability.

Solution formation is based on the assumption that the superposition of stable strategy designs produces a stable complete solution. An implied relationship between strategies and failure modes is that the number of strategies in a complete solution is never more than the number of failure modes because a single strategy may rehabilitate multiple failure modes. One way of considering the combination of strategy designs is by means of a morphological table (Finkelstein and Finkelstein 1983). Table 2 is the morphological table for the example in Figure 6. Following the morphological approach, a complete solution is a combination of compatible possible realizations of basic functional components. In Table 2, the functional components are the failures in the failure set and the possible realizations of each function are among the strategies in the strategy set. Table 2 is an alternative representation of a portion of Table A that can be used to visualize the solution formation. For example, gabion berm buttress is a complete solution because it contains both the sliding and bearing capacity functional components. If the strategy combinations of bracket pile underpin with shear pile and bracket pile underpin with ground anchor are compatible, then they form complete solutions because both the sliding and bearing capacity functional components are represented.

Table 2: Morphological Table for Solution Formation of Example in Figure 6

	Failure Set	
Strategy Set	sliding	bearing capacity
gabion berm buttress	•	•
bracket pile underpin		•
shear piles	•	
ground anchor	•	

The feasibility constraints used to prune the strategy set and strategy preliminary design procedures do not fall into patterns of organization that can be clearly represented in database tables. These phases of the design synthesis process typically require engineering rather than heuristic knowledge and can be efficiently represented as C language functions (Kernighan and Ritchie 1978). A unique set of feasibility constraints is defined for each possible strategy, with few constraints applying to more than one strategy. Soil constraints include strength limitations such as bearing capacity, shear strength, and sensitivity to vibration. Environmental constraints are related to aesthetics and possible impact such as noise or vibration on nearby people or structures and impact of ambient conditions such as heat, light or corrosive air or water on wall construction materials. Geometrical constraints are spatial, physical or access construction constraints including wall size, vicinity of neighboring structures, surface and subsurface utilities, right-of-way and equipment access. Other possible constraints include availability of equipment or materials and cost. A unique procedure is required for each strategy because preliminary design involves routine sizing of a predefined configuration. Strategy design parameters include, for example, height, width and profile (such as step-front or uniform) of gabion berms, or spacing, length and installation angle of ground anchors.

COMPUTER IMPLEMENTATION ISSUES

The components of the knowledge base described in this paper have been implemented in an integrated system for retaining wall failure diagnosis and rehabilitation design synthesis called RETAIN (Adams 1989). RETAIN integrates OPS83 production rules (Forgy 1986), C language functions and the INFORMIX database management system (Informix 1987). The network structure of the failure diagnosis knowledge base is a modified form of the inference network of assertions used by the PROSPECTOR system (Duda et al 1979) to identify ore deposits. The DIGR diagnostic inference engine (Rychener 1987) traverses the RETAIN failure network. While the RETAIN system was developed on a MicroVAX workstation, each of the software components have PC versions and the entire system has been successfully transported to a PC.

CONCLUSIONS

This paper describes the types of knowledge required and computer methods for representing and manipulating a heuristic knowledge base for retaining wall failure diagnosis and rehabilitation design synthesis. The knowledge base is organized in a manner that identifies gaps in existing knowledge and experience. Mechanisms for knowledge acquisition and system evolution are essential because the expertise for failure diagnosis and strategy design is scattered among practitioners whose personal experience may be limited to a few wall types, failure circumstances or strategies. For example, a specialty contractor may contribute knowledge regarding the application and design of a single strategy such as ground anchors. The failure diagnosis knowledge base is divided into models for diagnosing failure modes, organized according to wall type and represented as an inference network. The knowledge base organization permits evolution of the system as practitioners contribute new failure modes and wall type models by allowing models to be developed, tested, added, dropped or revised without influencing other models. The heuristic knowledge base for design synthesis is represented as relational database tables. One table stores relationships between failure modes and rehabilitation strategies, and another table stores incompatible relationships between strategies. The database tables may be easily revised as new failure modes and strategies are added to the system. Synthesis knowledge representation is simple, but manipulations are complex.

While the current version of RETAIN is not sufficiently complete for commercial use, the development of RETAIN has demonstrated that knowledge-based tools may be applied to solve a complex geotechnical engineering problem in the area of infrastructure management. The cost savings derived from the system would particularly benefit the agencies which are responsible for managing and maintaining numerous retaining wall, such as state departments of transportation and municipalities. The scope of knowledge and expertise required to complete the RETAIN system is very large. The expertise of even the most experienced engineer typically spans only a portion of the RETAIN knowledge base.

REFERENCES

Adams, T., Hendrickson, C., and Christiano, P. (1988). "An expert system architecture for retaining wall design." *Transportation Research Record 1187, Transportation Research Board, National Research Council*, pages 9–20.

Adams, T. (1989). "RETAIN: An integrated knowledge based system for retaining wall rehabilitation design." PhD thesis, Dept. of Civil Engineering, Carnegie Mellon University, Pittsburgh, PA, August.

Ciarico, A., Adams, T., and Hendrickson, C. (1989). "A cost estimating module to aid integrated knowledge-based preliminary design." In *Proceedings, Sixth Conference on Computing in Civil Engineering*, Atlanta, GA, September, ASCE.

Cooper, M. (1986). "Deflections and failure modes in dry-stone retaining walls." *Ground Engineering*, pages 28–33, November.

Duda, R., Hart, P., and Nilsson, N. (1976). "Subjective Bayesian methods for rule-based inference systems." In *Proceedings, AFIPS 1976 National Computer Conference*, pages 1075–1082.

Duda, R., Gaschnig, J., and Hart, P. (1979). *Expert Systems in the Micro-electronic Age*, chapter The Prospector Consultant System for Mineral Exploration, pages 153–167. Edinburgh University Press.

Finkelstein, L. and Finkelstein, A. (1983). "Review of design methodology." *IEE Proceedings*, Pt. A, Vol. 130(4):213–221, June.

Forgy, C. (1986). *OPS83 User's Manual, System Version 2.2*. Production Systems Technology, Inc.

Harty, H. and Peterson, G., editors (1984). *Guide to Selecting Maintenance Strategies for Captital Facilities*, volume 4 of *Guide to Managing Urban Capital Series (GMUC)*, chapter Maintenance Strategies and Options, pages 5–13. The Urban Institute Press, Washington D.C.

Informix Software, Inc. (1987). *INFORMIX User's Manual*.

Kernighan, B. and Ritchie, D. (1978). *The C Programming Language*, Prentice-Hall, Inc.

Leary, R. and Klinedinst, G. (1984). "Retaining wall alternatives." unpublished paper from U.S. Federal Hwy. Administration.

Rychener, M. (1987). "The DIGR rule-based diagnostic kernel." Unpublished user's manual, EDRC, Carnegie Mellon University, Pittsburgh, PA, April.

Schuster, R. and Krizek, R., editors (1978). *Landslides Analysis and Control, Special Report 176*, Transportation Research Board, Washington, D.C.

Terzaghi, K. and Peck, R. (1948). *Soil Mechanics in Engineering Practice*, John Wiley & Sons, Inc.

Tschebotarioff, G. (1973). *Foundations, Retaining and Earth Structures*, McGraw-Hill Book Co.

CONTRACTING PRACTICES FOR EARTH RETAINING STRUCTURES

Joe Nicholson, M.A.S.C.E.[1]

ABSTRACT: Recent publicity suggests that the U.S. construction industry is no longer cost effective. Lack of innovation is noted as a major cause. Contracting practices in this country tend to stifle innovation while those of Europe and Japan do not. To find out why, the competitive bid system is examined in detail. It is then compared to foreign practice since many of the recent improvements in contracting for earth retaining structures have been adapted from overseas. Various technical and commercial aspects of each of these improvements are discussed and a further recommendation is made for the future.

INTRODUCTION

As civil engineers, our careers depend heavily upon the health of the construction industry. Its problems affect each and every one of us since construction revenues are on the order of 10% of the nation's GNP, making it one of the largest components of that index. Uncertainty in the construction industry can have ripple effects throughout the entire U.S. economy. According to Carnegie Mellon University, our balance of payments is directly affected since construction and construction related sales are important U.S. exports[1]. Richard E. Heckert, the CEO of DuPont, said during testimony before Congress that "cost effectiveness in the construction industry is an important factor in assuring worldwide competitiveness for American industry[2]. Unfortunately, the U.S. construction industry is in trouble; it is no longer cost effective.

TELLTALE SIGNS

Most civil engineers are aware of recent publicity noting a decline in the competitiveness of U.S. contractors. The following indicators are telltale signs of this decline:

---

1 - Chief Executive, Nicholson Construction of America, P.O. Box 308, Bridgeville, PA 15017.

139

● Construction's share of the nation's GNP has slipped almost 30% since the 1960's[3].

● Labor productivity growth within the construction industry improved steadily between 1947 and 1968. Since 1969, has slipped 15 to 20% or back to pre-1957 levels[4,5]. And, according to the Bureau of Labor statistics, construction's productivity growth is by far the lowest of the 10 major U.S. industries[5].

● American companies are becoming non-competitive and are losing market share both here and abroad. During 1987, the Japanese alone accounted for almost 25% of the new construction in the United States[5]. Between 1981 and 1985 the market share for foreign firms doing business in the United States grew at a rate of 70% while overall construction in the U.S. was declining[5].

● In 1970, when the last Cornell conference was held, U.S. contractors had a firm hold on foreign construction work. However, in 1987 U.S. firms were outnumbered for the first time among Engineering News Record's "Top 250". In fact, U.S. firms now hold only a 25% share of the foreign market, while European and Japanese firms have 67%[6].

● Construction claims and litigation have increased at an alarming rate. The Department of the Navy's NAVFAC (Naval Facilities Engineering Command) reported an increase of more than 100% in its docket of litigation over a five year period[7].

● Other signs are all around us. Foreign firms are investing in local contractors and suppliers at an enormous rate. Large, well established U.S. firms are going out of business with little or no warning.

PARALLELS WITH THE STEEL INDUSTRY

The above signs augur poorly for the future of the American construction industry. Parallels may be drawn with the inability of many U.S. industries to compete in the global marketplace, particularly the steel industry. Just as steel failed to modernize after World War II, so has our construction industry. In the steel industry, research and development took a back seat to profits; innovation was instituted only too late; capital wasn't invested in the new plant and equipment necessary to remain competitive.

Construction has also failed to invest; not so much in assets, but more in the innovation required to improve its competitive position. Like steel, the U.S. construction industry stagnated and is realizing too late that it must compete globally by offering innovative services at competitive prices.

RESEARCH AND DEVELOPMENT

Research and development (R&D) are essential to the process of developing new, innovative techniques. U.S. contractors, architects and engineers as a group invest less than five cents out of every $100 in sales on R&D[8]. This compares unfavorably with Japanese contractors who spend ten to twenty times as much for R&D[1,9]. In fact, every

Japanese contractor is required to spend about ten cents of every $100 just to develop products which will be used five years in the future[5].

A report by the National Council on Public Work's Improvement addressing the infrastructure found the "... decline in R&D especially troubling in light of evidence that the nation is becoming more dependent on foreign producers..."[9]

Innovative techniques require advanced types of construction equipment. Certainly, U.S. contractors have invested heavily in equipment over the years. Unfortunately, it was left to equipment manufacturers to design the type of equipment which suited the contractor's needs. U.S. contractors almost never urged manufacturers to build equipment which could support new techniques developed through R&D efforts. One example of how this has impaired the industry is that with the end of the "baby boom" has come a severe labor shortage. It is reasonable to expect that appropriate R&D investment could have prepared the industry for just such a demographic fluctuation. Yet U.S. contractors are behind their Japanese counterparts in robotics and automation which could drastically reduce manpower requirements in the field[4].

Like most mature industries, construction is reluctant to try new techniques and procedures. It has erected rigid barriers to innovation by allowing custom, convention and long established practice to interfere with progress.

CONTRACTING PRACTICES

U.S. contracting practices are just such a barrier. To see why, it is necessary to examine them in detail. The contracting practice most unique to U.S. construction is the competitive bid. A legal cornerstone of the industry, competitive bidding was adopted in the mid-1800's to prevent corruption by government officials.
Another aim of competitive bidding was to inhibit the imprudent spending of tax dollars by guaranteeing the lowest possible price.

However, unqualified contractors soon obtained major public works projects. The structures they built sometimes collapsed with resulting property damage and loss of life. To alleviate this situation, language such as lowest "responsible" bidder and "if determined to be in the public interest" was added to existing laws on competitive bidding.

After these adjustments, competitive bidding supposedly mirrored the concept of American free enterprise by fostering healthy competition in the construction industry. It protected the small entrepreneur by permitting him to compete, so long as he could demonstrate the ability to perform. It also protected the owner by assuring the lowest possible price through what is referred to as "sealed" bidding.

The system has been in use for 150 years and is taken for granted by most within the industry. As to whether it is the best system available, one author put it, "the lower bidder system is like the democracy, it does not work but we do not have yet a better system."[10]

With today's level of competition, contractors focus only on their profit picture. And profits are low, averaging 3% of revenue (before tax)[11]. In 1989, specialty contractors reported a return on investment (ROI) of 5.2%[12]. Such performance is unheard of in any other industry so laden with risk.

Low margins have led to three major drawbacks with the competitive bid system:

```
┌─────────────────────────────────────────────────┐
│  DRAWBACKS TO THE COMPETITIVE BID SYSTEM         │
├─────────────────────────────────────────────────┤
│                                                  │
│     COLLUSION which limits competition           │
│                                                  │
│     CLAIMS and resulting litigation              │
│                                                  │
│     a lack of reward for QUALITY                 │
│                                                  │
└─────────────────────────────────────────────────┘
```

Collusion. Every few years, most of us hear of bid rigging scandals that receive state-wide or sometimes national attention. Contractors may be temporarily debarred from bidding government contracts but when firms are debarred, fewer bids are received and competition becomes limited. One way or another, collusion leads to price increases which undermine the foundation of the entire competitive process.

Litigation. In the current business climate, few bidders allow for contingencies against unforeseen conditions. Although it is prudent to do so, bidders understand that such contingencies will make them less competitive. Furthermore, many contractors admit to being unrealistic about their production forecasts to win an award. Other contractors make unrealistic bids intentionally if they believe their losses can be recovered through the claims process.

For all of these reasons, very few projects seem to be completed without a claim for extra work. The claim situation has degenerated to the point where many contractors begin a letter writing campaign, aimed at securing unjustified extras, immediately upon mobilization to the jobsite. Even reputable contractors find the claim process to be a "necessary evil" which protects them from the low bid award system.

Claims have become an inherent by-product of the competitive bidding process. Unfortunately, there is no value added by the litigation which often results. If the U.S. construction industry could devote as much time and energy to innovation as it wastes in litigation, there would not be a technology crisis facing us today.

Quality. For the purposes of this paper, the most serious drawback of the sealed bid system is the lack of emphasis it places on quality in the constructed product. With most products, low cost is synonymous with low quality. In construction, however, we are expected to believe that low cost and high quality go hand-in-hand.

And unlike a free marketplace where consumers are able to shop for
quality as well as price, the construction purchaser is prohibited by
law or convention from making a price versus quality decision.

There is no reward in our system for constructing durable products
which have low life-cycle costs. And low bids provide little margin
for the contractor to attempt new techniques or to upgrade the quality
of his current product.

THE CONTRACT DOCUMENTS

An equitable set of contract documents is essential if the sealed
bid system is to deliver any value to the owner. These documents must
ensure that price comparisons are based on a similar scope of work.
Oddly enough, it is these documents which stifle innovation and
inhibit quality in the constructed product.

Innovation. Construction owners try to build quality into the
project through the use of "cookbook" type technical specifications
(method specifications). Unfortunately, acceptable end results can
only be guaranteed if the owner is committed to costly inspection and
monitoring programs.

Construction technology can develop rapidly and no specifier can
stay abreast of all the new techniques, materials and equipment which
might produce acceptable results. For the contractor, it's difficult
to be innovative when he is required to construct work in a predetermined manner or by using prescribed materials and equipment. Yet
this is a consequence of the true method specification. The
contractor is actually discouraged from applying efficient techniques
because of a "Catch 22" situation where new methods can't be tried
until they have been specified; and they can't be specified until they
have been tried.

Minimal Quality. Proponents of the sealed bid argue that better
than minimum results are regularly achieved by drafting a good set of
specifications. Nevertheless, by inviting a low bid proposal, the
owner is encouraging the contractor to use methods and materials of
minimal quality. Rewards are obtained by rendering the absolute least
which is necessary to comply with the specifications.

The owner has only limited recourse in situations where the
contractor refuses to perform:

- add inspection personnel to ensure the quality of the product;
- terminate and replace the contractor;
- bar the contractor from further bidding.

All three methods are costly, and detract from the concept of value
for money. Anyone who's been involved in a project where the
contractor refuses to perform, knows the level of inspection required
to receive a quality product. Likewise, termination is a poor
substitute for obtaining the right contractor in the first place. And
finally, debarment is only effective as punishment when the owner
requires similar construction services in the future.

FOREIGN CONTRACTING PRACTICES

The sealed bid system is imperfect. The U.S. has begun to address this issue by comparing its contracting practices to those of Europe and Japan. It has been noted that the contractors in these countries behave differently with respect to R & D spending. It has also been noted that many of our latest innovative techniques for earth retaining structures (ERS) were imported from Europe and Japan. We should all be interested to know why foreign practices foster, rather than discourage, such innovation.

<u>End Result Approach</u>. Although many different practices are used overseas, the details are too complex for this discussion. However, there is a recurring theme contained in the contracting practices of Europe and Japan which is absent in U.S. practice. This theme is one which recognizes the important "end result" of any construction operation. Superlative end results are achieved by adopting plans and technical specifications developed jointly by a team consisting of the owner, his consultant and the contractor.

The contract documents almost never take a "cook book" approach to construction. Instead, they permit any solution which will yield the intended result. For example, the choice of method, equipment and product is left to the discretion of the contractor. The owner is commonly assured of quality by one or all of the following means:

QUALITY IN FOREIGN CONTRACTING PRACTICES

REFERENCES demonstrate success on previous projects

DURABILITY is secured by appropriate guarantees

NON-DESTRUCTIVE TESTS confirm results in the field

Unlike this country, owners typically confer with contractors during the planning stages of the job. The ERS contractor has in-house design capability, but for more complex projects he engages an outside designer to prepare preliminary plans. These plans are used for bid purposes, and of course, are based upon the contractor's preferred construction methods. When an outside consultant is retained, he is a recognized expert in his field, often with university status. His experience and reputation usually ensure immediate acceptance of his plans by the owner.

The contractor who can generate the most economies of design and production is usually the contractor with the lowest price. However, even public agencies are not always required to choose the lowest bidder. Especially for work of a critical nature, the caliber of the technical proposal and reputation of the design consultant are given

at least as much weight as price. The contractor's QA/QC and safety
programs and his experience are other determining factors. Price is
the final arbiter only when all other factors are perceived to be
equal.

Potential for Corruption and Abuse. In the U.S., the fear of
government corruption and abuse has prevented the widespread implementation of similar contracting practices. Certainly, corruption
exists in Europe and Japan. In one European country recently, several
top officials of a public railroad agency were fired for accepting
contractor's bribes[13]. Some countries have introduced bidding methods
to combat this behavior, such as choosing the median, rather than the
low bid from all those pre-qualified to perform the work. Another
typical method is to delay the award pending a strict review process.
Disinterested third parties conduct these reviews to ensure that the
owner's agent did not act for personal gain.

Another fear with foreign practice concerns the owner's ability to
ensure that the contractor performs as promised in his proposal. In
the U.S., it is routinely believed that contractors who prepare their
own designs and control their own methods routinely skimp on quality.
In Europe a unique solution has been devised to control the work
without relying on costly inspection services. In certain countries,
insurance companies actually become members of the project team and
serve as site QA manager. These firms still represent the contractor
but make financial guarantees to the owner. Unlike U.S. sureties who
only guarantee completion in the event of default,   European
"sureties" will guarantee the quality of the constructed product.
This guarantee is in the form of a fee payable to the owner in case
of latent defects in design, materials or workmanship which later
effect durability or performance.

Foreign Research and Development. Foreign countries view R&D as
more than a sideline to the business of construction. Instead, they
consider construction to be the commercial application of serious
research efforts. Only by investing heavily in such efforts can the
technologies evolve which are necessary to maintain competitive
advantage.

It is fair to point out that overseas R&D for construction is partly
the result of external influences. For instance, many retaining wall
techniques have been developed to maximize the use of expensive real
estate. Also, improvements in automated construction equipment were
necessitated by the fact that labor is not often cheap, nor can it be
employed on a temporary basis. However, the U.S. construction
industry is now facing the same external influences, and to compete
in a global marketplace, we must be as efficient as our foreign
counterparts.

EARTH RETAINING SYSTEMS IN THE U.S.

There are two notable differences between contracting practices for
earth retaining structures and most other construction projects:

● On the positive side, ERS represent new technologies which have been proven, both technically and commercially. Since ERS are relatively new to the U.S., the technology is still associated with a limited number of firms. This, coupled with the fact that ERS are installed in an unpredictable environment, i.e. the ground, leads to a heightened awareness of quality. It is this awareness which has provided an inroad for more flexible contracting practices.

● On the negative side, all of the drawbacks to the competitive bid system still exist within the realm of ERS. Since most ERS work is proposed by subcontractors or suppliers who have no direct link to the owner, certain defects even become magnified.

Qualification and Certification Clauses. Specifications which require a contractor to document his ability to perform the work always contain some type of qualification clause. Under such clauses, a limited review of a contractor's experience record is normally performed. On the other hand, certification refers to a process whereby suppliers of wall systems, e.g. mechanically stabilized earth, clarify design concepts and demonstrate prior experience after the award. In either case, the owner's approval is required before a subcontractor or supplier may be employed by the prime contractor.

Subcontractor qualification clauses for ERS projects have been employed repeatedly over the past ten to fifteen years but have rarely performed as intended. Remember that the prime contractor must be low bidder to get the job. In an atmosphere where price controls, the prime contractor feels he has a "right" to use the subcontractor of his choice. Furthermore, the prime contractor usually feels compelled to choose lower cost over experience for fear that his competitors will rely only on price in figuring their bids.

If the owner rejects the proposed ERS subcontractor after award, the prime contractor is responsible for any increased cost in obtaining a qualified firm. This places the prime contractor in an untenable position because he has neither the time nor the technical ability to ascertain the qualifications of every ERS subcontractor prior to the bid.

To combat this situation, prime contractors have devised numerous schemes which emasculate the intent of qualification clauses. First, specific contract language requiring firms to demonstrate prior experience has been circumvented by temporarily engaging the services of one person with the necessary expertise. Second, some contractors make arrangements with material suppliers to furnish technicians who witness only the more "critical" operations. And finally, prime contractors have mounted successful legal challenges to the owner's right to deny any subcontractor. The net effect of all these schemes is that experienced ERS organizations aren't in control of the work and quality usually suffers.

In contrast, certification of a mechanically stabilized system has been much more successful. Due to large development costs, only a handful of companies manufacture these systems. Provided the owner can recognize the proper wall system for his project, documented references can be used to weed out unacceptable proposals. In turn, the prime contractor can be forced to select from a group of possible alternates that the owner feels are viable.

However, one difficulty with the use of certification clauses involves a lack of expertise on the part of the excavation contractor installing the wall. Problems develop when this contractor isn't appraised of proper methods of foundation preparation or when inappropriate backfill material is substituted. To avoid such pitfalls, the wall manufacturer should be made responsible for the actions of the installer. Only adequate training and/or on-site control of the work will ensure an acceptable end result.

Pre-Qualification. A variation on the above theme called pre-qualification has met with recent success on ERS projects. Under this concept, contractors and suppliers demonstrate their abilities to the owner well before the bid date. Only approved firms may be considered by the prime contractor in preparing his proposal. In theory, this relieves the prime contractor from making "price versus quality" comparisons as the decision is more properly delegated to the owner.

Unfortunately, economic considerations and time constraints often invalidate this procedure. Prime contractors may still find themselves besieged with last minute proposals from subcontractors who claim the necessary qualifications. If such a proposal is competitive, the prime contractor is tempted to use it and then try to qualify the subcontractor after the award. Owners are often guilty of allowing this practice to continue in the interest of fair and open competition. If not controlled properly, pre-qualification clauses will produce results no better than the standard qualification process mentioned previously.

Furthermore, no national forum exists where standard guidelines can be established for the pre-qualification process. Nearly every owner uses a unique set of parameters upon which to base individual performance. Submitted references are rarely verified in the short time allotted between advertisement and bid. Unless and until some nationally recognized body sets standards for pre-qualification, the process will have only limited benefits.

Finally, the pre-qualification process doesn't guarantee involvement by specialty subcontractors during the project's conceptual phase. If an owner is to obtain true economies of design, foreign practice confirms that such early involvement is more important to the success of a project than any simple pre-qualification process.

RECENT U.S. IMPROVEMENTS - DESIGN/BUILD CONCEPT

It was previously stated that most current ERS technologies have been imported from Europe and Japan. Along with these technologies came new ideas in contracting practices. One of these already existed in the U.S. while others have been adapted to fit constraints imposed by the American sealed bid system.

Value Engineering. The "Value Engineering Proposal" is a form of alternate proposal long established in U.S. practice. Those who are familiar with this procedure know that it is of very limited use for ERS projects. This is not because the contractor must share his cost savings with the owner. Rather, it's because at bid time the prime contractor is unable to assess the following risks associated with using value engineering:

    a) the change may disrupt other work,

    b) the owner may not accept the scheme, and

    c) there may not be sufficient time for the approval process.

When presented with the above risks and little reward in return, most prime contractors simply choose never to use value engineering proposals.

    <u>Post-Bid Design</u>. This concept is used for the great majority of temporary ERS structures, i.e. shoring. For years, it's been recognized that construction costs are lower when the ERS contractor is permitted to act as designer and builder of temporary structures. Recently, the design/build idea has also gained acceptance for permanent work. The major contractural difference with the permanent case involves the owner's role in providing design guidelines and in his liability should a failure occur.

    Post-bid designs permit ERS contractors to submit technical proposals after the award. These proposals must be based upon parameters established in the contract documents beforehand. Sometimes, the owner prepares his own set of plans which any contractor may use in pricing a base bid. These plans establish the concepts which, as a minimum, the owner wants to incorporate in the project. Any contractor who believes he can produce a more efficient design may submit a price based on an alternate which, of course, is subject to the owner's approval.

    More typically the owner avoids the unnecessary expense of providing a base bid design. Instead, he provides only the wall envelope, an earth pressure diagram and relevant soil properties. After award, the contractor submits plans and supporting calculations based upon this information.

    The greatest difficulty with post-bid design surrounds the ability of the owner and contractor to agree on the design after award has been made. Since judgement plays a big part in any ERS design, the contractor's viewpoint doesn't always coincide with the owner's and disputes and delays can ensue.

    With post-bid designs, the prime contractor properly places total responsibility with his ERS subcontractor to obtain the owner's approval. This is one step better than the pre-qualification process because the prime truly avoids "price versus quality" comparisons, but the concept is still flawed. By the time the project is awarded, the ERS contractor has committed to a fixed price. Commercially, it is in his best interests to design to minimal standards since the wall will be less expensive to construct. To counter this, the owner feels compelled to over-specify the parameters upon which such designs are to be judged. In the extreme, this can stifle innovation in design as easily as method specifications inhibit the application of new construction techniques.

Pre-Bid Design. Another variation of the design/build concept has been devised to eliminate drawbacks with post-bid designs. Although very similar, pre-bid designs allow much more flexibility in ERS contracting. The pre-bid design has been used effectively for mechanically stabilized earth walls since their introduction into the U.S. market. Lately, this method has also been employed for design/build anchored walls.

The process begins when the owner calls upon specialty subcontractors or suppliers to submit designs in advance of the bid. Flexibility is derived from the fact that the owner and contractor are obliged to agree on all design issues prior to project award. This timing is crucial since the contractor is not under any financial pressure to dispute the owner's viewpoint. In most situations, the owner can easily make known the level of quality desired and the contractor need only comply.

Once the parties agree, the owner includes all acceptable design plans in the contract documents. Each design is labeled "proprietary" in favor of the contractor who proposed it. In other words, contractors may submit prices only on their own design/build scheme. Disputes which delay the post-bid design process are completely averted since contractors who fail to submit timely or acceptable plans are not allowed to construct the ERS project.

This method works best when the specialty contractor is permitted to prepare plans of a conceptual nature only. These conceptual plans don't include details which the contractor feels are unique to his design. So long as the supporting calculations address these details, the bid documents may include only enough information to make other contractors aware of the nature of the work.

LIABILITY

Liability is a an important issue with respect to any design/build proposal. For temporary structures, legal precedent has made the ERS contractor totally responsible for the adequacy of his design and installation. Exculpatory language in the contract documents usually relieves the owner of liability in case of failure, whether or not he reviewed the contractor's plans. Essentially unregulated, these practices can easily lead to failures of temporary earth retaining structures.

There are two reasons why similar problems are normally avoided in permanent design/build ERS. First, the documents establish strict guidelines for design and installation of the work. Second, and most importantly, the contractor's design is subject to a rigorous review and approval process.

This would appear to place at least some responsibility for adequacy of the design on the owner's shoulders. Unfortunately, this issue remains unresolved since not enough legal precedent exists to determine whether the owner, the ERS contractor, or possibly both are liable in the event of a failure.

TEAM APPROACH

For the system to work properly, it would appear that some type of risk sharing must be initiated. The ERS contractor should be responsible for the adequacy of his design and installation, but the owner must be responsible for the accuracy of project information upon which the design is based.

Risk sharing will come about when owners and contractors finally realize that they should act as a team with quality in construction as their goal. This team approach will lead to lower prices for the owner because the contractor will place fewer contingency costs in his bid. It will also lead to innovation since the contractor is rewarded for economies of design as well as installation. And finally, the team approach will lead to better quality since the owner and contractor will share similar goals.

Negotiated Bids. One final variation on the design/build concept involves the recent tendency towards negotiated, rather than sealed bid contracts. Private industry has used this system for years. The U.S. Postal Service, the Bureau of Reclamation and the Florida State Highway Department have just recently embraced the negotiated bid in one form or another.

The negotiated bid for ERS work requires a detailed technical proposal to be submitted separately from the financial proposal. Contractors are usually graded on a point system, weighted to account for critical features of the project. The point system is disclosed in advance and usually places more value on technical content than price. Criteria such as design technique, quality of materials and method of construction are typically used to judge technical content.

The Bureau of Reclamation has been pleased with the outcome of its negotiated bid program thus far. One official comments that the negotiation process allows the Bureau to question each contractor in detail to make certain he understands the complexity of the project. Such a process eliminates any controversy over the specifications, scope of work or quality level intended[14].

VESTED INTERESTS

In order for ERS contracting practice to change, attitudes in the U.S. industry as a whole must first be changed. This process will be controversial and hence very slow in coming. The key to change lies within industry groups who are in a position to recommend needed improvements. Three major industry groups are listed below along with their positions concerning changes in U.S. practice. This list was prepared from recent publicity statements and only represents the author's understanding of each group's position. It is interesting to note that the sheer volume of recent publicity indicates dissatisfaction with current U.S. practice. It also seems that battle lines are being drawn on both sides of the issue, often according to vested commercial interests.

The Insurance Industry. Insurance companies believe that design/
build jobs present unnecessary risk. This is not just due to design
liability issues, as one might expect. Instead, this is because they
feel that design/build contracts eliminate an important source of
possible relief from cost over-runs and delay damages for the
contractor.

AGC and ARTBA. The Associated General Contractors of America (AGC)
and the American Road and Transportation Builders Association (ARTBA),
two powerful lobby groups, oppose any change to the competitive,
sealed bid process. AGC is on record against a recently announced
FHWA proposal to experiment with the traditional low bid award by
saying it is "not in the public interest"[15]. ARTBA, which is remaining
"open minded" has reiterated its strong support for the competitive
bidding process but "will work with FHWA on this issue"[16]. The AGC also
strongly opposed changes to the Federal Acquisition Regulations which
permitted the Bureau of Reclamation, Army Corps of Engineers and
others wide flexibility in the use of negotiated bidding. They are
also opposed to design/build proposals in all but a few instances[17].

American Society of Civil Engineers. Our group has recognized that
changes must be made in the owner/engineer/contractor relationship if
quality is to be improved and failures are to be averted.   In their
1988 Manual of Professional Practice: Quality in the   Constructed
Project, ASCE points to this relationship as the key to any
improvement. The manual suggests that quality can be enhanced by
basing selection of the design professional on his ability and
reputation. For some reason, however, the manual states that
contractor selection should continue to be based on the lowest bid.

Unfortunately, the manual also defines quality as "conformance to
predetermined requirements". The author believes that conformance
implies minimal standards and that predetermined requirements inhibit
innovation. Instead, the author agrees with the consulting engineer
who defined quality as the "superlative result obtained from
comparative analysis"[18].

IMPROVEMENTS

U.S. contracting practice is slowly improving. The greatest
obstacles to change lie in our present laws and conventions. For
these to change, we must unite as an industry. Unfortunately, the
above indicates that our industry may be too fragmented for this to
occur.

With regard to earth retaining structures, the author feels that a
nationwide system of pre-qualification must be adopted for
design/build contractors and wall suppliers. If these firms had a
single venue for establishing their references, a meaningful list of
qualified firms could be compiled. This list could then be relied
upon by both public and private construction owners to solicit only
quality proposals for their projects.

For this to transpire, some nationally recognized group must
organize and regulate the system. Pre-qualification should be limited

to those subcontractors and suppliers willing to undergo a stringent "peer review" process. Reviews would be conducted by consulting engineers (such as ASCE members), with the intent of comparing the following three criteria for each candidate:

o Design Ability - examination of the credentials and references of the contractor's engineering staff (and/or outside consultants) together with an in-depth review of the design for one major project.

o Quality Assurance - inspection of the above mentioned project to determine whether all work was performed in accordance with the design.

o Commercial Endorsements - ten previous projects would be analyzed to ascertain owner satisfaction with the contractor's performance. Endorsements would be based upon both objective and subjective guidelines, including:

    1) meeting the project schedule;

    2) competence of the contractor's staff;

    3) durability of the constructed product.

At the completion of these reviews, each contractor would be issued a "pass/fail" grade in accordance with recognized standards for his particular category of work. Then, a list would be published indicating the organization's "seal of approval" for certain firms.

The above system would combine the most beneficial characteristics of pre-qualification and alternate bidding. Also, since either competitive or negotiated bidding could be used in conjunction with the system, no major changes to present U.S. contracting practices would be required.

CONCLUSIONS

The United States has the most economically stable construction market in the world. For this reason, it is very attractive to the overseas contractor hoping to expand. Until recently, we were impervious to such foreign intervention. Yet today, American contractors are losing market share as they try to compete with their foreign counterparts on a technical basis. This isn't unusual, given the disparity in associated R&D investment. Even within ASCE we see that nearly every geotechnical construction technique developed since the last Cornell conference was imported from Europe or Japan.

American contracting practices are the only remaining barrier to complete dominance of the industry by foreign firms. This is because our system is complicated and peculiar to foreign construction companies. Nevertheless, they are gaining experience every day and before long, U.S. contractors will be competing on a commercial as well as a technical basis.

The U.S. construction industry must recognize that the competitive bid system is largely responsible for its recent declines. Also, the industry must accept that there is more to learn from foreign companies than technique. Instead of focusing on the transfer of technology from overseas we should be looking for ways to import commercial practices that foster technology from within.

To do so, an overwhelming number of fragmented groups including owners, engineers and contractors must be united. Unfortunately, these groups are further fragmented into special interest groups of their own. Yet allying these diverse groups is the major challenge facing us today. We at ASCE must build upon our successful experience with recent developments in ERS contracting.

Unfortunately there is no single practice which provides a panacea. Design/build proposals, especially pre-bid designs, solve many problems related to earth retaining structures. Negotiated bids also seem promising. But if changes are to be made in a timely fashion, a national, fool-proof method of regulation is needed. Some sort of centralized oversight will be required to sway the many powerful critics of design/build and negotiated bidding. The peer review concept mentioned in this paper is possibly one such method.

Peer review would ensure that owners are protected from unscrupulous contractors and their bid practices. Quality in the constructed product could also be assured by rewarding contractors for superlative performance. Most importantly, peer review would allow the U.S. construction industry to take a team approach towards closing the innovation gap with its foreign competition.

REFERENCES

1. Construction at the Crossroads, Civil Engineering "Forum", December 1989, p.6.
2. The Bureau of National Affairs, Labor Reports, Vol. 133, No. 1620, p.35.
3. FMI Management Consultants, U.S. Markets, Construction Overview, 1989 Edition.
4. JA Pfeiffer, Construction Productivity Advancement Research Program Report, January 1990.
5. JH Wiggins, Construction's Critical Condition, Civil Engineering, October 1988, p.72
6. Engineering News Record, July 7, 1988.
7. Claims: The Navy's Better Way, Constructor, May 1989, p.20.
8. Building for Tomorrow, Global Enterprise and the U.S. Construction Industry, National Research Council, 1988.
9. Fragile Foundations, A Report on America's Public Work to the President and Congress by the National Council on Public Works Improvement, February 1988.
10. Prof. Zohar Herbsman, Paper (untitled) presented to the Transportation Research Board Task Force on Contracting Practices, August 1988.
11. Robert Morris Associates, Annual Statement Studies, 1989.
12. Construction Industry Annual Financial Survey, Construction Financial Management Association, 1989.
13. J. Nicholson, Nicholson Construction of America, Personal Conversation with Luigi Ginetti, Rodio SpA, Milan, Italy.
14. P.J. Nicholson, Nicholson Construction of America, Personal Conversation with a Bureau of Reclamation Contracting Officer,

   January 1990.
15. AGC National Newsletter, Vol. 42, No. 1, January 23, 1990.
16. ARTBA Newsletter, Vol. 34, No. 2, January 23, 1990.
17. Report on Construction Quality: Competitive Negotiation v. Sealed
    Bids in Federal Construction Procurement, AGC, September 1989.
18. O. Riley, Paper presented to the Transportation Research Board
    1990 Annual Meeting.

ALTERNATIVE WALL AND REINFORCED FILL EXPERIENCES ON FOREST ROADS

Gordon R. Keller[1], M.ASCE

ABSTRACT: Practices and experience with standard and nonstandard retaining structures and their selection by the USDA Forest Service on low and moderate standard rural roads are summarized. More detailed information is provided describing innovative and low cost alternative earth reinforced retaining structures, including welded wire walls, chainlink fencing walls, geotextile walls, timber and tire-faced walls, lightweight walls, and reinforced fills. Use of marginal quality backfill material is discussed.

INTRODUCTION

The U.S. Department of Agriculture, Forest Service, manages a vast system of roads which provide access to America's national forests. Over one quarter-million miles of roads exist with a wide variety of design standards ranging from major paved Federal Highways to local, narrow, single lane native surfaced roads. Most roads are in mountainous terrain and many resemble the county roads to which they are connected.

Because the Forest Service is governed by laws such as the National Forest Management Act (NFMA), the agency is directed to establish road standards appropriate for the intended use balanced between such factors as economics, environmental impacts, and public safety. A wide variety of retaining structure types and designs have been used in recent years to support the road system and best meet the intent of NFMA in steep terrain, unstable ground, for storm damage repairs and in narrow road sections.

Wall selection is initially indicated by site geometry and constraints, and is influenced by a wide variety of factors as discussed below. Many unique and moderately sophisticated designs and construction techniques have been used to fit specific site needs. A wide variety of standard and nonstandard structures are commonly used. Examples of this range of wall types will be discussed briefly along with some applications.

Most walls constructed by the Forest Service are less then 25 ft (7.5 m) high and are located in areas where site access is poor.

1 - Geotechnical Engineer, USDA, Forest Service, Plumas National
    Forest, PO Box 11500, Quincy, CA. 95971.

Use of lightweight or prefabricated wall components and on-site or local construction materials is highly desirable. Alternative earth reinforced walls using native backfill material and a variety of reinforcement materials and facing schemes satisfy these requirements and will be discussed in some detail as practical solutions to most forest roads needs. These walls also appear well suited for many private, local and county roads needs.

Considerable experience has been gained with reinforced fills which offer an economical alternative to retaining structures on moderately steep slopes where site geometry permits their use. Various techniques for slope face stabilization and varying slope angles have been tried and are discussed.

Typical soils developed in a mountain environment have a high friction angle, satisfying needed design strength criteria. However, fine grained, low quality material may be locally available which present other design and construction problems. Drainage of structures is nearly always incorporated into designs, and geocomposite drains, costing $2-$4 per ft^2, have proven to be very effective for drainage of most wall and reinforced fill sites.

WALL SELECTION CRITERIA

An appropriately chosen and well designed retaining structure should be stable, safe, practical to build, economical, and aesthetically and environmentally compatible with its surroundings. To achieve this given the wide variety of retaining structure options available to the designer, attention should be given to wall selection. A thorough treatment of wall selection factors considered in Forest Service applications has been published (5).

First the site objectives should be defined and a geotechnical site investigation should be conducted to properly define the problem and assess practical alternative solutions, particularly from a soils and foundation standpoint. Site conditions usually dictate the required wall size and often the basic wall type.

Gravity and reinforced backfill structures are most commonly used where site constraints are not severe. Gravity type structures typically require better foundation conditions and tolerate less deformation than earth reinforced structures, but also require slightly less foundation excavation. On steep sites and sites in massive bedrock, cantilever or anchored pile type walls are typically appropriate to minimize excavation, as well as to minimize site access problems and site disturbance. Wall selection in landslide terrain is often dictated by need for a lightweight structure, one that can tolerate considerable deformation, or by depth to competent material beneath the slide.

Economics typically influences wall selection once the basic appropriate wall types are chosen. Wall cost depends on a number of factors including: short term vs long term wall life; locally available or surplus construction materials; remoteness of the site and difficulty of terrain; need for standard vs nonstandard design solutions; construction timing constraints; size of the project; traffic needs during construction; equipment available; wall maintenance needs; and others.

Because the Forest Service is a multiple land use agency, environmental constraints identified by an interdisciplinary process can significantly affect wall selection. Visual constraints in a scenic mountain setting may dictate use of aesthetic wall facing materials or materials which are visually subtle or compatible with local surroundings. Site disturbance may need to be minimal and construction techniques needed to insure that sediment does not reach local waterways. Also a structure may need to minimize encroachment on a local stream or minimize impact on local wildlife habitat or migration routes.

Public safety must be insured during construction and the service life of any structure. However, varying degrees of risk can be incurred for some structures or types of structures depending on the remoteness of the site, cost savings involved, intended use, quality of engineering geologic input, and consequences of failure.

STANDARD AND NONSTANDARD WALLS USED

Nearly all types of existing walls have been used by the Forest Service nationwide. Most experience has been gained in the Pacific Northwest, California, and in the Rocky Mountains. These include both standard and nonstandard walls, where standard walls are defined as ones using "off-the-shelf" materials and commonly used, preexisting designs. Nonstandard walls are unique solutions or custom designs developed for specific applications. Standard structures used have included gabions, concrete gravity and concrete cantilever walls, rectangular and circular metal binwalls, timber and concrete cribwalls, Reinforced Earth Walls, etc. Internal structural stability is typically guaranteed by the manufacturer so the designer is principally concerned with site conditions and external stability factors of sliding, overturning, bearing capacity, and overall stability.

Nonstandard walls typically require detailed site investigation and specific geotechnical design analysis. Cost of these walls vary greatly depending on site specific conditions and materials used. Examples of a variety of nonstandard walls used and general applications are listed below:
- Driven H Pile walls are suited to sites on weathered bedrock or residual soil over bedrock and are used in steep terrain to minimize excavation. Timber lagging is commonly used, though sheet piling has been used in shallow applications. For walls over 8 to 12 ft (2.4-3.6 m) high tiebacks are typically required to resist earth pressures and minimize bending moments. Some 5/8 in. (15 mm) tieback anchor cables have corroded to failure within 15 years, so currently solid steel tie rods with corrosion protection are used.
- Drilled or Anchored H Pile walls are used on steep slopes on competent bedrock where pile driving is ineffective. Piles or piers are placed into footing holes drilled several feet into solid bedrock for toe support. Alternatively they can be supported by a footing and anchors at the toe and top of the piles tied to rockbolts or deadmen. Since alignment of the piles can be carefully controlled, precast concrete panels have been used as an alternative to timber lagging.

- Anchored Horizontal Sheetpiles and Vertical Culvert Pipe walls have been used where these materials were readily available or surplus. Support is achieved using passive earth anchor plates or strips buried in the backfill. Walls require minimal foundation preparation and can tolerate moderate settlement.
- Small nonstandard gravity structures have been constructed out of materials such as stacked used tires, stacked sack concrete, precast concrete blocks, GEOWEB cells, logs, and culvert pipe material. They are often used for support of roadway shoulders and low cut slopes.
- Vertical Sheetpile walls have been used along waterfronts and along boat ramps where site access is restricted. Also earth reinforced walls have often been used in Alaska for sea port facilities applications in areas with wide tidal fluctuations. Free draining, 8-inch minus (20 cm) shot rock is used for backfill material. Concrete panels and concrete blocks have been used as facing members in areas which are periodically submerged and are expected to be battered by drift debris and logs.

Because many standard and nonstandard walls have been built, because "as-built" modifications are common, and because varying factors of safety are used in design, depending on knowledge of site conditions and risk involved, a wall inventory and monitoring system is desirable. The Willamette National Forest in Oregon is instituting a prototype wall inventory data system to insure that monitoring is conducted by geotechnical personnel. Condition surveys are made, needed maintenance is scheduled, and evaluation of wall performance is done, particularly on unique and high risk walls.

ALTERNATIVE EARTH REINFORCED RETAINING WALLS

Many recent innovative designs have been developed utilizing earth reinforcement concepts. Forest Service walls have been built utilizing a variety of reinforcing and facing materials. Other alternative walls appropriate for high and moderate standard roads have been developed by the CALTRANS Transportation Laboratory, Sacramento, CA, and further discussions on the topic have been presented by the Federal Highway Administration (12).

Earth reinforced structures are the most commonly built type of wall used in the Forest Service today, mainly because of cost and ease of construction. Most utilize local or on-site backfill material and easily fabricated, flexible reinforcement elements. For walls under 25 ft (7.5 m) high, cost has typically ranged from $15.00 to $25.00 per ft^2 of face. Both frictional reinforcement systems, such as geotextiles, and passive resistance reinforcement systems, such as welded wire and geogrids, are commonly used. Lightweight geotextiles (i.e. 6 oz. per square yard) with nominal strength appear cost effective for reinforcement of walls to about 15 ft (4.5 m) high while stronger geotextile, welded wire or geogrid material is desirable for higher walls.

Wall sites are often in landslides or in areas having marginal foundation conditions so some allowable wall deformation is desirable. Also site investigation is often minimal for small walls. Earth reinforced structures which minimize foundation pressures, which have relatively wide foundations, and which tolerate deformation are desirable. Welded Wire walls, geotextile walls, chainlink fencing walls, timber and tire faced reinforced walls have all been used which satisfy these needs.

Welded Wire walls.   Welded Wire walls up to 30 ft (9 m) high have commonly been used by the Forest Service in Oregon, Washington, and California for several reasons. Many such walls have been built throughout this area so many contractors are familiar with their assembly. Local manufacturer's representatives typically provide excellent construction support services to contractors, both with technical advice and by providing on-the-job training to contractors unfamiliar with their product. Also, they are relatively easy to construct on grades and curves. Cost, ranging around $20-$29 per ft^2 of face, is competitive with other types of walls on medium size projects.

Either the wire wall can be specifically designed using published literature such as that presented in NCHRP 290 (8), or by using computer programs available from the manufacturer. Alternatively, for routine walls a conservative design, satisfying internal stability requirements, can be generated using the manufacturer's design tables for given wall heights and loading conditions. External stability of the structure can be focused upon, since this is the most common mode of wall failure, and time can be spent insuring that the structure has an adequate foundation.

Many wire walls have been constructed a little shorter than actually needed, perhaps because it is difficult to envision a 3-dimensional structure during design, because of the high unit cost of walls, or perhaps because of the 8-foot (2.4 m) increments in which wire walls are assembled. The end of the wall is often not adequately anchored into in-situ ground. Also material is commonly overexcavated at the ends of the wall and can only be placed back on a 1 1/2H:1V slope. In either case a debris chute may form at the end of the wall where backfill material will not catch. This can eventually result in a narrow road section, or it can lead to the additional expense of a wall extension. Plan on an adequate wall initially!

Use of fine backfill material has occasionally led to face settlement, both from poor compaction along the wall face and from fine soil migration through the wire face of the wall. Filter criteria must be satisfied. On several walls a layer of heavy, ultraviolet resistant geotextile has been placed against the wire mesh to contain the fine soil, thus minimizing face settlement. Use of tamped pea gravel or coarse material against the face will further minimize this problem and is generally recommended.

Reinforcing wire with a standard 0.4-ounce per square foot commercial galvanized coating has a 50-year design life in non-corrosive soils. A significant increase in design life can be gained by specifying 2.0-ounce per square foot hot-dip galvanizing, for a small increase in price. Alternatively, thicker wire can be specified to extend the design life.

These walls have been routinely used in a wide variety of applications. They have been particularly well suited for use around bridge abutments in conjunction with a gravel or rock backfill to minimize settlement.

Geotextile walls.   Geotextile-reinforced soil walls, pioneered by the Forest Service in Oregon and Washington, have perhaps the least expensive materials cost of any wall available today.   For reinforcement materials alone, the cost is as low as $1.25 per ft^2 of wall face.   Design procedures are widely published (7,11). Additionally, geotextile manufacturers now have design procedures available utilizing their products (1). These procedures have led to many successful, and perhaps very conservative designs for walls 10 to 20 ft (3-6 m) high.

Design is reasonably straightforward given specific site conditions, geometry, backfill properties, and chosen geotextile properties. Reinforcement lift thickness typically varies between 6 and 18 inches (15-45 cm), with thicker lifts being difficult to form. Base geotextile embedment length is typically nominal for pullout resistance and is dictated by the length required to resist sliding failure.

The constructed cost of geotextile walls should be minimal because of low materials costs. Experience, however, has shown that their cost is only slightly lower than most other walls, ranging between $15 and $25 per ft^2 of face.   Because of the flexibility of fabric, temporary forms must be used to support the wall face as each lift is constructed, making this process somewhat slow and labor intensive. Long 2"x12" dimension lumber and metal brackets are usually used for forms. One creative application of a semi-permanent form utilized hay bales placed along the face of the wall, with one layer of geotextile enveloping the soil and a second layer wrapped around the hay bale to hold it in place.

Most geotextiles must be protected from long term degradation by sunlight. Often a gunite layer is applied to the wall face. In remote forest applications the Forest Service may specify a protective coating of asphalt emulsion, which must be repeated several times during its service life. The final wall face itself usually has an irregular shape, but its appearance is acceptable in most rural settings (Figure 1). This type of wall is also ideal for temporary construction applications.

Choice of fabric does not appear to be critical for a geotextile type wall, provided that tensile strength requirements are adequate. A variety of woven and nonwoven fabrics have been used. Today with the "designer" oriented range of fabrics available, a slit-film woven fabric appears best for wall applications.   These fabrics typically have high tensile strength at low strain values, similar to soil stress-strain properties, at a cost typical of or lower than other geotextiles. The only reported disadvantage of slit-film woven fabrics is its tendancy to repel asphalt emulsion coatings.

Geogrid walls have been built in a design concept very similar to the geotextile wall, where additional materials strength and less creep are desired for a high wall, or to have a stronger, more durable facing material.

Lightweight geotextile walls.   Several walls, up to 28 ft (8.5 m), have been constructed in Oregon and Washington using wood chips or sawdust. This material, wrapped in a geotextile, produced a lightweight structure ideal for placement on an active slide deposit. Design and construction procedures for this wall were roughly similar to those of a normal geotextile wall since wood chips have a high friction angle (25 to 40 degrees, based on triaxial tests). Wood chips were spread and compacted in 18-inch (45 cm) lifts between the reinforcing layers. Compaction was difficult to measure or achieve so a procedural specification of several passes per lift was used. A final typical moist density of around 40 pcf (6.3 kN/m^3) was achieved.

Gradation of the wood backfill used has ranged from fairly clean 3-inch (75 mm) maximum size chips to a fairly dirty sawdust. Material within this range have had similar strength and performance properties. Long term permeability of the material was relatively low (10^{-2} to 10^{-3} cm/sec), due to swelling of the wood, so a drain should be placed behind such a structure (4).

Cost of one such structure was $12.25 per ft^2 of face in 1984. Performance of this wall has been satisfactory and settlement of the material after 9 years has been limited to about 5 percent of the structure height. Shredded used tires might also be used as a lightweight fill material in similar applications.

Chainlink fencing walls.   Several chainlink fencing walls to 22 ft (6.7 m) high have been constructed in Oregon and Washington using conventional 9-gage galvanized chainlink fencing material placed in 12 to 24 inch (30-60 cm) lifts in the backfill material for reinforcing (Figure 2).   This wall is a combination of procedures used for a conventional geotextile wall and a welded wire wall. Pullout resistance and strength parameters for the chainlink material are similar to those of a welded wire material, while the construction procedures for forming the face are similar to those used for a geotextile wall. A 1/4-inch (6 mm) galvanized hardware cloth is placed at the wall face to confine the backfill material.

The chainlink material can be uncoiled or spliced together to the exact length of material desired. Also the irregular mesh texture easily accomodates face settlement. This type of wall was found to be competitive with a welded wire wall, at a cost of about $22.00 per ft^2, but is less likely to be used because of the custom internal design needed for chainlink fencing.

Timber-faced walls.   An ideal type of reinforced soil wall appears to be one incorporating the ease and cost savings of

Figure 1. Geotextile Wall With   Figure 2. Chainlink Fencing
Typical Face Irregularity.   Wall Under Construction.

geotextiles or geogrid with durable and aesthetic timber facing members. A fabric-timber wall developed in Colorado (2) appears to be a nearly ideal combination of materials (fabric and railroad ties) which is easy to construct, aesthetic, and cost effective. Several such walls, to 18 ft (5.5 m) high, have been constructed by the Forest Service to date and appear pleasingly rustic and appropriate for a rural setting (Figure 3). Cost has varied between $14 and $19 per ft^2, depending on the price of treated timbers. Both a geotextile and geogrid have been used for reinforcement.

The connection detail of the reinforcing material to the timbers varies between designs. Techniques have included sandwiching the geotextile between the timbers, stapling the material to the timbers, wrapping it around the timbers and adding a face plate, and using an intermediate wrapped board sandwiched between the main timbers. This last technique appears best and strongest to date, though there is relatively little lateral load applied to the facing members when the reinforcement spacing is close. The lattice structure developed with this procedure can tolerate some deformation, it reduces timber needed by about 25 percent, and it is aesthetic. Timbers are often pinned together with spikes or rebar.

An interesting documented failure of a fabric-timber wall (9) shows the need for attention to the facing-connection detail. The timbers need to be tied to the reinforcing members every timber or two, at a 6 to 16-inch (15-40 cm) vertical spacing. Also the reinforcing material cost is only a small part of the overall project cost. Thus the vertical spacing of the geotextile or geogrid is not critical. Although a wide spacing may satisfy needed tensile reinforcement, narrower spacing needed for face connections and dictated by lift thickness often controls the design.

Tire-faced walls. A unique 10 foot (3 m) high reinforced wall recently designed and constructed on the Plumas National Forest in California used a slit film woven geotextile reinforcement combined with used tires for the facing members (Figure 4). Since used tires are abundant, geotextile cost is minimal, and since construction is simple, the cost of this type of low wall is minimal and could be under $13 per ft^2 of face, including installation of a drain. Actual cost was $17 per ft^2 because of unusually slow production rates for compacting the backfill.

Figure 3. Timber-Faced Wall
Showing Lattice Design.

Figure 4. Tire-Faced Wall
After Completion.

The design consisted of layers of geotextile on a 15-inch (38 cm) vertical spacing and two rows of tires, staggered on top of each other, placed to the front edge of the fabric (and wall face) between each fabric layer. Soil was compacted behind each layer of tires in 7-to 8-inch (18-20 cm) compacted lifts. This procedure was repeated for the 16 vertical rows of tires used. Local material was backfilled into each tire and hand compacted, filling effectively only the middle "hole" in the tire.

This type of wall needs to be built with a nearly 1H:4V face batter and tires staggered horizontally, one half tire diameter on each successive layer, to prevent the backfill soil from falling through the hole and space between tires on the next lower layer. Additional stagger and vertical offset of the tires could provide planting space in the tire holes for vegetation, adding long term biotechnical stabilization to the wall and improved appearance.

Tires formed the wall face. Placed 2 tires high between successive layers of geotextile reinforcement (Figure 5), they worked well and with minimal tire movement during compaction of the backfill with light construction equipment. Heavy compaction equipment or a higher wall might necessitate use of a layer of fabric between each row of tires. Also tires can be attached to the fabric to increase sliding resistance by threading a several foot long strip of fabric through the tire hole and burying the strips in the backfill.

Lateral movement of the "unattached" tires and long term settlement of the tire face are potential limitations for this type of wall. However, after over a year of monitoring wall settlement, deformation appears acceptable. As seen in Figure 6, face settlement on the top row of tires has been about 1 ft (.3 m), or 10 percent of the wall height. This is spread uniformly across the length of the wall, and is proportional to wall height. No deformation was observed in the road surface.

Aesthetics of tire walls may leave something to be desired, but they appear appropriate for low use, rural road repairs. The wall from a distance appears textured and visually acceptable. Also many walls supporting a road prism are rarely seen by the road user.

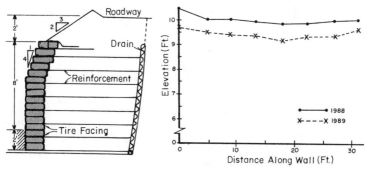

Figure 5. Section through Tire-Faced Wall.

Figure 6. Plot of Settlement Data for Uppermost Layer of Tires.

USE OF REINFORCED FILLS

Reinforced fills placed with a 1H:1V or steeper face slope can offer an economical alternative to retaining structures for those sites where the ground is too steep to catch a conventional fill slope yet is flat enough to catch an oversteep reinforced fill. This falls into an applicable ground slope range of 67 to over 150 percent. Reinforced fill heights have ranged from 15 to 50 ft (5-15 m) on forest projects.

A reinforced fill basically consists of a fill slope built with layers of a reinforcing member (geogrid or fabric) which add tensile resistance to possible shear failure through the soil. The spacing of the primary reinforcement is chosen to add the tensile strength needed to support the oversteepened fill slope and to prevent a deep seated slope failure. Spacing typically varies between 2 and 5 ft (.6-1.5 m) and depends on soil parameters, height of the fill, and strength of the geogrid. Intermediate reinforcement, placed between the primary reinforcement, typically consists of narrow (e.g. 5 ft (1.5 m) wide) strips of low strength geogrid placed along the fill face on a 1 ft (.3 m) vertical spacing. They prevent local failure on the oversteep face between the primary reinforcement layers and prevent failures due to construction equipment loading.

The cost for reinforced fills has been between that of a conventional fill and a retaining structure needed for that site. Reinforcing materials needed are roughly similar in quantity to those used in a wall so the principal cost savings comes from avoiding use of forms or installation of facing members, both typical in wall construction. These items are fairly labor intensive.

Installation of the geogrid is quite easy but care must be taken to insure that the geogrid is correctly oriented. Because of contractors' unfamiliarity with this design concept to date, constrution costs have not been as low as expected. Bid prices in 1987-1989 have averaged $10-$15 per yd^3 for controlled compaction of the material plus $4-$8 per yd^2 for the reinforcing geogrid, installed, including both the primary and intermediate reinforcement members. These same costs, expressed per square foot of vertical fill face, have ranged between $3.50 and $12. The lower cost can be realized on wide sites where equipment can operate efficiently.

Practical fill face slopes. With reinforcement, the final fill face typically achieved has been a 1H:1V slope, but this slope can vary between 1 1/4H:1V and 1/2H:1V, depending on the soil type used and extra measures taken. Approximately a 1H:1V slope is the steepest slope face achieved using a dominantly granular, low plasticity backfill material typical of mountainous terrain. Because the outer edge of the fill face is unsupported, good compaction in this area is very difficult to achieve. Without adequate density, soils placed on a steeper slope will typically not hold and local fill face instability will occur. Attempts to construct an unsupported 1/2H:1V fill face have failed, even using a slightly clayey soil, and a 1H:1V slope was the end result.

Use of a material with significant clay content may allow a somewhat steeper slope to be constructed but performance is probably still controlled by construction limitations at the fill face. In very granular, non-plastic material, such as decomposed granitic soil, experience has shown that a 1 1/4H:1V face is the most appropriate stable slope. A steeper slope will either ravel or needs some additional type of support, such as wrapping the reinforcement material around the face. If much wrapping is needed, a retaining structure may become more economical. Additional netting materials have been placed on the fill face to minimize face ravel, and the Siskiyou National Forest in Oregon has used 8-inch (20 cm) quarry rock on fill faces to achieve a stable 1:1 slope. Also a variety of prefabricated concrete blocks are available for facing material.

Some promise exists for achieving a 1/2H:1V slope face on low fills with use of biotechnical measures such as vegetal stabilization. Experience (3) with fills up to 11 ft (3.3 m) high on the San Juan National Forest in Colorado has shown that a 1/2H:1V slope can be constructed, without forms. They have placed a mixture of straw, clay-rich soil, manure and seed in 1 ft (.3 m) lifts along the outer couple feet of the fill face between geogrid reinforcing layers. The straw provides tensile strength to support the steep slope and erosion protection until the seed germinates which further adds root support. These measures can become labor intensive and to date cost has been $9.50 per ft^2. An aesthetic, well stabilized face has been produced.

A 3/4H:1V slope face has been achieved on the San Juan Forest by pneumatically shooting expanded fiberglass strands into the outer couple feet of the fill face as fill material was being dumped with a backhoe. This method, similar to but cheaper than methods developed in Europe, appears promising and very inexpensive. Only approximately 0.1 percent by weight of fiberglass reinforcing is added to the soil and cost has been about $3.50 per ft^2, including reinforcing geogrid within the fill.

Design procedure alternatives. Specific design procedures for reinforced fills have been well summarized (8) and simple step by step design charts have been developed by various manufacturers (10). Design charts, without detailed stability analysis, are commonly used to design projects. This approach appears appropriate for rural, non-critical applications on relatively small projects. Also the cost of reinforcing materials is a relatively small percentage of the total repair cost of a site, so optimization of a design is not critical and may not be cost effective.

Design procedures developed to date are typically suited to sites which have deep-seated, rotational failure surfaces or to new fill embankments where long reinforcing members can be used. However, the most common type of slope failure in mountainous terrain is a shallow debris slide, so the design procedure has been adapted to repair shallow fill failures over bedrock. Basically the reinforcement material and spacing is used as developed from published design procedures for the fill height needed, but the reinforcement length is shortened from the recommended value to fit the site geometry. Reinforcement length is determined to prevent sliding or overturning of the reinforced mass. Since reinforcement spacing is developed based on the most critical, typically deep seated failure surface, and since only a shallow failure is likely to occur on a debris slide, the design spacing is conservative. Internal stability of the reinforced mass is thus conservative.

External stability, however, may be marginal. The critical mode of failure for this type of design has been sliding along the fill-native material, or fill-bedrock interface. Limit equilibrium analysis of the fill as a sliding block has shown a factor of safety with respect to sliding of less than one unless the interface is stepped, forcing a failure through more competent in-place material or through the reinforcing members. If a geocomposite drain is used under the fill and placed along the fill-native material interface, as is commonly done, sliding failure is even more likely since the soil-to-geosynthetic friction angle is typically assumed to be only 60-90 percent of the friction angle of that soil itself (6). A graded aggregate drain can minimize this problem.

As determined on two fills designed to date with a benched soil-fill interface (Figure 7), adequate terracing of the interface can achieve a desired factor of safety of 1.5 against sliding. Terracing, however, produces additional construction difficulties, particularly for installation of a drain, which may increase the total project cost.

BACKFILL MATERIAL

On-site or local materials, often of marginal quality ("Marginal" soils are defined as fine grained, low plasticity materials which may be difficult to compact, have poor drainage, or have strength parameters sensitive to density), are consistently used by the Forest Service for backfill in retaining walls and fills. Coarse, granular, free draining backfill material, commonly specified by wall manufacturers, is very expensive to import to most construction sites. Coarse rockfill material, occasionally available, often has enough oversize material to make layer placement difficult and rocks can damage the reinforcement material. Local materials used to date have varied from silty sands to silts and clays (SM, SC, ML & CL Unified Soil Classifications) with over 50 percent fines (passing the No. 200 sieve). Marginal materials should be tested (i.e. triaxial tests) to determine their strength properties and strength-density relationships. As seen in Table 1, most soils used have had good frictional properties (typical of forest soils), usually exceeding a 30 degree effective friction angle (Phi) at the specified compaction density.

Since basic strength parameter requirements have been adequate and accounted for in the design, use of marginal backfill has been acceptable, but use can present subtle problems in construction and long term performance. Compaction of fine grained soils is sensitive to moisture content so close construction control is needed to insure that specified densities are achieved. Typically a density equal to 95 percent of the AASHTO T-99 maximum density has been specified. Occasional areas of poor compaction, particularly along the wall face, result in local areas of face settlement, as

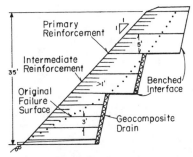

Figure 7. Reinforced Fill Section with Benched Soil-Fill Interface.

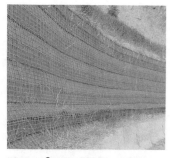

Figure 8. Welded Wire Wall Showing Local Face Settlement.

seen on the welded wire wall face in Figure 8. Unless settlement is severe it does not present a structural problem for the wall but it does make the face appear uneven, which may be an aesthetic issue. Granular backfill material is often used at the face to avoid this problem as mentioned previously. Extensive areas of poor compaction could affect the overall structural stability of the wall or overstress local reinforcing members.

Most walls built have had drainage provisions added, either to remove local groundwater or to insure that the backfill will remain in a drained condition (as assumed for the design). With fine grained, low permeability backfills, long term saturation of the

Table 1. Typical "local" soil used in structures.

Site	Wall Type	USC	% Minus 200	PI	Phi' deg	c' psf	Comments
Goat Hill	Welded Wire	SM	21	5	34	200	Some Face
		SC	20	8	31	300	Settlement.
L.North Fork	Reinforced	SM	38	2	34	100	Slight Slope
	Fill (1:1)	ML	55	3	33	150	Ravel.
B.Longville	Welded Wire	CL	50+	-	26	200	Poor Foundation.
Grave	Geotextile	SM	26	NP	35	850	Irregular Face.
Butt Valley	Tire-Faced	SC	38	8	26	400	10% Face Settlement.
Klamath	Timber-Faced	SM	27	NP	30	0	Minimal Settlement.
Willamette	Wood Chips+ Geotextile	GP	0	NP	32	0	5% Total Settlement.

Note: Phi' and c' are from Consolidated-Undrained tests @ 95% of T-99 Density. 1 psf = 0.0479 kN/m^2.

fill in a wet environment is possible, even with a typical chimney drain installed behind the structure. To prevent saturation, a layer of free draining gravel can be built into the wall. To prevent surface water from ever entering the fill, the backfill material may be "waterproofed" with a paved roadway surfacing, and positive surface drainage is incorporated into wall designs.

SUMMARY AND CONCLUSIONS

A variety of standard and nonstandard retaining walls, reinforced fills and alternative earth reinforced retaining walls have been constructed on USDA Forest Service roads in mountainous terrain. Choice of the structural measure depends on many factors including cost, site conditions, access, and experience. Reinforced fills on moderately steep slopes can offer an economical alternative to walls, and can be adapted to many site conditions. Many alternative low cost earth reinforced retaining structures have been constructed using on-site, often marginal backfill material. Table 2 below summarizes some options used and their merits.

Table 2. Summary of Alternative Earth Retaining Structures

Type of Structure	Height (feet)	Cost[*] (per ft^2)	Advantages/Disadvantages
Reinforced Fills	15-50	$4-$12	Less expensive than walls where they fit; Slope typically 1:1; 1/2:1 slope with extra measures.
Tire-Faced Walls	10	$12-$17	Potentially cheapest wall; Significant face settlement; Visually questionable.
Timber-Faced Walls	18	$14-$19	Optimum wall considering cost, durability and aesthetics; Easy to construct.
Geotextile Walls	10-20	$15-$25	Minimum materials costs; good temporary structures; irregular & non-durable face.
Lightweight Walls	28	$15-$20	Special geotextile walls suited for landslide terrain; Moderate settlement with sawdust.
Chainlink Fencing Walls	22	$20-$25	Similar to welded wire walls; Require a custom design; Accommodates face settlement.
Welded Wire Walls	6-30	$20-$29	Most commonly built FS wall; Good construction support from manuf.; Standard designs available.

Note: 1 ft = .3048 m; *1988 costs, typically including drainage, excavation and backfill. Total wall cost can increase significantly depending on wall size, site difficulty, and other road repair work.

Welded wire walls and geotextile or geogrid reinforced soil walls with various face materials appear ideal for rural, often remote applications. These walls and fills represent the low range of costs for road repair measures available today and are appropriate in the typical Forest Service setting.

ACKNOWLEDGMENT

The author wishes to thank Jim McKean, Ken Inouye, Michael Burke, Mark Truebe, Gerald Coghlan, Ken Buss, Don Porior, Robert Young, Bill Powell and Leslie Fisher for their assistance in providing information and suggestions, and for review of this paper.

REFERENCES

1. AMOCO Fabrics. Use of Construction Fabrics in Retaining Walls and Steep Embankments, AMOCO Fabrics Co, Atlanta, 1988, 22p.
2. Barrett, RK. Geotextiles in Earth Reinforcement, Geotechnical Fabrics Report, March/April, 1985, 15-19.
3. Burke, M. New Reinforced Soil Walls and Fills, Engineering Field Notes, ETIS, USDA, 20, Sept/October, 1988, 19-25.
4. Buss, K. Use of Sawdust on Forest Roads, 35th. Annual Road Builder's Clinic, Moscow, 1984, 19p.
5. Driscoll, D. Retaining Wall Design Guide-US Forest Service, Region 6, US Dept. of Agriculture, Portland, 1979, 302p.
6. Eigenbrod, KD & JG Locker. Determination of Friction Values for the Design of Side Slopes Lined or Protected with Geosynthetics, Can.Geotech.J, Vol. 24, 1987, 509-519.
7. Koerner, RM. Designing with Geosynthetics, Prentice-Hall, New Jersey, 1986, 424p.
8. Mitchell, JK & W. Villet. Reinforcement of Earth Slopes and Embankments, NCHRP 290, TRB, National Research Council, Washington, DC, 1987, 323p.
9. Richardson, GN & L. Behr. Geotextile-Reinforced Wall: Failure and Remedy, Geotechnical Fabrics Report, July/August 1988, 14-18.
10. Schmertmann, GR, VE Chouery-Curtis, RD Johnson & R. Bonepart, Design Charts for Geogrid-Reinforced Soil Slopes, Proc. Geosynthetics '87 Conf., New Orleans, 1987, 108-120.
11. Steward, J., R. Williamson, & J. Mohney. Guidelines for Use of Fabrics in Construction and Maintenance of Low-Volume Roads, FHWA-TS-78-205, Forest Service, USDA, Washington, DC, 1977, 153p.
12. Walkinshaw, JL. Handout on Retaining Wall Alternatives, Session Notes, Federal Highway Administration, San Francisco, 1986 and Effective Low Cost Retaining Walls, 41st Calif. Trans. & Public Works Conf., Garden Grove, 1989.

Up/Down Construction - Decision Making and Performance

Dr. James M. Becker, MASCE[1]
and
Mark X. Haley, MASCE[2]

ABSTRACT

This paper presents a discussion of the factors that are
involved with the planning and execution of Up/Down construction
projects.   Three case histories of recently completed projects
in Boston are presented, including a discussion of the decision
process and lessons learned.   Performance data for horizontal
wall   movements   and vertical settlements are included for
comparison   with   excavations   constructed with conventional
construction techniques.

INTRODUCTION

Over the past six years, the authors and their respective
organizations have had the opportunity to plan and execute three
major below-grade structures in the City of Boston. These three
below-grade structures represent a pioneering use of Up/Down
construction in the United States and have created space for the
development of a total of 1,250,000 square feet (116,000m^2) of
below-grade space.   This paper will trace the decision-making
process involved in each project and examine how the experience
gained on one project played a part in the planning and
development of the subsequent project.

The renewed interest in underground space in the United
States is due primarily to the continual increase in the value
of land in major urban areas. This increase in value places a
progressively greater importance on the ability to construct
underground space.   This need, when coupled with the problem of
site conditions and difficult subsurface conditions, has led to
the use of Up/Down construction.   These conditions have long
been common in Europe and the Far East; thus, these regions have
preceded the United States in the use of this construction
method.

---

1 - President, Beacon Construction Company, 50 Rowes Wharf,
    Boston, MA 02110; Chairman, Construction Committee, Friends
    of Post Office Square; and Senior Lecturer, Department of
    Civil Engineering, Massachusetts Institute of Technology,
    Cambridge, MA.

2 - Vice President, Haley & Aldrich, Inc., 58 Charles Street,
    Cambridge, MA 02141.

The recent renaissance of Boston as a major urban center, coupled with its unique subsurface conditions, the historic nature of many of its buildings, and the complexity of its utility and transit systems, makes it an ideal environment for the utilization of the Up/Down construction method.

This method allows for the simultaneous construction of a project's substructure and superstructure. The method is also referred to as Top/Down construction when a superstructure is not part of the project. This paper will deal primarily with the "Down" portion of this method.

Up/Down construction initially involves the installation of the substructure's wall, column and foundation systems, prior to excavation. The substructure's floor system is utilized as both the construction and permanent cross-lot braces, through their sequential installation as excavation moves downward. This method provides a secure, low-risk excavation support system that becomes progressively more attractive as the depth of the excavation increases. This attractiveness is derived from the potential for a shorter construction schedule, cost saving, as the price of a temporary support system becomes prohibitive, and technical feasibility as site conditions become more complex. Except for extremely deep excavations in excess of seven levels, the substructure portion of Up/Down construction usually takes longer to construct than the conventional method; however, the overall construction schedule is shorter due to the early start of the superstructure.

UP/DOWN CONSTRUCTION

Up/Down Construction evolved from the Milan method for subway construction which has been described as "cover, then cut" (14); that is, install parallel slurry walls, bridge between the slurry walls with a deck for traffic and then mine underneath. This concept lead to the development of Up/Down construction in Europe, the Far East, and even some early efforts in the United States (11,15). The use of Up/Down construction at the Olympia Center project in Chicago in the early 1980's was a stimulus to the introduction of the concept to Boston at the Rowes Wharf project (1,13).

Up/Down construction requires no radical changes in construction techniques, but rather a creative sequencing of techniques that have already been proven in either the building or heavy construction industry. Up/Down construction is best understood by dividing the process into four distinct, but highly interconnected, subsystems (see Figure 1):

Wall System - The wall system normally consists of a concrete diaphragm wall constructed by the slurry trench method. While it is the authors' opinion that other retaining walls may be used, they know of no use of Up/Down construction that did not make use of a concrete diaphragm wall.

Column/Foundation System - The installation of the permanent substructure columns and foundations prior to excavation is key to the Up/Down method. Although the permanent foundation elements do not necessarily need to be installed, as was done in

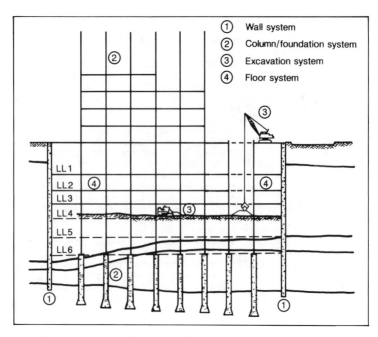

Figure 1. Up/Down Construction Process - Major Subsystems

a Los Angeles project (12), some form of foundation support is
required.   The foundation elements are individual high capacity
units designed to support the column loads by end bearing, side
friction or a combination of both. The column/foundation system
is normally constructed using either a caisson-auger drilling
technique or a slurry trench method (commonly called barrettes
or load-bearing elements). Columns can range from cast-in-place
concrete to structural steel. Key issues in this system are the
method of column placement in terms of dimensional accuracy and
the connection of the column to the foundation element. An
additional issue is the temporary stabilization of the column
through the depth of the substructure.   Normally, this is
accomplished by backfilling the annular space around the column
prior to mass excavation.

     Floor System - Floors are commonly cast-in-place flat slabs
or framed slabs.   The technique for forming each of the
progressively lower level slabs has ranged from standard form
work to casting the slab on a mud mat (8) to a drop-form system
(7).   In all cases care must be taken in designing the interface
between the floor and the diaphragm wall and columns; concrete
shrinkage and creep must be controlled, since they both can
actively contribute to wall movement.

Excavation System - The process of excavation in Up/Down construction is typically a horizontal mining operation below each previously completed floor slab. Special consideration must be given to the method and point of access to vertically remove the soil and supply materials for construction. Typically, the vertical access and soil removal is conducted through a series of "mucking" or "glory holes" constructed through each floor slab. The location and size of the "glory hole" must consider the logistics, equipment and sequencing of both excavation and slab construction. In addition, the location of these access point must be chosen so as not to interfere with the above-grade construction. Ventilation must also be considered to maintain viable working conditions. The careful planning, sequencing and coordination of excavation and floor construction is tantamount to successfully completing the down portion of the project on time and within budget.

CASE STUDIES

Three case studies will be presented for projects recently completed in Boston. They are presented in the sequence that they were constructed to provide a chronology of the decision process. Table 1 provides a summary of pertinent facts relative to those three projects.

In Boston, the soil conditions vary widely across the City due primarily to complex glacial and post-glacial activities. Within the colonial shoreline (See Figure 2 - Locus Map), three drumlins exist with the overburden soils typically consisting of glacially deposited sands or dense glacial till. Along the flanks of the drumlins, the glacial deposits are overlain by a

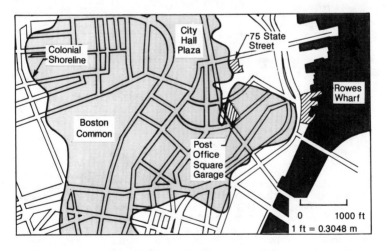

Figure 2. Locus Map of Boston

Table 1. Summary of Basic Project Facts

PROJECT	ROWES WHARF	75 STATE STREET	POST OFFICE SQUARE GARAGE
DEVELOPER	The Beacon Companies	The Beacon Companies	Friends of Post Office Square
GEOTECHNICAL	Haley & Aldrich	Haley & Aldrich	Haley & Aldrich
LEAD DESIGNER	Skidmore, Owings & Merrill Chicago	Skidmore, Owings & Merrill Chicago	Parsons, Brinckerhoff, Quade & Douglas
STRUCTURAL	Skidmore, Owings & Merrill	Skidmore, Owings & Merrill	Le Messurier Associates
CONSTRUCTION MANAGER	Beacon - O'Connell	Beacon Construction Company	Friends of Post Office Square
SUBSTRUCTURE	Perini Corp.	Turner Construction Company	J.F. White Company
NUMBER OF LEVELS	5	6	7
FOOT PRINT	70,000 SF	55,000 SF	73,000 SF
DEPTH	55 FT	65 FT	75 FT
NO. OF CARS	700	700	1400
COLUMN LOADS	Min/Max 1000/3800 kips	Min/Max 500/7000 kips	Min/Max 1400/3100 kips
WALL SYSTEM LENGTH/DEPTH/WIDTH	Slurry Wall 1300LF/65FT/30IN	Slurry Wall 1072LF/85FT/30IN	Slurry Wall 1150LF/90FT/36IN
COLUMN/ FOUNDATION SYSTEM	Precast Encased Structural Steel 112 Total Belled Caissons	Structural Steel with Concrete Encased 84 Total Belled Caissons (10) Straight-shaft Caissons (74)	Structural Steel with Structural Cast-in-Place Concrete 101 Total Load Bearing Elements
BEARING CAPACITY	Till - 40 KSF	Argillite - 20 KSF	Argillite - 20 KSF
FRICTION CAPACITY	Not Applicable	Argillite - 6 KSF	Argillite - 3 KSF
FLOOR SYSTEM	Cast-in-Place Reinforced Concrete Flat Slab Mud-Mat Form	Cast-in-Place Reinforced Concrete Flat Slab Panelized Drop Form	Cast-in-Place Reinforced Concrete Flat Slab Panelized Drop Form
EXCAVATION SYSTEM	1 Access Hole 3 Levels Mined Free-fall Clam Shell Bucket	2 Access Holes 5 Levels Mined Kelly Bar and Hydraulic Actuated Bucket	3 Access Holes 6 Levels Mined Kelly Bar and Hydraulic Actuated Bucket

(1ft. = 0.3048m)   (1kip = 17.794kN)   (1 KSF = 191.52 kN/m2)
(1in. = 25.4mm)   (1 SF = 0.929 m2)

medium stiff clay referred to as Boston Blue Clay that may be up
to 50 ft. (15m) in thickness. Beyond the colonial shoreline,
the soil conditions typically consist of fill and organic soils
overlying a marine deposit consisting primarily of clay varying
in thickness from 30 feet (9m) to in excess of 150 ft. (45m).
The clay, varying from stiff to soft, overlies a thin layer of
glacially deposited soils and bedrock. Groundwater is typically
measured throughout the Boston area within the tidal range.
Lowered water levels are measured in areas affected by utilities
or other below-grade structures.

Many of Boston's older buildings are unreinforced load-bearing masonry structures supported on either spread footings or short timber piles either of which are often founded in the clay stratum. These buildings, along with the City's aging utility systems and subway system are all easily impacted by any deep excavations associated with major substructures (10). Thus, the issue of technical feasibility can rapidly lead the project team to the use of a concrete diaphragm wall system ("slurry wall") for seepage cutoff and positive earth support to avoid problems associated with impact on adjacent structures. Once a slurry wall has been chosen for use on a project and the substructure extends in excess of three or four stories below-grade, Up/Down construction becomes a viable option.

ROWES WHARF

Rowes Wharf saw the introduction of Up/Down construction in the northeastern United States (1,2,4,8,13). This mixed-use project was built on the site of several old wharf structures and included 330,000 sq. ft. (37,000m^2) of office space, 100 residential condominium units, a 230-room hotel, a 700-car five-level below-grade garage, a ferry terminal facility, and a marina. Environmental considerations made it imperative to minimize the amount of harbor that would be filled, thus the designers developed a basic massing of a 15-story building on the land side of the parcel, along with three wharf buildings extending out over the water. This site organization required that the parking, along with a health club and the central service facilities, be located in the below-grade portion of the project.

The Rowes Wharf site is located on the easterly flank of Fort Hill, one of the three original Boston drumlins. The subsurface conditions primarily consist of glacially deposited soils with the Boston Blue Clay being absent at this site. See Figure 3 for a site plan and a below-grade section. In making decisions about the design and construction of the below-grade portion of the project, several major technical factors had to be taken into account. These factors included the type of foundation, methods of excavation and retention, problems associated with water, adjacent structures and existing subsurface conditions and the project's economic and schedule objectives.

The use of a concrete diaphragm wall would provide the necessary excavation support in addition to reducing the concern for water leakage and excavation dewatering. Another design issue that needed to be addressed was that the weight of the proposed building was not adequate to counterbalance the projected uplift force, which was in excess of 45 ft. (13.7m) of water head. The diaphragm wall could also solve this problem by extending the wall to below excavation level into the dense underlying till to provide a seepage cut-off and allow the lowest level floor to be designed as a fully relieved slab-on-grade. Therefore, the advantages of using a diaphragm wall far outweighed the cost premium associated with this wall

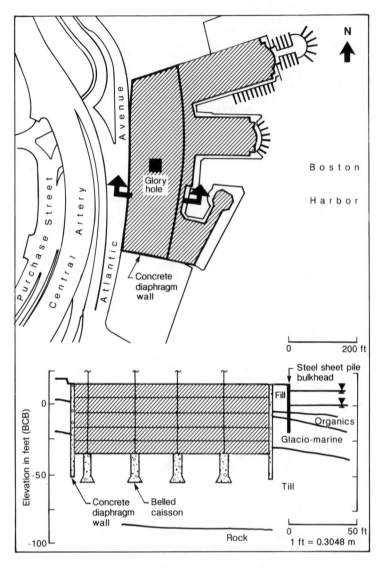

Figure 3.   Rowes Wharf – Plan and Section

system. During preliminary construction planning the use of tiebacks as a support system was evaluated. A concern was raised that water leakage through the tieback wall penetrations might make the permanent wall performance unsatisfactory. The designers suggested that the Up/Down construction method, which was recently used in their Olympia Center project in Chicago, be considered.

An early decision was made to use a heavy/marine contractor for the below-grade and foundation portions of the project. This contractor was managed by the construction manager, but contracted for directly by the developer. This decision was made to minimize the perceived risk of undertaking the below-grade and pier construction effort on the water's edge, by insuring appropriate experience and having the risk/reward relationship defined between the two directly affected parties. Another major consideration was that the contractor should be able to directly carry out a substantial portion of the work with their own forces, thus facilitating close coordination and sequencing of the work.

Based on the information gathered during interviews with heavy/marine contractors, and the design company's prior experience, scope drawings describing the Up/Down method were prepared. The evaluation of this alternative indicated an increase in direct construction costs of approximately $2 million, which was more than compensated for by a shortening of the overall project schedule by more than four months. While the overall project schedule was shortened, the schedule for the land-side subsurface work was actually extended.

The choice of the Up/Down method was prompted by more than simple economic and schedule incentives. The Up/Down procedure was also seen as minimizing certain construction risks, particularly:

Leakage - Wall leakage would be minimized, as tieback penetrations were eliminated and wall movement would be small, thus decreasing the working of the joints between slurry wall panels.

Delay - Once the initial excavation was completed, the superstructure work would be basically independent of subsurface problems. In addition, the excavation would be roofed over, providing weather protection as well as an acoustical barrier.

Tiebacks - Consistent concerns had been expressed as to the ability to use tiebacks on the water side of the project because of potential interference with the new sheet pile bulkhead and the new pile supported wharfs, as well as the potential lack of adequate subsurface conditions for grouting the tiebacks.

This qualitative evaluation was offset by one major issue; the risk of innovation. Short of retaining a heavy/marine firm from overseas, it was impossible to employ a local firm that had direct experience in the Up/Down construction method. Thus, the risks of innovation, delays from possible poor planning, unforeseen technical problems, lack of coordination, etc., had to be offset by the economic and schedule incentives, along with the minimization of risk.

In this particular case, once the concept was adopted, the design continued to evolve to the point where it became virtually impossible to build the final project using conventional subsurface construction techniques. Prior to making a final commitment, however, a testing program was carried out to ensure that caissons could be augered and belled at the site. The contractor proposed that the concrete floor slabs be cast on a form system created by a mud mat. The substructure columns were precast encased structural steel, where the precast was designed to carry substructure loads.

The actual construction activities experienced relatively minor difficulties in achieving an excavation rate of close to 1,000 cubic yards ($765m^3$) a day. Two problems, however, were encountered that were to affect future decision making. First, the ability to easily achieve bond breaking between the mud mat and the concrete floor slab proved difficult; and secondly, to achieve reasonable excavation efficiency, it was necessary to over excavate below the level of the next mud mat and then back fill to mud mat level. This was necessary due to the equipment size required to productively excavate the soils. Wall movements were small and the dense till that surrounded the site helped to ensure that there were virtually no water problems.

75 STATE STREET

The 31-story office tower development at 75 State Street includes six levels of below-grade space, utilized primarily for parking (7). The 1.37-acre ($5,500m^2$) site (Figure 4) is located near the original shoreline of the colonial Boston peninsula. The development includes 745,000 square feet ($69,000m^2$) of above-grade floor area and 350,000 square feet ($32,500m^2$) of below-grade space for a 700-car garage.

The subsurface conditions at 75 State Street are typical for sites in Boston located outside of the colonial shoreline, with the overburden soils consisting primarily of clay. The bedrock, 70 to 100 ft. (21 to 31m) below ground surface, is a highly altered argillite. A soil profile for the site, along with a substructure section, is included as Figure 4.

Early planning for the development considered, in addition to the subsurface conditions, the proximity to adjacent buildings supported on shallow foundations, the Massachusetts Bay Transit Authority Blue Line tunnel below State Street and existing foundations located within the site area (particularly a series of concrete-filled pipe piles from a previous above grade parking garage).

The requirements of the development, particularly the need to support both low-rise and high-rise building elements, in conjunction with existing site and subsurface conditions, dictated that a concrete diaphragm wall be utilized. The diaphragm wall, constructed by the slurry trench method, would provide a semi-rigid wall system eliminating the need to underpin adjacent structures and provide foundation support for the perimeter columns. In addition, the wall was incorporated into the permanent structure and provided a seepage cut off

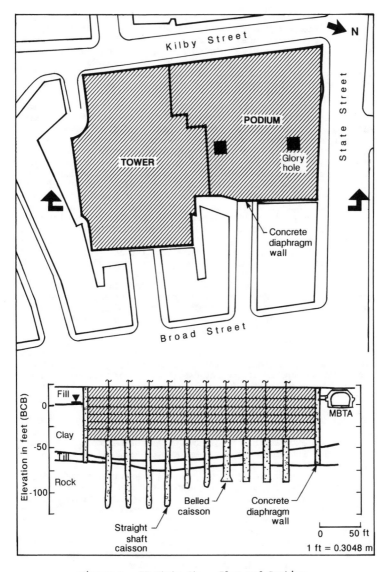

Figure 4. 75 State St. - Plan and Section

allowing the lowest floor level to be designed as a fully relieved slab-on-grade, eliminating problems associated with the variable building mass. Various temporary lateral support systems were evaluated during design including tiebacks, cross-lot bracing and inclined rakers. Tiebacks were deemed impractical at this site, if not impossible, due to the location of adjacent structures and to easements required to allow tiebacks installation.

The design team prepared conceptual designs for several alternative construction schemes. The evaluation of these alternative substructures showed that the time saved using Up/Down construction equalled approximately one to two months, at a cost premium of around $1 Million. The cost premium was essentially offset by a reduction of carrying costs for the shorter construction schedule. Therefore, the decision came down to one of technical feasibility and risk. The ability to locate several glory holes in the atrium of the building's low base, the inability to easily use tiebacks and the serious concerns for adjacent structures led to the ultimate decision to use the Up/Down method. Unlike Rowes Wharf, however, it was decided that a general contractor be engaged to manage both the substructure and the superstructure construction. In this case, the contractor eventually chosen subcontracted all of the work except for the cast-in-place floor concrete.

With the contractor selected, the final details of construction were developed. Concern as to the stability of the argillite during augering resulted in the design of alternative foundation types that included the preferred belled caissons and as an alternate, straight shaft caissons. This question of stability, extenuated by the inability to properly depressurize the rock through dewatering during construction, eventually led to the use of caissons drilled under slurry for placement of the core columns and their foundations (7). A major change was made in the forming of the floor slabs from the mud mat utilized at Rowes Wharf. A panelized drop form system supported off the permanent columns was utilized. This system was a direct result of experience at Rowes Wharf because of the previously mentioned need to over excavate below the mud mat level to achieve reasonable excavation efficiency. A further concern was the potential problem of upward seepage and soft soil conditions on which to cast a mud mat.

The construction process proceeded essentially as planned, with two exceptions. As noted above, the slurry caisson method was adopted for installation of the major core column foundations after several attempts to excavate and dewater the holes. The other problem was related to the fact that a large number of pipe piles had to be extracted within the site prior to excavation. The voids which were created in the clay (like Swiss cheese) plus pre-excavation along the slurry wall alignment, resulted in excessive inward movements of the wall before the floor systems could be placed. This fact helped to illustrate the critical nature of the excavation/floor slab construction sequence. This sequence must be carefully planned and coordinated to avoid unnecessary wall movement and delays in substructure completion.

POST OFFICE SQUARE GARAGE

The final case study is the Post Office Square Garage with a seven-level, 525,000 ft^2 (48,800m^2) below-grade structure. Located in the heart of the city's financial district, the new Post Office Square Garage, designed to provide parking for 1,400 vehicles, is being built completely underground and is to be covered with a 1.5-acre (6,000m^2) park (3). Prior to the construction of this project, one other Up/Down project was initiated in Boston by the contractor who had constructed 75 State Street. This project, 125 Summer Street, introduced the use of load-bearing elements constructed by the slurry trench method for the installation of the column/foundation system. These elements consisted of columns set directly into reinforced concrete that was tremied into either rectangular or cruciform slurry trenches. These columns were designed to transfer the loads to the foundation through a shear mechanism (shear studs) rather than end-bearing and base plates.

The Post Office Square Garage will be the deepest building excavation in Boston to date. The north end of the garage will bear directly on the area's bedrock - Cambridge Argillite. Many of the adjacent structures are founded on the thick clay layer that overlays the glacial till. The clay in this area is relatively soft compared to the other sites discussed. See Figure 5 for a site plan and subsurface section.

The constructability analysis carried out for this project indicated that the Up/Down construction method was the desired, if not the only, technical approach that would satisfy the concerns of minimizing risk and maximizing ground stability with regard to adjacency. Cost and schedule analysis indicated that at this depth, the Up/Down method would be no more costly or time consuming than conventional open-cut construction.

Another distinct advantage of the Up/Down method emerged as traffic management planning was carried out with the City's transportation department. This advantage, which had already been realized on the previous projects, was the ability to handle truck traffic during excavation directly on-site (on the ground floor slab) without the queuing normally required on city streets.

As the detailed planning progressed, decisions were made based upon the other Up/Down experience in the City. The key decision was to use the load-bearing element approach for the placement of the column/foundation system. While slightly more expensive than the caisson methods, the load-bearing elements could be installed by the slurry wall contractor, providing for a more efficient and coordinated operation that would save more than enough time to compensate for the added costs. Also, this approach did not require the completion of the slurry wall or site dewatering to proceed. Therefore, it eliminated much of the potential for ground loss during caisson installation and creation of the "Swiss Cheese Effect" noted at 75 State Street and thus reducing the potential for future wall movement. The floor system was also placed by the use of a panelized drop-form system.

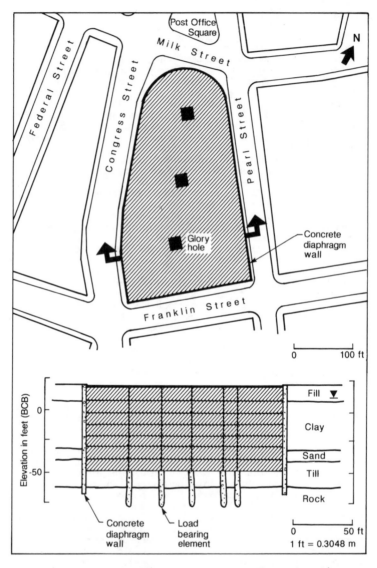

Figure 5.   Post Office Square Garage - Plan and Section

Because of the scale of this excavation, an extensive monitoring program was established to allow the project construction to be controlled by an observational approach, with proactive interaction of the design team with the contractor. In this case, a heavy contractor was chosen to carry out the construction, again emphasizing experience and the capacity and desire to carry out a substantial portion of the work with their own forces. Therefore, construction coordination was provided through direct control between the excavation and floor placement operation, which is a key to minimizing wall movement and thus potential impact on adjacent structures.

The Post Office Square Garage is not completed as of the writing of this paper; however, construction progress is on schedule and operations have run smoothly to date. Had it not been for the success of the projects prior to this one, it is questionable as to whether or not this garage project would have been attempted.

PERFORMANCE

In this section of the paper, the performance of the excavation support system as it relates to wall and adjacent ground or building movements will be discussed. This will be presented for only two of the cases as the third case, Post Office Square Garage, is in the early stages of construction. In addition, a comparison of performance for these projects with other excavations in Boston will be made.

At Rowes Wharf, where the soil conditions consist primarily of dense glacial deposits, horizontal wall movements were in the range of 1/4 to 3/4 of an inch (6 to 19mm). Refer to Figure 6. Adjacent building movements were not measured. The horizontal movements were well within predicted limits, with the ratio of maximum horizontal wall movement to depth of excavation was equal to 0.1%.

At 75 State Street, where up to 60 ft. of clay underlies the site, horizontal wall movements ranged from 1/2 to 2 inches (13 to 51mm). The largest movements were in the area where the clay was the thickest and a two level high mechanical room existed (Figure 6). Vertical movements of adjacent buildings that are located within 5 to 20 ft. (1.5 to 6m) of the excavation ranged from 1 to 4 inches (25 to 102mm), with the average movement in the range of 2 inches (51mm). However, 50 percent of this movement occurred prior to the start of excavation and was attributed to earlier pile extraction and caisson installation. The horizontal wall movements were within predicted ranges for an excavation of this depth. However, the vertical movements were significantly larger than predicted. The reason for these large movements is directly related to construction procedures and was one of the reasons for utilizing load-bearing elements at the Post Office Square garage.

A comparison of settlement data with respect to distance from the excavation was developed in Figure 7. The data from 75 State Street is compared to two other concrete diaphragm walls constructed in Boston in reasonably similar soil profiles, but

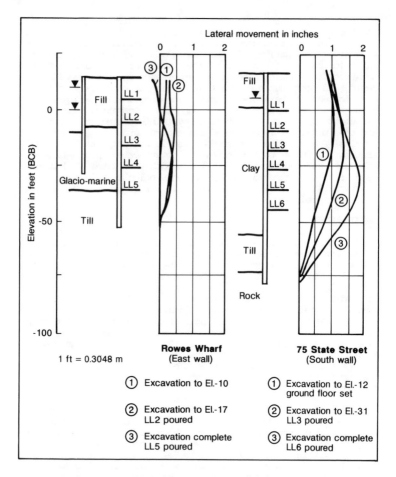

Figure 6.   Horizontal Wall Movements at Rowes Wharf
            and 75 State St.

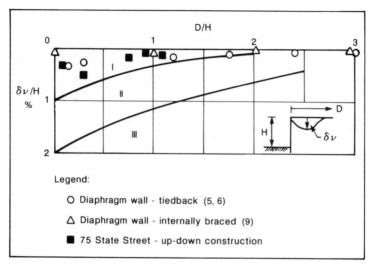

Figure 7.   Summary of Settlement Adjacent to Excavations as a
            Function of Distance from Edge of Excavation (Adapted
            from Peck (10), 1969, p. 266)

constructed with internal bracing and tiebacks, respectively
(5,6,9). The data indicate that the internally pre-stressed
braced excavation allowed less vertical movement to occur; the
tied-back and up-down construction indicated similar
performance. The ratio of settlement to excavation depth for 75
State Street was less than 0.5% and averaged 0.25%.

Table 2 summarizes the performance of numerous excavations
completed in the Boston area utilizing concrete diaphragm walls
and comparing them to three Boston projects utilizing up-down
construction. Although the depths of the excavations differed
and soil conditions varied between sites, a comparison of
performance can be made.

It can be concluded, based on these data, that Up-Down
construction can be completed at a level of performance equal to
or better than that experienced at excavations constructed using
conventional excavation and bracing methods. Thus, Up/Down
construction is considered a viable, technically feasible
alternative for construction of underground space. In addition,
Up/Down construction has the advantage of eliminating ground
loss and potential wall seepage typically associated with
tiebacks installation. Therefore, based on current experience
to date, if ground loss associated with column/foundation system
installation can be controlled Up/Down construction provides a
methodology whereby deep excavations can be completed
economically at relatively low risk.

Table 2. Summary of Horizontal and Vertical Movements
for Excavations in Boston Area

Project	Exca-vation Depth (ft.)	Wall Thick-ness (in.)	Bracing System	Summary of Maximum Movement (in.)			
				δhmax (in.)	δhmax/H %	δvmax (in.)	δvmax/H %
1. MBTA SOUTH COVE (5,6)	50	36	Cross-Lot	1.0	0.17	0.5	0.08
2. STATE TRANS-PORTATION BLDG.	30	24	Tied-Back	0.8	0.22	1.2	0.33
3. 60 STATE ST. (9)	30	24	Tied Back	0.75	0.21	1.2	0.33
4. MBTA DAVIS SQ.	56	24	Tied-Back	1.1	0.16	1.7	0.25
5. ONE MEMORIAL DR.	27	24	Tied-Back	1.0	0.32	1.2	0.37
6. ROWES WHARF	55	30	Up-Down	0.75	0.11	Not Measured	
7. 75 STATE ST.	65	30	Up-Down	2.0 1.1	0.26 0.14	4.0 2.4	0.51 0.31
8. 125 SUMMER ST.	60	30	Up-Down	0.6	0.08	0.38	0.05
9. POST OFFICE SQ. GARAGE	75	36	Up-Down	Data Not Available			

(1 ft. = 0.3048m)
(1 in. = 25.4mm)

SUMMARY AND CONCLUSIONS

The Up/Down construction method was introduced to the Boston
area when the economic and urban need for multiple levels of
underground space became apparent.   The method when first
utilized allowed a waterfront development project (Rowes Wharf)
with 5 levels of underground space to trim a minimum of four
months from the overall project construction schedule. In the
second   instance   (75   State Street), where six levels of
underground space was developed, it provided a method to solve
numerous   site   constraints   including   adjacency of existing
buildings and construction staging.   In the third case study
(Post Office Square Garage) where seven levels of underground
space are being developed, it proved to be the only way to
economically develop space of this depth at limited risk in a
congested urban area.   A summary figure of the three case

studies, indicating the soil conditions, depth of excavation and foundation system, are illustrated on Figure 8.

Each project, in sequence, incorporated the lessons learned during the previous development into the planning and construction for the following project. This was made possible because members of the development and design teams were participants in each subsequent development.

The Up/Down construction method adopts common construction procedures utilized in the heavy construction industry and involves four distinct systems: the wall system, the column/foundation system, the excavation system, and the floor system (refer to Figure 1). Each system is interrelated to the other systems, with the floor and excavation systems being very closely coordinated.

For multiple level below grade structures constructed in an urban area, the use of a concrete diaphragm wall has become common, especially when poor soil conditions and the need to protect adjacent structures exists. Once the decision has been made to incorporate a diaphragm wall into a project the use of Up/Down construction becomes economically and technically attractive if three to four levels of below grade space are to be built. Once the economic and technical issues of Up/Down have been evaluated the need to engage a qualified contractor must be recognized. In the case studies presented, various

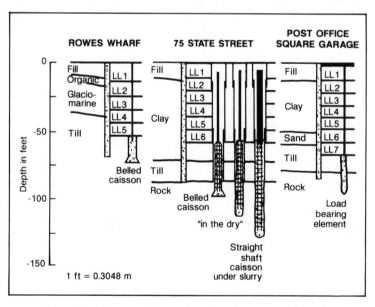

Figure 8. Summary of Geometry for Three Case Histories

contracting strategies were adopted. It is the authors' opinion
that utilizing a contractor with heavy construction experience
will best serve the project needs for the below grade portion of
the work.   This option provides the impetus for close
coordination of construction activities as the four key systems
of Up/Down construction are completed. In addition, it places
the risk and economic reward for completing the project directly
upon one responsible entity.

A summary of the decision process and other factors were
considered in the development of the three projects is included
on Table 3.   Also included are the time periods during which
construction was undertaken.

Based on performance data related to adjacent ground and
wall movements collected to date, Up/Down construction will
control movements to within acceptable limits. However, the
construction procedures utilized to install the wall and column
foundation system must preclude excessive loss of ground. The
excavation/floor construction sequence must be planned to
provide lateral support to the wall in a timely manner, to avoid
over-excavation at the wall, thus creating large unsupported
wall heights, and to allow a continuous sequence of construction
in order to maintain schedule.

In conclusion, Up/Down construction provides a relatively
low risk, technical and economic solution to the creation of
multiple levels of below grade space.   The method must be
undertaken in concert by the owner, designers and contractor for
it to be successfully employed. As more experience is gained in
its use, it is the authors' opinion that it will become the
primary excavation support system of choice for urban
excavations greater than 25 ft. when ground conditions are
difficult and the site is located within a congested urban
environment.

Table 3. Summary of Decision Factors and Schedule

PROJECT	ROWES WHARF	75 STATE STREET	POST OFFICE SQUARE GARAGE
COST PREMIUM	$2,000,000	$1,000,000	None
SCHEDULE ADVANTAGE	4 - 6 Months	1 - 2 Months	None
DECISION LOGIC	o Economic o Traffic Control o Stable Excavation o Water Related Risks	o Constrained Site o Traffic Control o Adjacency o Avoid Tieback Easements	o Constrained Site o Traffic Control o Technical Feasibility o Control Ground Movement
CONTRACT	Guaranteed Maximum Price	Guaranteed Maximum Price	Guaranteed Maximum Price
CONSTRUCTION START	March 1, 1985	Dec 1, 1986*	October 1, 1988
SUPERSTRUCTURE START	January 1, 1986	June 8, 1987	August 1, 1989**
BOTTOM OUT	August 15, 1986	December 1, 1987	June 1, 1990***

* Effective Start Date Following Demolition
** Effective Start - i.e. available to start a superstructure if one was required.
*** Scheduled

REFERENCE LIST

1. Becker, J.M., Rowes Wharf: A Case Study in Substructure
    Innovation, Tunnelling and Underground Space Technology,
    Great Britain, Pergamon Press, 1986.
2. Becker, J.M., Up/Down Construction Sails Into Boston,
    Concrete International, October 1987, Pp. 43-47.
3. Fournier, P., Park To Hide Deep Garage, New England
    Construction, May 8, 1989, Pp. 24, 29-30, 39, 46-47.
4. Fournier, P., Up/Down Speeds Pace at Boston's $180M Rowes
    Wharf, New England Construction, October 28, 1985, Pp.
    8-9, 12-13.
5. Lambe, T.W., Braced Excavations, Lateral Stresses in the
    Ground and Design of Earth-Retaining Structures, ASCE
    Speciality Conference Paper, Cornell University, 1970, Pp.
    1-27.
6. Lambe, T.W., Wolfskill, L.A. and Jaworski, W.E., The
    Performance of a Subway Excavation, Performance of Earth
    and Earth-Supported Structures, ASCE Speciality Conference
    Proceedings, Purdue University, 1972, Vol. 1-Part 2, Pp.
    1403-1424.
7. Lass, H. and S. Browne, Friction A Plus in Building
    High-Rise, Engineering News-Record, April 14, 1988, Pp.
    78-79.
8. Lass, H. and P. Green, Boston Project Grows Up and Down,
    Engineering News-Record, March 20, 1986, Pp. 64-65.
9. Johnson, E.G., Gifford, D.G., Haley, M.X., Behavior of
    Shallow Footings Near A Diaphragm Wall, ASCE Fall
    Convention and Exhibit, San Francisco, CA, 1972, Pp. 1-27.
10. Peck, R.B., Deep Excavations and Tunneling in Soft Ground,
    State-of-the-Art-Volume, 7th International Conference of
    Soil Mechanics and Foundation Engineering, Mexico City,
    1969, Pp. 225-290.
11. Sherman, P.J., Up/Down Construction: A Feasibility Study
    of Its Application in the U.S., Y. Rosenfeld,
    Massachusetts Institute of Technology, June 1986, Pp.
    1-167.
12. Sponseller, M.P., Deep Garage: A Shoe Horn Into Tight
    Site, Engineering News-Record, January 22, 1987, Pp.
    88-90.
13. Tatum, C.B., M.F. Bauer and A.W. Meade, Process For
    Up/Down Construction At Rowes Wharf, ASCE Journal of
    Construction Engineering and Management, Vol. 115, No. 2,
    June 1989, Pp. 179-195.
14. Toronto's Cover-Then-Cut Subway, Engineering News-Record,
    March 3, 1960, Pp. 37-39.
15. Upside-Down Basement Cuts Cost of 26-Story Los Angeles
    Building, Engineering News-Record, October 14, 1965, Pp.
    72-74.

CELLULAR SHEETPILE BULKHEADS

W.L. Schroeder,[1] F.ASCE

ABSTRACT: A comprehensive paper on cellular cofferdams was presented at the 1970 ASCE Conference on Lateral Stresses in the Ground and Earth Retaining Structures (14). Important field, laboratory and analytical studies on cofferdams and bulkheads have been made since that conference. This paper incorporates findings of these studies applicable to bulkheads. It is shown here that bulkheads behave differently than cofferdams, particularly because of the presence of an area fill which the bulkhead supports. The fill induces settlement due to its weight, unlike the water retained by a cofferdam, in addition to exerting lateral forces on the bulkhead. Seismic analysis is discussed, and construction procedures are outlined.

INTRODUCTION

A survey of 13 major U.S. coastal ports conducted for this paper suggests that cellular wharves are popular, but not nearly so popular as are anchored sheetpile bulkheads. Engineers at the ports surveyed indicate that the trend is currently away from full-height wharf retaining structures, toward pile supported wharf construction. Nonetheless, bulkheads for wharves and quays are often still designed to incorporate sheetpile cells as earth retaining structures. Unlike cellular cofferdams for dewatering projects, which are similar, cellular bulkheads usually support a structural deck and other surcharge loads. The earth fill in bulkhead cells is frequently densified to mitigate deck settlement and to enhance internal stability. The weight of the earth retained by a bulkhead may play a very important role in affecting the structure's foundation performance. Because bulkheads have a long design life, and are incorporated in permanent facilities, criteria for allowable foundation settlement, lateral deflection and seismic response can be more restrictive than are those for cofferdams. Figure 1 illustrates certain details which typify cellular wharf construction, and highlights differences between wharf bulkheads and cellular cofferdams, including: 1) the absence of connecting arcs between the backs of cells, 2) greater sheetpile length and penetration at the fronts of cells, 3) a concrete deck, and 4) a fendering system to protect the structure. The loaded (earth) side of a bulkhead may be referred to

---

[1]Prof., Dept. of Civil Engineering, Oregon State Univ., Corvallis, OR 97331.

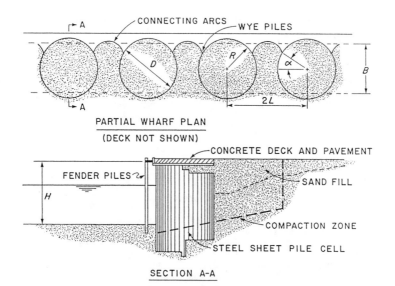

Figure 1. Typical Cellular Wharf.

as the back or outboard side, and the unloaded (water) side may be
referred to as the front or inboard side throughout this paper, and in
the various literature cited.   The outboard side of a cellular
cofferdam is loaded by water.   This terminology is noted here for
clarification.

The paper by Lacroix, Esrig and Luscher in the 1970 Cornell
conference (14) addressed cellular cofferdams specifically, yet much
of the topical coverage applies equally well to cellular bulkheads.
Since the 1970 conference there have been a number of field, laboratory and analytical studies which have further improved our knowledge of
cellular structures and their performance.   It is the intent of the
present paper to address certain issues related to bulkhead design and
construction that were not treated at the 1970 conference, and to
summarize field, laboratory and analytical research since that time
which is applicable to bulkheads.   Some of the new information
pertains to both bulkhead and cofferdam projects.

CELLULAR BULKHEAD DESIGN CONCERNS

A cellular bulkhead is a flexible gravity retaining structure. A
design needs to provide for (see Figure 2): 1) external stability, 2)
internal stability, and 3) tolerable vertical and horizontal movements. It is possible to show analytically that cellular bulkheads of
the usual proportion cannot fail by sliding (Figure 2a) or overturning
(Figure 2b), except for very special cases, because they are not rigid

EXTERNAL INSTABILITY   INTERNAL INSTABILITY

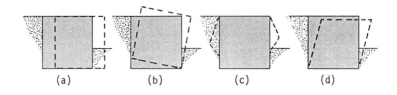

(a)   (b)   (c)   (d)

Figure 2.   Vertical Sections Through Sheetpile Cells Illustrating
            Failure Modes. a) Sliding, b) Overturning, c) Interlock
            Failure, d) Shear Distortion.

enough to do so. Forces on the cells must exceed levels which will
result in shear distortion (Figure 2d) before a sliding or overturning
failure can occur. Experience bears out analysis. There have been no
known failures of cofferdam or bulkhead cells by overturning (33), but
cells have been known to distort substantially (6,7,23,31,35).
Sliding failures which have been observed have been associated with
unanticipated, or unusual conditions. The sliding failure of the
Uniontown cofferdam on a clay seam is well-documented (32). An early
(1944) sliding failure occurred in the Portland, Oregon, ship repair
yards (see Figure 3) as dredged fill was being placed behind a
diaphragm cell bulkhead founded on sand. Though the embedded depth of
the cells below the dredge line was relatively large (30 ft [9.1 m]),
the dredge line sloped away from the immediate vicinity of the cells.
At the time of the failure there was a differential head of water (18
ft [5.5 m]), and a depth of submerged dredged sand fill 45 ft (13.7 m)
above the dredge line behind the bulkhead.

Terzaghi's classic paper (31) examines the possibility of failure
by penetration or pullout of sheetpiles below the dredge line, and
demonstrates that a bulkhead without penetration of the sheets below
the dredge line (a bulkhead on rock) has only slightly less distortion
resistance than one with penetration. The contribution of the
friction forces on the embedded portion of the sheets to distortion
resistance may, therefore, be safely disregarded. The analysis for
stability of cells on rock may be used for all cases. This finding
has been confirmed experimentally by pullover tests on cells with
differing sheetpile penetrations on the unloaded side (16).

Accordingly, to analyze a cellular bulkhead or cofferdam for
stability, it is only necessary to determine the adequacy of internal
stability, and to verify that there is sufficient resistance to
sliding. When these stability requirements have been met, other
matters such as lateral deflection and foundation performance may be
addressed.

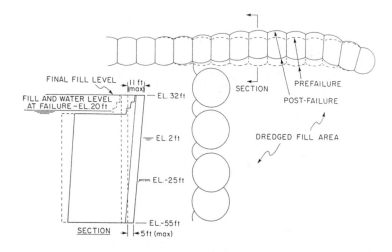

Figure 3. Sliding Failure of Diaphragm Cell Bulkhead (1 ft = 0.305 m).

TRIAL LAYOUT

Symbolic plan dimensions for a circular cell bulkhead are shown in Figure 1. For convenience of analysis the actual cell-arc combination is represented by an equivalent rectangular section of width B, and length 2L. The width, B, is found by requiring that the equivalent rectangular section and the actual section for a length 2L have either the same area or the same moment of inertia. For instance, α = 30° and B = 0.818D for 30° wye piles. To select a reasonable trial layout for analysis one usually determines the height of the cell above the dredge line, H, then estimates B = 0.85H, which leads to the result that the cell diameter, D, is about the same as the height above the dredge line. Given the approximate diameter of the main cells, the approximate circumference is calculated, and then, based on available sheetpile widths (actually driving distance) provided by manufacturers, the actual number of sheetpiles, and the actual circumference and diameter are determined. Wyes are then located and the number of sheets in an arc and actual arc dimensions are found.

INTERNAL STABILITY - INTERLOCK TENSION

To insure that the cells will not burst (see Figure 2c) it is necessary to determine that the forces in the interlocks and webs of the sheetpiles in horizontal planes do not exceed their strengths for a wide variety of conditions. Normally, sheetpiles are fabricated so that interlocks may be loaded to maximum allowable values without overstressing webs. However, in older cells considered for adoption in new work, and which have lost metal due to corrosion, webs may be weaker than interlocks, and the strengths of both should be evaluated. For U.S. manufactured sheetpiles the standard ultimate interlock strength, t_{ult}, is 16 k/in (28 kN/cm) and high strength interlocks have t_{ult} = 28 k/in (49 kN/cm). If calculated interlock force is t, then the factor of safety against interlock failure is

$$FS = t_{ult}/t \qquad\qquad\qquad (1)$$

A minimum value of FS = 2 is normally required for long-term loading conditions. The problem of evaluating the adequacy of interlock strength is, then, reduced to finding the interlock force, t, at key locations in the installation, for various service conditions. In selecting these locations, it is crucial to understand that unless the diameter of the cell or arc increases in service, with the consequent stretching of the interlocks after the sheetpiling are threaded and driven, there can be no interlock force. Thus, it is clear that restriction of sheetpile movement plays an important role in determining the actual level of interlock force in field installations.

There is no difference between a bulkhead cell and a cofferdam cell up to the time when the cell fill has been placed and compacted. Loading for this condition (referred to as the isolated cell) is axisymmetric. From this time on, bulkheads and cofferdams will typically differ structurally, and their behavior differs because of differing load conditions for the structures and their foundations. For instance, in a cofferdam there are typically front (unloaded side) and back (loaded side) connecting arcs. The arcs are filled after the cells, but the structure receives no unbalanced transverse load in the process. It is only when the cofferdam is unwatered that transverse loading occurs. For a bulkhead with only front arcs, placement of arc fill immediately subjects the structure to transverse load, which is gradually increased as the remainder of the backfill is completed. Placement of the backfill behind the bulkhead also subjects the foundation soils for the structure to a substantial vertical load, extending to far greater depth than does the influence of the weight of the cell fill in a cofferdam. Thus, while we may use observational data for the isolated cell case from either cofferdam or bulkhead projects to examine validity of design analysis methods, data on response of cofferdams and bulkheads to transverse loads must be compared with caution. The two cases differ substantially because of the important effect of backfill loads on bulkhead foundation response.

Interlock Forces Due to Main Cell Filling. Figure 4 shows a section of two cells and an arc, and a cross section which represents various stages in the service life of a cell. If p is taken as the net pressure tending to cause expansion of the cell, then for a given elevation, the interlock force is

$$t = pR \qquad\qquad\qquad (2)$$

Equation (2) may be used for appropriate combinations of elevation and service conditions to evaluate the factor of safety against a bursting

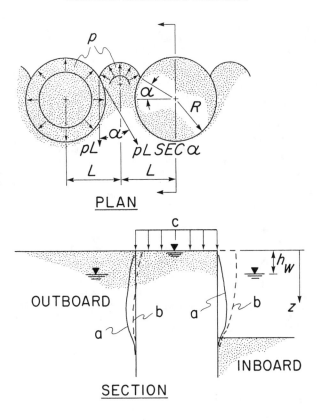

Figure 4. Interlock Forces in Sheetpile Cells.

failure. Designers must make two judgmental choices before doing the necessary calculations: 1) the elevation where interlock force is to be evaluated, and 2) the method of calculating net lateral pressure, p. Several design conditions usually need to be considered, as shown in Figure 4: a) during cell filling and during cell fill compaction, b) following placement of the fill behind the bulkhead, and c) during application of design surcharge loads. Special conditions may need to be examined; for instance, the case during filling where the water level inside the cell is higher than the outside, or where the same condition results from a large tidal range.

Terzaghi originally proposed that interlock force due to cell fill be calculated at the dredge line or rock surface (31). TVA later observed that the tips of sheetpiles were normally restrained, so that the maximum expansion of the cell (i.e., "the bulge"), and, therefore, maximum interlock force, occurred above the dredge line (1). Early observations at Terminal 4 in Portland, Oregon, (21) shown in Figure 5a agreed with TVA's recommendations concerning location of the bulge. A method (see Figure 5b) of determining penetration to provide fixed restraint and location of the bulge for sheetpiles driven in sands was developed later from available field and laboratory data (22). This approach has been confirmed in subsequent observations at Fulton Terminal 6 (23). It is important to note that the requirements for fixity, as determined by the method of Figure 5b, are based on actual penetration. Considerations for dredging, scour or conservatism will often govern the design penetration. Instrumentation of Lock and Dam 26(R) (see Figure 6a) has also provided field verification of the method of calculation for the magnitude and location of maximum interlock forces (28). Other work (27) shows that the maximum interlock force may occur very near the rock surface (see Figure 6b) for sheets which do not achieve any substantial penetration in the foundation.

Interlock force calculations for the effect of fill weight using Equation (2) require an "apparent" lateral earth pressure coefficient, K_i, for the cell fill. The coefficient is usually chosen based on interpretations of field observations of interlock forces. This "apparent" coefficient is not necessarily the same as would be chosen to calculate lateral earth pressures within the cell fill. For instance, near the dredgeline, where the sheetpile tips are restrained by embedment in the foundation, interlock forces are necessarily smaller than at the "bulge". Yet, the actual lateral earth pressure within the cell must certainly be greater. Field observations confirm

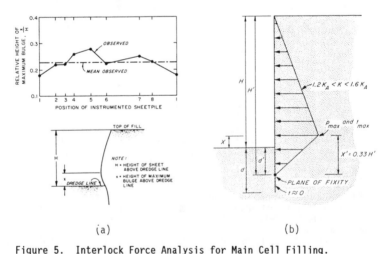

(a)   (b)

Figure 5. Interlock Force Analysis for Main Cell Filling.
           a) Location of Maximum Bulge in Sheetpiles (21),
           b) Method for Location of Maximum Interlock Force (22).

7   Schroeder

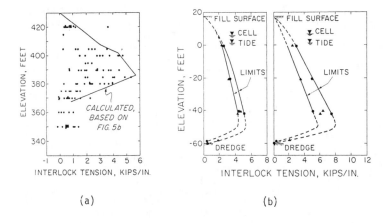

(a)   (b)

Figure 6.   Field Observations of Interlock Forces for the Isolated
            Cell Case. a) Lock and Dam 26(R) (28), b) Trident (27)
            (1 ft = 0.305 m, 1 k/in = 1.75 kN/cm).

Terzaghi's recommendation that for the isolated cell (Figure 4, case
a) K_i = 0.4 is a reasonable choice for the ratio of vertical to
lateral pressure for maximum interlock force calculations. K_i may
also be determined as a function of ϕ, the angle of internal friction
(22). It is usually nearly equal to the coefficient of earth pressure
at rest, K_o. During actual filling, values of K_i as great as 0.66 have
been calculated from interlock force observations (35). K_i is
properly taken as an effective stress value. Thus, for the construction (cell filling) case, where the cell is full of soil with
submerged unit weight γ' and water is overflowing the tops of the
sheetpiles a distance h_w above the water level outside the cell (see
Figure 4), at a depth z from the top of the sheetpiles

$$p = \gamma' z K_i + \gamma_w h_w \qquad\qquad\qquad (3)$$

and Equation (2) is used to compute interlock force. Selection of z
for calculation of maximum interlock force must be guided by evaluation of sheetpile tip restraint, as discussed earlier.
    <u>Interlock Forces in Common Wall and Wyes</u>. Most failures of sheet-
pile cells have been near or at the wye piles connecting the arcs to
the main cells (33). TVA proposed (1) that the interlock force at
this location be calculated as the tangential component of the total
force, pL, due to the projected lateral pressure, p, over half of a
typical section (see Figure 4). The "secant formula"

$$t_{max} = pL\sec\alpha \qquad\qquad\qquad (4)$$

resulting from this approach has been widely used. Both field and laboratory studies (16,22,28) show that Equation (4) is conservative, including the extensive data for Lock and Dam 26(R). The latter also showed that the difference between the interlock forces in the main cells and in the common walls was only about 30% of what Equation (4) would suggest. Swatek's (29) equation

$$t_{max} = pL \tag{5}$$

for the average interlock force in the common wall between the cell and the arc fills is less conservative than the secant formula, and bounds available field data for maximum common wall interlock force. It is recommended here for design use. Field studies frequently show (24) that placement of arc fill actually results in a decrease in interlock tension in the common wall behind the wyes.

Interlock Forces Due to Backfill and Surcharge. Following placement of backfill behind a bulkhead (Figure 4, case b) field and laboratory observations (16,21,35) show that interlock force on the back or loaded side is reduced, while it increases on the front, or unloaded side. NAVFAC (18), and the 1970 Cornell paper (14) suggest using a larger value of K_i to compute interlock force for loaded side (back) sheetpiles for this case, and a smaller value for the unloaded side sheetpiles, which is contrary to field observations. Field and laboratory measurements (16,21,27,35) suggest that interlock forces on the unloaded side will increase by about 25% of their isolated cell values. There are no accepted analytical methods for calculation of maximum interlock force on the unloaded side of a bulkhead. It is clear that the interlock forces are greater there than for the isolated cell case, and that they may be important for taller structures.

Surcharges may be applied to a cell, or to the area behind the cell. Uniform surcharges in the former case (see Figure 4, case c) simply increase vertical stress, and hence, interlock force. Any surcharge on the backfill behind the cell contributes to overturning moment, which mainly affects interlock force on the unloaded side of the bulkhead as described above. Interlock forces due to localized surcharge loads are usually largest in the upper part of a cell. Concentrated loads such as from crane wheels and loads on small areas of the deck above the cell ordinarily do not produce significant interlock forces. This is the case because 1) the upper part of the cell has the lowest interlock forces due to the weight of the cell fill, and 2) the stiffness of the deck structure is effective in distributing the load to a large portion of the cell area.

INTERNAL STABILITY - DISTORTION FAILURE

Bulkhead cells subjected to excessive moment tend to distort as shown in Figure 2d. Overturning (Figure 2b) is not a possibility because cells of the usual composition and proportions will distort before they can rotate as a rigid body, as explained earlier. Several methods for representing the possibility of a distortion failure analytically have been proposed, and each has been discussed in earlier papers and manuals (14,16,19,34). Current U.S. practice

(18,34) typically incorporates the vertical shear method of analysis originally published by Terzaghi (31), and only that method will be discussed here.

Justification for Vertical Shear Model. Figure 7a shows a transverse vertical section of a typical bulkhead. The center plane of this "beam" (neutral axis) is the plane of maximum shear, according to assumptions in simple beam bending theory. The vertical shear analysis method simply compares shearing forces on this plane to shearing resistance to derive a factor of safety against a distortion failure.

Experiments (16) on reasonably large (4 ft [1.2 m] diameter) model cells have demonstrated that vertical shear planes do form in cells at the time of a distortion failure. Figure 7b is a photograph of one of the test cells. Subsequent analytical work (12) using elastic theory showed that a vertical shear plane should form slightly ahead of the cell centerline in the lower part of the cell. Photoelastic studies (3) show that for double wall cofferdams, internal shear failure controls, but that the failure plane is most likely to be nearer the front sheetpiles than the centerline. The predominance of experimental and analytical evidence verifies that the vertical shear model adequately represents distortion of a sheetpile cell for purposes of stability analysis.

Vertical Shear Analysis. In Figure 7a an applied moment, M, results from the application of the forces ΣP_h and ΣP_v to a bulkhead. Bending stress distribution in the cell fill, according to conventional bending analysis, is triangularly distributed on either side of the neutral axis, with a resultant force, V. These bending stresses

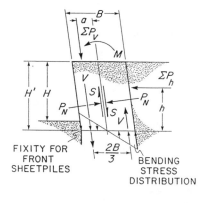

(a)   (b)

Figure 7.   Vertical Shear Distortion in Bulkhead Cells. a) Model for Analysis, b) Photograph of Model Cell Failure.

result in the couple 2VB/3 which is equal to the applied moment. The shear force $V = 3M/2B$ must be resisted by the shearing resistance, S, on the centerline of the bulkhead, which depends on the shearing resistance of the cell fill and the two interlocks (see Figure 1) in a typical section. The factor of safety against a shear failure will be the ratio of the shearing resistance on the centerline to the shearing force.

The relationships that follow are based on the presumption of a uniform effective unit weight of cell fill. In most actual cases, cell fill for a bulkhead is at least partially submerged, and as a result, effective unit weight is different above and below the water level. Also, various surcharge loads, such as deck weight or live load may be present on the cell fill. Consequently, the simple relationships derived will show symbolic results only. Actual analysis will require detailed representation of actual forces.

Due to the weight of the cell fill, the normal force on the centerline of the section in Figure 7a is $P_N = 0.5\ \gamma H^2 K_f$. K_f is the lateral earth pressure coefficient for the centerline of the cell fill. Shearing resistance per unit length due to the fill will be $S_F = P_N \tan\phi$, so that $S_F = 0.5\ \gamma H^2 K_f \tan\phi$. Interlock force distribution on the centerline of a cell driven in sand will generally be as shown in Figure 5b. If we define T as the total circumferential tension force in the interlock on centerline, then $T = \int_0^H pR\ dp$, which, for the case with no tip restraint (triangular distribution of interlock force over full height, H), becomes $T = 0.5\ \gamma H^2 K_i R$. Note that K_i is the apparent lateral earth pressure coefficient used for interlock force calculations, and is different from K_f. If we further define the shearing resistance in an interlock as S_I, and the friction coefficient for interlock sliding as f, then $S_I = Tf = 0.5\ \gamma H^2 K_i Rf$. Now, for a circular cell bulkhead like that shown in Figure 1, the factor of safety against shear failure on the centerline will be the ratio of resisting to driving forces on a typical section

$$FS = \frac{S_F(2L) + 2S_I}{V(2L)} = \frac{S_F(2L) + 2S_I}{\frac{3M}{2B}\ (2L)} \tag{6}$$

For the special case of uniform fill density, $2L = 2.2R$ and triangular distribution of interlock force

$$FS = \frac{B\gamma H^2 K_f \tan\phi + 0.9\gamma H^2 K_i fB}{3M} \tag{7}$$

Now, if K_i were taken equal to K_f (which it is not) and both were designated K, we have Terzaghi's equation

$$FS = \frac{B\gamma H^2 K(\tan\phi + 0.9f)}{3M} \tag{8}$$

Equations (6) and (8) represent the requirements for internal stability of a circular cell bulkhead. According to Equation (6), B, a function of R, must be large to provide resistance against tilting. There is an upper limit on R, however, according to Equation (2) because t is limited by available interlock strength. As a consequence, for current interlock strengths, it is difficult to produce an acceptable design when H > 70-80 ft (21-24 m), because the sheetpile interlocks will not support the cell fill with an adequate safety factor.

Consider now, certain details involved in the application of Equation (6) to evaluation of cell moment resistance. We must determine M, S_F and S_I. In usual design circumstances driving moment, M, is the result of external force applied to the cell. It has been shown (22) that the height of the cell above fixity, or the sheetpile tips if there is no fixity, as shown in Figures 5b and 7, should be used to represent the bulkhead height for moment and shearing resistance calculations. Figure 7a shows resultant vertical and horizontal external forces on a bulkhead, resulting in $M = \Sigma P_h(h) + \Sigma P_v(a)$. In Terzaghi's paper (31), these driving forces were not shown or considered in the foregoing derivation, probably because they complicate the calculation of S_F, which depends on the normal forces on the cell centerline. The original derivation included only the lateral forces due to the weight of the cell fill. Terzaghi also proposed the use of the Rankine value K_A for K_f for the calculation of S_F. Krynine's discussion (13) pointed out that the center plane is not a principal plane, and that the appropriate value of K_f [Equation (9)] should recognize that fact, and be calculated as

$$K_f = \frac{\cos^2\phi}{2-\cos^2\phi} \tag{9}$$

Krynine's K_f is widely used in practice (14,18,19,34), and is recommended here for calculation of the lateral earth force on the cell centerline. We must still account, however, for the effect of the force ΣP_h (Figure 7) on the shearing resistance of the cell fill. In the interpretation of model tests to failure (16), Equation (8) was used to back calculate K. It was determined that K was about 1.0. It was further suggested that the reason for this value exceeding that from Equation (9) had to do with the effect of lateral force ΣP_h being unaccounted for. A subsequent analytical study (12) using elastic theory showed that K for use in Equation (8) would depend on the magnitude and distribution of the applied lateral force, ΣP_h. Differentiating between bulkheads (applied lateral earth and surcharge forces) and cofferdams (applied lateral water forces) the analysis showed that for use in Equation (8) K = 0.6 (bulkheads) and 0.7 < K < 1.4 (cofferdams) would be appropriate. Thus, while interpretation of model test results suggested that K = 1 in Equation (8) and recommended its use, analytical work later showed that the recommendation may apply only to water loaded cofferdams. In the derivation of Equation (8) it was shown that K is assumed to be the same for determination of the lateral earth pressure on the cell centerline as for determination of interlock force. Further, it was assumed that interlock force has a particular (triangular) distribution over the height of the bulkhead. Finally, by making the assumption that the K's were identical, test data were used to evaluate how it was affected by lateral load. Not surprisingly, results of this approach turned out to apply only to the special cases from which they were derived. Their adoption to other circumstances can lead to error. However, the test data itself is still useful. It needs only an improved interpretation.

If Krynine's coefficient [Equation (9)] is used along with additional shearing resistance equal to $\Sigma P_h \tan\phi$ to evaluate S_F in Equation (6), if the actual interlock friction forces are used to calculate S_L, and if the shear force, V, is calculated for the circular cell cross section, new analysis of the model test (16) data produces the results shown in Table 1. The average calculated vertical shear factor of safety for the tests to failure is 0.99, which is very close to the actual value of 1.00 for each test. Therefore, the writer believes that use of Equation (8) should be discontinued, and that Equation (6) should be applied with M taken as the moment applied about the plane of fixity or the sheetpile tips, as appropriate, and Krynine's coefficient [Equation (9)] used to compute the effect of cell fill weight on shearing resistance. Additional shearing resistance equal to $\Sigma P_h \tan\phi$ should be added, so that $S_F = 0.5 \gamma H'^2 K_f \tan\phi + \Sigma P_p \tan\phi$ in Equation (6). Finally, it should be noted that calculation of the influence of interlock tension on centerline shearing resistance must incorporate the appropriate interlock force distribution. If it is assumed that maximum interlock tension is distributed as shown in Figure 5b, then $T = 1/3 \gamma K_i RH'^2$ and Equation (6) becomes, after some manipulation,

$$FS = \frac{B\tan\phi(\gamma H'^2 K_f + 2\Sigma P_h) + 0.6B\gamma H'^2 K_i f}{3M} \tag{10}$$

Equation (10) is recommended for vertical shear analysis of a bulkhead with a fill having uniform effective density and the "bulge" location as determined by the method of Figure 5b. In other cases, the more basic Equation (6) must be applied. A factor of safety of 1.50 is usually considered an acceptable minimum for static conditions.

Table 1. Re-evaluation of Model Test (16) Results.
(1 k = 4.45 kN, 1 k-ft = 1360 N-m)

Cell Number	P_{max} (k)	M_{max} (k-ft)	M'_{max} (k-ft)	Calculated FS
1	4.25	19.0	13.43	0.87
2	3.53	11.4*	10.25	1.00
3	4.45	19.8	12.77	0.92
4	2.65	6.5	6.49	1.18

M_{max} = Moment about sheetpile tips
M'_{max} = Moment about plane of fixity
* = Corrected errata

Bulkhead Deflections. It has been reported that cofferdams subjected to water loading will deflect 0.5-1% of the cell height above the dredge line, and that cofferdams loaded by soil backfill (bulkheads) will deflect more (6). The lower limit of deflection cited for cofferdams is based on observations at the Trident facility (27), where foundation conditions were especially good, and where the cell fill was vigorously densified and dewatered; and, at Lock and Dam 26(R) where there was an interior stabilizing berm in front of the cofferdam (5). The upper limit is based on Swatek's extensive experience (30). Data from Cell 66 at the Seagirt Marine Terminal and Cell 4 at Fulton Terminal 6 are cited to show that bulkheads have experienced lateral deflections from about 1% to an extreme value of about 13% of their free height (6). Early papers have shown very large deflections for cellular bulkheads (7,31).

At Terminal 4 (21), Fulton Terminal 6 (23) and at Long Beach (35), long-term settlements of front sheetpiles ranged between 10 and 22 in (25 and 56 cm) and deflections at the tops of the sheetpiles were from 15 to 33 in (38 to 84 cm). In each case, the cells were of substantially the same dimensions if the surcharge on the Long Beach cell is considered. The cell fill was not compacted at Long Beach. These long term deflections correspond to about 2-4% of the cell free heights, which is more than the 1% reported for cofferdams. End of construction deflections at Terminal 4 and at Fulton Terminal 6 were, however, only about 1% of the cell height, suggesting that the shorter life of cofferdams may account for much of the reported difference in observed deflections for cofferdams and bulkheads. It is of interest to note at this point that the deflections at failure in the model tests (16) correspond to 3-6% of the cell free height.

If Equation (10) is used to make calculations for hypothetical cofferdams (water load, submerged cell fill) and bulkheads (earth load, submerged or dry fill) it can be shown that structures with the same dimensions and the same fill and foundation have the same factor of safety, which suggests that they should experience the same deflection. This at first seems implausible, since a cofferdam typically carries a greater lateral load than does an earth bulkhead. However, according to Equation (10) the additional load results in additional shearing resistance on the centerline, with the result that the factors of safety for the structures are substantially the same. Thus, there is some analytical support for the reported observations that short term deflection of ordinary cofferdam and bulkhead cells both have been about 1% of their height. The analysis also shows that a cofferdam derives over 90% of its resistance to shear from the cell fill (less than 10% from the cross wall interlocks) while a bulkhead derives only about 70% of its resistance from the cell fill.

Figure 8 is a plot of available data comparing settlement and deflection for front sheetpiles on the Seagirt project (39 cells), the Fulton Terminal project (6 cells) and the Terminal 4 and Long Beach cells. Each of these projects involved cells with essentially the same diameter. The cells for Seagirt had about 15 ft (4.6 m) less free height than the other projects. The data does not represent long term observations in every case. It shows paired values of settlement and deflection that were available after construction was complete.

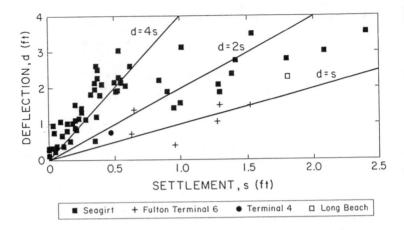

Figure 8. Comparison of Settlement and Deflection of Front Sheetpiles
           for Cellular Bulkheads (1 ft = 0.305 m).

It is apparent from Figure 8 that some relationship between settlement
and deflection certainly does exist.
     Many of the Seagirt cells, most of the cells of Terminal 6 and the
single instrumented cells at Terminal 4 and Long Beach experienced
transverse deflections between one and two times their settlements.
 Most of the Seagirt cells deflected from three to more than six times
their settlement, but these were unusual because of the presence of
soft clay in the lower third of the cells and in the backfill area,
which would contribute to larger than normal deflection.
     From Figure 8, it is clear that settlement and deflection of the
front (unloaded side) sheetpiles in a bulkhead are related.   If
settlement is small, and if good quality cell fill is used, deflections will be small.   Even for good quality fills, however, large
settlements lead to large deflections, as demonstrated at the Fulton
Terminal project. Therefore, it is of outstanding importance to make
good settlement estimates in planning a cellular bulkhead project. It
is even more important to keep weak compressible soils out of cell
fills, as demonstrated at Seagirt where there were large deflections
in spite of some rather small settlements.
     The Fulton Terminal and Long Beach projects appear to be the only
ones where field data are available to examine in detail the relationship between settlement and deflection.   In the initial measurement
program at Fulton Terminal it was planned to measure settlement at the
fronts and backs of seven cells. For various reasons, data was only
produced for three cells, and it did not show a consistent pattern of
relative magnitude of front and back sheetpile settlement. However,
data (see Figure 9b) developed after publication of the original paper
(23), clearly shows that the deck of the wharf has developed a regular
slope downward from back to front, along its entire length.   The

fronts of the cells clearly have settled more than the backs. The larger settlements near the east end of Berth 603 are attributed to the silt layer alluded to in the original paper and verified later by deep Dutch cone probing at Cells 2 and 4 (11). The silt layer was not present at or to the west of Cell 7. The front of the Long Beach cell settled about 1.4 ft (0.4 m) more than the back. The differential became more pronounced as settlement progressed.

Why do bulkheads tip increasingly forward with increasing settlement? Cells might logically be expected to tip back due to foundation compression settlement resulting from the weight of the cell fill and backfill. First, it is obvious that settlement and related deflection of the front sheetpiles could be the result of higher foundation stresses at the front of the bulkhead owing to the effects of moment as shown in Figure 7. Terzaghi explains the importance of higher toe pressures (31), especially if there is compressible material at the bottoms of the cells, as at Seagirt. A second, and less obvious reason may be associated with settlement differences between ends of a bulkhead and the middle. Figure 9 shows that at the Fulton Terminal project, settlement of the wharf was greatest near the middle, and likely would have been so even if the silt layer noted above was not present and the larger settlements under the east end of the structure had not occurred. This is in accord with expectations for the greater settlement on the side of an area load, with respect to the corners. Such a pattern of settlement would place the cells in compression in a direction along the wharf and induce an increase in cell diameter transverse to the wharf axis. Differential settlement, either back to front or front to back also promotes an increase in transverse

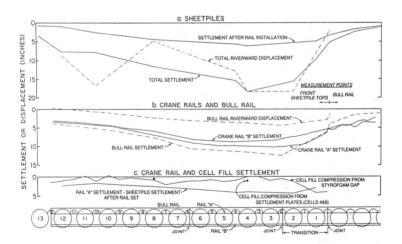

Figure 9. Movements at Fulton Terminal 6 Bulkhead (1 in = 2.54 cm).

diameter for a non-rigid cell. Because the backfill resists movement, this increase in diameter provides a third cause of tipping related to settlement. Measurements made at Long Beach (34) show that the instrumented cell deformed into an elliptical shape with the short axis parallel to the bulkhead. Compression resulting from settlement and the settlement itself thus had the effect of increasing front sheetpile deflection. Measurements at Lock and Dam 26(R) cofferdam also indicated an elliptical deformed shape (6).

Procedures are available for estimating deflection from bending theory (2) and by finite element analysis (5). Neither method would account for deflection due to compression along the bulkhead axis arising from settlement, as described above, due to the effects of differential settlement, or from the placement of arc fill, which has been observed (21,23). Deflection of the front sheets can be related to settlement empirically using Figure 8, but a good settlement estimate for the front sheets is required. Given the state of practice regarding settlement estimates for this type structure and the usual site conditions on variable alluvial deposits, great precision of a settlement estimate and, therefore, a deflection forecast is unlikely (11).

SEISMIC ANALYSIS

The 1970 Cornell conference addressed the subject of seismic analysis for retaining structures, particularly that aspect having to do with effects of earthquakes on earth and water forces (25). At the time, there was no widespread unanimity concerning approaches to seismic analysis, and there was less agreement about how to analyze quay (submerged) walls than ordinary retaining walls. Since 1970 only a single paper has appeared which deals directly with seismic analysis of cellular cofferdams and bulkheads (4). It proposed a pseudo-static analysis for stability evaluation. Matsuzawa, et al. (17) have re-examined external forces on submerged retaining structures.

The current state of affairs for seismic conditions may be summarized by referring to Figure 10, which shows loads recommended for analysis (4). To those loads shown, any additional loads due to surcharges and equipment would need to be added. A pseudo-static analysis is made, using Equation (6) or (10) as appropriate. It should be noted here that the body force, k_hW, makes no contribution to centerline shearing resistance.

The static factor of safety for most bulkhead cells is not large, typically being around 1.6 to 1.8. A minimum of 1.5 is usually recommended. The additional lateral forces shown in Figure 10 often cause the calculated factor of safety for a good static design to fall below one. Designers are sometimes tempted to disregard the analysis, arguing that the method is unproven, or overly conservative, that it does not address the real issue of tolerable movement, or that a large design earthquake is possibly not realistic. Usually, we accept a factor of safety of around 1.1.

The seismic analysis usually is done by applying the forces shown in Figure 10, and assuming that the response of the cell fill is governed by its drained strength. Bulkhead fills are, of course, usually compacted. Dense fills are often dilatant, and, as a

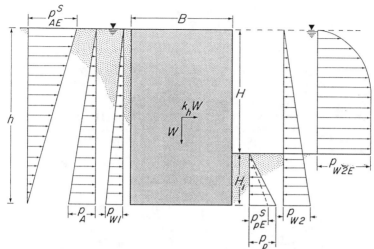

K_{AE}^S = Dynamic Active Pressure Coefficient for Submerged Soil
K_{PE}^S = Dynamic Passive Pressure Coefficient for Submerged Soil
W = Total Weight of Soil
k_h = Horizontal Seismic Coefficient
γ = Total Unit Weight of Soil
γ' = Submerged Unit Weight of Soil
γ_w = Unit Weight of Water

ACTIVE SIDE

Static Active Pressure = $p_A = K_A[\gamma'h]$
Static Water Pressure = $p_{W1} = \gamma_w h$
Dynamic Active Pressure =
$p_{AE}^S = [K_{AE}^S - K_A]\gamma'h$

INSIDE CELL

Horizontal Seismic Inertia = $k_h W$

PASSIVE SIDE

Static Passive Pressure = $p_P = K_P \gamma' H_1$
Note: For Dynamic Condition, Replace
K_P by K_{PE}^S
Static Water Pressure = $p_{W2} = \gamma_w(H + H_1)$
Dynamic Water Pressure = p_{W2E} =
7/8 $k_h \gamma_w H$; Force 7/12 $H^2 k_h \gamma_w$ Acts
at Height $(.4H + H_1)$ from the Base

Figure 10. Seismic Forces on a Cellular Sheetpile Bulkhead for Zero
Vertical Seismic Coefficient. Adapted from (4).

consequence, their short-term strength, especially for finer sands, is greater than their drained strength, which suggests that their dynamic factor of safety could be greater than the static value. It can be shown that the ratio of the vertical shear factor of safety calculated from undrained strength to that using drained strength, is

$$FS_{UD}/FS_D = \frac{\sin\phi[K_o+(1-K_o)A_f] + K_i f}{[1+(2A_f-1)\sin\phi][K_f\tan\phi+K_i f]} \tag{11}$$

where previous definitions apply and A_f is the pore pressure parameter (26).

The interesting result of this analysis is that for cells of the usual proportions $FS_{UD}/FS_D > 1$ for $A_f \leq 0.8$. In other words, for soils with $A_f \leq 0.8$, which would include most compacted cell fills, the factor of safety against shear failure for the seismic case should exceed that for the static case. While Equation (11) is unverified by experiment, it offers some hope that cellular bulkheads are not as vulnerable to seismic damage as drained strength analysis would suggest.

CONSTRUCTION

Certain considerations for construction were developed in the 1970 Cornell conference paper (14). Several are discussed here for emphasis, or to include the author's experience. The original paper is recommended reading, for completeness.

Sheetpile Selection. Both U.S. and foreign manufacturers supply sheetpiling that can be used in cells. The relative merits of these products will not be discussed here. There are, however, a number of general considerations that are important when sheetpiling are chosen for a particular job. Interlock strength is one of them. Standard strength interlocks are most commonly used, but lower strength and high strength interlocks are available for special applications. High strength sheetpile interlocks have, in some cases, made driving difficult because of excessive friction. The opening and the thumb of the interlocks for every sheetpile should be gauged on the job site to insure against driving difficulties and weaknesses. The webs of sheetpiling with a given interlock strength are available in different thicknesses. Web thickness should be chosen on the basis of working stress level for the expected interlock forces, and in some cases where corrosion may reduce metal thickness, to provide an adequate structural cross section over the project's design life. Special coatings, special steel alloys, and cathodic protection systems may be used to further enhance the durability of sheets.

Predredging. It has been shown that the presence of compressible fine grained soils in bulkhead cells results in large time-dependent lateral movements. It is, therefore, important to remove these materials by dredging before cell construction begins if at all possible. Removal after cells have been driven results in external compressive forces due to differential soil levels outside and inside the cell, and causes unbraced cells to buckle inward. Predredging is also important to limit driving distance for the sheetpiles. In most alluvial sands, sheets can easily be driven 20 or 30 ft (6 or 9 m). Swatek notes that 40 ft (12 m) is about the maximum practical driving limit (29), and one case where the total driven length was 65 ft (20 m) has been reported (20). The soils in this latter case were submerged loose to medium dense sands and soft silts. In soils containing gravel and cobbles there is a tendency for misalignment of sheetpiles and interlock binding which makes driving especially difficult.

   Sticking and Driving. Figure 11 shows the usual sequence of cell erection. Cells are formed on a double template, each consisting of two rings, about 6 in (15 cm) smaller radius than the inside of the cell. The template is positioned and supported on spud piles. The clearance provided between the template and the sheetpiles allows for blocking to set actual pile location, and helps during template removal. Wye piles are located and tacked to the template, then all of the sheetpiles are threaded between the wyes. [In at least one case (10) sheetpiles were threaded on a template and spud piles on land, then the entire assembly was lifted into position.] High winds make threading difficult, and often result in a job shutdown until they abate. Once the sheets are in place they are lifted and checked to be sure they run freely in the interlocks, and are plumb. Out of plumb piles promote toeing in or out of the sheetpiles during driving, and binding or splitting in the interlocks. Sheets are commonly driven in pairs (every other pair along the circumference), and only a short distance beyond adjacent sheets in order to minimize twisting.

Figure 11. Erection of Sheetpile Cells.

This distance, or "lead" is usually about 5 ft (1.5 m) where driving is easy, and may be limited to much less in difficult driving. By driving alternate pairs the total driving distance is twice the "lead." When adjacent cells have been driven, connecting arcs may be placed. Threading and driving of the arcs is best done before cells are filled, to eliminate driving difficulties resulting from expansion of the cells. Vibratory hammers have proven to be most effective for driving sheetpile cells in sands. Penetrating very dense soils and gravels is difficult and usually requires the use of impact hammers. An experienced, hard-working crew can usually stick and drive a cell of the usual proportions in two or three days. The first cell or two on a job often takes a week or more to complete. Figure 12 shows a bulkhead with all of the cells and arcs driven, and ready for filling.

Filling Cells and Arcs. Cells may be filled by a number of methods, including clam shell, hydraulic dredging, and conveyor. In every case, the fill should be dropped at the center of the cell to insure symmetrical expansion. Non-symmetrical placement results in severe cell distortion (35). Arc fill is sometimes placed before cell fill is compacted, particularly if there are back arcs. To restrict

Figure 12. Completed Sheetpile Cells Ready for Fill and Backfill.

transverse movement of the cells, however, it is best to compact the cell fill before arc fill (if there is no back arc) and backfill behind the cells is placed. If at all possible, cell fill should be clean granular material, amenable to compaction by various deep probing methods.

Densification. There is often some concern among engineers or contractors that densification of cell fills may increase interlock forces and split a sheetpile cell. This has indeed occurred, but for the cases in the writer's experience, the failures have been due to structural weaknesses in the sheetpiles. Measurements made at Terminal 4 (21) show that interlock force increases about 1 kip/in (1.8 kN/cm) as a result of vibratory probing. At Trident (27) the maximum interlock force increased about 2.5 kip/in (4.4 kN/cm) due to very vigorous deep compaction (see Figure 6). Drainage wells are sometimes used during compaction to enhance consolidation of the fill. There is no evidence which suggests that deep probing will cause general liquefaction of a loose cell fill.

Fill density is checked by standard penetration testing or Dutch cone probing, using standard correlations (9) between density or relative density, and penetration resistance. The writer's experience with clean medium to fine sand fills placed by all of the usual methods suggests that uncompacted fills have a relative density of around 40%. The fill at Long Beach (35) was not densified, and its surface settled about 1 ft (0.3 m) after placement, according to settlement plate observations. Figure 13 shows the result of

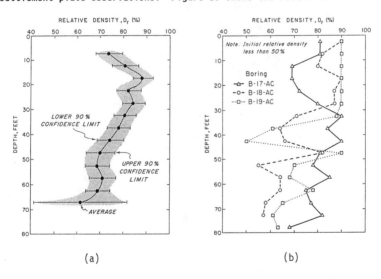

Figure 13. Densification Results for Sheetpile Cell Fills. a) Fulton Terminal (15), b) Trident (8) (1 ft = 0.305 m).

densification at the Fulton Terminal and Trident projects based on standard penetration testing (15,27). Settlement plate observations there showed a maximum of 1 in (2.54 cm) of settlement of the bulkhead deck at the Fulton Terminal, owing to compression of the cell fill. The specified relative density at Trident was 75%, and it was achieved or exceeded. Densification of backfill behind the bulkhead may proceed when cell fill densification is complete.

Deck Construction. If the bulkhead is to have a concrete deck, forming should be arranged to allow for fill settlement so that the concrete does not bear on the tops of the sheetpiles. Styrofoam blocking 4-6 in (10-15 cm) high above the sheets as shown in Figure 14 has proven adequate for this purpose.

If the deck is to support crane rails the rails may be set on ties in ballast, or supported directly on concrete beams cast integrally with the deck. The great length of most bulkheads (and the crane rails) can result in thermal expansion which tends to produce misalignment of tie-supported rails. Rails supported on monolithic deck beams have provided superior performance, in the writer's experience.

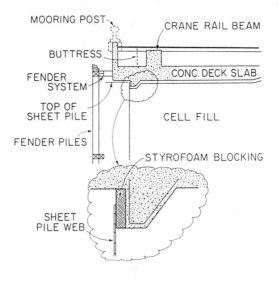

Figure 14. Detail for Isolating Deck Structure from Sheetpiles.

CONCLUSIONS

Cellular bulkheads are widely used to construct ship berthing facilities. They have proven to be reliable, economical and relatively maintenance free structures. Bulkheads have undergone relatively large movements without damage. A relatively rigid and continuous deck tends to minimize the effects of differential cell movements. Design methodology for cells recommended in this paper is semi-empirical and only slightly different from procedures reported for cofferdams in the 1970 Cornell conference. External stability (sliding) must be evaluated, and internal stability, which usually governs design, should be checked by the improved methods described herein. Methods for seismic analysis are not well-established, but indications are that in those instances involving compacted cell fills, seismic instability may be mitigated by mobilization of undrained cell fill strength.

Finite element methods are being developed to assess bulkhead deflections. Lateral deflections are strongly influenced by cell settlement. At this writing, the accuracy of deflection analysis depends very much on the careful selection of parameters to represent the cell and fill, and the experience and judgment of the designer doing the analysis. While our ability to forecast movements is still in the developing stage, we can take some comfort in the fact that there is a good experience base for estimating bulkhead deflection, and that many bulkheads have undergone movements which exceed those required to fully mobilize the strength of the cell fill, without impairment of function.

ACKNOWLEDGEMENTS

The author is indebted to the Port of Portland, Oregon, for its cooperation on field research projects and permission to publish data. Others who gave permission to use field data from project files include Shannon and Wilson, Inc. and the U.S. Army Corps of Engineers of St. Louis, Mueser Rutledge Consulting Engineers of New York, and the Kiewit Construction Group, Inc., of Omaha. Their cooperation is gratefully acknowledged. Elmer Richards, Jim Gould, and Larry Hansen made very helpful comments on the original manuscript.

REFERENCES

(1)   Cofferdams on Rock, TVA Technical Monograph 75, Tennessee Valley Authority, Vol. 1, Dec. 1957.

(2)   PP Brown, discussion of Field Study of a Cellular Bulkhead, by A White, JA Cheney & CM Duke, Transactions, ASCE, Vol. 128, Part I, 1963, 463-508.

(3)   NK Burki & R Richards, Photoelastic Analysis of a Cofferdam, J. Geot. Eng. (ASCE), 101(2), Feb 1975, 129-146.

(4)   S Chakrabarti, AD Husak, PP Christiano & DE Troxell, Seismic Design of Retaining Walls and Cellular Cofferdams, Proc. Conf. EESD (ASCE), Vol. 1, Pasadena, 1978, 325-341.

(5)   CW Clough & T Kuppusamy, Finite Element Analysis of Lock and Dam 26 Cofferdam, J. Geot. Eng. (ASCE), 111(4), 1985, 521-541.

(6)   CW Clough & JR Martin, Cellular Cofferdams-Developments in Design and Analysis, Proc. 2nd Intl. Conf. on Case Histories in Geot. Engr., St. Louis, 1988.

(7)   MM Fitz Hugh, JS Miller & K Terzaghi, Shipways with Cellular Walls on a Marl Foundation, Transactions, ASCE Vol. 112, Paper No. 2300, 1947, 298-324.

(8)   JB Forrest, Dewatering Cofferdam for the Trident Submarine Drydock, Proc. 10th Intl. Conf. on Soil Mech. and Found. Engr., Vol. 1, 1981, 123-126.

(9)   HJ Gibbs & WG Holtz, Research on Determining the Density of Sands by Spoon Penetration Testing, Proc. 4th Intl. Conf. on Soil Mech. and Fndn. Eng., 1957, 35-39.

(10)   SA Ghali, An Analysis Based on Full Scale Measurements of Interlock Piles of Cellular Structured Quays, Ocean Eng, 8(3), 1981, 259-294.

(11)   M Helal, Terminal 6, Berth 603, Settlement Analysis Using CPT Data, Report in Partial Fulfillment of MSCE Requirements, Oregon State Univ., Mar 1984, 66 p.

(12)   BH Ilhan, Theoretical K-Values for Vertical Shear Analysis, Report in Partial Fulfillment of MSCE Requirements, Oregon State Univ., Mar 1984, 31 p.

(13)   DP Krynine, discussion of Stability and Stiffness of Cellular Cofferdams, by K Terzaghi, Transactions, ASCE, Vol. 110, Paper No. 2253, 1945, 1175-1178.

(14)   Y Lacroix, MI Esrig & U Lusher, Design, Construction and Performance of Cellular Cofferdams, Proc. Conf. on Lateral Stresses in the Ground and Earth Retaining Structures, ASCE, Jun 1970, 271-328.

(15)   P Leycure & WL Schroeder, Slope Effects on Probe Densification of Sands, in Soil Improvement, A Ten Year Update, GSP No. 12, ASCE, 1987, 197-214.

(16)   JK Maitland & WL Schroeder, Model Study of Circular Sheetpile Cells, J. Geot. Eng. (ASCE), 105(GT7), Jul 1979, 805-821.

(17)   H Matsuzawa, I Ishibashi & M Kawamura, Dynamic Soil and Water
       Pressures of Submerged Soils, J. Geot. Eng. (ASCE), 111(10), Oct
       1985, 1161-1176.

(18)   Naval Facilities Engineering Command, Foundations and Earth
       Structures, Design Manual 7.2, May 1982, 119-127.

(19)   M Rossow, E Demsky & R Mosher, U.S. Army Corps of Engineers, TR
       ITL-87-5, Theoretical Manual for Design of Cellular Sheet Pile
       Structures, May 1987.

(20)   GM Lochhead, Waterfront Development for New Cement Plant, J.
       Waterways and Harbors (ASCE) 94 (WW2), May 1968, 149-157.

(21)   WL Schroeder, DK Marker & T Khuayjarernpanishk, Performance of
       a Cellular Wharf, J. Geot. Eng. (ASCE), 103(GT3), Mar 1977, 153-
       168.

(22)   WL Schroeder & JK Maitland, Cellular Bulkheads and Cofferdams,
       J. Geot. Eng. (ASCE), 105(GT7) Jul 1979, 823-837.

(23)   WL Schroeder, Wharf Bulkhead Behavior of Fulton Terminal 6, J.
       Geot. Eng. (ASCE), 113(6), 1987, 600-615. See also Discussion
       Closure, 115(7), 1989, 1029-1032.

(24)   WL Schroeder, Discussion of Sheetpile Interlock Tension in
       Cellular Cofferdams, by MP Rossow, J. Geot. Eng. (ASCE), 113(5),
       May 1987, 547-549.

(25)   HB Seed & RV Whitman, Design of Earth Retaining Structures for
       Dynamic Loads, Proc. Conference on Lateral Stresses in the
       Ground and Earth Retaining Structure, (ASCE) June 1970, 103-147.

(26)   AW Skempton, The Pore Pressure Coefficients A and B, Geotechni-
       que, Vol. 4, 1954, 143-147.

(27)   MD Sorota, EB Kinner & MX Haley, Cellular Cofferdam for Trident
       Drydock: Performance, J. Geot. Eng. (ASCE) 107(GT12), Dec 1981,
       1657-1676.

(28)   Shannon & Wilson, Inc., Lock and Dam 26(R), Analysis of
       Instrumentation Data, Vol. 1, Interim Report, St. Louis
       District, US Army Corps of Engrs, Oct. 1982.

(29)   EP Swatek, Cellular Cofferdam Design and Practice, J. Waterways
       and Harbors (ASCE) 93(WW3), Aug. 1967, 109-132.

(30)   EP Swatek, Summary, Cellular Structure Design and Installation,
       Design and Installation of Pile Foundations and Cellular
       Structures, Envo Pub. Co., Lehigh Valley, PA, 1970, 413-424.

(31)   K Terzaghi, Stability and Stiffness of Cellular Cofferdams,
       Transactions, ASCE, Vol. 110, Paper No. 2253, 1945, 1083-1202.

(32) HE Thomas, EJ Miller & JJ Speaker, Difficult Dam Problems-
      Cofferdam Failure, Civ. Eng., Aug 1975, 69-70.

(33) U.S. Army Corps of Engineers, An Analysis of Cellular Sheet Pile
      Cofferdam Failures, Office of the Chief of Engineers, April
      1974, 44 p.

(34) U.S. Army Corps of Engineers, EM 1110-2-2503, Design of Sheet
      Pile Cellular Structures.

(35) A White, JA Cheney & CM Duke, Field Study of a Cellular
      Bulkhead, Transactions, ASCE, Vol. 128, Part I, 1963, 463-508.

CELLULAR SHEET-PILE FLOODWALL IN COHESIVE MATERIAL
AT WILLIAMSON, WV

Robert E. Taylor[1] and David F. Meadows[2]

ABSTRACT: The U. S. Army Corps of Engineers' floodwall project currently under construction at Williamson, WV, is unusual in that: (1) the associated stream is fast-rising, (2) the riverbanks are unstable, and so the structure must contribute to the bank stability, and (3) usable land is scarce, and so the structure must be built as near to the river as possible.

Circular cellular sheet-pile walls have been selected as the principal structure because of their longevity, ease of maintenance, ductility, contributions to bank stability, and their adaptability to aesthetic treatment.

Because of the unstable bank conditions and planned retention of soft clays within the completed sheet-pile cells, the analyses and design considerations for the wall differed substantially from common practice. Conventional design methods had to be adapted to accommodate the soft clays, and finite-element analyses and prototype cells were used to confirm the design.

LOCATION AND HISTORY

Williamson, West Virginia, is a city of about 5000 people located on the Tug Fork of the Big Sandy River, a tributary of the Ohio River. The Tug Fork is about 140 miles (225.3 km) long and drains an area of 1,555 square miles (4027.45 square km). The middle of the stream marks the state border between West Virginia and Kentucky.

The Tug Fork Valley is extremely narrow and bounded on both sides by steep mountains. Only 10 percent of the land in this area has a grade of less than 12 percent, and the greater part of the land has slopes steeper than 35 percent. Most of the level land suitable for development is located along the stream channel. Growth in the area's coal industry resulted in intensive use of the valley floor for highways, railroads, commercial centers, and housing for the coal miners and their families with little regard for the associated flood hazards. Given its large share of the

1 - Structural Engineer, U.S. Army Corps of Engineers, Huntington District, 502 Eighth Street, Huntington, WV 25701-2070
2 - Geotechnical Engineer, U.S. Army Corps of Engineers, Huntington District, 502 Eighth Street, Huntington, WV 25701-2070

valley's level land and its location in the heart of the southern
West Virginia coal fields, Williamson quickly developed into the
center of commerce, education, government, and transportation activities for the Tug Fork Valley despite the ever-present threat of
flooding.

This threat of flooding has been realized time and again with 37
damaging floods since 1900. The flood of record occurred in April
1977 and caused damages in the Tug Fork Valley of about $300 mil-
lion by current day prices. The frequency of this flood exceeded
the 500-year event.

In the Water Resource and development Act of 1974, the U. S.
Army Corps of Engineers was tasked with developing a comprehensive
plan for the reduction of flood damage in the Tug Fork Valley. The
April 1977 flood later led Congress to pass Section 202 of the
Energy and Water Development Act of 1981. Williamson, West Vir-
ginia, which lies within the Huntington District of the Ohio River
Division of the Corps of Engineers, was one of the communities
named to receive flood protection.

UNUSUAL PROBLEMS ENCOUNTERED

Land use and riverbank instability caused some unusual design
and construction problems and required special criteria for the
flood-protection alternatives considered. Some of these problems
are characteristic to the Tug Fork Valley while others are isolated
to Williamson's central business district (CBD).

Extensive land use along the river was the cause of one problem.
The Tug Fork is dotted with communities along its length on both
sides, and there was much concern expressed that, if a structure
was built to protect one community and thus restrict the stream,
the flooding of the remaining communities would increase.   One
criterion for alternative selection was that the protection scheme
must not appreciably increase the hydraulic profile of the stream
and thereby would not increase the flooding of the other communities.

Also, intensive land use caused difficulties. Because usable
level land is scarce in the Tug Fork Valley, one goal of the
project is to develop as much flood-free land as possible. In considering alternatives for Williamson CBD project, the Corps sought
to preserve most of the city and provide usable land upon completion of construction. In addition, most of the land in Williamson
is already being used to its fullest, making it a challenge to work
in and around the city without disrupting and destroying the life
and economy of the city. An effort would be made in the design and
construction of the project to minimize the economic, ecologic, and
social impact on the city. This restricts the types of structures
and construction methods used.

Geotechnical investigation of the city revealed the project's
most difficult problem--an unstable river bank. Stability analysis
indicated that to provide flood protection on a stable riverbank
using a conventional flat base T-wall required laying back the
riverbank to slopes ranging from three horizontal to one vertical
to four horizontal to one vertical (12). In Williamson's case,

this would have required the taking of an unacceptable number of landward structures. A structure was therefore needed that would both contribute to the stability of the bank and allow maximum use of the land.

STRUCTURAL ALTERNATIVES AND SELECTION

Nearly 20 alternatives were investigated in addressing the geotechnical and hydrological problems. Conventional T-base walls, gravity walls, reinforced earth walls, anchored sheet-pile walls, and circular and diaphragm cellular sheet pile walls, singly and in combination were studied. Most of these alternatives were eliminated because they would have resulted in the hydraulic profile being raised, the riverbank remaining unstable, or landward structures being removed. It was concluded that the best solution was the construction of a circular cellular sheet-pile wall as illustrated in figure 1. This alternative not only meets the above three criteria but it also has a relatively long life, is easy to maintain by the local agency, has good ductility, and is easily adapted to aesthetic treatment.

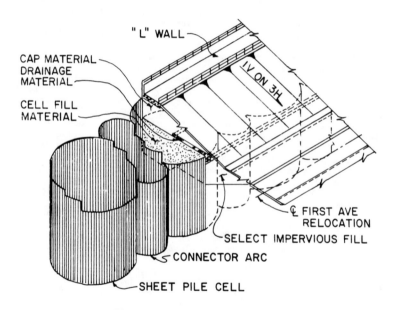

"L" WALL

CAP MATERIAL
DRAINAGE
MATERIAL

CELL FILL
MATERIAL

1V ON 3H

℄ FIRST AVE
RELOCATION

SELECT IMPERVIOUS FILL

CONNECTOR ARC

SHEET PILE CELL

Figure 1. Selected Alternative

The total length of the flood-protection alignment is 4,000 feet (1219.2 m) of which 2,833 feet (863.5 m) are cellular walls and the remaining length is conventional concrete wall, two pump stations, and five "quick closing"-type gate closures. (Quick-closing gates were chosen over conventional stop logs because of the Tug Fork River's fast-rising rate of 2.5 feet/hour--0.76 m/hr--and Williamson's limited man-power.) The length of cellular walls is made up of forty-one 53.27-foot (16.24-m) diameter cells with their connecting arcs. The design of the cell wall was complicated by soft in situ clayey material, the unstable bank, and the unusual dual purpose of the cells as a retaining wall for the bank during non-flood times and a floodwall when the river rises.

GEOTECHNICAL SITE CHARACTERISTICS AND DESIGN CONSIDERATIONS

The subsurface characteristics of the alignment for the project can be generally described as alluvial overburden underlaid by bedrock consisting of interbedded sandstone, siltstone, shale, and coal layers (12). The boring data indicated the presence of a 6-foot (1.83-m) to 25-foot (7.62-m)-thick blanket of artificial fill consisting of brick fragments, cinders, coal fragments, broken glass, wood fragments, and pockets of decayed vegetation for most of the floodwall alignment. The natural alluvial material consisted of sandy silty clays (CL) and clayey silty sands (SC) from the top of the ground (approximate elevation 660) and beneath the artificial fill to the approximate stream bed elevation (elevation 620) where clean sands and gravels begin and extend to bedrock (approximate elevation 594).

The sand content in the clay materials varies from 40 to 60 percent and the liquid and plastic limits are generally less than 35 and 15, respectively. The clayey soils were subdivided into layers based on their moisture contents, standard penetration (SPT) blow counts, and shear strength parameters. Clay materials of lower strength are generally found from a depth of 20 feet (6.10 m) to the interface of the sand and gravel substratum at a depth of 40 feet (12.19 m).

The soil profile along the floodwall alignment was developed using field exploration and laboratory test data. Analysis of this soil profile indicated six unique stratigraphic reaches existed along the project alignment. These stratigraphic reaches were selected based on variations in vertical extent of soil types present and their variations in strength.

Groundwater conditions along the alignment coincided with the river elevation of 625. Perched groundwater conditions were considered to be present at varying elevations in the fill materials.

A literature search addressing clays located in cellular sheet pile structures did not produce any relevant design criteria other than the recommendation that the cohesive material be replaced with granular material. As previously discussed, the removal of all of the cohesive material--approximately 40 feet (12.19 m)--would be impractical. It was determined, however, that 20 feet (6.10 m) of the material could be removed using a temporary cut slope of two horizontal to one vertical. This excavation would not significantly affect the construction work limits in the city. The bench created by removal of this material would reduce the quantity of cohesive material in the cell and provide a working platform for construction of the cells.

The depth of the cohesive material remaining in the cells averaged about 20 feet (6.10 m). For the Williamson project the material was classified as either a soft, medium, or stiff clay based on the SPT blow count. Q (unconsolidated-undrained), R (consolidated-undrained), and S (consolidated-drained) strengths were adopted for each material   type from the results of triaxial compression tests performed on undisturbed samples (13). Adopted Q strengths for the soft, medium, and stiff clays are   c=600 psf (28.73 kPa), c=800 psf (38.30 kPa), and c=1000 psf (47.88 kPa), respectively. R strengths are c=800 psf (38.30 kPa) and ϕ=16 degrees for the soft clay and c=1200 psf (57.46 kPa) and ϕ=14 degrees for the medium and stiff clays. Adopted S strengths for each of the materials are c=0 and ϕ=30 degrees.

PRELIMINARY DESIGN

The preliminary design considered the cells driven to rock through 60 feet (18.29 m) of overburden. In July 1984 a sheet-pile driving test program was conducted at the upstream end of the project to determine the feasibility of driving interlocked sheet piles 60 feet (18.29 m) and to compare various pile driving hammers (12). Twelve piles were driven at two locations with three different hammers. The hammers used were the L.B. Foster V-4000 vibratory hammer, McK-T 10b3 air hammer with 13,100 ft-lb (17,761 J) rated energy, and Kobe K13 diesel hammer with 24,400 ft-lb (33,082 J) rated energy. The piles were easily driven the 60 feet (18.29 m) to rock with all three hammers. The vibrations were measured with seismographs and found to be within limits for residential structures published by the U.S. Bureau of Mines (1), which recommends a maximum particle velocity of 0.5 inch/second (12.7 mm/s) for older structures with plaster-on-wood-lath interior construction. The maximum peak particle velocity recorded was 0.260 inch/second (6.6 mm/s) at 23 feet (7.01 m) from the piles using the diesel hammer with the pile tip depth at 43 feet (13.11 m). Test results showed that the piles could be driven through 60 feet (18.29 m) of overburden and that the diesel hammer generated higher particle velocities than the vibratory hammer for a given distance. It was therefore concluded that the cells could be driven to rock and that a vibratory hammer should be used. It was also recommended that a preconstruction inventory be made of the existing condition of structures within 200 feet (60.96 m) of the floodwall alignment.

It was later recognized that if the cells were driven to rock they would intercept the water table and cut the river off from the groundwater inside the city. This could have changed the existing groundwater condition beneath the city, with associated settlement and a reduction in bearing capacity. As the design evolved the founding elevation of the cells was raised 16 feet (4.88 m) off of rock (10 feet--3.05 m--below the stream bed) and the cells were designed as founded in sand rather than on rock.

ADAPTED CONVENTIONAL DESIGN METHODS

Even with a large portion of the unsuitable material excavated, a thick layer of clay would remain in some areas. The classical design methods which generally consider only granular cell fill had to be modified to address the cohesive soils inside the cells.

Cells at each of the six stratigraphic reaches were investigated as to their external and internal stability for the end-of-construction condition with and without the riverward wedge of in situ material (referred to as berms) and the long-term condition. External stability was checked by examining overturning, sliding, and bearing-capacity failure. Conventional methods of cell design were used since external stability is independent of the cell fill. Internal stability was analyzed for shear failure on the centerline of the cell (vertical shear), horizontal shear failure (tilting), and interlock tension. The internal-stability analysis was an adaptation of conventional methods.

Vertical Shear. The total centerline shear resistance per unit length of wall was computed as:

$$S_s = P_s \tan \phi \tag{1}$$

where S_s is the centerline shear resistance, P_s is the total lateral force at centerline of cell, per unit length of wall, due to cell fill, and ϕ is the angle of internal friction of the cell fill (10). This equation assumes a homogeneous granular cell fill. For a multi-layer fill the total centerline shear resistance is the sum of vertical shear resistances of each layer, and for cohesive soils the shear strength is equal to the cohesion, c. The following is a more general equation for S_s which is suitable for all cell fills:

$$S_s = \sum_{i=1}^{n} S_i = \sum_{i=1}^{n} (P_i \tan \phi_i + D_i c_i) \tag{2}$$

where S_i is the shear resistance of layer i, P_i is the lateral force of layer i, ϕ_i is the angle of internal friction of layer i, D_i is the thickness of layer i, c_i is the cohesion of layer i, and n is the number of layers. The frictional slippage resistance in the sheet-pile interlock is added to the centerline shear resistance to get the total shear resistance along the centerline of the cell. The interlock frictional resistance is equal to the total interlock tension force over the cell height multiplied by the coefficient of friction of steel on steel. The vertical shear factors of safety were found to be well above 1.25 for the end-of-construction condition and 1.5 for the long-term condition.

Interlock Tension. Two methods were used to calculate the maximum interlock tension. The TVA (9), the USS Steel Sheet Piling Design Manual (11), and EM 1110-2-2503 (10) approximate the interlock stress at the arc-cell connections using the "secant method":

$$t_{max} = pL \sec \theta \tag{3}$$

where t_{max} is the interlock stress at the connection, p is the maximum lateral inboard sheeting pressure, L is the distance from the arc centerline to the cell centerline, and Θ is the angle between the wall centerline and a line drawn from the connection to the cell center.   It was found that the secant method yielded unacceptably high interlock stresses.   Studies supported by model and field data indicate the secant formula predicts overly conservative interlock tension stresses (14).   Therefore, the stresses at the wye connection were recalculated as the sum of the main cell hoop tension and the tangential component of the connecting arc pull. The main cell interlock tension is equal to the maximum inboard sheeting pressure times the cell radius.   The tangential component of the connecting arc pull is equal to the maximum pressure in the arc times the arc radius multiplied by the cosine of the wye angle. This tangential method is given as:

$$t_{max} = pR + p_a r \cos \alpha \tag{4}$$

where t_{max} and p are as previously defined, R is the cell radius, p_a is the maximum lateral inboard pressure in the arc, r is the arc radius, and α is the wye angle.   Various locations of the maximum inboard sheeting pressure are recommended by TVA (9), Schroeder and Maitland (14), Terzaghi(6), and the Corps of Engineers (10).   The Corps recommendation was used and the maximum pressure was calculated at the point one fourth the height of the cell above the level at which cell expansion is fully restrained.   The tangential method gave 9 to 11 percent greater factors of safety than the secant method.   In most reaches berms were required for acceptable factors of safety for the end-of-construction condition.

It is noted that, after the Williamson CBD design, a third method of calculating interlock stress was recommended, known as the Swatek method (7).   The Swatek method computes the average interlock tension in the crosswall as:

$$t_{cw} = pL \tag{5}$$

where t_{cw} is the average interlock stress in the crosswall (rather than at the wye connection) and p and L are as defined before. This method increases the factors of safety of the secant method by secant Θ, or 15 percent when using a 30-degree (.524-rad) wye layout.   The Swatek method results in factors of safety 4 to 6 percent greater than those of the tangential method.   A further discussion of the Swatek method can be found in the Technical Report (7).

Horizontal Shear.   Cummings (1957) proposed a method for computing horizontal shear resistance based on his interior sliding theory, where failure by tilting is resisted by the horizontal shear resistance of the cell fill and the pile interlock friction resistance.   Using model tests, he concludes that shear resistance is developed only below a plane intercepting the base (or dredge line) of the inboard face at the angle of internal friction of the cell fill, and that the cell fill above the plane acts as a surcharge.   Again, the soft clays present a problem in the Williamson

case. When this plane intersects a cohesive soil layer, the angle
of internal friction equals zero and the plane is horizontal.
Since the passive resistance is developed only below the plane,
Cummings' method does not provide for any shear resistance from the
clays. The moments due to interlock friction and passive pressure
of the berms provided most or all of the resistance to tilting in
each reach. Adequate factors of safety were achieved, but horizontal shear through the clay layers was not addressed.

CONVENTIONAL STABILITY ANALYSIS

    Conventional limit-equilibrium stability analyses were performed
for end-of-construction and steady-seepage conditions for each of
the six reaches mentioned earlier (13). A wedge failure surface
was used to analyze horizontal sliding at the base of the sheetpile cell founded in the sand and gravel, a failure plane passing
through the cell at the base of the lower clay layer, the stability
of the riverside berm, and to address the need for the berm with
respect to the stability of the cell. The purpose of analyzing a
failure plane through the cell at the base of the lower clay layer
was to provide some idea of how the cell would behave with the clay
inside. For each analysis it was assumed that the sheet-pile cell
would act as a rigid structure. The horizontal dimension of the
cell was therefore considered as the neutral block and ignored the
presence of the steel. The active and passive failure planes were
not permitted to pass through any part of the cell.
    The end-of-construction and steady-seepage analyses performed on
the cell with a failure surface in the sands and gravels provided
factors of safety ranging from 2.4 to 2.6 with the berms in place.
When the berms were removed the factors of safety were reduced
slightly to the range of 2.1 to 2.4. Two of the reaches analyzed
had no riverside berms; their factors of safety ranged from 2.0 to
2.4. The required minimum factors of safety for the end-of-
construction and steady-seepage cases were 1.0 and 1.4, respectively.
    When the failure surface was analyzed at the base of the clay
layer through the cell, the results for the steady-seepage analysis
showed factors of safety ranging from 2.3 to 3.0 with the berms,
and from 2.4 to 3.2 without the berms. The results for the end-
of-construction analysis changed considerably. The factors of
safety decreased to the range of 1.0 to 1.4 with the berms and 0.7
to 1.3 without the berms. The reach at station 33+85 as shown in
figure 2 had a factor of safety below the required minimum of 1.0.
Thus the requirement for the berm to remain was determined for this
reach.
    The end-of-construction analysis was performed on the riverside
berms assuming no effects from the cells. The factors of safety
ranged from 2.3 to 2.9, which indicated that the berms would be internally stable against horizontal sliding.

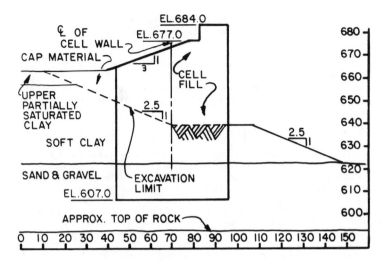

Figure 2. Section at Station 33+85

STRENGTH-GAIN-WITH-TIME ANALYSIS

When the cells are completed, it is assumed that the clay will have a Q (unconsolidated-undrained) shear strength. As time passes, the clay will consolidate, decreasing the void ratio and increasing the shear strength to the S (consolidated-drained) shear strength. The end-of-construction stability analysis indicates marginal factors of safety against horizontal sliding through the cell at the base of the clay layer and the steady-seepage analysis provided results greater than the minimum for each reach.   To determine the amount of time required for the material to change from the end-of-construction condition to the steady-seepage condition, a strength-gain-with-time analysis was performed (13).   The analysis was based on a correlation of the void ratio as determined from consolidation tests to the shear strength of the clay material.   Rather than perform this analysis for each reach, the reach at station 33+85 was selected since it had the lowest factors of safety against sliding for the end-of-construction condition.

The results of the analysis indicated that the primary settlement approximated 90 percent consolidation for the material in the cell.   Relating this settlement to the time-settlement curve, the clay material would achieve its consolidated-drained strength in about 1 year (13).

During this one-year period the cells in two of the six reaches would have a marginal factor of safety against sliding through the cell at the base of the clay layer.   The installation of wick drains was considered as a means of reducing the consolidation time

and improving the shear strength. This approach seemed workable
since the surcharge loading on the clay would be greater than the
preconsolidation stresses.   The e-log p curves showed that the
clays were normally consolidated.

Hansbo's procedure to determine drain spacing for a given degree
of consolidation, time, and surcharge was used to compute the drain
spacing (8).   Results indicated that ninety percent consolidation
could be achieved in one month using a drain spacing of 5 feet
(1.52 m).

FINITE ELEMENT ANALYSIS AND PROTOTYPE CELLS

Although adequate factors of safety were achieved by unique
methods, additional confirmation of this unconventional design was
judged appropriate.   Other questions were raised, also, such as
whether or not the relatively small amounts of fill to be placed in
the cells over the existing bank materials would be adequate to
develop the interlock forces, and what deflections could be expected.   The Waterways Experiment Station (WES) was contracted to
ensure the technical feasibility of the project's overall concept
by performing a finite element analysis (FEA).   For further assurance, the District drove two prototype cells and their connecting arcs at the downstream end of the project, instrumented to
measure their performance.   Task I of the FEA study addressed
specific details of the wall design such as construction details,
cell displacements, and sheet-pile stresses.   Task II analyzed the
field observations from the prototype cells.   The analyses were
performed using SOILSTRUC, a FEA computer code developed by Dr. G.
Wayne Clough of Virginia Polytechnic Institute and State University
for two dimensional (plane strain) soil-structure interaction
problems.

Task I.   In Task I (4), representative sections 18+00 and 33+85
were used as the analytical sections for the FEA study.   Station
18+00 as represented by figure 3 was selected because of its lack
of a riverside berm, and station 33+85 (see figure 2) had the
thickest layer of soft clays among the reaches and the worst
stability.   Five cases were considered in these analyses, including
combinations of drained and undrained clay layers, berms and no
berms, and flood loading and no flood loading.   The problem was
idealized using three types of elements: four-node quadrilateral
solid elements to model the soil and sheet pile materials; four-
node interface elements to model slippage between the sheet piles
and the soil; and spring (or bar) elements to model the hoop stiffness in the cell.

For station 33+85 the undrained condition with the berm removed
resulted in an estimated displacement of 12.9 inches (0.33 m) at
the top of the ground at elevation 666 and an interlock force of
8.2 kips/inch (1436 kN/m) at elevation 640.   In the drained condition these numbers reduced to 10.3 inches (0.26 m) and 5.3
kips/inch (928 kN/m), respectively.   During flooding the cells are
predicted to move landward 2.9 inches (0.074 m) and 3.8 inches
(0.097 m) riverward when unloaded.   Factors of safety against full
mobilization of the soil's strength were estimated and found to ap-

proach unity for the undrained condition and about 3 for the drained conditions. Obviously, as was also concluded by the conventional analysis, the stability of the cells is greatly enhanced if the clays inside the cells behave as drained materials and the riverside berm is left in place. It can also be seen by calculated movements during the flood loading and unloading cycle that the cells are flexible, which should be considered for any structure placed on top of the cells.

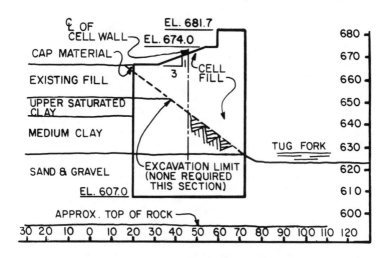

Figure 3. Section at Station 18+00

For station 18+00 the drained condition resulted in a predicted displacement of 3.8 inches (0.097 m) and an interlock force of 5.4 kips/inch (946 kN/m). Since there was no berm, the peak interlock forces occurred lower in the cell at elevation 620. There did not appear to be any severe stability problem.

A horizontal slice through the cell at station 33+85 was made using a generalized plane strain version of SOILSTRUC and analyzed to determine the pattern of interlock forces. The slice was taken at elevation 636, which corresponded to the approximate location of maximum computed interlock forces. These forces were calculated to occur within the common wall at the connecting wye where they reached a maximum of 8.2 kips/inch (1436 kN/m). The interlock stress averaged 5.0 kips/inch (876 kN/m) through the rest of the common wall, 3.3 kips/inch (578 kN/m) in the outer main cell wall, and 1.2 kips/inch (210 kN/m) for the arc.

The conventional analysis and finite element analysis yielded similar lateral stresses for corresponding conditions, but the conventional method computed much higher interlock stresses and lower elevation of peak interlock forces than did the FEA--that is, 11.35

kips/inch (1988 kN/m) at elevation 635 and 8 kips/inch (1401 kN/m) at elevation 645, respectively. The difference is related to the sequence of loading. Conventional analysis applies the entire lateral stress to the cell interlocks, while in the finite element analysis the interlocks take up only the stresses caused by the fill and surcharge (4).

Prototype Cells. In October 1985 a contract was let to construct two prototype cells together with their connecting arcs-- cell numbers 39 and 40 at the downstream end of the 41-cell alignment (2). An instrumentation program was developed that included strain gages, temperature sensors, inclinometers, surface-movement monitoring by surveying, settlement plates, piezometers, and vibration and noise monitoring. The strain gages measured hoop stresses, interlock stresses, and bending stresses and axial loading in the sheet piles. Inclinometer tubing was attached to the sheet piles to measure the deflected shapes of the cells. A system of movement monitoring points was located on top of the cells to monitor their distortion and the settlement of fill within the cells. The settlement plates founded at about elevation 640 measured settlement of the underlying natural clay. Piezometers were installed in the underlying clay stratum to measure both the increase in pore water pressure caused by installation of the piles and the filling of the cells and the decrease in pore pressure with time. Vibration and noise monitoring were performed during the pile driving to measure the ground vibrations and noise levels achieved, respectively.

Settlements ranging from 1.9 to 9.5 inches (48.3 to 241.3 mm) have been measured at the settlement plates in the cells. The largest settlement occurred in the first cell, most of which was attributed to the driving of the second cell and the connecting arcs during the filling operation of the first cell. The excess pore pressures had partly dissipated during construction and continued to dissipate relatively quickly after construction as measured by the piezometers. The inclinometer readings showed lateral riverward movements at the top of the cells of up to 8.0 inches (203.3 mm). The initial strain-gage readings at an arc pile suggested high interlock stress (15 kips/inch--2627 kN/m) prior to filling, possibly because of difficulties in aligning the arc piles. During the filling the interlock stresses increased slightly but began dropping in time. Vibrations measured during the pile driving resulted in a maximum peak particle velocity of .37 inch/second (9.4 mm/s) at 30 feet (9.14 m) from the pile. The maximum sound level measured was 115 db at a distance of 80 feet (24.38 m) from the vibratory hammer. The "threshold of pain" is considered to be 120 db.

Task II. Field observations of the prototype cells were used in Task II efforts to evaluate the overall FEA study (5). The data were used to improve the site characterization and cell-behavior idealization. The soil and structural representation and general behavior were checked for adequacy by comparing the predicted to observed performance. Task II was also used to determine whether or not cyclic flooding would increase the net movement of the cells toward the river.

The Task I model was modified to more accurately reflect actual conditions by using patch elements instead of horizontal springs to model the common wall, by using a sloping upper surface of the cell instead of applying a surcharge load, and by making adjustments in the hoop and shear stiffness of the sheet piles. The field observations and laboratory data indicated that the foundation soils behave like loose sand rather than soft clays and should be modeled accordingly.

Study Comments. Task II noted several important comments concerning this FEA and prototype cell study:

(1) Pile driving caused settlements up to 0.4 ft (0.12 m) and fill placement caused settlements up to 0.24 ft (0.07 m) after several days. Both field and laboratory data suggest that the soil is susceptible to disturbance by vibratory loading, so settlement around the project site should by monitored.

(2) The cells are clearly flexible structures that can be expected to move several inches riverward during construction and landward during flooding. Horizontal movements occurred primarily during cell fill placement, and all movements appeared to cease within 3 months. This must be considered for any structure built on top of the cells.

(3) The cyclic flood loading of the cells indicates that the net increase in lateral movements stabilized after two or three cycles. The lateral movements were elastic and fully recoverable with additional cycles.

OTHER CONSIDERATIONS

Parapet Wall. In order to minimize the height of the cell and provide an aesthetically pleasing structure, a reinforced concrete L-shaped parapet wall with a maximum height of 10 feet (3.05 m) will be built on top of the cells after the post-construction lateral movement has ceased. To withstand uplift, reinforcing bars will be welded to the outboard sheets and embedded in the wall base. The 1.5-inch (38.1-mm)-wide joints were designed with shear dowels to prevent horizontal shear movements at the monolith joints and allow rotation from differential lateral movements of the monoliths. The parapet wall was designed as a beam supported on the sheet piles alone in case the cell fill settles beneath the base. The FEA study found that the parapet wall will not significantly affect the lateral movements of the cells except near the top. The wall will have a decorative fractured-fin finish. The cell fill will be sloped and landscaped from the wall base down to the city-street elevation.

Construction Considerations. Since the internal stability of the cellular wall was found to depend greatly on the material in the cell, certain measures will be taken to improve the mechanical properties of the clays. As mentioned earlier, wick drains will be driven into certain clay areas to achieve a drained state in the clays as soon as possible in order to achieve higher factors of safety. Although the dissipation of pore pressure in the clay appears to be rapid, the wick drains will be used to assist the drainage. In order to reduce the fill settlement, which might in

turn reduce the long-term lateral deformation, the granular cell
fill will be densified by placing the material in 5-foot (1.52-m)
lifts and then flooding the cell. Weep holes, which will later be
covered, will be provided in the sheet piles to drain the material
and prevent overstressing of the interlocks. The cohesive cap
material will be compacted by mechanical means. After the cap
material has been placed, the material in the cell will be allowed
to consolidate for 9 months prior to construction of the parapet
wall. The final design is illustrated in figure 4.

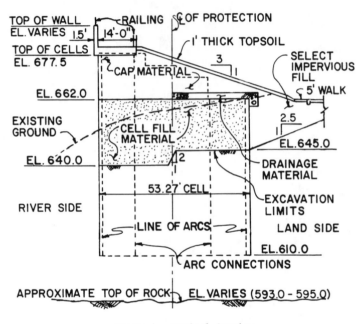

Figure 4. Typical Section

CONCLUSION

The local protection project for Williamson's central business
district has presented design challenges not normally encountered
in floodwall projects. The most unusual of these challenges is the
need for a flood-protecting structure that contributes to the
stability of the riverbank of low-strength clays. Because of the
unique application of the classical methods of sheet-pile cell
design to cohesive cell fill, a finite element analysis (FEA) was
employed to predict better the feasibility and overall performance
of the structure. In general, results of the FEA corresponded well
with the field observations of prototype cells, confirming that the
design was feasible and safe.

REFERENCES

1. DE Siskind, MS Stagg, JW Kopp, & CH Dowding, Structure Response and Damage Produced by Ground Vibration From Surface Mine Blasting, RI 8507, U.S. Dept. of the Interior, Bureau of Mines, Wash. DC, 1980, 74 p.
2. Dodson-Lindblom Associates, Williamson CBD Prototype Cells, U.S. Army Engineer District, Huntington,WV, Aug 1988.
3. EM Cummings, Cellular Cofferdams and Docks, J.Waterways and Harbor Division (ASCE), 83(WW3), Paper No.1366, Sept 1957, 13-45.
4. JF Peters, TL Holmes, and DA Leavell, Finite Element Analysis of Williamson CBD Sheetpile Cell Floodwall, Task I, U. S. Army Engineer District, Huntington WV, Oct 1986, 48 p.
5. JF Peters, TL Holmes, and DA Leavell, Finite Element Analysis of Williamson CBD Sheetpile Cell Floodwall, Task II, U. S. Army Engineer District, Huntington, WV, Feb 1987, 67 p.
6. K Terzaghi, Stability and Stiffness of Cellular Cofferdams, Transactions ASCE, 110 (2253), 1945, 1083-1119.
7. M Rossow, E Demsky, and R Mosher, Theoretical Manual for Design of Cellular Sheet Pile Structures, Technical Report ITL-87-5, U. S. Army Corps of Engi., Wash. DC, May 1987, 121 p.
8. S Hansbo, Consolidation of Clay by Band-shaped Prefabricated Drains, Ground Engineering, 12 (5), July 1979, 16-25.
9. Tennessee Valley Authority, Steel Sheet Piling Cellular Cofferdams on Rock, TVA Technical Monograph 75, Nashville, TN, 1957, 281 p.
10. U. S. Army Corps of Engineers, Design of Sheet Pile Cellular Structures, EM 1110-2-2503, Wash. DC, Draft.
11. United States Steel, USS Steel Sheet Piling Design Manual, United States, Sept 1970.
12. Williamson Flood Wall, Feature Design Memorandum No. 5, U.S. Army Engineer District, Huntington, WV, Oct 1984.
13. Williamson Flood Wall, Cell Supplement, Feature Design Memorandum No. 5b, U.S. Army Engineer District, Huntington, WV, Jan 1986.
14. WL Schroeder & JK Maitland, Cellular Bulkheads and Cofferdams, J. Geot. Eng. (ASCE), 105 (GT7), July 1979, 823-838.

EMPIRICAL SEISMIC DESIGN METHOD FOR WATERFRONT
ANCHORED SHEETPILE WALLS

George Gazetas[1] Panos Dakoulas[2], & Keith Dennehy[3]

ABSTRACT: An investigation of the documented field performance
of 75 anchored sheetpile quaywalls reveals that the anchoring
system is the most vulnerable component of the bulkhead.
Current pseudo-static design procedures often lead to deficient
anchoring, whose displacements and failure trigger excessive
permanent seaward displacement at the top of the bulkhead,
accompanied by cracking and settlement behind the anchor. The
results of the case histories lead to a Seismic Design Chart to
be used in conjunction with the pseudostatic procedure. The
Chart delineates between severe, moderate, and minor degrees of
damage, depending on the values of two dimensionless parameters
that are functions of the material and geometric
characteristics of the bulkhead, and the intensity of seismic
shaking. Before using the Chart, the engineer must ensure that
liquefaction-flow failure of cohesionless soils in the backfill
or the foundation is unlikely.

INTRODUCTION -- PAST SEISMIC PERFORMANCE OF ANCHORED BULKHEADS

     Port and harbor facilities are vulnerable to strong
earthquake shaking. Failures of such facilities have often
resulted in major disruptions of post-earthquake emergency
operations and have had serious economic consequences for the
stricken regions.
     Anchored bulkheads (Fig. 1), also called anchored (steel)
sheetpile walls, are often used as retaining structures in
wharves and quays, since they are relatively easy to construct
(requiring no dewatering, for example), while the soft or loose
soils that usually underline such waterfront structures can

---

1 - Prof. of Civ. Engrg., State Univ. of New York, Buffalo N.Y.
    14260; and National Tech. Univ., Athens, Greece.
2 - Asst. Prof. of Civ. Engrg., Rice Univ., Houston, Texas
3 - Formerly Grad. Stud., Rensselaer Polytech. Inst., Troy,
    N.Y.

hardly support the additional weight of gravity concrete walls. Anchored bulkheads are cheaper than supporting gravity walls on piles. Consequently, a large percentage of quaywalls are anchored bulkheads, and many of the reported quaywall seismic failures are therefore failures of anchored bulkheads.

Accounts of the earthquake performance of over a hundred anchored quaywalls in 26 harbors in Japan (mainly), in Alaska, in West   Indies, and in Chile have been published in the last decade [5,6,7,8,11]. Detailed listings of these reported case histories may be found in the theses of Abraham [1] and Dennehy [4], and in a report by Agbabian Associates [2]. The following conclusions emerge from a study of the performance of anchored bulkheads in these harbors:

1. Most of the observed major earthquake failures have resulted from large-scale liquefaction of loose, saturated, cohesionless soils in the backfill and/or in the supporting base (foundation). Such soils are not rare at port and harbor facility sites. Perhaps the most dramatic such failures have occurred in the Niigata, Japan, harbor during the 1964 earthquake.

2. Another frequent, although not as dramatic, type of anchored bulkhead damage takes the form of excessive permanent seaward tilting of the sheet-pile wall, accompanied by excessive seaward movement of the anchor block or plate relative to the surrounding soil; such an anchor movement manifests itself in the form of settlement of the soil and cracking of the concrete apron directly behind the anchor. Apparently, and in accord with the conclusions of pertinent detailed studies [2,4], such failures are the outcome of inadequate passive soil resistance against the anchor. Development of detrimental residual excess pore-water pressures in the backfill, leading to some soil strength degradation, cannot be excluded as having contributed to this type of failures in some of the reported cases.

3. Other, less frequent forms of damage include:
    - seaward permanent displacement of the sheetpile wall followed by soil subsidence and apron cracking between the wall and its anchor -- possibly the result of yielding/rupturing of the tie rod or breakage of its connection with the sheetpile, or perhaps the outcome of post-earthquake densification of loose soils;
    - seaward movement of the toe of the sheetpile due to inadequate passive resistance of the foundation soil against the embedded part of the bulkhead --

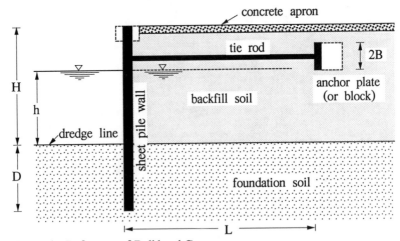

Figure 1. Definition of Bulkhead Geometry

resistance provided by the anchor
above the point of intersection of the
active failure surface of the wall and
the passive failure of the anchor is
neglected in the design

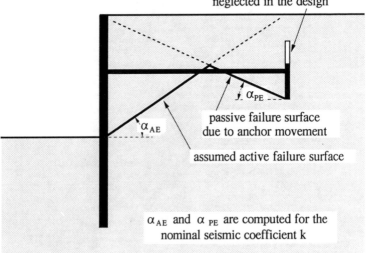

Figure 2. Code Procedure for Designing the Anchor

presumably due to insufficient depth of embedment, often made worse by a plausible soil shear strength degradation;

- localized seaward tilting of the cantilever upper part of walls carrying a heavy concrete block at their head (dashed lines in Fig. 1) -- apparently a result of flexural deformation/yielding of the sheetpile due to excessive inertia forces;

- general flexural failure of the sheetpile bulkhead (observed only once, and attributed to corrosion of the steel at dredge line elevation).

These observations suggest that anchored bulkheads are susceptible to earthquake damage and that the dynamic nature of the problem should be given sufficient attention during design and construction.

CURRENT SEISMIC DESIGN PROCEDURES

To the authors' knowledge, no comprehensive method of realistic dynamic analysis of anchored bulkhead systems subjected to strong shaking is well enough developed and validated to be used in practice. It is fair, however, to state that dynamic codes developed for site response (e.g. SHAKE, DESRA, CHARSOIL, etc.) or soil-structure interaction (e.g. FLUSH, ADINA, etc.) analyses have been utilized, albeit to study specific aspects (only) of the response of the system [7]. Simplified dynamic models specific for anchored bulkheads have also been developed [1, 4, 9], while research efforts for developing rational dynamic analysis models are underway [e.g., 24].

The difficulty of providing a comprehensive rigorous method arises from several factors, which include: the complicated wave diffraction pattern due to "ground-step" geometry; the presence of two different but interconnected structural elements in contact with the soil; the inevitably nonlinear hysteretic behavior of soil in strong shaking, including pore-pressure buildup and degradation, both in front and behind the sheetpile; the no-tension behavior of the soil-sheetpile interface; the presence of radiation damping effects due to stress waves propagating away from the wall in the backfill and in the foundation; and the hydrodynamic effects on both sides of the sheetpile wall. Until codes which can properly handle all these phenomena are developed, improving the pseudo-static procedures currently used in practice so that they can lead to safe and economic design merits our effort.

Pseudo-static procedures are of an empirical nature and determine dynamic lateral earth pressures with the Mononobe-Okabe seismic coefficient analysis [8,14,16,18]. Differences arise primarily with respect to the assumed point of application of the resultant active and passive forces P_{AE} and P_{PE} (on the two sides of the sheetpile wall), and the partial factors of safety introduced in the design.

The procedure developed and extensively used in Japan [8] is perhaps the most elaborate complete pseudo-static design procedure. It combines the use of the Mononobe-Okabe method with conventional static design procedures of anchored bulkheads. The vertical component of the ground acceleration is ignored, while the horizontal seismic coefficent, $k_h = k$, is chosen for a particular site as a product of three factors (according to the Japanese Code): a regional seismicity factor (0.10 ± 0.05) , a factor reflecting the subsoil conditions (1 ± 0.2), and a factor reflecting the importance of the structure. (1 ± 0.5). To account for the presence of water in the design procedure, and " apparent" seismic coefficient k' is used for soils below the water table:

$$k' = [\gamma_s/(\gamma_s - \gamma_w)] k \dotfill (1)$$

in which γ_s = the saturated unit weight of the soil, and γ_w = the unit weight of water.

Subsequently, the design proceeds as follows:

1. Estimation of the necessary length of the sheetpile embedment(D):
   This is computed by the free-earth support method. The safety factors usually required against the failure of embedment are 1.5 and 1.2 for static and seismic conditions respectively in sandy stratum. In cohesive soil strata, the usually required safety factor is 1.2 for both the static and seismic conditions.

2. Design of the tie rod:
   In the case of a sheet pile bulkhead constructed in sandy ground, tie rod tension is computed on the assumption that the bulkhead is a simple beam supported at the dredge line and the point of tie rod connection, and which carries the lateral earth pressure and the residual water pressure. In case of cohesive soil, tie rod tension is computed by the fixed-earth support method. Allowable stress of tie rods: 40% and 60% of the yield strength of steel for static and seismic conditions, respectively. These relatively low values of allowable stress are intended to account for bending moment in the tie rod due to surcharge, and for

concentration of lateral earth pressure at the point of tie-rod connection.

3. Design of the sheetpile cross-section:
   In sandy ground, the maximum bending moment is computed for the aforementioned simple beam. This maximum moment, which is about 40-50% of that computed by the free-earth support method, corresponds to the value computed by fully taking into account the moment reduction due to the flexibility of the sheetpile (Rowe [15]). The allowable stresses of the sheetpile for static and seismic conditions are 60% and 90% of the yield strength of steel, respectively.

4. Design of the anchor plate or block:
   Lateral resistance of an anchor plate should be 2.5 times the tie rod tension for both static and seismic conditions. Anchor plates should be placed behind the active failure wedge starting from the dredge line (Fig. 2). When the passive wedge of the anchor plate crosses the active wedge behind the sheet pile, the passive resistance of the soil above the point of intersection should be neglected in the computation of the lateral resistance of the anchor plate.

WEAKNESSES OF PSEUDO-STATIC DESIGN PROCEDURES

Kitajima & Uwabe [11] have compiled information on the seismic performance of 110 quaywalls (mostly anchored bulkheads) in Japan. Table 1 summarizes the conclusions of their study. The conveyed message for the adequacy of the pseudo-static design methods is negative: the percentage of bulkheads that suffered some degree of seismic damage did not decline following the adoption of the previously-described design procedure... (Year of construction seems also to have had little effect on damage statistics.)

Many of the "failures" included in the statistics of the foregoing Table were clearly due to extensive liquefaction of the backfill and/or the supporting base stratum; these cases will not be further addressed in this paper. Carefull study of the remaining "failures" [4] leads to the following conclusions regarding the major weaknesses of the psuedo-static procedure:

Table 1. Statistics of Seismic Damage to Anchored Bulkheads in Japan [11]

		Total Numbers	Number of Damaged Bulkheads	Percent of Damaged Bulkheads %
Total Numbers		110	70	64
Number of Bulkheads Designed According to the Japanese Procedure		45	29	64
Year Constructed	Before 1950	37	22	59
	1951 - 1960	11	6	55
	1961 - 1966	40	30	75
	after 1966	22	11	55

1. The values of the Code-specified seismic coefficient are not representative of the actual levels of acceleration that may develop in the backfill during moderate and strong earthquake shaking. Indeed there is little justification for the selected values. As noted by Seed[17]: "it is entirely possible that such empirical values of the seismic coefficient may lead to safe designs in many cases but until some means of judging their validity is developed, their use must be considered of questionable value". Futhermore, Wood [23] has observed that: "In general, seismic coefficients are chosen that are significantly less than the peak accelerations to be expected in a suitable design earthquake, apparently on the assumption that some permanent outward movement of the wall can be tolerated. There appears to be no rational basis for the magnitude of the reduction made."

Indeed, despite the increase of the design coefficient form k to $k' \cong 2k$ for soils under the water table, some of the failed bulkheads may have experienced greater "effective" peak accelerations than they were designed for. Strong ground shaking can induce accelerations in excess of 0.50g. On the other hand, moderately-strong ground shaking might be amplified by the (non-liquefiable) backfill and foundation stratum. Such an amplification could be substantial if a thick backfill-foundation

profile underlain by very stiff soil or rock is excited by an
earthquake motion rich in frequencies near its own natural
frequency(ies).   To demonstrate the possiblitiy for such an
amplification, theoretical, experimental and field evidence is
available [13,11,10].   Some examples: Nadim & Whitman [13] have
shown for rigid retaining walls that the permanent displacement
computed with a finite element model incorporating a Coulomb-type
sliding surface in the backfill is substantially greater than the
value obtained form rigid-plastic analysis, in which soil layer
response (and amplification) is ignored. Small-scale shaking-table
experiments conducted by the Japanese Port and Harbor Research
Institute [11] tend to confirm this behavior for anchored
bulkheads.   Although both the Nadim-Whitman and the shaking table
models may exaggerate such an amplification due to spurious wave
reflections at the lateral boundaries, some field evidence to this
effect is also available. Table 2 reports the peak accelerations
and the predominant periods of the motions recorded at the
Yamanoshita pier, in Niigata, during aftershocks of the 1964
earthquake. Measurements were taken atop the wall and at the free-
field ground surface, away from the wall. It is evident that the
peaks of the two components of motion which are parallel to the
wall are at a ratio of 1.3, while the predominant periods remain
the same.   However, the prependicular to the wall components of
motion reveal an amplification by a factor 2.6, with the
predominant period decreasing by a factor of 2.

Table 2.   Recorded Motions at Yamanosita Pier [10]

Direction of Component	Location of Record	Peak Acceleration (g)	Predominant Period (s)
perpendicular to quaywall	a	0.047	0.17
	b	0.018	0.37
parallel to quaywall	a	0.042	0.38
	b	0.031	0.36

a: top of quaywall; b: ground surface away from the wall

    Furthermore, the vertical component of the ground
acceleration, which is ignored by the method, increases the
"effective" acceleration that controls the seismic active and
passive pressures [3,16] by a factor of $(1-k_v)^{-1}$ [see Fig.3].

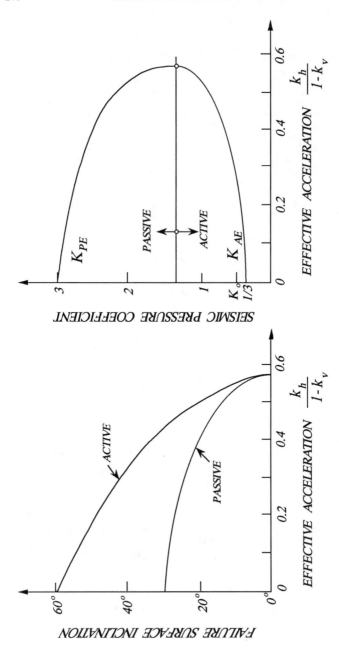

Figure 3. Effect of Horizontal and Vertical Seismic Coefficients on : (a) the Angle of the Active and Passive Sliding Wedges, and (b) the Active and Passive Earth Pressure Coefficients [3, 13]

On the other hand, the increase of k by a factor of about 2 for soils below the water table may only partially accomodate the detrimental effects of strength degradation due to pore-water pressure buildup. Also note that in the majority of the studied Japanese case histories the aforementioned increase in the seismic coefficient had little effect in the design of the anchor, as a significant part of the latter is located above the water table. And, finally, this increase of k was undermined by the unfortunate 20%-33% reduction in the required factors of safety, as outlined in the previous section.

In conclusion, it appears that many of the "failed" bulkheads experienced "effective" peak accelerations which were essentially 30% to 50% higher than what these walls had been designed for.

2. The available passive soil resistance against the anchor is often seriously overestimated by the Code procedure. While there is ample indirect empirical evidence supporting the above statement (recall the most frequent modes of failure), it is important to
develop a understanding of the causes of this inadequecy of the Codes.

To begin with, recall that the Japanese Code requires that the active sliding surface should start at the elevation of the dredge line. By contrast, even the static design of anchored bulkheads most often assumes that this surface originates at the point of contraflexure, or the point of zero moment in the sheetpile [21,22]. Tschebotarioff's "hinge at the dredge line" concept [19,20], useful as it may be for determining maximum bending moments in the sheetpile, is un-conservative for choosing the location and size of the anchor block/plate. In fact, it is more likely that the active failure surface originates at or near the "point of rotation" rather than at the points of contraflexure or zero-moment. The location of this point depends on the relative stiffness of the sheetpile wall and the overall rigidity of the anchoring system-- but, no doubt, is generally deeper than the points of contraflexure and zero moment.

Moreover, under seismic loading the "point of rotation" tends to move farther down, as repeatedly demonstrated in small-scale shaking-table tests [11,12]. The explanation is clear: when acceleration increases, the active soil pressures against the wall increase while the passive ones supporting the wall decrease (see Fig. 3). Hence, the effective "span" of the sheetpile beam (in the terminology of the free-earth support method) tends to increase, and the origin of the active sliding surface ($\cong$ "point of rotation") tends to be pushed downward. It appears that in several of the studied Japanese cases this "point" might have been located at a depth $f \gtreqqless D/2$, where the depth of embedment, D, usually takes values in the range of 50% to 80% of H. It is thus evident that the Code recommendation of placing the origin at the dredge line (Fig.2) would in most cases underestimate the required tie-rod length, L.

An additional factor in the Code procedure contributing to an overestimation of the available passive anchor resistance stems from the use of the seismic coefficient k rather than the "effective" peak acceleration in the backfill, or at least of the increased coefficient k'. As illustrated in Fig. 3, increased acceleration levels imply not only reduced passive forces, but also flatter failure surfaces. Recalling Fig. 2, it is obvious that a smaller in reality angle α_{PE} than that assumed in the Code design would (further) reduce the capacity of the anchor.

EMPIRICAL SEISMIC CHART AND PROPOSED DESIGN PROCEDURE

To arrive at a practical design chart (using the results of those case histories that did not involve liquefaction flow failures), two simple dimensionless indices have been selected. Their definition, significance, and methods of computation are explained below.

(a) The 'Effective Anchor Index' (EAI), representing the relative magnitude of the available passive anchor force: EAI is defined in Fig. 4, in terms of the horizontal distance d from the active failure surface to the tie-rod-anchor connecting point:

$$EAI = \frac{d}{H} \quad \dotfill \quad (2)$$

Note that the width of the anchor, 2B, does not appear directly in this index, despite its importance for the anchor resistance. This was a reluctant choice, out of necessity: in only a few of the analyzed case histories was this width reliably known! But, at least, 2B is indirectly reflected in Eqn 2 through the height H; indeed, according to the Code procedure, 2B depends chiefly on H and the backfill angle of shearing resistance ϕ.

The active failure surface is assumed to originate at the effective "point" of rotation, at depth f from the dredge line. While in actual design f could be estimated from a beam-on-Winlker- foundation analysis of sheetpile deformation, the following expression has been developed by the authors for a quick crude estimate (only):

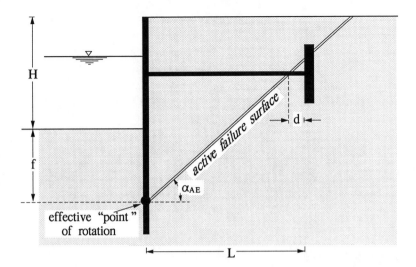

Figure 4. Definition of the Effective Anchor Index:
EAI = d / H

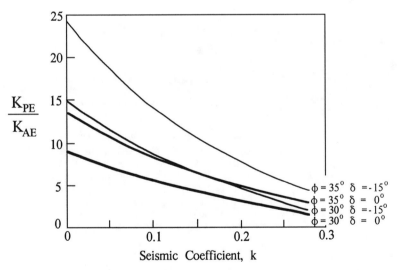

Figure 5. The Ratio of Active and Passive Earth Pressure
Coefficients as a Decreasing Function of k

$$f \simeq [0.50 \ (1 + k'_e) - 0.02 \ (\ \phi - 20^0) \ ] \ H \qquad \leq D \ \ldots (3)$$

In fact, taking f = D would lead to a slightly conservative length L.

The angle, α_{AE}, of inclination of the active sliding wedge is a decreasing function of the effective acceleration coefficient, k_e, as plotted in Fig. 4 for $\phi = 30^0$ and dry soil:

$$k_e = \frac{k_h}{1 - k_v} \ \ldots (4)$$

The horizontal and vertical seismic coefficients, k_h and k_v, should be taken as fractions of the anticipated peak ground acceleration components during the design earthquake shaking; e.g., as suggested by Seed,

$$k_h \simeq \frac{2}{3} \frac{max \ a_h}{g} \ \ldots (5)$$

— For cohesionless soils under the water table, to indirectly take into account both the potential strength degradation due to pore-water pressure buildup and the hydrodynamic effects, we suggest that k_e should increase to

$$k'_e \simeq 1.50 \ k_e \simeq \frac{max \ a_h}{1 - \frac{2}{3} max \ a_v} \ \ldots (6)$$

Having established k'_e, the angle α_{AE} can be approximated as:

$$\alpha_{AE} \simeq 45^0 + \frac{\phi^0}{2} - 135 \ (k'_e)^{1.75} \ \ldots (7)$$

for k'_e values between 0.10 and 0.50, ϕ values between 25^0 and 35^0, and regardless of the wall-soil friction angle, δ. Eqn. 7 was developed by Dennehy (1985) by curve fitting to the parametric results of the Monorobe-Okabe analysis. For k'_e and ϕ values outside these ranges, as well as for backfill and foundation consisting of several soil layers, the dynamic decrease of the inclination angle can be properly computed from the Coulomb-Mononobe-Okabe sliding-wedge analysis [18,16,14].

(b) The 'Embedment Participation Index' (EPI), provides a measure of the likely contribution of the embedment depth. If the wall were acting as a free cantilever (with no anchor), it would undergo horizontal displacement and rotation the magnitude of which would depend on the potential active and passive forces, F_{AE} and F_{PE}, and the respective moments of these forces about the "point" of rotation. In the interest of simplicity, and being restricted by the available data of the analyzed case histories, EPI is defined as:

$$EPI = \frac{F_{PE}}{F_{AE}} \left(1 + \frac{f}{f + H}\right) \dots\dots\dots\dots\dots\dots\dots\dots\dots \quad (8)$$

which for uniform backfill and foundation can be approximated as:

$$EPI \approx \frac{K_{PE}}{K_{AE}} r^2 (1 + r) \dots\dots\dots\dots\dots\dots\dots\dots\dots \quad (9a)$$

where

$$r = \frac{f}{f + H} \dots\dots\dots\dots\dots\dots\dots\dots\dots\dots\dots\dots \quad (9b)$$

The ratio K_{PE}/K_{AE} of the passive to the active earth pressure coefficient is, in general, obtained from a Coulomb-Mononobe-Okabe analysis. As illustrated in Fig. 5, K_{PE}/K_{AE} is a monotonically decaying function of the seismic coefficient and the angle of shearing resistance (assumed to be constant in this figure). Note that, for a wall-soil friction angle $\delta = 0$ and cohesionless soil,

$$1 \leq \frac{K_{PE}}{K_{AE}} \leq \tan^4(45^\circ + \phi/2) \dots\dots\dots\dots\dots\dots\dots \quad (10)$$

where the upper bound is the familiar ratio of the static ($k = 0$) Rankine earth pressure coefficients, whilst the lower bound is reached at a critical effective acceleration [3,16]

$$\frac{k_h}{1 - k_v} = \tan \phi \dots\dots\dots\dots\dots\dots\dots\dots\dots\dots \quad (11)$$

It is noted that the flexural rigidity $E_p I_p$, of the sheetpile does not appear explicitly in the above definition of EPI, despite its obvious importance on the magnitude and shape of the wall deformation. Again, this was done reluctantly since the sectional moment of inertia, I_p, was only rarely reported in the studied cases. Nonetheless $E_p I_p$ relates to the wall height H and the depth f, and hence it does affect (indirectly) EPI.

The two indices, EAI and EPI, computed for each one of the studied 75 anchored bulkheads, produce a point on the diagram of Fig. 6. The degree of damage of the particular bulkhead is reflected on the size and shading of the circle. Thus five different degrees of damage (0 - 4) are distinguished, as explained in Table 3, according to Kitajima & Uwake [11]. With justified reservation, in view of the rather crude way of characterizing the adequacy of the anchoring system and the effectiveness of embedment, and of the uncertainties regarding soil strength parameters and estimated ground acceleration, a clear picture emerges in Fig. 6. Three fairly distinct zones can be identified: Zone I, comprising mostly anchored bulkheads that suffered little or no damage (degrees of damage 1 or 0); Zone II, within which bulkheads suffered moderate

Table **3**. Qualitative and Quantitative Description of the Reported Degrees of Damage [11].

Degree of Damage	Description of Damage	Permanent Displacement of Top of Sheetpile (m)
0	no damage	< 0.02
1	neglible damage to the wall itself; noticeable damage to related structures (i.e. concrete apron)	0.10
2	noticeable damage to the wall	0.30
3	general shape of anchored sheetpile preserved, but significantly damaged	0.60
4	complete destruction, no recognizable shape of wall	1.20

* Average estimates

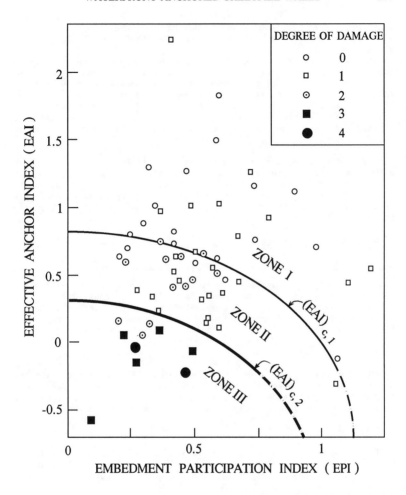

Figure 6. The Developed Seismic Design Chart

degrees of damage (up to degree of damage 2);   amd,   Zone III,   with
anchored bulkheads that suffered severe unacceptable damage (degrees of
damage 3 and 4).
    The shape of the lines delineating each zone does indeed suggest
that the degree of damage suffered by an anchored bulkhead is dependent
on both the adequacy of the anchoring system and the relative depth of
embedment. Notice, however, that the flat shape of these lines implies
that the importance of the Effective Anchor Index (EAI) is far greater
than that of the Embedment Participation Index (EPI), as one might have
anticipated from the earlier discussion on the types of observed
failures.

    Fig. 6 can serve as a Seismic Design Chart to be used in conjunction
with (and to rectify the inadequacies of) the aforementioned pseudo-
static design procedures. For instance, one can first follow the
design steps 1,2,and 3 that were outlined in the second section of this
article, but then determine the required length of the tie-rod form the
following geometric expression:

$$L \geq (h + f). \cot\alpha_{AE} + (EAI)_c \cdot H \quad \dots\dots\dots\dots\dots\dots (12)$$

where the critical value of the Effective Anchor Index, $(EAI)_c$, is
read from the "appropriate" delineating line of the Chart for the
specific value of the Embedment Participation Index (EPI),   as
illustrated in Fig. 6. "Appropriate" means the lower or the upper
curve depending on the acceptable degree of damage-- which in turn is a
function of the economic/social significance of the quaywall.

CONCLUSION

    An empirical chart has been developed (Fig. 6) for guiding the
design of anchored steel sheetpile bulkheads against strong earthquake
shaking.   Use of this Chart, along with Eqns 3, 5, 6, 7, 8, and 12,
would lead to safer (although not unduly conservative) bulkheads.

ACKNOWLEDGEMENT

    This research was funded by a grant (CEE-82006349) from the National
Science Foundation.   Thanks are due to Professor Rowland Richards Jr.
for his several thoughtful comments on the original draft of the paper.

REFERENCES

1. Abraham, R,."Dynamic Analysis of Anchored Bulkheads," Master's
   Thesis, Rensselaer Polytech. Inst., 1985.

2. Agbabian Associates, "Seismic Response of Port and Harbor Facilities," Report P80-109-499., E. Segundo, Calif., 1980.

3. T.G. Davies, R.R. Richards Jr., & K.H. Chen "Passive Pressure During Seismic Loading" Jnl. Geotech. Engrg. ASCE, Vol.112, No. 4, 1986, pp. 479-483.

4. K.T. Dennehy, (1985) "Seismic Vulnerability, Analysis and Design of Anchored Bulkheads," Ph.D. Thesis,Rensselaer Polytech Inst.

5. Duke, C.M. and Leeds, D.J. (1963), "Response of Soils, Foundations, and Earth Structures," Bull. Seism. Soc. Am., 53, 309-357.

6. Hayashi, S. and Katayama T. (1970), "Damage to Harbor Sturctures by the Tokachioki Earthquake,"Soils and Foundations, Vo. X, 83-102.

7. S.J. Hung & S.D. Werner, "An Assessment of Earthquake Response Characteristics and Design Procedures for Port and Harbor Facilities," Proc. 3rd Int. Earthq. Microzonation Conf., Seattle 1982.

8. Japan Society of Civil Engineers, "Earthquake Resistant Design of Quaywalls and Piers in Japan," Earthquake Resistant Design for Civil Engineering Structures, Earth Structures and Foundations in 1980, pp. 31-85.

9. Karkanias, S., "Seismic Behavior and Simplified Analysis of Anchored Sheetpile Bulkheads," Master's Thesis Rensselaer Polytech. Inst., 1983.

10.Kawakami, F. and Asada, A. (1966), "Damage to the Ground and Earth Structures by the Niigata Earthquake of June 16, 1964," Soils and Foundations, Vol. VI, 14-60.

11.Kitajima, S., and Uwabe, T., "Analysis on Seismic Damage in Anchored Sheet-piling Bulkheads, "Rep. of the Japanese Port and harbor Res. Inst., Vol. 18, 1978, pp. 67-130. (in Japanese).

12.Murphy, V.A., "The Effect of Ground Characteristics on the Aseismic Design of Structures," Proc. 2nd World Conf. on Earthq. Engrg., Tokyo, 1960, pp. 231-247.

13.Nadim, F. and Whitman, R.V., "Seismically Induced Movement of Retaining Walls, " Jnl. Geotech. Engrg. Div. ASCE, Vol. 109, No. 7, 1983, pp. 915-931.

14.Prakash, S., Analysis of Rigid Retaining Walls during Earthquakes," Int. Conf. on Recent Advances in Geotech. Earthq. Engrg. and Soil Dyn., Vol. 3, St. Louis, Mo., 1981, pp. 1-28.

15.Rowe, P.W. (1952), "Anchored Sheetpile Walls,"Proc. Inst. Civ. Engrs., 1, 27-70.

16.Richards, R. and Elms, D. (1979), "Seismic Behavior of Gravity Retaining Walls," J. Geotech. Engr. Div., ASCE, 105.

17.Seed, H.B., "Stability of Earth and Rockfill Dams during Earthquakes," Embankment-Dam Engineering, Casagrande Vol., John Wiley and Sons, 1973, pp. 239-269.

18. Seed, H.B. and Whitman, R.V., "Design of Earth Retaining Structures for of Dynamic Loads," ASCE Spec. Conf. Lateral Stresses in the Ground and Design of Earth Retaining Structures, Ithaca, N.Y., 1970, pp. 103-147.
19. Tschebotarioff, G.P. (1978), "Foundations, Retaining and Earth Structures," McGraw-Hill.
20. Tsinker, G.P., "Anchored Sheet Pile Bulkheads: Design Practice," Jrl. Geotech. Engrg. ASCE, Vol. 109, No. 8, 1983, pp 1021-1038.
21. U.S. Navy Bureau of Yards and Docks, Design Manual DM-7, U.S. Government Printing Office, Washington, D.C., 1962.
22. U.S. Steel, Steel Sheet Piling Design Manual, United States Steel Corporation, Pittsburg, PA., 1974.
23. Wood, J.H., "Earthquake-Induced Soil Pressures on Structures," Ph.D. Thesis, Calif. Inst. of Tech., 1973,.
24. Christiano, P.P. and Bielak J. (1985). "Nonlinear Earthquake Reponse of Tiedback Retaining Walls," Research Rep., Carnegie-Mellon Univ.

BEHAVIOR AND DESIGN OF
GRAVITY EARTH RETAINING STRUCTURES

J. Michael Duncan[1], F. ASCE, G. Wayne Clough[2], F. ASCE, and
Robert M. Ebeling[3], M. ASCE

ABSTRACT: The subject of gravity retaining structures is covered in
every elementary textbook on geotechnical engineering, and the basis
for design is typically described in terms of equilibrium analyses.
It is often assumed that the behavior of these systems are simple and
can be taken for granted. However, there are important soil-structure interaction effects in gravity walls, and more complete
understanding of these effects provides an improved basis for design.
Over the past 20 to 30 years, field studies and finite element
analyses have provided new insights into the behavior of gravity
walls.
    Almost all problems with gravity earth retaining walls have
occurred where clay backfills were used or where walls were founded
on clay foundations. Design for these conditions is appropriately
based on conservative empirical procedures that make allowance for
creep in clayey soils. On the other hand, currently used design procedures for gravity walls backfilled with sand or gravel and founded
on rock lead to excessive conservatism. In some cases current
methods indicate that a stable wall is on the verge of failure.

INTRODUCTION

    Gravity earth retaining structures are those that rely on their
weight to resist the forces exerted by the soil they retain. As
shown in Fig. 1, they can be constructed of reinforced concrete, mass
concrete, precast or metal cribs filled with earth or stone, gabions,
or corrugated metal cylinders filled with earth.
    Methods for estimating earth pressures on these structures, and
criteria for their stability, are discussed in most geotechnical

1   University Distinguished Professor, Dept. of Civil Engineering,
     Virginia Polytechnic Institute and State University, Blacks-
     burg, VA 24061
2   Harry C. Wyatt Professor and Chairman, Dept. of Civil Engineer-
     ing, Virginia Polytechnic Institute and State University,
     Blacksburg, VA 24061
3   Engineer, U. S. Army Corps of Engineers, Waterways Experiment
     Station, Vicksburg, MS

251

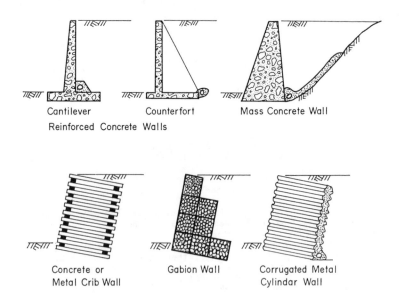

Cantilever   Counterfort   Mass Concrete Wall

Reinforced Concrete Walls

Concrete or   Gabion Wall   Corrugated Metal
Metal Crib Wall   Cylindar Wall

Fig. I. Gravity earth retaining walls

engineering textbooks and manuals. Few if any failures can be attributed to use of these conventional procedures in cases where the
backfill is a clean granular material and the wall is founded on
sand, gravel, or rock. Problems with gravity earth retaining structures have developed almost exclusively in cases where the backfill
and/or the foundation consists of clayey soil. Problems in these
cases have been due to improper assessment of the properties of the
backfill or foundation soils, or failure to make allowance for the
fact that clayey soils creep.

When conventional methods are used to analyze some existing walls
with clean granular backfills and rock foundations, the results indicate that the walls should be unstable, in spite of the fact that
they have performed well for many years. These findings suggest that
the assumptions underlying the conventional methods are erroneous for
walls that have granular backfills and are founded on rock, and that
they lead to excessively conservative results.

Even though the design of gravity earth retaining structures has
been considered by many geotechnical engineers to be an area little
in need of change or improvement, recent studies have shown that
there are important soil-structure interaction effects that are not
reflected in current design procedures. Better understanding of
these soil-structure interaction effects is the key to improvement in
design methods.

TERZAGHI'S DESIGN OF THE FIFTEEN-MILE FALLS DAM RETAINING WALL

An interesting and instructive case of a high gravity earth retaining wall founded on rock is that of the 150 ft (46 m) high mass concrete wall at Fifteen-Mile Falls Dam (now called Comerford Dam). The dam is on the Connecticut River, on the border between Vermont and New Hampshire. The wall, which retains the end of the earth embankment, was designed by Terzaghi in about 1928 (53). Although the wall should be unstable as judged by currently used methods, it has remained stable and performed well for nearly 60 years, and shows no signs of distress. It is interesting, therefore, to examine the basis for the design of the wall and to see how it differs from current practice.

A cross-section through the wall is shown in Fig. 2. In 1928, when this wall was being designed, the basic concepts required for rational design of earth retaining structures were still developing. Terzaghi therefore formulated a number of questions concerning the earth and water pressures that would act on the wall, and conducted large-scale retaining wall tests at MIT [(48) - (52)] to find answers to these questions. The answers not only provided the information needed for design of the Fifteen-Mile Falls Dam wall, they led to better understanding of the behavior of gravity walls. Unfortunately, some of these lessons have since been forgotten, or at least dropped from use in practice.

The Questions. The three questions posed by Terzaghi that are of interest here were concerned with the earth pressures that should be used in designing the wall:

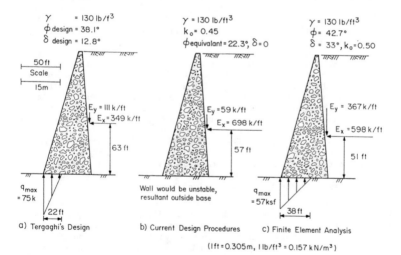

γ = 130 lb/f^3
ϕ design = 38.1°
δ design = 12.8°

50 ft
Scale
15 m

E_y = 111 k/ft
E_x = 349 k/ft

63 ft

q_{max} = 75 k

22 ft

a) Terzaghi's Design

γ = 130 lb/ft^3
k_o = 0.45
ϕ equivalent = 22.3°, δ = 0

E_y = 59 k/ft
E_x = 698 k/ft

57 ft

Wall would be unstable, resultant outside base

b) Current Design Procedures

γ = 130 lb/ft^3
ϕ = 42.7°
δ = 33°, k_o = 0.50

E_y = 367 k/ft
E_x = 598 k/ft

51 ft

q_{max} = 57 ksf

38 ft

c) Finite Element Analysis

(1 ft = 0.305 m, 1 lb/ft^3 = 0.157 kN/m^3)

Fig. 2. Analyses of the 150 ft high gravity wall at Fifteen-Mile Falls Dam.

(1)   "To what extent does the rigidity of the wall affect the inten-
      sity of the earth pressure?"
(2)   "What are the numerical values for the coefficients of internal
      friction and wall friction on which the design of the wall
      should be based?"
(3)   "Which of the known methods should be used for computing the
      forces acting on the wall?"

The Answers.   From the results of his retaining wall tests
Terzaghi determined the relationship between earth pressure and wall
movement used in the design of the Fifteen-Mile Falls Dam wall.   He
also measured the values of the angles of internal friction and wall
friction, and he was able to determine which earth pressure theory
was in best agreement with the measured behavior.   Based on this information he developed the following answers to the questions posed:

(1)   Even though the wall was mass concrete and was founded on a
      rock foundation (assumed by Terzaghi to be rigid), Terzaghi con-
      cluded that it would deflect enough to reduce the earth
      pressures below their at-rest values.   The deflection calculated
      by Terzaghi was thought to be sufficient to reduce the earth
      pressure coefficient to 0.25, as compared with an estimated at-
      rest value of about 0.40.   For comparison, the value of K_a was
      estimated to be about 0.21.

(2)   Based on his estimate of the possible long-term effects of
      stress relaxation due to high stresses at grain contacts and wa-
      ter percolating through the fill, Terzaghi recommended using
      reduced values of the angles of internal friction and wall fric-
      tion for the sand.   The values of ϕ and δ measured in high pres-
      sure tests were ϕ = 42.7 degrees and δ = 33 degrees.   Terzaghi
      used reduced values of ϕ = 38.1 degrees and δ = 12.8 degrees to
      calculate the forces shown in Fig. 2(a).

(3)   The retaining wall tests showed that a shear force of signifi-
      cant magnitude acted on the backs of retaining walls, even when
      the walls did not deflect.   The Rankine earth pressure theory,
      which indicates that the shear force should be zero for a level
      backfill surface, was therefore eliminated as a possibility for
      use in design.   The Coulomb theory was used, so that the effects
      of wall friction could be accounted for in estimating the forces
      exerted on the wall.   The resultant shear and normal forces on
      the back of the wall were assumed to act at a height of 0.42 H
      above the heel of the wall (H is the wall height), as indicated
      by the retaining wall tests, rather than at 0.33 H above the
      heel as inferred from the Rankine theory (53).

The Lessons.   Current procedures for design of the Fifteen-Mile
Falls Dam wall would be based on quite different assumptions than
those used by Terzaghi for its design.   Most practitioners today
would probably consider that a massive wall like the one shown in
Fig. 2, founded on a firm rock foundation, would be acted on by at-
rest pressure.   For example, the Corps of Engineers manual on engineering and design of retaining walls (59) states that gravity walls
on rock foundations should be designed for at-rest pressures (K_o =

0.45). For conditions where the ground surface is not horizontal or the back of the wall is not vertical, an equivalent value of ϕ (called ϕ_o) is calculated from the following expression:

$$K_o = \tan^2 (45 - \phi_o/2) \tag{1}$$

for $K_o = 0.45$, $\phi_o = 22.3$ degrees, as shown in Fig. 2(b). This value of ϕ_o is then used in the Coulomb equation with $\delta = 0$ to determine the horizontal and vertical forces on the wall. These forces are assumed to act at an elevation of 0.38 H above the heel of the wall. If subjected to these forces, the Fifteen-Mile Falls Dam wall would fail, as shown in Fig. 2(b). Yet the wall has not failed, and it has performed well for nearly 60 years.

Thus it clearly would not be necessary to design for at-rest pressure and $\delta = 0$. The example of the Fifteen-Mile Falls Dam wall therefore indicates that current conventional design procedures are excessively conservative when applied to large gravity walls founded on rock.

Similar examples have been found in recent studies by the Corps of Engineers [(34) - (36)]. These studies have shown that the earth-retaining walls founded on rock at Emsworth Lock, Montgomery Lock, and Troy Lock should be unstable as judged by current design criteria, in spite of the fact that they have performed well for many years. These cases indicate current conventional criteria for design of large gravity walls founded on rock result in incorrect indications regarding the stability and safety of existing walls. It can be inferred, therefore, that they also result in excessively costly designs for new walls.

FINITE ELEMENT ANALYSIS OF FIFTEEN-MILE FALLS DAM WALL

The results of a finite element analysis of the Fifteen-Mile Falls Dam retaining wall are shown in Fig. 2(c). It may be seen that this analysis results in a horizontal force not much less than the at-rest earth pressure force. The Poisson's ratio used in the analysis corresponds to $K_o = 0.5$, and the horizontal force calculated corresponds to $K_h = 0.41$. For comparison, the value of K_a computed by the Coulomb theory is 0.21.

Although the lateral earth pressure force calculated in the finite element analysis is nearly as large as the at-rest value, the analysis did not indicate that the wall should be unstable. As shown in Fig. 2(c), the portion of the base in contact with the foundation and the maximum bearing pressure at the toe of the wall are not much different from those estimated by Terzaghi. The forces used by Terzaghi correspond to a mobilized friction angle between the base of the wall and the foundation equal to 26.5 degrees, whereas the forces from the finite element analysis correspond to a mobilized base friction angle of 29 degrees.

There are, however, two very significant differences between the forces used by Terzaghi and the results of the finite element analyses: The horizontal force used by Terzaghi was much smaller than that computed in the finite element analysis, and the vertical force was also. These two differences have compensating effects, however,

and both Terzaghi's calculations and the finite element analysis
indicate that the wall should be stable.

The question naturally arises as to which set of forces shown in
Fig. 2 is the better estimate of the actual forces on the wall.
Given the fact that the wall is stable, the forces determined from
the conventional analysis (Fig. 2(b)) can be ruled out as a realistic
possibility. Since the wall would be stable if subjected to either
the forces estimated by Terzaghi (Fig. 2(a)), or those from the finite element analysis (Fig. 2(c)), the fact that the wall is stable
does not indicate which of these is more reasonable.

However, it seems likely that Terzaghi underestimated the horizontal force on the wall because he assumed that the variation of the
earth pressure coefficient with Δ/H would be the same for a 150 ft
(46 m) high wall as for a 5 ft (1.5 m) high wall (Δ = horizontal
movement at the top of the wall, H = wall height). As discussed in
subsequent sections of this paper, this is not the case. Measured in
terms of Δ/H, greater movement is required to reduce the earth pressure on a large wall a given amount below the at-rest value than is
required to reduce the earth pressure by the same fraction on a small
wall. This has been demonstrated by experiments with walls larger
than the 5 ft (1.5 m) high experimental wall used by Terzaghi, and by
finite element analyses. By scaling linearly from 5 ft to 150 ft
(1.5 to 46 m) to estimate the movements required to reduce the earth
pressures below their at-rest values, the writers believe that
Terzaghi underestimated the amount of movement required to reduce the
earth pressure, and thereby underestimated the pressures that would
act on the wall.

At the same time, however, he used a greatly reduced value of
mobilized friction angle on the back of the wall. As discussed subsequently, field measurements and finite element analyses indicate
that the mobilized wall friction angle increases as the depth of the
backfill increases, because the deeper the backfill, the more it
settles under its own weight, and therefore the greater the amount of
shear displacement between the backfill and the wall.

Thus, based on current understanding of the interaction between
high gravity walls on rock and their backfills, it appears that the
horizontal and vertical forces estimated by Terzaghi were smaller
than would be estimated today. However the effects of what now
appears to be an underestimation of these two forces were
compensating. The end result was an accurate evaluation of the stability of the wall, and a design that has resulted in 60 years of
good performance. Given the fact that the design was done in 1928,
when the necessary concepts were still developing, the design of this
very high wall has to be considered a marvelous achievement. It is
doubtful that modern engineers, with all the benefits of data accumu-
lated and analysis techniques developed since 1928, would be willing
to reduce the base width of the wall by a single foot.

MEASURED PERFORMANCE OF WALLS

Conventional earth pressure theories, like the Rankine and Coulomb
theories, are only capable of indicating what earth pressures would
act on walls in limiting conditions, when the strength has been fully

mobilized in the active or passive sense. In most service conditions, these pressures do not act on walls, because of factors of
safety incorporated in the design. Field observations help us understand the influence of the conservatism built into our designs, and
are needed to determine the effects of factors not considered in
earth pressure theories. These factors include soil creep, temperature effects, effects of rainfall and frost action, compaction loads,
surface loads, and the effects of vibrations. Fortunately, a considerable number of field measurement programs and experiments on large
walls have been undertaken in recent years. A number of the findings
of these studies are summarized in the following sections.

Pressure Measurements. Nearly all investigators have found that
earth pressures measured using earth pressure cells are scattered and
uncertain. Thus a single measurement of pressure has doubtful value,
because it is never clear whether a single measured value represents
the average behavior or an aberration. Carder et al. (9) and Weiler
and Kulhawy [(62), (63)] have discussed the difficulties involved in
earth pressure measurements, and the measures that are necessary to
achieve useful results. Even using the best procedures, however,
measured values of earth pressure often contain large scatter.

While a single measurement of pressure may never be fully
reliable, the results of multiple observations, when they are consistent, and when they have been verified by independent means, can
provide valuable insight into field performance. From the rather
large body of field and laboratory data available, some consistent
trends are discernable that are of considerable value in
understanding the range of applicability of earth pressure theories,
their limitations, and the importance of factors not considered by
theories.

Measured Pressures on Retaining Walls. Pressures on walls in service have been reported by Gould (23), Broms and Ingelson [(4), (5)]
Kany (30) and Rehnman and Broms (41). All observed that the measured
earth pressures were higher than active values. This is to be
expected as the normal situation even if active pressures were
employed in design, because a wall with a margin of safety against
sliding and overturning normally does not move far enough for the
earth pressure to decrease to the minimum active value.

Position of Resultant Earth Pressure Force. In cases where the
measurements permitted determination of the position of the resultant
earth pressure force, it has generally been found to be above the
lower third point. Terzaghi [(48) - (54)] found that the resultant
was located at 0.40 H to 0.45 H above the bottom of the wall, rather
than at 0.33 H as inferred from theory. Clausen and Johansen (11)
found the same range of locations for the resultant earth force on a
basement wall, and Sherif et al. (45) found that the at-rest earth
pressure resultant acted at 0.42 H. In Terzaghi's tests, as the wall
continued to move after the earth pressure force had reached its minimum active value, the position of the resultant eventually dropped
to 0.33 H. However, the movement required to achieve a triangular
pressure distribution (resultant at 0.33 H) was three to five times
as large as the movement required to reduce the earth pressure force
to its minimum active value. For pressures higher than the minimum

active value, it is to be expected that the earth pressure resultant may act at an elevation above the lower third point.

Pressures on Basement Walls. Although it is commonly presumed that basement walls are stiff and will be loaded by at-rest pressures, measurements indicate that this is not necessarily so. Clausen and Johansen (11) measured pressures on a basement wall backfilled with uncompacted sand that corresponded to K = 0.29. Similarly, Rehnman and Broms measured K = 0.35 for a basement wall backfilled with uncompacted gravelly sand, and K = 0.31 for a backfill of silty sand. The deflections measured by Rehnman and Broms were Δ/H about 0.0003 for the gravelly sand, and 0.0007 for the silty sand backfill. Roth et al. (42) measured pressures corresponding to K = 0.18 to 0.24 on a basement wall backfilled with silty sand.

Pressures Exerted by Narrow Backfill Zones. Some field measurements have shown smaller than normal active pressures for backfills placed in narrow trenches, and near the bottom of basement walls where excavations become narrow [(23), (30), (41)].

Friction Forces on Walls. A large number of field measurements shows that both free-standing retaining walls and basement walls are acted upon by downward-directed shear forces as well as normal forces [(3), (19), (22), (23), (30), (31), (48), (49)]. These shear forces, often neglected in design, have a very important stabilizing effect on retaining walls. The forces are generated by downward movements in the backfill as it settles relative to the wall (13). Only one or two tenths of an inch (2 to 5 mm) of relative shear movement between the backfill and the wall is needed to mobilize the full shear strength of the interface.

In his long-term tests on a 14 ft (4.3 m) high retaining wall backfilled with a sandy soil, Fukuoka (22) found that the shear force increased over a period of nine months after backfilling. It seems likely that this increase in wall friction was the result of gradual settlement of the backfill. Increases in the magnitudes of shear loads with time have also been observed in tests performed in the Instrumented Retaining Wall Facility at Virginia Tech.

Earth Pressures Due to Surface Loads and Compaction. When backfill is compacted behind a retaining wall or adjacent to a basement wall, the lateral pressures acting against the wall increase. The pressures reach a temporary peak when the compactor makes its closest approach to any point. As the compactor moves away, the lateral pressures decrease, but remain above their normal at-rest values. Broms and Ingelson (5), Rehnman and Broms (41), Vaughan and Kennard (60), Casagrande (10), Coyle and Bartoskewitz (16), Carder et al. (8), Carder et al. (7), and Sherif et al. (44) have all measured elevated pressures behind walls as a result of compacting the backfill.

Broms and Ingelson (5) and Carder et al. (7) found that small movements of the wall away from the backfill after compaction were enough to relieve the pressures induced by compaction, and reduce the pressures to their active values.

Rehnman and Broms found that vehicle loads acting on the surface of an uncompacted backfill induced significant lateral earth pressure on a wall, and that 40% to 80% of the increase remained when the loads were removed. In similar experiments where the backfill had been compacted, the same loads induced much smaller lateral pressures on the wall, and nearly all the load-induced pressure disappeared upon removal of the load. Thus it appears to be unnecessary to superimpose pressures due to surface loads with those induced by compaction, at least in the case where the magnitudes of the load-induced pressures would not exceed the residual lateral pressures from compaction.

Carder et al. (7) found that the earth pressures induced by compaction of a silty clay backfill decreased over a period of four months following compaction, eventually reaching normal at-rest pressure values.

Movements to Reach the Active Pressure Condition. Many different investigators, using laboratory tests on model retaining walls, or controlled field experiments on walls up to 33 ft (10 m) high, have measured the movements required to reduce earth pressure to their active values (Table 1). These data illustrate some significant points regarding active earth pressures:

(1) The movement (Δ/H) required to reach the minimum active pressure condition is about the same for rotation and translation, provided Δ represents the movement at the top of the wall.

(2) Two definitions of the active condition have been used. The first corresponds to the resultant active earth pressure force reaching its minimum value. The second corresponds to development of a triangular pressure distribution. In many experiments the first condition is achieved while the earth pressure has a

TABLE 1. MOVEMENTS REQUIRED TO ACHIEVE ACTIVE EARTH PRESSURE CONDITIONS

Investigators	Wall Height (ft)	Backfill	Compacted	Mode of Movement	Δ/H to Reach Active Pressure[1]
Broms & Ingelson (1971)	9.0	Sand	Yes	Rotate	0.0003[2]
Broms & Ingelson (1972)	28.4	Sandy gravel	Yes	Rotate	0.0009 to 0.0024 [2]
Carder et al. (1977)	6.6	Sand	Yes	Translate	0.0020[2]
Carder et al. (1980)	6.6	Silty Clay	Yes	Rotate	0.0009[2]
Matsumoto et al. (1978)	32.8	Silty Sand	Yes	Rotate	0.006 to 0.008[2]
Matsumoto et al. (1978)	32.8	Slag	Yes	Rotate	0.003 to 0.005[2]
Sherif et al. (1984)	4.0	Sand	Yes	Rotate	0.0005[2]
Sherif et al. (1984)	4.0	Sand	No	Rotate	0.0005[2]
Terzaghi (1934a)	4.9	Sand	Yes	Rotate	0.0011[2]
Terzaghi (1934a)	4.9	Sand	Yes	Translate	0.0011[2]
Terzaghi (1934a)	4.9	Sand	No	Rotate	0.0020[2]
Terzaghi (1936)	4.9	Sand	Yes	Rotate	0.002[2]
Terzaghi (1936)	4.9	Sand	Yes	Rotate	0.005[3]
Terzaghi (1936)	4.9	Sand	Yes	Translate	0.001[2]
Terzaghi (1936)	4.9	Sand	Yes	Translate	0.005[3]

(1) Δ = movement at top of wall, H = wall height
(2)   Movement to reach minimum total earth pressure force
(3)   Movement to develop triangular pressure distribution, with resultant at lower third point
(4)   1 ft = 0.305 m

nonlinear variation, with the resultant at an elevation higher
than 0.33 H. Continuing movement results in a triangular distribution of pressure and the resultant at 0.33 H.

(3) The larger the wall, the greater the value of Δ/H required to
reduce the active earth pressure resultant to its minimum value.
This finding is consistent with the fact that the stiffness of
sands does not increase in direct proportion to confining pressure, and therefore does not increase in proportion to wall
height. For large walls the movements required to reach the
active condition is on the order of $\Delta/H = 0.004$. This corresponds to about one inch of lateral movement for each 20 ft of
wall height (40 mm for each 10 m of wall height).

Movements to Reach the Passive Pressure Condition. The measured
movement required to reach the passive pressure condition varies over
a wide range, as shown in Table 2. Terzaghi (48) found that the
earth pressures in a compacted sand backfill were increased to K =
2.0 to 2.5 by movements of $\Delta/H = 0.001$; the maximum passive pressure
condition was not achieved in the tests. Broms and Ingelson (4)
found that for a bridge abutment in contact with compacted sand,
movements as small as $\Delta/H = 0.005$ increased the pressure to the theoretical maximum (K = $K_{pCoulomb}$). Tcheng and Iseux (47) and Carder et
al. (8) found that the movements required to mobilize the maximum
passive force in compacted sand ranged as high as $\Delta/H = 0.06$. A test
performed by Carder et al. (7) using a compacted silty clay showed
that $\Delta/H = 0.13$ was required to mobilize the full passive resistance.

TABLE 2. MOVEMENTS REQUIRED TO ACHIEVE PASSIVE PRESSURE CONDITIONS

Investigators	Wall Height (ft)	Backfill	Compacted	Mode of Movement	Δ/H to Reach Passive Pressure[1]
Broms & Ingelson (1971)	9.0	Sand	Yes	Rotate	0.005
Carder et al. (1977)	3.3	Sand	Yes	Translate	0.025
Carder et al. (1980)	3.3	Silty Clay	Yes	Rotate	0.132
Tcheng & Iseux (1972)	9.8	Sand	Yes	Rotate	0.02 to 0.06
Terzaghi (1934a)	4.9	Sand	Yes	Rotate	0.001[2]

(1) Δ = movement at top of wall, H = height of wall

(2)   Passive pressure not fully mobilized; k = 2 to 2.5

(3)   1 ft = 0.305 m

Changes in Earth Pressure With Time. In experiments and field
observations conducted over long periods, changes in pressure have
sometimes been observed that are not attributable to changes in environmental conditions such as temperature, rainfall, or earthquakes.
Hilmer (25) found that the pressures on a lock wall exerted by
compacted sandstone backfill increased by about 40 percent over a
period of two years, and then stayed essentially the same (except for
seasonal variations) for the next eight years. Matsuo (31) found
that the earth pressures on a 33 ft (10 m) high retaining wall backfilled with silty sand increased with time after reaching an active
condition. The increase was about 50 percent over a period of 20
days.

Rehnman and Broms (41) found that the distribution of the pressure exerted on a basement wall by a silty sand backfill changed with time, but the magnitude of the resultant force did not change much. Clausen and Johansen (11) found that the earth pressures exerted on a basement wall by a backfill of uncompacted sand remained constant over a period of 10 months.

Finally, as noted previously, Carder et al. (7) found that the pressures exerted on an experimental wall by compacted silty clay decreased over a period of four months from their high post-compaction values to normal at-rest values (K_o = 0.40).

These data are consistent in the following respect: If a wall does not move, the earth pressures tend toward at-rest pressure over time. If the pressures are initially below at-rest, they increase. If they are initially above at-rest, they decrease. If they are initially at-rest, they do not change. The fundamental reasons for this behavior are not understood, but it seems clear that at-rest pressures represent the long-term equilibrium condition for non-yielding walls.

Effect of Temperature on Earth Pressures. A number of investigators [(3), (4), (5), (15), (16), (17), (24), (25), (46)] have noted cyclic variations in measured earth pressures that were caused by temperature-induced wall movements.   Broms and Ingelson noted both daily and seasonal fluctuations.   In the other cases only seasonal variations were noted.

In most of these cases the earth pressures were seen to increase in the summer and to decrease in the winter. The reason for this is illustrated by the example shown in Fig. 3: In the summer, when the face of the wall is relatively warm, the face of the wall expands relative to the back. As a result, the wall deflects back against the fill, increasing the earth pressure. In the winter the wall deflects away from the fill, and the pressures decrease.

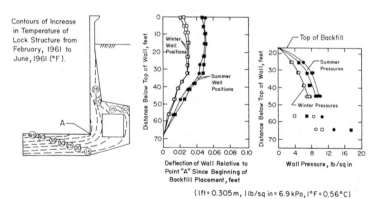

(1ft = 0.305 m, 1 lb/sq in = 6.9 kPa, 1°F = 0.56°C)

Fig. 3. Temperature-induced wall movements and changes in earth pressure at Port Allen Lock

The measurements made by Symons and Wilson (46) are an exception to the rule that earth pressures increase in the summer and decrease in the winter. Because the wall was connected at the top to a pavement slab extending across the top of the backfill, as the temperature of the slab increased in the summer, the wall was pushed away from the fill, and the earth pressures decreased. In the winter the slab contracted and the pressures increased.

Temperature-induced changes in earth pressure do not pose a significant problem in the sense that it is nearly inconceivable that they could cause failure of a wall. They can have a very important effect on measured earth pressures, however, and failure to understand the influence of temperature on earth pressures can lead to misinterpretation of field measurements.

Effects of Backfill Freezing. When the backfill behind a wall freezes, the water within the voids expands, and the pressure exerted on the wall increases. If extra water is drawn into the voids and clear ice lenses are formed, the tendency for expansion and the resulting pressures can be very large.

Sandegren et al.(43) described a case in which freezing caused failure of some of the tie rods anchoring a sheet-pile wall retaining a clay backfill. Vertical sand drains containing heating cables were used to prevent freezing of the backfill in subsequent winters.

Rehnman and Broms (41) conducted a field test in which a 6.6 ft (2.0 m) high cantilever wall backfilled with uncompacted silty sand was allowed to freeze. The earth pressures increased by 800 psf to 1000 psf (38 to 48 kPa) at some levels behind the wall, and decreased during the spring to levels slightly lower than the pressures before freezing. In a related experiment, 2.0 in (50 mm) thick insulation mats were used between the wall and the backfill. The insulation mats reduced the freezing rates, and their compressibility allowed the backfill to expand with only a small increase in the pressure on the wall.

The Corps of Engineers (59) has developed charts for determining the depth to which frost may penetrate. This depth (in feet) can be approximated as the air freezing index (in degree Fahrenheit-days) divided by 200. The air freezing index is the area below 32 degrees Fahrenheit (zero Celsius) on a plot of temperature vs time (In SI units, frost depth in meters = freezing index in degree Celcius-days divided by 360). For example, in a locale where the air freezing index is 1800 degree Fahrenheit-days (1000 degree Celcius-days) in a year, the combined thickness of wall and non-frost susceptible backfill should be 9.0 ft (2.7 m) to avoid problems with backfill freezing.

Effect of Rainfall Infiltrating the Backfill. Rehnman and Broms (41) performed a test to determine the effect of very heavy rainfall on the earth pressures exerted on their 6.6 ft (2.0 m) high experimental wall. The wall was backfilled with uncompacted gravelly sand, and was sprinkled at a rate of 5 inches (125 mm) per hour. After 4 hours the pressures on the wall had increased by about 50 percent, from K = 0.35 to K = 0.5. Subsequently, when the artificial rainfall was stopped, the pressures decreased to their initial values over a period of about 13 hours. Due to the heavy artificial rainfall, the surface of the backfill settled about 4 inches, or about 5 percent of its depth.

Effects of Simulated Earthquake Shaking. Sherif et al. (45) performed a series of shaking table experiments using a 4 ft (1.2 m) high retaining wall backfilled with Ottawa sand. The dynamic earth pressures measured during the tests were in agreement with values calculated using the Mononobe-Okabe theory [(32), (33)] for walls that had moved far enough to develop minimum active earth pressures. The measured dynamic pressures were about 30 percent larger than predicted by the Mononobe-Okabe theory in the cases of walls that had not moved far enough to reduce the earth pressures to their minimum active values.

RETAINING WALL FAILURES

Peck et al. (38) conducted a survey of railroad engineers to gather information about retaining walls and abutments that had performed in an unsatisfactory manner. Forty-eight percent of the engineers surveyed did not reply, and 31 percent reported no significant problems, indicating a low incidence of bad experience with retaining wall behavior.

The average height of the walls reported was 25 ft (7.6 m), and the majority of the walls described as having failed or having experienced problems moved 6 to 12 inches (150 to 300 mm). The results indicated that movements of 3 inches (75 mm) or less were not considered to be a problem.

As shown in Fig. 4, all of the walls that underwent progressive movement were backfilled with clay or were founded on clay. Fifty percent of the walls with problems were founded on pile foundations. Peck et al. (38) suggested that the factors of safety of walls founded on clay or backfilled with clay are probably lower than commonly believed. They recommended that more attention should be paid to evaluating the foundation conditions beneath walls, and to time-conditioned changes that occur in clayey backfills.

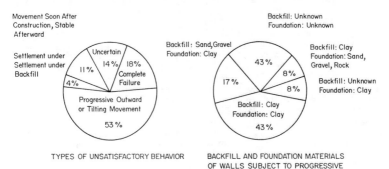

TYPES OF UNSATISFACTORY BEHAVIOR   BACKFILL AND FOUNDATION MATERIALS
                                          OF WALLS SUBJECT TO PROGRESSIVE
                                          OUTWARD MOVEMENT

Fig. 4. Failures and unsatisfactory behavior of retaining walls and abutments (after Peck, et al., 1948).

No failures were reported for walls backfilled with clean granular materials and foundations that did not contain clay. Peck and his colleagues interpreted this experience to indicate that the conventional methods of design for such walls are conservative.

Granger (24) described the failure of a 23 ft (7.0 m) high reservoir wall (retaining water) due to uplift and progressive softening of clay in the foundation, resulting from rainfall and leakage through joints in the reservoir floor. Brand and Krasaesin (2) reported a case of a deep-seated failure of a wall founded on piles; the authors concluded that the undrained shear strength of the soft clay in the foundation had been reduced by remoulding and disturbance due to pile driving. Peck, et al. (38) described two cases of large settlements and deep-seated movements in clay foundations beneath bridge abutments. In both cases the abutments were founded an piles driven into clay. Tschebotarioff (58) described a case of pronounced bulging in the upper portion of a 34 ft (10.4 m) high crib wall. The cribs were filled with uncompacted granular fill, and the backfill behind the cribs was compacted. Tschebotarioff suggested that the failure was due to the fact that the foundation beneath the heel of the wall was overloaded, and the wall settled more than the backfill. This caused the wall friction to act up on the wall and down on the backfill, opposite to the normal direction where the backfill settles relative to the wall. This unusual reversal of the shear force on the back of the wall had a very serious destabilizing effect.

FINITE ELEMENT ANALYSES - COMPARISONS TO FIELD MEASUREMENTS

Finite element analyses have proven useful for studying the interaction between retaining structures and the backfill they retain, particularly when they have been used in conjunction with field observations and instrumentation studies.

Clough and Duncan (13) showed that finite element analyses could be used to simulate the development of pressures on a wall and the movement of the wall during backfilling. In the case illustrated in Fig. 5, they showed that a wall can develop earth pressures lower than at-rest even when it tilts backward. As shown in Fig. 5, the backward tilt of the wall results from nonuniform foundation settlement, which is greater beneath the heel of the wall than beneath the toe. The normal to the foundation surface tilts backward even more than the wall. In a relative sense, the wall tilts forward from this line, away from the backfill. The wall also moves forward during placement of the backfill. Thus, even though the wall tilts backward, the pressures exerted on the wall are lower than at-rest values. This type of movement of a wall founded on clay has been described by Tschebotarioff (57).

Duncan and Clough (19) also used the finite element method to analyze the complex soil-structure interaction in Port Allen Lock. A curious aspect of the behavior of that structure was that the lock walls moved inward when the lock chamber was filled with water, rather than outward as intuition would suggest. At the same time that the walls moved away from the backfill, the earth pressures acting on the walls increased. The finite element analyses showed the same behavior, and provided the information needed to understand

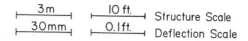

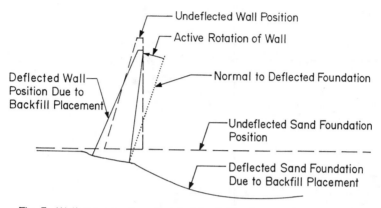

Fig. 5. Wall movements calculated by finite element analysis.

this seemingly anomalous occurrence. As shown in Fig. 6, the inward movement of the lock walls upon filling of the lock chamber was a result of a form of mass flow of the soil around the lock structure. The downward movements of the lock resulted from the increased weight of water in the lock chamber, and the upward movement of the backfill resulted from a rise in the water level within the backfill that accompanied the filling of the lock with water. It is clear that such a displacement pattern is consistent with an increase in wall pressures and simultaneous movement of the wall toward the center of the lock, in agreement with the measured behavior.

These results show how finite element analyses can be used to explain apparent anomalies in field instrumentation results. If this strange behavior was measured and not calculated, it would probably be attributed to some type of inaccuracy. If it was calculated and not measured, it would likely be rejected out of hand. Because it was observed and calculated, however, this behavior must be considered normal under the circumstances, and the finite element analyses provide a basis for understanding the phenomenon.

Matsuo, et al. (31) compared the results of finite element analyses with pressures measured on their large-scale retaining wall tests. They found, as have other investigators, that the calculated at-rest earth pressures were insensitive to the value of soil modulus used in the analyses, and were primarily determined by the value of Poisson's ratio. This finding illustrates the importance of using stress-strain formulations that are rational, and that use parameters whose values can be determined from the results of laboratory or field tests.

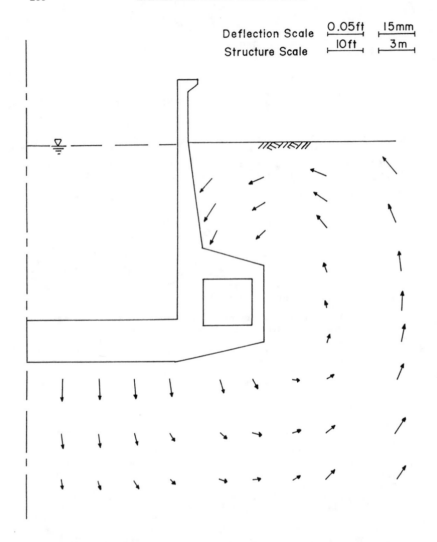

Fig. 6. Calculated movements due to filling Port Allen Lock
       with water.

Roth et al. (42) performed finite element analyses simulating the placement and compaction of a silty sand adjacent to a 110 ft (34 m) deep basement wall in Los Angeles. Earth pressure cells had been installed on the walls to measure the earth pressures during construction and later during possible earthquakes. Using nonlinear, stress-dependant modulus and Poisson's ratio values determined from laboratory tests on the backfill and the interface between the backfill and the basement wall, they calculated pressures that were, on average, about 15 percent larger than those measured.

FINITE ELEMENT ANALYSES - RETAINING WALLS FOUNDED ON ROCK

As mentioned previously, studies by the Corps of Engineers [(34) - (36)] have shown that walls at several older locks founded on rock foundations appear to be unstable as judged by current design procedures, in spite of the fact that the walls have performed well for many years. It was unclear whether these walls were actually in a precarious position or not, because it was considered possible that the stability of the walls depended on bond between the concrete and the rock foundation, a bond that might be susceptible to deterioration and brittle failure.

To develop an improved understanding of the interaction between gravity walls, their foundations, and their backfills, a program of finite element analyses was undertaken by the writers under the sponsorship of the Corps of Engineers. The primary challenge in these studies was to develop a means of allowing for the possibility of gradual loss of contact between the base of the wall and its foundation as the wall tips forward. The results showed that many aspects of conventional analysis procedures are correct, including the assumption that the base pressures vary approximately linearly across the foundation. However, the analyses also demonstrated that the backfill settles more than the wall, and develops downward shear loads on the wall, as shown by the field measurements described earlier.

One of the most important benefits of these studies has been to provide an explanation for the anomalous situation that walls judged unstable on the basis of current design methods are in fact performing well, without signs of instability. The difference between conventional analysis and actual performance is related to the significant shear loads that act on vertical planes through the heels of gravity walls, which are essentially unmoving and are acted upon by earth pressures close to at rest values. The stabilizing effect of these forces is neglected in most conventional equilibrium analyses of massive walls on rock, and the results of the analyses therefore do not reflect actual behavior.

Some examples are given in Fig. 7, for four walls on rock. In Fig. 7 the magnitude of the vertical shear force is expressed in terms of a vertical shear coefficient, K_v, which is related to the shear force on the vertical plane through the heel by the following equation:

$$T = 0.5 \; K_v \; \gamma \; H^2 \qquad\qquad (2)$$

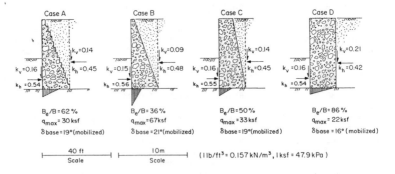

H = 40 ft, B = 16 ft for all cases, Foundation = Hard Rock (H = 12 m, B = 4.9 m for all cases)
Backfill and Toe Fill = Sand: γ = 135 lb/ft^3, ϕ = 39°, δ = 31° (maximum), k_a = 0.21, k_o = 0.51

Case A

k_v = 0.14
k_v = 0.16 k_h = 0.45
k_h = 0.54

B_e/B = 62 %
q_{max} = 30 ksf
δ_{base} = 19° (mobilized)

Case B

k_v = 0.09
k_v = 0.15 k_h = 0.48
k_h = 0.56

B_e/B = 36 %
q_{max} = 67 ksf
δ_{base} = 21° (mobilized)

Case C

k_v = 0.14
k_v = 0.16 k_h = 0.45
k_h = 0.55

B_e/B = 50 %
q_{max} = 33 ksf
δ_{base} = 19° (mobilized)

Case D

k_v = 0.21
k_v = 0.16 k_h = 0.42
k_h = 0.54

B_e/B = 86 %
q_{max} = 22 ksf
δ_{base} = 16° (mobilized)

|— 40 ft —| |— 10 m —| (1 lb/ft^3 = 0.157 kN/m^3, 1 ksf = 47.9 kPa)
   Scale   Scale

Fig. 7. Results of finite element analyses of four retaining walls on rock

where T = the shear force on the vertical plane through the heel of
the wall (force per unit length), K_v = vertical shear coefficient
(dimensionless), γ = unit weight of backfill, and H = wall height.

It may be seen that there are downward vertical shear forces on
both the fronts of the walls, due to the settlement of the toe fills,
and the backs of the walls, due to the settlement of the backfills.
The walls moved very little during placement of the toe fills and the
backfills; the values of Δ/H varied from 0.00003 for Case D to
0.00010 for Case B. As a result the earth pressures on the backs and
the fronts of the walls are close to at rest -- slightly lower than
at rest on the back and slightly higher than at rest on the front.
Even so, the settlement of the backfill relative to the wall as it is
placed behind the wall is sufficient to generate significant values
of shear force on the wall.

The values of the vertical shear coefficients on the backs of the
walls increase as the backs of the walls approach vertical. This
happens because the wedge of backfill between the back of the wall
and the vertical plane tends to settle and deform like the backfill.
As the wedge of backfill becomes narrower, there is a greater tendency for differential settlement between the wedge and the rest of
the backfill, and therefore larger shear forces develop on the
vertical plane through the heel of the wall.

Vertical shear also develops on the front of the wall during
placement of the toe fill. Because the settlement in this shallower
fill (18 ft, or 5.5 m deep) is less than the settlement in the backfill, the values of K_v for the toe fill are smaller. Parametric
studies have shown that the value of K_v depends on the stiffness of
the backfill, the depth of the backfill, the water level in the backfill, and the inclination of the back of the wall. Within the range
of parameters typical of sand and gravel backfills, and walls founded
on rock, values of K_v range from 0.09 to 0.21, as shown by the four
cases in Fig. 7.

SUMMARY AND CONCLUSIONS

Design Procedures and Stability Criteria. On the basis of laboratory studies, field studies, and analyses, it is clear that gravity
retaining walls should be considered in three categories for purposes
of design, as shown in Figs. 8 through 10. The first category is
those walls that are backfilled with clayey soils or that are founded
on clayey soils, as shown in Fig. 8. The second category is walls
that are backfilled with clean granular backfill and founded on sand
or gravel, as shown in Fig 9. The third category is walls that are
backfilled with clean granular backfill and founded on rock, as shown
in Fig. 10.

Walls with Clayey Soils in the Backfill or Foundation. Almost all
problems with retaining walls are associated with walls in this category, and the design of these walls should be undertaken following a
conservative approach. Because earth pressure theories do not
account for the effects of creep in clayey backfills, the earth pressures used for these backfills should be selected on the basis of
experience rather than theory. The design earth pressures developed
by Terzaghi and Peck (56) serve this purpose well. These have been
adopted in the NAVFAC Manual (18) and the Canadian Foundation Engineering Manual (6). Stability criteria for walls with clayey soils
in the backfill or foundation are shown in Fig. 8

Walls with Granular Backfills and Foundations of Sand or Gravel.
Few problems have been associated with walls in this category, and
none that are attributable to deficiencies in design methods. Sand
and gravel foundations are sufficiently compliant to allow the movement required to reduce earth pressures to values significantly
smaller than the at rest values. Active earth pressures calculated
by the Coulomb theory can be used in design, provided that the design
of the wall provides adequate safety against sliding and overturning.

While the value of K_a calculated by the Coulomb theory is insensitive to the value of δ used in the calculation, the value of δ is
nevertheless very important because it determines the magnitude of
the shear load on the back of the wall. As discussed in the previous
sections of this paper, shear loads of significant magnitude develop
whenever the backfill settles more than the wall (the normal situation), and these shear loads thus play an important part in enhancing
wall stability.

Two ways of analyzing wall stability are shown in Fig. 9. The
free-body considered can be the wall alone, as shown in Fig. 9(a), or
it can include the wedge of backfill between the back of the wall and
the vertical plane, as shown in Fig. 9(b). If the wall alone is
taken as the free-body, the value of δ used in design is determined
by the characteristics of the soil/wall interface. For concrete
walls, the value of δ is not likely to be less than 0.8 ϕ (39).

If the free-body includes the wedge of backfill between the wall
and the vertical plane through the heel, the shear load on the vertical plane can be estimated on the basis of the studies described
previously. For conditions where the backfill will settle relative
to the wall, it is unlikely that the magnitude of the vertical shear
load coefficient (K_v) will be less than 0.1. In cases where the cost

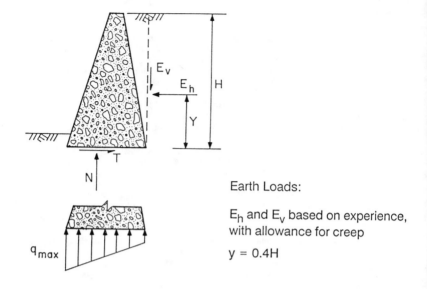

Earth Loads:

E_h and E_v based on experience, with allowance for creep

$y = 0.4H$

Stability Criteria:

(1)   N within middle third of base

(2) $q_{allowable} \geq q_{max}$

(3)   Safe against sliding

(4)   Settlement within tolerable limits

(5)   Safe against deep-seated foundation failure.

Fig. 8   Earth Loads and Stability Criteria for Walls with Clayey Soils in the Backfill or Foundation.

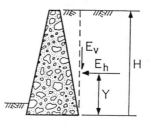

(a) Forces on Wall

(b) Forces on Vertical Plane
Through Heel of Wall

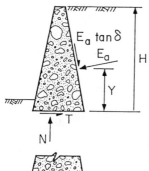

Earth Loads:

E_a or E_h calculated using Coulomb active earth pressure theory

δ or E_v estimated using judgment, with allowance for movement of backfill relative to wall.

$y = 0.4H$

Stability Criteria:

(1)   N within middle third of base

(2) $q_{allowable} \geq q_{max}$

(3)   Safe against sliding

(4)   Settlement within tolerable limits

Fig. 9   Earth Loads and Stability Criteria for Walls with Granular Backfills and Foundations of Sand or Gravel.

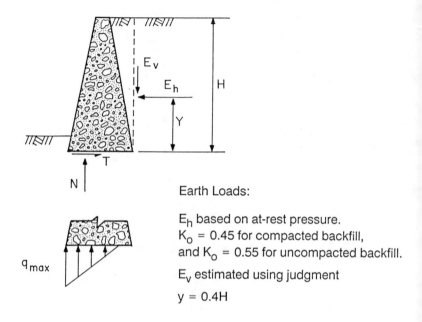

Earth Loads:

E_h based on at-rest pressure.
$K_o = 0.45$ for compacted backfill,
and $K_o = 0.55$ for uncompacted backfill.

E_v estimated using judgment

$y = 0.4H$

Stability Criteria:

(1)   N within middle half of base

(2) $q_{allowable} \geq q_{max}$ for foundation rock, and allowable compressive stress in concrete $\geq q_{max}$.

(3)   Safe against sliding

Fig. 10   Earth Loads and Stability Criteria for Walls with Granular Backfills and Foundations on Rock

is justified, finite element analyses can be used to make more
refined estimates of the shear load.

For low walls where laboratory tests are not used to measure the
properties of the backfill soil, the use of Coulomb theory is not
necessary. The design earth pressures developed by Terzaghi and Peck
provide a practical means of estimating the earth loads based on
knowledge of the type of soil to be used as backfill. For clean
granular fills these earth pressures are approximately the same as
would be calculated using the Coulomb earth pressure theory with reasonable estimates of unit weight and angle of internal friction for
the backfill.

The Terzaghi and Peck charts show the earth pressures on vertical
planes through the heels of the walls. For conditions of horizontal
backfill these charts show no shear load on the vertical plane. However, on the basis of the studies discussed previously, it appears
appropriate to include a shear load in the analyses even when the
backfill surface is level. On the basis of what is currently known
about the factors that control the magnitude of K_v, it seems likely
that K_v will not be smaller than 0.1, except in the unusual case
where the wall settles as much or more than the backfill.

Stability criteria for walls founded on sand or gravel foundations
are given in Fig. 9. Deep-seated foundation instability is not a
problem with walls on sand and gravel, with the exception of possible
liquefaction of loose deposits during earthquakes.

Walls with Granular Backfills and Foundations on Rock. Experience indicates that conventional design methods for walls in this
category are more conservative than necessary or appropriate. Conventionally these walls are designed for at-rest earth pressures,
with no shear on the vertical plane through the heel of the wall.
The studies discussed earlier indicate that, although these structures may deflect a little during backfilling, the amount is so small
that the earth pressures are reduced only slightly below at-rest
values. The same studies indicate, however, that settlement of the
backfill during placement results in development of shear forces on
the vertical plane through the heel of the wall, and that these
forces have a very beneficial effect on wall stability.

Based on these findings, it appears appropriate to use at-rest
horizontal pressures for design, and to include a vertical shear
force on the plane through the heel of the wall. As shown in Fig. 7,
the magnitudes of the shear loads depend on the orientation of the
back of the wall, whether the back of the wall is stepped or planer,
and, to some extent, on the depth of the fill. In any condition the
value of K_v is unlikely to be much smaller than 0.1, and it may be
considerably larger, as shown in Fig. 7. In cases where the cost is
justified, finite element analyses can be used to make more refined
estimates of the shear load.

Stability criteria for walls founded on rock are shown in Fig. 10.
These are similar to those for walls founded on sands and gravels,
except the resultant normal force on the base should fall in the
middle half of the base rather than the middle third. If the resultant falls at the extreme position from the heel of the wall, this
corresponds to the condition where contact between the base of the
wall and the underlying rock has been lost over one fourth of the

base width, and the effective base contact area (B_e) is equal to 75 percent of B.

Additional Design Requirements. In addition to the factors discussed in the previous paragraphs, design of gravity walls needs to address a number of other issues. These include provision of drainage, allowances for the effects of surcharge loads and increased earth pressures due to compaction, use of non-frost susceptible soils within the zone of frost penetration, and possible added earth pressures due to earthquake shaking. These issues are addressed in many textbooks and design manuals, and will not be considered in detail here.

Summary Remarks. The design of gravity earth retaining structures has not undergone very rapid changes in recent years, and has been considered by many geotechnical engineers to be an area where there was little need for additional studies or improvement in design procedures. It seems clear, however, that there are important soil-structure interaction effects even in gravity earth retaining structures, and that more complete understanding of these effects provides a basis for improved design procedures.

ACKNOWLEDGMENTS

The writers wish to express their appreciation to the many people who have contributed to the studies on which this paper is based. Levi Regalado performed the finite element analysis of the Fifteen-Mile Falls Dam wall, and Professor T. L. Brandon and Steven Winter of Virginia Tech contributed to the finite element research studies. Jeff Huffman assisted with the literature review. Reed Mosher, Don Dresler, Lucian Guthry and Carl Pace of the U. S. Army Corps of Engineers provided support and guidance for the research studies. Mary Duncan typed the text, and Beulah Prestrude drafted the figures. Grateful appreciation is expressed for all of this helpful assistance.

REFERENCES

(1) S Bang, Active Earth Pressure Behind Retaining Walls, J. Geot. Eng. Div. (ASCE), 111(3), 1986, 407-412.
(2) EW Brand & P Krasaesin, Investigation of an Embankment Failure in Soft Clay, Geot. Eng. J., SE Asian Soc. Soil Eng., 2(1), Bangkok, 1971, 53-66.
(3) H Brandl, Retaining Walls and Other Restraining Structures, Chapter 47 in Ground Engineer's Reference Book, Ed FG Bell, Butterworth's, Boston, 1987, 47/1-47/34.
(4) BB Broms & I Ingelson, Earth Pressure Against the Abutments of a Rigid Frame Bridge, Geotechnique, 21(1), 1971, 15-28.
(5) BB Broms & I Ingelson, Lateral Earth Pressure on a Bridge Abutment, Proc. 5th European Conf. on SMFE, (1), Madrid, 1972, 117-123.
(6) Canadian Geotechnical Society, Canadian Roundation Engineering Manual, 2nd Ed., 1985.

(7)   DR Carder, RT Murray & JV Krawczyk, Earth Pressures Against an
      Experimental Retaining Wall Backfilled with Silty Clay, Rpt 946,
      Transport and Road Research Laboratory, Crowthorne, Berkshire,
      1980.

(8)   DR Carder, RG Pocock & RT Murray, Experimental Retaining Wall
      Facility - Lateral Stress Measurements with Sand Backfill, Rpt
      766, Transport and Road Research Laboratory, Crowthorne,
      Berkshire, 1977.

(9)   DR Carder & JC Krawczyk, Performance of Cells Designed to
      Measure Soil Pressure on Earth Retaining Structures, Rpt 689,
      Transport and Road Research Laboratory, Crowthorne, Berkshire,
      1975.

(10) L Casagrande, Comments on Conventional Design of Earth Retaining
      Structures, J. SMFD(ASCE), 99(SM2), 1973, 181-198.

(11) CJF Clausen & S Johansen, Earth Pressures Measured Against a
      Section of a Basement Wall, Proc. 5th European Conf. on SMFE,
      (1), Madrid, 1972, 515-516.

(12) GW Clough & JM Duncan, Finite Element Analysis of Port Allen and
      Old River Locks, Report TE 69-3, University of California,
      Berkeley, 1969, 264 p.

(13) GW Clough & JM Duncan, Finite Element Analyses of Retaining Wall
      Behavior, J. SMFD (ASCE), 97(SM12), 1971, 1657-1674.

(14) GW Clough & JM Duncan, Temperature Effects on Behavior of Port
      Allen Lock, Proc. ASCE Specialty Conf. on the Performance of
      Earth and Earth-Supported Structures, Indiana, 1972, 1467-1479.

(15) HM Coyle, RE Bartoskewitz, LJ Milberger, & HD Butler, Field
      Measurements of Lateral Earth Pressures on a Cantilever
      Retaining Wall, TRR #517, 1974, 16-29.

(16) HM Coyle & RE Bartoskewitz, Earth Pressures on Precast Panel
      Retaining Wall, Journal, Geot. Eng. Div.(ASCE), 102(GT5), 1976,
      441-456.

(17) HM Coyle & RE Bartoskewitz, Field Measurements of Lateral Earth
      Pressures and Movements on Retaining Walls, Transp. Res. Rec.
      No. 640, 1, 39-49.

(18) Dept. of the Navy, Naval Facilities Engineering Command,
      Analysis of Walls and Retaining Structures, Chapter 3 in
      Foundations and Earth Structures, Design Manual 7.2, Virginia,
      1982, 59-127.

(19) JM Duncan & GW Clough, Finite Element Analyses of Port Allen
      Lock, J. SMFD(ASCE), 97(SM8), 1971, 1053-1068.

(20) RM Ebeling, GW Clough & JM Duncan, Methods of Evaluating the
      Stability and Safety of Gravity Earth Retaining Structures
      Founded on Rock, Report to U.S. Army Corps of Engineers,
      Waterways Experiment Station, (1), Vicksburg, 191 p.; (2),
      Vicksburg, 157 p.

(21) PD Evdokimov & DD Sapegin, Stability, Shear and Sliding
      Resistance, and Deformation of Rock Foundations, 1964,
      translated from the Russian, Israel Program for Scientific
      Translations, Jerusalem, 1967.

(22) M Fukuoka, Static and Dynamic Earth Pressures on Retaining
      Walls, Proc. Third Australia-New Zealand Conf. on Geomechanics,
      Wellington, (3), 1980, 337-346.

(23) JP Gould, Lateral Pressures on Rigid Permanent Structures, Proc. ASCE Specialty Conf. on Lateral Stresses in the Ground and Design of Earth-Retaining Structures, Cornell Univ., Ithaca, 1970, 219-269.

(24) VL Granger, Failure of a Reinforced Concrete Reservoir, Proc. 6th Int. Conf. SMFE (2), Montreal, 1965, 56-60.

(25) K Hilmer, Evaluation of a Ten-Year Measuring Program at the Eibach Lock, 1986

(26) TS Ingold, Retaining Wall Performance During Backfilling, J. Geot. Eng. Div. (ASCE), 105(5), 1979, 613-626.

(27) TS Ingold, A Retaining Wall Failure Induced by Compaction, Chapter in Failures in Earthworks, Ed T. Telford, Ltd., London, 1985.

(28) AM James, Low Friction and High Density of Fill Toppled Wall, Eng. News Record, July, 1965, 80-81.

(29) R Jarquio, Total Lateral Surcharge Pressure Due to Strip Load, J. Geot. Eng. Div. (ASCE), 107(10), 1981, 1424-1428.

(30) M Kany, Measurement of Earth Pressures on a Cylinder 30m in Diameter (Pump Storage Plant), Proc. Fifth European Conf. on Soil Mechanics, Madrid, 1972, 535-542.

(31) M Matsuo, S Kenmochi & H Yagi, Experimental Study on Earth Pressure of Retaining Wall by Field Tests, Soils and Foundations, 18(3), 1978, 27-41.

(32) N Mononobe, Earthquake-Proof Construction of Masonry Dams, Proc. of World Eng. Conf., (9), 1929, 275.

(33) S Okabe, General Theory of Earth Pressure, J. Japanese Soc. of Civil Eng., 12(1), 1924.

(34) CE Pace, Engineering Condition Survey and Structural Investigation of Emsworth Locks and Dam, Ohio River, U.S. Army Eng. Waterways Exper. Sta., Misc. Paper C-76-8, 1976, 67 pp.

(35) CE Pace, R Campbell & S Wong, Engineering Condition Survey and Evaluation of Troy Lock and Dam, Hudson River, U.S. Army Eng. Waterways Exper. Sta. New Misc. Paper C-78-6, 1981, 412 pp.

(36) CE Pace and JT Peatross, Engineering Condition Survey and Structural Investigation of Mongtomery Locks and Dam, Ohio River, U.S. Army Eng. Waterways Exper. Sta., Misc. Paper C-77-2, 1977, 179 pp.

(37) RB Peck, WE Hanson & TH Thornburn, Foundation Engineering, (2nd Ed), John Wiley & Sons, New York, 1974, 514 p.

(38) RB Peck, HO Ireland & CY Teng, A Study of Retaining Wall Failures, Proc. 2nd Int. Conf. SMFE, Rotterdam, 1948, 296-299.

(39) JG Potyondy, Skin Friction Between Various Soils and Construction Material, Geotechnique, (1), 1961, 339-353.

(40) DW Quigley & JM Duncan, Earth Pressures on Conduits and Retaining Walls, Rpt No. UCB/GT/78-06, University of California, Berkeley, 1978, 298 p.

(41) SE Rehnman & BB Broms, Lateral Pressures on Basement Wall. Results from Full-Scale Tests, Proc. Fifth European Conf. on Soil Mechanics (1), Madrid, 1972, 189-197.

(42) WH Roth, KL Lee & L. Crandall, Calculated and Measured Earth Pressure on a Deep Basement Wall, Proc. 3rd Int. Conf. on Numerical Methods in Geomechanics, Aachen, 1179-1191.

(43) E Sandegren, PO Sahlstrom & H Stille, Behavior of Anchored
Sheetpile Wall Exposed to Frost Action, Proc. Fifth European
Conf. on Soil Mechanics (1), Madrid, 1972, 285-291.

(44) MA Sherif, YS Fang & RI Sherif, KA and k0 Behind Rotating and
Non-Yielding Walls, J. Geot. Eng. Div. (ASCE), 110(1), 1984, 41-
56.

(45) MA Sherif, I Ishibashi & CD Lee, Earth Pressures Against Rigid
Retaining Walls, J. Geot. Eng. Div. (ASCE), 108(GT5), 1982, 679-
695.

(46) IF Symons & DS Wilson, Measurement of Earth Pressures in
Pulverized Fuel Ash Behind a Rigid Retaining Wall, Proc. Fifth
European Conf. on Soil Mechanics (1), Madrid, 1972, 569-574.

(47) Y Tcheng & J Iseux, Full Scale Passive Pressure Tests and
Stresses Induced on a Vertical Wall by a Rectangular Surcharge,
Proc. Fifth European Conf. on Soil Mechanics, Madrid, 1972, 207-
214.

(48) K Terzaghi, Large Retaining Wall Tests, I - Pressure of Dry
Sand, Eng. News Record, Feb. 1, 1934, 136-140.

(49) K Terzaghi, Large Retaining Wall Tests, II - Pressure of
Saturated Sand, Eng. News Record, Feb. 22, 1934, 259-262.

(50) K Terzaghi, Large Retaining Wall Tests, III - Action of Water
Pressures on Fine-Grained Soils, Eng. News Record, March 8,
1934, 316-318.

(51) K Terzaghi, Large Retaining Wall Tests, IV - Effect of Capillary
Forces in Partly Saturated Fill, Eng. News Record, March 29,
1934, 403-406.

(52) K Terzaghi, Large Retaining Wall Tests, V - Pressure of Glacial
Till, Eng. News Record, April 19, 1934, 503-508.

(53) K Terzaghi, Retaining-Wall Design for Fifteen-Mile Falls Dam,
Eng. News Record, May 17, 1934, 632-636.

(54) K Terzaghi, A Fundamental Fallacy in Earth Pressure
Computations, from Contributions to Soil Mechanics 1925-1940,
Boston Sociey of Civil Engineers, Boston, 1940, 71-88.

(55) K Terzaghi, General Wedge Theory of Earth Pressure, Trans. ASCE,
106, 1941, 68-97.

(56) K Terzaghi & RB Peck, Soil Mechanics in Engineering Practice
(2nd Ed), John Wiley & Sons, New York, 1967, 729 p.

(57) GP Tschebotarioff, Discussions, Proc. Bruss. Conf. 58 Earth
Press. Probl., 1958.

(58) GP Tschebotarioff, Foundations, Retaining and Earth Structures,
McGraw -Hill Book Company, New York, 1973.

(59) U. S. Army Corps of Engineers, Retaining Walls, Engineering and
Design, Manual EM 1110-2-2502, 1961, 13 p.

(60) J Vagneron & KL Lee, Design of Retaining Walls, Rpt to U. S.
Army Corps of Engineers, Waterways Experiment Station, 1975.

(61) PR Vaughan & MF Kennard, Earth Pressures at a Junction Between
an Embankment Dam and a Concrete Dam, Proc. Fifth European Conf.
on Soil Mechanics (1), Madrid, 1972, 215-221.

(62) WA Weiler, Jr. & FH Kulhawy, Behavior of Stress Cells in Soil,
Geot. Eng. Rpt. 78-2 to Niagara Mohawk Power Corp., Syracuse
Univ., Ithaca, New York, 1978, 289 p.

(63) WA Weiler, Jr. & FH Kulhawy, Factors Affecting Stress Cell
Measurements in Soil, Pres. ASCE Conf, Atlanta, Oct. 1979.

PERFORMANCE OF LARGE GRAVITY WALLS
AT EISENHOWER AND SNELL LOCKS

John G. Diviney,[1] M. ASCE

ABSTRACT:

This case history evaluates the performance of four large gravity retaining structures which were designed in the 1940s using then state-of-the-art practices. Design earth pressure assumptions, when compared with recently measured in-situ pressures, indicate that the wall design greatly underpredicted the earth pressures that would act on the walls. Because the earth pressures are greater than anticipated by the wall design, the gravity walls have exhibited extensive cracking which will require remediation.

INTRODUCTION

The Saint Lawrence Seaway project, a 190-mile (306 km) long system of canals, dams and locks, was designed in the 1940s and constructed in the early to mid-1950s. The project, a joint effort by the United States of America and Canadian governments, and one of the largest engineering undertakings of the 20th Century, had a construction cost of approximately $1.1 billion.

Included in the Saint Lawrence Seaway system are two locks near Massena, New York currently owned and operated by the United States Department of Transportation's Saint Lawrence Seaway Development Corporation (Corporation). These locks, the Dwight D. Eisenhower Lock and the Bertrand H. Snell Lock, are sister locks, each with chamber dimensions approximately 80 feet (24 m) wide by 800 feet (244 m) long. The lock walls at each lock consist of large, unreinforced concrete, gravity structures with base dimensions of approximately 65 feet (20 m) and heights of approximately 110 feet (34 m). All four lock walls are founded on massive dolomite bedrock and are backfilled with local glacial till which was excavated and stockpiled during lock construction. A typical cross section of the walls is shown on Fig. 1.

ORIGINAL DESIGN AND CONSTRUCTION OF LOCK WALLS

Subsurface conditions at both lock sites are similar, consisting of from 20 feet (6 m) to 110 feet (34 m) of glacial

---

1 - Vice President, Gannett Fleming, Inc., P.O. Box 1963, Harrisburg, PA 17105.

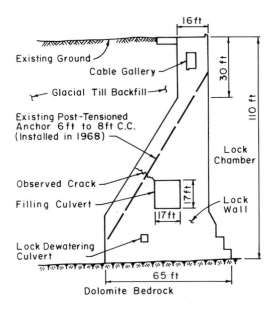

Existing Ground

Cable Gallery

Glacial Till Backfill

Existing Post-Tensioned
Anchor 6ft to 8ft C.C.
(Installed in 1968)

Observed Crack

Filling Culvert

Lock Dewatering
Culvert

16ft

30 ft

110 ft

Lock
Chamber

Lock
Wall

17ft

17ft

65 ft

Dolomite Bedrock

FIG. I. Typical Lock Wall Cross Section

till overlying bedrock. The till varies somewhat between
Eisenhower and Snell Locks with the till at Eisenhower Lock
consisting of brown and gray, fine to coarse gravel and fine to
coarse sand, with a trace of silt. The till at Snell Lock is more
fine grained consisting of brown and gray, fine to medium sand,
some silt, and some fine to coarse gravel. Gradation curves are
presented in Fig. 2.

The bedrock at both lock sites is Ordovician in age and belongs
to the upper part of the Beekmantown Formation. The rock consists
of massive dolomite for the most part but contains some interbeds
of shale and dolomitic shale.

The design of the locks, which required that the lock walls be
founded on the dolomite bedrock, necessitated large excavations to
expose the foundation bedrock. The top width of the excavations
at the locks ranged from 750 feet (229 m) to 1500 feet (457 m),
with 2H:1V slopes excavated in the glacial till. The excavated
glacial till was stockpiled at each site until wall construction
was completed and was then used as structural backfill. Because
of the large width of excavation the backfill was placed by
off-the-road dump trucks and dozers and was compacted with heavy,
self-propelled and dozer-drawn compactors to achieve a specified
minimum 95% standard proctor density. [Ref. 3,4]

According to United States Army Corps of Engineers construction
control reports dated 1956, in place wet unit weights of compacted
till backfill averaged approximately 145 pcf (2323 kg/m^3) at both

locks. [Ref. 2]   The design of the lock walls in 1942 had assumed
γw = 130 pcf (2082 kg/m³) for compacted backfill. [Ref. 1]
     The design of the lock walls in 1942 was performed by the
United States Army, prior to implementation of the current United
States Army Corps of Engineers' standard design practices for
gravity lock walls. The design assumed equivalent fluid pressure
values of 33 psf (1.6 kPa) and 93 psf (4.5 kPa) per foot of depth,
respectively, for the horizontal earth pressures of wet till
backfill and saturated (below water table) till backfill. [Ref. 1]
These equivalent fluid pressure values would correspond to a
horizontal earth pressure coefficient, K, of approximately 0.25,
although a coefficient was not explicitly used.

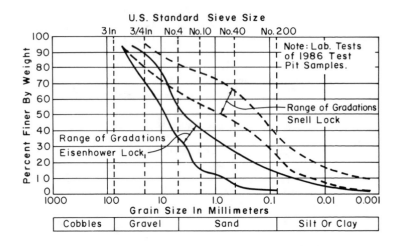

FIG. 2. Gradation Curves Of Glacial Till Used For Lock Wall Backfill

CONDITION OF LOCK WALLS

     Shortly after completion of the locks in 1958, extensive
cracking was observed in all four lock walls. Cracks, accompanied
by varying degrees of seepage, were noted in the upper, outside
corners of the 17' x 17' filling culverts. (Fig. 1) It was
presumed that these cracks had propagated between the filling
culverts and the backfaces of the walls, a distance of 12 feet,
and that the seepage was related to groundwater inflow. The
observation of almost 1600 lineal feet (489 m) of cracks at each
lock raised engineering concerns regarding the structural
stability of the lock walls.
     An internal system of post-tensioned anchors was installed in
the walls in 1968 to arrest observed crack enlargement and

propagation. Anchors, each with a 600 kip capacity and spaced at 6- to 8- foot intervals, were installed in all four lock walls at the location shown in Fig. 1. These anchors were designed and installed as an emergency measure, without any effort to measure either the state of stress in the wall backfill or horizontal earth pressures against the walls. [Ref. 4]

IN-SITU INVESTIGATIONS

Even with the installation of the internal anchors, there was concern that both the internal and external stability of the lock walls may be marginal due to horizontal earth pressures that exceed the pressures assumed during the wall design. In 1986 an in-situ testing program was undertaken by the Corporation at both locks to collect data to be used for detailed analyses of the lock walls. The in-situ testing program consisted of: (1) pressure-meter testing (PMT) of wall backfill to determine the state of horizontal stresses; (2) hydrofracture testing (HF) to verify the validity of pressuremeter data; (3) piezometer installations and measurements to determine the hydrostatic pressures acting on the lock walls; and (4) in-situ density testing to determine the unit weight of the wall backfill.

The PMT program was developed to sample in-situ horizontal stresses in the lock wall backfill to allow the determination of the horizontal earth pressures acting on the walls. Special equipment and field testing techniques were required for the PMT work due to the nature of the backfill material. Specific issues of concern that were addressed by the PMT program included: (1) protection of the rubber membrane (bladder) to prevent rupture by sharp edges on cobbles (resulting from drilling operations); (2) a wide pressure and expansion range in the probe was required due to large variations in till stiffness; (3) a short probe length was desirable since a high percentage of cobbles might allow only a short test length in cobble-free soils; (4) a probe capable of evaluating anisotropic stresses was required since compaction may have created an anisotropic stress condition. [Ref. 7]

These concerns were addressed by careful control of the preparation of the borehole test zone and by selection of a prototype pressuremeter with special capabilities. The pressuremeter used for the project was a Trimod Model III manufactured by Roc Test, Ltd. of Montreal, Canada. The probe, shown in Fig. 3, was designed to fit an NX-size borehole. The probe, approximately 36 inches (914 mm) in length, contains an 18-inch (457 mm) long single pressure cell, resulting in a length to diameter ratio of approximately 6. [Ref. 7]

A Chinese lantern-style stainless steel sheath was provided to protect the rubber membrane from rupture where sharp rock edges were encountered.

Membrane expansion was accomplished pneumatically by bottled nitrogen routed through a regulator. The regulator allowed operation of the probe at two pressure ranges, i.e., 1 kPa to

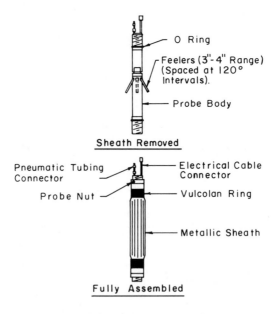

FIG. 3. Tri-Mod Model III Pressuremeter Probe (After Goldberg-Zoino Assoc., Inc., 1986)

2500 kPa, and 2 kPa to 10,000 kPa, to accommodate wide ranges in till stiffness. The applied pressure was monitored by one of three pressure gages: low pressures were monitored in 5 kPa increments on a 0 to 500 kPa gage, intermediate pressures were monitored in 25 kPa increments on a 0 to 2500 kPa gage, and high pressures were monitored in 100 kPa increments on a 0 to 10,000 kPa gage. The use of the three gages allowed precision in readings over a wide range of applied pressures. [Ref. 7]

To detect possible horizontal stress anisotrophy, the probe was designed with three independent strain gage transducers ("feelers"), spaced at 120° intervals around the circumference of the probe, to measure the cavity diameter. These "feelers" allowed precise readings of cavity diameter in 0.01 mm increments, but could also accommodate up to a 100 mm cavity diameter. An additional benefit of the probe design was that only borehole expansion in the vicinity of the "feelers" was necessary for accurate measurements. [Ref. 7]

Because the state of the horizontal stresses behind the lock walls was of primary concern a PMT testing procedure was developed which emphasized test increments at lower stress levels. This procedure provided data points and definition in the area of the PMT plots used to evaluate horizontal stress. [Ref. 7]

A total of 48 PMT tests were conducted behind the lock walls in eight boreholes, i.e. four boreholes at Eisenhower Lock and four at Snell Lock. Each borehole was offset 45 feet (14 m) from the backface of the upper portion of the lock wall. Six tests were performed in each borehole at approximately 10-foot (3 m) intervals to a depth of approximately 60 feet (18 m). The PMT testing was terminated at approximately 60 feet (18 m) to avoid the sloping portion of the lock wall at 70 feet (21 m).

An HF in-situ investigation method was employed, as a second in-situ technique, to verify the data obtained during PMT testing. HF is a test method involving the injection of fluid under pressure into a test zone until the soils in the test zone fracture, i.e., fail in tension. Because the HF method is a different technique from the PMT method, it served as an independent check of horizontal stress data. While not as sophisticated and accurate as the PMT testing, the HF method did allow for the approximate verification of PMT test data.

An HF set-up, consisting of a borehole packer, a volume measuring device, and a surcharge pressure application device, was assembled. Using this equipment, a thick drilling mud fluid was injected into the test zone until the soils failed in tension (fractured). The objective of the test was to develop the following relationships: (1) pressure versus cumulative borehole volume during the loading phase and (2) pressure versus time during the unloading phase. [Ref. 7]

A total of 20 HF tests were conducted behind the lock walls in seven of the boreholes that were used for the PMT testing. Three boreholes were tested at Eisenhower Lock, and four boreholes were tested at Snell Lock. Generally, three tests were performed in each borehole at 15 feet (5 m), 35 feet (11 m), and 55 feet (17 m), to distribute the HF tests equally among the pressuremeter test locations. [Ref. 7]

In addition to the PMT and HF testing, 28 vibrating wire piezometers were placed at selected depths, in the boreholes, to monitor in-situ piezometric head. The piezometric head data obtained from these piezometers was used to quantify: (1) the hydrostatic pressure on the backface of the lock walls, and (2) the hydrostatic uplift pressure beneath the lock walls. Also, sand cone density tests were performed in test pits excavated behind the lock walls to determine the in-situ density of the wall backfill.

RESULTS OF INVESTIGATIONS

PMT summary plots from all test locations were evaluated to determine the total in-situ horizontal pressure, Po. Tables 1 and 2 present summaries of interpreted pressuremeter results for each of the three measurement axes for Eisenhower and Snell Locks, respectively. As can be seen from these tables, no significant differences were found among the three axes. For this reason the average of the three values was taken as the best estimate of the horizontal stresses acting on the lock walls. [Ref. 9]

Tables 3 and 4 present summaries of interpreted HF results for Eisenhower and Snell Locks, respectively. A comparison with Tables 1 and 2 indicates that the HF test results substantiate the PMT results.

TABLE 1 - Summary of Interpreted Results from PMT Testing at Eisenhower Lock. (after Schmertmann, 1986)

Total In Situ Horizontal Pressure (kPa)

Boring	Depth (ft)	Axis U1*	U2	U3	Used	Quality** of Test
EN3	9.25	130	150	180	150	VG
	20	150	160	160	160	G
	30.5	180	150	140	160	VG
	40	220	230	200	210	VG
	51.6	190	150	150	170	VG
	62.4	270	310	260	280	VG
EN10	9	210	180	205	200	F
	19.5	350	350	270	320	F
	29.5	310	270	315	300	F
	39.5	310	310	300	310	G
	52	265	265	260	260	VG
	60	310	320	320	320	G
ES2	10.5	192	130	192	170	VG
	19.5	160	140	155	160	VG
	29	---	---	---	---	P
ES2	40.5	230	245	250	240	G
ES2	51.5	430	420	370	410	G
	61.5	200	200	190	200	VG
ES7	9.5	---	125	120	120	F
	22.5	---	---	---	---	P
	29.5	250	170	195	210	F
	39.5	240	240	240	240	VG
	51.5	270	300	270	280	VG
	60	260	270	220	250	G
ave. of 21 =	37.5	244	242	237	244	

Note: U1, U2, U3 spaced at 120° intervals
*Perpendicular to adjacent lock wall
**Qualitative evaluation from shape of PMT curve:
VG = very good G = good F = fair P = poor

TABLE 2 - Summary of Interpreted Results from PMT Testing at Snell
Lock. (after Schmertmann, 1986)

Total In Situ Horizontal Pressure (kPa)

Boring	Depth (ft)	Axis U1*	U2	U3	Used	Quality** of Test
SN1	10	100	100	100	100	G
	20	---	---	---	---	P
	30.5	220	200	220	215	VG
	40	140	160	210	170	G
	49.3	---	---	---	---	P
SN7	9.3	83	120	70	90	F
	20	220	150	160	180	G
	30	---	---	---	---	P
	43.6	230	190	240	220	VG
	50	303	270	270	280	VG
	60.5	280	260	300	280	VG
SS1	10	108	108	108	110	G
	20	183	173	153	175	G
	30.5	163	168	168	170	VG
	39.7	269	221	264	250	VG
	50	269	226	221	240	VG
	50	298	303	250	290	VG
SS7	10	140	230	280	210	F
	20	240	170	130	190	G
	30	220	230	225	225	VG
	40	145	170	205	180	VG
	50	210	230	240	225	G
	60	---	---	---	---	P
ave. of 19 =	34.4	201	194	201	200	

Note: U1, U2, U3 spaced at 120° intervals
*Perpendicular to adjacent lock wall
**Qualitative evaluation from shape of PMT curve:
VG = very good G = good F = fair P = poor

TABLE 3 - Summary of Interpreted Results from HF Testing at
Eisenhower Lock. (after Schmertmann, 1986)

			Total In Situ Pressure (kPa)		
Test No.	Boring	Depth (ft)	from Increase (Method 1)	from Decay (Method 2)	Used
1	ES7	38.1	258	242	250
2	EN10	55.3	no interpretation; questionable data		
3	ES7	62.8	299-375	306-378	310-380
4	EN3	12.5	118	119	120
5	EN3	33.5	155	150	150
Average of 4 =		36.7	217	213	216

TABLE 4 - Summary of Interpreted Results from HF Testing at Snell
Lock. (after Schmertmann, 1986)

			Total In Situ Pressure (kPa)		
Test No.	Boring	Depth (ft)	from increase (Method 1)	from decay (Method 2)	Used
1	SN1	55.0	343	355	350
2	SN1	13.0	138	127	130
3	SN1	33.0	207	207	210
4	SN1		packer blew		
5	SN1	63.5	351	367	360
6	SN7		packer blew		
7	SN7	23.0	162	137	140
8	SN7	53.3	271	269	270
9	SN7	63.5	395	389	390
10	SN7	12.8	108	114	110
11	SS7	33.1	---	141	140
12	SS7	53.8	251	222	230
13	SS1	13.0	135	---	130
14	SS1	33.3	184	158	160
15	SS1	53.0	331	343	340
Average of 12 =		392	240	236	236

Figs. 4 and 5 present plotted summaries of the interpreted total horizontal stress results from both PMT and HF tests for Eisenhower and Snell Locks, respectively. Included on these figures are plots of water pressure from piezometer data and plots of total vertical overburden stress based upon in-situ density testing.

By combining data from Eisenhower and Snell Locks, the average total vertical and horizontal stress and the average water pressure shown in Figs. 4 and 5 were used to determine the distribution of the effective horizontal earth pressure coefficient, Ko. This Ko profile is plotted in Fig. 6. The profile exhibits high Ko values near the ground surface which diminish with depth, reaching an asymptote of approximately Ko = 0.70. For comparison, the "equivalent" horizontal earth pressure coefficient consistent with the initial design of the lock walls, in 1942, is also shown in Fig. 6.

VALIDITY OF RESULTS

The horizontal earth pressure coefficients determined by in-situ testing are higher than might have been anticipated, especially when compared with K = 0.25 which corresponds to the initial design of the lock walls. However, based upon in-situ K

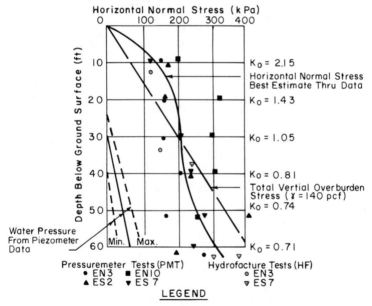

FIG. 4. Total Horizontal Stress From PMT And HF Testing At Eisenhower Lock
(After Schmertmann, 1986)

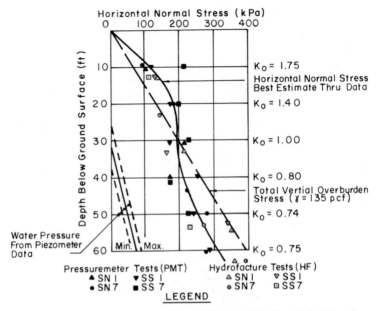

FIG. 5. Total Horizontal Stress From PMT And HF Testing At Snell Lock
(After Schmertmann, 1986)

measurements recently compiled by Schmertmann [Ref. 8] and observations made by D'Appolonia [Ref. 5] of stresses induced in sands by compaction, high earth pressure coefficients are not uncommon in compacted, coarse-grained backfill material where wall movement is restricted.

The validity of the in-situ test results has been substantiated by other observations; these include:

- K values calculated from in-situ test data decrease with depth as should be expected.
- Pressures measured by PMT testing were substantiated by HF test results.
- Results from testing at four lock walls are similar.
- K values were determined from calculations which incorporated a relatively large number of PMT tests and HF tests.
- Measurements recorded since the commissioning of the locks indicate that no wall movement has occurred.

Finite element analyses of the lock walls were performed to determine what effect the higher-than-anticipated horizontal earth pressures might have on internal wall stability. These analyses, which incorporated the horizontal earth pressures obtained from the in-situ testing, predicted tensile forces at various locations

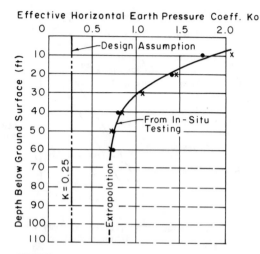

FIG. 6. Ko vs. Depth For Eisenhower And Snell Locks
(After Schmertmann, 1986)

within the gravity walls in excess of the tensile strength of the concrete. As is shown on Fig. 7, one of the tension zones predicted by the finite element analyses corresponds to the location of observed wall cracking and leakage. Other tension zones were predicted by analyses but could not be verified by field observation due to inaccessibility of the backface of the backfilled walls.

SUMMARY AND CONCLUSIONS

    In-situ pressuremeter and hydrofracture test measurements obtained in the backfill material behind the Eisenhower and Snell lock walls indicate high horizontal earth pressures. These pressures equate to horizontal earth pressure coefficients which range from $K_o = 0.70$ to $K_o > 2.0$. The high earth pressures are likely a result of: (1) compaction induced stresses which were introduced during placement of backfill material, and (2) lack of movement of the massive lock walls.
    The measured range of horizontal earth pressure coefficients is from 2.8 to more than 8 times greater than the earth pressure coefficient, $K = 0.25$, corresponding to the original lock wall design. The higher-than-anticipated earth pressures have resulted in extensive cracking and leaking in the gravity walls which will require remediation.

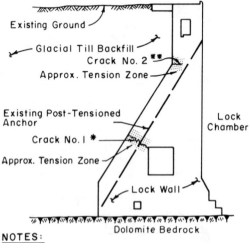

NOTES:
* Crack Location Predicted By Stress Analysis
  And Verified In The Field.
** Crack Location Predicted By Stress Analysis
   But Unverified.

FIG. 7. Typical Lock Wall Cross Section

ACKNOWLEDGMENT

    The studies described herein were sponsored by the United States
Department of Transportation, Saint Lawrence Seaway Development
Corporation, under Contract DTSL55-86-C-C0368. The writer
gratefully acknowledges the assistance of Corporation engineers
Steve Hung and Tom Lavigne during the studies.
    The studies were undertaken by an engineering team consisting
of: Gannett Fleming, Inc.; Clarkson University; Goldberg-Zoino &
Associates, Inc.; and Schmertmann & Crapps, Inc. Much of the
content of this paper was derived from reports prepared by this
team. The writer gratefully acknowledges the contributions from
all team members.

REFERENCES

1. Corps of Engineers, US Army, Robinson Bay Lock/Analysis of Design, Final Report, Massena, New York, 1942.
2. Corps of Engineers, US Army, Unit Weight of Compacted Till, St. Lawrence Seaway, Memorandum, January 6, 1956.
3. Corps of Engineers, US Army, St. Lawrence Seaway International Rapids Section/Foundation Report, Dwight D. Eisenhower Lock, Buffalo, New York, October 1958.
4. Corps of Engineers, US Army, St. Lawrence Seaway International Rapids Section Works Solely for Navigation and Thousand Islands Section/Design Memorandum No. 4, Part A/Appendix C - Structural Design Data/I-Design Criteria for Lock Walls and Sills, Revised May 27, 1985.
5. DJ D'Appolonia, RV Whitman, & E D'Appolonia, Sand Compaction with Vibratory Rollers, ASCE Journal of the Soil Mechanics & Foundation Division, Vol. 95, No. SM1, January 1969.
6. Gannett Fleming Geotechnical Engineers, Inc., Data Collection and Evaluation for Engineering Study of Lock Stability, Eisenhower and Snell Locks, Massena, New York, Harrisburg, Pennsylvania, November 1986.
7. Goldberg-Zoino & Associates, Inc., Eisenhower/Snell Locks, Contract DTSL55-86-C-C0368, In-Situ Testing - Final Report, Newton Upper Falls, Massachusetts, October 1986.
8. JH Schmertmann, Measure and Use of In-Situ Lateral Stress, The Practice of Foundation Engineering, Northwestern University, 1985.
9. JH Schmertmann, Final Report: Horizontal Pressures on Eisenhower & Snell Lock Walls, Schmertmann & Crapps, Inc., Gainesville, Florida, October 10, 1986.

THE BEHAVIOR OF BRIDGE ABUTMENTS ON CLAY

Malcolm D. Bolton[1], Sarah M. Springman[2] and H. Wing Sun[3]

ABSTRACT: Aspects of the behavior of bridge abutments on clay have
been simulated in small-scale centrifuge models. Data of two
contrasting tests are presented. One shows the behavior of a full-
height abutment with a spread base over firm to stiff clay, during
backfilling, deck loading, and subsequent clay consolidation. The
other investigates the lateral loading effect of embankment
surcharge on pre-driven piles, such as might be used to support an
abutment on softer clay. A simplified deformation mechanism captures
approximately the observed patterns of undrained lateral movements.
This can form the basis of a design method which accounts properly
for soil-structure interaction.

INTRODUCTION

A recent statistical study for FHWA (3) of bridge abutments in
North America concluded that lateral movements at the bridge deck
connection were much more significant than vertical movements in
determining whether the structure remained serviceable. Limits of
100 mm vertical movement and 50 mm horizontal were thought to be
applicable to a wide range of bridge deck and abutment types.
Unfortunately, there is a lack of fully instrumented tests from
which calculation methods for lateral movements could have been
derived. Furthermore, the most typical scenario leading to
unserviceability involved both horizontal and vertical movements of
bridge deck supports. Clay soils were most frequently cited as the
seat of damaging movements, but the reports of damage involved
shallow and deep foundations in equal numbers. Apparently, the
likelihood of lateral movement for piled bridge abutments over
weaker soils is similar to that for abutments on spread footings
over stiffer soils.
The source of these movements, and the means for their
prediction, is explored below with respect to full-height abutments
of the type shown in figure 1, (a) with a spread base, (b) on piles.

1 - Lecturer, 2 - Research Fellow, 3 - Research Student,
Cambridge University Engineering Department, Trumpington Street,
Cambridge, CB2 1PZ, England.

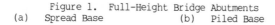

Figure 1.   Full-Height Bridge Abutments
(a)   Spread Base   (b)   Piled Base

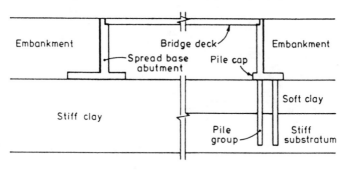

CENTRIFUGE TEST METHOD

The principles of centrifuge testing are now well known (6). When the scale of a model is reduced by a factor n and its "self-weight" increased by the same factor in a centrifuge, the distribution of vertical stresses in an equivalent field scale prototype, made of identical materials, is correctly replicated. If, in addition, the model boundaries are sufficiently remote from the focus of interest, or themselves simulate field boundaries, then the complete response to some desired stimulus can be determined from the behavior of the model. Applicable scaling factors are listed in table 1.

Table 1.   Scaling Factors Multiplying Centrifuge Model Parameters

Length		n		
Stress, pressure, strength		1		
Strain, rotation		1		
Displacement		n		
Consolidation time		n^2		
Bending stiffness	per pile n^4		per m	n^3
Bending moment	per pile n^3		per m	n^2

Two sets of criteria must additionally be satisfied before the modelling can be accepted as valid. First, deviations from the ideally uniform body force field must be reduced to acceptable levels. In the tests to be described the variation in "gravity" at different centrifuge radii causes stress deviations of less than 3%, and the inclination of the body force due to curvature is less than 5% in the zones of soil which interact with the structures. These favourably small errors are due to the relatively large working radius (4 m) of the Cambridge Geotechnical Centrifuge: figure 2.

Figure 2. Outline of a Model on the Cambridge Beam Centrifuge

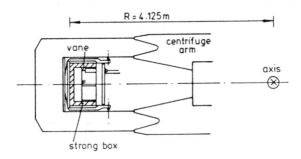

Second, the stress history of the soil in the model should replicate that in the desired prototype. Our approach to the ideal was to aim for realistic soil densities and strengths, but to accept that geological processes and construction methods could not be modelled directly. The equipment consisted of a liner (200 x 675 mm in plan) in which the pre-existing soil strata (clay over sand, for example) could be prepared, a consolidometer which could receive the liner and subject these soils to any desired overconsolidation cycle, and a strong box with a stiff, lubricated Perspex window, which could hold the liner in a centrifuge test up to 100g. The soil could then be brought into equilibrium with the desired water table set by an overflowing standpipe, permitting swelling or re-consolidation to take place. Finally, it was possible to conduct a model site investigation, in flight, to demonstrate that the required soil strength profile had been achieved. Figure 3 shows a typical vane test and undrained strength profile for 200 mm of kaolin clay in test HWS7 (referred to later), representing 20 m depth at 100g.

Figure 3. Vane Strength Profiles at 100g

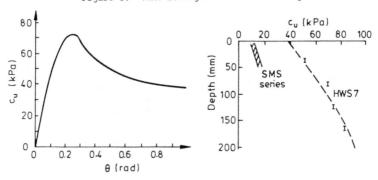

ABUTMENT ON SPREAD FOUNDATION

Test HWS7 is one of a sequence aimed at investigating the soil-structure interaction of bridge abutments founded directly over firm to stiff clay. Kaolin slurry was consolidated, as described earlier, over a period of about 1 month. A staged load-unload cycle was used to create a firm clay block into which pore pressure transducers could be inserted and backfilled. The block was then taken to its maximum consolidation pressure of 660 kPa before swelling in stages to 66 kPa. Excess water was then removed, the consolidometer opened, and the soil was trimmed to shape inside the liner which was then slid into the strongbox. The back surface of the clay was forced against the greased box, while the front surface was marked with a matrix of black plastic bullets within a painted grid, which were used to measure subsoil displacements from photographs taken through the Perspex window in flight.

The aluminium alloy wall was intended to model at 1/100 scale a plane strain section of abutment wall retaining 8 m of granular backfill. The 5.5 mm thickness modelled the bending rigidity of a 1 m thick reinforced concrete prototype. Figure 4 details the locations of wall instrumentation: in (a) the 13 fully active strain gauge bridges used as bending moment transducers (bmts) on both stem and base, and in (b) the 7 linear variable differential transformers (lvdts) used to indicate wall movement, together with 1 to show any soil heave in front of the base. These measurements are projected in terms of the vertical and horizontal displacement, and rotation, of a reference point (Figure 4 b) taken at the stem-base junction.

A deck-load simulator was designed to investigate the effects of the installation of a bridge deck resting on the top of the abutment, and the subsequent passage of heavy vehicles. Two Rolafram jacks bore on vertical, strain-gauged, brass rods which were fitted at both ends with spherical bearings to minimise the inclination and eccentricity of vertical thrust transmitted down to the wall crest.

Figure 5 shows a view of the model wall in place, relative to the sand hopper which was used to deposit dense granular fill behind the wall in flight. This construction was performed once the clay had come fully into equilibrium at 100 g with its imposed piezometric level, and following the taking of an in-flight vane probe such as that shown in figure 3. A medium and uniformly sized, sub-rounded, dry silica sand (Leighton Buzzard 30/52) was poured in 11 layers to form an 80 mm high embankment behind the wall. Each pour lasted 0.3 s and the interval between layers was 20 s, so the whole construction period of 200 s corresponded to 23 days at full scale. The high relative velocity of the sand grains as they struck the accumulating fill created a relative compaction of 97% (modified Proctor), with a dry unit weight of 16.5 kN/m^3 taken from mass and profile measurements after the test. Overall height was controlled within ±5%, and the depth of soil above the top surface of the wall base in HWS7 was 75 mm ± 4 mm.

Figure 4. Instrumentation Around the Wall

a) Bending Moment Transducers

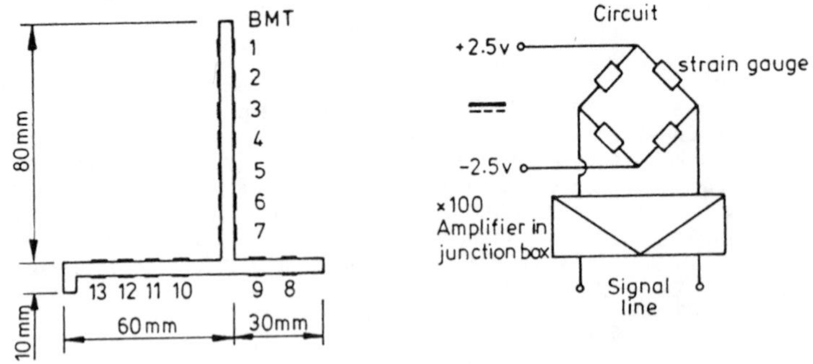

b) Displacement Transducers

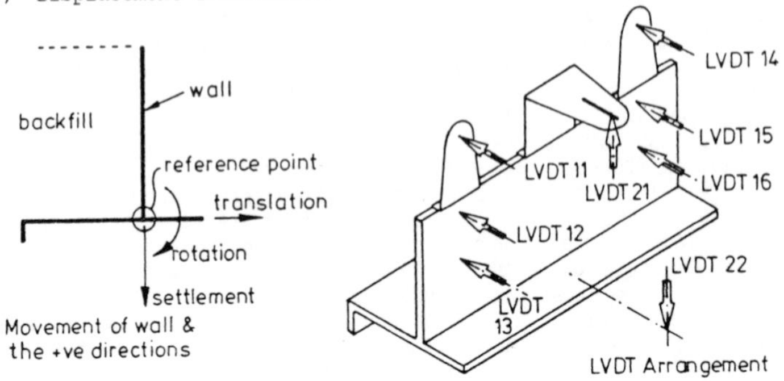

Figure 5. General Arrangement of Model HWS7

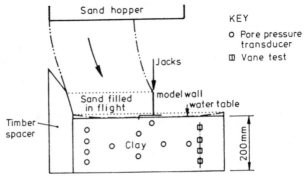

Embankment construction caused an immediate heave, forward translation and backward rotation of the reference point on the stem-base junction, corresponding to an instantaneous rotation about a high center, see figure 6a. For the next 36 minutes (250 days, prototype) consolidation of the underlying clay magnified the lateral displacement mode and permitted general settlement: see figure 6b. The time histories of base translation, settlement and rotation, and bending moments near the junction, 10 mm up the wall stem, and 8 mm along the toe, are shown in figure 7. The initial period of embankment building led to a tendency for outward movement of the wall base, permitting pressures in the accumulating fill to remain close to fully active. Subsequent consolidation with the fill in place led to a tendency for inward rotation which was strongly resisted by the fill: earth pressure increments high on the stem caused bending moments to rise, but the stiffness of the fill kept the point of rotation high, so the base continued to move outwards.

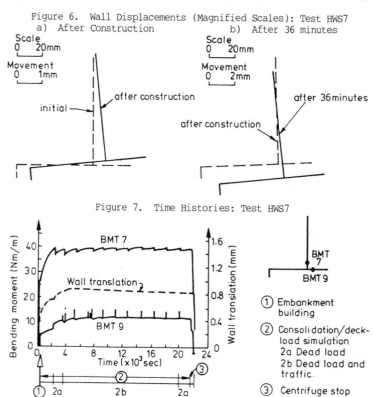

Figure 6. Wall Displacements (Magnified Scales): Test HWS7
a) After Construction   b) After 36 minutes

Figure 7. Time Histories: Test HWS7

Figure 7 shows that following construction and the first 36 minutes of consolidation, certain bridge loading effects were simulated. A vertical load of 240 N (120 kN/m, prototype) was first applied to represent the dead weight of the deck. This was left to consolidate for a further 24 minutes (168 days, prototype). A second increment of load, equal to the first, was then successively applied and removed in 15 cycles with a duration of about 10 s (1 day prototype) to simulate traffic effects. As the vertical load exceeded its past maximum, the rates of backward rotation, forward translation and settlement all increased, the net effect of which was apparently to permit lateral earth pressures to fall slightly. Subsequent unloading-reloading cycles had little effect on the wall stem, though the bending moments in the base continued to reflect the cyclic changes of thrust in the stem.

Through all these stages it became clear that the changes of lateral earth pressure due to consolidation in the underlying clay could be at least as significant as those induced on backfilling and by deck loading. In terms of prototype displacements, however, the ultimate settlement of 93 mm, coupled with a base translation of 93 mm, and sufficient backward rotation to exactly eliminate horizontal movement at the elevation of the deck, fall within FHWA limits (3). If earth pressures can be safely predicted, spread bases could be acceptable for abutments on firm to stiff clays.

LATERAL LOADING OF PILES DUE TO EMBANKMENT CONSTRUCTION

If unacceptable lateral displacements have been observed in piled abutments of the form shown in figure 1b, it presumably follows that greater attention should have been paid to lateral loading of the piles, including the effects of softer superficial clays being squeezed against piles under the influence of the embankment surcharge. This mechanism has been investigated (7) in a series of centrifuge tests conducted at 100g on 1/100 scale models of idealised prototypes consisting of one or two rows of piles. The particular test described here is SMS7 in which a single row of 5 free-headed piles, diameter d = 12.7 mm (1.27 m, prototype), spacing s = 40 mm (4 m, prototype) s/d = 3.15, which have been driven through a 60 mm deep soft clay layer into a 100 mm deep sand layer (6 and 10 m prototype, respectively): figure 8. To simplify the embankment loading condition, surcharge was applied adjacent to the piles using a tailored latex rubber bag in contact with the clay, constrained on its other five sides, and capable of being pressurised to a measured air pressure.

The piles were made from 12.7 mm diameter 18SWG aluminium alloy tubing and were 300 mm long, fully penetrating the model soils, into which they were driven with the aid of a conical shoe. The bending stiffness of the pile was comparable to a solid reinforced concrete pile of the same diameter. Each pile was externally strain gauged with 8 half-bridge bending moment transducers, which were protected by an acrylic moisture barrier and two layers of shrinkfit plastic tubing. Signals were balanced and amplified on the model package.

Figure 8. General Arrangement of Model SMS7

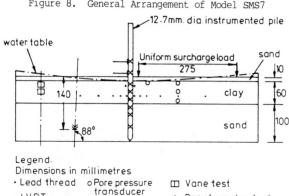

Legend.
Dimensions in millimetres
· Lead thread   o Pore pressure   ⊞ Vane test
⊢ LVDT   transducer   ⅋ Penetrometer test
                  x Strain gauge

The centrifuge package was identical to that described earlier. The same soils were also used, but with a different consistency. Preparation in the liner began with the dry pouring of a 100 mm layer of medium loose Leighton Buzzard sand, followed by its slow saturation by upward flow of water. Weighings before and after this process indicated an average relative density of about 60%. The kaolin slurry was then placed in the liner which had been inserted inside the consolidometer. After an initial, staged, consolidation cycle to facilitate the insertion of pore pressure transducers, the consolidation pressure was taken to its maximum value of 86 kPa. The liner was removed three days before the centrifuge flight and trimmed to a total soil depth of 160 mm.

The liner was then slid inside the greased strongbox and the window was attached. The model piles were installed at 1g, before the loading apparatus and site investigation gear were assembled. This did not exactly replicate stress-strain conditions at full scale but comparisons between piles inserted at 1g and in flight have shown lateral capacity variations of only about 10% (2).

The final operation was temporarily to remove the window, so that markers could be placed as described earlier. In addition, horizontal lead threads were inserted using a hypodermic needle, parallel to the row of piles, so that internal deformations could be obtained from radiography after the test. The package was then re-assembled and a 10 mm layer of sand was placed over the soft clay, so as better to define its surface.

The pore pressures, which are initially negative after removal from the consolidometer and storage during model making, increased on testing at 100 g due to the increase in total vertical stress. After 2 hours (2.3 years, prototype) the pore pressures had come into equilibrium with the imposed groundwater level indicated on figure 8. The investigation of surcharge loading effects could then begin.

The surcharge was increased to 19 kPa and then by nominal 20 kPa intervals up to 93 kPa. A complete unload-reload loop was then followed. Each loading stage was held for about 2 minutes (14 days, prototype) during which a photograph was taken : the re-attainment of 93 kPa was held constant for 30 minutes (0.57 years, prototype) so that vane tests could be carried out.

Vane tests were conducted on the far side of the piles from the surcharge. The range of data from all the tests with a similar stress history is shown on figure 3. The peak shear strength increased from about 11.5 kPa at a depth of 10 mm (1 m, prototype), up to about 16.5 kPa at 50 mm (5m, prototype).

On completion of the vane tests, surcharge loading was continued in nominal 20 kPa increments with unload-reload loops at 152 and 189 kPa, by which time the clay was beginning to be squeezed under the edge of the bag housing and past the piles. At this point the test was stopped. Evidence from the period of constant surcharge during vane testing indicated that the changes of pile bending moments during loading were much more significant than subsequent changes as the clay consolidated under a load increment. Ideally, the interpretation of the effect of a load increment should be in terms of the undrained strength of the clay operative when the load was applied, taking consolidation into account.

A best-fit polynomial was drawn through the bending moment data points, imposing zero values at the pile's free ends: figure 9a. Datum values were taken as those immediately before surcharging started. By double differentiation and double integration of the bending moment polynomial, the net lateral pile pressure (figure 9b) and the relative pile deflections (figure 9c) were calculated. The lateral pressures, derived by double differentiation of fitted data points, should be taken as approximate. The distinction between action on the pile in the soft layer and reaction provided for the pile in the stiff layer is, however, clear. The lvdts at the pile head could only fix position (not rotation). Absolute pile displacements were plotted by assuming that the pile tips did not displace. The inference from figure 9c is that although the embedded length of the piles in the sand was too small for them to be considered fully fixed, the majority of pile head displacement was still due to pile curvature rather than tip rotation. It will be seen that the maximum bending moment occurred about 25 mm or two diameters below the clay-sand interface, and that it increased roughly in proportion to the surcharge.

Equivalent full-scale piles would deflect 56 mm at their crest, and have a maximum bending moment of 2.2 MNm, at a surcharge of 152 kPa corresponding approximately to an 8 m high embankment. This would generate acceptable bending strains, up to 0.26×10^{-3} in the 1.27 m diameter piles. Of course, abutment designers must also allow for the effects of a stiff pile cap, and the lateral thrust of the backfill, each of which will modify the bending moment profile compared with that observed in the experiment. The objective of the model test was to investigate lateral loading on the piles due to soil squeeze, under a simple set of boundary conditions.

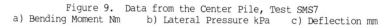

Figure 9.   Data from the Center Pile, Test SMS7
a) Bending Moment Nm   b) Lateral Pressure kPa   c) Deflection mm

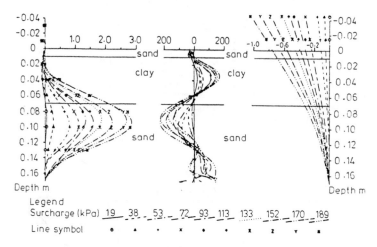

GEO-STRUCTURAL MECHANISMS

Three conditions are to be applied in the solution of problems in solid mechanics: equilibrium, compatibility and material deformability. When applied in terms of stresses and strains at a point, some means has to be found of integrating these expressions to satisfy arbitrary boundary conditions of force or displacement. This can be achieved using the principles of continuum mechanics as embodied in the finite element method. The FE program CRISP has been used (8) to analyse the transient response of an abutment wall on clay due to sequential backfilling. Computed wall displacements were qualitatively similar to those of HWS7 in figure 6, but great attention had to be paid to the parameters describing the non-linear stress-strain relations if correct magnitudes were to be derived. Practising engineers often reject FE analyses for routine design, when they recognise that the results are highly sensitive to the values assumed for input parameters with which they are unfamiliar. A similar rejection may follow if the crucial behavior mechanisms are obscured by unnecessary detail.

An alternative approach, used widely in structural mechanics, is to apply equilibrium and compatibility conditions not at every point but in a global idealization for a particular element. For example, engineers' beam theory usually ignores stress concentration at supports and beneath loads, and neglects deflection due to shear.

This structural mechanics approach has been effective in enabling engineers to design structures which deform tolerably. A similar approach will be used here to shed some light on the design of serviceable bridge abutments.

We will consider just one aspect of the larger problem: the immediate undrained lateral soil movements which must occur beneath the edge of a block of vertical surcharge. Figure 10 shows in (a) the stress increments and in (b) the soil displacements, which comprise a simplified geo-structural mechanism. The vertical plane OV is considered frictionless, and displacements beneath the planes AV and PV (inclined at 45°) are neglected. These assumptions are both consistent with the adoption throughout the deforming region of vertical and horizontal principal directions.

Figure 10. A Geo-Structural Mechanism for Undrained Surcharge

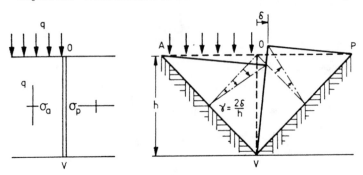

A further simplification is to imagine that the real soil is replaced by a homogeneous isotropic material with properties similar to those at mid-depth of the actual stratum, and possessing an initial earth pressure coefficient of unity. Let the increments of horizontal stress σ_h in AOV and POV be σ_a and σ_p: if the plane OV is to remain in equilibrium with zero external forces, then these horizontal stress increments must be equal. The increase in mobilized shear strength is half the increase in deviatoric stress, that is $\frac{1}{2}(q - \sigma_h)$ in active zone AOV and $\frac{1}{2}\sigma_h$ in passive zone POV.

If point O is taken to displace by δ horizontally and δ vertically then the shear strain in both deforming triangles must be $\gamma = 2\delta/h$, where h is the depth of the clay. If the shear strains are the same then so must be the shear stresses, if the soil is isotropic, so that $c_{mob} = q/4$. Figure 11 shows data of normalised undrained shear stress versus strain for kaolin tested in plane compression at OCRs of 10 and 3.3, corresponding to one third full depth in HWS7 and SMS7 respectively. Tests at OCRs of 7 and 2 for mid-depth would have been more ideal. Applying this data to test HWS7 we get q = 7.5 x 16.5 = 124 kPa, so c_{mob} = 31 kPa. Now c_u = 70 kPa at mid-depth, so

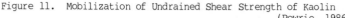

Figure 11. Mobilization of Undrained Shear Strength of Kaolin
(Powrie, 1986)

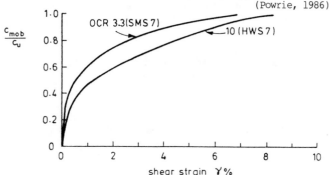

$c_{mob}/c_u = 0.44$. This requires $\gamma = 8.5 \times 10^{-3}$, so the prediction is
that $\delta = 8.5 \times 10^{-3} \times 20 / 2 = 85$ mm at prototype scale. Now it will be
recalled that the outward movement of the reference point was only
about 66 mm directly after construction in test HWS7, so the
calculation has been slightly conservative in this case.

It might, of course, be argued that the base of the wall can not
be dislocated in the fashion of figure 10: indeed the suppression of
the "step" at the reference point O could be taken to be the cause
of the backward rotation which accompanied translation in the model.
The proposed mechanism is approximate, and intended mainly to enable
the designer to decide whether the order of magnitude of soil
movements will be acceptable. Adaptations for non-homogeneous,
anisotropic soils, and which account more carefully for the normal
and shear stresses applied to the clay, are being developed.

EXTENSION TO LATERAL PRESSURE ON PILES

Suppose that the vertical plane OV in figure 10 contains piles of
diameter d and center spacing s, which develop average lateral
pressure p_{av} due to relative soil-pile displacement. For global
equilibrium of plane OV it must follow that

$$\sigma_a - \sigma_p = p_{av} \, d/s = p \, d/s \qquad (1)$$

if the pressure p on the pile were taken to be uniformly distributed
with a value calculated at mid-depth. The shear strain in the
active and passive triangles will be

$$\gamma = \tfrac{1}{2}(q - \sigma_a)/G \ = \tfrac{1}{2}\sigma_p \,/G \qquad (2)$$

This permits the calculation of the mid-depth soil displacement

$$\delta_s = h \, \gamma \, /4 \qquad (3)$$

However, elastic analysis (1) of the plane strain displacement of a rigid, adherent disc moving through a medium with shear modulus G_r provides that, for a relative displacement δ_r,

$$p = 5.33\ G_r\ \delta_r/d \tag{4}$$

Taking the same uniform pressure profile on the pile, considered to be perfectly fixed at the clay-sand interface, the pile displacement at mid-depth would be

$$\delta_p = 0.044\ p\ d\ h^4/EI \tag{5}$$

Then the relative displacement of the soil against the pile at mid-depth will be

$$\delta_r = \delta_s - \delta_p \tag{6}$$

Equations 4, 5 and 6 allow an independent assessment of mid-plane lateral soil displacement, for comparison with (3):

$$\delta_s = 0.19\ p\ d/G_r\ +\ 0.044\ p\ d\ h^4/EI \tag{7}$$

Substituting (2) into (1) we obtain

$$q = 4\ \gamma\ G +\ p\ d/s \tag{8}$$

Making a string substitution: (3) for γ, and (7) for δ_s, we get:

$$\frac{p}{q} = \cfrac{1}{\cfrac{d}{s} + \cfrac{0.71\ Gdh^3}{EI} + \cfrac{3\ G\ d}{G_r\overline{h}}} \tag{9}$$

This provides a quasi-elastic estimate of lateral pressure on a row of piles, due to the undrained imposition of surcharge. For a uniform elastic material, clearly $G_r = G$. However, for soil which has the non-linear stress-strain curve shown in figure 11, the larger strains near the pile will lead to a secant modulus reduced by a factor of about 2. Taking other values appropriate to the first two increments of load in the prototype of SMS7: $d = 1.27$ m, $s = 4$ m, $h = 6$ m, $G = 800$ kPa, $G_r = 400$ kPa, $EI = 5.13 \times 10^6$ kNm^2, equation 9 gives (retaining the same order of terms):

$$p/q = 1\ /\ [0.32 + 0.03 + 1.27] = 0.62$$

This compares with the evidence of the deduced lateral pile pressure from SMS7 in figure 9b, where p appeared to be parabolically distributed with a peak value of p/q very close to unity, and therefore a similar average value to that calculated.

Further model studies and FE analyses are in hand, so that an optimum calculation method can be devised. However, it is already clear that the reduction in soil stiffness due to enhanced strains around the pile is much more significant than its absolute value.

When undrained soil strength c_u is fully mobilized everywhere, plasticity analyses show (5) that the limiting lateral pressure $p_u \simeq 10.5 \ c_u \pm 1.35 \ c_u$, over the possible range of soil-pile adherence factors. Replacing equation 3 by the plastic alternative

$$q - \sigma_a = \sigma_p = 2c_u \qquad\qquad (10)$$

and substituting in equation 1, with $p = 10.5c_u$, we get

$$q = c_u \ (4 + 10.5 \ d/s) \qquad\qquad (11)$$

The initial quasi-undrained loading phase of test SMS7 is compared in figure 12 with the predictions of equations (9) and (11). The data are compared on normalized axes of q/c_u and p/c_u, where p is the mean pressure inferred to act on the piles in the soft layer, and c_u is the shear strength at the centre of that layer. The calculation is sufficient, in this case, to allow the approximate prediction of lateral pressures in the loading range $q < 4c_u$, within which the undrained settlement might be tolerable.

CONCLUSIONS

1. Lateral displacements of bridge abutments are strongly related to the soil-structure interaction which occurs as backfill is placed. Centrifuge tests of abutments on spread foundations over firm to stiff clay showed an outward displacement during backfilling, which tended to reduce lateral earth pressures towards fully active values. On consolidation, however, the tendency for backward rotation led to a significant increase in earth pressure, albeit with negligible movement at the deck support.

2. Lateral displacements of piled bridge abutments may be significantly underestimated if the lateral loading of the piles themselves, due to squeeze of soft clays under the embankment, is not accounted for. Centrifuge tests of piles subject to surcharge loading showed that lateral pressures could be of the same order of magnitude as the surcharge. These came into effect as soon as the surcharge was placed: subsequent consolidation hardly affected them.

3. It was shown that a geo-structural mechanism comprising two deforming triangles could offer a rationale for the estimation of lateral displacements and pressures, which was sufficiently accurate in the two cases examined to form a basis for design. The feature of the method was its approximate treatment of equilibrium and compatibility which permitted non-linear stress-strain data from appropriate soil tests to be incorporated directly in simple calculations.

Figure 12.   Elasto–Plastic Interaction Diagram for Surcharge–Induced
Lateral Pressure on a Pile

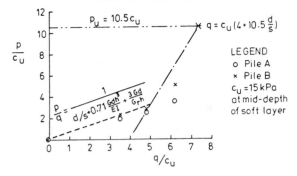

REFERENCES

1.   FJ Baguelin, RA Frank & Y Said,   Theoretical Study of Reaction
Mechanism of Piles, Geotechnique, 27 (3), 1977, 405–434.

2.   WH Craig,   Installation Studies for Model Piles, Proc.   Symp.
Applications of Centrifuge Modelling to Geotechnical   Design,
Univ. of Manchester (UK), 1984, Ed. Craig, Balkema, 440–455.

3.   LK Moulton,   HVS Ganga Rao & GT Halvorsen,   Tolerable Movement
Criteria for Highway Bridges, RD–85/107, FHWA, Wash. DC, 1985.

4.   W Powrie,   The Behaviour of Diaphragm Walls in Clay,   Ph.D.
Thesis, University of Cambridge (UK), 1986.

5.   MF Randolph & GT Houlsby,   The Limiting Pressure on a Circular
Pile Loaded Laterally in Cohesive Soil,   Geotechnique, 34 (4),
1984, 613–623.

6.   AN Schofield,   Cambridge University   Geotechnical   Centrifuge
Operations, Geotechnique, 30 (3), 1980, 227–268.

7.   SM Springman,   Lateral   Loading   on Piles   Due   to   Simulated
Embankment Construction, Ph.D. Thesis, University of Cambridge
(UK), 1989.

8.   HW Sun,   Soil Structure Interaction Problem of Retaining Wall
on Compressible Foundation, M.Phil Thesis, University of
Cambridge (UK), 1987.

ACKNOWLEDGEMENTS

The   work described here was supported by research contracts   let
to the first author by the Transport and Road Research Laboratory of
the UK Department of Transport.   The opinions expressed here are the
authors' and do not necessarily coincide with those of   the
Laboratory or the Department.

Wing   Sun   is grateful for the financial support of the   Croucher
Foundation of Hong Kong, and the Overseas Research Student   Award
from   the   CVCP of   British Universities.

CEMENT STABILIZED SOIL RETAINING WALLS

Derek V. Morris[1], M.ASCE and William W. Crockford[2], A.M.ASCE

ABSTRACT: A new design of retaining wall is proposed, utilizing facing panel units anchored into a cement stabilized backfill. Only short anchors are required and as long as the intact strength of the stabilized soil has been sufficiently improved by cement addition, the structure becomes a conventional mass gravity structure. Strength properties are specified, as well as design criteria on the basis of finite element analysis. A field case history is described, and costs of the system are potentially significantly less than other designs of a medium height retaining wall.

INTRODUCTION

The need for economical methods of retaining earth fill or supporting the sides of an excavation, has resulted in many alternative designs of retaining structure, even of the traditional gravity type walls. Mass concrete walls have long been superseded by designs that make more efficient use of the retained earth fill, to assist in providing overall stability.

A common commercial example uses galvanized steel straps and select backfill to form the retaining wall mass behind a precast concrete facing. The overall strength of the fill is provided by friction between the soil and the reinforcement. Although very successful, and substantially cheaper than either traditional cantilever or mass concrete walls, such a design is still labor intensive.

This paper describes a new and potentially even more economical design of retaining wall, investigated at the Texas Transportation Institute of Texas A&M University for the Texas Highway Department, utilizing facing panel units anchored into a cement stabilized soil backfill.

Overall stability is provided by the self-weight of the stabilized soil, artificially given an intact strength by the relatively inexpensive addition of between 4 to 8% cement by mass. Moreover the design is non-proprietary, not subject to any licensing restrictions. A generic diagram is shown in Figure 1.

---

1 Assistant Professor, Department of Civil Engineering, Texas A&M University, College Station, Tx 77843-3136.

2 Engineering Research Associate, Texas Transportation Institute, College Station, Tx 77843-3135

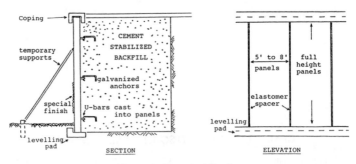

Figure 1. Proposed Design

PRINCIPLE OF DESIGN

The basic principle underlying the design of cement stabilized soil retaining walls is that the stabilized backfill behind the facing panels essentially forms the retaining wall. This is in contrast to other commercial designs, where the resistance to shear failure is produced by reinforcing strips or bars or fabric in the backfill. If the strength of stabilized soil is sufficiently improved by the addition of cement, then stability analysis is straightforward, as the stabilized cross-section can be considered as an integral monolithic structure. There are also significant advantages in construction, as many practical difficulties are substantially reduced. Large lifts of stabilized soil can be easily compacted to required densities, the likelihood of soil wash-out between units is reduced, and trafficability of the compacted fill is improved

These in fact are the reasons that this design was suggested originally, as it was found that many contractors were voluntarily electing to cement-stabilize earth fill for earth reinforced type walls, even though there was no engineering requirement for them to do so. The contractors did so at their own expense, because of improved constructability, as mentioned above. Additional advantages were less concern for long-term internal erosion or seepage of fine sand backfill, and less concern for corrosion of steel earth reinforcement or anchors. This made the backfill into a structure that was inherently stable, without the need for any reinforcement.

In certain circumstances it may also be advantageous to use a lightweight fill (such as gypsum or calcium sulphate) in situations where low unit weight is desirable, such as soft soil foundations unable to withstand high bearing pressures. Tests to investigate the suitability of fill materials are described in the next sections, but in general it appears that a wide variety of soils is capable of stabilization for this purpose.

Anchor Design. The facing panels can be tied into the stabilized soil by means of short anchors whose main function is to retain the individual facing units. They are not required to carry primary loads or to prevent structural failure. The main design criteria is that the pull-out resistance should be sufficient to counteract the outward forces that act on facing units. These forces are quite small in nature, and derive primarily from incidental wind loading and from eccentricities between the self-weight of a panel and its foundation. To verify the performance of short anchors in stabilized soil, tests were conducted on 6 mm diameter bolts embedded to a length of 30 mm and 45 mm in sand stabilized with 5 % cement. These were performed in accordance with ASTM standard C900 for pull-out of anchor bolts in concrete, so that a direct comparison could be made between behavior of anchors in stabilized soil and behavior in weak concrete. Curing times of 7, 14, and 28 days were used, so that the results could be compared to the known strength properties of the cement stabilized soil mix under these conditions.

Resistance to pullout was observed to be significant even for small anchors. Comparison of the capacity in cement stabilized soil with standard predictive formulae (and incorporating the known compressive strength characteristics of the stabilized material) have indicated that measured capacities exceed theoretical predictions, once due allowance is made for the actual strength.

As a result, the following conservative formula is suggested for predicting the capacity of a rough anchor in stabilized soil. It is analogous to pullout formulae for concrete, (although the coefficents are different) but assumes that the capacity is the sum of shear and frictional components.

$$P = A_s(7d\sqrt{f_c} + \gamma H f) \tag{1}$$

where:

A_s = surface area of anchor (ignoring outermost 0.2 m) in m^2
d = equivalent circular diameter in mm only.
f_c = compressive strength of cement stabilized soil in Pa only.
γ = unit weight of cement stabilized soil.
H = overburden depth of anchor
f = soil friction coefficient, which may be assumed to equal the tangent of the residual friction angle of the soil cement matrix.

It is also desirable to provide a design that can mobilize full capacity with the soil after only a minimal amount of wall movement. Figure 2 shows a possible anchor design, where a rigid galvanized steel bar is held to the wall with a nut and bolt connected to a pair of eyes in the facing panel. The bent bar can be kept in the upright position until the level of the compacted stabilized soil reaches desired elevation and is then rotated down to the soil and the point is forced in. This geometry would be easy to install and easy for the contractor to compact around. A threaded section could be incorporated, that would allow for a pretensioning load to be applied, if desired.

ENGINEERING PROPERTIES OF STABILIZED SOILS

Since the stabilized backfill forms probably the most important element of the proposed design, careful consideration is necessary as to the structural suitability of such materials. In general it is found that the fill material, mix

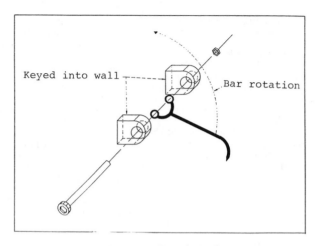

Figure 2. A Generic Anchor

proportions, construction methods, and environmental conditions all have an influence on the engineering properties of cement stabilized soil. These factors can be divided into the following groups:

Nature of Materials and Proportions of Mix. The type and nature of the soil affect the properties of compacted and hydrated soil-cement mixtures. As the clay content increases, there is an increase in the percentage of cement required to produce a given strength and modulus of elasticity (9), (7). Also, for soils of the A-4 group, the cement requirement increases with the liquid limit (4). The trend is more marked for soils of the A-6 and A-7 groups, for which cement requirements will typically be much higher (between 8 to 16%). The presence of organic matter, sulfates and cations associated with clay sized minerals in the soil, influences the strength and setting time of soil-cement, although the pH - value may not have an effect (9). Experiments on the effect of moisture content show that the compressive strength increases to a maximum at slightly less than the optimum moisture content for sandy and silty soil (AASHTO classifications A-2 and A-3), and at greater than optimum for the clayey soil (6). Compressive strength and resistance to wetting and drying, and freezing and thawing also increases as cement content is increased.

Mixing and Compaction. Studies on the effects of delayed compaction on stabilized soil cement have concluded that a) the durability, compressive strength, and density of soil cement decreases considerably and uniformly after a delay of two hours or more in compaction of the soil cement after mixing. b) the loss of compressive strength, durability and density due to delayed com-paction can become so great that any physical improvements derived from the addition of portland cement are nullified, c) compaction in soil cement mixes should not be delayed beyond the initial setting time of cement gel, and d) if a delay between mixing and compaction cannot be avoided, then retarding agents such as calcium lignosulfonate and hydroxylated carboxylic acid should be added in trace amounts to slow down the cementation process (3).

Some studies show that, for granular soil, specimens molded by impact compaction give higher cohesion values than the corresponding specimens molded by kneading compaction. For silty soil, specimens molded at optimum moisture content by kneading compaction give higher cohesion values than the corresponding specimens modeled by impact compaction. Friction values appear not to be influenced by the method of compaction, cement content or age.

Curing Conditions and Age. Moist soil, water proof paper, asphalt, tar and asphalt emulsion have all been found effective in retaining the moisture of soil cement, which is vital for the development of strength during curing (8). Studies conducted on the influence of temperature on mix strength indicate that the 7 - day compressive strength increases by about 2% per degree Celsius at normal temperatures. As with concrete, the strength also increases with age.

Effect of Admixtures. Soil admixtures and additives can be used to improve the reaction between the soil and cement. Poorly reacting sandy soils can be improved by addition of normally reacting soils (5). Addition of 4 to 6% flyash can significantly improve the strength of certain stabilized sands and reduce shrinkage cracking during curing of certain cement stabilized clay soils. Additives like calcium lignosulphonate and hydroxylated carboxylic acid, can be used to delay the initial setting time of cement, thus providing a good bond between subsequent layers of soil cement (2).

STRENGTH PROPERTIES

In order to address specific technical questions on the suitability of marginal, light weight and clayey soil fills, a substantial laboratory program was undertaken, to investigate strength and failure characteristics. In addition to unconfined compression testing, triaxial tests were conducted on two kinds of materials, a poorly graded sand from Houston and calcium sulphate (fluorogypsum), to provide a basis for the prediction of stress - strain characteristics at different confining pressures, and to obtain information about cohesion and internal friction. Type I Portland cement was used as a stabilizer for sand, whereas Type II Portland cement was used as a stabilizer for gypsum. Sand cement specimens were compacted at the Texas compaction energy (TEX-113) whereas gypsum cement specimens were compacted at the ASTM D-1557 energy. Impact compaction was used for all specimens. Seven percent cement content by weight of dry soil was chosen for both materials and all specimens were molded at optimum moisture contents. Sand cement specimens were cured at 95% and 50% humidity and tested at an age of 75 days, and, in some cases, at 14 days. Sand samples with 5% cement were cured at 95% humidity and tested at 7, 21 and 28 days. Just before testing, the cylindrical samples were taken out of curing and were placed in a Texas triaxial cell (TEX-117E, ASTM D3397) with porous stones top and bottom. Load was applied at a rate of 0.02 mm/s until failure. Table 1 summarizes the geotechnical properties of the fills tested.

Results Obtained. Typical stress-strain curves for the specimens are shown in Figure 3 for various confining pressures and curing conditions. As would be expected, increasing confining pressure required increased applied stresses to cause failure. Failure strains also increased as confining pressure was increased. Stress - strain curves were linear up to about one-third of the ultimate stress.

Table 1. Geotechnical Properties of Sand and Gypsum

	SAND	GYPSUM
% Passing sieve no. 2	98	82
% Passing sieve no. 40	87	46
% Passing sieve no. 200	4	0.67
Plasticity Index	NP	
AASHTO Classification	A-3	
Unified Classification	SP	
Optimum moisture content (for 7% cement)	13.5%	6.0%
Maximum dry density (for 7% cement)	1745 kg/m^3	1650 kg/m^3

Table 2. Elastic and Strength Properties of Cement treated Materials

Material	Cement Content (%)	Age (days)	Curing Humidity (%)	E (MPa)	C (kPa)	φ (deg)
Sand	7%	14	95%	730	286	44
Sand	7	75	95	875	489	42
Sand	7	75	50	875	737	45
Sand	5	7	95	262	184	32
Sand	5	21	95	441	270	35
Sand	5	28	95	689		
Gypsum	7	50	95	262	375	57

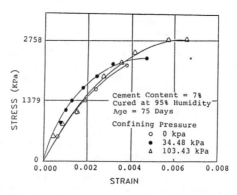

Figure 3. Stress-Strain Curve for Stabilized Sand

Table 2 presents a summary of tests conducted on sand cement samples and gypsum cement samples at different confining pressures. Specimens cured at 50% humidity had higher strengths and failure strains than those cured at 95% humidity. It is evident from typical Mohr - Coulomb envelopes (Figure 4) that reducing curing humidity appears to increase cohesion.

However, like concrete, there was a noticeable increase of soil cement strength with age. It is also clear from the table that as the cement content is decreased, the strength decreases as well as the modulus of elasticity, cohesion and internal friction.

Cement treated gypsum samples exhibit similar characteristics. Strength as well as failure strains increase with confining pressure. However, the stress-strain curve is linear up to 60% of the ultimate stress. The material proved to be about 25% weaker than the sand equivalent, in return for a density reduction of about 20% compared to normal fill. In comparison, it was noticed that gypsum cement has a higher angle of internal friction and lower cohesion than cement treated sand. However, the modulus of elasticity is almost half that of cement treated sand.

In general, the results indicate that a wide range of sandy and silty backfills are capable of being stabilized in this way, with unconfined compressive strengths in the range of 1 to 4 MPa, for cement contents of 3.5 to 7%. As shown in the next section, these strengths are likely to be adequate for low to medium height retaining walls.

NUMERICAL ANALYSIS OF STRENGTH REQUIRED

Finite element numerical modelling was carried out to predict the response of the proposed wall system under different geometric and loading conditions. Particular attention was paid to investigating the distributions of displacement, stresses, plastic deformation (or failure mode), and ultimate load bearing capacity of the system. Only 2-dimensional finite elements (plane strain) were required and nonlinear plastic material characteristics were used. The concrete-panel facing and anchors were not incorporated in the analysis because of their minor influence on the overall stability of the wall system.

A commercially available finite element program ("ABAQUS") (1) was utilized. This was capable of modelling different types of materials including metals, soils, rock and concrete, and performing linear as well as non-linear static and dynamic analysis with pre- and post - processing facilities.

Input Data. The finite element mesh for a typical analysis (modelling a 10 m x 8 m cement-stabilized wall) is shown in Figure 5. The model consists of two-dimensional quadrilateral/triangular solid elements of plane strain condition. The lower boundary is assumed to be rigid, and the vertical boundaries at the left- and right-hand sides are assumed to be on rollers to allow downward movement due to construction of the embankment. The total model consists of 142 nodal points and 150 elements. In order to predict failure of the unstabilized soil constituting the backfill adjoining the retaining wall mass as well as the foundation, the Drucker -Prager model was adopted in preference to the Mohr - Coulomb failure criterion. Cement stabilized soil (block A) was assumed to behave in a similar fashion to plain concrete with suitably reduced values.

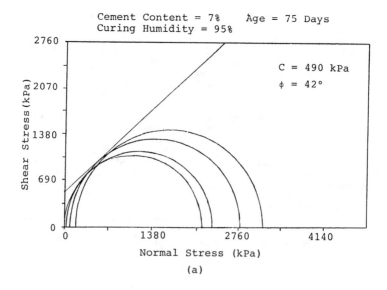

Cement Content = 7%   Age = 75 Days
Curing Humidity = 95%

C = 490 kPa
ϕ = 42°

(a)

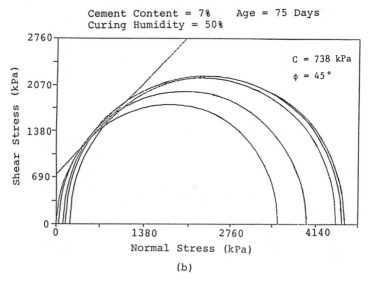

Cement Content = 7%   Age = 75 Days
Curing Humidity = 50%

C = 738 kPa
ϕ = 45°

(b)

Figure 4. Mohr-Coulomb Envelopes at Failure for Cement Stabilized Sand

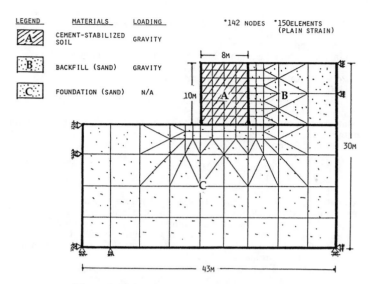

Figure 5. Finite Element Model (10M x 8M wall)

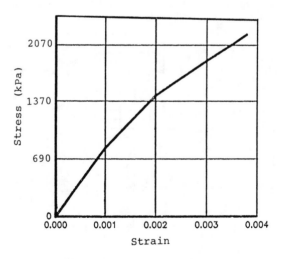

Figure 6. Stress-Strain Curve Utilized

Material properties of the cement-stabilized soil were obtained through laboratory testing, as follows.

Unconfined Compressive Strength 2.2 MPa
Uniaxial Tensile Strength 0.20 MPa (9% of Unconfined Strength)
Initial Young's Modulus = 830 MPa
Poissons' Ratio = 0.14 (from literature)
Unit Weight = 20 kN/m³
Uniaxial Stress-Strain Relation used was an idealized tri-linear relation, shown in Figure 6.

Material properties of typical sandy soil were used for the granular backfill and foundation soil, as follows:

Internal Friction Angle: 35 degrees
Cohesion: 5 kPa
Initial Young's Modulus: 100 MPa
Poisson's Ratio: 0.3
Unit Weight: 17.5 kN/m³

Gravity was used as the primary source of loading on the system. The cement-stabilized wall and soil backfill were treated as if they were constructed on top of the existing grade without excavation. The applied load consisted only of the self-weight of the cement-soil wall and granular backfill. Ultimate load bearing capacity of each model was assessed in terms of the multiple of gravity each model could withstand.

Typical Results. The displacement field is shown at an exaggerated scale in Figure 7. In general, the wall moves downward and rotates clockwise. The foundation soils below the embankment compress from the weight of overburden, with some small corresponding heave in front of the wall. Maximum distortion occurs at the toe of the wall. For normal aspect ratios, for which a satisfactory factor of safety exists against outward toppling, the wall rotates toward the backfill instead of overturning toward the face of wall.

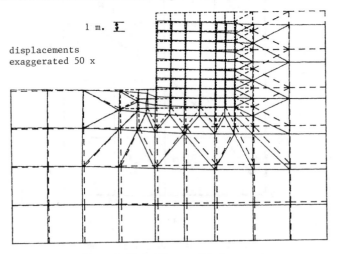

1 m.

displacements
exaggerated 50 x

Figure 7. Finite Element Displacements

The distribution of maximum compressive stresses is shown in Figure 8. As might be expected, a significant stress concentration was observed around the toe of the wall. The maximum compressive stress at the toe was about 1.5 times the overburden pressure. For a 10 m high wall this value represents approximately 13% of the unconfined compressive strength of the cement-stabilized soil.

The distribution of tension (not shown) shows maximum tensile stresses near the toe and the top of the cement-stabilized wall. The maximum tensile stress observed within a wall of these dimensions, represents approximately 3% of the uniaxial tensile strength of the cement-stabilized soil implying that failure will occur primarily in compression, or crushing.

Maximum shear stresses (not shown) occurred at the toe of the wall. Results indicate that shear stresses drop sharply along the boundary between the wall and soil backfill, due to low interface friction between the two different materials.

Of more direct interest are the contours of plastic strain, also shown in Figure 8. These indicate that there will be a localized plastic zone within the soil beneath the toe of the wall. Possible ultimate failure modes may be established by carefully inspecting the contours of plastic deformations computed under the ultimate loading conditions. The figure shows contours under the ultimate loading of 7.8 g (which corresponds theoretically to a 78-meter tall wall), and suggests that the wall would fail by rotating around its toe, due to bearing capacity failure of the foundation soils. Ultimate loading was determined by the multiple of gravity at which the program no longer converged.

Alternative Designs. Other aspect ratios were also analyzed in this fashion, to examine the effect of differing wall widths in relation to the height. However, the overall behavior was similar, certainly for cross-sections that would be considered to be acceptable in engineering practice. For example, a wall 10 m high and only 4 m wide (generally regarded as unacceptably narrow) produced a compressive stress concentration in the toe of 1.8 times the overburden pressure, and failed at a maximum height of 58 m, representing an approximately 20% reduction in safety factors with respect to a wall of double the width.

Alternative shapes for the cross-section were examined including an indented rear face, and triangular cross-sections tapering either towards the top or the bottom. The only one that appeared to offer any advantages, was a triangular section tapering towards the top. This effectively offered the same performance as a rectangular cross-section, but with major potential savings in stabilized fill volumes (and therefore cost). In principle it should also be possible to reduce the cement content selectively towards the top of a wall, but this has not been investigated at this stage.

Recommendations. The results of these numerical analyses support the concept of the design described. They indicate that the system will perform satisfactorily under typical field conditions to heights well in excess of those normally required for highway construction (typically 10 m to 20 m). However, in view of the stress concentration factors predicted by the finite element analyses, and in view of the brittle nature of the material, a factor of safety of at least 3 (and preferably 5) against crushing of the stabilized fill is recommended, defined as:

$$F_s = f_c / \gamma H \qquad (2)$$

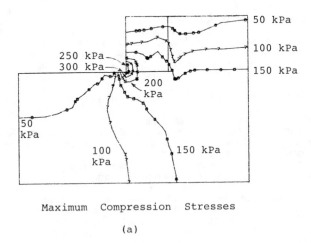

Maximum   Compression   Stresses

(a)

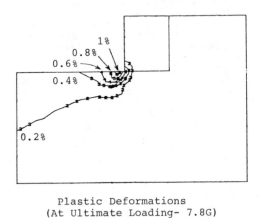

Plastic Deformations
(At Ultimate Loading- 7.8G)

(b)

Figure 8. Typical Results of Numerical Analysis

where f_c is the unconfined compression strength of the fill, H is the wall height, and γ the unit weight of stabilized soil.

FIELD CASE HISTORY

A cement stabilized soil retaining wall has recently been constructed in California, based on similar design concepts. Photographs of the wall, which had a maximum height of 10 m, and a width of approximately 5 m are shown in Figure 9. The design used a proprietary interlocking concrete block facing, without any separate tie backs. Cementitious bond between the facing blocks and the stabilized backfill (which was carefully compacted manually in this region) was sufficient to retain the facing. In addition, the wall was constructed with a slight backward slope of 1 horizontal to 8 vertical (7°). Some terraces were incorporated into the side sections, but not in the main body.

Design and supervision were carried out by 2R Engineering, of San Marcos, California who specified stabilized backfill with a minimum cement content of 4% by weight and at least 1.4 MPa cylinder crushing strength. Sandy soil was used (which gave an average strength of 4 MPa), but some fines content was permitted. Overburden stresses in the wall were limited to 1/6 of the crushing strength, treating the toe as the critical point, and allowing for any eccentricity of the wall center-of-gravity by simple elastic theory. In other respects, design followed conventional gravity wall principles.

As well as sloping the front face backward, it was also possible to produce a rearward slope on the rear face of the wall (1 horizontal to 4 vertical or 15°), which was incorporated into construction against original earth, by using the cut slope as a rear form and compacting stabilized fill directly against it. This had the advantage of moving the wall center of gravity rearwards, so that the resulting eccentricity of the self-weight could to some extent neutralize the overturning moments on the wall due to soil pressures, and relieve high stresses at the toe.

Anticipated Costs. Historical costs of the wall shown were $250,000 in 1985 for 1,500 m² of wall, giving a unit price of about 170$/m². This compares with costs at that time of 250 to 300$/m² for a "soft wall".

Subsequently, unit costs have slightly decreased, to about 220$/m² in 1989 for "soft wall" in Texas, and bulk cement costs have dropped from about 0.07$/kg to 0.06$/kg. Current estimates of construction costs of similar cement stabilized soil walls (with non-union labor) range from 120 to 140 $/m², so that significant economies are indicated if such designs achieve wide acceptance.

ACKNOWLEDGEMENTS

The study on which this paper is based, has been funded as a research project to the Texas Transportation Institute by the State Department of Highways and Public Transportation, and is prepared in cooperation with the U.S. Department of Transportation, Federal Highway Administration (SDHPT contact representative - M.P. McClelland).

The help of M.R. Bollam in preparing this manuscript is gratefully acknowledged.

Figure 9. Photographs of Field Retaining Walls

DISCLAIMER

The contents of this paper reflect the views of the authors who are responsible for the facts and the accuracy of the data presented herein. The contents do not necessarily reflect the official views or policies of the Federal Highway Administration or the State Department of Highways and Public Transportation. This paper does not constitute a standard, specification, or regulation.

REFERENCES

1. ABAQUS User's and Theory Manuals, Version 4.6, Hibbitt, Karlsson & Sorensen, Inc., 1987.
2. A. Arman and T.J. Dantin, The Effect of Admixtures on Layered Systems Constructed with Soil-Cement, Engineering Research Bulletin 86, Louisiana State University, 1965.
3. A. Arman and R.S. Saifan, Effect of Delayed Compaction on Stabilized Soil-Cement, Engineering Research Bulletin 88 , Louisiana State University, 1965.
4. M.D. Catton, Research on Physical Relations of Soil and Soil-Cement Mixtures, Highway Research Board Bulletin 20, 1940.
5. M.D. Catton and E.J. Felt, Effect of Soil and Calcium Chloride Admixtures on Soil-Cement Mixtures. HRB Proceeding 23, 1943.
6. E.J. Felt, Factors Influencing Physical Properties of Soil-Cement Mixtures, Highway Research Board Bulletin 108, 1955.
7. R.L. Handy, D.T. Davidson, and T.Y. Chou, Effect of Petrographic Variations of Southwestern Iowa Loess on Stabilization With Portland Cement, HRB Bulletin 98, 1955.
8. A.W. Maner, Curing Soil Cement Bases, HRB Proceedings 31, 1952.
9. F. Reinhold, Elastic Behavior of Soil Cement Mixtures, Highway Research Board Bulletin 108, 1955.
10. E.G. Robbins and P.E. Mueller, Development of a Test for Identifying Poorly Reacting Sandy Soils Encountered in Soil-Cement Construction, Highway Research Board Bulletin 267, 1960.

NORTH AMERICAN PRACTICE IN REINFORCED SOIL SYSTEMS

James K. Mitchell[1], F.ASCE and Barry R. Christopher[2], M.ASCE

ABSTRACT: Soil reinforcement is now widely used for the construction of retaining walls, embankment slopes, and natural or cut slopes. Several types of reinforcement and proprietary systems are available for these purposes. Based on past experience, the results of extensive laboratory model tests, observations on instrumented full-scale structures, and analytical and finite element studies, safe and economical design of reinforced soil structures is now possible. A generic design approach is summarized which takes into account considerations of relative stiffnesses of the reinforcement and soil, durability of the reinforcement, and deformations of the reinforced structure. Analysis of the seismic stability is included.

INTRODUCTION

The reinforcement of soil is one of the most significant advances in Civil Engineering construction since the development of reinforced concrete. In fact, reinforced soil may well replace reinforced concrete as the conventional design material for earth retaining structures.

"Reinforced soil" is a generic name that is applied to combinations of soil and distributed linear or planar inclusions; e.g., steel strips, steel or polymeric grids, geotextile sheets, steel nails, that are capable of withstanding tensile loadings and, in some cases, bending and shear stresses as well. "Mechanically stabilized earth" refers principally to reinforced soil placed as fill and also includes anchored systems. The composite material has found its greatest applications for earth retaining structures of various types and in the reinforcement of embankments so that steeper side slopes may be used. Reinforcement is also used at the base of embankments over soft ground; however, these systems are beyond the scope of this paper. Reinforcement of an in-situ soil by the insertion of long metal rods or "nails" is an additional form of mechanical stabilization of soil that is becoming increasingly used for support of excavations and for stabilization of slide masses. Schematic diagrams of a reinforced soil wall, reinforced embankment slopes and soil nailing are given in Figure 1.

Since the introduction of Reinforced Earth into the United States in the early 1970's, soil reinforcement for construction of earth retaining structures and embankments has become very widely used. The rapid acceptance of various types of reinforced soil systems for a variety of applications can be

---

[1]Prof., Dept. of Civil Engineering, University of California, Berkeley, CA 94720.
[2]Technical Manager, Polyfelt, Inc., Atlanta, GA 30328.

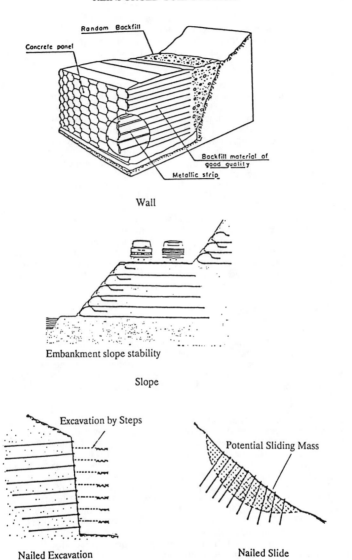

Wall

Embankment slope stability

Slope

Nailed Excavation

Nailed Slide

Figure 1. Mechanically Stabilized Soil Systems: Wall, Slope, Nailed Excavation, Nailed Slide

attributed to a number of factors, including low cost, aesthetics, reliability, simple construction techniques, the ability of reinforced soil structures to adapt to different site conditions and to withstand substantial deformations without distress, and the keen competition between suppliers of the different systems that are used.

The purpose of this paper is to provide, in condensed form, an overview of the applications, analysis, design, and construction of the earth reinforcement systems that are used in the United States at present. It draws heavily on the results of two recently completed studies (Mitchell and Villet, 1987 and Christopher et al. 1990) which provide comprehensive state-of-knowledge, state-of-practice, and design information. In view of the extensive use of earth reinforcement over the past 15 to 20 years and the very extensive literature already available, neither the mechanism of earth reinforcement nor soil-reinforcement interaction are analyzed in detail in this paper.

As previously indicated, multi-anchor systems are sometimes also considered a form of mechanically stabilized soil. These structures contain a large number of anchors distributed in a regular manner throughout the soil mass, as do reinforced soil and nailed soil structures. However, anchor systems rely on passive resistance against the anchors and tendon action between the anchors and wall facing; whereas, soil reinforcement relies on continuous stress transfer along the full length of the reinforcements. In anchored systems the facing resists and retains the fill; in reinforced systems the facing serves a local role in preventing soil sloughing and erosion. Detailed consideration of multi-anchor systems is beyond the scope of this paper.

DESCRIPTION OF SYSTEMS

The essential components of earth reinforcement systems are the reinforcements, the backfill or in-situ soil with which the reinforcements interact, and the facing. They can be described by the reinforcement geometry, the stress transfer mechanism, the reinforcement material, the extensibility of the reinforcements, and the method of placement, as shown in Table 1. The primary systems in use in North America at the present time are described briefly below.

Placed Reinforced Soil Systems. A summary of the available systems in terms of the reinforcement and facing panel details is given in Table 2. The general arrangement of the different elements is shown in Figure 2. In each case the layers of reinforcement are spaced vertically at distances typically from one to three feet (0.3 to 1.0 meter). A granular soil is usually specified for use in the reinforced soil volume, and specifications for this material do not vary significantly among the different systems.

The geometries and some mechanical properties of the different reinforcement types are shown in Figure 3. Reinforcement systems are generally characterized as extensible or inextensible due to their somewhat different behavior (Christopher et al. 1990). Extensible reinforcements can deform without rupture to deformations greater than can the soil in which they are included. Inextensible reinforcements cannot. Generally, the polymeric geotextiles and geogrids are extensible; whereas the steel strip and bar mat systems are inextensible. The stresses that must be carried by the reinforcements depend both on the type of reinforcement and the density of reinforcements in the soil (Adib, 1988). The total load carried by extensible reinforcements can equal that in inextensible systems if the density of extensible reinforcements is high enough so that the soil cannot yield.

Table 1.   Comparison of Reinforced Soil System
(Adapted from Jewell, 1984 and NCHRP 290)

APPLICATION	REINFORCEMENT TYPE		ALLOWABLE SLOPE ANGLE (30° 60° 90°)	RECOMMENDED SOIL TYPE* (Clay Silt Sand Gravel; .002 .02 .2 2mm)	STRESS TRANSFER MECHANISM		REINFORCEMENT MATERIAL		EXTENSIBILITY		PROPRIETARY SYSTEM / PRODUCT NAMES
					Surface Friction	Passive Resistance	Metal	Non-Metal	Extensible	Inextensible	
IMPORTED EMBANKMENT TYPE APPLICATION	STRIP	Smooth	I	I	•		•			•	Reinforced Earth
		Ribbed	I	I	•		•	•	(I)	•	Reinforced Earth / Paraweb
	GRID		I	I		•	• / •	•		• / •	VSL, MSE, GAE, RSE, and Welded Wire Wall / Maccaferri Gabion / Tensar, Mirafi and Tenax Geogrids
	SHEET		I	I				•	I		Geotextiles
	BENT ROD		I	I		•	•			•	Anchored Earth
	FIBER		I	I	•		•	•	I		
IN SITU GROUND IMPROVEMENT APPLICATION	FLEXIBLE, SMALL DIAMETER NAILS		↓	IN SITU SOILS	•		•			•	
	RIGID, LARGE DIAMETER PILES		↓	IN SITU SOILS		•	•	•		•	

*Based on stress transfer between soil reinforcement. Other Criteria may preclude use of soils for specific applications.

Table 2. Reinforcement and Face Panel Details for Several Reinforced Soil Systems Used in North America.

System Name	Reinforcement Detail	Typical Face Panel Detail[1]
Reinforced Earth: (The Reinforced Earth Company 1700 N. Moore St. Arlington, VA 22209-1960)	Galvanized Ribbed Steel Strips: 0.16 in (4 mm) thick; 2 in (50 mm) wide. Epoxy coated strips also available.	Facing panels are cruciform shaped precast concrete 4.9 ft x 4.9 ft x 5.5 in (1.5 m x 1.5 m x 14 cm). Half size panels used at top and bottom.
VSL Retained Earth (VSL Corporation, 101 Albright Way, Los Gatos, CA 95030)	Rectangular grid of W11 or W20 plain steel bars, 24 in x 6 in (61 cm x 15 cm) grid. Each mesh may have 4, 5 or 6 longitudinal bars. Epoxy coated meshes also available.	Precast concrete panel. Hexagon shaped, (59-1/2 in high, 68-3/8 in wide between apex points, 6.5 in thick (1.5 m x 1.75 m x 16.5 cm).
Mechanically Stabilized Embankment. (Dept. of Transportation, Div. of Engineering Services, 5900 Folsom Blvd., PO Box 19128 Sacramento, CA 95819	Rectangular grid, nine 3/8 in (9.5 mm) diameter plain steel bars on 24 in x 6 in (61 cm x 15 cm) grid. Two bar mats per panel. (connected to the panel at four points).	Precast concrete; rectangular 12.5 ft (3.81 m) long, 2 ft (61 cm) high and 8 in (20 cm) thick.
Georgia Stabilized Embankment (Dept. of Transportation, State of Georgia, No. 2 Capitol Square Atlanta, GA 30334-1002) Arlington, VA 22209-1960	Rectangular grid of five 3/8 in diameter (9.5 mm) plain steel bars on 24 in x 6 in (61 cm x 15 cm) grid 4 bar mats per panel	Precast concrete panel; rectangular 6 ft (1.83 m) wide, 4 ft (1.22 m) high with offsets for interlocking.
Hilfiker Retaining Wall: (Hilfiker Retaining Walls, PO Drawer L Eureka, CA 95501)	Welded wire mesh, 2 in x 6 in grid (5 cm x 15 cm) of W4.5 x W3.5 (.24 in x .21 in diameter), W7 x W3.5 (.3 in x .21 in), W9.5 x W4 (.34 in x .23 in), and W12 x W5 (.39 in x .25 in) in 8 ft wide mats.	Welded wire mesh, wrap around with additional backing mat and 1.4 in (6.35 mm) wire screen at the soil face (with geotextile or shotcrete, if desired).
Reinforced Soil Embankment (The Hilfiker Company 3900 Broadway Eureka, CA 95501)	6 in x 24 in (15 cm x 61 cm) welded wire mesh: W9.5 to W20 - .34 in to .505 in (8.8 mm to 12.8 mm) diameter.	Precast concrete unit 12 ft 6 in (3.8 m) long, 2 ft (61 cm) high. Cast in place concrete facing also used.
Tensar Geogrid System (The Tensar Corporation 1210 Citizens Parkway, Morrow, GA 30260)	Non-metallic polymeric grid mat made from high density polyethylene of polypropylene	Non-metallic polymeric grid mat (wrap around of the soil reinforcement grid with shotcrete finish, if desired), precast concrete units.
Miragrid System (Mirafi, Inc. PO Box 240967 Charlotte, NC 28224)	Non-metallic polymeric grid made of polyester multifilament yarns coated with latex acrylic.	Precast concrete units or grid wrap around soil.
Maccaferri Terramesh System (Maccaferri Gabions, Inc. 43A Governor Lane Blvd. Williamsport, MD 21795)	Continuous sheets of galvanized double twisted woven wire mesh with PVC coating.	Rock fill gabion baskets laced to reinforcement.
Geotextile Reinforced System	Continuous sheets of geotextiles at various vertical spacings.	Continuous sheets of geotextiles wrapped around (with shotcrete or gunite facing). Others possible.

[1]Many other facing types as compared to those listed, are possible with any specific system.

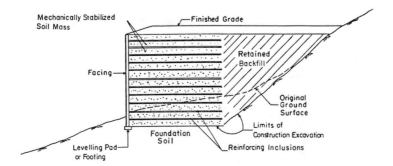

REINFORCED SOIL STRUCTURE — PRINCIPAL ELEMENTS

Figure 2.   Generic Cross Section of a Placed Soil Reinforced Structure and its
             Geotechnical Environment

In-Situ Reinforced Systems. The reinforcement generally consists of steel
bars, metal tubes, or other metal elements which resist not only tensile stresses,
but also shear stresses and bending moments. The reinforcements are installed
at relatively close spacings; e.g., one nail for each 10 to 60 sq.ft. (1 to 6 sq.m.).
They are either placed in drilled bore holes and grouted full length or driven
into the ground. They are not usually prestressed. The ground surface around
the nails is usually covered by shotcrete reinforced by wire mesh or by
intermittent plates or panels that act similarly to large washers on a bolt.
    Micro-piles, also called root piles, have been used for underpinning
structures, reinforcement of foundation soils and stabilization of slopes, with
the latter being a more recent application. The root piles are cast-in-place
concrete piles with diameters from 3 to 12 inches (75 to 300 mm). The smaller
diameter piles contain a centrally located steel bar or pipe; the larger diameter
piles may contain a reinforcing cage. Root pile systems differ from soil nailing
in that the piles are arranged in a three-dimensional pattern to form a network
that encompasses the soil and in which the piles interact with each other. In a
nailed system the nails act as independent resisting elements.

U.S. APPLICATIONS OF REINFORCED SOIL SYSTEMS

    Over the past 15 to 20 years reinforced soil systems have become very
widely used as retaining structures of various types, in embankments, and for
the support of open excavations. Reinforced soil is used also for bridge
abutments and wing walls, containment structures for water and waste
impoundments, dams and dikes, and for seawalls. They are particularly well
suited for economical construction in steep terrain, on unstable ground, and in
situations where ground deformation is likely. Reinforced in-situ ground (soil
nailing) is an economical means for excavation support, and stability is
achieved without the need for internal bracing.

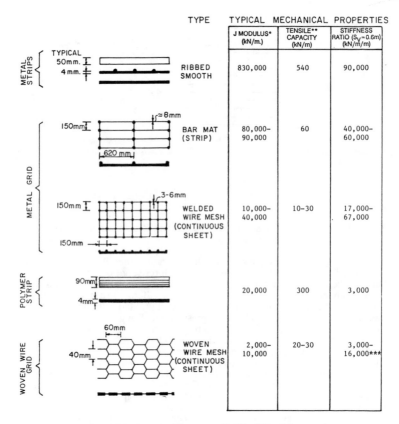

J represents the modulus in terms of force per unit width of the reinforcement.

* J = E(A$_C$/b) where: A$_C$ = total cross section of reinforcement material and

    b = width of reinforcement

    E = modulus of material

** Allowable values with no reduction for durability considerations

*** Confined

Figure 3.   Types and Mechanical Properties of Reinforcement Types
(Extended from Schlosser and Delage, 1987)

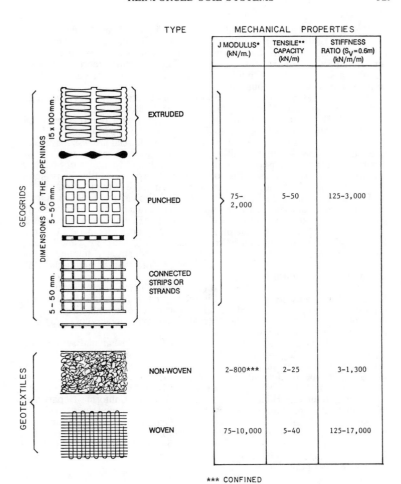

Figure 3 Continued

The first reinforced soil wall constructed in the United States was a Reinforced Earth system built in 1972 on California State Highway 39 northeast of Los Angeles. Since then, more than 4,500 walls have been built in the United States. Over 12000 Reinforced Earth structures representing over 50 million square feet (4.6 million square meters) of wall facing have been completed in 37 different countries.

As was indicated in Table 2 several other proprietary and non-proprietary systems have been used since the introduction of Reinforced Earth.

The Hilfiker Retaining Wall, which uses welded wire reinforcement and facing, was developed in the mid-1970's, and the first experimental wall was built in 1975 to confirm its feasibility. The first commercial use was made on a wall built for the Southern California Edison Power Company in 1977 for repair of roads along a power line in the San Gabriel Mountains of Southern California. In 1980, the use of welded wire wall expanded to larger projects, and to date about 1600 walls have been completed in the United States.

Hilfiker also developed the Reinforced Soil Embankment (RSE) system, which uses continuous welded wire reinforcement and a precast concrete facing system. The first experimental Reinforced Soil Embankment system was constructed in 1982. Its first use on a commercial project was in 1983 on State Highway 475 near the Hyde Park ski area northeast of Santa Fe, New Mexico. At that site, four reinforced soil structures were constructed totaling 17,400 square feet (1600 square meters) of wall face. More than 50 additional RSE systems have been constructed since then.

A system using strips of steel grid (or "bar mat") type reinforcement, VSL Retained Earth, was first constructed in the United States in 1981 in Hayward, California. Since then, 150 VSL Retained Earth projects containing over 600 walls totaling some 5 million square feet (465,000 square meters) of facing have been built in the United States.

The Mechanically Stabilized Embankment (MSE), a bar mat system, was developed by the California Department of Transportation based on their research studies started in 1973 on Reinforced Earth walls. The first wall using this bar mat type reinforcement system was built near Dunsmuir, California about two years later. Here, two walls were built for the re-alignment and widening of Interstate Highway No. 5. Since then, the California Department of Transportation has built numerous reinforced soil walls of various types.

Another "bar mat" reinforcing system, the Georgia Stabilized Embankment System was developed recently by the Georgia Department of Transportation, and the first wall using their technology was built for abutments at the I-85 and I-285 Interchange in southwest Atlanta. Many additional walls have been constructed using this system.

Polymeric geogrids for soil reinforcement were developed after 1980. The first use of geogrid in earth reinforcement started in 1981. Extensive marketing of geogrid products in the United States started in about 1983 by the Tensar Corporation. Since then over 300 wall and slope projects have been constructed using this type of reinforcement.

The use of geotextiles in reinforced soil walls started after the beneficial effect of reinforcement with geotextiles was noticed in highway embankments over weak subgrades. The first geotextile reinforced wall was constructed in France in 1971, and the first structure of this type in the United States was constructed in 1974. Since about 1980, the use of geotextiles in reinforced soil has increased significantly, with over 80 projects completed in North America (Yako and Christopher, 1988).

The highest vertical reinforced soil wall constructed in the U.S. to date is about 100 ft (30 m). Polymeric reinforced soil walls have been constructed to a height of 40 ft (13 m).

Soil nailing was first used in North America in Vancouver, B.C. in the late 1960's for temporary excavation support for industrial and residential buildings. Advancements in the construction of these systems has continued over the past several years, and it is now a widely used method for excavation support.

DURABILITY CONSIDERATIONS

As the required service life of reinforced soil structures may exceed 75 to 100 years for permanent structures, the durability of the reinforcements and, to some extent the facings, is an important factor. The life of structures with metallic reinforcements will depend on the corrosion resistance of the reinforcement. Practically all the metallic reinforcements used for construction of embankments and walls are made of galvanized mild steel. Epoxy coating can be used for corrosion protection, but it increases cost and is susceptible to construction damage, which can significantly reduce its effectiveness. Methods for the estimation of the required corrosion allowance under normal and aggressive soil conditions for galvanized reinforcements are given by Mitchell and Villet (1987) and Christopher et al. (1990).

Soil nails for permanent applications are generally protected against corrosion by the grout used during placement and by electrostatically applied resin bond epoxy.

Polymeric reinforcement, although not susceptible to corrosion, may degrade due to physico-chemical activity in the soil, such as hydrolysis, oxidation, and environmental stress cracking. In addition, it is susceptible to construction damage, and some forms may be adversely affected by prolonged exposure to ultraviolet light. Susceptibility to degradation varies widely among different polymer formulations. The durability of geosynthetics is a complex subject, and research is ongoing to develop reliable procedures for quantification of degradation effects. Moderate strength geosynthetics have tensile strengths of about 100 lbs/in (17.5 kN/m); some are now available that have strengths well over an order of magnitude higher. Current procedure to account for strength loss due to construction damage and as a result of aging and chemical and biological attack is to decrease the initial strength of the intact, unaged material for design. A minimum strength reduction of 10 percent is recommended for chemical and biological effects, and it may be up to 50 percent for construction damage (Christopher et al. 1990).

REINFORCED FILL MATERIALS

Well-graded, free draining granular material is usually specified for permanent placed soil reinforced walls. Lower quality materials are sometimes used in reinforced embankment slopes. Experience with cohesive backfills is limited; however, their low strength, creep properties and poor drainage characteristics make their use undesirable. Some current research is focused on the use of cohesive soil backfills.

The following gradation and plasticity limits have been established by AASHTO-AGC-ARTBA Joint Committee Task Force 27 for mechanically stabilized embankment:

U.S. Sieve Size	Percent Passing
4 inch	100
No. 40	0-60
No. 200	0-15

Plasticity Index (PI) less than 6 percent

It is recommended that the maximum particle size be limited to 3/4 inch (19 mm) for geosynthetics and epoxy coated reinforcements unless tests show that there is minimal construction damage if larger particle sizes are used.

SOILS SUITABLE FOR SOIL NAILING

In situ soil reinforcement using nails is most effective in dense granular and stiff, low plasticity silty clay soils. Soil nailing is not cost effective in loose granular soils with SPT N-values less than about 10 or a relative density less than 30 percent. In poorly-graded soils with a uniformity coefficient less than 2 nailing is not practical because of the necessity to stabilize the cut face prior to excavation by grouting or slurry wall construction. Soil nails will generally not be used in soft cohesive soils with undrained shear strength less than 0.5 tsf (48 kPa) because of the inability to develop adequate pullout resistance. Highly plastic clays and clays with a plasticity index greater than 20 percent may be unsuitable owing to excessive creep deformations.

SOIL-REINFORCEMENT INTERACTION

Soil reinforcement interaction is a key parameter in the design of reinforced soil systems and is usually evaluated by measuring pull-out resistance. An evaluation of the long-term pull-out resistance of the reinforcement; i.e., the ultimate tensile load to cause sliding of the reinforcement through the reinforced soil mass, is needed relative to three criteria:

1. Pull-out capacity - the pull-out resistance of each reinforcement should exceed the design tensile force by a specified factor of safety.
2. Allowable displacement - the relative soil to reinforcement displacement needed to mobilize the design tensile force should be less than the allowable displacement.
3. Long-term displacement - the pull-out load should be less than the critical creep load for the reinforcement or the soil, in the case of cohesive soils.

Reinforcement pull-out resistance is mobilized through one or a combination of two basic soil-reinforcement interaction mechanisms: interface friction and passive soil resistance against transverse elements of composite reinforcements such as bar mats, wire meshes, and geogrids, as shown schematically in Figure 4. The basic characteristics of pull-out performance in terms of load transfer mechanism, relative soil-to-reinforcement displacement required to fully mobilize the pull-out resistance, and creep potential of the reinforcement in granular or non-plastic soils are summarized in Table 3.

Several methods and design equations have been developed to estimate the pull-out resistance (Mitchell and Villet, 1987). They are difficult to compare for different reinforcements because they are based on different interaction parameters. To provide a common basis for different types of reinforcements,

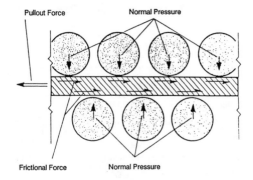

Pullout Force

Normal Pressure

Frictional Force

Normal Pressure

A) FRICTIONAL STRESS TRANSFER BETWEEN SOIL AND REINFORCEMENT

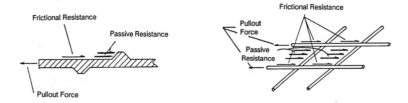

Frictional Resistance

Passive Resistance

Pullout Force

Frictional Resistance

Pullout Force

Passive Resistance

B) SOIL PASSIVE (BEARING) RESISTANCE ON REINFORCEMENT SURFACES

Figure 4. Stress Transfer Mechanisms for Soil Reinforcement

the following general relationship is proposed for the pull-out resistance per unit width of reinforcement, P_r:

$$P_r = F^* \cdot \alpha \cdot \sigma'_v \cdot L_e \cdot C \qquad (1)$$

in which: $L_e \cdot C$ = the total surface area of the reinforcement in the resistant zone behind the potential failure surface

L_e = the embedment or adherence length on the resisting zone behind the failure surface

C = the reinforcement effective unit perimeter: 2 for strips, grids, and sheets; π for nails

F^* = pull-out resistance or friction-bearing interaction factor

α = a scale effect correction factor

σ'_v = effective vertical stress at the soil-reinforcement interfaces

Table 3. Basic Aspects of Reinforcement Pull-out Performance in Granular Soils

	Generic Reinforcement Type	Major Load Transfer Mechanism	Displacement to Pull-out	Long Term Performance in Granular Soils
Mechanically Stabilized Embankments	Inextensible strips	Friction		
	Smooth	L.D.	0.05 in	Non-creeping
	Ribbed	H.D.	0.5 in	
	Extensible composite plastic strips (Paraweb)	Frictional	Dependent on extensibility	Dependent on structural or material creep
	Extensible sheets	Frictional (interlocking)	Dependent on extensibility	Dependent on material creep
	Geotextiles	L.D.		
	Inextensible grids			
	Bar mats	Passive H.D.	2 to 4 in	Non-creeping
	Welded wire meshes	Frictional +	0.5 to 0.8 in	
	Woven meshes	Passive H.D.		
	Extensible grids			
	Geogrids	Frictional + passive H.D.	Dependent on extensibility	Dependent on material creep
	Anchors	Passive	0.2 to 0.4 in	Non-creeping
Ground Reinforcement	Nails (ground reinforcement)	Frictional H.D.	0.08 to 0.12 in	Non-creeping

Note: **L.D. - low dilatancy**   H.D. - high dilatancy effect   1 in = 25.4 mm

The pull-out resistance factor F^* is most accurately obtained from pull-out tests. Alternatively, it can often be obtained from empirical or theoretical relationships given by the reinforcement supplier or in the literature. It may also be estimated from the general equation:

F^* = passive resistance + frictional resistance

$$= F_q \cdot \alpha_\beta + K \cdot \mu^* \cdot \alpha_f \qquad (2)$$

in which: F_q = embedment or surcharge bearing capacity factor
α_β = a structural geometric factor for passive resistance
K = ratio of the actual normal stress to the effective vertical stress, dependent on the reinforcement geometry
μ^* = apparent friction coefficient
α_f = structural geometric factor for frictional resistance

The scale effect correction factor α reflects the non-linearity of the P_r - L_e relationship and depends primarily on the extensibility of the reinforcement material. For inextensible reinforcements it is approximately 1; however, it can be significantly less than 1 for extensible reinforcements. It can be obtained from pull-out tests using reinforcements of different lengths or by calibrated analytical or numerical load-transfer models. Detailed presentation of the pull-out capacity design parameters is beyond the scope of this paper; however, they are given in the FHWA report (Christopher et al. 1990), as are analytical procedures for evaluating displacement and creep potential from pull-out tests.

DESIGN OF REINFORCED FILL WALLS

The specific design procedures used for various proprietary wall systems and for different reinforcement materials, as well as a number of design examples, are given in the NCHRP report (Mitchell and Villet, 1987). Detailed and simplified generic design guidelines common to all reinforced soil wall systems are given in the FHWA design manual (Christopher et al. 1990). Herein it is possible only to overview the generic design approach and design steps.

Reinforced soil wall design consists of determining the geometric and reinforcement requirements to prevent internal and external failure. The complete design of a reinforced soil wall involves:
  (1) the selection of the backfill soil and its placement conditions,
  (2) selection of the type of reinforcement,
  (3) selection of the facing material,
  (4) determination of the lengths and spacing of the reinforcements, and
  (5) specification of construction procedures as necessary.

Evaluation of both the external and internal stability of the reinforced structure are essential analyses that form a part of the design process.

The methodology in the FHWA design manual enables preliminary evaluations to assess the suitability of reinforced soil for a given project, the design of simple wall systems, and the checking of designs provided by others. It is limited based on experience to walls with near vertical faces with granular backfills that are up to 100 ft (30m) high for inextensible steel reinforcement and 50 ft (15 m) high for extensible polymer reinforcements. Either segmented panels or flexible facings are assumed, as well as adequate drainage of the backfill to prevent the development of hydrostatic pressures. Excluded are rigid, full-height wall facings, as they have been found to significantly alter the stress regime within the reinforced fill. The methodology has a significant empirical component drawn from exisiting proprietary design methods, performance evaluations on full scale structures, model test results, and theoretical and numerical analyses.

Extensible and Inextensible Reinforcements. The reinforcements restrain the lateral displacement of the backfill, and the extensibility of the reinforcements compared to the deformabilty of the fill controls the magnitude of the horizontal stress in the backfill (Mitchell, 1987; Adib, 1988). Closely spaced, inextensible reinforcements produce an unyielding mass which can retain essentially a K_0 state of stress. In fact, owing to compaction stresses against reinforcements with passive interaction components the horizontal stress may be greater than the conventionally calculated K_0 stress in the upper few feet of the structure. On the other hand, when extensible reinforcements are used, the fill yields laterally so that an active condition can develop before failure of the reinforcement.

Reinforcement Tension. The distribution of tensile forces in the reinforcements is generally as shown in Fig. 5(a). The locus of maximum tensile forces divides the reinforced fill into two zones, as shown in Figs. 5(b) and 5(c): the active zone which tends to pull away from the backfill and the resistant zone which holds the active zone in place. For analysis purposes the boundaries between the active and resistant zones are somewhat different for extensible and inextensible systems as shown.

Horizontal earth pressure. Current practice is to design the reinforcements to resist the active earth pressure in the case of extensible reinforcement systems; i.e., systems reinforced using geosynthetics. For inextensible reinforcements; i.e., ribbed steel strips and bar mats, a K_0 state of stress is assumed at the ground surface, with K decreasing to the active (K_a) value at a depth of 20 ft (6m). In the new FHWA design manual (Christopher et al. 1990) a procedure for selection of the appropriate horizontal earth pressure coefficient is proposed which is based on the composite system stiffness. This stiffness factor depends on both the stiffness of the reinforcing material and the amount of reinforcement per unit width of wall. Thus it reflects both the type and the amount of reinforcement in the wall.

Design Steps. The design of a reinforced soil wall with a near-vertical face and uniform reinforcement lengths involves the following steps. Complete details and an example are given in the FHWA design manual.

1. Establish design limits and external loadings as shown in Figure 6.
2. Determine the properties of the foundation soil.
3. Determine properties of both the backfill soil and the retained soil.
4. Establish design factors of safety and performance criteria.
   Recommended minimum factors of safety and deformation limits are:
   External Stability:

Sliding	1.5
Bearing capacity	2.0
Overturning	2.0
Deep seated	1.5
Settlement	project requirements
Horizontal displacements	face batter and project requirements
Seismic stability	> 75% static F.S.

   Internal Stability

Rupture strength	allowable reinforcement tension
Pull-out resistance	1.5
Seismic stability	1.1
Durability	depends on design life

5. Determine preliminary wall dimensions (Fig. 6).
6. Determine lateral earth pressures on back of wall.
7. Check external wall stability; adjust reinforcement length as necessary.
8. Estimate settlements using conventional methods.
9. Calculate maximum horizontal stresses in reinforcements at each level. The vertical stress in the soil at each level, composed of overburden fill, surcharge loads and stresses from any concentrated loads, is multiplied by the appropriate K coefficient. The resulting horizontal stress in the soil is then multiplied by the tributary area to a reinforcement to get the force to be resisted by the reinforcement.
10. Check internal static stability and determine required reinforcement for each layer.
11. Check internal seismic stability.
12. Estimate anticipated lateral displacements for construction control.
13. Prepare specifications.

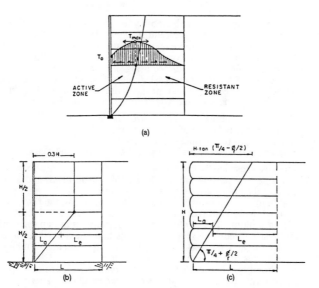

(a)

(b)   (c)

Figure 5.   Tensile Forces in the Reinforcements and Schematic Maximum
             Tensile Force Lines: (b) inextensible reinforcements,
             (c) extensible reinforcements

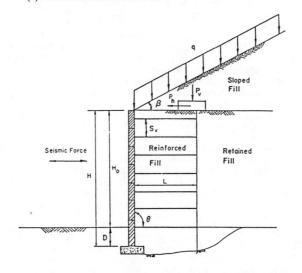

Figure 6.   Geometric and Loading Characteristics of a Reinforced Soil Wall

External Stability. External stability of the reinforced soil mass will generally control the length of the reinforcement; therefore, it is advantageous to perform this evaluation before evaluating the internal stability requirements. External stability involves the overall stability of the stabilized soil mass considered as a whole and is evaluated using slip surfaces outside the stabilized soil mass. Factors of safety for external stability are analyzed based on classical analysis of reinforced concrete and gravity wall type systems. External failure of the reinforced soil mass is generally assumed to be possible by:

- sliding of the stabilized soil mass over the foundation soil
- bearing capacity failure of the foundation soil
- overturning of the stabilized soil mass
- slip surface failures entirely outside the stabilized soil mass

In some cases, especially for complex geometrics, the critical slip surface is partially outside and partially inside the stabilized soil mass, and a combined external/internal stability analysis may be required.

Internal Stability. To provide an internally stable mass, the reinforcement is required to resist horizontal stresses so that it will not break, elongate excessively, or pullout. The reinforcement requirements are analyzed using the maximum tensile forces line (Figure 5). This line is assumed to be the most critical potential slip surface. The length of reinforcement extending beyond this line will thus be the available pullout length.

Seismic Stability. The performance of properly designed and constructed reinforced soil walls during earthquakes has been excellent. There are no cases of seismically induced damage or failure known to the authors. The bases for the seismic design recommendations in the new FHWA design manual are an unpublished report to the Reinforced Earth Company by H. B. Seed and J. K. Mitchell in 1981 and an extensive finite element study and half-scale shaking table model tests reported by Segrestin and Bastick (1988).

The results of this research have shown that the maximum tensile forces line is essentially the same as under static loading. The lower reinforcements are required to withstand the greatest dynamic increment, and the use of a pseudo-static dynamic thrust P_{AE} as proposed by Seed and Whitman (1970) for conventional retaining walls is appropriate. The design takes into account the tensile forces from the vertical overstresses due to P_{AE} and from an inertia force P_I acting on the active zone. The forces are then distributed among the different layers of reinforcements based on the tributary areas in the resistant zone.

The force and pressure diagrams and the equations necessary for analysis of the seismic stability of walls with inextensible reinforcements are given in Figures 7(a) and 7(b) for external and internal stability, respectively. In the relationships shown, α_{og} is the peak horizontal ground acceleration and α_{mg} is the maximum acceleration developed in the wall. The analysis for walls with extensible reinforcements would be based on an active zone geometry similar to that on Figure 5(c). The method should be safe for extensible reinforcement systems because the greater damping in less stiff systems and the high factor of safety on tensile stresses in the reinforcements to allow for creep under long-term static loads should compensate for any amplification of accelerations relative to those in stiffer walls.

Lateral Wall Displacement Evaluation. There is no standard method for prediction of the lateral displacements in reinforced soil walls, most of which occur during construction. The horizontal movements depend on compaction effects, reinforcement extensibility, reinforcement length, reinforcement to panel connection details, and deformability of the facing system. A crude

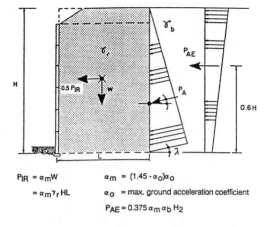

$$P_{IR} = \alpha_m W \qquad \alpha_m = (1.45 - \alpha_o)\alpha_o$$

$$= \alpha_m \gamma_r HL \qquad \alpha_o = \text{max. ground acceleration coefficient}$$

$$P_{AE} = 0.375 \, \alpha_m \, \alpha_b \, H_2$$

SAFETY FACTORS FOR SEISMIC - 75% STATIC

(a) External Stability Analysis

(1) Determine T_{m1}

$T_{m1} = S_v \sigma_h$

(2) Determine: Dynamic Increment

$T_{m1} = P_{IA} \dfrac{(R_c \cdot L_e)i}{\Sigma (R_c \cdot L_e)i}$

R_c = reinforcement coverage ratio = b/S_H

b = strip, grid, or sheet width

S_H = horizontal spacing

$T_m = T_{m1} + T_{m2} =$ max. tensile force

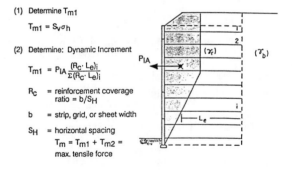

(b) Internal Stability Analysis

Figure 7. Seismic Stability Analysis of Reinforced Soil Wall

estimate of the probable lateral displacements of simple structures that may occur during construction can be made based on the reinforcement length to wall height ratio and reinforcement extensibility, as shown in Figure 8. Recent finite element analyses (Schmertmann et al, 1989) have shown that while reinforcement length has only little effect on the maximum tensions in the reinforcement, its effect on lateral deformation are large. For example decreasing the reinforcement length from 0.7H to 0.5H could increase the horizontal deformation by 50 percent.

The preceding design considerations are intended for simple wall geometries and typical soil conditions. Guidelines for more complex walls are given in the FHWA design manual.

Complex Stabilized Structures. The basic design methods generally consider simple stabilized structures with horizontal reinforcement layers having approximately the same length (Figure 2) throughout the full height of the structure. Although most structures fall into this category, more complex structures are sometimes built or considered at the design stage, including:

- Structures with inclusion layers of different lengths.
- Structures with inclusion layers of different inclinations.
- Structures with inclusion layers of different strengths.
- Structures with full height facing panels.
- Structures with multiple facings (or "stacked" wall designs).
- Structures supporting a sloping soil surcharge.
- Composite structures such as a Reinforced Earth wall constructed above a slope stabilized by soil nailing.

It is important to realize that the applicability of usual methods may be questionable in some cases. For example:

- Traditional soil mechanics methods used for evaluating external stability may not be applicable to complex structures because such structures may be far from being rigid bodies, or because their shape may make it difficult to use traditional methods.
- Semi-empirical methods developed in the past two decades to evaluate the internal stability of stabilized soil structures may not be appropriate for complex structures, because these methods have been established using measurements made on simple reinforced soil structures.

Consequently, in addition to the recommended design approach for simple structures, a global limit equilibrium analysis should be considered. The limit equilibrium analysis should consider three types of slip surfaces: (1) slip surfaces entirely inside the structure; (2) slip surfaces entirely outside the structure; and (3) slip surfaces partially inside and partially outside the structure. Other special design considerations for the specific complex structures listed above and guidance for each system are provided in the FHWA design manual (Christopher et al. 1990).

DESIGN OF REINFORCED SOIL SLOPES

The complete design of reinforced embankment slopes requires (1) selection of fill and determination of fill properties, (2) selection of reinforcement type, (3) determination of reinforcement lengths and spacings, and (4) preparation of specifications.

Reinforced slopes are analyzed using modified versions of classical limit equilibrium slope stability methods. A circular or wedge shaped potential failure surface is assumed, and the relationship between resisting and driving forces or moments determines the factor of safety. Reinforcement layers that intersect the potential failure surface are assumed to increase the resisting forces and moments based on their tensile capacity and orientation. The tensile capacity is taken as the minimum of the allowable pull-out resistance behind the failure surface and its allowable design strength. The shear and bending resistances of linear and planar reinforcements used for embankment reinforcement are neglected. Computer programs are available to carry out the needed calculations.

The orientation assumed for the reinforcement tensile force influences the calculated safety factor. The assumption that the tensile forces act always in the horizontal direction, as shown in Figure 9, is conservative in terms of moment equilibrium. Near failure, however, extensible reinforcements may elongate along the failure surface. The following inclination assumptions are recommended:

Inextensible reinforcements:	T parallel to reinforcements
Extensible reinforcements:	T tangent to the sliding surface

Reinforced Slope Design Steps. The steps for design of a reinforced soil slope, such as shown in Figure 9, are:
1. Establish the geometric and loading conditions.
2. Determine the properties of the natural soils.
3. Determine the properties of the available fills.
4. Establish the performance criteria: safety factors, allowable stress in reinforcement, durability criteria.
5. Evaluate the stability of the unreinforced slope.
6. Choose reinforcement lengths and distribution to provide a stable slope: stability of the reinforced mass against sliding, deep-seated failure, or excessive settlement.

Detailed guidance for these steps is given in the FHWA Design Manual. Several alternative approaches are also given in the FHWA Geotextile Engineering Manual (Christopher and Holtz, 1985). Recommended minimum factors of safety are:

External Stability	
Sliding	1.5
Deep-seated	1.3
Compound	1.3 for failure through reinforced zone
Dynamic loading	1.1
Settlement	project requirements
Internal Stability	
Slope stability	1.3
Allowable tension	see FHWA Design Manual
Pull-out resistance	1.5 for granular soils
	2.0 for cohesive soils
	3 ft (1 m) minimum length in resisting zone

Seismic Stability. Under seismic loading a reinforced soil slope is subjected to two additional forces that must be taken into account in the evaluation of external stability, as shown in Figure 10: (1) an inertia force P_I acting on the active zone, and (2) a dynamic thrust P_{AE} calculated according to the pseudo-

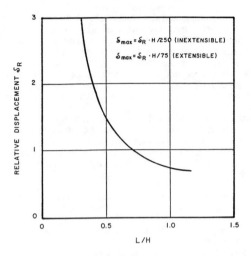

NOTE: Based on a 20 ft high wall, the relative displacement
increases approximately 25% for every 400 psf of
surcharge. Experience indicates that for higher walls
the surcharge effect may be greater.

Figure 8.   Curve for Estimation of Lateral Displacements Anticipated
at the End of Construction

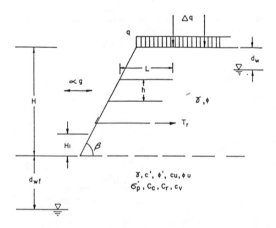

Figure 9.   Geometry and Parameters for Design of Reinforced Slopes

static dynamic earth pressure (Mononobe-Okabe). Only 60 percent of the inertia force is taken into account, because P_{AE} and P_I are unlikely to peak simultaneously. The internal stability is analyzed using only a horizontal pseudo-static force $(\alpha \cdot g \cdot m)$ in which α is the seismic coefficient, g is the acceleration due to gravity, and m is the mass of the sliding zone.

DESIGN OF NAILED SOIL RETAINING STRUCTURES

The stability of a nailed soil structure relies upon (1) the transfer of resisting tensile forces generated in the inclusions in the active zone into the ground in the resistant zone through friction or adhesion mobilized at the soil-nail interface and (2) passive resistance developed against the face of the nail. A detailed analysis of soil nailing and procedures for design are given by Elias and Juran (1988).

Ground nailing using closely spaced inclusions produces a composite coherent material. As shown in Figure 11 the tensile forces generated in the nails are considerably greater that those transmitted to the facing. The significant difference between the concept and behavior of nailed and anchored structures is also shown in the figure.

The design procedure for a nailed retaining structure includes (1) estimation of nail forces and location of the potential sliding surface, (2) selection of the reinforcement type, cross-sectional area, length, inclination, and spacing, and (3) verification that stability is maintained during and after excavation with an adequate factor of safety.

Methods for determination of the tensile, bending, and shear stresses in the nails are given in the FHWA design manual that are based on a limit equilibrium analysis.

Evaluation of a global safety factor that includes the nailed soil and the surrounding ground requires determination of the critical sliding surface. This surface may be located totally inside, totally outside, or partially inside and partially outside the nailed zone. Limit equilibrium methods are usually used, and the Davis method (Shen et al. 1981) is recommended because of its simplicity and availability in the public domain. The Davis method has been modified by Elias and Juran (1988) to permit input of interface limit lateral shear forces obtained from pull-out tests, separate geometric and strength data for each nail, facing inclination, and a ground slope at the top of the wall. The concrete facing elements (shotcrete, cast in place concrete, or prefabricated panels are considered for design to be analogous to a beam or raft of a unit width equal to the nail spacing supported by the nails.

CONSTRUCTION CONSIDERATIONS

The construction of reinforced soil systems is both simple and rapid. The construction sequence for reinforced soil walls consists of:
(1) preparing the subgrade,
(2) placing and compacting backfill in normal lift operations,
(3) laying the reinforcement layers into position, and
(4) installation of the facing elements.

Special skills or equipment are generally not needed, and locally available labor can usually be used. The vendors of proprietary systems often provide training in construction techniques for their systems.

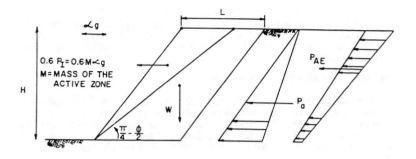

Figure 10. Forces for external stability analysis under seismic loading

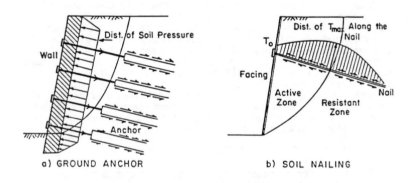

Figure 11. Comparison Between Ground Anchors and Soil Nailing

The installation of soil nails is equally simple, and follows a sequence of excavation, reinforcement installation, and facing placement at successive levels until the desired depth is reached.

The different construction steps for the various types of reinforcement systems are described in some detail in the FHWA design manual. Of particular importance is that facing panels be accurately positioned and aligned to insure both a proper fit and a proper appearance.

Large, smooth drum vibratory rollers are best for backfill compaction. Sheepsfoot rollers should not be used because of possible damage to reinforcements. Heavy rollers should not be used within about three feet of the wall face, however, because they may distort the wall face, overstress the reinforcements at face connections, and damage the facing panels. In this zone small, walk behind vibratory rollers or plate compactors are preferred.

Wrap-around facings of geotextile, geogrid, and wire mesh require careful installation to preserve a neat and uniform appearance. Temporary forms or blocking may be needed to assist in holding the material until the backfill has been placed and compacted. In slopes where no facing is used and when open meshes or grids are used as facings for slopes and walls an erosion control mesh or geotextile may be required. Such faces should be seeded as soon as practical.

CONCLUSION

From field observations, model tests, and analytical and numerical studies a great deal is now known about the behavior, design, construction, and performance of reinforced soil systems. The state of practice in mechanically stabilized soil systems has been reviewed briefly in this paper, with particular emphasis on U.S. procedures. Comprehensive reports and manuals are now available that contain information on the types and characteristics of different reinforcement systems and on their design and construction. With the aid of the information in this paper and in the cited reports and manuals, adequate information is at hand for any engineer to independently assess the feasibility of the use of earth reinforcement, to evaluate existing and proposed structures, and to carry out simple designs.

Areas that require continued research include the prediction and control of deformations, the analysis and design of structures of complex geometry, the effects of unusual loading configurations, the influences of ground, backfill, and reinforcement conditions on seismic response, the durability of different reinforcement types, the use of lower quality soils and the optimization of reinforcement dimensions.

REFERENCES

1. Adib, M.E., "Internal Lateral Earth Pressure in Earth Walls," Ph.D. Dissertation in Civil Engineering, University of California, Berkeley, 1988.
2. Christopher, B.R., Gill, S. A., Giroud, J.P., Juran, I., Mitchell, J.K., Schlosser, F. and Dunnicliff, J., "Design and Construction Guidelines for Reinforced Soil Structures - Volume I," and "Summary of Research - Volume II," Report No. FHWA-RD-89-043, Federal Highway Administration, U.S. Department of Transportation, 1990.

3. Christopher, B.R. and Holtz,R.D., "Geotextile Engineering Manual", Prepared for FHWA, National Highway Institute, Washington, DC, Contract No. DTFH61-80-C-00094, 1985.
4. Elias, V. and Juran, I., "Soil Nailing", Report for FHWA Contract No. DTFH-61-85-C-00142, 1988.
5. Jewell, R.A., "Material Requirements for Geotextiles and Geogrids in Reinforced Slope Applications", Proc. 23rd Int. Man-Made Fibres Congress, Dornbirn, Austria, Sept. 1984.
6. Mitchell, J.K., "Reinforcement for Earthwork Construction and Ground Stabilization", Proc. VIII Pan American Conference on Soil Mechanics and Foundation Engineering (1), Cartagena, Columbia, 1987, 349-380.
7. Mitchell, J.K. and Villet, W.C.B., "Reinforcement of Earth Slopes and Embankments", National Cooperative Highway Research Program Report No. 290, Transportation Research Board, Washington, DC, 1987.
8. Schlosser, F. and Delage, P., "Reinforced Soil Retaining Structures", in The Application of Polymeric Reinforcement in Soil Retaining Structures, P.M. Jarrett and A. McGown, ed., 1987, pp. 3-65.
9. Schmertmann, G.R., Chew, S.H., and Mitchell, J.K., "Finite Element Modeling of Reinforced Soil Wall Behavior", Report No. UCB/GT/89-01, Department of Civil Engineering, University of California, Berkeley, 1989.
10. Seed, H.B. and Whitman, R.V., "Design of Earth Retaining Structures for Dynamic Loads", Proc. ASCE Specialty Conference on Lateral Stresses and Earth Retaining Structures, Cornell University, Ithaca, NY, 1970, 103-147.
11. Segrestin, P. and Bastick, M.J., "Seismic Design of Reinforced Earth Retaining Walls: The Contribution of Finite Element Analysis", Proc. Int. Symposium on Theory and Practice of Earth Reinforcement, Kyushu, Japan 1988, 577-582.
12. Shen, C.K., Bang, S., Romstad, J.M., Kulchin, L. and Denatale, J.S., "Field Measurements of an Earth Support System", J. Geot. Eng., ASCE, 107 (12), 1981.
13. Yako, M.A. and Christopher, B.R., "Polymerically Reinforced Retaining Walls and Slopes in North America", in The Application of Polymeric Reinforcement in Soil Retaining Structures, P.M. Jarrett and A. McGown, ed., 1987, 239-282.

ACKNOWLEDGMENTS

The information in this paper is condensed from Report No. FHWA-RD-89-043, "Design and Construction Guidelines for Reinforced Soil Structures", 1989, prepared for the Federal Highway Administration under Contract DTFH61-84-C-00073. The second author of this paper was Principal Investigator and the first author was Technical Director for this project. Gary R. Schmertmann reviewed the manuscript and contributed useful suggestions.

MECHANICALLY STABILIZED EARTH RETAINING STRUCTURES IN EUROPE

BY F. SCHLOSSER [1]

ABSTRACT: the focus of this paper is to review the history and development of Mechanically Stabilized Earth Retaining Structures in Europe, starting from the ancient concept of crib walls to the recent technology of Texsol. As a result of the complex soil/reinforcement interactions occurring in the various systems, significant research, including reduced and full scale experiments, instrumentation of actual walls, and numerical analysis, has been performed to provide a better understanding of the behavior and aid in the development of appropriate design methods. This report will deal primarily with retaining walls constructed of reinforced backfill material, but will also briefly cover the in situ technique of soil nailing.

INTRODUCTION AND CLASSIFICATION

The term "mechanically stabilized earth retaining structure" (MSERS) incorporates a large number of different techniques, ranging from crib walls to multi-anchored walls, and more recently, reinforced soil structures. Although the idea to associate structural elements with soil in the construction of retaining walls is very ancient, during the last decades the European countries have played a major role in the development of this area, and for certain techniques a pioneering role, as in Reinforced Earth invented by Henri Vidal in the early 60's. As a result of an increase in the number of retaining walls in which the fill material contributes at least partially to the behavior of the wall, it is becoming more difficult to define exactly what is incorporated in the term "mechanically stabilized earth retaining structure". The scope of this paper is limited primarily to works utilizing fill material. However, there is an increasing tendency to construct earth retaining walls using a "mixed" structure, composed of a lower in-situ part and supporting an upper section built of fill material. For this reason, this state of knowledge report will touch on certain in-situ techniques, such as soil nailing.

---

[1]Professor, Ecole Nationale des Ponts et Chaussées, Paris, France; President-Director, Terresol, Puteaux, France.

It is possible to classify the MSERS into three general groups (Figure 1): crib walls; multi-anchored walls; and reinforced soil walls.

In a MSER wall, the reinforced soil mass behaves like a monolith, while still remaining relatively flexible. All of these systems are constructed of soil and more rigid structural elements (anchors, reinforcements, prefabricated wall elements), in which the soil/inclusion interaction constitutes one of the aspects characterizing the behavior. However, the behavior of these three systems is significantly different, with respect to the mechanism by which the soil within the reinforced mass participates in the internal stability.

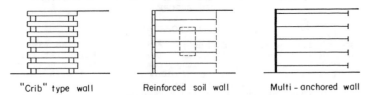

"Crib" type wall   Reinforced soil wall   Multi - anchored wall

Fig. 1: Main Types of Mechanically Stabilized Earth Retaining Structures.

This paper will review the European developments and experience in the field of MSERS. It will deal primarily with reinforced soil walls utilizing fill material and will encompass the following areas: European history and development; soil/inclusion interaction mechanisms; behavior and design of reinforced fill walls; and the recent technology of Texsol.

HISTORY AND DEVELOPMENT

Crib walls - Although crib walls have been utilized in the U.S. for more than 50 years, the first walls of this type were built centuries ago in the Alpine areas of Austria. The structural elements were generally made of wood (branches or tree trunks) with layers alternately oriented in the longitudinal and transversal directions. These cells were then filled with soil. Presently, they are commonly used, but also restricted to the construction of forest roads or temporary structures. In 1965, the technique of constructing crib walls using concrete prefabricated elements was begun in Austria, which lead to a very rapid and large development of this type of MSER wall in Austria (Figure 2). By 1980, more than 150,000 m^2 of crib walls had been built in Austria. The main advantages of this technique are: rapid construction and providing drainage for the structure.

A large research program was initiated in the 70's by the Austrian Ministry of Roads and Constructions in order to investigate the behavior of concrete crib walls and to produce appropriate design methods. The results and

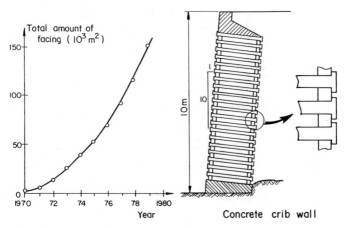

Fig. 2: Development of Crib Walls in Austria during 1970–
        1980 (Brandl, 1980).

information related to reduced scale models, full scale
experiments, observations, and failures have been published
by Brandl (12, 13). One of the main conclusions obtained
from this research is that the soil inside the cells
behaves similar to a silo, with primarily a convex distribution of vertical stresses acting on the horizontal planes.
The state of stresses in the cells are the at-rest earth
pressure (K_O). In addition, it was determined that the
pressure at the base is more uniform than predicted by the
gravity wall theory, as a result of the internal
deformation of the structure.
    For practical purposes, the most important factor is the
horizontal displacements of the structure. A vertical
concrete facing crib wall produces a maximum lateral
displacement at the middle of the wall, which is equal to
approximately 2% of the wall height (H) when the wall is
constructed simultaneously with the retained backfill, and
approximately 5% when the backfill is placed after
construction of the crib wall. An inclination of the
facing of 1V/10H reduces these lateral displacements by a
factor of 2. However, the displacements are still
relatively large, which may explain why this technique,
largely widespread in Austria, has until now not been
frequently utilized in the other European countries.
    There has been a large number of different types of
concrete crib walls produced during the two last decades.
The variation is related primarily to the type of structural
elements utilized, which consequently affects the behavior
and the design. However, it is possible to separate the
various types into two general groups, specifically, normal
crib walls with closed cells, in which silo pressure
conditions prevail, and the more recent crib walls, in which

the rear longitudinal elements are omitted for economical reasons. In this latter type, the silo effect is reduced and Brandl recommends that the design should incorporate the monolith theory or consider the analogy with a Reinforced Earth wall (13).

Multi-anchored walls – The first example of a multi-anchored technique in the field of MSERS was the Ladder Wall, invented and patented in 1929 by the famous French dam engineer André Coyne. In this system (Figure 3) the facing, generally inclined, consists of prefabricated panels placed one above the other, to allow some vertical deformation of the structure. The anchors ties are relatively short and consequently, the reinforcements (discrete or continuous) in the upper part of the wall are located within the Coulomb failure wedge. The Brest harbor wall (Figure 3), built in 1935, was the first major application of the system, and utilized anchor ties 1.7 m long for a wall height of 4.5 m. The inclination of the facing was 65° with respect to the horizontal.

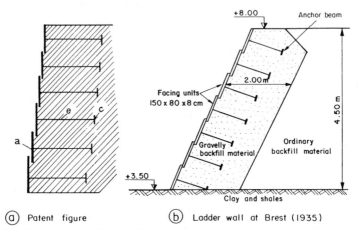

(a) Patent figure   (b) Ladder wall at Brest (1935)

Fig. 3: Coyne Ladder Wall System (Coyne, 1945).

Despite research on reduced scale models, Coyne did not fully explain the actual mechanism occuring in a Ladder Wall (16). He mentioned essentially that the structure formed by the facing, the anchors and the reinforced soil appeared to behave similarly to a monolith, sustaining small internal deformations, compared to the global displacements of the structure.

During World War II, the development of this technique was halted and despite Coyne's wishes, it was not frequently utilized during the reconstruction of France following the war. Following a long period in which the Ladder Wall system was not used, several procedures, related to this type of MSERS have been proposed, including the British

technique of Anchored Earth (37). In this technique, the
facing is made of vertical concrete panels and the anchor is
realized by twisting the ties at their free extremities,
thereby increasing the passive resistance to displacements.
A new multianchorage system, Actimur, has been developed in
France since 1984. It combines a vertical sheet-pile facing
and horizontal tie-rods fitted with metallic anchor disks.
However, despite interest in these new versions of the Coyne
Ladder Wall, this technique has not been very well
developed. Economically, it appears that it is more
advantageous to use the entire length of the reinforcements
in the mobilization of the soil-inclusion interaction, than
to concentrate this interaction only at the extremities.

Reinforced Soil Walls - The invention of Reinforced Earth
(R.E.) by the French architect and engineer Henri Vidal in
1963 and the rapid development of this new technique during
the late 60's (first wall built in France in 1965) was the
starting point for soil reinforcement techniques, especially
those relating to soil retaining systems where the
reinforcement is periodical and the soil/reinforcement
interation is mobilized along the entire reinforcement.

Figure 4 shows the amount (surface area) of facing
utilized in R.E. structures built around the world, per
year. It can be noted that the usage of R.E. is increasing
relatively linearly. Geographically, Europe represents 33%
of the total surface area, while the USA and Canada comprise
34% of the total market. Considering the distribution by
project type, 48% are related to urban highways and 21% to
mountainous highways, including essentially retaining walls
and bridge abutments.

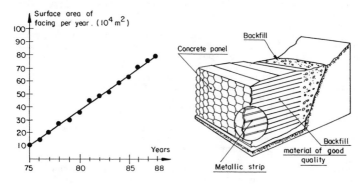

Fig. 4: Development of Reinforced Earth Walls Worldwide
        Since 1975.

Three events have had a significant effect on the
development of the R.E. technology. The first was the
choice to use galvanized steel for the strips and facing
(initially the facing was made of U-shaped metallic
elements). At the beginning, the use of fiber glass

reinforced polymers was proposed by Vidal. In 1966, an
experimental wall using fiberglass reinforced plastic failed
after only 10 months, apparently as the result of bacteria
attack. This resulted in the use of stainless steel and
aluminum strips for R.E. structures built in France.
However, currently (10 to 15 years after construction), in a
large amount of the walls built with stainless steel and
some built with aluminum strips, the inclusions are
corroded, indicating that these metals are not adequately
resistant against corrosion when embedded in soil.
Therefore, all R.E. structures currently utilize galvanized
steel strips for the reinforcements. The second event, was
the development of a standard cruciform facing panel in
1971. This type of facing allows various architectural
possibilities, including curved facings, and is presently
the worldwide trademark of the Reinforced Earth development.
The third event affecting the development of R.E.
structureswas the introduction of ribbed strips in 1975.
This new technological aspect was directly the result of
research on the soil/reinforcement frictional interaction.
The ribs increase the restrained dilatancy effect, which is
the main phenomenon in the three-dimensional friction
mechanism.

Two large research projects on Reinforced Earth were
performed in Europe from 1967 to 1977, including reduced
scale models and full scale experiments. The first project
was performed in France at the Laboratoire Central des Ponts
et Chaussées (LCPC), in connection with the inventor H.
Vidal (2,9). It provided an understanding of the fundamental behavior of R.E. and elaborated on the first design
methods and specifications. A second project was performed
some time later, at the Transport Road Research Laboratory,
in the U.K. (10,37). Although the Department of Transport
in the U.K. was convinced by the advantages of this
technique, they wanted to develop their own method.
Therefore, in addition to fundamental research, a large
technological study was performed relating to the facing,
strips, and metal corrosion.

From 1975 until the present, the Reinforced Earth
Companies around the world have performed a lot of applied
research, to understand the fundamental behavior of R.E.
structures, and the technological aspects related to
corrosion and degradability of the strips, different kinds
of facings and walls, and new types of R.E. abutments, dams
and other structures. Some of the main results from this
applied and fundamental research will be presented
hereafter.

After more than 10 years, the major developments in the
R.E. technique appear to be related partly to the following
features: good behavior, even under critical situations
(foundation soil movements, seismic events, etc.);
attractive and aesthetic facings; good durability when using
galvanized steel; and economy.

Polymeric Sheet and Grid Reinforced Soil Walls - The rapid
development and acceptance of the R.E. technique at the
beginning of the 70's gave motivations for finding
alternative reinforcing materials. This resulted in the use

of geotextiles in multi-layered reinforced soil retaining systems. In this technique, geotextile sheets provide simultaneously the planar reinforcements and the facing, which is constructed by wrapping the end of the sheet upwards in a "U" shape. The first geotextile reinforced wall was built in France in 1971 by the LCPC (42). It was an experimental wall using non-woven fabric (Bidim) and a very poor backfill material (wet clayey and sensitive soil). The wall was 4 m high and was founded on a very compressible soil (peat layer 3 m thick). Since this first application, geotextiles have been frequently used in reinforced soil retaining walls, as a consequence of their low cost, drainage properties, and the possibility of using poor backfill material. However, until now, their utilization has been rather limited, because of their high deformability (particularly in the case of non-woven geotextiles) and relatively unaesthetic appearance of the facing. In 1987, the French laboratory LCPC patented a technique of geotextile reinforced soil walls, called Ebal Wall, in which a precast concrete facing is placed in front of the soil reinforced wall in order to facilitate the erection of the wall, improve the aesthetic aspect of the facing, and to protect the geotextile facing from degradation by ultraviolet rays.

Metallic and polymeric grids have also been used as reinforcements, although Europe was not a pioneer in this area. The first grid reinforced soil retaining structure (welded wire bar mat) was constructed in the USA in 1974, along Interstate highway 5, near Dunsmuir, California (20). The bar mat/soil interaction is complex and involves both friction along the longitudinal bars and passive resistance against the transversal elements. Because of the mobilized passive resistance, bar mats are more resistant in pull-out than strips, but only for large displacements (5 to 10 cm). If the lateral displacements necessary to generate the passive resistance are acceptable for the structure, bar mat reinforcement permits the use of poor quality backfill material, with a relatively large fine-grain portion. In the early 80's, Netlon, in the UK, developed and manufactured a plastic grid reinforcement, called Tensar. This product consists of a high strength oriented polymer grid structure obtained from punched and stretched polymer sheets. Tensar has become rapidly accepted in a large variety of soil reinforcement applications (embankment reinforcement, retaining walls, rafts, repairs of slope failures, and gabions) and has resulted in a new type of two-dimensional reinforcement, called geogrids. Compared with non-woven geotextiles, geogrids exhibit a larger deformation modulus and tensile resistance. Up to now, geogrids have been widely used in all areas of soil reinforcement, except retaining walls. Similar to geotextiles, the problem related to the facing must still be solved, specifically, the unaesthetic appearance of the geogrid facing, difficulties in construction, and the method for attaching the geogrid to the prefabricated facing panels.

Three-dimensional Reinforcement: Texsol – The principle of the Texsol reinforcement method consists of mixing granular soil with several randomly placed continuous filaments to produce a composite material (32). The filaments are made of polyester with a diameter of 0.1 mm and a tensile strength of approximately 10kN. The first wall, similar in shape to a masonry wall, was built in France in 1983, with an external slope angle of 60°, with respect to the horizontal. This recent and interesting technique will be further discussed in a following section of this paper.

INTERACTION MECHANISMS

Review of the Mechanisms – The interaction between the soil and the structural elements of MSER walls is comprised of various mechanisms and can be different for the various techniques. In crib walls the main phenomenon is the development of the silo effect within the cells, composed of alternating transversal and longitudinal horizontal beams. The soil located inside the cell is initially at the K_o state of stress. However, as a result of the applied thrust at the back of the wall, the state of stress can change to the K_p state at the rear of the cell and the K_a state at the front. The thrust induces forces at the connections and bending in the elements, as a consequence of the action of lateral shear stresses and lateral pressures. These various forces result in the development of a complex interaction mechanism.

In multi-anchored walls, the primary interaction mechanism is the passive resistance mobilized between the soil and the anchor. Although studies have been performed to investigate the interaction mechanism generated in these two MSER systems, most of the research on interaction mechanisms has focused on reinforced soil walls. In this method, the primary interaction phenomenon is soil/inclusion interface friction, which generates tensile forces in the reinforcements. For planar and smooth reinforcements the friction along the inclusion has a relatively classical behavior. On the other hand, the soil/reinforcement interaction of linear inclusions, such as metallic strips, placed in dense granular soils involves a 3-D friction mechanism, as a consequence of the restrained dilatancy effect (44).

Geotextiles also mobilize their resistance by interface shear along their planar surface, which depends on the amount of soil interlocking within the fabric. However, unlike metal strip reinforcements, geotextiles cannot mobilize any restrained dilatancy effect. In addition to the generated friction, passive resistance against transversal elements in grids, and lateral pressure against longitudinal reinforcement exhibiting bending stiffness, can affect the resistance. Often, several different mechanisms are generated simultaneously, as in the case of the pull-out resistance of a bar mat (mobilized friction and passive resistance), or in the failure mechanism developed in the region where the failure surface intersects the reinforcement (friction and lateral resistance).

This latter mechanism is dependent on the orientation of the reinforcement. The effect of reinforcement orientation was initially studied in England by Jewell, who performed direct shear tests on reinforced soil samples, utilizing a plate inclusion placed at different orientations (24). Jewell demonstrated that the development of tensile forces in the reinforcement depends primarily on the inclination of the reinforcements, with respect to the sliding surface. When an inclusion is oriented in a direction of compressive strains, the reinforcement becomes ineffective, and can result in a decrease in the shear strength of the soil. Palmeira and Milligan (1989) further investigated this effect using a large scale direct shear box, and found that when sufficient bond was available to prevent pull-out failure, tensile failure of the rein-forcement could occur at very small bending deformations (bending angle β less than 5°).

Friction and Restrained Dilatancy Effect - The mobilized resistance along a longitudinal reinforcement can be influenced by the effect of the restrained dilatancy, resulting in the development of a 3-D friction mechanism. Pull-out of a linear inclusion (Figure 5a) induces shear displacements in the zone of soil surrounding the reinforcement. In a compacted granular soil, this zone tends to dilate, but the volume change is restrained by the surrounding soil, inducing an increase in the applied normal stresses on the inclusion. This behavior was first observed from the interpretation of pull-out tests on R.E. strips, which gave values for the coefficient of friction significantly greater than the actual values measured in the laboratory (44). This has led to the concept of an apparent coefficient of friction μ^*, which is related to the overburden pressure γh ($\mu^* = \tau/\gamma h$). The values for μ^* obtained from pull-out tests on actual structures are presented in Figure 5a, in addition to those obtained from shear tests performed with no volume change, which represents the extreme case of restrained dilatancy. It can be observed that μ^* (or tan ϕ^*), measured in the pull-out or direct shear tests conducted at constant volume, decreases with the initial normal stress ($\sigma_0 = \gamma h$) similarly to the well-known phenomenon of the decrease of soil dilatancy with the confining pressure.

The increase in the normal stress on the inclusion during pull-out was initially measured in West Germany by Wernick, using a special cell located on the pulled-out bar to measure simultaneously the applied shear and normal stresses (54). Wernick demonstrated that the stress path followed during pull-out rapidly attained the Mohr-Coulomb curve ($\tau = \sigma \tan\phi$), with a maximum increase in the normal stress of approximately four times the initial value. In France, Plumelle was able to determine the approximate volume of the sheared zone surrounding the inclusion during pull-out (40). He used a tie-rod placed in a compacted Fontainebleau sand embankment and pressure cells located at

different distances from the tie rods. As shown in Figure
5b, the influence of restrained dilatancy above the tie-rod
has an effect on the applied normal stress for a relatively
large distance from the inclusion (more than 3 diameters
above the tie-rod).

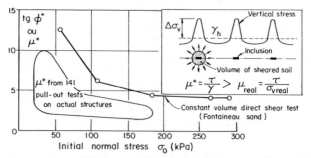

a)   Restrained Dilatancy Effect (Guilloux et al., 1979).

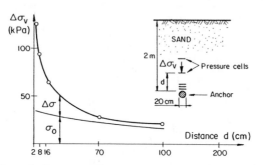

b)   Normal Pressure Measurements around a Tensioned
     Inclusion (Plumelle, 1984).

Fig. 5: Restrained Dilatancy Effect on the Soil-Linear
        Inclusion Friction.

    In soil nailing, the determination of the shear stresses
acting along the nails is not made by considering an
apparent coefficient of friction, but rather by predicting
the limit frictional shear stress τ_l, similar to the skin
friction developed along piles. It is interesting to note
that τ_l is practically independant of the depth, since
there is compensation between the decrease in dilatancy and
the increase of overburden pressure with depth. For
backfilled reinforced soil walls, in accordance with
specifications on Reinforced Earth, the apparent coefficient
of friction μ^* is taken to decrease linearly with depth
until 6 m, after which it becomes constant (35).

One parameter which influences the value of the apparent coefficient of friction μ^* is the compressibility of the soil surrounding the dilatant zone along a tensioned linear inclusion. This aspect has been theoretically and experimentally studied in France by Boulon et al. (11), by considering an analogy between the tensioned inclusion in the soil and special direct shear tests performed on plates at various controlled normal stiffnesses (Figure 6a). They assumed that the thickness (e) of the sheared zone (interface) along the inclusion is small compared to the radius (R) of the inclusion. By introducing the pressure-meter modulus E_M, for relating the normal stress to the soil volume change, the following formula can be obtained:

$$\sigma/u = 2E_M/R = k \qquad (1)$$

in which σ and u are the normal stress rate and radial displacement rate respectively. The expression $k = 2E_p/R$ is called the normal stiffness of the soil. Depending on the value of this parameter, the value of the limit shear stress τ_l is determined as a function of the initial normal stress $(\sigma_o = \gamma z)$ by charts derived from the results

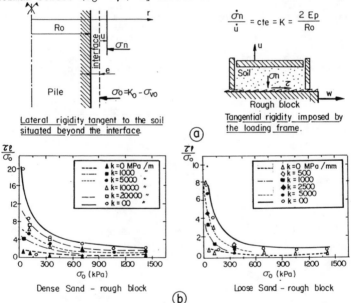

Fig. 6: a) Interface Shear Developed in an Axially Loaded Inclusion and one Subjected to Direct Shear; b) Influence of the Soil Stiffness on the Limit Lateral Friction (Boulon et al., 1986).

of direct shear tests on various plates. Figure 6b presents charts related to a rough plate, for both loose and dense sand samples.

One of the main features of the soil/inclusion frictional interaction is the small relative displacement required to completely mobilize the friction, generally on the order of magnitude of 1-5 mm. It explains why a small deformation of the reinforced soil mass can mobilize tensile forces in the reinforcements.

Passive Resistance - Passive resistance of the soil is mobilized in the soil/inclusion interaction when the inclusion is tensioned and includes protuding elements (large ribs, discs) and/or transversal members (bar mats), or when the inclusion has some bending stiffness and is sheared or bent along the zone of potential failure. Generally, in these cases, the overall resistance is mobilized by a combination of both friction and passive resistance. To analyze the respective importance of the two phenomena, Bacot (3) and Schlosser et al. (48) performed pull-out tests on bar mats. In addition, Morbois and Long (36) measured the mobilization of the pull-out resistance for rods equipped with circular transversal passive anchor plates (Figure 7). It is interesting to note that while the resistance mobilized by friction reaches a maximum for very small displacements at the head of the rod (≈ 5 mm), the displacements required to mobilize passive earth pressure are much greater (over 25 mm). As a consequence, when both mechanisms are generated, the friction is always completely mobilized before the passive resistance, indicating that the

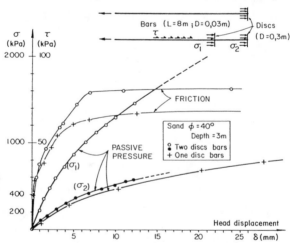

Fig. 7: Mobilization of Friction and Passive Pressure on Bars Equipped with Anchoring Discs (Morbois and Long, 1984).

soil   reinforcement   friction   is   the   most   important
interaction phenomenon.

REINFORCED SOIL SYSTEMS

General Behavior – In a reinforced soil retaining wall,
the earth mass is subdivided into two zones, the active zone
and the resistant zone, as shown for the first time by
Schlosser in 1971 (43). In the active zone, the soil tries
to move away from the structure, but is restrained by
friction developed along the inclusions. The mobilized shear
forces are directed toward the front of the wall, which
results in an increase of the tensile force with distance
from the facing. Consequently, the maximum tensile force in
the reinforcement does not occur at the wall facing, but
rather at some distance away from the facing. In the
resistant zone, the shear stresses are oriented away from
the facing and prevent slippage of the reinforcement at the
soil/inclusion interface.

From the results of reduced scale (9, 10, 33) and full
scale experiments (6, 7), in addition to instrumented R.E.
walls it can be observed that the maximum tensile force line
is significantly different from that predicted by the Mohr-
Coulomb failure wedge for classical retaining structures
(49). The presence of the horizontal and practically
inextensible metallic strips restrains lateral deformations
and consequently, completely changes the stress and strain
patterns in the soil (5). In a R.E. mass, the first family
of zero extension lines (β) coincides with the strip
direction, while the second family (α) is vertical and
corresponds to the direction of the potential failure plane
(Figure 8). This is quite different than the behavior
occuring in a classical retaining wall where the zero
extension lines are oriented at $\pm$ ($\pi/4 + \phi/2$) with respect

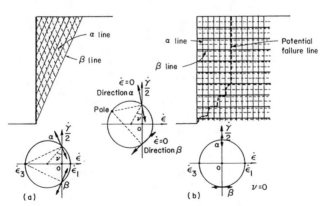

Fig. 8: Influence of Inclusions on the Potential Failure
        Lines (Bassett and Last, 1978).

element analysis seem to demonstrate that the maximum tensile force line moves back towards the Coulomb failure wedge for more extensible reinforcements (50).

Full scale experiments on reinforced soil walls, utilizing both inextensible and extensible reinforcements demonstrate differences in the structural behavior. Extensible reinforcements induce lateral displacements, especially at the top of the structure, which lead to the mobilization of an active state of stress (K_a). On the other hand, inextensible reinforcements seem to develop an at-rest (K_o) state of stress in the upper part of the structure, related to the small lateral displacements of the facing (Figure 9). Futhermore, the mechanism leading to an at-limit equilibrium state is more complex than occuring behind a classical vertical retaining wall. In a reinforced soil wall, large horizontal shear stresses can be mobilized in the vicinity of the reinforcements, even under very small lateral displacements (46). The K_o state of stress, observed in the upper part of a wall constructed with inextensible inclusions is probably the result of overstresses induced by compaction. In the lower part of the wall, the tensile forces in the reinforcements correspond to a state of stress below K_a, which is caused by the arching effect developed between the base and the upper section of the wall.

Failure of a reinforced soil mass can occur by two mechanisms, slippage of the reinforcement at the soil/inclusion interface, or breakage of the reinforcements. Slippage failure normally occurs for short inclusion length to wall height ratios (L/H), when the soil/inclusion interaction is not adequate to prevent pull-out. If the reinforcement has sufficient length beyond the active zone to prevent pull-out, failure will occur by breakage of the inclusions when the generated tensile force exceeds the tensile strength of the reinforcements. In the latter case, the failure surface will correspond to the location of the maximum tensile forces in the inclusions. Unlike classical retaining structures, which typically fail by the Mohr-Coulomb's failure wedge, reinforced soil walls develop a curved failure surface, that is vertical at the top of the backfill and intersects the top of the fill at a distance of 0.3 of the wall height H, (Figure 9). Theoretical calculations and actual field observations have shown that the failure surface in reinforced soil structures can be accurately represented by a log spiral curve (27).

Compaction and Arching Effects - The compaction process is a parameter which can affect the generated tensile forces in the inclusions, especially when they are inextensible. Compaction affects the tensile forces in the reinforcements by inducing additional overburden stresses and consequently lateral stresses in the soil, which are in turn transferred to the reinforcements. Since these stresses are largely plastic they are unrecoverable, leaving a high residual lateral pressure. The additional overburden stresses were found to decrease with depth and are a function of the type of compactor utilized (Figure 10). The increase in overburden pressure measured in an actual fill

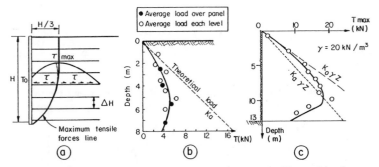

Fig. 9: Tensile Forces in the Inclusions; a) Theoretical
        Distribution for Reinforced Earth; b) Geotextile
        Reinforced Wall (John et al., 1983); c) Reinforced
        Earth Wall, Thionville, France, (T.A.I., 1989).

was found to correspond fairly well to those obtained from
finite element calculations and the theoretical method
proposed by Ingold in 1987 (23). At the top of the wall,
the measured values of the tensile forces are approximately
equal to the at-rest state of stress. However,
measurements of the facing displacements have demonstrated
that some horizontal movements do occur, even for
inextensible reinforcements, and therefore the tensile
forces in the reinforcements should correspond to a state of
stress less than K_O. These large tensile forces are the
result of compaction, which generates overstresses in the
fill and consequently induces additional tensile forces in
the inclusions (Figure 11). This effect is especially
significant in the upper reinforcement layers, since in the
lower inclusions there is a release of the overstresses as a
result of the lateral displacements generated by the
extension of the reinforcements, under the increase of the
overburden pressure. As a result of the low rigidity of
extensible reinforcements (geotextiles), the amount of
overstress is small and rapidly released. One limitation of
the finite element method is that in general, it does not
take into account the increase in tensile load within the
reinforcements resulting from the compaction process.
However, finite element analysis procedures have been
developed which allows the induced lateral stress generated
by compaction to be taken into account (15, 51), and the
results show relatively good agreement with the experimental
values.

    The high tensile stresses in the inclusions, at the top of
the wall, may also be related to the arching effect
developed in the reinforced soil mass. The arching of the
soil results in a stress transfer into the upper and lower
reinforcement levels, thereby increasing the tensile force.
However, at the base of the wall, the stresses will be
transfered into the foundation material, and will not be

felt by the lower inclusions. It is more likely that the high tensile stresses are a result of a combination of the compaction and the arching effects, rather than the inextensibility of the inclusions resulting in complete lateral restraint of the horizontal displacement at the top of the wall.

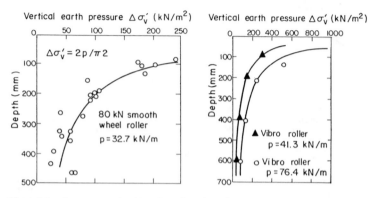

Fig. 10: Increase in the Overburden Pressure as a Result of Compaction (Ingold, 1987).

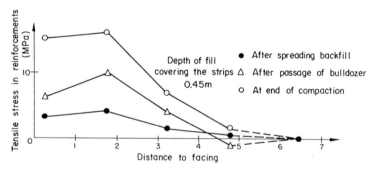

Fig. 11:   Effect of Compaction on the Tensile Forces in the Reinforcements: Granton Wall Great Britain, 1973 (T.A.I., 1989).

Facing Rigidity – The rigidity of the wall facing affects the development of the lateral pressure and displacement at the facing. To determine this effect, reduced scale tests were performed using different facing stiffnesses (34). The stiffness of the facing affects the behavior in two opposing fashions. Increasing the stiffness of the wall facing decreases the deformation of the wall while increasing the lateral pressure applied to the facing (Figure 12). It can be observed that the lateral pressure is reduced with a decrease in the facing stiffness, while the facing deformation only slightly increased for a decrease in stiffness. It therefore seems that an optimal stiffness exists in which the tensile forces in the inclusions can be reduced without exceeding the deformation criteria. McGown et al. have proposed the placement of a small compressible layer behind the wall facing of actual structures for the purpose of reducing the lateral pressure without increasing the facing deformation, providing that the thickness of the compressible layer is greater than the horizontal deformation generated in the reinforced mass (34).

From finite element calculations (6), it was determined that the ratio between the tensile force at the facing and the maximum tensile force (T_0/T_{max}) increases with the rigidity of the facing (Figure 13). For standard reinforced

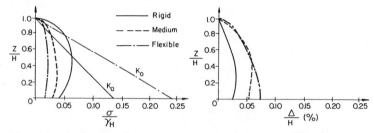

Fig. 12: Variation in the Lateral Facing Pressure and Facing Displacements for Different Facing Stiffnesses (McGown et al., 1987).

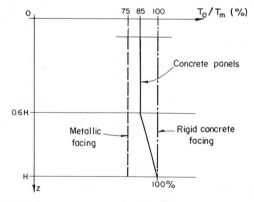

Fig. 13:   Ratio of the Facing Tensile Force to the Maximum Tensile Force for Various Types of Facings (Results from Bastick, 1984).

concrete facing panels, and for depths greater than 0.6H, it was found that the ratio can approach one.

Design Methods – There have been many design methods proposed for the design of reinforced soil structures (19, 22, 26, 45). The majority of these methods are based on limit equilibrium analysis and differ primarily in the shape of the assumed failure surface. Recently a new method has been proposed based on yield design by a homogenization method (17). In this method, the soil is considered as a homogeneous material with anisotropic properties. This theory provides an upper bound solution and has demonstrated a good correlation to the results obtained from reduced scale models (8). The currently utilized method to design R.E. walls is an approach proposed by Schlosser and Segrestin, which uses a working stress equilibrium analysis to check the stability along the potential failure surface, represented by the maximum tensile force line (45). A complete limit equilibrium method was further developed by Schlosser, resulting in the development of the TALREN program (47). This method is based on a multi-criteria analysis which takes into account the four following failure criteria: the tensile and bending strength of the reinforcement, the shear strength of the soil, the maximum soil/inclusion lateral friction, and the maximum lateral earth pressure on the reinforcements. This program is a very general program that is capable of taking into account all kinds of inclusions, and is not restricted to retaining structures built of reinforced soil.

To aid in the development of design methods for the correct dimensioning of reinforced soil structures, there has been significant research into the behavior of reinforced soil. Recently, this research has focused on full-scale experiments, instrumentation of actual structures, and numerical analysis using the finite element method. The behavior of R.E. and geosynthetic reinforced soil structures has been extensively studied and are therefore relatively well known. On the other hand, limited full scale research has been performed on determining the behavior of soil nailed walls. To provide a better understanding of the behavior, determine the limitations of the method, improve the design procedures, and elaborate on the design specifications, a research project, entitled "CLOUTERRE" was undertaken in France to study the behavior of soil nailed walls, relating to the three principal failure modes: breakage of the inclusions, slippage of the inclusions, or soil failure resulting from an excessive excavation phase without reinforcements (41).

BEHAVIOR OF REINFORCED SOIL WALLS

Reinforced Earth – The behavior of R.E. walls have been extensively studied over the past 20 years, with both reduced and full scale models, and numerical analysis utilizing the finite element method. Results of 3-D reduced scale models show that for L/H ratios less than 0.3, failure will occur as a result of a loss in adherence

between the soil and the strip (27). For larger L/H values
the scale walls failed by breakage, and the failure zone
corresponds to the line of maximum tensile force. From
reduced scale tests (2, 9, 31), it was determined that the
wall height to generate failure was greater than that
predicted by the Mohr-Coulomb failure wedge, and the failure
surface intersected the top of the backfill at a distance
from the facing, closer than that predicted by the Rankine
theory (27).

Based on the results obtained from reduced and full scale
tests, and finite element calculations, Terre Armee Int. has
recently postulated the bi-linear distribution for the
maximum tensile forces, presented in Figure 14. This
surface is applicable to most wall geometries, and for
surcharges located near the facing. For walls having a
surcharge at a large distance from the front of the wall,
the maximum tensile force line tends to migrate to the
surcharge. The proposed failure surface is seen to
accurately correlate to the results obtained from reduced
scale experiments, finite element calculations, and actual
R.E. walls, especially those obtained from real structures
(Figure 15). At the bottom of the wall, the results from
finite element and reduced scale models tend to
underpredict the location of the maximum tensile force.

A limitation of both reduced and full scale experiments
is the difficultly in evaluating the thrust applied to
the R.E. wall by the soil located behind the reinforced
mass. To evaluate the thrust effect, finite element
calculations have been performed by Terre Armée Int. (1, 52)
and demonstrate that the value of the maximum tensile force
and the location of T_{max} did not significantly vary for
L/H ratios ranging from 0.4 to 1.0 (Figure 16). Therefore it
could be concluded that the thrust had little effect on the

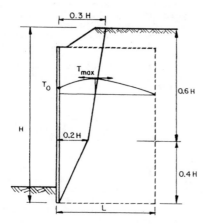

Fig. 14: Proposed Distribution of the Maximum Tensile Forces
        for Reinforced Earth Walls (Bastick, 1984).

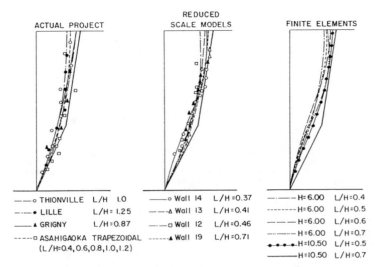

Fig. 15: Position of the Maximum Tensile Forces with Depth
for Various Test Methods (Bastick, 1984).

tensile force and consequently, the walls could be designed
using a vertical pressure equal to the applied overburden
pressure (γz). On the other hand, the results obtained by
Terre Armée Int. demonstrated that the maximum tensile force
was influenced by the slope angle of the embankment
supported by the reinforced mass, especially in the lower
reinforcement layers (Figure 17). Thus the thrust has an
effect on the mobilized tensile force, especially for walls
supporting sloped embankments. For this reason, the French
Specifications on Reinforced Earth have proposed using the
thrust at the back of the wall equal to the vertical
pressure determined from the Meyerhof distribution (35).
Recently, Terre Armée Int., as a result of a finite element
study, has proposed to take into account an inclination of
the thrust (β) given by the following formula:

$$\beta = (1.2 - L/H)\phi \tag{2}$$

where: ϕ is the soil internal friction angle.

As the wall becomes less slender (L/H increasing), the
inclination of the thrust is reduced, and becomes horizontal
for an L/H ratio of 1.2. This equation shows relatively
good correspondence to actual structures in which the
vertical pressure at the base of the wall was measured and
the thrust inclination could therefore be determined (6).
In addition to the finite element study, Terre Armée Int.
performed two full-scale experiments on walls reinforced

with short strips (0.4 < L/H < 0.45), and utilizing two
different cross-sections (7). The results (Figure 18)
demonstrated a good correlation to the proposed theoretical
distribution for the maximum tensile force. A comparison of
the results obtained from the full scale tests with the

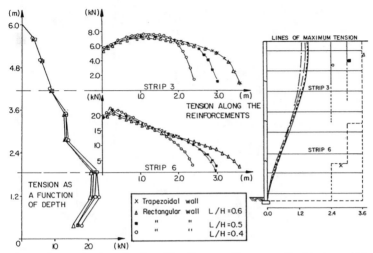

Fig. 16: Variation in Tensile Forces Along an Inclusion with
         Depth for Various L/H Ratios (T.A.I., 1989).

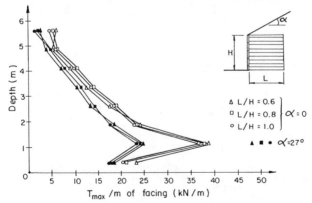

Fig. 17: Variation of the Maximum Tensile Forces with Depth
         Obtained from Finite Element Calculations (Results
         from T.A.I., 1988).

results obtained by finite element calculations, and dimensional analysis using the Meyerhof distribution is presented in Figure 19. It can be observed that the measured values are approximately equal to the calculated values, especially at the bottom of the wall. The best correlation exists between the experimental values and the numerical analysis considering the Meyerhof distribution and a horizontal thrust. In the full scale experiments, the vertical pressure was measured at the base of the wall. It can be noted that the actual distribution does not exactly correspond to that proposed by Meyerhof and would probably be better represented by a trapezoidal or triangular shape.

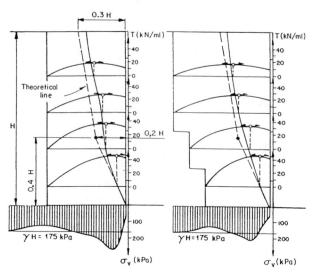

Fig. 18: Distribution of the Maximum Tensile Forces and Vertical Pressure at the Base, for Two Full-scale Reinforced Earth Walls (Bastick et al., 1989)

Soil Nailing - In both soil nailing and Reinforced Earth, the main phenomenon in the soil/inclusion interactions is the mobilized interface friction along the reinforcement, generating tensile forces in the inclusion. However, there are three principal differences in the behavior of these two methods. The first fundamental difference is the construction procedure, which can yield differences in the stress and strain patterns, especially those developed during the construction phase. The second principal difference is related to the stiffness of the reinforcement. In a soil nailed wall, the inclusions can resist both tensile forces and bending moments, while the R.E. strips, as a result of their low lateral rigidity, can only resist tensile forces. The third difference is related to the soil

type, specifically the cohesion, water content, and
heterogeneity of the soil which are utilized by these two
methods. As a result, the design method used for soil
nailing is presently quite different from that utilized for
Reinforced Earth.

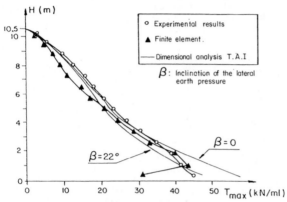

Fig. 19:   Theoretical and Experimental Distributions of the
           Maximum Tensile Forces in the Inclusions (Bastick
           et al., 1989).

    To compare the behavior of walls reinforced by both soil
nailing and R.E., Juran et al. in 1985 performed tests with
reduced scale models and compared the measured results to
the calculated values obtained from the finite element
method (28). By measuring the horizontal displacements and
the maximum tensile forces in the inclusions it was
determined that the horizontal deformation was significantly
greater for the soil nailed structure, especially after the
final construction phase. In addition, the maximum tensile
forces generated in the inclusion is larger for the soil
nailed wall. This behavior is related to the construction
process, which can significantly influence the behavior of
the structure. In R.E., placement of the successive,
compacted embankment layers results in settlement of the
structure. On the other hand, the excavation process
utilized in soil nailing results in horizontal movements at
the front of the wall, generating larger tensile stresses in
the inclusions, than those occuring in Reinforced Earth.
    From the numerical models, it was observed that for soil
nailing, the tensile forces corresponded to the at rest
state of stress at the top of the wall and decreased to
approximately the active stress state at the base of the
wall. Conversely, the calculated forces for R.E. were
approximately equal to the active state over the entire
height of the wall. In the reduced scale models, the
experimental behavior for the soil nailed structure was
similar to the calculated behavior. On the other hand, for

the R.E. wall, the stress state was equal to K_o at the top and decreased to approximately K_a at the bottom. This demonstrated that the compaction process has a significant effect on the behavior of reinforced soil walls which are constructed by compacted fill layers. The compaction results in increased lateral stresses generated in the reinforcement, and is especially important at the top of the wall. In the finite element calculations, the lateral stresses generated by compaction cannot usually be taken into account, resulting in an under-prediction of the tensile forces.

To aid in the proper dimensioning of soil nailed walls, three full scale tests were performed in France, within the framework of the CLOUTERRE project, to investigate the three principal failure modes of soil nailed walls. The results from the main full scale experiment, within the framework of the CLOUTERRE project are described in detail, in an accompanying paper to this conference (41). In the main full scale experiment, by measuring the horizontal displacement at various locations from the wall facing, it was determined that there was a zone extending between two and four meters from the facing, which is moving as a rigid block. This zone corresponds to the active zone of the soil nailed wall. The vertical displacements were of the same order of magnitude as those obtained for the horizontal displacements. The maximum horizontal ($\delta_h$max) and vertical displacements ($\delta_v$max) occured at the top of the wall, and the ratio ($\delta_v$max/$\delta_h$max) was approximately one for the final excavation phases, and greater than two for the early phases. During construction, the first resisting force mobilized, as a result of the excavation process, is the tensile forces in the nails. These forces are the primary component in the resistance of the inclusion and can be mobilized even under small displacements. Shortly prior to failure and under large deformations, the bending stiffness of the nails is mobilized, which prevented the soil nailed wall from completely failing. This result is consistent with that obtained from laboratory tests performed by Palmeira and Milligan (38), which were previously discussed.

In the main CLOUTERRE full scale experiment, the soil mass was instrumented with bands of black sand, which allowed the failure zone to be visualized after failure of the soil mass. It was observed that the line of maximum tensile forces and the maximum bending generated in the nails coincide with the actual failure zone observed in the soil.

Another factor which can affect the behavior of soil nailed structures is the orientation of the nails. It was determined that the horizontal displacement doubled when the nail was oriented at an angle of 30° to the horizontal axis (28). In addition, the tensile force in the reinforcement was less for the non-horizontal nail. In reinforced soil walls the zero extension line is horizontal, and therefore, the maximum tensile force will be generated when the reinforcement is oriented in the horizontal direction. This large tensile force will resist horizontal movements of the

wall facing, resulting in a reduction in the displacement
with respect to inclined nails.

Although the techniques of soil nailing and R.E. are
normally utilized alone, it is possible to construct a
"mixed structure" which incorporates a soil nailed mass as a
support for a reinforced backfill. In this process, an in-
situ soil nailed wall is constructed by excavation phases
and then once it is complete, a reinforced soil wall is
constructed by layers above the nailed wall.

Geosynthetic Reinforced Walls - Polymeric materials
(geosynthetics) have been considered as an alternative to
metallic reinforcements. However, a limitation of these
materials is that they posess a low stiffness, relative to
steel, and consequently, the amount of deformation required
to attain the maximum shear strength can exceed the
allowable displacement of the structure. In addition,
experience and research have shown that these materials can
also deteriorate with time, depending on the soil
environment. Because of the extensibility of the fabric,
the behavior of the inclusion is significantly different
than that for steel reinforcements.

The principal difference is in the developed strain
pattern. For steel reinforcements, the high stiffness of
the inclusion results in only small deformations in the soil
mass, particularly along the reinforcements. As a result,
the maximum tensile force line is vertical in the upper part
of the wall and intersects the ground surface at a distance
of approximately 0.3H from the wall. On the other hand,
geosynthetics, especially non-woven geotextiles, have a
relatively low stiffness and therefore undergo sufficient
deformation to attain an active state, in a manner similar
to that developed behind a translating retaining wall. As a
consequence, the maximum tensile force line for the
reinforcements shifts to the Mohr-Coulomb failure plane.
Furthermore, as a result of geosynthetics' extensibility,
the shear stress is concentrated at the front of the
reinforcement and decreases with length. Therefore, the
results of reduced scale tests cannot be accurately
interpolated to full scale walls with longer strips. In the
design of geosynthetic reinforced soil structures, it is
necessary to meet the strain compatability requirements.
Therefore the critical state friction angle (ϕ_{cv}) should
be used in the design since the strain necessary to reach
the maximum strength may be greater than the allowable
deformation, and the critical state value will still be
valid under large displacements (22).

Another factor which can affect the behavior of reinforced
soil walls using extensible reinforcements, is the creep
deformations which can occur with time. In walls utilizing
steel inclusions (R.E., soil nailing, steel grids, etc.),
there is no creep of the reinforcements. On the other hand,
for some fabrics, the creep developed in the inclusion can
be significant and the fabric may develop strains in the
long term above the allowable level.

In Europe, geotextiles and geogrids have also been
extensively utilized to reinforce soil. To provide a better
understanding of the behavior of reinforced soil utilizing

extensible reinforcements, many of these structures where
instrumented to provide measurements on the strains
developed during construction and with time. Similar to
inextensible reinforcements, the tensile forces in
geosynthetics at the top of the wall can approach (or
exceed) the at-rest state of stress as a result of the prestress generated in the inclusions by the compaction
process. Because of the low rigidity of geosynthetics, the
maximum tensile forces rapidly decrease with depth and can
approach (or be less) than that corresponding to the active
state of stress.

In geosynthetic reinforced soil walls, the design is
usually controlled by the allowable deformation of the
reinforcements. Therefore, the main aim of instrumenting
the reinforcements in actual reinforced soil structures
constructed in France was to determine the effect of
construction on the measured strains along the reinforcement
and measure the developed creep strains. From the strain
measurements, it was determined that the strain generated in
the reinforcement during the construction of a layer was
normally between 70 and 95% of the total strain developed
during the construction process. The compaction effort near
the facing results in the straining of the layers where the
forms had already been removed. In addition, if the soil is
not adequately compacted near the facing, the soil mass can
deform upon removal of the forms, resulting in the
generation of strains near the facing which are
approximately equal to those developed in the layer by the
end of construction. The construction of a slope on top on
the reinforced wall resulted in additional strains being
mobilized along the inclusion, away from the facing.

The surface resistance of polymers has necessitated
investigation of the effect of damage occuring in the
inclusion during the compaction process. The damage is
normally assessed by visual inspection and tensile strength
tests on virgin samples and samples taken from the
constructed wall after construction. These tests
demonstrated that the majority of the damage was surface
abrasion of the inclusion, which had a negligable effect on
the tensile and creep strength and peak strain in the
inclusion. On the other hand, the strain required to reach
failure increased by 25% for the samples tested after
construction (14). Tests reported by Delmas et al. in 1988
also demonstrated that in general, the construction
procedure does not significantly change the tensile strength
of the fabric (18). However, it must be noted that in
certain cases, especially for sharp, angular soils the
tensile strength and the elongation at failure can be
reduced. Perrier and Lozach found that the tensile strength
of various geofabric could be reduced by up to 35% as a
result of physical damage resulting from compaction (39).

TEXSOL

Since the development of the R.E. technique, extensive
research has continued in the area of soil reinforcement.
During the past decade, this research has resulted in the

development of new techniques, specifically Texsol. Texsol is a three-dimensional reinforcement technique, which was developed in France at the Central Laboratory for Bridges and Roads (LCPC) in 1980. The procedure consists of mixing continuous polymer fibers, usually polyester with a diameter of approximately 0.1 mm, with granular soil, thus producing a composite material capable of resisting tensile forces. By the end of 1988, 85 projects had been built in France, utilizing 100,000 m^3 of Texsol reinforced soil.

Continuous fiber reinforcement (Texsol) demonstrates some similar behaviors to that obtained for soils reinforced with individual fibers. The placement of the fibers in the soil increases the soil's strength by the development of an artificial cohesion and also by increasing the friction angle. The shear resistance of Texsol results from soil/fiber interaction generated at the contacts between the soil grains and the fibers, and by interlocking of the soil particles in the fiber framework. From triaxial tests, the increase in the friction angle ranges from 0 to 10°, while the apparent cohesion was equal to 100 kPa/0.1% fiber content. The placement method for Texsol produces an anistropic material, and therefore, the shear strength is affected by the orientation of the fiber placement with respect to the shear plane. The maximum apparent cohesion was measured when the thread was oriented perpendicular to the failure plane. On the other hand, there was almost no apparent cohesion mobilized when the fiber or threads were placed parallel to the shear surface (Figure 20).

One advantage of the Texsol procedure is that in general, it has the possibility of utilizing soil on or near the job site as its base material. The method is efficient for use with silty sands and gravels. However, if there is a possibility of the reinforced mass becoming saturated, it is essential to utilize clean sands and gravels and consequently, all Texsol reinforced walls are presently built utilizing clean sand (4).

To acquire a better understanding of the behavior of Texsol reinforced structures a series of 6 full-scale experiments were built in France by the Centre d'Experimentation Routiere de Rouen between 1987 and 1988, and conducted until failure by either an applied surcharge or saturation of the wall (29, 30). Five walls (3 m high) were utilized to retain a backfill and were loaded to failure by an applied surcharge. The sixth wall (5 m) was constructed as a vertical free-standing wall, which allowed the load to be applied by increasing water pressure at the back of the wall. The method of surcharge, the density, and the densification procedure were varied for the walls.

The five full scale experiments utilizing an applied surcharge demonstrated that the surcharge method, soil density, and the compaction procedure had little effect on the ultimate load required to generate failure, with the five walls failing under an applied surcharge of approximately 75-78 kPa (29). On the other hand, the behavior of the walls was significantly affected by the various construction and test conditions. Two of the walls were tested under identical conditions except for the

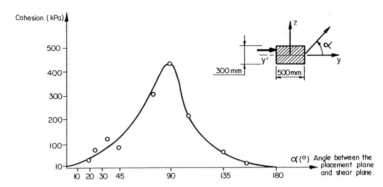

Fig. 20:   Anisotropic Cohesion of Texsol as a Function of the Angle between the Placement and the Shear Planes (Baguelin, 1989).

orientation of the deposit plane for the Texsol fiber. It was found that the wall in which the fibers were deposited horizontally exhibited three times more horizontal displacement than when the fibers were placed at a 10° angle with the horizontal. In all the walls, collapse occured by overturning of the upper section of the wall, resulting in complete failure of the structure.

The sixth wall was loaded by applying water pressure at the back of the wall (30). The water level was raised by increments until failure occured. The wall displayed monolithic tilting about its toe, with the maximum horizontal displacement occurring at the top of the wall. Failure of this wall occured when the water level reached 2.96 m, which was approximately 2/3 of the wall height.

In two of the walls, the vertical displacements at the top of the wall was measured in addition to the horizontal displacements. In the wall loaded by increasing water pressure, the maximum vertical and horizontal displacement measured just prior to failure gave a displacement ratio ($\delta_v$max/$\delta_h$max) of 0.8. For the other wall, it was determined that the maximum displacement occured at a distance of 0.5 m from the top of the wall and the displacement ratio was equal to 0.2. However, it must be noted that in this wall the load was applied by a surcharge, therefore resulting in more significant settlements of the structure.

Based on the results of laboratory and full scale experiments, Texsol walls are generally constructed with polyester fiber, approximately 0.1 mm in diameter, placed in clean sand in the amount of 0.2% of the soil weight. The fiber reinforced soil is compacted at a density corresponding to the Standard Proctor value. In the design of Texsol walls, both the internal staility, related to rotation of the wall, and the external stability, related to bearing capacity failure, sliding along the base, and rotation about the toe, must be verified. The design safety factor for Texsol walls ranges from 1.3-1.5, depending on the structure type. For walls embedded in the foundation soil, the embedment depth must be taken into account in the internal stability (4).

CONCLUSION

As demonstrated by this paper, the European experience in MSERS is relatively extensive. From the development of crib walls in Austria two centuries ago, to the invention of R.E., with the rapid development of reinforced soil structures, European engineers have contributed to numerous new soil reinforcement techniques. It is interesting to note that generally with these techniques, actual structures were built before the mechanisms were well understood. As a result, a large effort has been made during the last two decades to better understand the fundamental mechanisms and interactions between the structural elements and soil of MSERS. The silo effect in crib walls, the 3-D friction with the restrained dilatancy effect in strip reinforcement, and the durability and material properties of the reinforcements are some of the important aspects currently studied. New techniques are presently being developed, which demonstrates that the European effort to develop new reinforcement methods will continue.

APPENDIX. REFERENCES

1. Anderson, P.L. and Bastick, M.J. (1983). Finite Element Study of Reinforced Earth Structures using the Rosalie Program (10.5 m walls with short strips), Internal Report #R31, Terre Armée Int., December, 320 p.
2. Bacot, J. and Lareal, P. (1973). Etude sur modèles rèduits tridimensionnels de la rupture de massifs en Terre Armée, Revue Travaux, No. 463.
3. Bacot, J. (1981). Contribution à l'étude de frottement entre une inclusion souple et un matériau pluvérulent: cas de la Terre Armée, Doctorate thesis, University of Lyon, France.
4. Baguelin, F. (1989). Texsol: Retaining Works, Technical Report for the Laboratoire Central des Ponts et Chaussées (LCPC), 52 p.
5. Bassett, R.H. and Last, H.C. (1978). Reinforced Earth below Footings and Embankments. ASCE Symp. Earth Reinforcement, Pittsburgh, 222-231.
6. Bastick, M.J. (1984). Behavior of Reinforced Earth Walls using Short Strips, Internal Report #R34, Terre Armée Int., July, 94 p.
7. Bastick, M., Schlosser, F., Amar, S. and Canepa, Y. (1989). Monitoring of a Prototype Reinforced Earth Wall with Short Strips, Proc. 12th I.C.S.M.F.E, Rio de Janiero, 1221-1222.
8. Ben Assila, A. and El Amri, M. (1984). Comportement à la rupture des murs en Terre Armée et clouée, Thesis, Ecole Nationale des Ponts and Chaussées, Paris.

9. Binquet, J. and Carlier, P. (1973). Etude expérimentale de la rupture de murs en Terre Armée sur modèles Tridimensionnels, Internal Report, LCPC.

10. Boden, J.B., Irwin, M.J. and Pocock, R.G. (1977). Construction of Experimental Reinforced Earth Wall, TRRL Symp. Reinforced Earth and Other Composite Soil Techniques, Edinburgh, Supplementary Report No. 457, Transport and Road Research Laboratory, U.K.

11. Boulon, M., Plytas, C. and Foray, P. (1986) Comportement d'Interface et prévision du frottement latéral le long des pieux dt tirants d'ancrage, Revue Française de Géotechnique, Vol. 2, 31-48.

12. Brandl, H. (1980). Behavior and Design of Crib Walls, Report #141 for the Bundesministerium Fur Bauten und Technik, Vienna, Austria, 222 p.

13. Brandl, H. (1982) Design and Construction of Crib Walls, Report #208 for the Bundesministerium Fur Bauten und Technik, Vienna, Austria, 287 p.

14. Bush, D.I. and Swan, D.G. (1987). An Assessment of the Resistance of Tensar SR2 to Physical Damage during the Construction and Testing of a Reinforced Soil Wall, NATO Adv. Research Workhop Appl. Polymeric Reinf. in Soil Retaining Structures, Royal College of Canada, Kingston, Ontario, 8 p.

15. Collin, J. (1986) Earth Wall Design, Ph.D Dissertation, University of California, Department of Civil Engineering, 440 p.

16. Coyne, A. (1945) Murs de Soutènement et Murs de Quai "à Echelle", Le Génie Civil, May.

17. De Buhan, P., Mangiavacchi, R., Nova, R., Pellegrini, G. and Salençon, J. (1989). Yield Design of Reinforced Earth Walls by a Homogenization Method, Géotechnique, 39(2), 189-201.

18. Delmas, P., Gourc, J.P., Blivet, J.C. and Matichard, Y. (1988). Geotextile-Reinforced Retaining Structures: A Few Instrumented Examples, Proc. Int. Symp. Theory and Practice of Earth Reinforcement, Japan, 511-516.

19. Department of Transport, U.K. (1978). Reinforced Earth Retaining Wall and Bridge Abutments for Embankments, Technical Memorandum BE 3/78, Highway Procedures and Legislative Division, London.

20. Forsyth, A. (1978). Alternative Earth Reinforcements, ASCE Symp. Earth Reinforcement, Pittsburgh, 358-370.

21. Guilloux, A., Schlosser, F. and Long, N.T. (1979). Laboratory Study of Friction between Soil and Reinforcements, Proc. Int. Conf. Soil Reinforcement, Vol 1, Paris, 35-40.

22. Ingold, T.S. (1982). An Analytical Study of Geotextile Reinforced Embankments, Proc. 2nd Int. Conf. on Geotextiles, Las Vegas.

23. Ingold, T.S. (1987). Compaction-Induced Earth Pressures under K_O - Conditions, discussion submitted to ASCE, Journal of the Geotech. Eng. Div., ASCE, 113(11), 1403-1405.

24. Jewell, R.A. (1980) Some Effects of Reinforcement on the Mechanical Behavior of Soil, Ph.D thesis, University of Cambridge, England.

25. John, N.W.M., Ritson, R., Johnson, P.B., Petley, D.J. (1983). Insturmentation of Reinforced Soil Walls, Proc. 8th E.C.S.M.F.E., Vol. 2, Helsinki, 509-512.
26. Juran, I. (1977) Dimensionnement interne des ouvrages en Terre Armée, Thesis for Doctorate of Engineering, Laboratoire Central des Ponts et Chaussées, Paris.
27. Juran, I. and Schlosser, F. (1979). Theoretical Study of Traction Forces in Slabs of Reinforced Earth Structures, Proc. Int. Conf. Soil Reinforcement, Vol. 1, 77-82.
28. Juran, I, Shafiee, S. and Schlosser, F. (1985) Numerical Study of Nailed Soil Retaining Sturctures, Proc. 11th I.C.S.M.F.E., Vol. 3, San Francisco, 1713-1716.
29. Khay, M., Mascre, D., Trufley, H. and Vinceslas, G. (1987). Texsol Application: Retaining Wall. Experimental Works C.E.R., Internal Report for the Centre d'Experimentation Routiere, 80 p.
30. Khay, M., Mascre, D., Trufley, H. and Vinceslas, G. (1988). Texsol Experimental Wall No. 6 with a Vertical Facing. Test Until Failure by Hydrostatic Loading, Internal Report for the Centre d'Experimentation Routiere, 80 p.
31. Lee, K.L., Adams, B. and Vagneron, J. (1973). Reinforced Earth Retaining Walls, Journal of the Soil Mech. and Found. Div., ASCE, Vol. 99.
32. Leflaive, E. (1988). TEXSOL: Already more than 50 Successful Applications, Proc. Int. Symp. Theory and Practice of Earth Reinforcement, Japan, 541-545.
33. Legeay, G. (1978). Etude sur modèles réduits de murs en Terre Armée, Thesis, Université Pierre et Marie Curie.
34. McGown, A., Andrawes, K.Z., and Murray, R.T. (1987) The Influence of Lateral Boundary Yielding on the Stresses Exerted by Backfills, Soil-Structure Interactions, a collection organized by the Ecole Natonale des Ponts et Chaussées, Paris, 585-592.
35. Ministere des Transport (SETRA). (1979). Les Ouvrages en Terre Armée: Recommandandations et Régles de l'Art, Internal Report for the Direction des Routes et de la Circulation Routière, 194 p.
36. Morbois, A. and Long N. (1984) Etude de procédé Actimur: Rapport de recherche. Laboratoire Central des Ponts et Chaussées, Paris.
37. Murray, R.T. and Irwin, M.J. (1981). A Preliminary Study of TRRL Anchored Earth, TRRL Supp. Report #674.
38. Palmeira, E.M. and Milligan, G.E. (1989). Large Scale Direct Shear Tests on Reinforced Soil, Japanese Society of Soil Mech. and Found. Eng., 29(1), 18-31.
39. Perrier, H. and Lozach, D. (1986) Essai de poinconnement sur geotextile - Influence des sollicitations de compactage. Compte rendu de mesures, Internal Report for the Laboratoire Centrale des Ponts et Chaussées.
40. Plumelle, C. (1984). Improvement of the Bearing Capacity of Soil by Inserts of Group and Reticulated Micropiles, Proc. Int. Conf. on In situ Soil and Rock Reinforcement, Paris, 83-89.

41. Plumelle, C., Schlosser, F., Delage, P. and Knochenmus, G. (1990). French National Research Project on Soil Nailing: Clouterre, paper accepted for publication to the 1990 Specialty Conf. on Design and Performance of Retaining Structures, Cornell, 16 p.
42. Puig, J., Blivet J.C. and Pasquet, P. (1977). Remblai armé avec un textile synthétique, Int. Conf. on the Use of Fabrics in Geotechnics, Ecole Nationale des Ponts et Chaussées, Vol. 1, Paris, 85-89.
43. Schlosser, F. (1971). Terre Armée: recherches et réalisations, Bulletin de Liaison des LCPC, No. 62).
44. Schlosser, F. and Elias, V. (1978). Friction in Reinforced Earth, ASCE Symp. Earth Reinf., Pittsburgh, 735-763.
45. Schlosser, F. and Segrestin, P. (1979). Dimensionnement des ouvrages en Terre Armée par la methode de l'equilibre local, Proc. Int. Conf. Soil Reinforcement Vol. 1, Paris.
46. Schlosser, F. and Mitchell, J.K. (1979). General Report, Proc. Int. Conf. Soil Reinf., Vol. 3, Paris, 25-62.
47. Schlosser, F. (1980). Internal Report on the TALREN Design Method, Ecole Nationale des Ponts et Chaussées, 20 p.
48. Schlosser, F., Jacobsen, H. and Juran, I. (1983). Soil Reinforcement, General Report for Specialty Session No. 5, Proc. 8th European Conf. on Soil Mechanics and Foundation Engineering, Vol. 3, Helsinki, 83-104.
49. Schlosser, F. and Juran, I. (1983). Behavior of Reinforced Earth Retaining Walls from Model Studies, Chapter 6 in Developments in Soil Mechanics and Foundaton Engineering, Vol. 1, Ed. P. Banerjee and R. Butterfield, Applied Science Pub., London, 197-229.
50. Schlosser, F. and Bastick, M. (1985). Reinforced Earth: New Aspects and New Applications, 3rd Int. Geotech.Seminar Soil Improvement Methods, Singapore, 273-284.
51. Seed, R. and Duncan, J. (1986). FEM Analyses: Compaction Induced Stresses and Deformations, Journal of Geotechnical Eng., ASCE, 112(1), 23-43.
52. Terre Armée Int. (1988). Finite Element Study of Reinforced Earth Structures Supporting a High Slope, using the CESAR Program, Internal Report #R46, July, 280 p.
53. Terre Armée Int. (1989) Reinforced Earth Retaining Walls Published documentation, 20 p.
54. Wernick, E. (1978). Stress and Strains on the Surface of Anchors, Revue Française de Géotechnique, Vol. 3, 113-119.

LARGE CENTRIFUGE MODELING OF
FULL SCALE REINFORCED SOIL WALLS

Makram Jaber[1], A.M. ASCE, James K. Mitchell[2], F. ASCE
Barry R. Christopher[3], A.M. ASCE, and Bruce L. Kutter[4], M. ASCE

ABSTRACT: A recently completed research program on "Behavior of Reinforced Soil" done for the Federal highway Administration included large scale centrifuge tests on several models of instrumented full scale walls. The prototype walls were 20-ft (6.1 m) high and were reinforced using steel strips, bar mats, geogrids, and non-woven geotextiles. The model walls were built at a geometric scaling of 1:12. Details of model design and instrumentation are described in this paper, and the main results are presented. Good agreement was obtained between predicted and measured tensions in the reinforcements, lending credibility to the centrifuge modeling technique for the study of reinforced soil structures.

INTRODUCTION

The usefulness of small-scale physical modeling of engineered earth structures is often limited because the stress levels in the models are much smaller than in the full scale structures, thus leading to different soil properties. The centrifuge, however, provides a tool for geotechnical modeling in which prototype structures can be studied as scaled-down models while preserving the stress states required to develop the appropriate soil properties.

A recently completed research program on "Behavior of Reinforced Soil" performed for the Federal Highway Administration included large scale centrifuge tests on several models of full scale, instrumented reinforced soil walls. These tests constituted one of the first series of tests on the new large National Geotechnical Centrifuge, located at the University of California at Davis. The centrifuge testing had two objectives; namely, (1) to determine how well the scaled-up measurements of the centrifuge models agreed with the behavior of the full-scale walls, and (2) to evaluate the suitability of a numerical finite element code for prediction of the centrifuge model behavior. The present paper is concerned with the first of these two objectives.

[1]   Assistant Project Engineer, GeoServices Inc., 5950 Live Oak Parkway, Suite 330, Norcross, Georgia 30093.

[2]   Professor, Department of Civil Engineering, University of California, Berkeley, California 94720.

[3]   Principal Engineer, STS Consultants, Ltd., 111 Pfingsten Road, Northbrook, Illinois 60062.

[4]   Associate Professor, Department of Civil Engineering, University of California, Davis, California 95616.

FULL-SCALE WALLS

As part of the FHWA project mentioned above, four full-scale reinforced earth walls were modeled using the National Geotechnical Centrifuge. The prototype walls were 20-ft (6.1 m) high and were reinforced using 14-ft (4.3 m) long Reinforced Earth steel strips, VSL bar mats, Tensar geogrids and non-woven geotextile. Precast concrete Reinforced Earth cruciform facing panels were used in the first three walls. The geotextile was folded up to form the facing of the fourth wall. The main components of the full-scale walls are listed in Table 1. A gravelly sand from Algonquin, Illinois, was used as backfill and retained soil. The backfill was compacted with walk-behind vibratory roller compaction equipment to 95 % of ASTM Standard Proctor Method which resulted in a dry density of about 129 pcf (20.3 kN/m^3).

Facing	Reinforcement	Horizontal Spacing
Cruciform Concrete Panels	Steel Strips 50 mm x 4 mm Ribbed	2.42 ft (0.74 m)
Cruciform Concrete Panels	Bar Mats 6in. x 24 in. VSL 3/8 in. diameter of bars	4.48 ft (1.48 m)
Cruciform Concrete Panels	Geogrids Tensar SR2	Continuous Sheet
Wrapped Geotextile	Non-Woven Geotextile 16 oz / yd^2	Continuous Sheet

Note: All walls were 20-ft (6 m) high with 14 ft (4.3 m long reinforcements spaced vertically at 2.5-ft (0.76 m).

Table 1. Full-Scale Walls Modeled in the National Geotechnical Centrifuge.

CENTRIFUGE TESTS

The main purpose of the large centrifuge model tests was to obtain quantitative comparisons between the behavior of centrifuge models and that of the corresponding prototypes. A relatively small geometric scaling factor of 1:12 was therefore chosen so that the model walls would be large enough to enable meaningful instrumentation to be included. The main components of the model walls are listed in Table 2. A schematic representation of the reinforced soil model walls is shown in Figure 1.

The similarity in the overall geometry of the walls could be easily satisfied. Although certain components of the wall such as the ribs on Reinforced Earth strips could not be modeled, the similarity between models and prototypes was conserved for the properties that affect most the behavior of reinforced soil walls, namely soil strength and reinforcement axial stiffness.

Model No.	Facing	Reinforcement	Horizontal Spacing
1	Cruciform Aluminum Panels	Galvanized Steel Mesh 0.5 in x 2 in. 0.032 in. diameter of bars	4.48 in. (114 mm)
2	Cruciform Aluminum Panels	Steel Strips 0.5 in x 0.004 in.	2.42 in. (57 mm)
3	Cruciform Aluminum Panels	Miniature Geogrid	Continuous sheet
4	Wrapped Geotextile	Non-Woven Geotextile Phillips 2oz / yd^2	Continuous Sheet

Note: All model walls were 20-in (508 mm) high with 14-in (356-mm) long reinforcements spaced vertically at 2.5 in (63.5 mm).

Table 2. Large-Scale Centrifuge Model Walls.

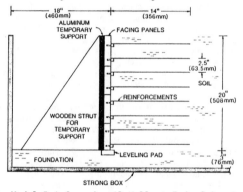

Figure 1. Model Reinforced soil Wall at End of Construction

Soil Properties. The soil that was used in the prototype walls was a local gravelly sand. This soil could not be used "as-is" to build the centrifuge models because of the large particle size of the gravel with respect to the size of the model. The prototype soil was therefore scalped by passing it through a #4 sieve. The resulting model soil was pluviated through air to a dry density of 117 pcf (18.4 kN/m^3).

Triaxial compression tests on both prototype and model soils at dry densities of 129 pcf (20.3 kN/m^3) and 117 pcf (18.4 kN/m^3), respectively, gave similar stiffnesses and strengths (friction angle of 43° to 44°). The friction angles were not affected in any appreciable way by changes in confining pressures within the range expected in the walls. The tensions developed in the model reinforcements were multiplied by a factor $\gamma_{prototype}/\gamma_{model}$ before comparing them with the measured values in the prototypes, to account for the difference in dry densities between the model and prototype soils.

Facing Panels. Standard Reinforced Earth facing panels were used in three of the full-scale walls. These panels have a cruciform shape and are made of 5.5-in (140 mm) thick concrete. The facing panels used in the model had the same shape and a scaled-down geometry. Due to the difficulty of manufacturing miniature concrete panels, aluminum panels were used. The thickness of these model facing panels was designed so as to yield a section with proper scaled down bending stiffness (Jaber, 1989).

Reinforcements. The reinforcements affect the behavior of reinforced soil walls through two basic properties, namely, their axial stiffness and their surface interaction with the soil (Adib, 1988). To model correctly the behavior of reinforced soil walls, at the design scaling factor, n, the following two similarity equations have to be satisfied in which the subscript m refers to the model and p refers to the prototype:

$$(EA)_m = \frac{(EA)_p}{n^2} \tag{1}$$

and,

$$(b \tan\delta)_m = \frac{(b \tan\delta)_p}{n} \tag{2}$$

where: EA = axial stiffness of reinforcement; E = Young's Modulus; A = Cross-sectional area of reinforcement; b = width of reinforcement; δ = apparent angle of friction between reinforcement and soil; and n = geometric scaling factor (i.e., 12) (Jaber, 1989).

Steel Strips: Ribbed Steel strips, 50 mm wide and 4 mm thick, were used in the full-scale Reinforced Earth Wall. The model strips, in Model 2, were 0.5-in (12.7 mm) wide and were cut from 4 thousands of an inch (0.1 mm) thick steel shim. They have a cross sectional area of 0.002 sq in (1.3 mm^2), 1/144 times smaller than that of the full-scale strips. Since they are made of the same material, namely steel, they satisfy the axial stiffness similarity requirement (Equation 1).

The reason for using model strips with the correct scaled-down area, but modified width and thickness was an attempt to satisfy equation 2. The Reinforced Earth strips had a measured apparent friction angle, δ, of 60° (FHWA research report), and the 0.5-in (12.7 mm) wide shim had a δ of about 17°, as measured in laboratory pullout tests. The desired and used (b tanδ)$_m$ values are therefore 0.284 in. (7.2 mm) and 0.153 in. (3.9 mm), respectively. The agreement is not very good. To increase the values of (b tanδ) for the model to the same value as that for the prototype would require increasing b while decreasing the thickness. The 0.5-in (12.7 mm) wide model strip was adopted, however, because of concerns on the dependability of attaching strain-gauges on strips thinner than 0.004-in (0.1 mm).

Bar Mats: 3/8-in (9.5 mm) diameter bars were used in the full-scale VSL bar mat wall. Each bar mat had 4 longitudinal bars spaced at 0.5-ft (0.15 m), center-to-center, and transverse bars at 2-ft (0.6 m) spacing. The model bar mats were scaled-down by a factor of 1:12. 1/2-in by 1/2-in (12.7 mm by 12.7 mm) galvanized steel mesh was cut to form exact models of the bar mats. Every three out of four transverse bars were cut out to obtain the correct transverse spacing in the model mat. Galvanized steel mesh was used because the galvanization process gave added stiffness to the connections between transverse and longitudinal bars not unlike the welds in the full-scale mats. The model bars had a diameter of 0.032-in (0.8 mm).

An apparent friction angle of 50° at a confining pressure of 50 kN/m^2 (1020 lb/ft^2) was measured in laboratory pullout tests on the model bar mats and soil. The friction angle obtained from laboratory tests on the full-scale bar mats under the same confining pressure was about 40° (FHWA research report).

Geogrid Reinforcement: Continuous sheets of geogrid (Tensar SR2) were used as reinforcements in one of the full-scale walls. The Tensar Corporation, Morrow, Georgia, provided the miniature geogrids that were used in Model 3. These grids were formed in the laboratory, using the same manufacturing process of SR2 grids, i.e., a sheet of polypropylene was punched and stretched in one direction in order to obtain a scaled-down axial stiffness. Laboratory pullout tests were not felt to be necessary, since the pullout capacity of the continuous geogrids was sufficiently large that pullout was not a concern in the wall design.

Non-Woven Geotextile: The full-scale geotextile was a polyester, continuous filament, needle-punched non-woven geotextile ("Quline 160" provided by the Wellman Quline Company, Charlotte, North Carolina). It had an unconfined axial stiffness (defined as force per unit length divided by strain) of about 250 lb/in (44 N/mm) at 10 % strain and a confined axial stiffness of 1,100 lb/in (193 N/mm) at 5 % strain, as provided by STS Consultants, Northbrook, Illinois. The desired model geotextile would therefore have an unconfined axial stiffness of 21 lb/in (3.7 N/mm). The lightest non-woven geotextile that could be found, a 2-oz. heat-bonded non-woven geotextile with an unconfined axial stiffness of 54 lb/in (9.5 N/mm) manufactured by Phillips Fabrics, was chosen as the model geotextile.

The non-woven geotextile wall was, therefore, not properly modeled, as the axial stiffness of the model fabric is more than twice that of the required axial stiffness to correctly model the full-scale fabric. Nonetheless, the axial stiffness of the model geotextile is much smaller than that of the other model materials. The results of this test were considered valuable in investigating the effect of reinforcement axial stiffness on the behavior of reinforced soil walls. Furthermore, analytical studies carried out by Adib (1988) suggest that when the axial stiffness of reinforcements is as low as that of common geotextiles (i.e. several orders of magnitude less than that of other types of reinforcements), the tension in the reinforcements is not significantly affected by variations in the axial stiffness of the reinforcement.

The axial stiffness of the reinforcements used in the FHWA prototype walls and the centrifuge models are listed in Table 3. The values reported in Table 3 are normalized with repect to the vertical and horizontal spacing between reinforcements, (i.e., reported values = $EA/(S_v x S_h)$, in order to provide a basis for comparison between various types of reinforcements.

Type of Reinforcement	Normalized Axial Stiffness, $EA/(S_v x S_h)$	
	Prototype	Model
Steel Strips	1,480 K/ft^2 (71,000 kN/m^2)	1,400 K/ft^2 (67,200 kN/m^2)
Steel Bar Mats	1,150 K/ft^2 (55,200 kN/m^2)	1,200 K/ft^2 (57,600 kN/m^2)
Geogrid	20 K/ft^2 (960 kN/m^2)	20 K/ft^2 (960 kN/m^2)
Geotextile	1.2 K/ft^2 (57.6 kN/m^2)	3.1 K/ft^2 (150 kN/m^2)

Table 3.   Axial Stiffness of Reinforcements.

Instrumentation. The instrumentation of the full-scale reinforced soil walls included strain-gauges on the reinforcements, and inclinometers near the wall face and further back in the walls. The instrumentation of the large centrifuge model walls included strain-gauges on the reinforcements and LVDT's for measuring the outward movement at two points on the face. A general layout of the instrumentation of the model walls is shown in Figure 2.

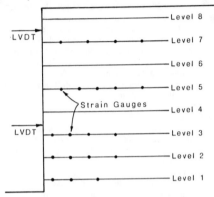

Figure 2. Layout of Instruments in the Centrifuge Model Tests

RESULTS OF LARGE CENTRIFUGE TESTS

Measured tensions rather than measured strains in the reinforcements are reported, because the strain-gauge arrangements were directly calibrated for load. Moreover, the strains that were measured in the geogrid and the geotextile walls do not have any important physical meaning, as the stiffness of the strain-gauge arrangements is different from the stiffness of the material itself. It should be noted, also, that only the increase in tensions during the tests were actually reported, i.e. 1g was taken as the initial condition. These values were not corrected for "true zero" because of the difficulty of taking reliable measurements of tension before the reinforcement was laid down. Furthermore, a similar problem was found in the field; i.e., the initial conditions were taken after one half to a full layer of backfill was constructed above the instrumented reinforcements.

Predicted prototype tensions based on the centrifuge model results are presented in the form of equivalent lateral earth pressure, i.e. $T/(S_v \times S_h)$, where T = tension in the reinforcement; S_v = vertical spacing between reinforcement; and S_h = horizontal spacing between reinforcement. This is the most meaningful value to compare, because at 12 g, the pressures in the centrifuge should correspond directly to the values in the full scale structures (i.e. scaling factor 1:1). The predicted values are adjusted for the different densities of prototype and model soils, as described earlier.

Model 1: Bar Mat Reinforcement. Model 1 was an exact 1:12 model of a full-scale 20-ft (6.1 m) high VSL wall. A typical example of measured tensions in the reinforcements during the centrifuge test is shown in Figure 3.

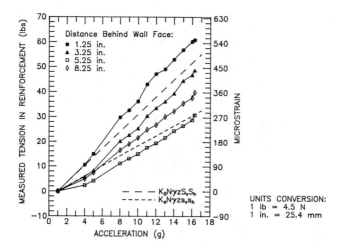

Figure 3. Tension in Reinforcement vs Acceleration
Level 3 Above Bottom of VSL Model Wall

At all levels, the tension in the reinforcements increased with a basically linear trend, as the acceleration was increased from 1g to 16.5g. Some of the traces were slightly curved upwards between 1g and 6g. This may be due to initial slack in the reinforcement. Another explanation may be that the soil is stronger under very low stress levels than under higher levels due to increased dilatancy, therefore less stress is transferred to the reinforcements.

Figure 4 shows the measured maximum tensions at each instrumented level for the four centrifuge model walls at 12 g. In this figure, the tensions are reported in units of force per unit length (lb/in) as they are normalized with respect to the horizontal spacing between the reinforcements (i.e., reported values = T/S_h). The normalized values provide a better basis for comparisons between the different reinforcement systems. In model 1, the measured maximum tensions above the first layer of reinforcement plot above the K_a line, with most of them close to the K_o line. This result seems to agree with previous results (Collin, 1986; Adib, 1988). The measured value at the lowest level of reinforcement is well below the K_a line, a phenomenon commonly observed both in instrumented reinforced soil walls and in finite element analysis of such walls, and usually attributed to base friction at the foundation level.

Figure 5 shows a comparison between the measured tensions from the full-scale VSL wall and the predicted values, based on the results of Model 1. The agreement is reasonably good. More detailed results of the field tests can be found in the FHWA research report.

The increase in outward movement of the facing panels at two points is shown in Figure 6. The maximum displacement, near the top of the model wall, amounted to about 0.2 percent of the total height of the wall at 17g.

The outward movement of the face of the full-scale VSL wall, measured by an inclinometer adjacent to the wall face, is compared in Figure 7 to that of the corresponding model. The movements are normalized with respect to the walls height and therefore scale at 1:1. The face movement of the centrifuge model is less by a factor of about 2.5 than the face movement of the prototype. The difference between model and prototype face deformations is probably due to the difference in construction sequence between the two structures: the facings of the model walls were braced during construction, and were therefore not allowed to move as is usually the case in field construction.

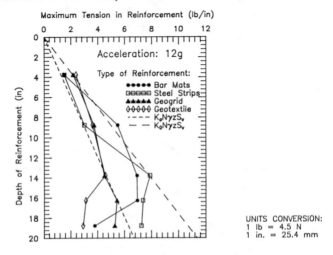

Figure 4. Maximum Tension vs Depth for Centrifuge Model Walls

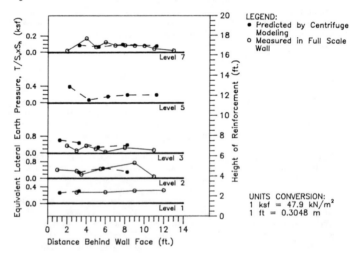

Figure 5. Comparison of Predicted and Measured Tension for VSL Full Scale Wall

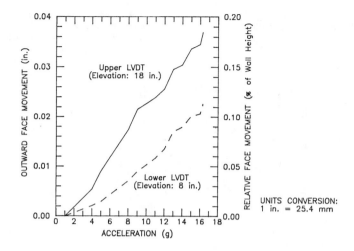

Figure 6. Outward Face Movement of VSL Model Wall

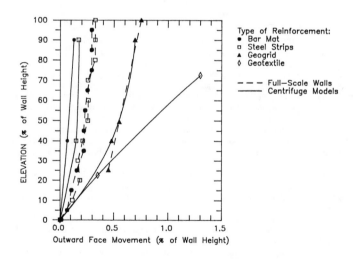

Figure 7. Comparison of Predicted and Measured Face Movements

Model 2: Steel Strip Reinforcement.   Model 2 was a scaled-down model of the full-scale Reinforced Earth Wall, with steel shim reinforcements as described earlier.

Typical measured tensions in the reinforcements during the centrifuge test are shown in Figure 8.

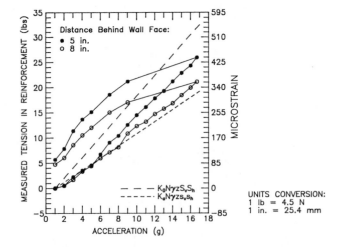

Figure 8. Tension in Reinforcement vs Acceleration Level 2 Above Bottom of Reinforced Earth Model Wall

Measurements were taken during both loading and unloading of this model.   The development of stress in the instrumented reinforcements seem to follow the same linear trend found in Model 1 during the loading portion of the test. The unloading curves were non-linear, however, and the reinforcements retained a residual tension at the end of the test. This behavior may be due to a higher coefficient of lateral earth pressure at-rest in the over-consolidated soil mass. Another reason may be that because the outward movement of the wall face cannot be recovered at the end of the test, the reinforcement cannot fully rebound as the acceleration is decreased back to 1g unless slip can occur at its interface with the soil.

The distribution with depth of maximum measured tensions in Model 2 is shown in Figure 4. Although all the values plot as expected between the K_a and K_o lines, it was surprising that the maximum tensions in the top half of the wall were very close to the K_a line.

The predicted prototype reinforcement tension distribution are presented in Figure 9. Unfortunately, limited field measurements are currently available due to instrumentation drift. The data is still under evaluation for refinement. The limited comparisons that could be made between model and prototype seem to indicate, however, that the the tensions measured in the prototype are larger than those measured in the centrifuge model.

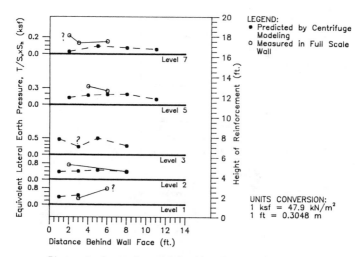

Figure 9. Comparison of Predicted and Measured Tension
for Reinforced Earth Full Scale Wall

The outward face movement of Model 2 increased linearly with increase in acceleration, and reached about 30% more than the movements in Model 1.

The outward movement of the face of the full-scale Reinforced Earth wall, measured by an inclinometer adjacent to the wall face, is compared in Figure 7 to that of the corresponding model. The face movement of the centrifuge model is 1.5 times smaller than that of the prototype. Again, the difference can probably be attributed to construction differences.

Model 3: Geogrid Reinforcement. Model 3 was a scaled-down model of the full-scale geogrid wall in which Tensar geogrid was used as reinforcement.

Typical measured tensions in the reinforcements during the centrifuge test are shown in Figure 10. The development of stress in the instrumented reinforcements seems to follow the same nearly linear trend observed in previous models. At the end of the test, the residual tensions retained in the geogrid reinforcements were higher than those retained in the smooth steel strips used in model 2. This is attributed to the larger geogrid-soil friction angle.

The predicted and measured prototype reinforcement tensions are shown in Figure 11. The agreement is very good for levels 2, 3, 5 and 7. Sufficient data is not available for meaningful comparisons at level 1.

The distribution of maximum tension with depth for Model 3 at 12g is shown in Figure 4. The maximum tension is close to the K_o line near the top of the wall, and drops off to the K_a line at mid-depth of the wall. The values for one strain-gauge point, 2.95-in (75 mm) behind the wall face at level 5, was dropped because it does not seem to correspond to the general distribution of tension in this type of wall, i.e. gradual increase in tension from the wall face to a distance of about 4-in to 6-in (102 mm to 152 mm) behind the wall face, and then gradual decrease.

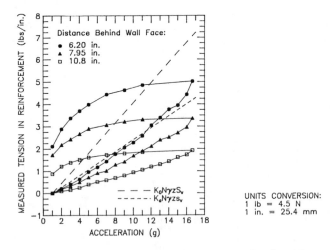

Figure 10. Tension in Reinforcement vs Acceleration
Level 5 Above Bottom of Geogrid Model Wall

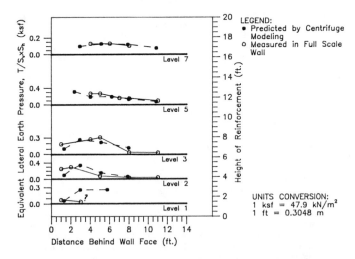

Figure 11. Comparison of Predicted and Measured
Tension for Geogrid Full Scale Wall

No record of face movement was obtained for this model because of an error in mounting the LVDT's. However, an initial and a final reading were taken, and showed that the outward wall face movement for Model 3 was appreciably larger than in previous models, because of the higher ductility of the geogrids. The outward movements 2-in (51 mm) and 12-in (305 mm) below the top of the wall reached 0.7 and 1 percent of the height of the wall at the end of the test, respectively.

The outward face movement of the full scale geogrid wall, measured by survey techniques, is compared to the face movement predicted from the centrifuge test. The agreement is good.

Model 4: Geotextile Wall. Model 4 was a scaled-down model of a geotextile wall. The full-scale wall was originally under-designed to obtain appreciable deformations in the structure. As discussed earlier, a weak enough model geotextile was not found to truly model that wall.

Typical measured tensions in the reinforcements during the centrifuge test are shown in Figure 12.

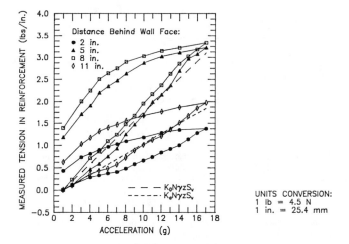

Figure 12. Tension in Reinforcement vs Acceleration Level 7 Above Bottom of Geotextile Model Wall

The predicted prototype reinforcement tensions are presented in Figure 13. Unfortunately no data was available from the field, at full height of the geotextile wall, because the full-scale reinforcement was severely under-designed and most of the strain-gauges were disabled before the wall was topped off. The distribution of stress obtained in the centrifuge seems reasonable, and shows clearly increasing tension from the face to about 5-in (127 mm) behind it, beyond which it slowly decreases.

The distribution with depth of the measured maximum tensions in Model 4 at 12g is shown in Figure 4. The values obtained are close to the K_o line near the top of the wall, and near K_a below the middle of the wall. It is interesting to note that the absolute value of stress from the top of the wall to the third level of reinforcement from the bottom of the wall is almost constant, a behavior predicted by finite element analysis of this type of wall by Collin (1986). The maximum tension for the geotextile model drops to well below the

K_a line, two levels above the bottom of the wall. This is an unusual behavior that may be due to the effect of friction at the foundation which would be larger in the case of geotextile walls, because of the greater deformation of these structures.

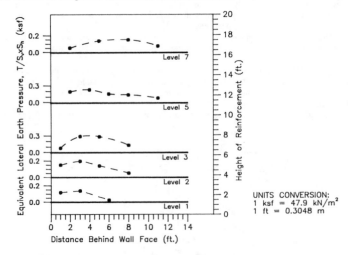

Figure 13. Predicted Tension for Geotextile Full Scale
Wall, based on Centrifuge test

The outward movement at the face reached 0.5 percent and 1.75 percent of the height of the wall, at elevations 4.5-in (114 mm) and 14.5-in (368 mm), respectively. Face deformation data for the full-scale geotextile wall was not available at the time this paper was written.

Comparison of the Behavior of the Four Centrifuge Model Walls.   Four centrifuge tests were performed on models with vastly different types of reinforcements. The reinforcements included steel bar mats, steel strips, geogrids, and non-woven geotextile.

The results of these four tests, when put together (Figure 4), seem to confirm the conclusions reached by Adib (1988), namely that the maximum tension and the distribution of tensions in the reinforcements are affected by both the reinforcement axial stiffness and interaction with the adjacent soil. Stiffer reinforcement results in generally larger tension in the bottom half of the wall. In the upper half of the wall, however, the relative movement between reinforcement and soil required to develop the pullout capacity of the reinforcement plays an important role. Smooth steel strips which have a lower pullout capacity than geogrids or geotextiles exhibit lower stress in the upper half of the wall, although the axial stiffness of steel reinforcement is appreciably higher than that of the geosynthetics. This trend is reversed in the lower half of the walls, where the factor of safety against pullout is large for all systems, and reinforcement axial stiffness becomes the dominant factor.

The relative movements of the faces of the four models, at the design acceleration of 12g, are compared in Figure 7. The outward movement seems also to be affected by the axial stiffness of the reinforcement: the geosynthetic reinforced systems exhibit face movements that are more than one order of magnitude larger that the steel reinforced systems.

CONCLUSIONS

Four full-scale reinforced soil walls were modeled and tested using the National Geotechnical Centrifuge. The reinforcements included VSL bar mats, Reinforced Earth steel strips, Tensar SR2 geogrids and non-woven geotextile.

Good agreement was obtained between predicted and measured tensions in the reinforcements, lending credibility to the centrifuge modeling technique for the study of reinforced soil structures. Investigation of the behavior of reinforced soil walls with unusual reinforcement patterns, such as walls with different length reinforcements, or narrow walls, or walls under high surcharge loads, for which little field data is available would be especially amenable to study by this means.

Differences between full scale walls and centrifuge models may occur due construction techniques and compaction-induced stresses. If compaction-induced reinforcement tensions and deformations are important (Seed et al., 1986; Collin, 1986), their effect cannot be determined in the centrifuge unless techniques are developed for in-flight model building.

The outward face movements of the centrifuge model walls were smaller than those of the corresponding prototypes for walls reinforced with steel strips or bar mats. The differences have been attributed to differences in construction details between the centrifuge and prototype walls. Better agreement was obtained for the geogrid reinforced wall. A possible explanation may be that when the deformations due to reinforcement strain are relatively large, as is the case for geogrid walls, the effects of construction-induced deformations are minimized.

Comparison of the maximum tensions developed in the different model walls seemed to suggest that both the reinforcement axial stiffness and the relative movement between soil and reinforcement affect their magnitude at working stress. The higher the stiffness, the higher the tension in the bottom half of the wall; and the higher the pullout capacity, the higher the tension in the upper half of the wall. The outward movement of the wall face depends mainly on the reinforcement axial stiffness, shown by the fact that the deformations of geosynthetic reinforced model walls were about one order of magnitude larger than those of steel reinforced model walls.

ACKNOWLEDGMENTS

This research was done under subcontract to STS Consultants, Northbrook, Illinois, as a part of Federal Highway Administration Contract DTFH61-84-C-0073, entitled "Behavior of Reinforced Soil". Professor C.K. Shen provided valuable guidance to the centrifuge testing program. Mr. Bill Sluis' help in building the large centrifuge models and operating the National Geotechnical Centrifuge was critical for the success of the testing program. Messrs. L. Chang, S-H. Chew, G. Schmertmann and M-H. Chang assisted in building the model walls.

REFERENCES

1. M.E. Adib, Internal Lateral Pressures in Earth Walls, Ph.D. Thesis, University of California, Berkeley, 1988.

2. J.G. Collin, Earth Wall Design, Ph.D. Thesis, University of California, Berkeley, 1986.

3. M. Jaber, Behavior of Reinforced Soil Walls in Centrifuge Model Tests, Ph.D. Thesis, University of California, Berkeley, 1989.

4. R.B. Seed, J.G. Collin, and J.K. Mitchell, FEM Analyses of Compacted Reinforced Soil Walls, Proc. Second International Conference on Numerical Methods in Geomechanics, Ghent, Belgium, 1986.

EXTERNAL STABILITY OF A REINFORCED EARTH WALL
CONSTRUCTED OVER SOFT CLAY

By Nelson N.S. Chou[1], M.ASCE and C.K. Su[2], A.M.ASCE.

ABSTRACT: A Reinforced Earth (RE) wall, approximately 1000 feet (304.8 m) long and up to 40 feet (12.2 m) in height, was constructed over a 10 foot (3.0 m) thick layer of very soft clay east of Julesburg, Colorado. Several possible wall failure modes were analyzed. The safety factors indicated the wall would be unstable if no precautions were taken. The following remedial measures were then recommended: (1) staged construction, (2) a temporary berm built in front of the wall, (3) a select cohesive material for embankment behind the wall, and (4) an instrumented monitoring system. To date, the field measurements have shown a close agreement with the predicted movements. The RE wall has remained stable during and after construction.

INTRODUCTION

The Colorado Department of Highways (CDOH) constructed a grade separation structure over the Union Pacific Railroad tracks on State Highway 385 near Julesburg, Colorado (Figure 1). To allow for future widening of the railroad roadbed and to provide sufficent clearance from the tracks, a Mechanically Stabilized Earth (MSE) wall was proposed with a Reinforced Earth (RE) wall design chosen. The wall was approximately 1000 feet (304.8 m) long, with a height ranging from 5 feet (1.5 m) at both ends to 40 feet (12.2 m) near the center, as shown in Figure 2. The reinforced earth wall was a propietary design submitted by the Reinforced Earth Company. The original geotechnical investigation was performed by a local consulting firm. The CDOH Geotechnical Section reviewed the design and decided to perform an independent study on the external stability of the wall.

1, 2 - Senior Geotechnical Engineer and Geotechnical Engineer, respectively, Colorado Department of Highways, 4340 East Louisiana Avenue, Denver, CO 80222.

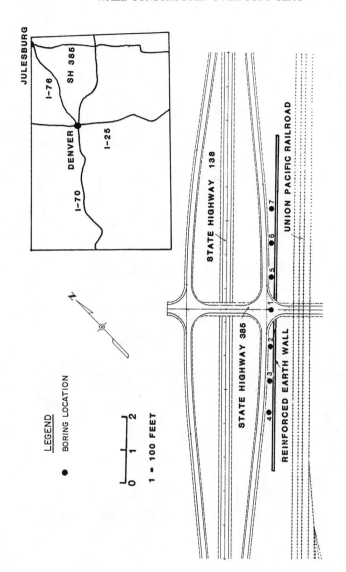

Figure 1: Site Location of Reinforced Earth Wall

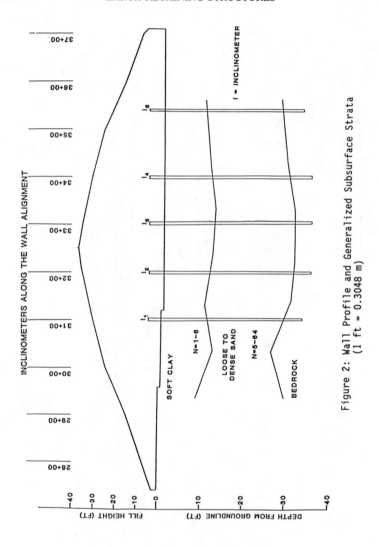

Figure 2: Wall Profile and Generalized Subsurface Strata
(1 ft = 0.3048 m)

SITE EXPLORATION

The Julesburg area is located in the lower South Platte River Valley. The alluvium in the South Platte River Valley consists mainly of clay, sand, and gravel. Seven test holes were drilled by CDOH along the proposed RE wall alignment (Figure 1). The cross section of the RE wall is shown in Figure 3. Generally, the subsurface at the proposed location of the wall consists of soft to very soft silty clay overlying loose to dense sand and gravel. Figure 2 shows the generalized subsurface strata and wall profile. Bedrock was encountered at depths between 28 to 70 feet (8.5 to 21.3 m). The water table was encountered approximately 4 feet (1.2 m) below the ground surface, and "pumping" of the ground surface occurred when construction vehicles passed. The subsoils in the area were divided into the following strata shown in Table 1.

Table 1: Subsoil Properties of RE Wall Site

Depth (ft)	Soil Description	Blow Counts[1] (per ft)
0 - 4	CLAY, silty, medium stiff (crust)	3-6
4 - 14	CLAY, silty, very soft	1-4
14 - 21	SAND, loose	5-9
21 - 31	SAND, medium dense to dense	20-64

(1 foot = 0.3048 m)
1 - Standard Penetration Test (per AASHTO 206-74)

LABORATORY TESTING

Laboratory tests were conducted to primarily study the engineering characteristics of the soft clay stratum. The tests included classification, vane shear, direct shear, unconfined compression, triaxial UU, CIU, CAU, and consolidation. The silty clay was classified as CL by the Unified Soil Classification and as A-7-6 by the AASHTO Classification system. The following soil properties were obtained from the laboratory tests, as shown in Table 2.

Table 2: Laboratory Test Results

Depth (ft)	Soil Description	W(%)	LL	Pl	S_u[1](psf)	ϕ[2](deg.)
0 - 4	Silty Clay (Crust)	21.0	26	10	480-1000 822 (Avg.)	0
4 - 14	Soft Silty Clay	39.6	45	23	40-790 415 (Avg.)	0
14 - 21	Loose Sand	12.5	-	-	-	35
21 - 31	Dense Sand	10.5	-	-	-	40

(1 foot = 0.3048 m, 1 psf = 47.88 Pa = 0.04788 kPa)
1 - Obtained from vane shear tests
2 - Obtained from direct shear tests

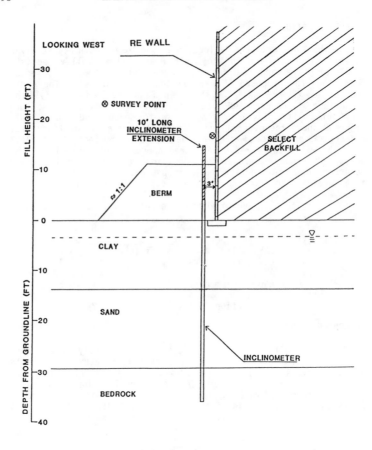

Figure 3: Typical Cross-Section of Reinforced Earth Wall
(1 ft = 0.3048 m)

STABILITY ANALYSES

The RE wall system must satisfy the requirements of both external and internal stability. After review of the design plan, the internal stability of the wall (designed by the manufacturer) was judged adequate, however, the wall external stability was questionable. Several possible failure modes in external stability, evaluated by considering the reinforced earth mass as an equivalent semi-rigid composite gravity block, were analyzed and the safety factors calculated as shown in Table 3. The average shear strength parameters shown in Table 2 were used in the analyses.

Table 3: Calculated Safety Factors Based on
Various Failure Modes

Failure Mode	Safety Factor	
	Calculated	Required
Bearing capacity	0.86	2.0
Overturning	4.34	1.5
Sliding	1.30	1.5
Slope stability		
(at the end of construction		
and undrained condition)		
- Janbu Method	0.93	1.25 - 1.50
- Ordinary Method	0.85	1.25 - 1.50
- Modified Bishop Method	1.08	1.25 - 1.50

The overall slope stability (Figure 4) and bearing capacity were the major concerns. A low safety factor of 0.86 for bearing capacity was obtained, based on the undrained shear strength of 415 psf for the soft clay. The following three approaches: Janbu, Ordinary, and Modified Bishop methods were used in the slope stability analysis with safety factors of 0.85 to 1.08 obtained. At the time of design, the source of the embankment (random) fill material was not finalized. Calculations were based on the assumption that the embankment fill would be borrowed from a nearby pit. This borrow material would be well compacted, silty sand, with the following properties: C (cohesion)= 150 psf (7.2 kPa), $\emptyset$ (friction angle)= 39 degrees, and γ_t (total unit weight) = 130 pcf (20.4 kN/m^3).

Embankment on soft foundation may fail progressively because of differences in the stress-strain characteristics of the embankment and foundation. The strength of both the embankment and foundation material should be reduced to allow for progressive failure, using the reduction factors recommended by Chirapuntu and Duncan [1]. A reduction factor (Re) of 0.67 was obtained in our analysis, which reduced the cohesion (C), and internal friction (tanϕ) of the soils by one third. The safety factors calculated by the Janbu method, based on various borrow site soil properties, are listed in Table 4. When this reduction factor was considered, the safety factors obtained from all the soils, except soil No. 5, were close to 1. These values were smaller than 1.25, the minimum safety factor

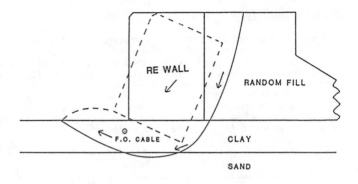

Figure 4: Possible Slope Stability Failure Mode

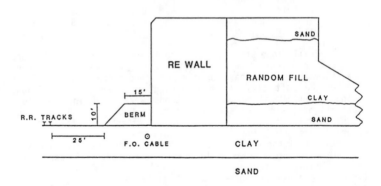

Figure 5: Location of Fiber Optic Cable and Placement
           of Cohesive Soil
           (1 ft. = 0.3048 m)

recommended by Task Force 27, AASHTO-AGC-ARTBA [2].

Based on the above analyses, borrow soil No. 5 was chosen. The material was silty clay, with AASHTO classification A-7-6(11). The probability of catastrophic failure calculated for this soil based on the mode of sums method (Standardized normal distribution) [3], was about 50 percent.

Table 4: Overall Slope Stability Analyses by Using
Different Embankment Fills

Borrow Soil No.	Embankment Soil Strength Properties				Factor of Safety by Janbu Method,[1]
	C(psf)	ϕ(deg.)	C'(psf)	ϕ'(deg.)	
1	150	39	100	28	0.934
2	300	35	200	25	0.961
3	450	28	300	20	0.988
4	750	22	500	15	1.106
5	1500	14	1000	9	1.401

(1 psf = 47.88 Pa = 0.04788 kPa)
1 - Values After Duncan's Strength Reduction

FIBER OPTIC CABLE

The U.S. Telecommunication Company had previously buried a fiber optic cable in a concrete box 4 inches (101.6 mm) below the ground and 1.5 to 2.0 feet (0.5 to 0.6 m) away from the face of the proposed wall (See Figure 5). The cable location raised a concern because: (1) a stability failure due to any of the above reasons might cause catastrophic damage to this very expensive cable, and (2) under the influence of the embankment and wall loads, the cable could be subjected to both the vertical and horizontal deformation, even if stability was not a problem. According to U. S. Telecommunication, the cable could tolerate up to 18 inches (457.2 mm) of deformation without distress. It was calculated that the magnitude of possible cable movement was in the range of 9 to 15 inches (228.6 to 381.0 mm), provided the wall and the foundation soils were in stable condition. If instability of the wall did occur, the cable would definitely move beyond the tolerance magnitude.

REMEDIAL MEASURES

The following remedial measures were considered for use to strengthen the stability of the RE wall:

(1) Use of cohesive soil for embankment - The cohesion would contribute to a significant   increase of safety factor, as indicated by Table 4. Since the quantity of this cohesive soil was limited, it was recommended that this material be sandwiched in the middle portion of the embankment shown in Figure 5.

(2) Reinforcement - Multiple layers of geogrid or high-strength

geofabric were considered for embankment reinforcement. This
option was abandoned because of the high cost and large
quantities of geotextiles needed.

(3) Excavation of the soft clay - Due to the high cost and possible
environmental impact, this alternative was not recommended.

(4) Extension of the reinforcing strips in the lower sections of
the RE wall - Making the extensions longer would increase the
safety factor for slope stability, however, since this
modification deviated from the standard design, the Reinforced
Earth Company would no longer be responsible for their product.
It was therefore decided not to use this alternative.

(5) Relocation of the fiber optic cable - The safety of the cable
was threatened if a wall stability problem developed. Because
the cost of repair and possible litigation resulting from
damage would undoubtedly be high, it was recommended that the
cable be moved southward, a sufficent distance from the wall
footing area.

(6) Construction of a 10 foot (3.0 m) high berm - This option was
considered one of the most cost-effective measures to increase
the safety factor for both slope stability and bearing
capacity. In addition, stability for sliding would also be
improved.

(7) Staged Construction - It was recommended to build the wall in
4 phases, each 10 feet (3.0 m) in height, with at least a 1
month waiting period between each phase.

(8) Instrumentation - An instrumented monitoring system was
proposed to measure the performance of the wall during and
after construction.

The measures (1), (5), (6), (7), and (8) were adopted by CDOH.

STAGED CONSTRUCTION

During construction of the RE wall, the undrained shear strength
(S_u) will increase by consolidation. The gain of undrained shear
strength, dS_u, can be approximated from the following simple
equation [4]:

$$dS_u = m \times I \times d\sigma_c \times U\%$$

Where   m = Slope between S_u and σ_c, I = Influence factor of
loading, $d\sigma_c$ = Increase in mean value of consolidation pressure =
$d(\sigma_1 + \sigma_2 + \sigma_3)/3$, and U (%) = Percentage of consolidation.

The value of m can be obtained from either of the following
methods:

(i) Based on the results of a typical triaxial CU test (Figure 6),
by increasing mean consolidation pressure, the undrained shear
strength can be improved. In Figure 6, the undrained shear strength
(A and B) are moved horizontally to the corresponding σ_c values (A'
and B'). M(m) is then represented by the slope of σ_c and S_u.

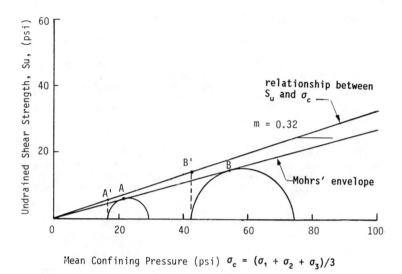

Figure 6: Undrained Shear Strength Increase with Consolidation
from a CAU (Total Stress) Test
(1 psi = 6.895 kpa)

(ii) Skempton's empirical equation [5], where:

$$S_u/P' = 0.11 + 0.0037 \times (PI)$$

P' = effective overburden pressure and PI = Plasticity Index. With
a representative PI of 23, a S_u/P' value of 0.20 was obtained. It
should be noted that in (i) the mean consolidation pressure is used,
while in (ii), the overburden pressure is adopted. If a K_o of 0.5
is assumed, the S_u/P' ratio obtained from (ii) is very close to m
obtained from (i).

    Total shear strength gain with resulting bearing capacity safety
factor improvements are presented in Table 5. The Table shows the
undrained shear strength of the clay between the 4-14 foot (1.2-4.3
m) interval (Su=415 psf) along with the shear strength improvement
(dSu) at each 10 foot (3.0 m) increment of staged construction.

Table 5: Comparison of RE Wall Bearing Capacity With and
Without Staged Construction Considerations

Fill Height (feet)	dS$_u$ (psf)	S$_u$+dS$_u$ (psf)	Without Stage Construction	With Stage Construction
			Safety Factor	
10	0	415	1.78	1.78
20	180	595	1.05	1.27
30	205	800	0.89	1.14
40	230	1030	0.86	1.10
Long term	409	1439	--	1.54

(1 foot = 0.3048 m, 1 psf = 47.88 Pa = 0.04788 kPa)

    From Table 5, it can be seen that the safety factors of bearing
capacity, during each stage of construction, increased significantly
when the increased strength was taken into account.

    Table 6 shows the total stability safety factor improvement when
*all of the remedial measures were taken into account. The drastic
increases of safety factor in overturning and sliding were
essentially due to the use of cohesive embankment (random) fill.
The active thrust was significantly reduced when cohesion was
introduced, even if the long term lateral perssure (Ko condition)
against the RE wall was considered in analyses.
    *Following removal of the berm.

Table 6: Comparison of Safety Factors Before
and After the Remedial Measures

Failure Mode	Before	After
	Safety Factor	
Bearing capacity	0.86	1.10
Overturning	4.34	8.88
Sliding	1.30	7.09
Slope stability	0.93	1.40

INSTRUMENTATION, MONITORING AND PERFORMANCE DATA OF THE RE WALL

During the period of construction, the behavior of the wall was monitored by various instrumentation. The instruments used for the monitoring program are listed in Table 7.

Table 7: Summary of Instrumentation for the RE wall

Item	Number Installed	Function
Piezometer	9	Monitor excess pore water in soft clay during staged construction
Inclinometer	5	Monitor horizontal deformation of RE wall foundation
Liquid Settlement Transducer	2	Monitor settlement at the base of RE wall
Survey Point	8	Monitor deformation of RE wall panels
Stand Pipe	2	Monitor the local ground water elevation

The monitoring system was designed to detect early stages of potential wall failure. In the event of an impending failure, construction forces were to be alerted and work on the wall was to be temporarily suspended. In the case of an emergency, the protective berm was to be immediately increased in height. As it turned out, the wall was constructed on schedule to a maximum height of 40 feet and the instrumentation operated normally during the 6 month construction period. Instrumentation data showed a close correlation with the predicted movements (See Figures 7 and 8). Maximum settlement of the wall footing was 10.8 inches (274.3 mm) near Station 32+95. The predicted value was between 10-11 inches (254.0-279.4 mm). The vertical inclinometers measured between 2.5-3.0 inches (63.5-76.2 mm) of lateral deformation which was within the 3-4 inches (76.2-101.6 mm) predicted.

The wall facing did not show any visible signs of distress since the differential settlement value of 2.5 inches (63.5 mm) per 100 feet (25.4 m) of length was within the limits of the wall design. Analysis of the stress behavior within the panels and reinforcing strips was beyond the scope of this paper. However, a finite element analysis studying the internal behavior of the RE wall, including the effects due to foundation settlement, will be addressed in a forthcoming paper.

SUMMARY AND RECOMMENDATIONS

The external stability, i.e., bearing capacity, overturning, sliding, and slope stability of the RE wall were all analyzed. The overall slope stability and bearing capacity were considered to be the most critical problems. Consequently, several remedial measures were taken to improve the stability of the wall. These included:

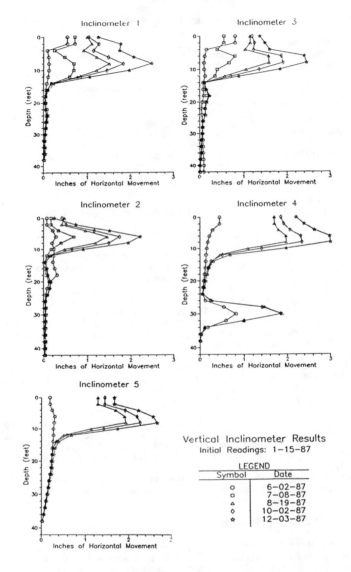

Figure 7: Lateral Movements vs Depth from Inclinometers
(1 ft = 0.3048 m; 1 in = 25.4 mm)

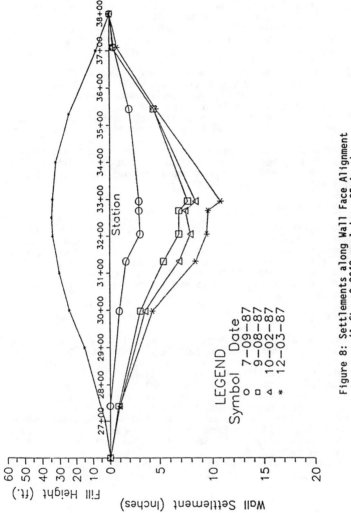

Figure 8: Settlements along Wall Face Alignment
(1 ft = 0.3048 m; 1 in = 25.4 mm)

(1) Constructing a temporary berm, 10 feet (3.0 m) high and 15 feet (4.6 m) wide, in front of the wall face.
(2) Using cohesive soil for the embankment material behind the structural backfill of the wall. This resulted in the cohesion significantly increasing the safety factors for slope stability, overturning and sliding.
(3) Relocating the fiber optic cable away from the wall.
(4) Adopting a staged construction procedure. The safety factor of the bearing capacity was increased when the undrained shear strength gain was taken into account.
(5) Instrumenting and monitoring the wall performance to detect early signs of potential failure during and after construction.

The field instruments performed as planned and their measurements show a close agreement with the predicted movements. The adoption of the remedial measures proved successful and have resulted in a stable wall as of the date of this paper.

ACKNOWLEDGEMENTS

The authors are indebted to numerous Colorado Department of Highways personnel for administrative and technical support, including: Duane L. Muller, Frank Lopez, and Ken Wood of District 4; Ed Belknap, Larry Bong, Paul Macklin, John B. Gilmore, Lorraine Castro, Johneen Fitz, and George Pavlick of Staff Materials Branch. Mr. Ed Belknap also assisted in editing and reviewing the finished manuscript.

REFERENCES

1. S Chirapuntu and JM Duncan, The Role Fill Strength in the Stability of Embankment on Soft Clay Foundations, Geotechnical Engineering Research Report, Department of Civil Engineering, University of California, Berkeley, 1975.
2. Task Force 27, AASHTO-AGC-ARTBA, Draft Copy, July, 1986.
3. JR Benjamin and CA Cornell, Probability, Statistics, and Decision for Civil Engineers, McGraw-Hill Book Company, 1970.
4. N Chou, KT Chou, CC Lee, and KW Tsai, Preloading by Water Testing Eliminated Sand Drains for a 65,000 Ton Raw Water Tank in Taiwan, Proceedings of the 6th Southeast Asian Conference on Soil Engineering, 1980.
5. AW Skempton, Discussion on The Planning and Design of New Hong Kong Airport, Proc. Inst. Civil Eng., vol. 7, 1957.

FINITE ELEMENT MODELING OF REINFORCED SOIL WALLS AND EMBANKMENTS

Mazen Adib[1], M. ASCE, James K. Mitchell[2], F. ASCE, and
Barry Christopher[3], M. ASCE

ABSTRACT: Five reinforced soil walls and four reinforced soil
embankments were constructed as part of a Federal Highway
Administration sponsored project on the behavior of reinforced
soil structures. This paper contains the results of finite
element modeling of these structures, comparisons between FEM
predictions and measured field behavior, and conclusions
concerning factors influencing stresses and deformations in
reinforced soil structures and their implications for design.

INTRODUCTION

As part of a Federal Highway Administration (FHWA) project to
study the behavior of reinforced soil walls and engineered slopes,
five walls and four embankments were constructed by STS
Consultants Ltd. in Illinois. Several types of soil reinforcement
were used to construct these structures, and each structure was
instrumented. A description of each wall and embankment is given
in Table 1. (Several other walls were constructed but are not
reviewed in this paper.)

Computer program SSCOMP (Seed, 1983) was used to predict
stresses and deformations in these walls and embankments. In this
paper, the actual performance of these structures is presented and
compared with the predicted values. Comparisons are presented for
the most important factors that control the internal stability of
the structures; namely:

1. Location of maximum tension in the reinforcement,
2. Distribution of tension along the reinforcement,
3. Distribution of maximum tension in the reinforcement with
   depth, and
4. Lateral movement of wall face

---

1 - Staff Engr., Dames & Moore, 221 Main Street, Suite 600, San
    Francisco, CA 94105.
2 - Prof. of Civil Eng., University of California, Berkeley, CA
    94720.
3 - Formerly Principal, STS Consultants, now Polyfelt, Inc., 1000
    Abernathy Rd., Atlanta, GA 30328.

Wall	Soil Type	Reinforcement	Horizontal Spacing (ft)	Height (ft)
1	Sand with Gravel (SW)	Steel strips	2.4	20
2	Sand with Gravel (SW)	Geogrid	Continuous	20
3	Sand with Gravel (SW)	Bar Mat	4.92	20
4	Cobbles	Bar Mat	4.92	20
5	Silt (ML)	Bar Mat	4.92	20

TABLE 1A: TYPES OF SOILS AND REINFORCEMENT USED IN THE
            INSTRUMENTED FHWA PROJECT WALLS (1 ft = 0.305 m)

Embankment	Soil Type	Reinforcement	Height (ft)	Slope
1	Silt (ML)	Geogrid	20	1H:1-1/2V
2	Silt (ML)	Woven Geotextile	20	1H:1-1/2V
3	Silt (ML)	Geogrid	25	1H:1V
4	Silt (ML)	Woven Geotextile	25	1H:1V

TABLE 1B: TYPES OF SOILS AND REINFORCEMENT USED IN THE
            INSTRUMENTED FHWA PROJECT EMBANKMENTS (1 ft = 0.305 m)

FINITE ELEMENT PROGRAM SSCOMP

    The capacity for finite element modeling of compaction induced
stresses was developed by Seed (1983) in program SSCOMP. The soil
is modeled using quasi lateral isoparametric elements and the
hyperbolic stress strain constitutive relationships developed by
Duncan et al. (1980). The reinforcement is modeled using elastic
bar elements. The soil-structure interaction is modeled using
interface elements capable of transferring shear stresses between
the soil and the reinforcement. These elements have zero
thickness and follow a hyperbolic constitutive relation.
Compaction induced stresses are modeled using a hysteretic model
for stresses resulting from cyclic loading under K_0 conditions
(Seed, 1983).

    Material Properties. The soil, reinforcement, and interface
element properties used in the finite element analyses of the FHWA
structures are shown in Tables 2, 3, and 4. As finite element
program SSCOMP is a plane strain code, the reinforcements must be
represented by continuous sheets. This assumption imposes
additional constraints when modeling discrete reinforcements. For
walls 1, 3, 4, and 5, which have discrete reinforcements, the
cross sectional area of the reinforcement in SSCOMP was reduced to
maintain the same reinforcement density (reinforcement cross
section area to soil cross section area) as in the actual wall.

A special finite difference code, PULL, (Adib, 1988) was developed to analyze the reinforcement pull-out tests done as part of the FHWA project. The interface element behavior in this code was modeled by the same hyperbolic constitutive relation as used in modeling the interface element behavior in SSCOMP. The finite difference code was used to estimate the interface element properties needed for the finite element analysis.

Finite Element Meshes. The meshes used for modeling the structures are shown in Figure 1. The mesh boundaries were sufficiently far from the reinforced zone that their influence on the computed results was small.

COMPARISON BETWEEN FINITE ELEMENT MODELING AND FIELD MEASUREMENT

A detailed comparison between the finite element modeling results and field measurements for each structure was presented by Adib (1988). As noted earlier, emphasis in this paper is on the influence of soil type and reinforcement type on the factors that control the internal stability of these structures.

1) Influence of Reinforcement Type. This influence can be best shown by comparing the behavior of walls 1, 2 and 3, which were constructed using steel strips, geogrids, and bar mats, respectively, as reinforcement.

a) Distribution of Tension Along the Reinforcement.

The distribution along the different levels of reinforcement in walls 1, 2, and 3 is shown in Figure 2. The comparison between the finite element prediction and the measured behavior is good for walls 1 and 3. In wall 2, however, the comparison is reasonable only for the distance between the wall face and the point of maximum tension. Beyond this point, the finite element analysis predicts higher values than were actually measured. There is a significant drop in the measured tension after the maximum value is reached which is not shown by the finite element analysis. The distribution of tension in walls 1 and 3 on the other hand, decreased smoothly from the location of the maximum tension to the free end, which is more consistent with prior analyses and measurements.

b. Location of Maximum Tension. The locus of maximum tension in the reinforcement is shown in Figure 3. The finite element prediction of this locus is reasonable for wall 3. For wall 1, there were too few recovered data to reliably locate the locus for the full-scale wall. The finite element prediction of this location in wall 2 fell about 2 ft behind the measured one. There is, however, a qualitative similarity between the measured and the

Material Property Description	Sand with Gravel (SW)	Silt (ML)	Cobbles	Foundation Soil (SW)
a) Hyperbolic Parameters				
- Unit Weight (K/ft^3)	0.130	0.130	0.105	0.130
- Young's Modulus Number, K	460	200	700	600
- Young's Modulus Exponent, n	0.5	0.6	0.4	0.25
- Failure Ratio, R_f	0.7	0.7	0.7	0.7
- Bulk Modulus Number, K_b	230	100	225	450
- Bulk Modulus Exponent, m	0.5	0.5	0.3	0.0
- Cohesion, C (Ksf)	0.0	0.05	0.0	0.0
- Angle of Internal Friction (deg)	40.	35.	42.	36.
- Reduction in Friction Angle	0.	0.	0.	1.0
- At-rest Lateral Earth Pressure coefficient, Ko $(1-\sin \Phi)$	0.357	0.426	0.33	0.412
- Unloading/Reloading Modulus Number, K_{ur}	690	300	1050	300
b) Compaction Parameters				
- Frictional Component of Limiting Lateral Pressure Coefficient	3.07	2.46	3.02	2.57
- Cohesion Component of Limiting Lateral Pressure Coefficient	0.0	0.04	0.0	0.0
- Unloading/Reloading Earth Pressure Coefficient K2 and K3	0.157	0.244	0.15	0.22
- Fraction of Peak Lateral Stress, F	0.56	0.43	0.55	0.466

Note: Refer to Seed (1983) for definition of parameters.

TABLE 2: SOIL PROPERTIES AND PARAMETERS USED IN THE FINITE
ELEMENT ANALYSIS OF FHWA WALLS AND EMBANKMENTS

($1 \ K/ft^3 = 157.2 \ KN/m^3$, 1 ksf = 47.82 KN/m^2)

Structure	Reinforcement Type	Axial Stiffness (K/ft)
Wall 1	RECO (Steel Strips)	3745.
Wall 2	Tensar SR2 (Geogrid)	140.*
Walls 3, 4, and 5	VSL 6"x24" (Bar Mat)	2593.
Embankments 1 and 3	Signode TNX 250 (Geogrid)	90.*
Embankments 2 and 4	Amoco 2006 (woven Geotextile)	24.**

*   Secant Stiffness at 2% Strain
** Secant Stiffness at 10% Strain

TABLE 3: REINFORCEMENT PROPERTIES USED IN THE FINITE ELEMENT
          MODELING OF FHWA WALLS AND EMBANKMENTS
          (1 K/ft = 14.585 KN/m)

Structure	Friction Angle (deg)	Reduction+ in Friction Angle (deg)	Shear Spring Coefficient	Modulus Exponent	Failure Ratio
Wall 1**	5.0	4.0	660	1.0	0.9
Wall 2*	50.0	3.0	15000	0.9	0.8
Wall 3*	11.0	1.5	500	1.0	0.9
Wall 4*	11.0	5.0	330	0.76	0.8
Wall 5*	5.7	1.0	183	1.0	0.83
Embank 1&3*	50.0	3.0	15000	.9	0.8
Embank 2&4**	30.0	3.0	3500	1.0	0.9

+   Reduction in Friction Angle for increase in pressure of one
    atmosphere.
*   Properties determined from Pull-out tests
** Properties determined from Collin (1986)

Notes: Refer to Seed (1983) for definition of parameters.
        Friction angle reduced for discrete modeling in Walls 1,
        3, 4, and 5.

TABLE 4: INTERFACE ELEMENT PROPERTIES USED IN THE FINITE ELEMENT
          ANALYSIS OF FHWA WALLS AND EMBANKMENTS

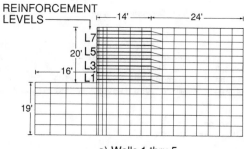

a) Walls 1 thru 5

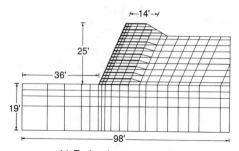

b) Embankments 1 and 2

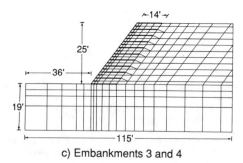

c) Embankments 3 and 4

Figure 1.   Finite Element Meshes Used in the Analysis of the In-
             strumented FHWA Project Walls and Embankments.
             (1 ft = 0.305 m)

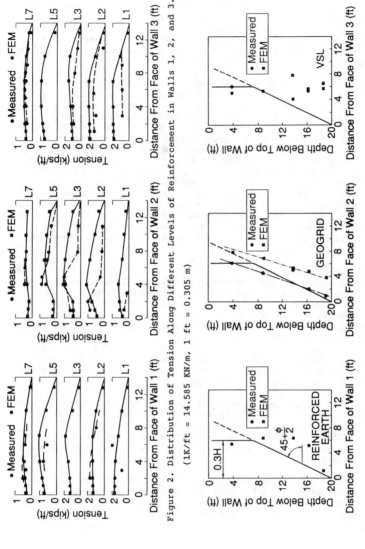

Figure 2. Distribution of Tension Along Different Levels of Reinforcement in Walls 1, 2, and 3.

(1K/ft = 14.585 KN/m, 1 ft = 0.305 m)

Figure 3. Location of Maximum Tension in Walls 1, 2, and 3. (1 ft = 0.305 m)

computed loci: the line that connects the measured locations is
inclined at 72 degrees from the horizontal, which is 7 degrees
steeper than the angle of the Rankine plane of 65 degrees.
Similarly, the line that connects the predicted locations is
inclined at the same angle. The locus of the maximum tension in
walls 1 and 3 is almost parallel to the wall face.

c) <u>Increase of Maximum Tension with Depth</u>. This increase of
maximum tension with depth is shown in Figure 4. Also shown in
this figure are the values of tension based on both the K_0 and
K_a values (the at rest and active lateral earth pressure
coefficients). Reasonable agreement between the measured and
predicted distribution was obtained for all three walls. The
influence of the friction between the reinforced mass and the
underlying rigid foundation is clearly shown in Figure 4. This
friction reduces the amount of tension in the bottom reinforcement
layer. From a practical point of view, this means that the most
critical reinforcement level is the one above the bottom
reinforcement layer. Caution should be taken, however, when the
wall rests on soft foundation because the bending strains due to
the settlement of foundation may cause larger than expected
tension. In this case, the bottom reinforcement layer may become
the most critical.

The measured values also reflect the influence of reinforcement
stiffness. As shown in Figure 4, the magnitude of the measured
values decreased in the order: wall 1, wall 3, wall 2. The
reinforcement stiffness in these walls, as shown in Table 3,
decreased in the same order. The importance of reinforcement
stiffness has been shown in previous studies; e.g. Collin (1986),
Mitchell (1987).

d) <u>Deformation at Wall Face</u>. The inclinometer in wall 2 did
not behave satisfactorily, therefore, only the deformations for
walls 1 and 3 are shown in Figure 5.

The finite element analysis overpredicted the magnitude of the
displacements, possibly because the stiffness of the soil in
Table 2 was underpredicted; nevertheless, the comparison is
qualitatively reasonable. The actual amount of deformation in
both walls was on the order of 1 inch, or 0.4% of the height of
the wall.

2) <u>Influence of Soil Type</u>. Walls 3, 4, and 5 had the same
reinforcement (VSL Bar Mat), but each wall was constructed using a
different soil type. Therefore, the comparison of the behavior of
these walls shows the influence of soil type.

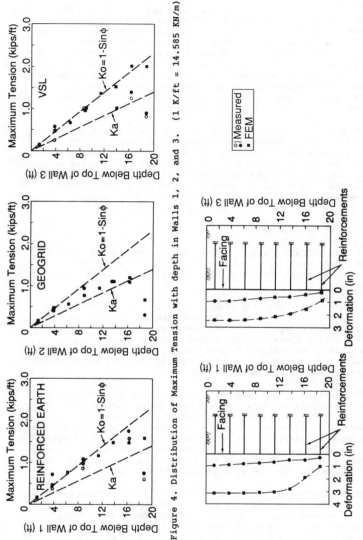

Figure 4. Distribution of Maximum Tension with depth in Walls 1, 2, and 3. (1 K/ft = 14.585 KN/m)

Figure 5. Deformation at the Face of Walls 1 and 3. (1 ft = 0.305 m, 1 in = 0.0254 m)

a) <u>Increase of Maximum Tension with Depth</u>. This increase is
shown in Figure 6. The comparison between the finite element
prediction and the measured values is good. In wall 4, where the
soil is the stiffest of the three soil types, the measured and
predicted values fell close to the at-rest values. In wall 3 the
measured tensions are close to the K_0 values in the top half of
the wall but decrease to near the active values in the bottom half
of the wall. Apparently, the stiffness of the soil has an
influence on the maximum tension in a reinforced soil wall. The
stiffer the soil, the higher the maximum tension.

b) <u>Deformation at the Wall Face</u>. The inclinometer at the face
of wall 5 did not behave satisfactorily therefore only the
deformations at the faces of wall 3 and 4 are shown in Figure 7.
The finite element analysis overpredicted the deformations in each
case. Qualitatively, however, the comparison is reasonable. Both
the finite element and the measured deformations show larger
deformations in wall 3 than in wall 4. This is consistent with
smaller deformations in stiffer walls.

BEHAVIOR OF REINFORCED EMBANKMENTS

1) <u>Embankments 1 and 2</u>. These two embankments had the same
slope (1-1/2 vertical to 1 horizontal) but were built using
different types of ductile reinforcement; i.e., geogrid and woven
geotextile.

<u>Distribution of Tension Along the Reinforcement</u>. The finite
element analysis predicted rather well the distribution of tension
along different levels of reinforcement in embankment 1, Figure
8. The finite element analysis prediction of the tension in
embankment 2, however, is 3 times larger than the tensions
computed from strain measurements, possibly because the stiffness
of the geotextile was underpredicted.

<u>Distribution of the Maximum Tension with Depth</u>. The finite
element analysis prediction of the distribution of maximum tension
with depth is very close to the measured distribution in
embankment 1, as shown in Figure 9, the tension increases linearly
with depth in the top 10 ft of the embankment and remains almost
constant below that level. For embankment 2, the maximum tension
increases with depth, whereas the finite element prediction shows
similar behavior to that in embankment 1.

2) <u>Embankments 3 and 4</u>. The distribution of tension along the
reinforcement in embankment 3 and 4 is presented in Figure 10.
The tension was relatively small starting from zero at the
embankment face and reached maximum at the middle of the
reinforcement. The finite element analysis predicted higher
maximum tension values than were actually measured.

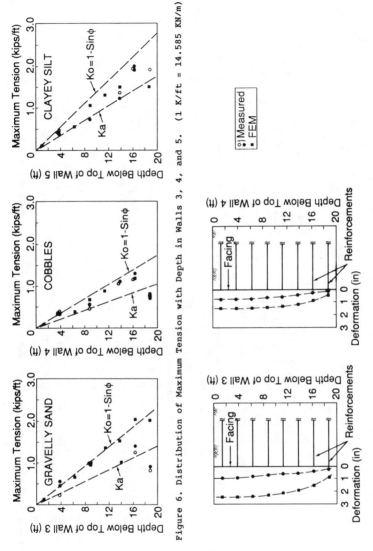

Figure 6. Distribution of Maximum Tension with Depth in Walls 3, 4, and 5.   (1 K/ft = 14.585 KN/m)

Figure 7. Deformation at the Face of Walls 3 and 4.   (1 ft = 0.305 m, 1 in = 0.0254 m)

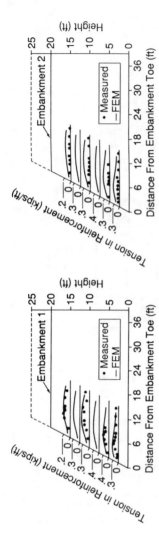

Figure 8. Distribution of Tension Along Different Levels of Reinforcements in Embankments 1 and 2.
(1 K/ft = 14.585 KN/m, 1 ft = 0.305 m)

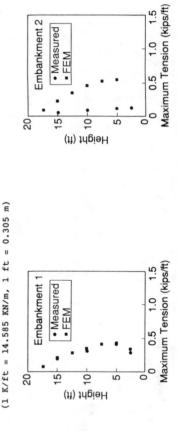

Figure 9. Distribution of Maximum Tension with Depth in Embankments 1 and 2. (1 K/ft = 14.585 KN/m)

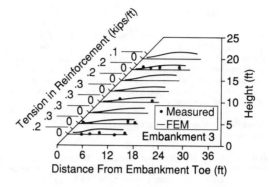

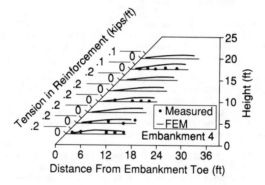

Figure 10: Distribution of Maximum Tension with depth in Em-
            bankments 3 and 4.
            (1 ft = 0.305 m, 1 K/ft = 14.585 KN/m)

SUMMARY AND CONCLUSIONS

Finite element analyses were conducted for five reinforced soil walls and four reinforced soil embankments. Comparisons were made between the finite element prediction and the measured values of the soil stresses, deformations, location of the maximum tension, and the distribution of tensile forces along the reinforcement. Three types of soil and five types of reinforcement were used in constructing the structures.

The observed behavior and comparisons have resulted in the following conclusions:

1) The locus of the maximum tensile forces in walls with stiff reinforcement is almost parallel to the wall face at a distance of 0.25 H. The finite element prediction of this locus is very close to the measured one.

2) The locus of the maximum tensile forces in walls with ductile reinforcement is inclined at 72 degrees from the horizontal. The finite element prediction of this locus fell about 2 ft behind the measured one.

3) The finite element prediction of the deformation in the walls is qualitatively reasonable.

4) The finite element analysis predicted reasonably well the distribution of tensile forces in the reinforcement, both along the reinforcements and with depth.

In walls with stiff reinforcement the distribution is characterized by a steady but slow increase in the tension from the wall face to a distance of about 0.3 L behind the face. Beyond this point, the tension decreases until it becomes zero at the free end of the reinforcement. The finite element analysis predicted slightly higher reinforcement tensions than actually measured between the wall face and the location of the maximum tension. Beyond this point, the finite element prediction matched very well the measured tensions.

In walls with ductile reinforcement, the tension near the wall face is only about one half the maximum tension. The distribution is characterized by a fast increase in the tension from near the wall face to the location of the maximum tension. The finite element prediction of this distribution matched relatively well the measured one up to the location of the maximum tension. Beyond this point, the finite element analysis predicted higher tension values than actually were measured.

5) The finite element analysis predicted the reinforcement tensions in embankments with Geogrid reinforcement reasonably well. In embankments reinforced with Geotextile, the finite element prediction was not as good, possibly because the geotextile stiffness was underpredicted.

ACKNOWLEDGMENT

The research described in this paper was financially supported in part by grants from the Institute of Transportation Studies at the University of California at Berkeley and from the California Department of Transportation, Contract RTA-54G462 titled: "Stresses and Deformations in Earth Reinforcement Systems." The construction of the full-scale walls and embankments, their instrumentation, and the monitoring of that instrumentation formed a part of FHWA contract DTFH61-84-C-00073, "Behavior of Reinforced Soil Structures." The reduction of the raw data was done by Mr. S. H. Chew, graduate student at U.C. Berkeley. The finite element program SSCOMP was debugged and made available on MicroVax system by Dr. Makram Jaber, GeoServices, Atlanta, GA.

REFERENCES

1. Adib, M.E. (1988), Lateral Earth Pressures in Reinforced Soil Walls, Thesis Submitted in Partial Satisfaction of the Requirements for the Degree of Doctor of Philosophy, Department of Civil Engineering, University of California, Berkeley, California.

2. Collin, J.G. (1986), Earth Wall Design, Thesis Submitted in Partial Satisfaction of the Requirements for the Degree of Doctor of Philosophy, Department of Civil Engineering, University of California, Berkeley, California.

3. Duncan, J.M., Byrne P., Wong, K.S., and Mabry P. (1980), Strength, Stress-Strain, and Bulk Modules Parameters for Finite Element Analysis of Stresses and Movements in Soil Masses, Report No. UCB/GT/80-01, Department of Civil Engineering, University of California, Berkeley, California.

4. Mitchell, J.K., Reinforcement for Earthwork Construction and Ground Stabilization, Proceedings of the VIII Pan American Conference, Cartagena, Vol. 1, pp. 349-380, August 16-21, 1987.

5. Seed, R.B. (1983), Soil-Structure Interaction Effects of Compaction-Induced Stresses and Deflections, Thesis Submitted in Partial Satisfaction of the Requirements for the Degree of Doctor of Philosophy, Department of Civil Engineering, University of California, Berkeley, California.

FAILURE OF BLUE HERON ROAD EMBANKMENT SUPPORTED BY REINFORCED EARTH
WALL

Utpalendu Bhattacharya[1], M.ASCE
and Frank B. Couch, Jr.,[2] M.ASCE

ABSTRACT: Blue Heron road in the Big South Fork National River
and Recreation Area in south-central Kentucky, provides access to a
restored coal mining and loading area.   The road follows steep
topography characteristic of the Cumberland Plateau and is built on
extensive colluvial deposits.   For much of its length, the
colluvium is underlain by the Mississippian Pennington Formation.
The upper part of this formation in contact with the colluvium
consists of highly weathered Pennington Shale.   The road embankment
is supported in several locations by reinforced earth walls founded
on the colluvium.   After heavy rains in the fall of 1986, a section
of the road embankment supported by one of the reinforced earth
walls moved significantly towards the river damaging the wall and
the road surface.   Gravel was used as backfill in the reinforced
earth area behind this wall.   No underground drain system was
installed behind the wall.   Since the road surface was unpaved
during the rains, runoff and underground water from the adjacent
bluff entered the weathered Pennington Shale through the overlying
gravel and colluvium.   The shale structure, in the presence of
water, broke down to weak Pennington Clay.   The embankment and the
riverbank slope started sliding on this weak clay under the load of
the embankment fill.   With the progress of this movement, the clay
strength gradually approached its residual value.   This reduction
in the clay strength accelerated the embankment and the slope
movement damaging the road and the reinforced earth wall.   An
emergency remedial rock berm constructed on the riverbank slope
essentially stopped the movement of the Blue Heron road embankment.
A long-term stability analysis of the remedial berm based on
residual soil strength parameters of the Pennington Clay gave a
minimum factor of safety (F.S.) of 1.3.   Two years of
instrumentation data, collected after the berm construction, showed
only negligible movements.   This indicates that the emergency berm

1 - Civil Engineer, Geotechnical Br., US Army Corps of Engrs,
    Nashville Dist, Nashville, TN 37202-1070
2 - Chief, Geotechnical Br., US Army Corps of Engrs, Nashville
    Dist, Nashville, TN 37202-1070.

is adequate to provide long-term stability to this section of the road.

INTRODUCTION

Big South Fork National River and Recreation Area (BSFNRRA) has been developed in Kentucky and Tennessee on the eastern side of the Cumberland Plateau by the U.S. Army Corps of Engineers, Nashville District (7). The Big South Fork of the Cumberland River runs through the mountainous area of this region. The improvement of Blue Heron road located near Stearns in McCreary County, Kentucky is a part of the BSFNRRA project. The last stretch of 2.5 miles (4 kilometers) of the road runs between a bluff and the Big South Fork River, and provides access to various recreational facilities developed near an abandoned coal mine in a gorge area. The original 15 ft (4.6 m) to 20 ft (6.1 m) wide undeveloped road was constructed in the late 1930's in the colluvium consisting of sands, silty and clayey sands, and silty and sandy clays with

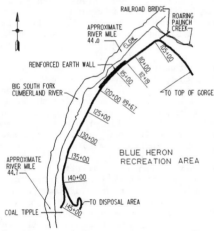

Figure 1. Blue Heron Road in Recreation Area (1 ft = 0.3048 m)

cobbles and boulders (5,6,7). These materials are the products of the weathering of the shales, sandstones, siltstones and coals which have moved downslope, mainly under gravity, over a long period of geologic time (4,5). The current road improvement was accomplished by using cut-and-fill operation. The embankment fills on some steep grades are supported by reinforced earth walls (RE walls).

The RE wall, about 4 ft (1.2m) to 25 ft (7.6 m) high, between Sta. 112+19 and Sta. 119+66.8 in Fig. 1 is on the riverbank near River Mile 44. In the fall of 1986, after heavy rainfall the wall between Sta. 116+00 and Sta. 118+50 moved significantly towards the river causing extensive damage to the embankment surface and the

wall panels. The failure analysis and the design and performance
of the remedial measures to restore the support system are
discussed in this paper. Under similar subsurface conditions,
several other embankment and RE wall slides occurred along Blue
Heron road between 1986 and 1989. These failures are not described
in this paper because of space limitations.

DESCRIPTION OF EMBANKMENT FAILURE

Four borings drilled during the wall design indicated 8 ft
(2.4 m) to 12 ft (3.7 m) of colluvial materials and coal of varying
consistencies underlain by very stiff to hard clay and weathered
shale (5). A depressed area near Sta. 118+00 served as a natural
drain for the runoff from the bluff. No borings were located in
this area. The installation of the RE wall in the colluvium was
started in the fall of 1985. The base of the affected wall was
between Elev. 755 ft (230.1 m) and Elev. 761 ft (232 m). About 7
ft (2.1 m) to 10 ft (3 m) of the colluvium was removed behind the
wall to lay the reinforcing steel straps which were then covered
with the gravel backfill. The area behind the reinforced earth was
filled with compacted on-site clayey soils. No subsurface drain
system was installed behind the wall. The slope adjacent to the
wall base was protected by riprap. By the end of April 1986 the
backfill was 2 ft (0.6 m) to 3 ft (0.9 m) below the top of wall.
The surface drain at the foot of the bluff to carry runoff from the
bluff area was not then constructed. The coping for the northern
part of the wall was completed to its finish elevation of 779.4 ft
(237.6 m). During installation of coping for the remaining wall in
May 1986, the wall height was found to be less than the height
originally installed. No visible distortion of the wall was,
however, observed. To monitor any possible wall settlement, the
elevation of the top of wall was measured by survey method at 49
points marked on the top of wall. The elevations, generally
recorded on weekly basis are plotted in Fig. 2. The measurement of
lateral movement of the top of wall was started on 1 August 1986
with reference to a baseline near the bluff away from any disturbed
soil behind the wall. The lateral movements, generally recorded
once in every one to two weeks are shown in Fig. 3.

The wall settlement was insignificant until September 1986.
After several heavy showers, the settlement on 29 October 1986 was
about 1.5 inches (38 mm) near Sta. 117+00. Except for some minor
cracks and openings at the panel joints, the overall structural
integrity of the wall appeared to be unaffected. The settlement
increased to 6 inches (152 mm) on 12 November 1986. This greatly
increased the panel damage. On 14 November, following additional
rains, extensive cracks, 6 to 24 inches (152 to 610 mm) wide and
parallel to the wall, were observed on the clayfill surface about
25 feet (7.6 m) behind the wall between Sta. 116+00 and Sta.
118+50. The road surface near the cracks subsided by about 3 feet
(0.9 m). The wall base appeared to have moved towards the river.
The above observations indicated deep-seated sliding of the
embankment below the reinforced earth structure. Fig. 4 shows a

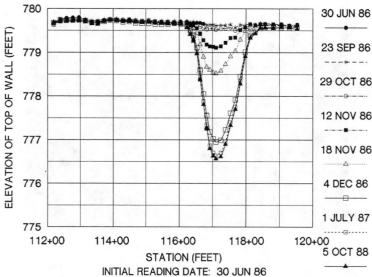

Figure 2. Settlement of Reinforced Earth Wall (1 ft=0.3048 m)

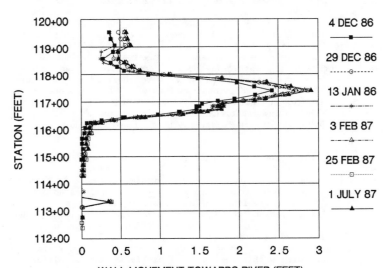

Figure 3. Lateral Movement of Reinforced Earth Wall
(1 ft = 0.3048 m)

Figure 4.   Failed Embankment of Blue Heron Road

photograph of the damaged embankment on 14 November 1986.   To
prevent further movement, construction of an emergency rock berm on
the riverbank slope was started on 5 December 1986.   By that time
the maximum wall settlement was 2.66 ft (0.8 m) and the lateral
movement measured for the first time since 1 August 1986 was 2.4 ft
(0.7 m).   The gravel fill spilled through several large gaps at the
panel joints causing large depressions on the backfill surface.
Numerous cracks were observed on the riverbank slope below the
wall.   During site preparation for the remedial berm, two springs
with steady discharges of clear water were observed on the slope
near Sta. 117+00.   Two additional springs were also encountered in
an excavation for the berm rock materials on the hill slope above
the road.   The spring locations and a cross section at Sta. 117+20
developed for designing the emergency berm system are shown in Fig.
5.

GEOLOGY AND PRELIMINARY SUBSURFACE INVESTIGATION

     The Blue Heron road site is in the Cumberland Plateau
physiographic province.   The bedrock of the site is composed of the
Pennington Formation, which is Upper Mississippian in age, and the
overlying Lee Formation, which is Lower Pennsylvanian in age.   The
Pennington Formation consists of grayish-red or olive-green shale,
with sandstone and limestone beds.   The Lee Formation consists of
shale, usually carbonaceous or sandy, interbedded with siltstone
and sandstone with a thin coal bed.   These formations of differing
ages are exposed at the site as the result of the Big South Fork of
the Cumberland river partially dissecting the area.   The surface of
this area is covered by colluvial materials (4,5).

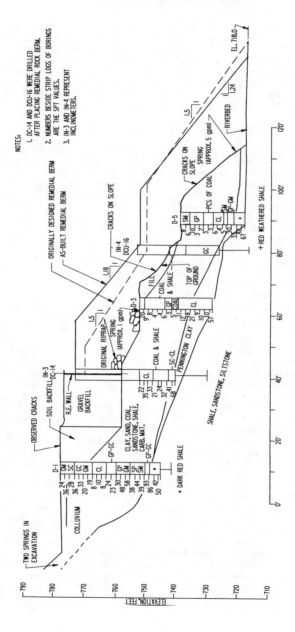

Figure 5. Blue Heron Road Cross Section at Sta. 117+20 with Strip Logs of Borings (1 ft = 0.3048 m)

The information from the design stage borings was found inadequate to explain the settlement first recognized in June 1986. Hence, six SPT borings, D-1 through D-6, including three (D-1, D-3 and D-5) near Sta. 117 were drilled in August 1986 by the Corps. The borings indicated about 16 to 25 ft (4.9 to 7.6 m) of generally compressible colluvial materials underlain by Pennington Clay and weathered Pennington Shale. The boring locations are shown in Fig. 6. The strip logs of D-1, D-3 and D-5 are shown in Fig. 5

ANALYSIS OF EMBANKMENT FAILURE

A significant property of the weathered Pennington Shale is its very rapid breakdown to a weak, highly plastic impervious Pennington Clay in the presence of water. The Cumberland Plateau area in east-central Tennessee and adjacent areas of Kentucky are generally underlain by colluvium on weathered Pennington Shale. In many places, both surface and subsurface water percolated through the highly permeable colluvium, and became trapped on the relatively impermeable weathered shale. The shale then broke down to weak Pennington Clay in the presence of water (4). Many natural slopes underlain by this material are on the verge of instability. Many embankment failures along Interstate highways I-24, I-40 and I-75 in the Cumberland Plateau area were attributed to the embankment load disturbing the equilibrium at the interface between the colluvium and the weak Pennington Clay (4).

The Blue Heron road embankment was constructed in the colluvium overlying the weathered Pennington Shale. The breakdown of the weathered Pennington Shale samples from Blue Heron road site was observed during slaking tests performed in the laboratory. Such softening of the weathered shale was, however, not recognized during the embankment design. At the RE wall site, the structure of the weathered shale, generally remained undisturbed before the major rainstorm that occurred in October 1986. Hence the wall settlement was insignificant till September 1986, as shown in Fig. 2. Because of lack of any drain system, runoff from the bluff during heavy rains in the fall of 1986 entered the exposed gravel fill and the relatively permeable colluvial materials, and encountered the impermeable weathered Pennington Shale. The discovery of several springs in the colluvium across the height of the slope in the failed area, as shown in Fig. 5, indicates that the spring water also flowed to the underlying weathered shale. The shale in the presence of water then broke down to weak Pennington Clay. The embankment and the riverbank slope started sliding on this weak clay under the load of the embankment fill. With the progress of this movement, the clay strength gradually approached its residual value. Progressive failures of various natural slopes on clay shales due to gradual reduction of shear strength to its residual value were reported by Bjerrum (1). Such reduction of the shear strength from the peak to the residual shear strength value was observed in the laboratory direct shear tests conducted on undisturbed Pennington Clay samples from Blue Heron road site. This reduction in the clay strength accelerated the RE

wall and the embankment movement damaging the wall panels and the
road, as observed on 14 November 1986. The movement continued to
increase until the remedial rock berm on the riverbank slope was
installed in December 1986.

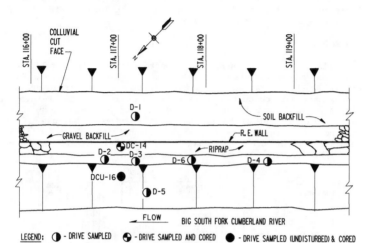

Figure 6.   Boring Locations on Blue Heron Road Wall Area
            (1 ft = 0.3048 m)

EMERGENCY REMEDIAL ROCK BERM

     The remedial system consists of a rockfill berm with its toe
extended to 20 ft (6.1 m) onto the riverbed rock surface. Since no
reliable strength information was available, the design was based
on cohesion, c=0 and the angle of shearing resistance, $\phi=10^{\circ}$ for
the Pennington Clay. The actual geometry of the failed slope could
not be determined by survey at that time. Hence an approximate
cross section of the failed slope was analyzed for a F.S.=1 to back
figure the above strength parameters. Using these parameters, the
rockfill berm was designed for a F.S. =1.5. The design cross
section is shown in Fig. 5. The berm was constructed of sandstone
from roadway excavation supplemented by a nearby quarry. To
minimize erosion, the surface was sealed with fine stone and
covered with topsoil to support vegetation. After construction,
the effective wall height was reduced to 4 to 7 ft (1.2 to 2.1 m).
The as-built cross section of the remedial berm is shown in Fig. 5.
The berm construction was completed between 5 December and 28
December 1986.
     To determine if this emergency berm is capable of providing
long-term stability to the embankment, a stability analysis of the
as-constructed berm based on undisturbed Pennington Clay strengths
was conducted. Borings DC-14 and DCU-16 on Fig. 6 were drilled in

March 1986 on the upper and the lower berms, respectively. Undisturbed samples were collected from DCU-16, and also from U-15, which was drilled at another slide area near Sta. 132+00. Since the two slide areas are underlain by similar materials, the U-15 soil strengths were also used in the analysis. The boring locations are shown in Fig. 6, and the boring logs in Fig. 7. The strip logs are included in Fig. 5.

LABORATORY TESTS

Triaxial, direct shear and repeated direct shear tests were conducted on undisturbed samples from the Blue Heron road site. Consolidated undrained triaxial tests with pore pressure measurement were conducted on five samples including two Pennington Clay samples from U-15 and two samples from DCU-16. Most of the test specimens had to be patched during trimming because of the presence of gravel in the samples. The specimens, about 1.3 to 1.4 inches (33 to 36 mm) in diameter and 2.6 to 3 inches (66 to 76 mm) in height were consolidated under the confining pressure (σ_3)

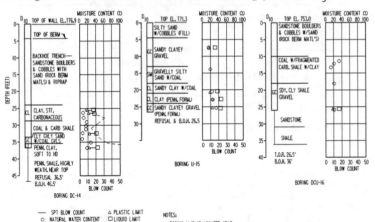

Figure 7. Logs of Borings at Blue Heron Road Slide Area
          (1 ft = 0.3048 m)

ranging from 0.5 TSF (0.024 kN/m^2) to 4 TSF (0.192 kN/m^2), as shown in Tables 1 and 2. Time to apply the deviator stress for a maximum axial strain of 15% to 20% ranged from 440 to 1080 minutes. The deviator stress at failure, $(\sigma_1 - \sigma_3)_f$, for each specimen is also included in Tables 1 and 2. Pore pressure was measured during the application of the deviator stress. The strength parameters determined from the total stress envelope are called the R parameters, and those from the effective stress envelope are the R-bar parameters (2). Since the Pennington Clay strength used for the stability analysis was determined from the repeated direct

shear test, as described below, further details of the triaxial
test data are not discussed. The R and R-bar parameters are,
however, included in Tables 1 and 2 to compare with those from the
direct shear and the repeated direct shear tests. Details of all
soils test data for the slide area were published elsewhere (8).

The direct shear and the repeated direct shear tests were
conducted on two undisturbed Pennington clay samples from boring U-
15 and one from DCU-16. Each test was conducted on a nominal 3
inch (76 mm) square and 1/2 inch (13 mm) thick sample consolidated
under a normal load of 0.75 to 4 tons per square foot (72 to 383
kN/m^2) and then sheared at a strain rate of 0.00033 to 0.00046
inch/minute (0.00838 to 0.01168 mm/minute). In the repeated direct
shear test the sample was repeatedly sheared by reversal of the
direction of shear until a minimum shear stress was determined (2).
The shear strength parameters are given in Tables 1 and 2. The
ratio of the residual shear stress to the peak shear stress for the
two Pennington Clay samples ranged from 0.30 to 0.74. The strength
and the deformation data for these two samples are plotted in Fig.
8.

A mineralogical composition test by X-ray diffraction analysis
was run on a Pennington Clay sample from boring U-15. The
resulting diffraction pattern indicated that the sample consisted
predominantly of quartz and dolomite with 17% illite, 8% kaolinite,
and a trace (less than 3%) of mixed layer clay minerals. Slaking
tests were performed on two cored samples of the weathered
Pennington Shale, about 1-7/8 in (48 mm) diameter and 1 in (25 mm)
in height, obtained from a boring, DC-10, in the slide area near
Sta. 132+00. The air-dried samples crumbled to very soft clay in 1
to 2 minutes after being immersed in water.

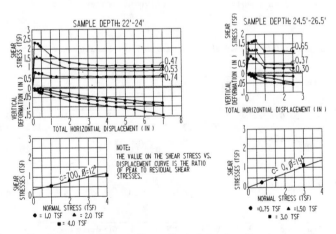

Figure 8. Repeated Direct Shear Test Data for Pennington Clay
           Samples from Boring U-15
           (1 ft=0.3048 m, 1 PSF=0.04788 kN/m^2, 1 TSF=95.76 kN/m^2)

Table 1. Summary of Shear Strength Test Results for Boring U-15

Depth (Ft)	Materials	σ_3 TSF	$(\sigma_1-\sigma_3)_f$ TSF	Type of Str Envel	c PSF	ϕ DEG	γ_{sat} PCF	γ_{moist} PCF
6.5-8.5	Sandy	0.5	1.9	R*	800	26	141	133
	clayey	1.0	3.0	R-bar**	500	30	141	133
	gravel	2.0	3.9					
13.0-14.8	Gravelly	0.5	2.8	R	600	29	128	125
	silty snd	1.0	3.6	R-bar	0	38	128	125
	with coal	2.0	6.1					
20.0-22.0	Sandy	2.0	2.7	R	1400	12	129	122
	clay with	4.0	3.2	R-bar	1100	24	129	122
	coal							
22.0-24.0	Clay	1.0	2.4	R	1800	10	131	128
	(Penn)	2.0	2.7	R-bar	1000	19	131	128
		3.0	3.1	Dir sh	500	31	135	135
				Res dir sh	700	12	135	135
24.5-26.5	Sandy cl	1.0	1.9	R	1200	14	122	122
	gravel	2.0	3.1	R-bar	1100	25	122	122
	(Penn)	4.0	4.1	Dir sh	1100	23	118	118
				Res dir sh	0	19	118	118

1 ft=0.3048 m, 1 PSF=0.04788 kN/m^2, 1 PCF=0.1571 kN/m^3
*Total stress envelope   **Effective stress envelope

Table 2. Summary of Shear Strength Test Results for Boring DCU-16

Depth (Ft)	Materials	σ_3 TSF	$(\sigma_1-\sigma_3)_f$ TSF	Type of Str Envel	c PSF	ϕ DEG	γ_{sat} PCF	γ_{moist} PCF
18.5-21.0	Gravelly	1.0	2.0	R*	1000	19	133	133
	sandy	2.0	3.2	R-bar*	0	34	133	133
	clay	3.0	4.0					
				Dir sh	0	43	134	134
				Res dir sh	100	28	134	134
21.5-23.5	Sandy	1.0	3.1	R	2200	11	128	124
	clay	2.0	3.6	R-bar	1000	27	128	124
		3.0	3.9					

1 ft=0.3048 m, 1 PSF=0.04788 kN/m^2, 1 PCF=0.1571 kN/m^3
*Total stress envelope   **Effective stress envelope

LONG-TERM STABILITY ANALYSIS

For the long-term stability analysis, the minimum values of the Pennington Clay strength parameters (c=0, ϕ=19°) and of the colluvium (c=500 psf, ϕ=30°) were selected from Tables 1 and 2. The remedial rock berm was designed on an emergency basis considering a very approximate geometry of the failed slope. Hence the laboratory strength values of the Pennington Clay are somewhat different from the back-calculated values (c=0, ϕ=10°). The

γ = MOIST OR SATURATED UNIT WEIGHT OF SOIL

c = COHESION

ϕ = ANGLE OF SHEARING RESISTANCE

STRENGTH PARAMETERS

SOIL NO.	SOIL TYPE	γ PCF	C PSF	ϕ DEG
①	SOIL BACKFILL	120	2000	30
②	GRAVEL BACKFILL	130	0	40
③	BERM ROCKFILL ABOVE WATER	135	0	45
④	BERM ROCKFILL BELOW WATER	145	0	45
⑤	COLLUVIUM ABOVE WATER	122	500	30
⑥	COLLUVIUM BELOW WATER	128	500	30
⑦	PENNINGTON CLAY	118	0	19
⑧	ROCK	160	40000	0

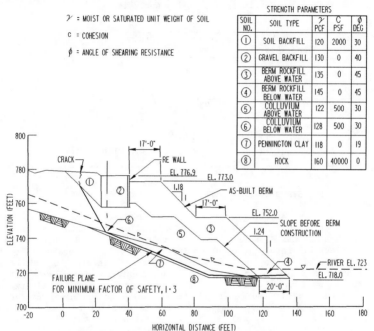

Figure 9. Cross Section of Slope at Sta. 117+20
(1 ft=0.3048 m, 1 PSF=0.04788 kN/m^2,1 PCF=0.1571 kN/m^3)

strength values for other materials in the slope were assumed based on previous experience with similar materials. The strength parameters and the as-built cross section of the remedial berm at Sta. 117+20 are shown in Fig. 9. The stability program UTEXAS2 was used for running the analysis by the Spencer method (3). Five cases were studied for various changes in the Pennington Clay strength and the river level keeping the strength of other materials the same. The results are summarized in Table 3.

Based on the minimum strength (c=0 and ϕ=19°) of the Pennington Clay, the minimum F.S. is 1.3, as shown in Case 1. This appears to be adequate for the long term stability of the embankment system.

The F.S. increases to 1.4 when the river level rises to Elev. 750 (Case 2). Case 3 was studied for the residual strength (c=700 psf, $\phi=12°$) of the other Pennington Clay sample given in Table 1. This gives a F.S.=1.5. Case 4 shows an example of the sensitivity of the F.S. with reduction in the Pennington Clay strength. The F.S. reduces to 1.2 for c=0 and $\phi=17°$. The F.S. of 0.9 in Case 5 indicates the instability of the slope before construction of the remedial berm.

Table 3. Results of Stability Analysis

Case	Strength of Penn. Clay		River Elev. (Ft)	Minimum F.S.
	c (PSF)	ϕ (DEGREE)		
1	0	19	723	1.3
2	0	19	750	1.4
3	700	12	723	1.5
4	0	17	723	1.2
5	0	19	723	0.9 without berm

1 ft = 0.3048 m,   1 PSF = 0.04788 kN/m^2

STUDY OF INSTRUMENTATION DATA

Two inclinometers, IN-3 and IN-4 were installed in the boreholes of DC-14 and DCU-16, respectively. The inclinometer data are plotted in Fig. 10 and Fig. 11. Pennington Clay is located approximately at Elev. 741 in DC-14, as shown in Fig. 5. A sudden break at about the same elevation in the inclinometer curves in Fig. 10 indicates that some sliding continued along the Pennington Clay layer after the construction of the remedial berm. The maximum riverward movements within two years of the initial readings never exceeded 0.25 inch (6 mm). The maximum movements parallel to the wall are within 0.6 inch (15 mm). Some inconsistency in the direction of the movements appears to be due to alternate wetting and drying of the riverbank materials caused by the fluctuating river level. The wall settlement between the last two observations in Fig. 2 is less than 1 inch (25 mm), and the lateral movement in Fig. 3 less than 0.5 inch (13 mm). The above settlement and the movements are small for the size of the structure.

CONCLUSIONS

The large movement of the failed Blue Heron Road embankment was due to its deep-seated sliding on the underlying weak Pennington Clay. The inclinometer data of IN-3 indicates movement on this clay. Runoff and underground water from the bluff area entered the permeable backfill and colluvium, and encountered the Pennington Shale. The shale in the presence of water broke down to weak

Pennington Clay. The softening of the shale after being immersed
in water was observed in the laboratory. The embankment started

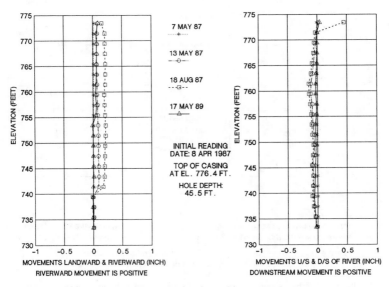

Figure 10. Blue Heron Road IN-3 Inclinometer Data
(1 ft = 0.3048 m, 1 inch = 25.4 mm)

moving on the weak Pennington Clay. With the progressive movement,
the clay strength was gradually reduced to its residual value
causing the large movement due to sliding. Repeated direct shear
tests on the Pennington Clay indicated reduction of the peak shear
stresses to the residual values. The emergency remedial berm on
the river slope effectively stopped the embankment movement. A
reanalysis of the slope based on undisturbed soil strengths
indicated a F.S. of 1.3. Two years after the berm installation,
the lateral movements of the wall and the riverbank slope and the
wall settlement were found to be less than 1 inch (25 mm). The
above slope and the wall movements since the construction of the
remedial system are small for the size of the structure. Based on
this observation, the remedial berm, without any modification,
appears to have stabilized the embankment structure on a long term
basis.

REFERENCES

1. L Bjerrum, Progressive Failure in Slopes of Overconsolidated
     Plastic Clay and Clay Shales, Terzaghi Lectures, 1963-1972,
     ASCE, 1974, 139-187.

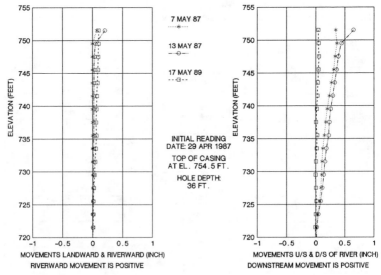

Figure 11.   Blue Heron Road IN-4 Inclinometer Data
(1 ft = 0.3048 m, 1 inch = 25.4 mm)

2. Dept of the Army, Corps of Engrs, Laboratory Soils Testing,
    Engineering Manual, EM 1110-2-1906, Washington, DC, Nov 1970
    (Rev. May 1980).
3. EV Edris and SG Wright, User's Guide, UTEXAS2 Slope Stability
    Package, Vol 1, Dept of the Army, US Army Corps of Engrs,
    Washington, DC, Aug 1987.
4. DL Royster, Highway Landslide Problems Along the Cumberland
    Plateau in Tennessee, Bulletin of the Assn. of Engineering
    Geologists, Vol. X, No. 4, 1973.
5. US Army Corps of Engrs, Nashville District, Specifications for
    Blue Heron Road Part II and Recreation Part I (Appendix A),
    Big South Fork Natl. River and Recreation Area, KY & TN, May
    1985.
6. US Army Corps of Engrs, Nashville Dist, Master Plan, Feature
    Design Memo No. 7, Vol I, Chap 5, Big South Fork Natl. River
    and Recreation Area, KY & TN, Jun 1981.
7. US Army Corps of Engrs, Nashville Dist, General Design
    Memorandum, Big South Fork Natl. River and Recreation Area, KY
    & TN, Chap 2, Dec1976.
8. US Army Corps of Engrs, Nashville Dist, Stability Analysis of
    Blue Heron Road (Recreation Area, Part 1), Sta. 132+00 & Sta.
    117+00, Big South Fork Natl. River and Recreation Area, KY &
    TN, Nov 1987.

CONSTRUCTION INDUCED MOVEMENTS OF INSITU WALLS

By G. Wayne Clough [1] F. ASCE and Thomas D. O'Rourke [2] M. ASCE

ABSTRACT: The issue of movements of insitu walls has become more important with the growth of new technology in this area, and the increase in litigation associated with damages caused by the movements to adjacent facilities. New insights into the subject are possible given the increasing numbers of instrumented case histories, and the ability to model the problem using the finite element method. In this paper movements of insitu walls are examined by updating the existing data base using information on both conventional and new systems. The effort differs from previous work in that the movements are divided so that effects of the basic excavation and support process can be separated from those caused by factors such as ancillary construction activities. The influence of movements on adjacent structures are considered. The results allow refinements in trends of maximum movement and displacement profiles relative to those given in previous literature. Information is given in a form to provide tools that can be used for design predictions.

INTRODUCTION

Twenty five years ago the subject of insitu walls involved a relatively simple technology which included mainly temporary sheet-pile and soldier pile walls with crosslot bracing or rakers and earth berms. In the intervening years innovations have had a major impact on this field (Table 1). With the change in technology has come a growing interest in the movements of insitu wall systems, reflecting the increasing litigation over damages caused by excavations constructed within insitu wall systems, and applications in more critical situations.

---

1 - Professor and Head, Department of Civil Engineering, Virginia Polytechnic Institute and State University, Blacksburg, VA 24061-0105.
2 - Professor, School of Civil & Envr. Engineering, Cornell University, Ithaca, NY 14853.

TABLE 1. Innovations In Insitu Wall Technology

Tie Back Supports	Shotcrete and Soil Nail Walls
Soil Nail Supports	Jet Grouted Walls
Diaphragm Walls	Lime Column Walls
Secant and Tangent Pile Walls	Root Pile Walls
Chemically Grouted Walls	Soil Cement Walls

The first practical approach for estimating movements for insitu wall systems was proposed by Peck (42) (Fig. 1). Data were compiled on settlements of the ground adjacent to temporary braced sheetpile and soldier pile walls. Peck's chart gave the settlement, divided by excavation depth, H, plotted against the distance from the insitu wall also divided by the distance H. Three categories of behavior

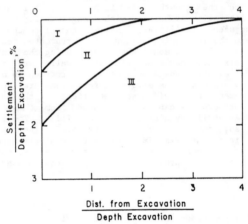

I - Sand and Soft to Hard Clay, Avg. Workmanship

II - Very Soft to Soft Clay
    1. Limited Depth of Clay Below Bott. Exc.
    2. Significant Depth of Clay Below Bott. Exc.,
       But $N_b < N_{cb}$*

III - Very Soft to Soft Clay to a Significant Depth
    Below Exc. Bott. and $N_b > N_{cb}$

    * N_b = Stability No. Using C "Below Base Level" = $\dfrac{\gamma H}{C_b}$
    N_{cb} = Critical Stab No. for Basal Heave

Figure 1.   Summary of Soil Settlements Behind Insitu Walls
           (Peck, 42).

were defined, with the smallest movements indicated for sands, stiff
clays, and soft clays of small thickness (Category I). The maximum
movements in the Category I conditions near the wall were 1% of the
excavation depth. Recent performance with well designed and
constructed insitu walls shows improvement over this standard, an
example being the 36m excavation in stiff clay for the Columbia
Center in Seattle, Washington, where the maximum movements were less
than 0.1% H (19). This reflects progress in control of movements
that has come through the use of newer design and construction
technologies.

Peck's chart also included data for excavations in soft clays
where basal stability was an issue and the thickness of the clay
below the excavation was large (Categories II and III). Movements
in these conditions exceed those in Category I because plastic
yielding occurs beneath the excavations. Control of movements for
this type of behavior is more difficult than for the Category I
condition, but progress is being made.

Since the publication of the Peck (42) paper, many workers have
contributed to our knowledge of this subject (3, 12, 18, 27, 39,
57). This paper is intended to be a logical extension of the
previous work. The goals are: (1) Update the available data in
terms of the new wall technologies and the enhanced base of
information on movements; (2) Clarify ground movement patterns and
refine current methods for estimating wall movements and settlement
distributions adjacent to deep excavations; and, (3) Relate the
likely ground movements patterns for different soil types with the
potential severity of building damage. Various construction
activities which often are performed concurrently with the
excavation and support process also are reviewed, and methods for
evaluating ground movements from these sources are proposed.

BASIC MOVEMENT TRENDS

Movements of insitu walls are a function of many factors,
including the soil and groundwater conditions, changes in groundwater level, depth and shape of excavation, type and stiffness of
the wall and its supports, methods of construction of the wall and
adjacent facilities, surcharge loads, and duration of wall exposure
among others. Estimating wall movements requires that all possible
factors be considered. In this section of the paper the discussion
concentrates on those movements due to the basic excavation and
support process.

Maximum Movements - Stiff Clays, Residual Soils and Sands. It
is characteristic of stiff clays, residual soils and sands that
basal stability is not an issue except in unusual cases (described
later in this paper). Peck's 1969 data suggested that in normal
circumstances movements of excavation support systems in these
soils were limited to 1% H. Later, using case histories from the
literature, Goldberg et al. (18) showed that the maximum horizontal
movements for insitu walls and settlements of the retained soil
masses in such materials were usually less than 0.5% H. To test
this finding, Figs. 2 and 3 were prepared to show maximum horizontal
movements and soil settlements respectively as a function of H.

These plots use the Goldberg et al. (18) data plus new information on conventional walls and new systems including soil nailed walls. Some points on the plots exhibit very large movements relative to the main portions of the data. These special cases are identified by the reference number from which they were obtained, and are unique in that they were influenced by factors outside the basic excavation and support process. Excluding points indicated as special cases, the following conclusions may be drawn:

(1) The horizontal movements tend to average about 0.2% of H.

(2) The vertical movements tend to average about 0.15% H.

(3) There is ample scatter in the data, with the horizontal move-ments showing more than the vertical movements.

(4) There is no significant difference between trends of the maxi-mum movements of different types of walls, and this includes even the new soil nail and soil cement walls.

Figs. 2 and 3 are useful to understand movement patterns and also can be used as design tools to estimate maximum wall and soil movements. However, the question to be answered for any particular project is, "Will it be one of those which falls on the average trend line, or will it experience larger or smaller movements?" This issue can be addressed in two parts. The first concerns the case where the movement is within the general scatter. The second addresses the special cases where movements are 0.5% H and above, and is considered later in the paper.

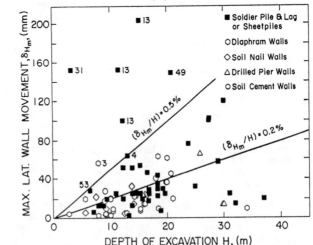

Figure 2. Observed Maximum Lateral Movements for Insitu Walls in Stiff Clays, Residual Soils and Sands.

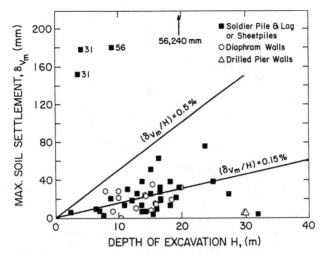

Figure 3.   Observed Maximum Soil Settlements in the Soil
Retained by Insitu Walls.

For those cases with movements that fall within the scatter around the 0.2% H trend line, it is notable that the data roughly follow a linear relationship with depth. This suggests that the soil masses are behaving approximately as elastic materials. Using the assumption of elastic soil behavior, a series of finite element analyses were performed with modulus values typical of those for stiff soils. Within the analysis framework, several parameters important to insitu wall performance were varied, including soil stiffness, wall stiffness, support spacing, and the coefficient of lateral earth pressure. In Fig. 4, the predicted maximum lateral wall movements are plotted against H, and it can be seen that the data follow an approximately linear response with excavation depth, centered around a trend line of 0.2% H. The predicted response is consistent with the average behavior observed in Fig. 2. The parameters wall stiffness and strut spacing were found to have only a small influence on predicted movements because the soil in these circumstances is stiff enough to minimize the need for the structure. However, soil modulus and coefficient of lateral earth pressure had a more significant impact (Fig. 4). The variations in predicted behavior from the finite element analyses illustrate factors that could easily lead to the scatter found in the observed behaviors in Figs. 2 and 3. The data also suggest that in a stiff soil environment variations in soil stiffness have a more profound effect on wall behavior than system stiffness.

Maximum Movements - Soft and Medium Clays. As opposed to stiffer soils, basal stability in soft and medium clays may be at issue, and as a result, movement patterns in these conditions can be dominated by deflections beneath the excavation. Peck (42) recognized this

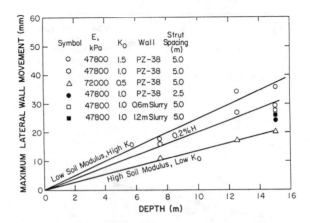

Figure 4.   Predicted Maximum Lateral Wall Movements by Finite
Element Analysis Modeling Stiff Soil Conditions.

and qualified his movement data on the basis of stability number, N_b, defined as $\gamma H/c_b$, where γ is the unit weight of the soil above the excavation, and c_b is the undrained shear strength of the clay beneath the excavation. When the magnitude of the stability number exceeds the bearing capacity factor for failure of the base of the excavation (6 - 9), then movements can become large. The acceleration of the movements comes about as a result of plastic yielding in the soils at and beneath the base of the excavation. Mana and Clough (27) and more recently Clough et al. (10) defined movements in terms of the factor of safety against basal heave (FS). Fig. 5, a plot of maximum lateral movement of the wall versus FS, shows that as the FS falls below 1.5, movements increase rapidly. Fig. 5 also illustrates the influence that wall stiffness and support spacing can have on movements. These factors are most important when FS is low. Of course, there are many other factors that affect movements beyond those included in Fig. 5. Allowances for other parameters such as soil anisotropy, support stiffness, and excavation dimensions are given in References 8 and 27.

In circumstances where movements are primarily due to the excavation and support process, Fig. 5 can be used to assist in predicting maximum lateral wall movements in clay soils. The figure can also be used to estimate maximum soil settlement since this parameter is approximately equal to maximum horizontal wall movement. It is notable that the chart shows that as the FS becomes over 2, and base stability is guaranteed, the maximum movements decrease below 0.5%. This is consistent with the data in Fig. 2 and 3 for stiff soils where the FS is usually well above 2. On the other hand, if the FS is close to one, movements can exceed 2% H even with good construction.

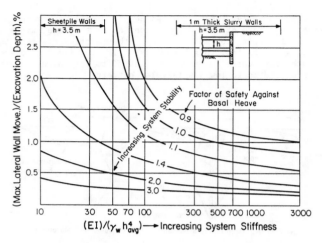

Figure 5.   Design Curves to Obtain Maximum Lateral Wall Movement
            (or Soil Settlement) for Soft to Medium Clays
            (Clough et al., 10).

Charts such as those in Fig. 5 have to be used with caution,
especially where the FS value is below 1.5.   In these conditions
construction variables can cause significant increases in movements.
Cases of this type are illustrated later in this paper.

General Patterns of Ground Movement. Inclinometer and settlement
measurements for braced and tied-back excavations have disclosed a
general pattern of wall movement and adjacent ground deformation
which is illustrated in Fig. 6.   During the initial stages of
construction, soil may be excavated before the installation of
support.   In some cases, soil is excavated when the upper levels of
support are not preloaded or lack sufficient stiffness to restrict
inward movement.   The wall deforms as a cantilever, and the adjacent
soil settles such that vertical surface movements increase in
inverse proportion to distance from the edge of excavation.   Settlements during this stage of construction may be bounded within
a triangular distribution of displacement, as shown in Fig. 6a.
Cantilever displacement profiles are characteristic of flexible
systems that have zero to low preloads, flexible walls where the
supports are installed only after considerable excavation has been
made, and cantilever walls.

When the excavation advances to deeper elevations, upper wall
movement is restrained by installation of new support or   stiffening
of existing support members.   Deep inward movement of the wall
occurs, which is shown as an   incremental component of the total
displacement in Fig. 6b.   The combination of cantilever and deep
inward components results in the cumulative wall and ground surface

displacement profiles shown in Fig. 6c. If deep inward movement is the predominant form of wall deformation, as is the case with many deep cuts in soft to medium clay, then settlements tend to be bounded by a trapezoidal displacements. If cantilever movements predominate for cuts in sand and stiff to very hard clay, then settlements tend to follow a triangular pattern.

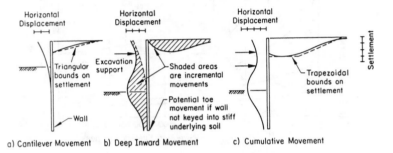

a) Cantilever Movement   b) Deep Inward Movement   c) Cumulative Movement

Figure 6.   Typical Profiles of Movement for Braced and Tied-Back Walls.

There is considerable justification in theory and measurement for these patterns of movement. Theoretical and experimental studies by Milligan and coworkers (29, 30) have shown that incremental plastic deformations of the wall will generate deformations at the ground surface consistent with those for the cantilever and deep inward movements modes delineated in Fig. 6. Field measurements of horizontal strains at excavations in different types of soil (16, 39, 52) show similar patterns, with triangular contours of strain caused by cantilever wall deformation and deep concentric contours of strain bounding zones of maximum settlement caused by walls subject to deep inward displacement.

Recent developments such as the soil nailed wall are shown in Fig. 2 to have maximum wall movements in line with those of conventional systems. Wall deflection profiles for four soil nail case histories in Fig. 7 show that these systems have the largest movements at the top of the wall, and a cantilever shape. This is consistent with the fact that the nails usually are not pre-loaded and the wall system is relatively flexible.

Displacements Adjacent to Excavations. Observations of movements of soil adjacent to excavations provide the best way to define ground deformations likely to occur in the field. However, it is well known that field measurements can involve significant movement components caused by activities such as dewatering or deep foundation construction within the excavation. It is logical to treat these components separately and to focus on the movements caused by excavations and support systems. In the past this has not always been the case, with the result that the conventional approaches are often biased, reflecting activities not directly tied to the excavation and support process.

For the following discussion, each set of field measurements was
screened   to   preclude   movements   not   primarily   related   to   the
excavation and support of the cut. This involved discounting some
case history information,   and in other cases, breaking out the

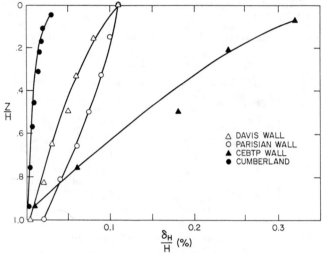

Figure 7.   Non-Dimensionalized Wall Movements for Soil Nailed
            Walls (Adapted from Juran and Elais, 23).

incremental movements pertaining to the main excavation and bracing
stages of construction.   Subsequent sections of the paper provide
data pertinent to the effects of factors beyond those considered at
this juncture.

Excavations   in   Sand.   Fig. 8 summarizes settlements for
excavations in predominantly sand and granular soil profiles.   The
data involve various types of walls and supports, including soldier
pile and lagging with cross-lot struts (40), soldier pile and
laggng with tiebacks (55), sheetpiles with tiebacks (47, 48), and a
concrete diaphragm wall with crosslot struts (14).   The excavations
were in granular soils above the water table, which in most cases
was lowered by dewatering.   The Charter Station excavation was an
exception in that a groundwater recharge system was operated to
reduce settlement of adjacent property.   Settlements are expressed
as a percentage of the maximum excavation depth and plotted as a
function of the ratio of distance from excavation to maximum
excavation depth.

The   maximum   settlements are typically less than 0.3%, a finding
consistent with the data given in Fig. 3, and as indicated in other
references (18, 39).   The settlements show a consistent pattern of
increasing vertical displacement with decreasing distance from the
edge of the excavation. As shown on the figure, a triangular bound
applies to the data.

Excavations in Stiff to Very Hard Clays. Fig. 9 summarizes settlements for excavation sites in stiff to very hard clays, involving horizontally supported concrete diaphragm walls (15, 21, 48, 52), tied-back concrete diaphragm and soldier pile walls (19, 22, 45, 46), and other support systems (53, 54). As discussed previously, displacements caused by ancillary construction activities were removed, if present, from reported measurements to represent more accurately the displacements caused by the excavation

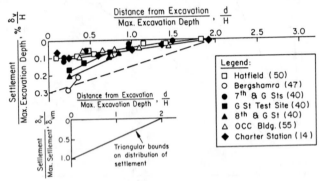

Figure 8.   Summary of Measured Settlements Adjacent to Excavations in Sand.

and bracing process. For example, vertical and horizontal movements observed during the installation of secant piles at Bell Common (52) were screened from the data summarized in Fig. 9. Both settlement and horizontal displacements, which are expressed as percentages of maximum excavation depth, are plotted versus the ratio of distance from the cut to maximum excavation depth.

The settlements are only a small percentage of excavation depth, with maximums less than 0.3%, but are distributed over three times the excavation depth from the edge of the cut. Small amounts of heave are also indicated in some cases from Houston, where the heave has been as much as 0.05 to 0.1% the excavation depth (54).

Records of horizontal ground displacements are more variable than those for settlements, and show two distinct zones of movement (Fig. 9). The majority of horizontal displacements fall within a triangular boundary with the same dimensions as those pertaining to the observed settlements. This zone corresponds to excavations which have been braced with relatively stiff supports. A second zone also is drawn in the figure. This zone contains measurements from the Neasden (45) and Bell Common (52) highway excavations in London Clay. Both excavations were affected by their support systems. At Bell Common, movements immediately adjacent to the edge of construction were influenced by a 3.5-m-deep cut in which temporary sheetpiles were propped against the permanent secant pile wall. At the Neasden Underpass, block movement of London Clay led to horizontal displacement in the zone of tieback anchor support.

An interesting feature is that, even when relatively high horizontal displacements were observed, settlements did not exceed the general triangular bounds shown in the upper figure. For cuts in stiff to very hard clays, it appears that horizontal movements will tend to equal or exceed their vertical counterparts. As an upper bound, the ratio of horizontal to vertical movement can be as high as 2.5:1.

It should be recognized that stiff to very hard clay represents an approximate classification, based on consistency and undrained shear strength. Performance in stiff to very hard clays is influenc-

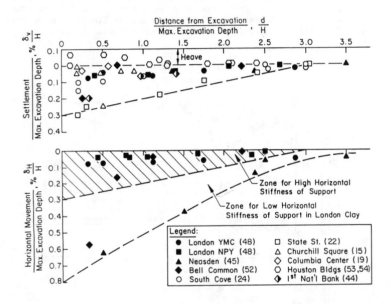

Figure 9.   Summary of Measured Settlements and Horizontal Displace-
            ments Adjacent to Excavations in Stiff to Very Hard Clay.

ed as a minimum by in-situ horizontal stress, degree of fissuring, degree of weathering and plasticity, and variations in behavior are likely. Inclinometer measurements at most sites represented in Fig. 9 show that substantial cantilever components of wall deformation occurred (15, 19, 21, 22, 45, 48, 52). Moreover, the surface surveys showed that settlements, when present, decreased in approximately direct proportion to distance from the edge of excavation. For these reasons, it is recommended that a triangular bounds on the settlement profile be used as a first order estimate of the distribution of movement.

Excavations in Soft to Medium Clays. Fig. 10 summarizes settlements for excavations in soft to medium clay, involving cross-lot struts supporting sheetpile (16, 17, 26, 33-37, 41), soldier pile and lagging (41), and concrete diaphragm walls (41). In some cases, berms and rakers were used either as full or supplemental support (41) for the excavations. Settlement as a percentage of maximum excavation depth is plotted versus distance from the cut as a fraction or multiple of maximum excavation depth. Zones pertaining to various levels of workmanship and soil conditions as described by Peck (42) also are shown in the figure. Although the zones are helpful in delineating broad trends of performance, it is evident that the scatter in settlement magnitude places limitations on this type of empirical approach for making predictions about movement.

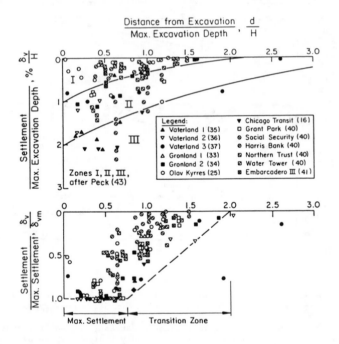

Figure 10.   Summary of Measured Settlements Adjacent to Excavations in Soft to Medium Clay.

When the settlements are plotted as fractions of maximum settlement, a relatively well-defined grouping of the data is evident. The settlement distribution is bounded by a trapezoidal envelope in which two zones of movement can be identified. At $0 \leq d/H \leq 0.75$, there is a zone in which the maximum settlement occurs.

At $0.75 < d/H \leq 2.0$, there is a transition zone in which settlements decrease from maximum to negligible values.

The largest differential soil movements occur near or within the transition zone. In Fig. 11, the maximum angular distortion, β, is plotted relative to the maximum settlement, δ_{vm}, as measured at excavations represented in Fig. 10. The angular distortion is defined as a differential settlement between two points divided by the distance separating them. It is a local slope of the settlement profile, and has been used as an index of potential building damage (e.g., 2, 35). In each case, the maximum angular distortion was calculated for a single line of settlement points perpendicular to the edge of excavation and plotted relative to the maximum settlement in that line. In almost all cases, the maximum angular distortion was located between 0.5 and 1.25 times the maximum excavation depth from the edge of cut.

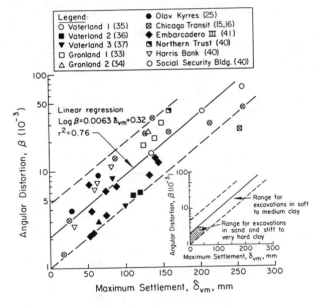

Figure 11. Angular Distortion Plotted Relative to Maximum Settlement for Excavations in Soft to Medium Clay

The linear regression of the data is plotted, for which a relatively high coefficient of determination, $r^2 = 0.76$, was obtained. For maximum settlement less than 50 mm, angular distortion is relatively small. Angular distortion, however, appears to be an exponential function of maximum settlement so that the degree of deformation increases rapidly for movements exceeding 50 mm. In the inset diagram, the range of angular distortion versus settlement for excavations in sand and stiff to very hard clay is shown. The relatively small values of settlement anticipated for cuts in sand and stiff to very hard clay are reflected in low angular distortions, compared with those for excavations in soft to medium clay.

Summary of Settlement Profiles. Fig. 12 presents dimensionless settlement profiles recommended as a basis for estimating vertical movement patterns adjacent to excavations in sand, stiff to very hard clays, and soft to medium clays. With a knowledge of the maximum settlement, the dimensionless diagrams in Fig. 12 can be used to obtain an estimate of the actual surface settlement.

In using these diagrams, it should be recognized that they pertain to settlements caused during the excavation and bracing stages of construction. Movements associated with other activities, such as dewatering, deep foundation removal or construction, and wall installation, should be estimated separately. As previously discussed, excavations in stiff to very hard clays show variable behavior, with heave possible for some conditions. For these materials, the dimensionless diagram in Fig. 12 should be used as a conservative estimate, provided that the wall is stable and not affected by poor construction practice. In making judgments about stiff to very hard clays, it is often valuable to refer to local construction experience.

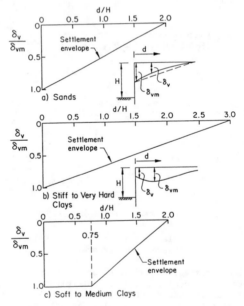

Figure 12.   Dimensionless Settlement Profiles Recommended for
              Estimating the Distribution of Settlement Adjacent
              to Excavations in Different Soil Types.

INTER-RELATIONSHIPS BETWEEN MOVEMENTS AND CONSTRUCTION

The charts given to this point relate to conditions where the movements can largely be attributed to the basic excavation and support process. However, in many cases movements of insitu walls are caused by parameters outside of this category. One principal source of movements is related to the construction of the wall itself as well as other facilities inside the excavation. In most cases, the events that are described in the following paragraphs lead to movements in addition to those due to normal conditions.

Wall Installation Processes. In estimating movements for an insitu wall project it is common to envision the wall in place, and consider what occurs after this point. However, the placement of wall can generate movement.

Sheetpiles are installed by driving, and unless the driving is particularly hard, it is usually done with a vibratory hammer. Vibrations from the driving process are commonly monitored, and they tend to follow a regular pattern (10). The vibrations can cause problems for an excavation project, ranging from complaints from persons about perceived nuisance, to settlements of the ground in the presence of loose or medium sands. Using measured settlements from vibratory sheetpile driving in loose and medium sands, Fig. 13 was developed to provide a plot of vertical strain in the ground caused by pile driving vibrations. The chart takes into account that the sheetpiles will be driven in a long line on either side of the point of interest. To use the figure, first, the level of vibrations to be caused by the driving needs to be selected based on the ground conditions. This determines which line is to be used on the chart. Then, the shortest distance from the point of interest to the sheetpiles is used to find the appropriate strain. To obtain the settlement estimate, the selected strain value is multiplied by the thickness of the sand through which the piles will be driven.

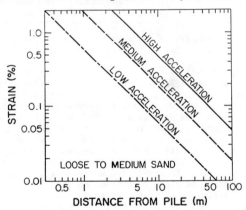

Figure 13.   Vertical Strain Induced in Loose to Medium Sands by Vibratory Sheetpile Driving (Clough, et al., 10).

The settlements concentrate near the sheetpile and decrease rapidly with distance from the sheetpile. Importantly, significant settlements can occur near a sheetpile even though conventional criteria for structural damage due to vibrations show no damage can occur.

Fig. 14 summarizes vertical surface displacements observed during concrete diaphragm wall construction at four slurry walls in granular soil (11), soft to medium clay (25), and stiff to very hard clay (21, 48). The surface displacement and distance from the edge of the wall are expressed as a ratio and fraction, respectively, of the maximum wall depth. The data show that the settlements are a very small percentage of the maximum slurry panel or secant pile depth. Even though settlement is a small percentage of the wall depth, there is still the potential for significant movement, especially when the walls are deep. Slurry wall construction in Hong Kong, for example, has involved panels as deep as 37 m, and generated settlements exceeding 50 mm (14). Maximum settlements elsewhere have been in the range of 5 to 15 mm.

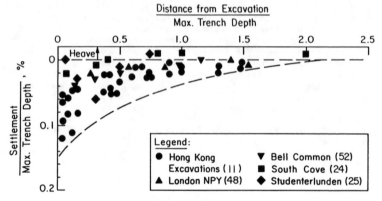

Figure 14.   Summary of Measured Settlements Caused by the Installation of Concrete Diaphragm Walls.

Construction Technique. Poor construction can obviously account for large movements of an insitu wall. Cases from reference 13 as plotted in Figs. 2 and 3 illustrate this simple idea, with each instance showing far above average movements. Problems in these cases were derived from poor installation of soldier piles, late placement of lagging, and improper anchor grouting. Many more large movement cases have obviously occurred due to poor construction than the data would indicate, but the movements were not reported or measured. The quality of construction for an insitu wall project depends upon many factors, including the experience of the contractor in the subsurface conditions at the site and with the insitu wall system being used. However, the geotechnical engineer cannot escape responsibility for poor construction if he does not properly identify subsurface conditions that can cause problems during construction. For example, if there is a layer of loose sand at a site that will settle under sheetpile driving, then this has to be pointed out in the geotechnical report. Thus, it behooves the geotechnical engineer to know how insitu walls are constructed since otherwise he cannot properly predict how the soil will behave.

In soft or stiff clays insitu walls will exhibit creep and contribute extra movement if supports are not promptly installed. Burland et al. (3) cite a case history in London Clay where movements were observed due to undrained creep, leading the writers to conclude: "Deformations following excavation in London Clay are time-dependent and if movements are to be kept to a minimum the provision of positive support is required as quickly as possible." Mana and Clough (27) provided a diagram showing the rate of movement for walls in San Francisco Bay Mud as a function of time required for support installation. The rate of movements was highest in the first 12 hours after an excavation step was made, illustrating the need for prompt installation of supports if movements are to be minimized.

Construction and Removal of Deep Foundations. Deep excavations often are performed concurrently with the removal of existing piles and the installation of new deep foundations. Measurements summarized by O'Rourke (39) adjacent to a 14m deep braced cut in San Francisco show that substantial ground movements were caused when existing timber piles were removed from the site and prestressed concrete piles were driven. Pile extraction in soft clays results in the lost volume of the pile and any attached soil as the pile is pulled from the ground. Conversely, driving of displacement piles tends to heave the adjacent ground.

Deep seated movements have been observed during drilled shaft construction due to dewatering difficulties and squeezing of clay into the shafts or annular voids surrounding steel linings. At the Sears Tower in Chicago, as much as 250mm of settlement occurred in an adjacent street as pumping of the shaft to remove water from the shaft bottom caused sand and silt to flow with the water (39). Fortunately, measures can be taken to reduce and eliminate ground loss associated with drilled shaft construction, such as augering under slurry, tremie placement of concrete, and steel casings sized for a tight fit with the augered holes.

Depth of Excavation Below Support Placement. There are often incentives for a contractor to excavate soil below the specified location for a support. However, to do so will increase the wall movements as has been documented in a number of case histories (7, 13, 40, 42).

In Fig. 15 the effects of excavation well below a support level are illustrated by measurements of lateral movements for a braced excavation in soft San Francisco Bay Mud. In the average case the supports were installed shortly after the support level was reached, while in one location the excavation was taken 10m below the support level before the brace was placed. The lateral wall movements were increased by over 100% for the case which included over-excavation over that where the supports were installed without over-excavation. This problem can be minimized by specifying the allowable distance that soil can be removed below the design support levels, and enforcing the limit.

SPECIAL GEOTECHNICAL FACTORS THAT INFLUENCE WALL MOVEMENTS

In the earlier discussion in this paper, geotechnical conditions were included in a general sense by classifying movements in terms of soil types, and in the case of soft and medium clays, considering the possibility of basal shear. However, there are special conditions that deserve mention which can have a significant effect on insitu wall performance.

Wall Settlement. Insitu wall designs are predicated on the assumption that the wall itself will not settle significantly. Settlement of an excavation support wall can de-stress a tieback system, and cause racking of a brace system. In excavations in soft clays with low FS values and a wall that does not bear on firm material, relative settlements can occur from one side of an excavation to the other due to differences in soil conditions or surcharge loads (9). To prevent this problem, it is advisable to penetrate the wall to a bearing layer where possible. If the clay

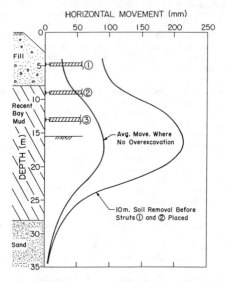

Figure 15.   Effects of Over-Excavation Below Support Levels in Soft Clays - Phase II Excavation, San Francisco (Adapted from Davidson, 13).

extends to a great depth below the excavation, it may be uneconomical to extend the entire wall perimeter to a hard layer to prevent settlement. In one case history involving sheetpiles, every fifth sheetpile was extended to the underlying hard layer to provide control over possible settlement, and this solution was apparently successful (51).

In conditions where a soil is underlain by rock, wall settlement can occur where the wall bearing is obtained at the top of the rock and the excavation extends below this point. White (56) has described several failures and near failures due to the exposure of slip planes in the rock bearing layer. The large movements in these cases exceeded 0.5% (Fig. 3). Corrective measures such as the installation of rock bolts need to be taken before the excavation is extended below the bearing surface.

Movements in the Anchorage Zone of an Anchored Wall. Highly overconsolidated clays are often characterized by the presence of high lateral stresses. Fig. 4 showed that for a braced wall, the movements are larger when the excavation takes place in a soil with a higher coefficient of lateral earth pressure. This effect can be more prominent in the case of an anchored wall where the movements can extend to include the anchors themselves. Burland et al. (3) reported observed movements in London Clay that extended several excavation depths behind an excavation supported by an anchored diaphragm wall, and this case produced movements above 0.5% H (Fig. 2). The behavior occurred in spite of the use of long anchors with a no-load zone defined by a line sloped at 45° from the bottom of the excavation. Other large movements reported for anchored walls in overconsolidated cohesive materials occurred for cases in Los Angeles and Frankfurt. The Los Angeles situation can be explained from the use of relatively short anchors where the no-load zone was defined by a line sloped at 60° from the horizontal (26). The Frankfurt wall movements were attributed to the lack of three dimensional arching effects since the excavation was planar, having been opened for a subway (49).

In spite of reports of anchored walls in stiff clays and shales with large movements, there are at least an equal number of anchored wall projects in these soils that report small movements (5, 19, 28). The key to differences in behavior appears to lie in the level of the lateral stress in the clays. Even local variations in the stress history of London Clay are noted to have an influence on system movements (3). In design, it behooves the engineer to rely on local experience and properly conducted consolidation tests to determine the degree of importance of the lateral stress relief factor. It is also apparent that it is important to use a conservative choice for the depth of the anchorage zone in stiff clays and shales. Even in highly stressed clays, the movements diminish with distance from the wall.

Use of Earth Berms. Earth berms are used in some cases to provide temporary support for a wall before structural supports can be placed. Experience with earth berms is mixed. In stiff soils, experience is usually positive (3). In such soils earth berms are reasonably effective because the soil can mobilize its passive resistance with relatively small movement, and the lateral movements of the wall are not produced by deep seated soil strains. Experience in soft clays is not so positive (6, 7). In these soils, considerable movement is needed to mobilize the passive restraint, and in the presence of a low FS, the excavation induced movements occur below the berm and carry it with the rest of the soil mass (6).

Water Movement and Piezometric Pressures. Water movements in and around an excavation can occur through flaws in a supposedly impervious wall, by flow under the excavation wall, by flow along boundaries between soils of different permeability, by flow along the wall itself if the wall penetrates an underlying aquifer, and due to groundwater lowering by dewatering or flow through the wall (Fig. 16). Also, if piezometric pressures in an aquifer underlying an excavation are not properly reduced, heaving of the base can occur, leading to loss of passive restraint for the wall. Of the 12 cases in Fig. 2 and 3 that show large movements, five were related to problems with water and water control (4, 7, 13, 31).

There are a range of techniques for minimizing the problems associated with groundwater. One step involves using the correct wall type and wall installation procedures for the ground conditions. For example, hard driving can lead to a loss of interlock in a sheetpile wall. In the presence of the dense sands, gravels or rubble, the alternatives to solve the potential problems include: (1) Preaugering holes for alternative sheetpiles; (2) Predriving with a spud pile to open a hole and move rubble aside; (3) Backhoe excavation of upper rubble layers and back filling with sand to clear the way for sheetpile driving; or, (4) Using a different wall type such as a diaphragm wall. The first three choices minimize the risk of splitting of interlocks, while the last one, albeit an expensive approach, eliminates the problem.

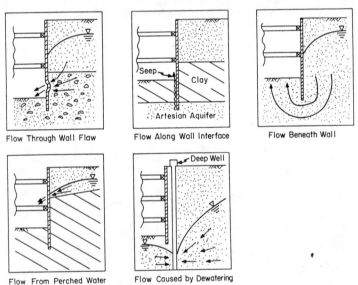

Flow Through Wall Flow   Flow Along Wall Interface   Flow Beneath Wall

Flow From Perched Water   Flow Caused by Dewatering

Figure 16.   Potential Water Flow Situations Which Can Lead to Ground Movement.

The issue of consolidation due to water table lowering is complicated in the urban environment because the soils are often preconsolidated to some degree by natural causes as well as water table lowering in previous construction. The preconsolidation effects found in many urban environments helps to control settlements if the water table drops. Reference 44 presents guidelines that are useful to the designer when dewatering is used.

Heaving and piping in an excavation bottom can occur where an aquifer underlies the excavation. Milligan and Lo (31) describe these types of problems and the solutions associated with them.

SUPPORT SYSTEM DESIGN CONSIDERATIONS

In addition to the geotechnical and construction influences on insitu wall movements, the structural support system is also an important consideration. This element is particularly significant because it can be controlled by the designer. It is at once important to be prepared to utilize this element to improve the wall behavior while at the same time to be able to realize the limitations of its influence.

Wall Stiffness. Both theory and intuition lead to the conclusion that increasing the stiffness of the insitu wall helps to reduce movements. Unfortunately, the influence of this factor is often not well understood, and its effects misinterpreted. Fig. 5 shows that in clays the effect of wall stiffness is a function of the degree of stability against basal heave. In conditions where the clay is inherently stable, a stiff wall is much less effective in reducing movements than in conditions where there is a potential problem with basal stability. In the latter environment a properly designed wall can serve as an important element in carrying load that the soil cannot. However, it is notable that in the presence of a low FS condition a stiff wall is just as subject to movements due to extraneous construction factors as a flexible wall (39). Thus, the potentially positive effects of a stiff wall can be negated if all conditions are not considered.

In cohesionless soils there is no problem with basal stability unless special conditions are encountered. In the presence of a stable base increasing the wall stiffness theoretically does not significantly reduce wall system movements. However, there are subsidiary advantages for using a stiff cast-in-place concrete wall in cohesionless soils. First, the construction of this type of wall is less subject to risk than others such as the sheetpile wall where the interlocks can be lost during driving. Second the cast-in-place wall helps control water movement since it is relatively impervious. Finally, during the placement of the concrete for the cast-in-place wall, the concrete forms intimate contact with the soil surrounding the wall, and eliminates problems that might develop with closure of voids that can exist behind lagged walls. Thus, while the stiffness of the cast-in-place wall may not be that effective in controlling movements in cohesionless soils, when combined with the influence of susidiary factors, its effects are magnified.

Support Spacing and System Stiffness. Wall stiffness is only one element in the stiffness of the support system. As pointed out by Goldberg et al. (18), another, and perhaps more important factor, is the spacing of the supports. In Fig. 5, the system stiffness is defined in terms of both the flexural stiffness of the wall and the support spacing, and the support spacing term is raised to the fourth power. The smaller the support spacing, the stiffer the support system, and this applies to either the vertical or horizontal support spacing.

The interplay between wall stiffness and support spacing can be illustrated by comparing the lateral movements of two excavations which were constructed in San Francisco within several blocks of each other (Fig. 17). Both used diaphragm walls and the soil conditions were similar, consisting of 6m of rubble fill over about 20m of soft Recent Bay Mud. Of interest is the fact that while the excavation for the One Market Plaza site was one half as deep as that for the Embarcardero BART station, the One Market Plaza support system had a theoretical system stiffness factor of only 10, while that for the Embarcardero BART station was 2600. The large difference in support system stiffnesses derives in large part from the use of wide brace spacings for the One Market Plaza wall of 7.5m as opposed to an average of 3.3m for the Embarcadero BART Station wall. As is seen in Fig. 16, in spite of the fact that the

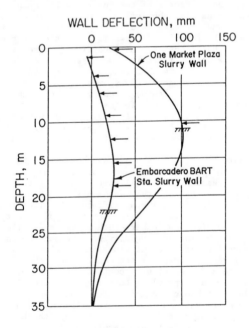

Figure 17.   Comparative Braced Wall Case Histories in Soft Clay
             Illustrating Effects of System Stiffness.

Embarcadero BART station excavation was more than twice the depth of that for the One Market Plaza wall, the horizontal movements of the former wall were one fourth those of the latter (101 vs. 28mm). A primary factor in the difference in response was the much smaller support spacing used in the Embarcardero BART excavation as opposed to that of the One Market Plaza excavation. This pattern of response is predicted by the data in Fig. 5.

The One Market Plaza wall behavior is all the more interesting in that its system stiffness is lower than many sheetpile wall systems, and comparisons of movements shows that it moves more than a sheet-pile wall with smaller support spacings. This behavior confirms that support spacing is a major parameter in defining the system stiffness.

Support Stiffness. In the interest of simplicity, the stiffness of a support system is defined in terms of as few parameters as possible. Thus, in Fig. 5 only the wall stiffness and the support spacing are included in system stiffness. However, this does not mean that other factors do not have some influence on the support system behavior. Clearly the supports themselves in the form of the braces, tiebacks, rakers, and nails also will influence the overall system stiffness to some degree. The theoretical stiffness of the supports can be defined in terms of AE/L, where A is the area of steel, E is the modulus of the steel, and L is the unsupported length of the support. Because tiebacks are preloaded in tension, their theoretical stiffness is close to their actual stiffness. Braces and rakers on the other hand are compression members and their actual stiffness is significantly affected by the nature of the connections used to link them to the wall and the use of preloading (20, 39). Soil nails also have an effective stiffness less than their theoretical value in that they are commonly not preloaded, and require movements to mobilize the stiffness.

Finite element analyses have shown that within the normal range of system parameters, support stiffness is not as important a factor as either wall stiffness or support spacing (27). The range of effect of support stiffness on predicted movement is on the order of plus or minus 20% relative to the values given in the charts earlier in the paper. However, should an unusually stiff or flexible support be employed, the effect can be greater than 20%.

Preloading. Preloading of an insitu wall generally improves performance of the wall from two standpoints. Most importantly, preloading takes the slack out of a support system that otherwise would have to be taken up by movements of the wall. As noted, for a compression member like a brace, this increases the effective stiffness of the brace. Secondarily, preloading reduces the stress levels in the soil that are induced by the excavation process. This latter effect allows the soil to follow an unloading-reloading response instead of the softer primary loading response. Quantifying the effects of preloading is difficult using only observed data. Finite element analyses have been used to determine the influence of preloading, although the results would not account for the effects of removing slack from the system (27). Reasonable values of preloads can be chosen from diagrams used to determine brace loads, e.g. Peck (42).

GROUND MOVEMENT EFFECTS ON ADJACENT STRUCTURES

Building response to ground movements has been viewed tradition-
ly as deformation imposed by differential settlement. Braced and
tied-back excavations, however, generate ground movements which
involve horizontal as well as vertical components. The key to
evaluating effects on adjacent buildings is to recognize the
influence of horizontal movement and to account for how it modifies
differential vertical displacements in setting levels of potential
damage for neighboring structures. Various researchers have
investigated the interaction of differential horizontal and vertical
movements on structures (e.g., 32, 40). Most recently, Boscardin
and Cording (1) have proposed a two-dimensional plot of horizontal
strain versus angular distortion in which zones of potential damage
to buildings are mapped. The zones were established by theoretical
considerations of structural response to deformation, field
observations of building damage, and measurements of differential
horizontal and vertical displacements associated with the damage.
The plot offers a convenient reference space in which to evaluate
the damage potential of excavations in different types of soil.

Figs. 18a and 18b show the range of horizontal strain and angular
distortion most likely for excavations in sand, stiff to very hard
clay, and soft to medium clay, superimposed on the zones of potential building damage mapped by Boscardin and Cording. Ranges of the
ratios of horizontal to vertical surface movements adjacent to
excavations in sand and soft to medium clay are given by O'Rourke
(40). The range of the ratio of horizontal to vertical surface
movement adjacent to excavations in stiff to very hard clay was
established on the basis of data summarized in this paper. It is
assumed that the ratio of horizontal and vertical movement between
two points affected by the excavation will be in the same behind the
excavation is approximately constant for points separated by
distances of $0.25H$ to $0.5H$. The severity of damage is related to
magnitude of displacement.

For excavations in sand and stiff to very hard clay, settlements
typically are a small percentage of the excavation depth. Accordingly, the range of horizontal strain and angular distortion associated with excavation in these soils is likely to be relatively low.
Fig. 18a shows that the worst damage is likely to be bounded by a
moderate to severe level, and that controls can diminish the
severity of movement to negligible values. On the other hand, if
horizontal movements are allowed to increase because of insufficient
support stiffness for excavations in stiff and very hard clay, then
moderate to severe levels of damage may be possible even if angular
distortion is about 1×10^{-3}.

Excavations in soft to medium clay can result in large vertical
and horizontal movements so that there is a chance of severe to very
severe damage, as illustrated in Fig. 18b. Settlements tend to
exceed horizontal displacements adjacent to excavations in soft to
medium clays, primarily as a result of deep inward movement of the
excavation wall and time-dependent consolidation effects. When
braces are insufficiently stiff such that large cantilever wall
movements occur, Fig. 18b shows that the deformation severity may

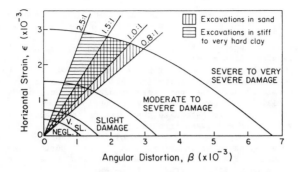

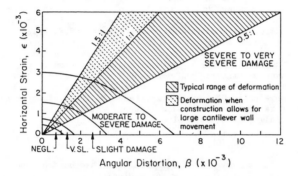

Figure 18. Range of Deformations Typical of Excavations in Various
            Soils Relative to Building Damage Potential
            [After Boscardin and Cording (1)]

increase from levels of slight to very severe damage, primarily as
the result of an increased horizontal component of movement.

    Although the trends shown in Figs. 18a and 18b can help establish
approximate bounds on performance, the response of a real building
must be judged according its characteristics in terms of its
construction, use, and current state of repair. In assessing the
potential for damage, a preconstruction survey and review of
structural drawings are necessary. The preconstruction survey
establishes the current state of repair, allows for verification of
the assumed structural configuration, and helps to identify
potentially weak or troublesome parts of the facility.

    In some cases, the weakest part of a building is located below
ground. In many urban locations, underground vaults connect building
basements with delivery locations near the curb line. As a result
of the vault, the building may be substantially closer to the

excavation than is implied by the building line shown on a site survey map. Differential ground movements imposed on the vault can be transmitted directly to its juncture with the building, where concentrated damage in the form of cracks and separations along the front facade will appear. More importantly, a gas or water line connection with a building may represent a point of local restraint where deformation from differential soil movement will concentrate. Unless a careful survey is performed before construction, sensitive and potentially weak connections may be overlooked.

It is important to recognize that distinctions between levels of severity in building damage frequently are unclear, or at least sufficiently ambiguous to be changed by the environment or affected by what may appear to be an unimportant detail. During deep excavation for Gallery Place Station on the Washington, D.C. Metro, architectural damage manifested itself at the National Portrait Gallery in the form of thin cracks and separations (38). Some of these cracks caused masonry tiles to fall, thereby endangering the public and threatening to damage the art inside. In this case, estimating movements from the soil displacements typical of excavations in sand would have led to a correct prediction that architectural damage was a likely outcome of such deformation. The loss of building function which actually did occur would not have been obvious.

Old masonry, historic and monumental structures offer some of the most challenging problems when evaluating building response to ground movement. Of great importance in these structures are the architectural ornamentation and statuary which may be weakly secured in potentially troublesome places.

SUMMARY AND CONCLUSIONS

The issue of movements of insitu walls has become more important over the past decade with new technologies being introduced, and increased concern about litigation over possible damages to adjacent facilities. This paper presents information intended to update previously available data, clarify ground movement patterns, and relate likely ground movements with the potential severity of building damage. Attention is paid to sorting out observed data so that effects of the basic excavation and support process are separated from those related to construction events, and unusual geotechnical and support system parameters.

Principal conclusions are:

The Basic Excavation And Support Process

1)   In stiff clays, residual soils and sands, maximum lateral wall movements and settlements of the retained soil average about 0.2% to 0.3% H with a scattering of case history data up to 0.5% H. This appears to apply for new soil nail and soil cement walls as well as for conventional systems. System stiffness has only a marginal influence on movements in these soils.

2) In soft to medium clays maximum movements are a function of
    the FS value with values ranging from over 2% H at FS's below
    1.2 to less than 0.5% H at FS's above 2. Where the FS value
    is below 1.5, the system stiffness can have a significant
    influence on movements.

3) Wall displacement profiles with the maximum movement at the top
    are characteristic of cantilever walls and walls with minimal
    preload and flexible supports. The soil nailed wall is typical
    of this category.

4) In stiff clays and sands, settlement profiles of the ground
    surface behind an insitu wall are shown to be triangular in
    shape, with the maximum settlement occurring at the wall
    position.

5) In soft to medium clays, settlement profiles of the ground
    surface behind an insitu wall are shown to be trapezoidal in
    shape, with the largest movements occurring near the wall.

The Influence Of Construction Activities

1) Wall installation can induce additional movements to those of
    the basic support process by vibratory pile driving in loose
    sands and closure of the trench opened for a diaphragm wall.
    Both of these sources of movements tend to be localized, but
    can be significant within 5 to 10m of the wall.

2) Additional movements can also be generated by (a) poor con-
    struction technique (b) slow installation of supports after the
    excavation level is reached (c) construction and removal of
    foundations within the excavation and, (d) removal of soil
    below the design level of a support. Any of these factors can
    cause wall movements to become much larger than expected.

Effects of Geotechnical Factors

1) Settlements of the insitu wall, per se, can occur in soft clays
    where bearing for the wall is not obtained, or where the wall
    bears on a layer that is exposed by the excavation which con-
    tains a flaw such as a relict joint or slickenside.

2) In an overconsolidated clay with high insitu lateral stresses,
    movements induced by the excavation will penetrate further from
    the wall than in other soils. For an anchored wall, this can
    lead to movements of the tieback anchorages.

3) An earth berm may be an effective form of supplementary support
    for stiff soils, but is much less so in soft to medium clays.

4) Proper treatment and control of groundwater and piezometric
    pressures is essential for restraining insitu wall movements.

Role of Support System

1) Increasing wall stiffness tends to reduce system movements, but this is most effective in soft to medium clays. The positive impact of a high wall stiffness can be undercut by ancillary construction activities, improper water control, and use of large support spacings.

2) The support spacings are more important than wall stiffness in defining system stiffness and helping control movements.

3) Support stiffness per se influences system stiffness, but within normal ranges of this parameter, its has less influence on movements than wall stiffness or support spacing.

4) Preloading helps limit movements, and is recommended for braced or anchored walls where movements are of concern.

General

1) Excavations in front of insitu walls induce both horizontal and vertical movements, and evaluation of potential damage to structures requires consideration of the contribution of each. In soils where settlements are small, lateral movements caused by cantilever effects can damage a structures, even when settlements are small.

2) Building response to ground movements is a function of the nature and condition of the building, and a preconstruction survey will be very useful, if not critical, for assessing the potential for damage. Of special importance are subsurface structures connected to buildings, such as vaults and water and gas services.

3) Movements of insitu walls can be predicted within bounds so long as principal sources of displacement are considered.

ACKNOWLEDGEMENTS

The information presented herein is derived from the work of many contributors, including collegues and former students of the writers who worked with them in the past. The writers would like to express their appreciation to all whose efforts made this paper possible. The help of Ms. Judith Brown in preparing the manuscript is sincerely appreciated. Mr. Nick Harman helped in production of Figures 2, 3 and 4.

REFERENCES

1. MD Boscardin & EJ Cording, Building Response to Excavation-Induced Settlement, Journal of Geotechnical Engineering, ASCE, Vol. 115, No. 1, Jan. 1989, pp. 1-21.

2. JB Burland & CP Wroth, Settlement of Buildings and Associated
    Damage, Proceedings, Conference on Settlement of Structures,
    Pentech Press, London, U.K., 1974, pp. 611-654.
3. JB Burland, B Simpson & HD St. John, Movements Around Excava-
    tions in London Clay, Seventh Eurpoean Conf. on Soil Mech, and
    Fdn. Engr., Vol. 1, Brighton, 1979, pp. 13-29.
4. KR Chapman, Performance of Braced Excavations in Washington,
    D.C., Thesis Presented in Partial Fulfillment of M.S. Degree,
    University of Illinois, Urbana, Illinois, 1970.
5. GW Clough, PR Weber & J Lamont, Design and Observation of a
    Tied-Back Wall, Proceedings, ASCE Conference on Performance of
    Earth and Earth-Supported Structures, Purdue University, Vol.
    I, 1972, pp. 1467-1480.
6. GW Clough & GM Denby, Stabilizing Berm Design for Temporary
    Walls in Clay, Journal of the Geotechnical Engineering
    Division, ASCE, Vol. 103, No. GT2, February 1977, pp. 75-90.
7. GW Clough & RR Davidson, Effects of Construction on Geotech-
    nical Performance, Proceedings, Specialty Session III, Ninth
    International Conference on Soil Mechanics and Foundation
    Engineering, Tokyo, 1977, pp. 15-53.
8. GW Clough & LA Hansen, Clay Anisotropy and Braced Wall Behav-
    ior, Journal of the Geotechnical Engineering Division, ASCE,
    Vol. 107, No. GT7, July 1981, pp. 893-914.
9. GW Clough & MW Reed, Measured Behavior of Braced Wall in Very
    Soft Clay, Journal of the Geotechnical Engineering Division,
    ASCE, Vol. 110, No. 1, Jan. 1984, pp. 1-19.
10. GW Clough, EM Smith & BP Sweeney, Movement Control of Excava-
    tion Support Systems by Iterative Design, Proceedings, ASCE,
    Foundation Engineering: Current Principles and Practices, Vol.
    2, 1989, pp. 869-884.
11. JW Cowland & CBB Thorley, Ground and Building Settlement Assoc-
    iated with Adjacent Slurry Trench Excavation, Ground Movements
    and Structures, Pentech Press, London, U.K., 1985, pp. 723-738.
12. RR Davidson, Effects of Construction on Deep Excavation Behav-
    ior, Thesis Presented in Partial Fulfillment of Degree of
    Engineer, Stanford University, Stanford, California, July 1977,
    184 pp.
13. DJ D'Appolonia, Effects of Foundation Construction on Nearby
    Structures, Proceedings, 4th Pan American Conference on Soil
    Mechanics and Foundation Engineering, Puerto Rico, Vol. 1,
    1971, pp. 189-236.
14. RV Davies & DJ Henkel, Personal communication on Charter Sta-
    tion, Hong Kong, Ove Arup and Partners, 1981.
15. Z Eisenstein & LV Medeiroz, A Deep Retaining Structure in Till
    and Sand. Part II: Performance and Analysis, Canadian
    Geotechnical Journal, Vol. 20, No. 1, 1983, pp. 131-140.
16. RJ Finno, Observed Performance of a Deep Excavation in Clay,
    Journal of Geotechnical Engineering, ASCE, Vol. 115, No. 8,
    Aug. 1989, pp. 1045-1064.
17. RJ Finno, Saturated Clay Response During Barced Cut Construc-
    tion, Journal of Geotechnical Engineering, ASCE, Vol. 115, No.
    8, Aug. 1989, pp. 1065-1084.

18. DT Goldberg, WE Jaworski & MD Gordon, Lateral Support Systems and Underpinning, Report FHWA-RD-75-128, Vol. 1, Federal Highway Administration, Washington, D.C., Apr. 1976.

19. WP Grant, Performance of Columbia Center Shoring Wall, Proc., 11th International Conference on Soil Mechanics and Foundation Engineering, Vol. 4, San Francisco, CA, 1985, pp. 2079-2082.

20. LA Hansen, Discussion - Ground Movements Caused by Braced Excavations, Journal of Geotechnical Engineering, ASCE, Vol. 109, No. 3, March 1983, pp. 485-487.

21. WE Jaworski, An Evaluation of the Performance of a Braced Excavation, Ph.D. Thesis, Massachusetts Institute of Technology, Cambridge, MA, June 1973.

22. EB Johnson, DG Gifford & MX Haley, Behavior of Shallow Footings Near a Diaphragm Wall, Preprint 3112, ASCE Fall Convention, San Francisco, CA, Oct. 1977.

23. I Juran & V Elais, Soil Nailed Retaining Structures: Analysis of Case Histories, Geotechnical Special Publication No. 12, ASCE, 1987, pp. 232-244.

24. K Karlsrud, Performance and Design of Slurry Walls in Soft Clay, Preprint 81-047, ASCE Spring Convention, New York, NY, May 1981.

25. K Karlsrud & F Myrvoll, Performance of a Strutted Excavation in Quick Clay, Proceedings, 6th European Conference on Soil Mechanics and Foundation Engr., Vienna, Vol. 1, 1976, pp. 157-164.

26. PA Maljian & JL VanBeveren, Tied-Back Deep Excavation in Los Angeles Area, Journal of the Construction Division, ASCE, Vol. 100, September 1974, pp. 337-356.

27. AI Mana & GW Clough, Prediction of Movements of Braced Cuts in Clay, Journal of the Geotechnical Division, ASCE, Vol. 107, No. GT6, June 1981, pp. 759-778.

28. CI Mansur & M Alizadeh, Tie-Backs in Clay to Support Sheeted Excavation, ASCE, Journal of the Soil Mechanics and Foundations Division, Vol. 96, No. GM2, March 1970, pp. 495-509.

29. GWE Milligan, Soil Deformations Behind Retaining Walls, Ground Movements and Structures, Pentech Press, London, U.K., 1985, pp. 707-722.

30. GWE Milligan, Soil Deformations Near Anchored Sheet-Pile Walls, Geotechnique, Vol. 33, No. 1, Mar. 1983, pp. 41-55.

31. V Milligan & KY Lo, Observations on Some Basal Failures in Sheeted Excavations, Canadian Geotechnical Journal, Vol. 7, 1970, pp. 136-144.

32. National Coal Board, Subsidence Engineers Handbook, National Coal Board Production Department, London, U.K., 1975.

33. Norwegian Geotechnical Institute, Measurements at a Strutted Excavation, Oslo Subway, Gronland 1, Technical Report 1, Blindern, Norway, 1962.

34. Norwegian Geotechnical Institute, Measurements at a Strutted Excavation, Oslo Subway, Gronland 2, Technical Report 5, Blindern, Norway, 1965.

35. Norwegian Geotechnical Institute, Measurements at a Strutted Excavation, Oslo Subway, Vaterland 1, Technical Report 6, Blindern, Norway, 1962.

36. Norwegian Geotechnical Institute, Measurements at a Strutted Excavation, Oslo Subway, Vaterland 2, Technical Report 7, Blindern, Norway, 1962.

37. Norwegian Geotechnical Institute, Measurements at a Strutted Excavation, Oslo Subway, Vaterland 3, Technical Report 8, Blindern, Norway, 1962.

38. TD O'Rourke, Ground Movements from a Deep Excavation in Sands and Interbedded Clay, Failures in Earthworks, Thomas Telford, London, U.K., 1985, pp. 371-383.

39. TD O'Rourke, Ground Movements Caused by Braced Excavations, Journal of the Geotechnica Engineering Division, ASCE, Vol. 107, No. GT9, Sept. 1981, pp. 1159-1178.

40. TD O'Rourke, EJ Cording & MD Boscardin, The Ground Movements Related to Braced Excavations and Their Influence on Adjacent Buildings, Report DOT-TST 76T-23, U.S. Department of Transportation, Washington, D.C., Aug. 1976.

41. TD O'Rourke, Personal files on Embarcadero III Project, San Francisco, CA.

42. RB Peck, Deep Excavations and Tunneling in Soft Ground, State-of-the-Art Report, 7th International Conference on Soil Mechanics and Foundation Engineering, Mexico City, State-of-the-Art Volume, 1969, pp. 225-290.

43. JP Powers (Ed), Dewatering-Avoiding its Unwanted Side Effects, ASCE Manual, 1985, 69 pp.

44. WL Shannon & RJ Strazer, Tied-Back Excavation Wall for Seattle First national Bank, Civil Engineering, ASCE, Vol. 40, Mar. 1970, pp. 62-64.

45. GC Sills, JB Burland & MK Czechowski, Behavior of an Anchored Diaphragm Wall in Stiff Clay, Proceedings, 9th International Conference on Soil Mechanics and Foundation Engineering, Tokyo, Vol. 2, 1977, pp. 147-154.

46. AW Skempton & DH MacDonald, The Allowable Settlement of Buildings, Proceedings, Institution of Civil Engineers, Part III, 5, 1956, pp. 727-784.

47. H Stille, Behavior of Anchored Sheet Pile Walls, Royal Institute of Technology, Stockholm, 1976.

48. HD St. John, Field and Theoretical Studies of the Behavior of Ground Around Deep Excavations in London Clay, Ph.D. Thesis, University of Cambridge, U.K., Nov. 1975.

49. D Stroh & H Breth, Deformation of Deep Excavation, Proceedings, Conference on Numerical Methods in Geomechanics, Vol. II, June 1976, pp. 686-700.

50. IF Symons, JA Little, TA McNulty, DR Carder & SGO Williams, Behavior of a Temporary Anchored Sheet Pile Wall on Al (14) at Hatfield, Research Report 99, Transport and Road Research Laboratory, U.K., 1987, 40 p.

51. RG Tait & HT Taylor, Design, Construction and Performance of Rigid and Flexible Bracing Systems for Deep Excavations in San Francisco Bay Mud, Preprint No. 2162, ASCE, Jan. 1974, 25 pp.

52. P Tedd, BM Chard, JA Charles & IF Symons, Behavior of a Propped Embedded Retaining Wall in Stiff Clay at Bell Common Tunnel, Geotechnique, Vol. 34, No. 4, 1984, pp. 513-532.

53. EJ Ulrich, Internally Braced Cuts in Overconsolidated Soils, Journal of Geotechnical Engineering, ASCE, Vol. 115, No. 4, Apr. 1989, pp. 504-520.
54. EJ Ulrich, Tieback Supported Cuts in Overconsolidated Soils, Journal of Geotechnical Engineering, ASCE, Vol. 115, No. 4, Apr. 1989, pp. 521-545.
55. KR Ware, M Mirsky & WE Leuniz, Tieback Wall Construction Results and Controls, Journal of the Soil Mechanics and Foundations Division, ASCE, Vol. 99, No. SM12, 1973, pp. 1135-1152.
56. RE White, Anchored Walls Adjacent to Vertical Rock Cuts, Proc., Conference on Diaphragm Walls, Session VI, Institution of Civil Engineers, London, September 1974, pp. 35-42.
57. KS Wong & BB Broms, Lateral Wall Deflections of Braced Excavations in Clay, Journal of Geotechnical Engineering, ASCE, Vol. 115, No. 6, June 1989, pp. 853-870.

GROUND MOVEMENT ADJACENT TO BRACED CUTS
by

Safdar A. Gill[1] , F-ASCE and Robert G. Lukas[2] , M-ASCE

ABSTRACT: This paper discusses ground movements at eight building sites in the downtown Chicago area, where excavations were made for two or more basement levels that were stabilized by retention systems consisting of either steel sheeting, soldier piles, or slurry walls.

At each of these sites, measurements of ground displacement were obtained. At some sites, lateral deformations were obtained using an inclinometer while at other sites, lateral and vertical movements were obtained from survey monuments. Ground displacement for each of these sites was predicted and the predicted movements compared to the recorded movements.

INTRODUCTION

One of the earliest methods of predicting settlements adjacent to basement excavations in downtown Chicago was established by Peck (4). This method of prediction is based upon many settlement observations obtained throughout the Chicago area with excavations extending from one to three basements. Since the soil conditions are essentially the same throughout the downtown Chicago area, the diagrams can also be correlated with strength of soil. However, the procedure does not take into account the embedment of the vertical retention member, and the type and stiffness of the bracing system which are also important factors affecting offsite ground movements.

Since 1969 numerous investigators have monitored movements adjacent to braced cuts and have used finite element methods for predicting these movements. One method was developed by Clough (2) which takes into account some of the more important factors influencing offsite movements, namely:

1. The factor of safety against base heave of the bottom of the cut which is dependent upon the depth of cut, surface loading and shear strength of the soil.

[1,2] Senior Principal Engineer, STS Consultants, Ltd, 111 Pfingsten Road, Northbrook, Illinois 60062

2. The stiffness of the retention system which depends upon the spacing between walers and the properties of the vertical retention member.
3. The distance from the edge of the excavation.

The authors of this paper have used the Clough procedure, using the Bjerrum and Eide (1) method of calculating the base heave factor of safety and adjusting the base heave factor of safety calculation to take into account the embedment of the perimeter retention system into stiffer soil deposits.

BASE HEAVE FACTOR OF SAFETY

At all the project sites, a soft to medium strength clay is present below a depth of about 17 feet (5.2m) below ground surface, which means that excavations for two or more basements will extend into these weak deposits. Thus, the factors of safety against base heave are generally quite low. Even minor changes in factors of safety can result in significant variations in predicted off-site movements, so it is important that this factor of safety be calculated accurately.

Analysis for base heave can be performed by several methods, all of which are based on the analogy of a uniformly loaded footing outside the sheeting at the level of the cut. No heaving can occur unless the load due to the weight of the soil (plus any surcharge thereon) exceeds the bearing capacity of the soil located below the level of the bottom of the cut. According to the procedures developed by Terzaghi (5) and Tschebotarioff (7), this load is reduced by the shearing stresses along the vertical back face of the block of soil overlying that plane. Terzaghi gives a bearing capacity factor of 5.7 and Tschebotarioff 5.14. The failure is characterized as a slip on a well defined rotational surface which is approximately circular below the base of the cut and extends vertically to ground surface above this level, as shown on the following sketch.

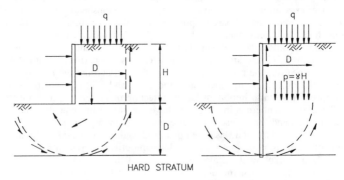

HARD STRATUM

FROM TERZAGHI (5)   FROM BJERRUM AND EIDE (1)

The authors are suggesting that a more realistic method of analysis is to use the Bjerrum and Eide (1) procedure which considers a localized failure at the base of the excavation without mobilizing the shear strength up to ground surface. In this approach, the soil behind the retention system above cut level acts as a surcharge. Charts are available for determining the stability number, N_c, as a function of the geometry of the excavation and depth of cut.

The authors used equation (1) to calculate the factor of safety against base heave for all the projects described in this paper.

$$F.S. = \frac{c_b \, N_c}{\gamma H + q - c_a H/D} \qquad (1)$$

where: N_c = bearing capacity factor determined from charts prepared by Bjerrum and Eide (1), which takes into account the geometry of the excavation and the depth of cut.

c_b = undrained shear strength of the soft to medium clays at and below cut level.

γ = unit weight of soil at and above cut level

H = depth of cut

q = surcharge pressure, if any, from traffic or adjacent buildings. The foundation pressures and the reduction in loads from adjoining basements must be taken into account.

c_a = adhesion at the back face of the sheeting above cut level

D = depth below cut to a firm layer

According to this equation, the driving force for base heave is the weight of the soil above the cut level, plus the surcharge, if any. This driving force is reduced by the adhesion which develops along the back face of the sheeting, provided the sheeting extends sufficiently deep so as to be embedded into a firm soil. The available resistance of the embedded portions of the sheeting must exceed the downward force from the adhesion. In calculating the available adhesion in fill and granular soils, the adhesion was based upon lateral pressures used in the design of the sheeting and bracing system multiplied by tangent ϕ. In cohesive soils, the undrained shear strength was multiplied by an adhesion factor.

The resisting forces are based upon the undrained shear strength of the soils below cut level and the stability factor determined from Bjerrum and Eide charts.

The advantage of using equation (1) is that the benefits that arise from extending the sheeting or slurry wall into a stiff deposit are at least partially taken into account in reducing the driving forces and thereby increasing the basal heave factor of safety. This method of adjusting the factor of safety for the longer sheeting was originally suggested by others (5,6).

PROJECT SITES

The typical profile at each of the eight project sites is shown in Figures 1 to 8. These profiles indicate the locations of the walers and the types of vertical retention members that were used, the depths of the cut, and the typical soil conditions. The undrained shear strengths are also shown for the clayey soil deposits and these were used in the calculation of the factor of safety against base heave.

At sites A and C through H, the construction sequence was similar. The vertical portion of the retention system was installed around the perimeter of the site. This was then followed by a shallow excavation within the interior from which foundation caissons (drilled piers) were installed. At most sites, the center portion of the site was then excavated to the final grade, leaving berms of soil along the perimeter to stabilize the soil mass until the interior grade beams could be constructed. Afterwards, walers and inclined rakers were installed to the interior grade beams and then the berms removed for the placement of the lower rakers. At each corner, corner braces were installed in lieu of the rakers. At a few of the sites, crosslot bracing was used for some or all of the bracing. Construction generally proceeded relatively uninterrupted from the start of the installation of the vertical portion of the retention system to completion of the excavation. In most cases, the struts were prestressed to about 50 to 100 tons (445 to 890 kN).

Five of the sites had certain peculiarities, such as geometry of site, offsite surcharge loads, construction methods, etc. which affected ground movements. These features were considered in the prediction of movements as discussed below.

Site B. The construction sequence at this site consisted of excavating the site from street grade of +13.5 ft (+4.1m) Chicago City Datum (CCD) to +6 CCD (1.8m) from which pile foundations and the perimeter soldier beams and upper waler were installed. The basement excavation was started on March 1, at the south end of the site where the basement grade was -16.67 ft (-5.1m) CCD. As the excavation proceeded to the north, a slope failure took place within the interior of the site on March 18, deflecting the previously driven piles. Since the pile caps could not be formed until the piles were repaired, there was no place to brace the internal rakers, so the excavation stood open without bracing until October, when the upper rakers were installed. During this time interval, settlement and lateral displacement readings were taken on the tops of the soldier piles and the lateral displacement readings along one street are shown on Figure 10.

Since the average slope of the berm was approximately 2H:1V, the relatively large lateral movements before installation of the rakers were

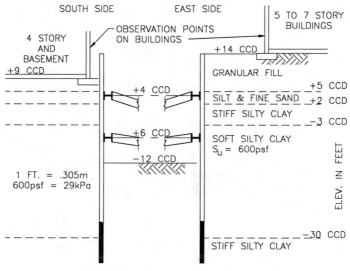

SOUTH SIDE, PZ27 TOE @ −35 CCD
EAST SIDE, FN 2N , ALTERNATE SHEETS
AT −26 & −36 CCD

FIGURE 1 − PROFILE AT SITE A

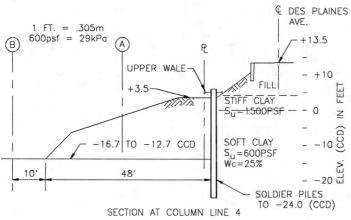

SECTION AT COLUMN LINE 4

FIGURE 2 − TYPICAL PROFILE, SITE B

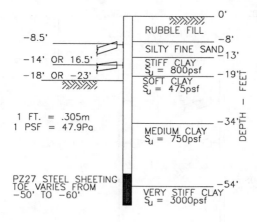

FIGURE 3 — SITE C, TYPICAL PROFILE

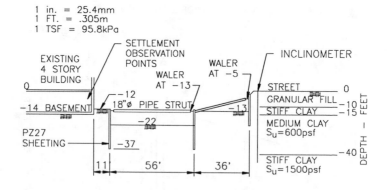

FIGURE 4. SITE D TYPICAL PROFILE

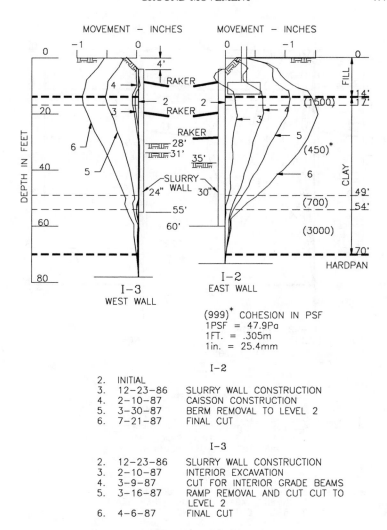

FIGURE 5: PROFILE AND INCLINOMETER DATA FOR SITE E

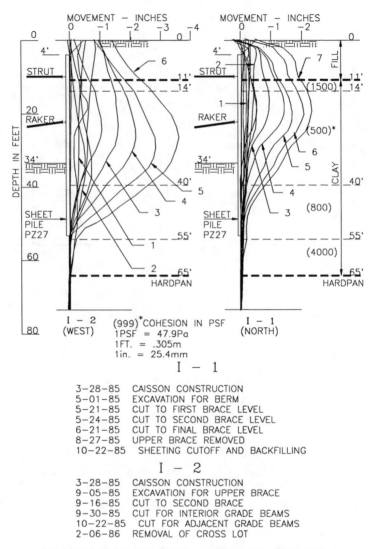

FIGURE 6: PROFILE AND INCLINOMETER DATA FOR SITE F

I – 1

3–28–85	CAISSON CONSTRUCTION
5–01–85	EXCAVATION FOR BERM
5–21–85	CUT TO FIRST BRACE LEVEL
5–24–85	CUT TO SECOND BRACE LEVEL
6–21–85	CUT TO FINAL BRACE LEVEL
8–27–85	UPPER BRACE REMOVED
10–22–85	SHEETING CUTOFF AND BACKFILLING

I – 2

3–28–85	CAISSON CONSTRUCTION
9–05–85	EXCAVATION FOR UPPER BRACE
9–16–85	CUT TO SECOND BRACE
9–30–85	CUT FOR INTERIOR GRADE BEAMS
10–22–85	CUT FOR ADJACENT GRADE BEAMS
2–06–86	REMOVAL OF CROSS LOT

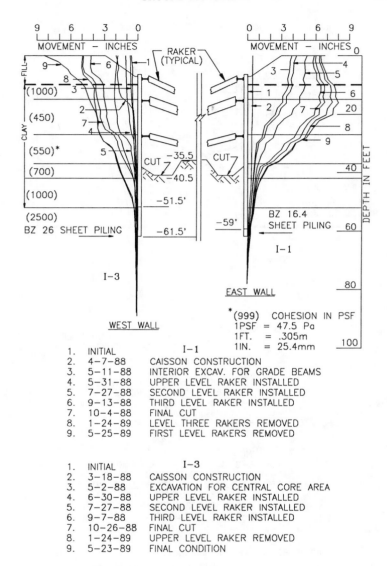

FIGURE 7: PROFILE AND INCLINOMETER DATA FOR SITE G

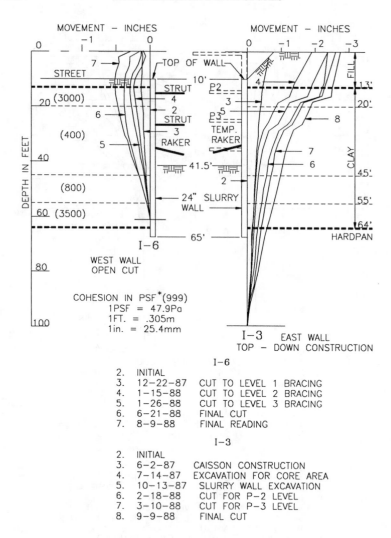

FIGURE 8: PROFILE AND INCLINOMETER DATA FOR SITE H

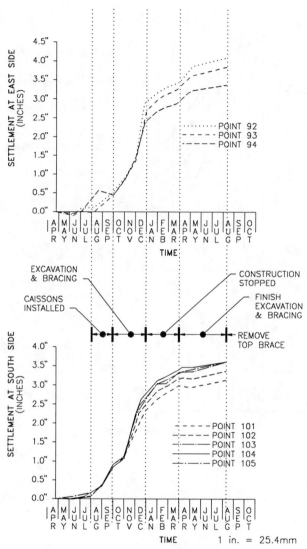

FIGURE 9 — SITE A SETTLEMENT

1 in. = 25.4mm

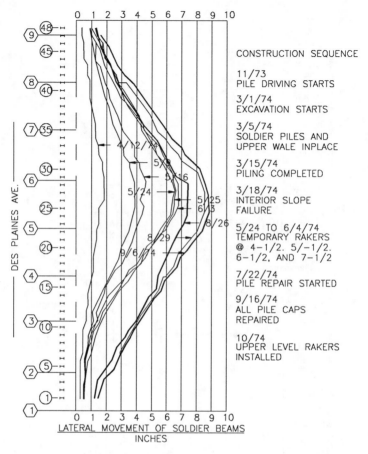

FIGURE 10 — LATERAL MOVEMENTS OF SOLDIER PILES — SITE B

somewhat more than anticipated. However, using a procedure developed by Clough and Denby (3), movements on the order of 7 inches (178mm) are predicted for a free end wall because the factor of safety against base heave at this site is relatively low. In the deeper portion of the basement excavation, the stability number, $\gamma H/s_u$, equals 5.55 and for the shallower portion of the cut the stability number computes to be 4.70. These high values of stability number result in relatively large predicted movements for the berm, even though the berm had a relatively flat slope. Another factor that undoubtedly influenced the magnitude of the lateral deflections is that some excavations within the interior of the site were made as part of the pile repair scheme and for construction of the pile caps. However, none of these repair operations were started until July 22, and at that point, the soldier piles had already displaced up to 7 inches (178mm) near the center of the site.

These findings indicate that berms are not very effective in restraining the soil mass when the stability number is high, especially if there is a time delay between making the cut and installing the rakers.

Site D. Prediction of ground movements in advance of construction was complicated by the stepped excavation as shown in Figure 4. The procedure that was followed was to assume that all of the soil and live load above the first basement level, -13 ft (-4m), consisted of a surcharge. The stress induced at the lower basement level, -22 ft (-6.7m), from this surcharge was then calculated and converted to an equivalent height of soil. At the building side, the pressure induced from the building foundations calculates to approximately 5.4 ft (1.6m) of additional equivalent soil height, thereby making the depth of cut 18.4 ft (5.61m). Along the street side of the cut, the equivalent depth of cut was found to be 20.6 ft (6.3m), because the pressure from the soil mass and the street surcharge induced higher pressures at the second basement cut level than from the structure.

Site E. This site has on one side a 10 story building with one basement situated about 10 ft (3m) from the excavation. This building was built in the 1890s and is supported on a grillage type footing at a depth of 13 ft (4m) from street grade. Based on the structural details, it was estimated that the bearing pressure on the footings was in the range of 3.5 to 4 ksf (168 to 192 kN/m^2) and the factor of safety against bearing capacity failure in the underlying clay was not far above 1. The retention system was designed to obtain a high system stiffness using a slurry wall and three levels of braces at a maximum vertical spacing of 9 ft (2.7m). A 30 inch (762mm) thick slurry wall was constructed in short panels about 7 to 9 ft (2.1 to 2.7m) long each, for the retention wall and it was extended into very stiff clays at a depth of 60 ft (18.3m).

Calculation of the base heave factor of safety included allowance for the footing surcharge and relief due to the existing basement.

Work at this site proceeded at a fast rate and thus ground movements due to creep were minimal.

Site F.   This site was only 120 ft (36.6m) wide and did not provide sufficient berm width with adequate slopes to dig the interior for constructing heel blocks for rakers.   Hence, one level of crosslot struts, with a central pile support, was used at the upper level.   The two short ends were supported by three corner braces at each corner.   The corner braces and crosslot struts were installed when the excavation was shallow and helped to support the ground before the cut extended into the soft clays.   The beneficial effects of these braces were partly lost when a relatively long delay occurred in constructing the interior grade beams due to a labor strike.   Inclinometer readings taken during this work stoppage showed continual movement due to creep although no excavation was going on.

Site G.   This site had the deepest excavation and considerable time was required for interior excavation and for constructing grade beams for raker heels.   About 3 to 4 inches (70 to 102mm) of movements occurred before any raker could be installed.   The topmost raker served as a temporary brace and was removed after the second level raker was installed.   The excavation was essentially supported by two levels of braces which would be considered minimal bracing for this depth of cut.   Significant movements also occurred during the removal stages because the backfill between the sheetpiles and the basement walls was several feet below a waler level when that level's braces were removed.

PREDICTED AND MEASURED MOVEMENTS

The measured movements at seven of the eight project sites are shown in Figures 5 to 11.   No offsite movements are shown for Site D because the measured movements were 1/2 inch (12.7mm) or less, so the deformation with various levels of construction activity is not possible to discern.

The predicted versus measured movements are summarized in Table 1. Also shown in this Table are the calculated factor of safety against base heave, plus the calculated stiffness of the retention system.

For many of the sites, there are two observations of predicted and calculated deflections.   This is because many of the sites had varying levels of cuts or some sites had buildings adjacent to one side of the site which added additional surcharge.

The observed and predicted movements appear to be reasonably in close agreement for those sites where inclined rakers were used in the bracing system.   However, the measured movements are less than predicted at Sites

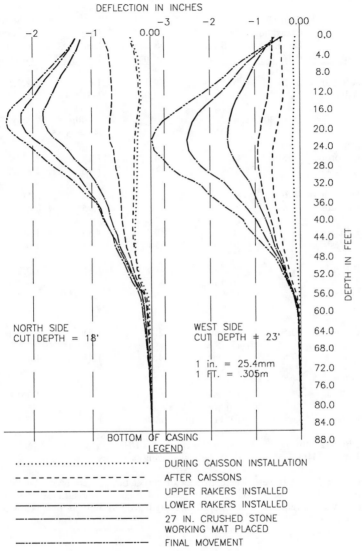

FIGURE 11 - INCLINOMETER READINGS, SITE C

TABLE 1

Site	Equivalent Excavation Depth (ft)	Vertical Spacing of Braces (ft)	System* Stiffness	Base Heave F.S.	Predicted/ Measured Movement (inches)
A-South	27.5	10	59.4	1.67	2.2/2.5
A-East	31	10	32.0	1.51	2.7/3.3
B	26 to 30	--	--	stability No = 5.55	7/5 to 7
C-North	28.8	6.5	247	1.18	2.1/2.2
C-West	23	8.5	91	1.16	2.8/2.7
D-North	18.6	8	142	1.55	0.5/0.5
D-South	17.4	8	142	1.95	0.6/0.4
E-East	35	10	1085	1.55	1.9/1.5
E-West	28	10	556	1.23	2.0/0.9
F-North	31	12.5 max 10.5 avg	48.8	1.35	3.7/2.3
F-West	35	12.5 max 12.0 avg	28.7	1.25	5.4/3.5
G-East	40.5	13.5 max 12.75 avg	22.0	1.41	5.9/8.1
G-West	40.5	13 max 12.5 avg	44.0	1.41	4.4/7.0
H-West	35	11 max 9.0 avg	847	1.4	2.0/0.6
H-East	35	10' between floors	556	1.4	2.0/2.7

1 in = 25.4mm
1 ft = .304m
*System stiffness = EI/H^2
H = Vertical spacing of braces.

E-west, F, and H-west, where crosslot or corner braces were used. At Sites E and H, the use of slurry walls provided high system stiffness and the observed movements. At Site D, the predicted, as well as the measured movements are small, due to high system stiffness and the use of crosslot bracing. The stiffness of the crosslot brace associated with the installation of the brace before any cut is made below that level appear to definitely minimize the offsite movements. The crosslot and corner braces provide lateral restraint to the perimeter walls before the excavation is made deeper than the brace level. This procedure also eliminates creep in the soft clays which continually occurs during the long period required for the interior deep excavation and construction of the heel beams. Limited size of berms provide limited passive resistance to the unsupported vertical walls and the slope of the berm continues to creep.

Other sources of offsite movements include the installation of the caissons and removal of the rakers. Within Figures 5 to 9 and 11, the monitoring points indicate that movements on the order of 1/2 inch (12.7mm) occur during the installation of the drilled piers for the support of the building, which is before excavation of the interior of the site. On those projects where backfill is placed between the vertical retention member and the basement walls, an additional 1/2 inch (12.7mm) of movement appears to occur when the rakers at each level are removed. This is attributed to compression of the backfill within this space. Generally, it is difficult to obtain compaction in this area because of the narrow working space and the presence of the walers, so compaction is generally quite poor. Walers are often removed when the backfill is several feet below the design level as opposed to design requirements of backfill level just below the walers. At those sites where the basement wall was poured tight to the sheeting, this movement following removal of rakers was not observed.

CONCLUSIONS

Based upon the data generated for the eight project sites in the Chicago area, it is concluded:

1.   The procedure for calculating offsite movements developed by Clough (2), compares favorably with measured movements when the retention system consists of inclined rakers that are extended to grade beams or a mat within the center of the site.

2.   Where crosslot braces are used, the movements appear to be on the order of 1/2 of the values predicted by Clough (2).

3.   A suggested procedure for calculating the base heave factor of safety is presented. This procedure which takes into account adhesion on the back face of the vertical portion of the retention system appears reasonable based upon the close agreement between predicted and measured movements.

4.   Other factors contribute to offsite ground movements. One factor is movements associated with the installation of drilled piers. Another source of movement is compression of the fill which is generally not

well compacted in the space between the basement wall and the vertical portion of the retention system.

5. Expeditious construction and preloading of braces result in smaller movements. Movement due to creep continues until the excavation is backfilled.

6. The most important parameters in minimizing ground movements are the stiffness of the system and basal heave factors of safety. Stiffness is increased very significantly by reducing vertical spacing of the braces. To obtain acceptable factors of safety against base heave, the vertical members must be anchored into underlying stiff soils.

REFERENCES

1. L. Bjerrum & O. Eide, Stability of Strutted Excavation in Clay, Geotechnique, Vol VI, Mar 1956, pp 32 to 47

2. G. Clough, Effects of Excavation Induced Movements in Clays on Adjacent Structures, Chicago Geot Lect Series (ASCE), 1986

3. G. Clough & G. Denby, Stabilizing Berm Design for Temporary Walls in Clay, J Geot Eng (ASCE) 130(2), Feb 1977, 75-90

4. R. Peck, Deep Excavations and Tunneling in Soft Ground, Seventh Int Conf on Soil Mech and Fdn Eng, State of the Art Volume, Mexico 1969, pp 261 to 270

5. K. Terzaghi, Theoretical Soil Mechanics, 5th Edition, John Wiley, New York, June 1948, Article 69

6. K. Terzaghi & R. Peck, Soil Mechanics in Engineering Practice, 2nd Edition, John Wiley, New York, Jan 1967, pp 265-266

7. G. Tschebotarioff, Soil Mechanics, Foundations and Earth Structures, McGraw Hill, New York, 1951 pp 412-417

ESTIMATING LATERAL EARTH PRESSURES FOR
DESIGN OF EXCAVATION SUPPORT

Sam S.C. Liao[1], M.ASCE and Thom L. Neff[2], M.ASCE

ABSTRACT: This paper presents a rational framework for the
estimation of actual (rather than apparent) lateral earth pressures
for the design of excavation support, developed after a
re-examination of Peck's (1969) method and various other
theoretical and practical considerations. The proposed approach
first requires an estimation of the at-rest and the active earth
pressures. The earth pressure diagram for design is then specified
to be intermediate between these two conditions. The specified
pressures would depend on the amount of excavation movements which
are anticipated and/or which are acceptable in the design.
Insights into the behavior of cohesive soils, during undrained
excavation unloading and subsequent long-term drainage, are
obtained through the application of the stress path method. A
previously published case study is re-examined in the context of
the proposed approach, and indicates that the approach is
reasonable.

INTRODUCTION

The method presented by Peck (1969) for specifying apparent
lateral earth pressures forms the basis of excavation support
design still used by the geotechnical profession today. Though the
apparent earth pressure concept predates the specific
recommendations by Peck (e.g., see Flaate, 1966 and Terzaghi and
Peck, 1967), the use of this concept in the United States has
become commonly referred to as Peck's method.

Though Peck's method and its derivatives have been used for the
past two decades, there are difficulties associated with the
method. These include: (1) The method was developed for soil
profiles consisting predominantly of either sand or clay; the
extension of the method to mixed profiles is not straightforward.
(2) For sands, the method implicitly assumes that the excavation is
dewatered and that there is a significant lowering of the
groundwater table behind the excavation support wall; in cases
where relatively impermeable concrete diaphragm walls are used,

1 - Sr. Geotech. Engr., Bechtel/Parsons Brinckerhoff, One South
    Station, Boston, MA 02110
2 - Sr. Prof. Assoc., Parsons Brinckerhoff, 120 Boylston St.,
    Boston, MA 02116

this assumption may not be valid. There is a related concern for clays with respect to longer term conditions, where partial drainage may occur. (3) The method was developed based on data from H-pile and lagging and sheetpile walls, and may not be applicable to concrete diaphragm walls (which are more rigid) and the top-down construction method. (4) The method is based on apparent rather than actual earth pressures, and this is a problem in cases where the actual pressures are desired for input to an analysis.

In addition to the above problems, the specification of long-term soil pressures is often required for diaphragm walls intended for use as both temporary excavation support and as part of the permanent structure. Since Peck's method is intended only for use as temporary excavation support, its relationship to the long-term lateral pressures needs to be clarified.

This paper presents a re-examination of the problems related to the use of Peck's method, and suggests a rational approach to the specification of actual lateral earth pressures for design. It is important to note that the approach presented in this paper is not intended to replace the use of Peck's method in specifying apparent pressures for strut or tieback design using the tributary area method of analysis.

Not all the elements of the methodology proposed in this paper are innovative or new. Several elements have been proposed by others and have been used in practice, and as a result of repeated use, have become standards to a certain extent. The contribution of this paper is to systematically synthesize the various empirical and theoretical elements that relate to the specification of lateral earth pressure diagrams for design.

REVIEW OF PRESENT PRACTICE

Design Conditions. Design specifications or recommendations by geotechnical consultants often include provisions for flexible versus rigid wall systems, and dewatered versus not-dewatered excavations. The inclusion of these various provisions is usually intended to give the contractor several options for his design and methods in performing the excavation. In terms of soil behavior, there are three conditions that need to be considered in design. They are: (1) construction short-term (undrained) condition; (2) construction long-term (drained or partially drained) condition; and (3) post-construction/permanent (drained) condition.

For granular soils, the construction short-term and long-term conditions are usually synonymous since drainage in these soils occurs very quickly. Differences in the construction short- and long-term conditions are generally only significant for cohesive soils. Peck's method is strictly only applicable for the construction short-term condition. Whether the specification of a construction long-term condition is necessary, will depend on the construction schedule and the permeability of the actual soil. In practice, the specification of post-construction conditions is frequently different from the specification of construction

long-term conditions. Often, the construction long-term condition
is based on the active earth pressure condition using effective
stress parameters (c' and ϕ'), while the post-construction lateral
stresses are assumed to revert back to the pre-construction at-rest
(K_O) condition. Whether this practice is logical will be
discussed later in the paper.

Peck's Method. The use of Peck's method results in apparent
earth pressure diagrams that are rectangular or trapezoidal in
shape. A summary of the method is shown in Fig. 1, where Peck's
recommendations have been generalized to accommodate similar
methods. The coefficients b_1, b_2, and b_3 shown in Fig. 1 can
be varied to produce differing proportions for the trapezoid, and
can also be used to produce triangular-shaped diagrams. The width
of the diagram in Fig. 1(a) is denoted as p and is expressed as a
multiple β of the excavation depth H. The expressions for
evaluating β as a function of the earth pressure coefficient and
the total unit weight γ are also shown in Fig. 1. For sand, the
value of the active earth pressure coefficient K_A is given by

$$K_A = \tan^2(45 - \frac{\phi'}{2}) \tag{1}$$

For clays, the specified pressure is a function of the stability
number N_s, which is defined (Peck, 1969) as:

$$N_s = \gamma H/S_u \tag{2}$$

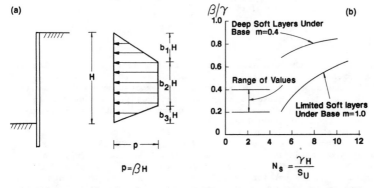

SOIL TYPE	b_1	b_2	b_3	Equation for β	Typical Value of β
SAND	0	1.0	0	$\beta = 0.65 K_A \gamma$	0.2γ
SOFT TO MEDIUM CLAY (N_s>5 or 6)	0.25	0.75	0	$\beta = K_{AT}\gamma$	0.4γ to 0.8γ
STIFF FISSURED CLAY (N_s<4)	0.25	0.50	0.25	$\beta = 0.2\gamma$ to 0.4γ	0.30γ

Fig. 1. Summary of Peck's (1969) Recommendations for Apparent
Earth Pressures

where γ is the total soil unit weight, and S_u is the average undrained shear strength on the side of the excavation. A similar stability number N_{sb} is defined using the shear strength of the soil below the base of the excavation.

For values of $N_s \leq 4$, which Peck (1969) associated with "stiff-fissured" clay, Peck's expression for β is independent of the shear strength and is simply specified to equal a value between 0.2γ to 0.3γ. For $N_s > 4$, Peck's method specifies β equal to $K_{AT}\gamma$, where K_{AT} is defined as:

$$K_{AT} = 1 - m\frac{4}{N_s} = 1 - m\frac{4Su}{\gamma H} \tag{3}$$

The subscript "T" indicates that the value of K_{AT} is the total stress rather than the effective stress coefficient. The coefficient m is an empirical factor that takes into account the effects of potential base stability problems.

Having expressed K_{AT} in this fashion, Peck's recommendations for determining the values of β for soft to medium clays and for stiff-fissured clays can be conveniently summarized as shown on Fig. 1(b). Note that Peck (1969) recommended using values of the coefficient m = 0.4, where deep soft layers exist under the base of the excavation and when $N_s > 6$ to 8. Otherwise, for excavations with limited thickness of a soft layer, Peck recommended using m = 1.0. A more theoretically rigorous expression for K_{AT} was developed by Henkel (1971) to take into account the base failure mode effects on lateral stresses, is given by:

$$K_{AT} = (1 - \frac{4Su}{\gamma H}) + \frac{2\sqrt{2}D}{H}(1 - \frac{5.14\,Sub}{\gamma H}) \tag{4}$$

where S_u is the undrained shear strength above the bottom of the excavation, S_{ub} is the shear strength below the excavation, and D is the depth from the bottom of the excavation to an underlying hard stratum. Henkel's work was not readily accessible in the United States, but was recently brought to light and is described in more detail by Christian (1989).

Typical values of β as a function of γ are shown in Fig. 1, with the values of p ranging usually from p = $0.2\gamma H$ to p = $0.8\gamma H$. For sand, the typical value is calculated based on a friction angle $\phi'=30°$. For soft to medium clay, the average or typical value is based on an assumption of $N_s = 8$. In practice, the width of the diagram p is often expressed solely as a function of the excavation depth H, and usually varies from p = 20H to p = 40H, where H has units of feet and p has units of psf (Note: 1 psf = 47.9 N/m^2).

Peck's Method versus K_A and K_O Conditions. Some designers in practice base the specified earth pressure distributions on the active or at-rest pressures, calculating the resultant of an assumed triangular diagram and redistributing it as a rectangular or trapezoidal distribution. On the Washington, DC subway project, a criterion was used where the total active resultant was

multiplied by factors of 1.1 to 1.4 depending on the stiffness of
the wall and the corresponding amount of tolerable excavation
movement. Then the resultant factored load was redistributed in a
trapezoidal pressure diagram (Parish, 1989). A similar approach
using a redistributed pressure equal to 1.3 times the Rankine
pressure resultant, was used on the Central Artery North Area
(CANA) Project (I-93 and Route 1 Interchange), presently under
construction in Boston. A method based on evaluating the earth
pressure diagram as intermediate between K_O and K_A conditions
was used by J. Gould, as described in Golder et al (1970). Thus,
it is of interest to compare the total resultant of Peck's apparent
pressure diagrams to the resultant based on roughly equivalent
active and at-rest conditions, where triangular stress
distributions are assumed.

For sands, assuming a typical value of $\phi' = 30°$ and a dewatered
excavation, the total resultant using Peck's envelope is denoted as
$P_{Peck} = 0.217 \gamma H^2$. If the sand is normally consolidated
($K_O = 1 - \sin \phi' = 0.5$), the resultant of a triangular K_O
stress distribution would be $P_{Ko} = 0.250 \gamma H^2$. Similarly, for
active conditions one can calculate $P_{Ka} = 0.167 \gamma H^2$. Thus it can
be observed that $P_{Peck} = 0.87 P_{Ko}$ and $P_{Peck} = 1.3 P_{Ka}$,
indicating that the total resultant using Peck's recommendations
for sand is intermediate between the total resultants for the
active and at-rest conditions. For clays the comparison is more
difficult because pore pressure and drainage conditions during
excavation are not easy to define.

The comparison of total resultants as described above is
strictly not valid. Peck (1990) has indicated that "As soon as you
think of the apparent pressure as having a total resultant, you
depart from the essential concept that the apparent pressure
diagram is an envelope. At no cross section in an open cut should
the strut loads add up to the equivalent of the apparent pressure
diagram. One strut may have a total corresponding to the diagram,
but if it does, the others should be appreciably less."

Nevertheless, the comparison above does indicate that there is a
basis for defining the pressure on a wall during excavation to be
between the K_O and K_A conditions. If P_{Peck} were less than
P_{Ka} or greater than P_{Ko}, this hypothesis would have less
empirical validity.

Use of Buoyant Unit Weight. In excavations with relatively
impermeable support walls and/or where significant drawdown of the
water table does not occur, geotechnical engineers often adapt
Peck's recommendations using the buoyant unit weight (γ') rather
than the total unit weight (γ) of the soil (or possibly a pro-rated
unit weight in cases where the water table is at an intermediate
depth between the ground surface and the excavation bottom). Then
the lateral pressure due to hydrostatic pressure is superimposed
onto the earth pressure diagram. Though this approach follows a
logic of sorts, the validity of this method is questioned by
engineers who view Peck's recommendations as a purely empirical
method based on total stresses. To cast Peck's method into an
effective stress framework by using buoyant unit weights would, in

this view, be equivalent to unnecessarily tampering with a successful method of design. The proponents of this view would advocate using Peck's method with total unit weights, even when the water table behind excavation wall is not perturbed, and not account for the hydrostatic pressures. Generally, this results in more severe loading conditions than the approach adapted using buoyant weights.

COMMENTS ON SHAPE OF PRESSURE DISTRIBUTIONS

General. The shapes of the apparent earth pressure distributions recommended by Peck were based on empirical observations. The fact that the distributions are trapezoidal or rectangular, rather than the triangular shape expected from geostatic or Rankine active conditions, has commonly been attributed to the construction sequence, the stiffnesses of the excavation wall and bracing, and arching in the soil. The purpose of this section is to examine more fundamental phenomena that may explain the expected shape of lateral pressure distributions.

Apparent Versus Actual Pressure. The concept of apparent earth pressures, in addition to being an empirical method of approximating soil behavior, represents a simplified procedure for solving an indeterminate structural analysis problem. In both its derivation and application, it is assumed that the strut (or tieback) loads are a summation of pressure behind a tributary area of wall, as shown on Fig. 2. The assumed tributary area extends from the mid-point between the strut and the strut immediately above it to the mid-point between the strut and the strut immediately below. The tributary areas of the top and bottom struts in an excavation are adapted to account for the appropriate boundary conditions. As described by Peck (1969), "the envelopes, or apparent earth pressure diagrams..., were not intended to represent the real distribution of earth pressure..., but instead constituted hypothetical pressures from which there could be calculated strut loads that might be approached but would not be

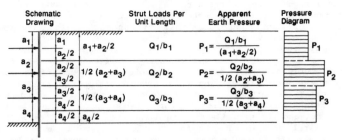

Fig. 2. Schematic of the Apparent Earth Pressure Concept (after Flaate, 1966).

exceeded in the actual cut."

To illustrate the differences between actual earth pressures and apparent earth pressures, a structural analysis was made for a hypothetical excavation with a depth H = 40 ft. and an assumed triangular soil lateral stress distribution behind the wall, as shown in Fig. 3(a). The struts were assumed to be rigid, and the strut loads were calculated using STAAD-III, a widely available microcomputer structural analysis program. The apparent earth pressures were then backcalculated from the strut loads, and are shown in Fig. 3(c). Thus, where the imposed load has a smooth distribution, the backcalculated apparent pressure is a stepped distribution, generally increasing with depth.

It was on the basis of backcalculated distributions similar in concept to that in Fig. 3(c) that Peck's envelopes were derived. In fact, many of the backcalculated distributions are similar in shape, as shown in Fig. 4 which is excerpted from Peck (1969). A comparison of Fig. 3(c) with Fig. 4 indicates that in many cases, a triangular distribution of actual soil pressure may be representative of soil behavior. Consequently, in some cases, strut loads based on Peck's envelopes may be excessively conservative, especially for the top struts, and especially if the apparent pressures are used in place of actual pressures in design calculations. However, not all case studies of backcalculated apparent pressures indicate a pattern similar to Figs. 3(c) and 4 (see Flaate, 1966).

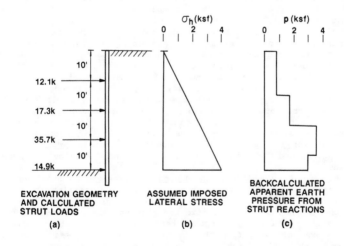

Fig. 3. Calculated Loads and Apparent Earth Pressure for a Hypothetical Case with a Triangular Imposed Pressure Distribution. (Note: 1 ft = 0.305 m; 1 ksf = 47.9 kN/m²)

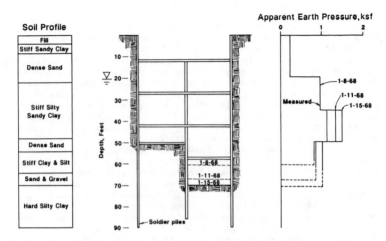

Fig. 4. Example of Apparent Earth Pressures (after Peck, 1969).
(Note: 1 ft = 0.305 m; 1 ksf = 47.9 kN/m²)

In practice, the specified apparent earth pressure diagrams are
sometimes not distinguished from actual earth pressures, and
structural designers often treat them equivalently. Given the
present availability of microcomputers and structural analysis
programs, and the ease of solving indeterminate structural
problems, the original intended application of Peck's apparent
earth pressure diagrams have fallen into disuse. Instead,
structural designers often assume that the apparent earth pressures
are the actual earth pressures, using them as input to analysis,
i.e. they avoid the use of the tributary area method of calculating
strut loads. This procedure is clearly improper, considering the
derivation of apparent earth pressures, although it is not clear
whether the procedure leads to overly conservative or unsafe
designs. It is the writers' opinion that, in order to rectify this
situation, geotechnical engineers should indicate clearly when they
are specifying <u>actual</u> earth pressures versus <u>apparent</u> earth
pressures, and indicate the applicable method for calculating strut
loads.

K_O and K_A <u>Conditions as Bounds on Pressure</u>. Starting from
at-rest (K_O) conditions in the ground, the insertion of a soil
retaining element and subsequent excavation ultimately causes
movements that tend to mobilize the internal shearing resistance of
the soil. Mobilization of the shear strength leads to a reduction
of lateral stresses from the K_O condition, and if enough movement
occurs, an active Rankine failure (K_A) condition results in the
soils retained in back of the wall. For clays, of course, there is
the added intermediate step of buildup of excess (usually negative)
pore water pressures, resulting in undrained active (K_{AT})

conditions.   However, for purposes of simplifying the discussion,
consideration of undrained conditions will be discussed
separately.   It is thus recognized that the K_O and K_A
conditions form the upper and lower bounds on the values of lateral
earth pressure (except in cases where the preloads in struts or
tiebacks are excessive).   Since both the K_A and K_O conditions
are usually associated with triangular distributions, the purpose
of this section is to show how intermediate conditions between K_A
and K_O can result in non-triangular distributions.

Fig. 5 shows a hypothetical pattern of movement of an excavation
wall and the inferred corresponding perturbations caused by such
movements on the lateral soil pressure. It is hypothesized that
the greater the lateral movement at any depth level, the closer the
soil will tend towards the active condition.   Fig. 5(b) shows the
distribution of lateral earth pressure for a normally consolidated
soil, and Fig. 5(c) shows the corresponding distribution for an
overconsolidated soil.   The non-triangular shape of the
intermediate distribution is much more evident for the
overconsolidated case because the associated K_O condition is
non-triangular to begin with.

The above exercise suggests that the trapezoidal earth pressure
distributions proposed by Peck (1969) for dense sands and stiff
fissured clays can theoretically reflect the shapes of actual
(rather than apparent) soil pressure distributions behind a wall.
However, it is more likely that non-triangular distributions will
occur in overconsolidated dense sands and stiff to hard clays. For
loose sands and soft clays, the actual pressure distributions will
tend to be more of a triangular shape similar to that shown in
Fig. 5(b).

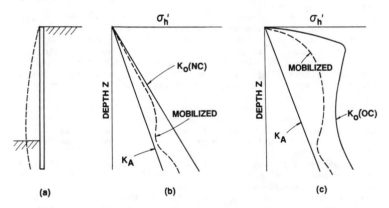

Fig. 5.   Schematic Model for Redistribution of Lateral Stress Due
          to Wall Movement: (a) Hypothesized Movement; (b)
          Hypothesized Stresses - Normally Consolidated Soil; (c)
          Hypothesized Stresses - Overconsolidated Soil.

STRESS PATH METHOD

   Basic Concepts.   The stress path method is presented here as a
useful way of visualizing the soil behavior behind an excavation
wall, particularly for undrained conditions in clay.   The concepts
of the stress path method were proposed by Lambe (1967), and have
been updated by Lambe and Marr (1979).   Lambe (1970) has also used
the concept in the analysis of excavation support walls.   A brief
review and an application of the concepts, specifically related to
long-term drained behavior of the excavations in clay, is presented
below.
   In the stress path method, the state of stress for a soil
element in the ground is represented by a point in a stress space,
i.e., a plot of q versus p or p', where the values of q, p' and p
are defined as:

$$q \ = \ 1/2 \ (\sigma_v \ - \ \sigma_h) \tag{5}$$

$$p' \ = \ 1/2 \ (\sigma_v' \ + \ \sigma_h') \tag{6}$$

$$p \ = \ 1/2 \ (\sigma_v \ + \ \sigma_h) \tag{7}$$

where σ_v' and σ_h' are respectively the effective vertical and
horizontal stresses, and where σ_v and σ_h are total stresses.
Note that p' is an effective stress parameter and that p is a total
stress parameter, and that q can be both an effective stress or a
total stress parameter, since the pore pressure is cancelled out in
the subtraction of σ_h from σ_v.   Given these definitions, we can
trace the path that a soil element follows as the state of stress
changes in the ground.
   In a plot of q versus p' or p, it can be shown that straight
lines extending from the origin represent equal values of the ratio
K of σ_h' to σ_v' or the corresponding total stress ratio K_T.
This is illustrated in Fig. 6 for values of K or K_T ranging from
0.25 to 3.0.   Of course, soil elements cannot exist at all
arbitrary values of q and p', but are restricted to be within the
boundaries formed by values of K corresponding to the active and
passive failure conditions.
   Effective stresses are important in terms of understanding soil
behavior.   However, it should be noted that only total stresses
that can be directly measured by stress cells, and total stresses
are the stresses which must be resisted by the bracing or tie-backs
of an excavation support system.
   Stress Paths During Excavation.   For purposes of simplicity of
illustration, it will be assumed in the following discussion that
the initial conditions in the ground corresponds to an isotropic
state of stress, i.e., $K_O = 1$.   It can then be shown that the
total stress path followed by a typical soil element (at
approximately mid-depth) behind a wall experiencing unloading due
to excavation is a 45-degree line extending upward and to the
left.   This is illustrated in Fig. 7(a), where the total stress
path is marked with the label TSP.   The equivalent drained

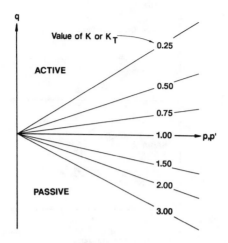

Fig. 6.   Lines of Equal K or K_T in q-p space.

effective stress path is labeled DSP in the same figure. The DSP
and the TSP lines are parallel and are consistently separated by a
distance u_s, which denotes the static pore water pressure.

The drained stress path would be the effective stress path
experienced by the soil, if there were no generation of excess
positive or negative pore pressures due to soil loading. This
generally occurs for granular soils. However, for relatively
impermeable soils, the effective state of stress would deviate from
the DSP line, following an undrained effective stress path (ESP)
similar to that illustrated schematically in Fig. 7(a). The
deviation of the ESP from the DSP is the excess pore water
pressure Δu, which in Fig. 7(a) is shown as having a negative
value.

Consistent with the notion of K_O and K_A being bounds on the
lateral pressure distribution, the stress paths experienced by the
soil element during excavation unloading reduces from K_O,
crossing the lines of equal K towards the lower bound value of
K_A. This is true whether one is considering effective or total
stress.

Drainage of Excess Pore Pressure. After excavation is
completed, dissipation of the excess (usually negative) pore
pressures occurs with time. The stress paths followed by the soil
during drainage of excess pore pressures will depend on the amount
of lateral constraint imposed by the wall on the soil. In an
unconstrained situation, where the total stress σ_h is kept
constant and the soil is "free" to deform, the effective stress
path (ESP) followed by a soil element during dissipation of pore
pressures is a horizontal line. The ESP during dissipation
terminates when it reaches the corresponding DSP of excavation

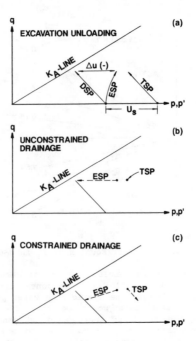

Fig. 7. Schematic of Stress Paths for a Representative Soil Element Behind An Excavation Support Wall.

unloading. The total stresses, however, do not change during unconstrained drainage, and the total stress path (TSP) is represented in Fig. 7(b) as simply a point. This is analogous to stress conditions that could easily be imposed in a triaxial test by simply allowing drainage after undrained unloading of a soil sample.

However, if a wall has been installed in the ground that constrains the deformations of the soil, then it is hypothesized that the effective stress path during drainage will not be horizontal, but will follow the path show in Fig. 7(c). In general, it is hypothesized that the path will be non-linear, and will depend on the stress-strain properties of the soil. However, as in the care of unconstrained drainage, the ESP will terminate when it reaches the DSP of excavation unloading. The corresponding total stress path (TSP) will then move down along a 45 degree line, as shown in Fig. 7(c). Thus, during constrained drainage, it is obvious that the K_T value corresponding to total stresses would increase with time. This is consistent with observations that the strut or tieback loads in an excavation tend to increase with time, e.g. see DiBiagio and Roti (1972) and Ulrich (1989a, 1989b).

An analogy to constrained drainage can be drawn to imposing undrained loading on a soil sample in a triaxial cell, and then adjusting the cell pressure and axial load so that there are zero horizontal strains while allowing the excess pore water pressures to dissipate. Figure 8 shows a comparison of unconstrained versus constrained drainage, and also offers an intuitive explanation of the rationale for total stress increases during constrained drainage.

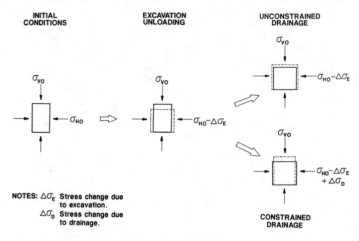

NOTES: $\Delta\sigma_E$ Stress change due to excavation.
$\Delta\sigma_D$ Stress change due to drainage.

Fig. 8. Schematic of Stress Changes and Deformations for Unconstrained and Constrained Drainage. Note: Dashed Lines Indicate Deformed Shape After Stress Changes.

These observations lead to an interesting practical conclusion regarding long-term lateral earth pressures. As discussed previously, it is usually assumed in design for the permanent condition that the stresses in the soil ultimately revert to pre-construction K_0 conditions. From the analysis presented, it is clear that this is not the case. In the context of the simple illustration described here, the effective K could increase slightly, depending on the constraint conditions and the soil properties, but the analysis indicates (though it is not a proof) that there is no logical mechanism for the soil to completely revert back to the K_0 condition. Long-term measurements of a retaining wall in stiff clay reported by Carder and Symons (1989) are consistent with this conclusion.

The effective stress K value decreases during the excavation process, and with dissipation of pore pressure, and may increase slightly with time. Its ultimate value will depend on the amount of stress relief that has been caused by movements during the construction process and the degree of lateral constraint during drainage of excess pore pressures. The apparent increase of strut or tieback loads with time can probably, in most cases, be explained in terms of negative pore water pressure dissipation leading to increases in the total stress imposed on the excavation wall.

Additional Complications. To illustrate the basic conclusions regarding excavation support that can be derived using the stress path method, some simplifying assumptions regarding the construction process were made. These include the assumption that the initial installation of the wall element (e.g., slurry diaphragm walls) did not substantially affect the in situ stress conditions, and that there was no friction or adhesion effects between the soil and the wall. It was also assumed that the pre-existing static water table was maintained during the construction, either because of the low permeability of soil (i.e., slow response time for equilibration despite dewatering of the excavation) or that the groundwater level was maintained by the wall element and/or groundwater recharge. For illustrative purposes, Fig. 7 showed the stress paths starting from a K_O condition equal to 1.0, and it was assumed that soils will generally exhibit negative excess pore pressures during excavation unloading.

It should be emphasized that the above assumptions were made solely for clarity in presentation. Soil elements at different locations behind the wall will follow different stress paths. In practice, additional complications may have to be considered on a case by case basis. Though most of the conclusions described in this section will continue to be valid, this may not be so in many instances. For example, lowering of the groundwater table will affect the stress paths significantly. It is also easy to envision a condition where groundwater lowering continues long enough to cause additional consolidation of the soil due to increased effective stresses. This may perhaps alter the conclusion that the soil in the long term will not revert to K_O conditions. Also, for very soft clays, it may be possible that the excess pore water pressure during excavation will be positive rather than negative, which would change the conclusion that the total stresses would always increase with time.

PROPOSED APPROACH

A rational framework for the estimation of actual lateral earth pressures for design is proposed, based on a synthesis of the observations presented in the previous sections of this paper. The proposed approach results in actual rather than apparent earth pressures. It is not advocated that his approach be used to replace the use of Peck's pressure envelopes for calculating strut or tie-back loads using the simple tributary area method of analysis. Rather, the proposed approach is intended for use in conjunction with rigorous methods of indeterminate structural analysis. The steps in the approach are as follows:

1. Estimate the at-rest (K_O) effective horizontal stress prior

to excavation, as best as practical, using either measurements based on in-situ tests, laboratory tests, or empirical correlations.

2. Estimate the active (K_A) earth pressure based on the effective stress friction angle and cohesion.

3. Determine the amount of movement which is tolerable in the design, or estimate the movements which are expected (e.g., whether average or better than average workmanship is expected of the contractor).

4. Specify the design lateral pressure distribution to be intermediate between the K_O and K_A conditions, with lower pressures associated with greater excavation movements. Using laboratory or field measurements combined with judgement, estimate the degree of soil strength mobilization which would be associated with the movement, and estimate the degree of lateral stress reduction from the K_O to the K_A condition at various depths along the excavation.

5. Use the stress path method, preferably combined with laboratory testing, to estimate the long term lateral earth pressures. For long-term construction conditions, and for the permanent conditions, the earth pressure will generally not revert back to the K_O condition. However, especially for clays, it should be recognized that the total lateral pressure will increase with time due to the dissipation of the usually negative excess pore water pressures developed during excavation.

6. Superimpose the water pressures from hydrostatic conditions or from flow conditions (due to dewatering) onto the estimated earth pressure to obtain the total (soil plus water) lateral pressure diagram for design.

The proposed approach has several advantages over the Peck (1969) method. The first is that the method can easily accommodate soil profiles consisting of a several different strata with different K_O and K_A values. The second is that that proposed approach utilizes only effective stresses, so there is no question as to how the pressures due to groundwater should be considered in the design. Thirdly, the proposed approach allows the use of one's judgement or additional analyses (such as use of a finite element model in an iterative fashion) to account for different stiffnesses of the wall and strut or tieback system or different construction sequencing (e.g., top-down construction).

Admittedly, there are many difficulties in the actual implementation of the proposed approach. The most problematic, of course, involves Step 4, where one must estimate the amount of stress relief or mobilization of shear strength that occurs with movement. For a wall displacing in a rotational mode, the amount of wall movement (at the top) required to fully mobilize the Rankine active state is given by various authors to range from 0.0005 to 0.001 times the depth of the excavation for a dense sand, but would be substantially higher for loose sand. For cohesive soils the equivalent movement required is generally stated to range from 0.004 to 0.02 times the depth of excavation (see Peck, Hanson, and Thornburn, 1974; Wu, 1975; NAVFAC DM 7.2, 1982). It is

believed that this is an area where further research would be
especially beneficial.

A conservative approach would be to assume that no wall movement
occurs, so that the K_O condition would govern the design.
However, it is recommended that in no case should the earth
pressure coefficient be lower than $1.3K_A$, unless very large
movements of the wall are expected and acceptable in design. This
is believed to be consistent with Peck's (1969) recommendations,
except for soft to medium clays where Peck's method is considered
by the writers to be excessively conservative in the particular
case where relatively stiff soils or rock exists under the base of
the excavation.

CASE STUDY

To evaluate the approach proposed in this paper, the case study
reported by DiBiagio and Roti (1972) of the Telefonhuset Building
in Oslo, Norway was re-examined. The case study involved the
construction of a building with a basement that required excavation
to a depth of 19 m. The excavation wall was a concrete diaphragm
wall, 1 m thick, constructed by the slurry method. The wall was
keyed into bedrock, of which the surface was at the bottom of the
excavation.

The soil profile consists of approximately 2 meters of fill
overlying a clay deposit which appears to be normally consolidated,
except for a desiccated crust near the top of the clay stratum.
The clay is approximately 17 m thick and overlies the bedrock. The

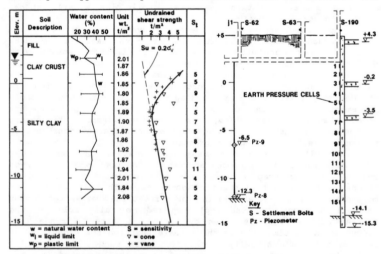

Fig. 9. Soil Profile and Instrumentation for Telefonhuset
        Building, from DiBiagio and Roti (1972).

soil profile and a schematic of the instrumentation for the wall is shown in Fig. 9.

The Telefonhuset Building study was of particular interest because it is one of the very few case studies reported where total pressure cells were used to measure the lateral stress. In most of the case studies reported in the literature, measurements of lateral pressures are apparent pressures derived from strut and tieback loads. The Telefonhuset case study is also of interest because it involves a generally soft clay and a situation where the base of the excavation is a hard stratum. As described above, this is a situation for which the writers believe Peck's method is excessively conservative.

Some assumptions had to be made in re-analyzing the case study in the context of the approach proposed in this paper. For purposes of simplicity, it was assumed that the total unit weight of the soil was constant and had a value of 1.9 t/m^3 (18.6 kN/m^2). It was also assumed that the building shown immediately adjacent to the excavation was constructed with a compensated "floating" foundation, and hence contributed little additional stress other than the stress which existed originally in the ground.

Based on the plasticity index (PI), which ranged from 15 to 26, it may be assumed that ϕ' = 30° and that the normally consolidated $(K_O)_{NC}$ = 0.5 (Ladd et al, 1977). Based on the Atterberg limits data given for the clay, and assuming that the shear strength of the clay could could be described by the SHANSEP approach (Ladd and Foott, 1974; Ladd et al, 1977), the undrained shear strength S_u was assumed to be:

$$S_u = 0.2\ \sigma_v'\ (OCR)^{0.8} \tag{8}$$

The values of OCR were back-calculated using the above equation and the shear strength profile shown in Fig. 9. Then, the overconsolidated value of K_O was calculated assuming:

$$(K_O)_{OC} = (K_O)_{NC}(OCR)^{0.5} \tag{9}$$

An implicit assumption in applying the above equations to this case study is that the K_O values resulting from apparent overconsolidation of the soil by desiccation are the same as those which would result from stress relief following compression by mechanical loading.

The results of the calculations of K_O and K_A lateral stresses are shown on Fig. 10, superimposed on the data measured by DiBiagio and Roti (1972). Note that both the measurements and the calculations are for total stresses that include the pore water pressure. The comparison indicates that the measured stresses behave in a manner consistent with the concepts described in this paper. As the excavation progresses to deeper stages, and greater movements occur, the stresses move away from the K_O towards the K_A condition.

Near the bottom the excavation, the stresses are less than that

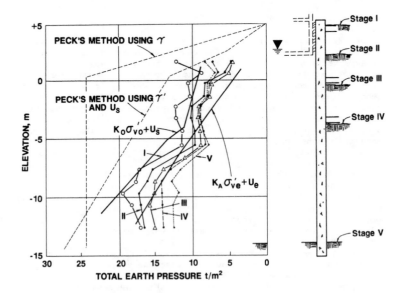

Fig. 10. Comparison of Calculated Earth Pressures with Measured Data from Telefonhuset Building, adapted from DiBiagio and Roti (1972). (Notes: Subscript "e" Indicates Stresses Calculated Based on Actual Measurements of Pore Pressure From Excavation and Dewatering. 1 t/m² = 9.81 kN/m².)

calculated based on the K_A condition. This is because of decreases in the pore pressure behind the wall due to dewatering. If the effective lateral stresses were back-calculated from the measured stresses, assuming hydrostatic conditions (u_s), they would actually be negative. The stresses marked as "$K_A \sigma_{ve}' + u_e$" in Fig. 10 are based on measured pore pressures, where the subscript "e" indicates excavation conditions. However, the pore pressures were measured in piezometers which are at a horizontal distance of approximately 12 m from the wall, and hence would not accurately reflect the pore pressures at the actual locations of the stress cells on the wall. Thus, there are 2 components to the decrease of total pressure shown in Fig. 10, the first being that due to the mobilization of soil shear strength, and the second being that due to reduction of pore pressure by dewatering. In many cases, as it has occurred near the bottom of the Telefonhuset excavation, the dewatering effect is clearly more significant.

Fig. 10 also shows a comparison of the range of apparent earth pressures calculated using Peck's method with either total or buoyant unit weight. It is clear that Peck's method tends to overestimate the lateral pressures in this case, especially above the mid-height of the excavation.

In the long term, DiBaggio and Roti observed total stress distributions which increased with time to an average value 25% greater than the minimum stresses shown in Figure 10, but which were 15% lower than the maximum measured stresses.

Other case studies which indicate phenomenon similar to that found in the Telehonhuset Building excavation include those presented by Tedd et al (1984) and Clarke and Wroth (1984). However, these cases involved embedded cantilevered walls in stiff overconsolidated clays.

SUMMARY AND CONCLUSIONS

A rational framework for estimating actual lateral earth pressure diagrams for the design of excavation support has been developed after a re-examination of Peck's (1969) method and various other theoretical and practical considerations. It is important to note that the approach presented in this paper is not intended to replace the use of Peck's method in specifying apparent pressures for strut or tieback design using the tributary area method of analysis.

The proposed approach first requires an estimation of the at-rest and the active earth pressures. The earth pressure diagram for design is then specified to be intermediate between these two conditions. The specified pressures will depend on the amount of excavation movements which are anticipated and/or which are acceptable in the design. The pressure diagram derived in this fashion is an effective stress distribution, on which must then be superimposed the lateral stress due to pore water pressure.

Admittedly, some specifics of the implementation of the proposed approach may be difficult, and judgement is required in its application. Further research (especially involving instrumented excavations) to define these specifics is recommended. However, in terms of the general applicability to complex soil profiles, varying groundwater conditions, and differing wall stiffness and methods of construction, the proposed approach provides a rational basis for incorporating all these factors.

An interesting observation, derived from application of the stress path method to undrained excavation in clay, is that long term increase of total lateral pressure on a wall with time can be explained in terms of excess pore pressure dissipation and the lateral restraint caused by the wall. It has also been indicated that except possibly in the case where dewatering causes significant consolidation of the soil, it is unlikely that the post-construction lateral stress can completely revert back to its original K_o condition.

With regard to Peck's (1969) method, it has been shown that there are theoretical bases for the trapezoidal or rectangular shape of apparent pressure diagrams other than the common attribution to the effects of construction sequencing, structural stiffness factors, and soil arching. It is hypothesized that

trapezoidal or rectangular distributions are more likely to occur
for overconsolidated soils, i.e., dense sands and stiff to hard
clays. For normally consolidated sands and clays, the pressure is
more likely to be closer to a triangular distribution. However,
future research involving in situ measurements are needed to verify
this hypothesis.

It is the writers' opinion that Peck's method is significantly
conservative for soft to medium clay in the case where there is a
relatively hard stratum directly beneath the base of the
excavation. This may also be true in the case where the excavation
wall is keyed into a hard stratum that may be at a significant
depth below the bottom of the excavation. Also, when specifying
pressure diagrams for design, it is important to indicate whether
apparent or actual pressures are given, since this should affect
the designer's method of calculating strut or tieback loads.

ACKNOWLEDGMENTS

This paper is the by-product of various studies conducted by the
writers as part of the Central Artery/Tunnel Project, currently
under design in Boston. The writers wish to thank the
Massachusetts Department of Public Works (MDPW) and the management
of Bechtel/Parsons Brinckerhoff (B/PB) for their encouragement in
developing the concepts presented in this paper, and for granting
permission for its publication. Specifically, the writers wish to
acknowledge W. Twomey, L. Barbieri, A. Ricci, and L. Bedingfield of
the MDPW; and D. Marshall, C. Carlson, L. Silano and K.K. See-Tho
of B/PB. The writers also wish to thank C.C. Ladd, W.A. Marr, and
R.B. Peck for providing comments on this paper. However, any
opinions, errors or omissions in this paper are solely the
responsibility of the writers, and any explicit or implicit
opinions stated by the writers should not be inferred to be
official policies or procedures of the MDPW or B/PB.

REFERENCES

1. Carder, D.R. and Symons, I.F. (1989). "Long-Term Performance
    of an Embedded Cantilever Retaining Wall in Stiff Clay."
    Geotechnique, 39 (1), 55-75.
2. Christian, J.T. (1989). "Design of Lateral Support Systems."
    Proceedings, Seminar on Design, Construction, and Performance
    of Deep Excavations in Urban Areas, BSCE Section, ASCE, held
    at MIT, Cambridge, MA.
3. Clarke, B.G. and Wroth, C.D. (1984). "Analysis of Dunton Green
    Retaining Wall Based on Results of Pressuremeter Tests."
    Geotechnique, 34(4), 549-561.
4. DiBiagio, E. and Roti, J.A. (1972). "Earth Pressure
    Measurements on a Braced Slurry-Trench Wall in Soft Clay."
    Proc. 5th Europ. Conf. on Soil Mech. and Found. Engr.,
    Madrid, Vol. 1, 473-483.

5. Flaate, K.S. (1966) "Stresses and Movements in Connection with Braced Cuts in Sand and Clay." Ph.D. Thesis, Univ. of Illinois.
6. Golder, H.Q., Gould, J.P., Lambe, T.W., Tschebotarioff, G.P., and Wilson, S.D. (1970). "Predicted Performance of Braced Excavation." J. of Soil Mech. and Found. Div., ASCE, 96 (SM3) 801-815.
7. Henkel, D.J. (1971). "The Calculation of Earth Pressures in Open Cuts in Soft Clays." The Arup Journal, 6(4), 14-15.
8. Ladd, C.C. and Foott, R. (1974). "New Design Procedures for Stability of Soft Clays." J. of Geotech. Engr. Div., ASCE, 100 (GT7), 763-789.
9. Ladd, C.C. et al (1977). "Stress Deformation and Strength Characteristics." Proc., IX Int'l Conf. on Soil Mech. and Found. Engr., Tokyo.
10. Lambe, T.W. (1967). "The Stress Path Method." J. of Soil Mech. and Found. Div., ASCE, 93(SM6), 309-331.
11. Lambe, T.W. and Marr, W.A. (1979). "Stress Path Method: Second Edition." J. of Geotech. Engr. Div., ASCE, 105(6), 727-738.
12. Lambe, T.W. (1970). "Braced Excavations." Proc. ASCE Specialty Conf. on Lateral Stresses in the Ground and Design of Earth Retaining Structures, Cornell Univ., Ithaca, N.Y., 149-218.
13. NAVFAC DM 7.2 (1982). Foundations and Earth Structures, U.S. Govt. Printing Office, Washington, D.C.
14. Parish, W.C. (1989). Personal Communication.
15. Peck, R.B. (1969). "Deep Excavation and Tunnelling in Soft Ground." Proc. 7th Int'l Conf. on Soil Mech. and Found. Engr., Mexico City, SOA Volume, 225-290.
16. Peck, R.B. (1990). Personal Communication.
17. Peck, R.B. Hanson, W.E., and Thornburn, T.H. (1984). Foundation Engineering, 2nd Edition, John Wiley & Sons, Inc. New York.
18. Tedd, P. Chard, B.M., Charles, J.A., and Symons, I.F. (1984). "Behavior of a Propped Embedded Retaining Wall in Stiff Clay at Bell Common Tunnel." Geotechnique, 34(4), 513-532.
19. Terzaghi, K. and Peck, R.B. (1967). Soil Mechanics in Engineering Practice, 2nd Ed., John Wiley and Sons, Inc., New York.
20. Ulrich, E.J. (1989a). "Internally Braced Cuts in Overconsolidated Soils." J. of Geotech. Engr., ASCE, 115 (4), 504-520.
21. Ulrich, E.J. (1989b). "Tieback Supported Cuts in Overconsolidated Soils." J. of Geotech. Engr., ASCE, 115 (4), 521-545.
22. Wu, T.H. (1975). "Retaining Walls," in Foundation Engineering Handbook, Edited by H.F. Winterborn and H.-Y. Fang, Van Nostrand Reinhold Co., New York.

SUBWAY DESIGN AND CONSTRUCTION FOR THE DOWNTOWN SEATTLE TRANSIT PROJECT

Arthur J. Borst[1], M.ASCE; Thomas L. Conley[2], M.ASCE;
Daniel P. Russell[3] and Ralph N. Boirum[4], M.ASCE

ABSTRACT: Design, construction and monitoring of three subway stations and a cut-and-cover tunnel in Seattle are presented. The design and construction addressed short and long-term pressures, seismic considerations and stiffness requirements for mitigating settlements of adjacent buildings. The monitoring program addressed the actual pressures on the support of excavation, the shoring wall movements and associated building settlements.

INTRODUCTION:

The Downtown Seattle Transit Project (DSTP) is a 1.8 mile subway through Seattle's Central Business District. The subway is part of the Municipality of Metropolitan Seattle (Metro) regional transit facility. The project consists of three underground stations, 1,080 feet of cut-and-cover tunnel, two retained cut staging area stations and 10,200 feet of double tube shield driven tunnel. The subway also crosses an 80 year old active railroad tunnel in two locations. The subway, which is adjacent to 90 existing buildings, is designed to initially accommodate dual-mode (electric and diesel) busses. Provisions for conversion to light-rail are included in the design and construction of all elements. A general site plan is shown in Figure 1.

This paper addresses the design considerations, construction techniques and construction monitoring for the cut-and-cover tunnel and the three subway stations. Design considerations included strength and stiffness requirements to minimize settlements of adjacent buildings, streets and utilities. Construction techniques included flexible and rigid wall systems which were internally braced and/or temporarily or permanently tied back. Construction monitoring of both the support of excavation and adjacent structures included the use of survey markers, inclinometers, strain gauges, load cells, tilt-meters and tape extensometers.

---

[1] - Associate, Andersen Bjornstad Kane Jacobs, Inc., 220 West Harrison, Seattle, WA 98119
[2] - Lead Engineer, Parsons Brinckerhoff Quade & Douglas, Inc., 999 3rd Avenue, Suite 801, Seattle, WA 98104
[3] - Associate, INCA Engineers, Inc., 11820 Northup Way, Suite 210, Bellevue, WA 98005
[4] - Associate, Shannon & Wilson, Inc., 400 North 34th Street, Suite 100, Seattle, WA 98103

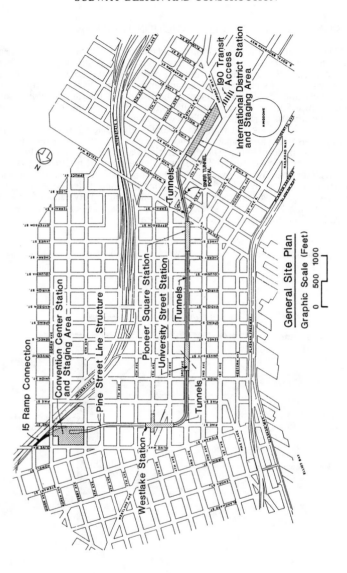

Figure 1. Site Plan

DESIGN CRITERIA AND CONSTRUCTION TECHNIQUES:

The selection of the design criteria and construction techniques for the earth retaining structures was based on numerous considerations. These factors included: size of the station or tunnel, proximity of the excavation to existing structures, types of soil(s) anticipated, presence of groundwater, anticipated wall movements and building settlements, and estimated tolerable building settlements. Since Seattle is in a seismically active area, earthquake effects on both the temporary and permanent facilities were addressed.

In general, where excavations were directly adjacent to buildings, rigid shoring walls with closely spaced supports were used to minimize settlements in lieu of direct underpinning. The rigid shoring walls typically consisted of drilled shafts tangent to each other. These shafts were filled with structural concrete and steel shapes. Where excavations were located such that wall movements would not significantly affect building settlements, more flexible shoring walls were used. These walls consisted of soldier piles spaced between six feet and 10 feet. Lagging between the soldier piles was either timber or shotcrete. A general station cross-section is shown in Figure 2.

Lateral wall movements were predicted using a two-dimensional, elastic finite element analyses. These results were tempered based on empirical data.

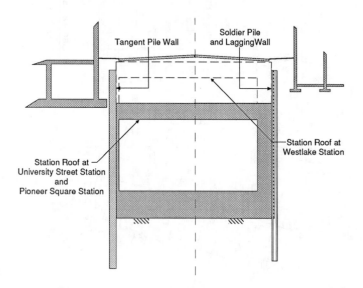

Figure 2. General Station Section

The design pressures for the support of excavation systems were based primarily on the type of soil and whether the system was required to be rigid or flexible. Spacing of the struts or tiebacks was also determined based on the rigidity requirements. Cross-lot struts along Pine Street and at Westlake Station were preloaded to 25 percent of their design loads. At Pioneer Square Station, the struts were preloaded to 50 percent. The design pressures are shown in Figure 3. The specific construction techniques for the stations and the cut-and-cover tunnel are addressed in the following description of each project element.

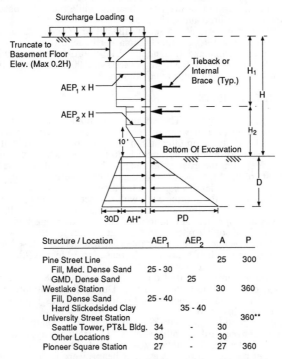

Structure / Location	AEP$_1$	AEP$_2$	A	P
Pine Street Line			25	300
Fill, Med. Dense Sand	25 - 30			
GMD, Dense Sand		25		
Westlake Station			30	360
Fill, Dense Sand	25 - 40			
Hard Slickedsided Clay		35 - 40		
University Street Station				360**
Seattle Tower, PT&L Bldg.	34	-	30	
Other Locations	30	-	30	
Pioneer Square Station	27	-	27	360

* AH Used only to determine pile embeddment to resist kickout.
** 240 Above and 40' either side of the BNRR Tunnel.
D = 12 Feet, minimum.
Ignore passive resistance in upper 2 Feet of excavation bottom.
Pressures in pounds per square foot (psf).

Figure 3. Recommended Earth Pressures for Temporary Braced Wall Design

GEOLOGY:

The geology of downtown Seattle, which is located in the central Puget Sound area, generally consists of very dense to hard soils which have been deposited by glaciers and subsequently consolidated by the weight of glacial ice. These soils consist of glacial outwash sand and gravels, glacial-marine drift, till, and glacio-lacustrine silts and clays. Along portions of the project corridor, the consolidated soils have been covered by fill placed during numerous regrading projects through the past century. Fill depths of 20 feet are not uncommon. Most of the silts and sands are water bearing, while the over-consolidated clays are frequently fractured and slickensided.

Seattle is also located in a seismically active (Zone 3) area. Major earthquakes include the 1965 Seattle-Tacoma earthquake (magnitude 6.5) and the 1949 Olympia earthquake (magnitude 7.1), approximately 60 miles south of Seattle. Bedrock is estimated to be in excess of 2,000 feet deep in the downtown Seattle area and active faults are not known.

PROJECT ELEMENTS:

Cut-and-cover Tunnel. The cut-and-cover tunnel on Pine Street is approximately 1,080 feet long between 6th and 9th Avenues. The tunnel width varies from 38 feet wide between 6th and 8th Avenues to 70 feet wide at 9th Avenue. Since the right-of-way on Pine Street is 80 feet, the excavation line is approximately 21 feet from the existing buildings. The geology along Pine Street generally consists of hard till and/or consolidated clay overlain by fill. Groundwater conditions vary considerably based on the time of year and weather conditions.

The location of the shoring walls with respect to the adjacent buildings, and the competent (when dry) soils, indicated that the use of a flexible shoring system was appropriate. An internally braced soldier pile and lagging system was chosen. The soldier piles were space at six feet in the 50 foot deep area of the cut at 6th Avenue. The soldier pile spacings were increased to a maximum of 10 feet as the depth of excavation decreased to 30 feet at 9th Avenue. Three and four inch thick timber lagging was used throughout.

The excavation was started during the spring of 1987. The cut-and-cover tunnel was completed in October, 1987. During this six-month period, Seattle experienced its driest weather in 100 years. There was little, if any, need to dewater during this period.

Westlake Station. Westlake Station is located in Seattle's retail center on Pine Street between 3rd and 6th Avenues. The station is approximately 684 feet long and 74 feet wide. The station was constructed by the cover-and-cut method since the permanent station precast roof beams were at street level. As previously mentioned, the right-of-way along Pine Street is 80 feet. Three feet remained between the station structure and the adjacent buildings. Since many of the adjacent buildings are older, their foundations encroached into the right-of-way. In most cases, the outside of the station structure was within inches of existing foundations. This space constraint, along with the need to minimize settlements of Seattle's major department stores, dictated that the shoring wall be rigid and be incorporated into the permanent station walls. Tangent piles were used to form the shoring walls which then became the permanent station walls when married with a shotcrete liner wall.

The soils encountered in this area of Pine Street were more varied than at the cut-and-cover tunnel. Loose sand, medium dense sand and silt, till, and over consolidated clay were encountered. The sand and silt were also water bearing which required that many of the drilled tangent piles be cased or drilled using bentonite slurry. Casing was used primarily due to Metro's and the city's concerns of the extensive use of slurry in the retail core.

The tangent pile walls were supported using a combination of cross-lot struts and temporary tiebacks. The tangent piles were typically 42-inch drilled shafts filled with steel wide flanges and structural concrete. The excavation extended to 50 feet below the existing street.

University Street Station.   University Street Station is located on 3rd Avenue which has an 84-foot right-of-way. The support of excavation north of University Street was a more flexible soldier pile and shotcrete lagging system with temporary tiebacks. The soldier pile wall was used here since the major adjacent building, the Cobb Garage, was underpinned to allow for a major entrance into the station. The support of excavation south of University Street became one of the most sensitive areas of the entire project.

During the final design of this station, a major new development was under construction on the southwest side of the station. The basement of this 55-story building extends 60 feet below street level to below the station invert. Directly across the street are the Seattle Tower and the Pacific Northwest Bell Buildings. The Seattle Tower is a 28-story reinforced concrete frame building with a brick facade. The building, which was built in the 1920's and is on the National Register of Historic Places, is generally considered to be one of the finest examples of architecture in the city. The building had previously experienced approximately 1/2 inch of settlement during excavation of the double tube shield driven tunnel for this project. To further aggravate matters, both the subway station and the existing buildings were directly above an active twin-track railroad tunnel built between 1905 and 1910. This condition is illustrated in Figure 4. It became apparent that this area required special consideration.

A very rigid permanently tied-back tangent pile wall was chosen to control additional building settlements. The tangent piles were 36 inch diameter drilled shafts filled with structural concrete and steel wide flanges. The permanent tiebacks were six inch diameter pressure grouted anchors spaced at approximately six foot by eight foot centers. The anchors were tensioned and locked-off at an average load of 120 kips.

Pioneer Square Station. Pioneer Square Station is located on 3rd Avenue in the southern portion of the alignments. The station is flanked by some of the oldest and heaviest buildings along the project corridor. Many of the existing foundations encroach several feet into the 84-foot right-of-way. Like Westlake Station, the Pioneer Square Station walls come within inches of many of the existing footings. For these reasons, a rigid tangent pile wall was chosen for the support of excavation. This wall would also be incorporated into the permanent station walls.

The tangent piles used were 36 inch diameter filled with structural concrete and steel wide flanges. The pile walls were supported by cross-lot struts since the presence of numerous utilities in the upper 30 feet of the excavation and depths of adjacent basements and footings essentially eliminated the use of tiebacks. The tangent piles were drilled using bentonite slurry in order to minimize sloughing of the water bearing sand and silt.

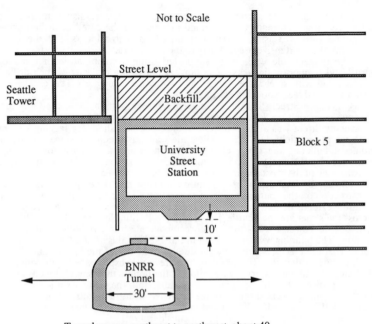

Tunnel crosses southeast to northwest, about 40
degrees to the axis of the station

Figure 4. Maximum Wall Deflections at Each Inclinometer Casing Pine Street
Line Structure and Westlake Station

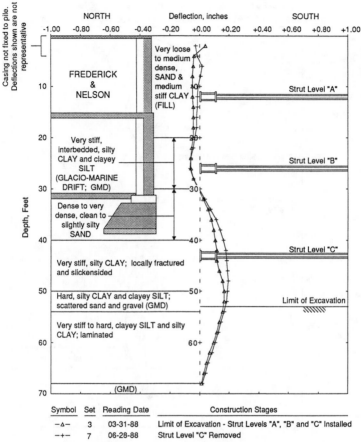

Figure 5. Wall Deflections at Critical Construction Stages
Westlake Station - Casing P1-6

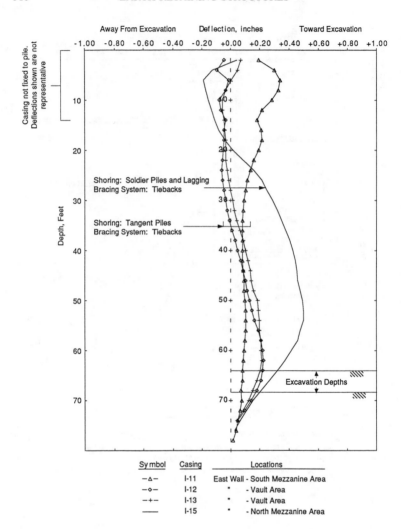

Figure 6. Maximum Wall Deflections at Each Inclinometer Casing
University Street Station

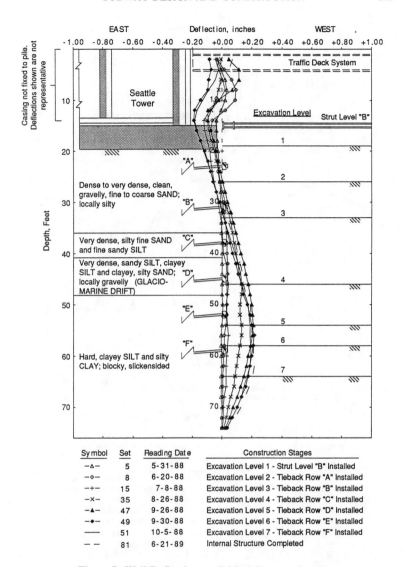

Figure 7. Wall Deflections at Critical Construction Stages
University Street Station

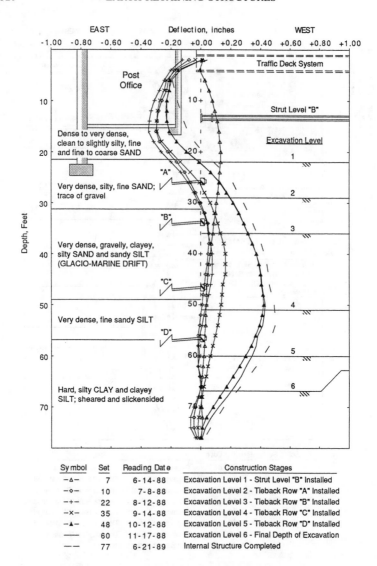

Symbol	Set	Reading Date	Construction Stages
—△—	7	6-14-88	Excavation Level 1 - Strut Level "B" Installed
—◇—	10	7-8-88	Excavation Level 2 - Tieback Row "A" Installed
—+—	22	8-12-88	Excavation Level 3 - Tieback Row "B" Installed
—×—	35	9-14-88	Excavation Level 4 - Tieback Row "C" Installed
—▲—	48	10-12-88	Excavation Level 5 - Tieback Row "D" Installed
——	60	11-17-88	Excavation Level 6 - Final Depth of Excavation
— —	77	6-21-89	Internal Structure Completed

Figure 8. Wall Deflections at Critical Construction Stages
University Street Station

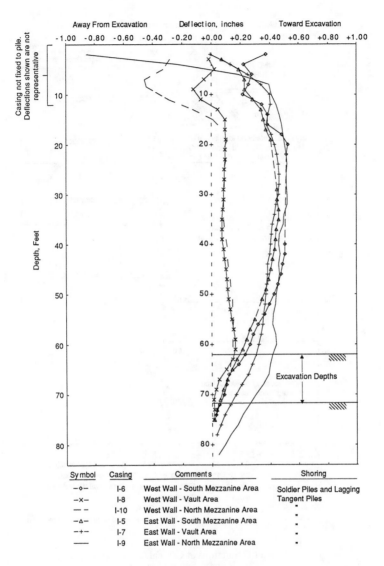

Figure 9. Maximum Wall Deflections at Each Inclinometer Casing
Pioneer Square Station

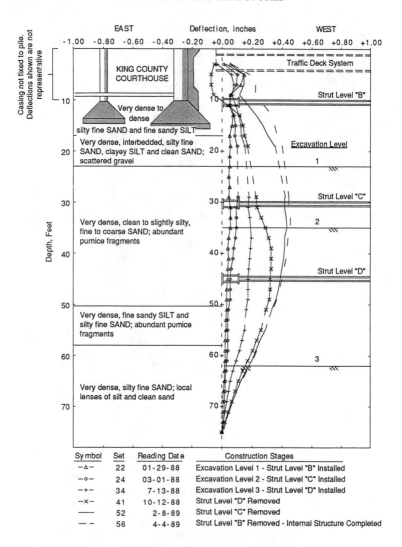

Figure 10. Wall Deflections at Critical Construction Stages
Pioneer Square Station

Casing #	Location	Shoring Wall	Bracing System	Soil Conditions	Design Strut Stress ksi	Measured Strut Stress ksi *	Max. Lateral Deflection inches	Adjacent Settlement inches
PI-3	Pine Street Line	Soldier Piles & Lagging	Struts	Very dense (v.d.) Till to 50'; hard CLAY/SILT below	11.0	NI	0.19**	0.18
PI-5	Pine Street Line	Soldier Piles & Lagging	Struts	Fill to 20'; very stiff to hard CLAY/SILT to 30'; v.d. SAND to 40'; very stiff to hard CLAY/SILT below	11.0	NI	-0.2, 0.09**	0.10
PI-6	Pine Street Line	Soldier Piles & Lagging	Struts	Fill to 20'; very stiff to hard CLAY/SILT to 30'; v.d. SAND to 40'; very stiff to hard CLAY/SILT below	11.0	NI	0.19**	0.12
I-5	Pioneer Square Station	Soldier Piles & Lagging	Struts	v.d. SAND to 51'; v.d. silty fine SAND and fine sandy SILT below	12.0 12.4	B = 2.5 C = 5.0	0.46	0.29
I-6	Pioneer Square Station	Soldier Piles & Lagging	Struts	v.d. SAND/SILT to 24'; v.d. fine to coarse SAND below, occasional hard SILT	12.0 11.4	B = 2.2 C = 8.2	0.53	0.19
I-7	Pioneer Square Station	Soldier Piles & Lagging	Struts	hard clayey SILT to 21'; v.d. fine SAND to 29'; hard SILT to 53'; v.d. fine SAND/SILT below	12.0	NI	0.47	0.41
I-8	Pioneer Square Station	Soldier Piles & Lagging	Struts	v.d. SAND to 29'; hard CLAY to 37'; v.d. TILL to 42'; v.d. fine SAND/SILT below	12.0	NI	0.16	0.26
I-9	Pioneer Square Station	Soldier Piles & Lagging	Struts	v.d. TILL/OUTWASH to 28'; v.d. fine SAND and SILT below	12.0 12.3	B = 5.5 C = 5.9	0.53	N/A
I-10	Pioneer Square Station	Soldier Piles & Lagging	Struts	v.d. TILL to 17'; v.d. fine SAND/SILT to 45'; hard CLAY to 50'; v.d. TILL to 65'; v.d. fine SAND below	12.0 12.3	B = 5.5 C = 5.9	0.15	0.20
I-11	University Street Station	36" Dia. Tangent Piles	Tiebacks	v.d. SAND/SILT to 27'; v.d. TILL to 47'; v.d. SAND to 61'; v.d. fine SAND/SILT to 77'; hard CLAY below	5.8	NI	0.22	0.03
I-12	University Street Station	36" Dia. Tangent Piles	Tiebacks	v.d. SAND to 41'; v.d. TILL to 48'; hard CLAY/SILT below	5.8	NI	0.23	0.35
I-13	University Street Station	36" Dia. Tangent Piles	Tiebacks	v.d. SAND to 41'; v.d. TILL to 48'; hard CLAY/SILT below	5.8	B = 3.2	0.22	0.50
I-15	University Street Station	Soldier Piles & Lagging	Tiebacks Struts	v.d. SAND to 31'; v.d. TILL to 49'; hard SILT to 56'; hard CLAY below	5.8	B = 4.5	0.50	0.07

* 'B' and 'C' indicate the strut level, 'A' level being the street, which was not instrumented.
** Inclinometer readings were discontinued on casings PI-3, PI-5 and PI-6 after the excavation was completed and the floor was cast, but before the struts were removed.
NI Indicates struts were not instrumented.
N/A Indicates that instruments were not accessible for monitoring
   Inclinometers were installed within tangent piles and soldier piles.

Table 1. Summary of Shoring Wall Performance

SUMMARY OF OBSERVED PERFORMANCE:

Results of the instrumentation program are illustrated in Figures 5 through 10. Westlake Station wall movements are summarized in Figure 5. University Street Station wall movements are summarized in Figures 6, 7 and 8. Wall movements at Pioneer Square Station are summarized in Figures 9 and 10. Table 1 summarizes, at each location, the types of shoring wall and bracing systems, predominant soil conditions, strut stress, maximum wall deflections and adjacent building settlements.

At the Pine Street line structure, the wall movements were generally less than one-quarter inch even though a more flexible soldier pile and lagging system was used.

The Westlake Station tangent pile wall movements were also less than one-quarter inch. Settlements of adjacent buildings were limited to one-eighth inch.

The rigid tangent pile shoring wall in the southern portion of University Street Station typically moved less than one-quarter inch. The more flexible soldier pile wall movements averaged one-half inch. Building settlements adjacent to the rigid wall were between one-eighth inch and one-half inch.

The movements of the rigid tangent pile walls at Pioneer Square Station varied between one-eighth inch and one-half inch. Adjacent settlements were generally between one-quarter and three-eighths inch.

CONCLUSIONS:

The Pine Street soldier pile wall moved less than anticipated during design. This probably is due to the fact that the excavation was generally in till during one of the driest summers on record.

The Westlake Station tangent pile walls performed as anticipated while the adjacent building settlements remained well within a tolerable range.

Both the flexible and rigid walls at University Street Station performed essentially as predicted. Settlement of the Seattle Tower varied between one-eighth inch and one-half inch while settlement of the PT&T Building was virtually eliminated during station construction.

The majority of the tangent pile walls at Pioneer Square Station, while performing fairly well, did not perform as well as anticipated. However, although some wall movements approached one-half inch, adjacent building settlements did not exceed one-quarter inch.

APPENDIX. UNITS

To convert	To	Multiply by
in	mm	25.4
ft	m	0.305
miles	Km	1.609
kips	kN	4.45
lbs/sq ft	Pa	47.9
kips/sq ft	kPa	47.9

ANCHORED EARTH RETAINING STRUCTURES
-DESIGN PROCEDURES AND CASE STUDIES

Seth L. Pearlman, P.E., M. ASCE[1]
John R. Wolosick, P.E., A.M. ASCE[2]

ABSTRACT: Current design practice for anchored earth retaining structures (AERS) is reviewed. This practice may vary due to personal experience or regional influences. A soil/structure interaction design methodology consisting of a beam on elastic foundation analysis is proposed to structurally design and model AERS. Two case studies are presented to demonstrate the applicability of the proposed method. The good correlation of predicted and actual measured deflection is discussed. Recommendations for use and possible future refinement of the beam on elastic foundation method are suggested.

INTRODUCTION

Permanently anchored (tied-back) earth retaining structures (AERS) have been constructed in the United States since at least 1969[1]. The principal structural elements are commonly driven or pre-drilled steel soldier piles, reinforced concrete soldier piles, concrete diaphragm walls (slurry walls) or reinforced concrete slabs. Ancillary items such as lagging, facing, wales or horizontal distribution reinforcement completes these structures. Several design procedures for AERS have been published in the literature [2, 3, 10, 11, 14, 15]. The geotechnical and structural assumptions that are employed may vary significantly due to the designers' personal experience, regional or geographic considerations. AERS are designed globally to prevent failure of the supported soil mass, and structurally to resist assumed earth pressure distributions. The sophistication of designs for these structures has increased markedly in recent years. Many of the procedures currently used by engineers to design AERS have not been widely published, since they have often been advanced by specialty contractors to gain a competitive advantage.

Data from instrumented AERS projects suggest that the performance of these structures during and after construction does not always correlate well with that predicted during design. This disparity

---

1 - Chief Design Engineer, Nicholson Construction of America, PO Box 308, Bridgeville, Pa. 15017
2 - Chief Engineer, Nicholson Construction, Inc., 4070 Nine/McFarland Dr., Alpharetta, GA 30201

suggests that more refined design methods are necessary. A methodology is proposed for the structural design of AERS using a soil/structure interaction approach. The method utilizes a convenient computer program, which facilitates the practical application of beam on elastic foundation (BEF) analyses to more accurately model the performance of these structures. The proposed method is appropriate when a soil secant modulus for the range of anticipated lateral stresses can be determined from test data. The authors' experience indicates that the model is appropriate up to soil strain levels of about 0.007 in/in (0.007 mm/mm). Deflected retaining wall profiles from two case histories are compared to the shapes predicted by the beam on elastic foundation model to demonstrate the applicability of the method. The analytical method is shown to correlate well with the case histories.

STATE OF PRACTICE OF DESIGN

AERS are designed for overall stability of the retaining wall system and for internal stability of the structural elements. For the purpose of this paper, overall and internal stability design requirements are referred to as global and structural design requirements, respectively. A proper design must satisfy both criteria, and these criteria may interact in the design process. In the authors' experience, AERS are typically designed according to the following general procedures:

1) A geotechnical investigation is performed to determine the nature and strength of the materials that are to be supported.

2) An appropriate earth pressure distribution (Rankine, Coulomb, or empirical apparent earth pressure) is selected. The choice of the most appropriate distribution has been reviewed in various papers and design manuals [7, 16, 17]. The publications are often augmented by the designer's experience and the specific site constraints, such as sensitivity of the supported surrounding area to movements.

3) Horizontal reactions are determined at the anchor locations and at the embedment points to resist the selected earth pressure.

4) Moments and shears for the vertical wall elements are computed. Most designers use a continuous beam analysis (stiffness method) but some assume pins (i.e., zero moments) at all anchor supports except the top. The continuous beam method is recommended by the authors.

5) Global stability is checked using limit equilibrium slope stability methods, graphical methods or a simple wedge method [13, 14, 15, 16]. In each case, the intent is to provide a system where the anchor bond zones and/or the structural embedded elements are located beyond possible failure planes.

6) If the analysis in procedure 5 above indicates that additional support forces are required to satisfy global stability requirements, anchor forces or vertical element embedments are increased, and the structural design is modified accordingly.

Geotechnical Appraisal. All AERS require a geotechnical appraisal of the site which should include: 1) borings a maximum of every 100 feet along the wall; 2) borings behind the wall to identify anchor bond zone materials; 3) standard penetration testing in soils; 4) tests to determine soil strength and modulus parameters, such as triaxial shear, direct shear, consolidation, dilatometer and Menard pressuremeter; and 5) cores in rock where appropriate with tests to determine the shear strength of the rock.

Unfortunately, these procedures are not always implemented, which may lead to the inaccurate assumption of geotechnical parameters. The anchor forces are derived from the assumed earth pressure distributions. If the selected soil shear strength parameters (angle of shearing resistance and cohesion) are much lower than actual, then higher than necessary design anchor forces will result. These forces may also produce inward deflections of the wall face with time and reduction of the anchor load due to shortening of the anchor free length. Several authors have reported net inward wall deflections of AERS[6, 8, 9]. Conversely, if the selected soil parameters are higher than actual, or if low anchor lock-off loads are employed, outward deflections of the wall will occur with time. Larger than predicted anchor loads will result with any post construction outward movement of the wall. Other factors such as vertical distance between the anchors, structural wall stiffness, and soil response when laterally relieved may produce outward wall movements.

For the components of an AERS to work in harmony, all design constraints and parameters must be selected reasonably, i.e., conservative assumptions may prove to be as detrimental to the process as non-conservative assumptions. The choice of geotechnical parameters has the greatest effect on the design.

Often, owners or engineers place too much emphasis on the structural requirements and too little on proper determination of appropriate soil parameters. This is due to the relative comfort level engineers have with deterministic structural analyses and the discomfort level associated with the accurate prediction of geotechnical parameters. These soil and rock parameters (e.g. shear strength) affect the global and local design constraints, and the resulting design can have performance or design inconsistencies if the parameters are not properly chosen. For example, Pearlman et.al.[9] discussed a case history for an anchored bridge abutment in a known slide area, with a residual soil shear strength developed from a direct shear test. The resulting anchor loads from a STABL4 analysis forced the structure back into the soil. The resulting bending moments in the wall were not the same as those predicted by design.

Global Analysis. All AERS should be checked for global stability to evaluate the potential for deep failure surfaces outside the anchor and soldier pile system. Often, engineers ignore this check due to high factors of safety obtained on previous work in similar settings.

Other global stability checks include considering the anchored zone of soil as a contiguous mass, and determining the factor of safety based upon resistance against sliding and overturning. Graphical methods may require anchor bond zone locations to be founded outside potentially unstable wedges of supported soil[16]. However, this type of analysis does not dictate the anchor load required to satisfy structural equilibrium or to provide a desired global safety factor.

Limit equilibrium analyses are often used to check global stability [19,10]. The program STABL4/5[14,15] developed at Purdue University is a general purpose slope stability limit equilibrium program that computes factors of safety for failure surfaces of any shape. The program allows the addition of ground anchors into the computation of the factor of safety for the system. This method has been employed by the Pennsylvania Department of Transportation[13] and gives reasonable results if certain cautions are exercised[22]. Specifically, the program interprets the anchor load as a negative driving force rather than a positive resisting force, thereby influencing the safety factor calculation. Factors of safety may also be unreliable for steep failure surfaces.[23]

Structural Analysis. Structural design of AERS consists of sizing the elements that transfer the anchor reactions at the face of the retaining wall into the soil. These elements generally are chosen by the designer to resist empirical apparent earth pressure distributions. The structures are often analyzed as continuous beams with certain end conditions and with the anchors acting as supports. The anchor loads generated through this type of analysis are applied to the structure during construction when the anchors are locked off. The most common approach is to lock off the anchors at 100 percent of the computed design loads. However, some specifiers prefer lock-off loads other than 100 percent. The resulting designs can be classified into one of the following general categories depending on the accuracy of both the chosen geotechnical parameters and the empirical apparent earth pressure diagram:

1) Design anchor reactions are less than the actual soil pressures resulting in a net outward movement of the retained earth mass.

2) Design anchor reactions are exactly equal to the actual soil pressures producing an equilibrium condition with almost zero net wall movement.

3) Design anchor reactions are greater than the actual soil pressures, but equal to the reactions gained when analyzing the wall to resist apparent earth pressure diagrams. This case illustrates the application of over-conservative geotechnical parameters and/or the inherent conservatism that is built into the recognized apparent earth pressure diagrams. This case results in an inward movement of the wall during anchor stressing.

4) Design anchor reactions are greater than those dictated by the appropriate apparent earth pressure distributions. In this case, the anchor loads may be increased due to other factors such as anticipated deep seated instability or anticipated future surcharges. Thus, the anchors are designed to resist forces in addition to the apparent earth pressures, and are locked-off at loads greater than indicated by apparent pressure diagrams. This case generally results in a significant inward movement of the wall during anchor stressing[9], unless special structural details are included to resist these anchor loads.

Category 1 may be considered unconservative since most designs require that the anchors be minimally sized to resist the appropriate apparent earth pressure distribution. This type of design is

sometimes practiced in the construction of temporary shoring
structures where large wall movements are allowed, and the risk
associated with the performance of this type of structure during its
short service life is evaluated. This category may also arise when
the Rankine triangular earth pressure distribution is used in the wall
design[16].

Category 2 is an ideal design solution where a balance between
applied anchor load and earth pressure is obtained. AERS designs
generated for this category theoretically include correctly sized wall
elements since the applied anchor loads would be equal to the earth
pressures.

Category 3 may be considered conservative by some, but in reality
could result in unconservative designs if the anchors are loaded to
such a high degree that the resulting deflected shape and moment
diagram are completely different than assumed during design. This
tendency to require greater than required anchor loads is recognized
and discouraged by Trow[8].

Category 4 results from the use of AERS for reasons other than
simple earth retention. AERS are a recognized method for landslide
repair. In this special case, the structural elements are oversized
to transmit the required landslide resisting forces into the ground.
Primary to the structural design in these cases is recognition that
the ground is acting as a structural reaction element to the
anchor/structure system.

Other Structural Design Considerations. Typical structural design
for AERS utilizes continuous beam analysis with supports at the
anchors and at the base of the excavation. The modelling of the end
conditions of the vertical elements is more complicated. There are
several design assumptions available in choosing the appropriate end
condition at the base of the excavation. These include:

1) Free end: when the vertical element is not able to penetrate
a layer capable of lateral support. For example, a driven pile
stopped at a hard rock with no penetration at a location where the
excavation is planned to extend to or below the rock line.

2) Fixed end: when the vertical element is drilled or driven
deeply into very stiff materials, and near or complete fixity is
assumed.

3) Pinned support: the embedded portion of the vertical element
is located in soil or rock and a hinge or point of zero moment is
assumed to develop (Figure 1).

4) Structural springs: the embedded portion of the vertical
element is located in soil or rock and supported laterally by a series
of spring elements. The use of this assumption is limited to soil and
rock that exhibit elastic response at low strain levels (i.e. up to
about 0.007 in/in (0.007 mm/mm)) which includes many soils and rocks
but is not appropriate for soft clay (Figure 1).

The fixed end assumption should be rarely used when designing AERS
since no condition is ever truly fixed. A preferable method of
modelling a high degree of structural fixity is to use stiff spring
elements as the end condition for the beam. The pinned condition is
probably the most commonly used design assumption. This method was
initiated and advanced because it facilitated hand calculations before

computers were readily available for design[3].   Many AERS have been
successfully designed and constructed using this assumption.

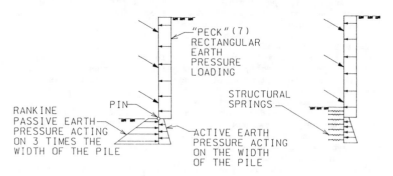

FIGURE 1

SCHEMATICS OF THE PINNED AND SPRING END CONDITIONS

The structural spring assumption was suggested by Lambe in 1970[5],
but practical application of these models effectively began with the
widespread use of personal computers in design.   The use of a spring
model in the design of AERS requires the development of a stiffness
based computer solution.   The spring model can be used to accurately
model the end condition of the vertical wall element at low strain
levels in elastic soils.   The spring model can be used to verify
Teng's[3] pinned condition assumption by identifying the point at which
a pin (zero moment) occurs.   The spring model can also indicate to the
designer whether sufficient embedment has been provided to confirm the
fixed end assumption.

SOIL/STRUCTURE INTERACTION APPROACH

As a further extension to the structural spring model for the
embedded portion of the vertical element, a soil/structure interaction
computer program based on beam on elastic foundation (BEF) theory has
been developed by Nicholson Construction[21].   The program automatically
inputs springs on 1 foot centers along the entire vertical wall
element as shown schematically in Figure 2.

Program inputs include the stiffnesses of the springs which are
input as an elastic secant modulus, tributary width of the vertical
element, and allowable soil capacity to avoid overloading of the
spring elements.   Pile stiffness is also input.   Shear, moment, and
deflected structure diagrams are generated.   Spring forces are
calculated and compared with passive soil capacities.   Any springs
that exhibit tension above grade are set to zero.   The program offers
a practical design tool that does not require long set-up times to
generate the model.

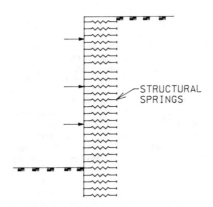

FIGURE 2
SCHEMATIC OF SOIL/STRUCTURE INTERACTION MODEL

As with any geotechnical analysis, the input of reasonable soil properties is important to the successful application of this program. The elastic secant modulus data can be derived from stress-strain curves generated from Menard pressuremeter or dilatometer data. These data can be particularly useful, since the soils are tested laterally, which is the direction most appropriate for AERS. Triaxial shear and consolidation test data can also yield stress-strain curves, and secant moduli can then be determined. Judgement and experience must be applied to these data to adjust for the possible effects of anisotropy. The moduli can be calculated within the range of expected stress levels that will be encountered by the structure. When these types of field or laboratory tests are not available, published correlations based on standard penetration testing may be used to determine approximate soil secant moduli[12,18]. Experience has shown that the model is appropriate up to soil strain levels of about 0.007 in/in (0.007 mm/mm).

The relative stiffness of the vertical wall elements and the soil directly in contact with these elements has an effect on the performance of the system. A very stiff and unyielding vertical wall element will spread the load more evenly than a very flexible vertical wall element, which produces more deflection in the vicinity of the anchor. Very stiff vertical wall elements can be designed for much higher bending moments than flexible wall elements. The designer must recognize that the total system stiffness is a function of the structural element stiffness, the soil elastic secant modulus and the distances between anchors.

In the following sections, two case histories of AERS are presented and the performance of the retaining walls is compared to that predicted by the soil/structure interaction approach.

BAD CREEK PUMPED STORAGE PROJECT

Project Background. The Bad Creek pumped storage facility is currently being constructed near Salem, South Carolina. The owner, overall designer and construction manager of the project is Duke Power Company. The facility is located adjacent to Lake Jocassee, which acts as the lower reservoir for the pumping scheme. The project also includes a very large underground powerhouse, a large upper reservoir with associated dams and dikes, a large vertical shaft, and several tunnels. Construction of the intake structure at Lake Jocassee required the construction of several permanent and temporary AERS. These retaining walls were bid as design/build packages. Nicholson Construction designed and built four of the walls during 1988.

The soil and rock at the retaining wall locations consisted of residual silty sands and biotite gneisses typical of the Piedmont Physiographic Province. The soils at the case study wall location exhibited very high relative densities and were classified as partially weathered rock. In general, standard penetration resistances at this location were on the order of 50 blows per 6 inches (150 mm) of penetration.

Wall Design. The AERS consisted of soldier piles and wood lagging. The soldier piles consisted of double wide-flange steel beams that were spaced 8 ft (2.44 m) center to center (Figure 3). The piles were designed as continuous beams resisting Peck[7] apparent earth pressure distributions for cohesionless soils. The angle of shearing resistance was specified at 34 degrees by Duke Power company.

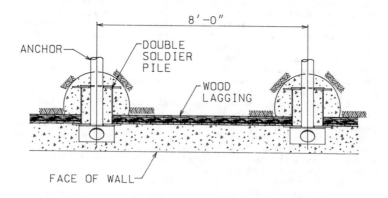

FIGURE 3
SCHEMATIC PLAN VIEW OF BAD CREEK
RETAINING WALLS
(1 ft = 0.3048 m)

Design excavation depths for the Intake/Portal structure were about 95 ft (29.0 m). This depth was well into the bedrock, down to the invert level of the power tunnels. The AERS were designed to retain the soil portions of the excavation, down to the top of rock. Due to anticipated variations in the depth to rock shown by the soil borings, wall design heights varying from 15 to 50 ft (4.6 to 15.2 m) were generated. The soldier piles were socketed 1 ft (0.3 m) into the rock. A 10 ft to 20 ft (3 m to 6 m) wide bench was provided at the top of rock to protect the rock sockets from the blasting operations performed to extend the excavation to final depth. The final invert of the excavation varied from 5 ft (1.5 m) to 75 ft (22.9 m) below top of rock. Even though the soils were very dense, the design heights at some sections of the wall required anchor loads as high as 370 kips (1646 kN).

Many of the retaining wall sections were designed using both the stiffness method and the BEF analysis for comparison purposes. The design moments from the two analyses were comparable for negative moments. However, the positive moments were significantly smaller with the BEF analysis since the method allows pressure redistribution in the soil behind the soldier piles and anchors (Figure 4). The constructed wall section shown on Figure 4 was designed using the BEF analysis since it was considered more representative of field conditions. The stiffness method assumes no relative movement between the soil and the structure, and therefore cannot account for the pressure redistribution and resulting bending moment reduction in the soldier piles.

The soil and rock stiffnesses for the BEF model were chosen based on calculations from triaxial shear and consolidation testing in similar materials, and from the results of Menard pressuremeter data from other sites within the Piedmont Physiographic Province. These data were correlated based on standard penetration resistances obtained adjacent to undisturbed samples and pressuremeter tests in boreholes.

Wall Performance. Instrumentation of the walls located in residual materials consisted of two inclinometers attached to soldier piles numbered 32 and 23. The inclinometer casings were installed in steel pipes that were welded to the soldier piles. The plastic inclinometer casings were installed to depths of about 6 ft (2 m) below the bottom of the piles by drilling through plugs at the bottoms of the steel pipes after the soldier piles were placed into the ground. The inclinometer data showed that the soldier piles generally deflected into the soil during construction. This type of performance indicated that the earth pressures used in the design were conservative.

The actual deflected profile of Pile 32 from the inclinometer data is compared on Figure 4 with the deflected profiles generated during design using the stiffness method and the beam on elastic foundation analysis. The measured deflections were very small, with calculated elastic soil strains of about 0.002 to 0.003, which indicated that the BEF analysis was appropriate for this case. The predicted deflections from the stiffness method were not at all similar to the measured deflections. The actual deflections compare favorably with those predicted by the beam on elastic foundation analysis.

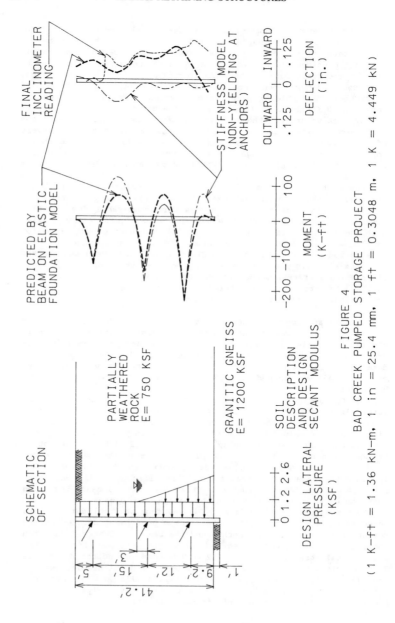

FIGURE 4
BAD CREEK PUMPED STORAGE PROJECT

(1 K-ft = 1.36 kN-m, 1 in = 25.4 mm, 1 ft = 0.3048 m, 1 K = 4.449 kN)

Pile 23 was designed and constructed using the stiffness method. At the end of construction, a calculation of the deflected shape of the soldier pile was made using the BEF analysis. Although the overall deflected shape measured by the inclinometer was similar to the calculated profile, the actual deflections (maximum = 0.4 in (10 mm)) were greater by about a factor of five. It is theorized that the soils adjacent to this pile were not as dense as indicated by the closest soil boring. Interestingly, the maximum bending moments computed with the BEF analysis were only 9 percent less than generated by the stiffness method. The results from this case are not presented in a figure.

BRADY'S RUN PARK DIAPHRAGM WALL

Project Background. A two-lane roadway in Brady's Run Park, Beaver County, Pennsylvania is located parallel to and 40 ft (12 m) above a man-made lake. The roadway had shown evidence of underlying slope instability for many years prior to a major slide in the fall of 1987. The slide was most likely triggered by a major lake maintenance project that consisted of drawdown of the lake and dredging. In September of 1987, the roadway owner (Pennsylvania Department of Transportation) let an emergency contract for the construction of an AERS to restore the damaged roadway. The contract documents included a design that consisted of predrilled steel soldier piles, temporary wood lagging and permanent concrete lagging. A provision for a contractor's design alternate was included. For economic and site-specific construction reasons, Nicholson Construction proposed and constructed an alternate design consisting of about 17,000 sq. ft. (1580 sq. m) of 30 in (762 mm) thick reinforced concrete diaphragm wall (structural slurry wall) with an additional cast-in-place top section of about 5,000 sq. ft. (465 sq. m). The alternate design cost approximately one-half million dollars less than the owner's design. The construction of the wall was complete by the summer of 1988.

It was determined by a geotechnical investigation, including the installation of inclinometers in the failure zone, that the slide was occurring at the interface of shaley colluvial soils located approximately 30 to 45 ft (9 to 14 m) below the surface with a zone of weathered shale whose surface sloped towards the lake.

Wall Design. The owner specified a dual design criteria to determine the required anchor forces for the wall. One criterion was to apply a Peck[7] rectangular apparent earth pressure (Figure 5) using an active earth pressure coefficient of 0.577 and a soil density of 125 pcf (19.6 kN/m^3). The second criterion was to analyze the system globally using the limit equilibrium slope stability program STABL4[14, 15] at each stage of construction.

The anchor loads used in the design were governed by the rectangular apparent earth pressure diagram. The owner also required a change in grade for the final roadway configuration, raising the road up to 10 ft (3 m). To attain this, an 18 in (450 mm) section of cast-in-place wall was extended up from the diaphragm wall and backfilled. During design, a beam on elastic foundation analysis was used to generate shear and moment distributions for the structural design of

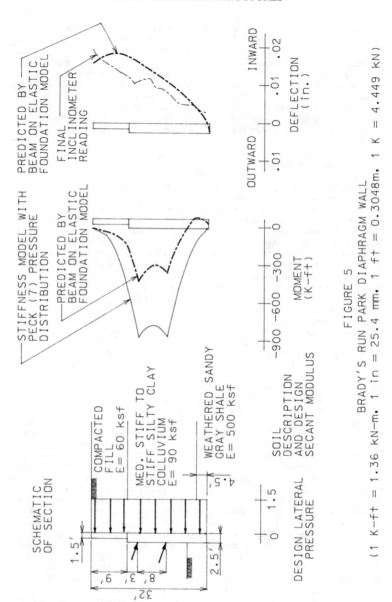

FIGURE 5

BRADY'S RUN PARK DIAPHRAGM WALL

(1 K-ft = 1.36 kN-m, 1 in = 25.4 mm, 1 ft = 0.3048m, 1 K = 4.449 kN)

the diaphragm wall and to predict the lateral deflection of the wall (Figure 5). Analysis of the structure using the Peck[7] rectangular apparent earth pressure distribution was considered, but deemed too conservative for this purpose due to the flexibility of the upper section of wall. Figure 5 illustrates a 32 ft. (9.8 m) high section of the structure and the bending moments obtained from a hypothetical design using the Peck[7] rectangular apparent earth pressure diagram.

Wall Performance. Instrumentation of the diaphragm wall consisted of inclinometers installed within the wall at various locations along the structure. Generally, the inclinometer data showed that the diaphragm wall deflected into the soil during construction due to the stressing of the anchors. This type of performance indicated that the earth pressures used in the design were conservative. One inclinometer was placed in a panel representative of a design case. The actual deflection of this wall section as measured by the inclinometer is shown in Figure 5. Predicted deflections[22] from the BEF analysis and the actual wall deflections are compared in Figure 5. The soil and rock stiffnesses for the model were chosen based on data published by Terzaghi[12]. These data were correlated based on standard penetration resistances. The measured deflections were very small, which indicated that the BEF analysis was appropriate for this case. The deflected profile shows good agreement with the actual deflections and the BEF analysis. The calculated strain was on the order of 0.002 in/in (0.002 mm/mm). The results from this project show that wall design utilizing an empirical apparent earth pressure distribution acting over the entire wall height would have been overconservative.

RECOMMENDATIONS FOR FUTURE STUDY

The BEF analysis provides a practical tool for the design of the structural elements in an AERS. The computer program developed can be effectively utilized for designs on projects with a wide variety of soil conditions. The method should be verified with more case studies. Some situations may require more refined soil modeling techniques. Soils with non-linear response or cases with high soil strain levels might best be modelled as elasto-plastic or piecewise linear.

The key to the design program described is the relative ease and short set-up time for the generation of the model and the analysis. Analysis times on personal computers are on the order of a few seconds. The inclusion of refined models will lengthen input times and significantly increase the calculation times. The modification of the beam on elastic foundation analysis to include more refined soil models should only be applied where known soil conditions or further research indicates it to be necessary.

SUMMARY AND CONCLUSIONS

The state of practice for design of anchored earth retaining structures (AERS) in the authors' experience has been reviewed. A soil/structure interaction method consisting of a beam on elastic foundation (BEF) analysis is proposed. Two case histories designed using this method are presented to demonstrate its viability. Design predictions using the BEF analyses correlate well with the actual performance of the case histories. Hence, the following conclusions are offered:

o The method is practical, with fast input and calculation times.

o Immediate turnaround of parameter studies allow for rapid convergence on optimized designs.

o Results from the analysis method correlate well with observed performance of completed AERS indicating that the analysis method is realistic.

o For the cases shown, analysis using the BEF method resulted in more economical but sufficiently conservative designs.

o It is recommended that the proposed BEF design method be applied to the structural design of AERS.

o It is recommended that additional AERS projects are instrumented to further confirm the applicability of the BEF design method.

REFERENCES

1. Nicholson, P.J., "Rock Anchor System for Securing Failing Wall," Transportation Engineering Journal, ASCE, March, 1979, pp. 199-209.
2. Clough, G.W., D'Appolonia, E., et al., Workshop #5, Proc. of the Conf. on Analysis and Design in Geotechnical Engineering, June 9-12, 1974, U. of Texas, Austin, Tex., pp 100-102.
3. Teng, W.C., Foundation Design, 1972, Prentice-Hall, Inc.
4. Clough, G.W., Weber, P.R. & Lamont, J., "Design and Observation of a Tied-Back Wall," Proc. of the Spec. Conf. on Performance of Earth and Earth-Supported Structures, June 11-14, 1972, Purdue U., Lafayette, In., pp. 1367-1389.
5. Lambe, T.W., "Braced Excavations," 1970 Spec. Conf., Lateral Stresses in the Ground & Design of Earth Retaining Structures, June 22-24, 1970, Cornell U., Ithaca, N.Y., pp 149-217.
6. Gould, J.P., "Lateral Pressures on Rigid Permanent Strucs.," 1970 Spe. Conf., Lateral Stresses in the Ground & Design of Earth Retaining Structures, June, 1970, Cornell U., Ithaca, N.Y., pp 219-269.
7. Peck, R.B., "Deep Excavations & Tunneling in Soft Ground," Seventh International Conference on Soil Mechanics and Foundation Engineering, Mexico, 1969, pp 225-290.
8. Trow, W.A., "Experiences With Shored Excavations," Canadian Geotechnical Journal, Vol. 24, January, 1987, pp. 267-278.
9. Pearlman, S.L., et al., "Instrumenting a Permanently Tied-Back Bridge Abutment-Planning Installation, and Performance,"

Proceedings of the 5th International Bridge Conference, Pittsburgh, Pa., June 13-15, 1988, pp 40-50.

10. Peck, R.B., Hanson, W.H., Thornburn, T.L., Foundation Engineering, 1974, Wiley.

11. Cheney, R.S., "Permanent Ground Anchors", Report No. FHWA-DP-68-1R, November 1984, FHWA, Washington, D.C.

12. Terzaghi, K., "Evaluation of Coefficients of Subgrade Reaction", Geotechnique, Vol. V, Dec., 1956.

13. Boghrat, A., "Use of STABL Program in Tied-Back Wall Design", J. of Geotechnical Engineering, ASCE, April, 1989, pp. 546-552.

14. Carpenter, J.R., "Slope Stability Analysis Considering Tiebacks and Other Concentrated Loads", Joint Highway Research Project No. 86-21, School of Civil Engineering, Purdue U., W. Lafayette, Ind.

15. Achilleos, E., "User Guide for PC STABL 5M", Joint Highway Research Project No. 88/19, School of Civil Engineering, Purdue U. W. Lafayette, Ind.

16. Canadian Foundation Engineering Manual, 2nd Ed., Canadian Geotechnical Society, 1985.

17. Winterkorn, H.F., Fang, H.Y., Foundation Engineering Handbook, Van-Nostrand Reinhold, 1975.

18. Bowles, J.E., Foundation Analysis and Design, 4th Ed., McGraw Hill, 1988.

19. Morgenstern, N.R., "The Analysis of Wall Supports to Stabilize Slopes", Proceedings of ASCE Conven. April 29, 1982, Application of Walls to Landslide Control Problems, pp. 19-29.

20. Sangrey, D.A., "Evaluation of Landslide Properties", Proceedings of ASCE Conven. April 29, 1982, Application of Walls to Landslide Control Problems, pp. 30-43.

21. Hetenyi, M., Beams on Elastic Foundation, McGraw Hill, 1960.

22. Gannett Fleming Transportation Engineers, Calculations for Beaver County S.R. 4012 S-17159, Sept. 25, 1987.

23. Correspondence, Gannett Fleming Corddry & Carpenter Inc. to District 11-0 Engineer, Pennsylvania Department of Transportation, October 16, 1986.

SLURRY WALL DESIGN AND CONSTRUCTION

George J. Tamaro , F.ASCE[1]

ABSTRACT: This paper is concerned with design and construction developments in structural slurry wall (diaphragm wall) construction within the United States during the period 1970 to 1990. Major improvements have been made in the method of analysis of the wall/support system, in the development of wall reinforcing methods and in the development of better tools for the execution of the work. However, construction tools and methods used effectively elsewhere in the world remain either untried or little used here.

INTRODUCTION

The first structural slurry wall was installed in New York City in 1962, more than a decade after the method's first use in Europe. The work was of small scale, a 7 meter diameter by 24 meter deep shaft, and went almost unnoticed. Other works followed in Boston[12] and San Francisco[1,6] with the World Trade Center slurry wall, constructed in New York in 1967[10], putting all on notice that slurry wall construction was here to stay.

Early slurry walls were installed using lightweight, rudimentary tripod rigs mounted on rails and operating cable hung clamshell buckets. Obstructions were removed by clamshell bucket, lightweight chisels and on occasion by dynamite! Rock was removed by chisel and on occasion by special cutting devices.

SLURRY MATERIALS

Slurry materials have remained essentially unchanged over the last two decades, with slurries continuing to be blended from powdered bentonite and water. Higher yield, treated bentonites have been developed but offer nominal advantage, except for the reduced cost of transportation over great distance. The main disadvantage of treated bentonites lies in the sensitivity of these materials to ground water contaminants, particularly organics (broken sewers) and chlorides (sea water). The use of additives in the blend remains limited, except for a few alchemists who concentrate on the slurry and ignore other aspects of the work. Major improvements have occurred in the testing, cleaning and reuse of the slurry. Contrac

---

1- Partner, Mueser Rutledge Consulting Engineers, 708 Third Avenue, New York, NY 10017

tors and engineers provide field testing facilities to test and
verify bentonite quality. Desanding devices, consisting fine
screens and cyclones, are capable of removing sufficient suspended
soil to permit two or more reuses of the slurry prior to disposal.
Offsite disposal remains an unsolved problem, with greater and
greater resistance to acceptance of the material at landfills or
"dumps".

EXCAVATION TOOLS AND THEIR PERFORMANCE CHARACTERISTICS

Soil excavating buckets have improved in weight, endurance and
digging characteristics, with some buckets being used effectively as
chisels for removal of boulders, hardpan and soft rock. Buckets are
raised and lowered by cable or kelly bar and opened and closed by
cable or hydraulics. The crane hung, cable operated clamshell
bucket, weighing in excess of 10 tons, is most used by U.S. contractors due to its versatility, relatively low operating cost and
effectiveness in almost all U.S. geologic environments.

A Japanese manufacturer designed a drilling tool consisting of a
line of vertical rotary drills which dislodged the soil. The drill
spoil was then removed by suction from the bottom of the bentonite
filled trench. Sand separators were used to remove the soil from
the bentonite. The tool was not capable of removing hard obstructions, boulders or rock. The tool saw limited use in the '70's and
is not currently in use by any contractor in the U.S.

French and Italian equipment manufacturers are currently producing a "hydromill", a grinding device consisting of two milling heads
mounted parallel to each other and rotating on an axis perpendicular
to the trench. The milling heads rotate in opposite directions and
can dislodge soil and soft rock from the bottom of the trench. The
drill spoil is lifted by suction and removed from the bentonite by a
series of sand separators. The machine was first used in Baltimore
on a structural wall application, and is reported to produce a
better finished wall to tighter tolerances in uniform soils, but at
a higher cost than a clamshell bucket operation.

Rock drilling tools range from heavy drop chisels to track
mounted chisel drills to large diameter roller bits driven by oil
well drilling equipment and, as a last resort, to drilling and
blasting. Occasionally auger rigs are used to predrill hardpan or
soft rock in order to facilitate clamshell advancement. An attempt
to slice rock with a hydraulic ram driven steel blade went un-
rewarded at the World Trade Center project, discouraging further
attempts at innovation in that direction. The basic U.S. rock
removal tool remains the drop chisel, weighing in excess of 10 tons
and stretched to lengths of 10 meters (Figure 1 presents a summary
of performance characteristics of each tool).

PLAN CONFIGURATION AND GEOMETRIC SHAPES

Panel lengths and widths remain determined by the dimensions of
the excavating tool, however the tool can be manipulated to produce
panel lengths varying from 2 to 10 meters and elements in the shape
of most letters of the alphabet. "H" and "cross shaped" elements

		CABLE HUNG CLAMSHELL		KELLY BAR CLAMSHELL		REVERSE CIRCULATION ROTARY DRILL			PERCUSSION	
		MECH	HYDR	MECH	HYDR	ROLL BIT	GANG DRILL	HYDRO MILL	DROP CHISEL	REVER CIRC.
NATURE OF SOIL	SOFT COHESSIVE	GOOD	GOOD	GOOD	GOOD	----	POOR	POOR	----	----
	MED GRANULAR	GOOD	GOOD	GOOD	GOOD	----	GOOD	GOOD	----	----
	HARD PAN	GOOD	FAIR	FAIR	POOR	GOOD	POOR	GOOD	GOOD	GOOD
	BOULDER	FAIR	----	POOR	POOR	FAIR	----	FAIR	GOOD	GOOD
	ROCK SOFT	POOR	----	----	----	GOOD	----	FAIR	GOOD	GOOD
	ROCK HARD	----	----	----	----	GOOD	----	----	FAIR	GOOD
DEPTH OF EXCAVATION	0 - 15M	GOOD	GOOD	GOOD	GOOD	GOOD	GOOD	----	GOOD	GOOD
	15 - 30M	GOOD	FAIR	FAIR	FAIR	GOOD	GOOD	GOOD	GOOD	GOOD
	> 30M	GOOD	POOR	POOR	POOR	GOOD	POOR	GOOD	GOOD	GOOD
SITE CONSTRAINT	URBANIZED	GOOD	GOOD	GOOD	GOOD	FAIR	FAIR	FAIR	FAIR	GOOD
	LOW HEAD ROOM	GOOD	FAIR	----	----	----	FAIR	----	FAIR	GOOD
LABOR CONDITIONS	STRICT RULES	GOOD	FAIR	GOOD	FAIR	FAIR	POOR	FAIR	GOOD	GOOD
	LACK OF SKILL	GOOD	FAIR	FAIR	POOR	FAIR	POOR	POOR	GOOD	FAIR

FIG. 1 PERFORMANCE OF EXCAVATION TOOLS

remain the most popular for load bearing elements while "T" shaped elements remain popular for bending elements. The actual shape achieved in the field remains dependent upon ground conditions; good ground yields well defined shapes while poor ground yields amorphous shapes which barely resemble the intended shape.

SLURRY WALL REINFORCEMENT

Slurry walls were first reinforced using conventional reinforcing steel bars assembled into cages.   The bars were proportioned to resist bending in the vertical direction with supplemental bars serving as "internal walers" or to distribute forces around inserts or openings. Panel end joints must be formed by stop end pipes or other suitable forming devices.   This method of reinforcement is currently most popular and is used extensively throughout the U.S., particularly for permanent walls and where tieback anchors are to be used (Fig. 2).

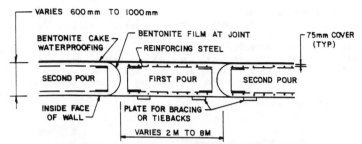

FIG.2 CONVENTIONAL REINFORCED CONCRETE WALL

"Soldier beam and concrete lagging walls" are becoming increasingly more popular, particularly in deep narrow excavations requiring high bending resistance and internal bracing. These walls were installed first in San Francisco (1963)[4,5], using augers to predrill closely spaced holes for installation of a series of vertical steel beams which served as vertical reinforcing and as panel joints. The space between the beams was excavated under slurry and filled with concrete. The concrete was strengthened with reinforcing bars in those cases where the beams were spaced beyond the bending capacity of the unreinforced concrete. Later (1973)[15], these walls were excavated directly by clamshell bucket and a pair of steel beams and a reinforcing cage were assembled together and installed into the excavation as a unit prior to concreting. Secondary panels required only the installation of the reinforcing steel cage (Fig. 3).

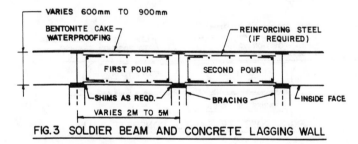

FIG.3 SOLDIER BEAM AND CONCRETE LAGGING WALL

The excavation must be close to correct alignment, otherwise the rigid beam/cage assembly will not fit into the trench. Occasionally, trimming devices fabricated of two beams and a bottom cutting edge are used to cut the trench true to shape.

Consideration has been given to the substitution of fiber reinforced concrete for the reinforcing bars and conventional concrete however economic and logistical considerations have mitigated against this use.

On several occasions in the late 70's precast concrete wall elements were used to reinforce slurry walls. This method produces the best quality of finished wall but is limited in use by cost and specific site constraints. This method requires that the panel excavation be supported by a cement-bentonite slurry which will

eventually harden.   Wall panels are best cast on site adjacent to
the work and installed directly into an oversized panel excavation.
The precast panels are temporarily suspended within the excavation
until the cement-bentonite hardens. Excavation usually proceeds in a
linear fashion with precast panel installation following. The
method becomes less practical in those environments where utility
crossings or obstructions can be expected.   Rubber waterstops are
usually required since the cement-bentonite joints are subject to
drying and shrinkage cracking (Fig. 4).

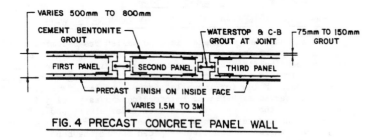

FIG. 4 PRECAST CONCRETE PANEL WALL

    Post-tensioned concrete walls as illustrated in Figure 5, have
been used successfully in Europe but have not yet been used in the
U.S.   High bending resistance can be obtained, but, cost and labor
considerations mitigate against their use.

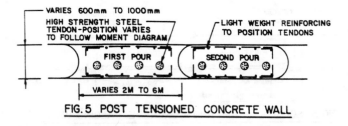

FIG. 5 POST TENSIONED CONCRETE WALL

WALL SUPPORT SYSTEMS

    Slurry walls are temporarily laterally supported during con-
struction by raker bracing, cross-lot bracing, tieback anchors,
berms, or permanent floor construction.   At some excavations,
attempts have been made to mix these various systems on one particular wall segment. The behavior of a mixed system cannot be
easily evaluated and can result in overloading of the supports or
overstressing of the wall system as a result of different behavior
of the mixed system. Mixed support systems are not recommended for
slurry wall construction.

Lateral supports are usually spaced vertically and horizontally at 3 to 6 meter spacing, with soil conditions usually dictating the spacing rather than wall bending capacity. Cross-lot bracing has been installed utilizing tubular or wide flange shapes in narrow cuts. In larger excavations, cross lot bracing has required special treatment to prevent increases in stresses resulting from temperature increases or inward wall movements due to temperature decreases. Preload ranges upwards to 100% of the design load. In general, preloads have been limited to 50% to 80% of the design load, since design loads, in many cases, exceed actual loads in the field[8]. Preloads in excess of these limits would result in movements of the wall away from the excavation and/or overloading of the bracing. Sloping raker bracing and/or corner bracing is used in wider cuts where cross-lot bracing is not practical.

Where appropriate, tiebacks have proven to be the most competitive and advantageous support system since they provide a clean open site, and permit preloading of the wall and pretesting of the anchor system. Soil anchors are increasingly used for both temporary and permanent support, but must be monitored for creep affects.

On occasion, the top down (Milan method) method has been used to support perimeter walls during excavation. The first U.S. example was constructed in Chicago in 1971[2]. Interior columns and foundations are installed concurrently with the perimeter slurry wall. The superstructure construction then proceeds upward while basement construction proceeds downward at the same time. Excavation and floor slab construction advances downward, level by level, until final subgrade is reached. Excavation by this method is more expensive; however, savings are realized by elimination of temporary bracing and by advancing the start of construction of the superstructure.

A variation on this technique has been used successfully in Europe where an annulus of floor construction has been installed around the perimeter leaving the interior of the site open. The "ring" floor system is more practical for wide, shallow excavations, while the "full floor" system is practical in narrow, deep excavations.

The self-supporting wall, constructed to a circular or elliptical plan configuration, has been used to construct deep, large diameter shafts, with or without liner walls or other interior supports. These segmental "thin shell" structures carry significant lateral soil and water loads through primary compression with some secondary bending affects, but are highly sensitive to unsymmetrical loading. Walls constructed to an incorrect plan alignment require ring beams or a liner wall to provide ring bending capacity and buckling stability. Most engineers utilizing this technique have made provisions for temporary ring walers in the event the plan alignment has not been maintained and supplemental support is required. In the case where the wall is to be used only as a temporary support, ring walers can be removed as the permanent construction advances upward. The permanent structure can also be modified to incorporate these walers within the permanent structure. In almost all cases where uplift forces dictate structural dimensions, the slurry walls have been connected to the permanent structure permitting some savings in permanent structure concrete[3].

WALLS WITH COMBINED LOADS

Slurry walls have been used primarily as temporary or permanent lateral earth retention systems, but increasingly are being used to also support vertical loads from the superstructure or the temporary vertical components of tieback loads. In those cases where the wall is founded in rock, there is more than adequate vertical bearing capacity available to support vertical loads directly. In many cases where the wall is founded in soils and tieback anchors are used, the anchors must be placed at a flat angle to the horizontal in order to minimize the vertical reaction. Where vertical super-structure loads are to be carried by the wall, it is necessary to extend either part or all of the wall down to a suitable bearing strata or to support the walls on piles. This latter technique was successfully used in 1972. Few examples of vertical support of a wall by steel piling are available at this time.

Because of the rather irregular shape of the wall and uncertain cross-sectional properties of the wall (usually greater than those assumed for the design). Most engineers design the wall for bending and ignore the very small effects of either the tieback load or the vertical wall load. On the few occasions where it becomes necessary to check the wall for vertical loading as well as bending, it has been found that the vertical loads provide a nominal uniform compression and improve the bending capacity by reducing critical tensile stresses.

WALL FINISHES, TOLERANCES AND WORKMANSHIP

Slurry walls have proven to be most appropriate in difficult ground conditions where fills, obstructions and soft ground are expected to be encountered. In disadvantageous soil conditions, the walls usually are constructed to dimensions significantly beyond the neat line dimension shown on the plans. In many cases obstructions encountered in the excavation are broken free from the side walls of the excavation and large overpours are experienced. The only techniques that prove practical in overcoming these problems are either the pre-excavation and replacement of material with lean concrete or the full excavation of the panel, replacement with lean concrete and re-excavation of the panel. In Italy, grouting of open granular materials has been used successfully to stabilize side walls of the slurry filled excavation. U.S. contractors have not yet considered grouting necessary nor economical and have yet to use this technique.

In most cases contractors have not elected to use lean concrete for the redefinition of the slurry wall excavation, as a result massive overpours have been experienced in certain ground condi-tions. Several contractors have attempted to overcome the overpour problem by attaching forms to the reinforcing cage. These forms are set at the neat line dimension and provide a break to which the overpour concrete can be removed. Unfortunately, in most cases where this method is used, the form work does not remain in its precise alignment and some irregularity of the surface occurs.

Other contractors have elected to cast the wall and then cut back the wall using hydraulic and pneumatic breaker equipment. Regardless of which method is used a "smooth finished wall" is never achieved.

Verticality measurements have remained rather crude, with the primary method of verifying the position of the panel remaining the measurement of the position of the clamshell bucket during excavation. This method does not identify the location of overbreaks, but at least it can assure that a sufficient width and length of trench exists for the full excavation, and that the reinforcing cage can be installed and hung vertically within the panel. On some occasions the cages move laterally or vertically during concrete placement.

Where necessary to install a wall thinner than the smallest available width of clamshell bucket, cages have been installed with styrofoam and wood blockouts to reduce the overall thickness of the cast concrete wall. Two problems occur using this method. First, the reinforcing cage may not be sufficiently heavy to overcome the buoyancy of the cage blockout assembly, and the cage may have to be forced down into the slurry filled excavation rather than held suspended from the guide walls. Second. buoyant cages prove to be a problem during the placement of concrete since they tend to rise with the placement of the concrete.

Japanese slurry wall contractors have used sonic devices to measure trench alignment. A graph of the width of the trench is obtained permitting an evaluation of the condition of the trench as excavated. The problem remains that once a misalignment or bulge is discovered, the contractor must either backfill the trench with lean concrete and re-excavate or he must cast the panel to the condition discovered and later remove the concrete by conventional methods.

No resolution to the tolerances and finish problem, other than the use of precast concrete panel construction, is foreseen in the immediate future.

METHODS OF ANALYSIS

Four methods of analysis are currently used for the design of slurry walls in the United States. The first method, the time-tested use of Terzaghi-Peck apparent pressure diagrams[13], is the most well known and has been used successfully for temporary walls with uniform and reasonably spaced supports.

However, structural designers have preferred to use a staged excavation analysis which assumes that the slurry wall acts as a continuous beam spanning from support to support and to a soil "subgrade reaction". This method assumes nondeforming supports and ignores overall wall movements. When compared to methods which attempt to acknowledge wall movements, the method has been found to yield higher negative moments behind the supports and lower positive bending moments between supports, however, calculated reactions and shears appear reasonable.

The staged excavation procedure has been improved through the use of the beam on elastic foundations method (finite difference method)[9] to estimate a) the movement of the wall during excavation, b) the effect of the installation of deformable supports on a deformed wall and c) the behavior of the subgrade soil.

Attempts to utilize finite element methods in the design of slurry walls rather than to just estimate the behavior of the soil behind the wall have met with mixed success. Site specific, detailed soil properties, wall properties and geometry are required prior to use of this method as a design tool.

All of the foregoing methods are discussed in detail in a paper prepared by Kerr and Tamaro[11] for inclusion in these proceedings.

The ultimate strength method of design has almost replaced the working stress method, with many engineers fully automating the design process and their operations. Computer programs have been developed to analyze, design and detail typical walls, but methods of accurately estimating the behavior of the usually oversized wall during all stages of construction remain elusive.

DESIGN RESPONSIBILITY

For the most part the final design of slurry walls, bracing or tieback anchors, and the proportioning of reinforcing steel remains the responsibility of the contractor. Consulting engineers generally prepare construction documents showing the geometry of the wall, a finished design for the permanent condition, a criteria for the design of the wall for temporary conditions and a requirement that the contractor prepare an independent design which considers excavation staging and all temporary wall loading conditions.

A few specialized consultants prepare full design documents for all load cases rather that leave any phase of the design of the wall to others. In many cases contractors offer alternative designs which reflect special equipment and/or methodology. The alternative designs are usually accepted in lieu of the original design if found equal in performance to the original design.

COST

Slurry walls prove to be cost competitive to conventional wall systems when at least two of the following conditions can be met[7]:

    a) Slurry wall can be used as the permanent wall.
    b) Slurry wall eliminates the need for underpinning.
    c) Slurry wall eliminates the need for dewatering outside the
       site.
    d) Natural or man made obstructions restrict the use of conven-
       tional sheeting systems.
    e) Slurry wall can be used to support vertical loads.

Slurry walls remain a construction bargain, doubling in cost over the two decade period while the ENR Cost Index quadrupled. Figure 6 presents estimated costs of slurry wall construction in 1970 and 1990[14].

SUMMARY

Over the past 20 years an increasing quantities of slurry walls have been installed in the United States. Walls have been installed for lateral or vertical load retention and to a variety of plan

SLURRY WALL COSTS ($ PER SQUARE METER OF WALL)

	DEPTH < 15M		15M TO 30M		DEPTH > 30M	
	1970	1990	1970	1990	1970	1990
SOFT SOILS SPT < 50	$ 150	$ 400	$ 350	$ 600	$ 350	$ 800
MEDIUM SOILS 50 <SPT<100	$ 250	$ 500	$ 450	$ 700	$ 550	$ 900
HARD SOILS SPT > 100	$ 350	$ 600	$ 550	$ 800	$ 650	$1000
SOFT ROCK* SPT > 200	$ 500	$ 700	$ 700	$ 900	$ 900	$1500
HARD ROCK** 600mm KEY	$1000	$1500	$1200	$1800	$1500	$2000
HARD ROCK** 6M SOCKET	$1500	$2500	$1800	$3000	$2100	$3500

FOR 75Kg/M^2 REINFORCING STEEL ADD $50/M^2, (1970) $1030/M^2,(1990)
ADD 25% TO 50% FOR CONSTRUCTION IN URBAN AND/OR UNION AREA
* SANDSTONE OR SHALE   ** GRANITE, GNEISS OR SCHIST

FIG. 6 SLURRY WALL COSTS.

configurations.   Four different wall reinforcement techniques have
been used with the conventional reinforcing steel cage method
remaining most popular. Several million square feet of this type of
wall has been successfully installed.   The satisfactory structural
behavior of the walls is unquestioned, but the quality of the
finishes remains a constant problem. Tools and techniques have been
improved over the 20 year period with almost all of the work in the
United States being executed by cable hung clamshells operated by
conventional crawler crane equipment. Occasional attempts have been
made to use other more sophisticated tools in specific ground conditions. The use of these tools remains the exception rather than the
norm in slurry wall construction.
     Analytical design tools have improved over the 20 year period.
However, the most contemporary, the finite element method remains
untested and unproven as a design tool.   More conventional methods,
including the finite difference methods (with Winkler springs
representing soil behavior) have been used successfully for production design and are becoming increasingly more easy to use.
     Lateral support systems have also improved over the past two
decades with the increasing use of tiebacks and the top down method
in preference to conventional bracing using cross lot or raker brace
schemes.
     Lastly, slurry wall construction has remained a specialists
method in this country.   A handful of skilled specialized contrac-

tors are available for the execution of this work throughout the United States. A few general contractors have undertaken slurry wall construction in conjunction with their more conventional construction. These general contractors do not usually compete as subcontractors on slurry wall construction.

REFERENCES

1. W.Armento, Cofferdam for BARTD Embarcadero Subway Station, J SM&FO (ASCE), 99, (10) 1973, 727-744.
2. J.Cunningham & J. Fernandes, Performance of Two Slurry Wall Systems in Chicago, ASCE Conf. Performance of Earth and Earth Supported Structures, Purdue Univ., Lafayette IN, 1972, 1425-1437.
3. Engineering News Record, Pump Pit Goes Deep into Wet Soil, December 3, 1987, 29.
4. B.Gerwick, Slurry Trench Techniques for Diaphragm Walls in Deep Foundation Construction, Civil Engineering, December 1967, 70-72.
5. B.Gerwick, The Use of Slurry Trench (Diaphragm Wall) Techniques in Deep Foundation Construction, Conference Preprint 496, ASCE Structural Engineering Conference, Seattle, May 1967.
6. B.Gerwick, Personal Communication, March 15, 1989.
7. K.Godfrey, Subsiding Problems with Slurry Walls, Civil Engineering, January 1987, 56-59.
8. J.Gould, Lateral Pressures on Rigid Permanent Structures, Lateral Stresses in the Ground and Design of Earth Retaining Structures, ASCE Conf. Cornell University, Ithaca, NY, June 1970.
9. T.Halliburton, Numerical Analysis of Flexible Retaining Structures, J SM & FD, (ASCE), 94, (6), 1968, 1233-1251.
10. M.Kapp, Slurry Trench Construction for Basement Wall of World Trade Center, Civil Engineering, April 1969, 36-40.
11. W.Kerr & G. Tamaro, Diaphragm Walls-Update on Design and Performance, ASCE Conf. Design and Performance of Earth Retaining Structures, Cornell Univ., Ithaca, N.Y. 1990
12. P.O'Neill, Personal Communication, August 8, 1989.
13. R.Peck, Deep Excavations and Tunneling in Soft Ground,State-of-the-Art Report, Proc. 7th Int. Conf. ISSMFE,Vol. III, Mexico, 1969, 225-281.
14. A.Ressi, Personal Communication, January 24, 1989.
15. Roads & Streets, Slurry Wall Special Equipment Solve No Room Excavation Problem, October 1973.

SHORING FAILURE IN SOFT CLAY

Paul G. Swanson[1] and Thomas W. Larson[2]

ABSTRACT: In March of 1988, a shoring failure occurred along a
section of the Washington Metropolitan Area Transit Authority
(WMATA) subway station excavation in the Anacostia area of
Washington, D.C. Lateral and vertical movements on the order of
several feet occurred prior to excavating to the required depth
necessitating a corrective design. A summary of the soil design
parameters as interpreted from the pre-construction studies and
studies performed after the failure indicate the stability number
$N = \gamma H/S_u$ may vary from 5 under design conditions to as high as 12
for S_u interpreted from DMT values taken in the failed area.

INTRODUCTION

WMATA is the governing authority for the operation, maintenance
and development of the subway system servicing Washington, D.C.
The system will eventually include 100 miles (160 km) of surface
and subsurface railway through two major geologic provinces and a
wide variety of soil types, groundwater, and foundation conditions.

The site of the proposed station is located within the Coastal
Plain Geologic Province. The subsurface stratigraphy consists of
8 to 12 feet (2.4 to 3.7 m) of sand/clay fill overlying
approximately 40 feet (12 m) of very soft organic clay. Beneath
the soft clay soils are layers of firm to dense clayey sand and
stiff to hard clays of Cretaceous age.

The station excavation required the full depth of the soft clay
stratum to be shored temporarily. Project documents for
construction of the station were based on geotechnical data
derived from the original exploration studies for the Metro
System. The temporary shoring system proposed by the contractor
consisted of steel H-piling (usually W21 x 111 section) driven on
8 foot (2.4 m) centers. The piling was driven 5 to 10 feet (1.5
to 3 m) into the dense sands and clays approximately 40 to 50 feet
(12 to 15 m) below the existing ground surface. The piling was

---

1 - Chief Engineer, Law Engineering, 4465 Brookfield Corporate
    Drive, Chantilly, VA 22021
2 - Senior Geotechnical Engineer, Law Engineering,
    4465 Brookfield Corporate Drive, Chantilly, VA 22021

braced at three levels using 20 to 35-inch (50 to 90 cm) diameter steel casing extending approximately 60 feet (18 m) across the excavation.

At the time of the failure, excavation was complete and shoring in place within the approach tunnels to the south of the station. Partial excavation had been completed within a 400-foot (120 m) section of the station to a depth of approximately 30 feet (9 m) when significant lateral "squeezing" of the soft clay occurred resulting in a lateral soil movement of 10 feet (3 m) (reference Figure 1). Along the east side of the excavation, the ground surface subsided on the order of 5 feet (1.5 m) causing lagging boards to break and dislodge from the pile flanges. Soldier piling was bowed as much as 18 inches (4 cm) into the excavation and twisted 45 degrees away from the axis of the excavation. Further excavation was halted and a detailed field investigative study was undertaken to determine the cause of failure and a cost effective re-design of the bracing system.

Figure 1. Failed Section of Bracing System

SUBSURFACE CONDITIONS

Pre-construction Exploration. The pre-construction exploration studies for this portion of the subway system were performed in 1980. Standard soil test borings were performed using wash boring techniques, standard penetration testing and thin-wall Shelby tube sampling. A plan view of the site area with boring locations is shown in Figure 2. The stratum of primary concern within the subsurface stratigraphy was the soft, organic clay designated as the A1 soils. The A1 clays typically have a liquid limit of 80, plastic limit of 40, and in situ moisture contents at or above the liquid limit. Design strength parameters for this stratum were determined from laboratory index testing, unconfined compression tests, unconsolidated undrained triaxial compression tests, and consolidation testing. A summary of the undrained strength determined by the testing is shown in Table 1. A summary of the post failure testing is given for comparison.

Table 1. Undrained Strength Summary of A1 Soil

	Undrained Strength psf (kPa)		
	-5 to -20	-20 to -30	-30 to -40*
Pre-Construction	500 (24)	650 (31)	900 (43)
Post-Construction			
UU tests; failed area	400 (19)	355 (17)	315 (15)
Dilatometer; failed area	250 (12)	270 (13)	270 (13)
CU tests; ext. & comp.	460 (22)	525 (25)	670 (32)
Dilatometer	545 (26)	480 (23)	525 (25)

* Elevation, ft, msl

Post-Failure Exploration. Post-failure testing included an emphasis on in-situ field testing. Three undisturbed sample borings were drilled to confirm the pre-construction testing. In addition, six dilatometer probe borings were drilled to assess strength and stress history within the A1 clays. The dilatometer probe used was similar to that developed by Marchetti, 1980. Estimations of undrained shear strength, S_u, have been made using the relationship:

$$S_u = 0.22 \sigma'_v (0.5K_d)^{1.25}$$

where: σ'_v = effective vertical pressure
K_d = horizontal stress index

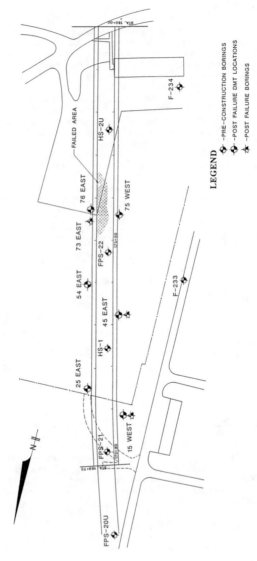

Figure 2. Test Locations

Dilatometer measurements were obtained every two feet within the clay stratum. Rotary drilling techniques were used to advance the boring through the overlying fill soils. Upon completion of the dilatometer probe, a standard soil test boring was drilled immediately adjacent to the dilatometer probe. A combination of standard penetration tests (ASTM D 1586) and undisturbed tube sampling (ASTM D 1587) were used to obtain samples. Three-inch (8 cm) diameter steel Shelby tubes were used for both standard push and piston samples.

Laboratory strength measurements were obtained from controlled strain undrained triaxial compression and extension tests with pore pressure measurements. Undrained strength was taken as the peak value of deviator stress. The results summarized in Table 1 indicate good correlation between extension and compression triaxial testing and dilatometer data both within and outside the failed area. Overconsolidation Ratio, (OCR), as interpreted from the dilatometer testing, ranged between 1.0 and 1.6. A plot of the strength estimations from the dilatometer probes is shown in Figure 3.

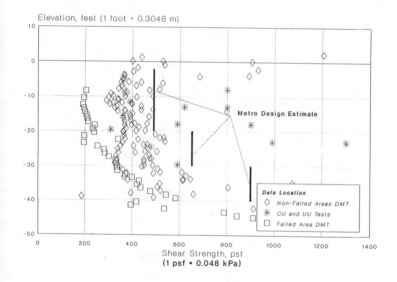

Figure 3.   Strength Estimations from Dilatometer

ANALYSIS OF THE FAILURE

   <u>Design Conditions</u>. The original shoring design was undertaken
by the general contractor. Lateral pressures were defined by the
owner's geotechnical consultant using guidelines published by
Ralph Peck at the 7th ICSMFE (Peck, 1969). The design lateral
pressure diagram is shown in Figure 4.
   The original pressure diagram included in the design drawings
for this portion of the excavation was a trapezoidal distribution
with a maximum value of 2300 psf (110 kPa) specified for the
middle half of an excavation depth of approximately 45 feet (14
m). This maximum pressure was used for design of the struts. A
similar distribution with a maximum pressure of 1500 psf (72 kPa)
was used for design of the beams.
   Strength information obtained during the pre-construction
exploration studies came from a few borings well removed from the
station area. Much reliance was placed on previous testing of the
A1 soils at various other locations along the subway route. The
results of this testing indicated a soft soil with a highly
variable S_u value. To establish a design standard, a characteristic strength increase with increasing effective overburden
pressure was assumed.

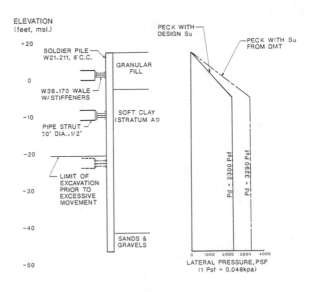

Figure 4.   Retaining System Design and Lateral Pressure Diagram

Failure Conditions. - Evidence of impending failure was noticeable early on in the excavation process. Survey measurements
taken on the piling at the second strut level (approximately
elevation -9 feet (-2.7 m)) showed horizontal movement into the
excavation on the order of 4 to 6 inches (10 to 15 cm). The
piling had encroached upon the excavation "neat" line 12 to 48
inches (30 to 45 cm) upon excavation to the third strut level 30
feet (9 m) deep. Soil was clearly moving laterally past the piles
below the lagging and at some places breaking or dislodging the
lagging from the pile flanges.

Visual inspections of the shoring within the failed area
indicate that the shoring failed as a result of base shear. The
lateral movement, twisting of the pile sections and creep of the
clay into the excavation appear to have occurred below the depth
of lagging as the excavation was made. The calculation of safety
factor related to this failure mechanism is obviously controlled
by the estimation of soil shear strength. Using the shear strength
parameters specified for the design, a minimum safety factor of
less than 1.0 is obtained for failure at the -20 foot (-6 m)
level. However, based on the presumption that the undrained
strength of the clay increases with depth, similar base failures
would not be predicted at deeper excavation depths.

The results of the dilatometer testing of these soils suggest a
different interpretation of the strength profile. Referring to
Figure 3, very little if any relative strength increase with depth
was observed. This may be due in part to the presence of the 10
to 13 feet (3 to 4 m) of fill above the A1 soils and the construction dewatering along the alignment of the excavation. It is
believed that some degree of consolidation may still be occurring
within these soils from the change in effective stress state.

The dilatometer interpretation of the clay strength is further
substantiated by analyzing the buckling potential of the soldier
piles. Lateral soil displacement below the base of the excavation
would load the beam between the strut and the bearing stratum,
resulting in a potential buckling failure. An analysis of this
condition predicts a buckling failure of the pile for a lateral
soil pressure on the order of 2760 psf (130 kPa), approximately 20
percent higher than the maximum predicted pressure from the Peck
relationship using the design shear strength values. However, the
soil was observed to have failed between the piles and the piles
did not buckle, which indicates that the soils had a lower
strength than the design value. A lateral bearing analysis of the
piles indicates the strength of the soil below elevation -20 feet
(-6 m) would be on the order of 500 psf (25 kPa) to meet both
these criteria.

Wide disparities in strength interpretation between field and
laboratory testing resulted in a correspondingly large variation
in the predicted lateral pressures, making redesign of the shoring
very difficult. Depending on the strength selected, the predicted
maximum lateral pressure could vary from 2300 to 3550 psf (70 to

170 kPa). Based on the observed deflection of the soldier piles, it was clear that the design pressure of 2300 psf (70 kPa) had been exceeded. It was also clear from the dilatometer data that some lower strength value (i.e.. higher lateral pressure) would be applicable within the failure area relative to areas outside the failed area.

Shoring Redesign. Owing to the many constraints associated with an ongoing construction schedule, redesign of the shoring system had to be generally consistent with the previous construction. Maximum lateral pressures of 3350 psf (160 kPa) within the failed area and 2950 psf (140 kPa) outside the failed area were used for redesign of the soldier beams. It was recognized that a more compact pile section was desirable to reduce the alignment sensitivity of the design. The new design consisted of heavier pile sections driven slightly deeper than the previous design and an extra level of bracing. The spacing of the piles was reduced to 4 feet (1.2 m) on center to eliminate the shear failure of soil between piling below the lagging boards.

Within the failed area, a number of vibrating wire strain gages were placed on the struts to determine actual load applied to the shoring system during excavation. The results of that instrumentation indicate the lateral load on the shoring system to correspond to a maximum pressure of approximately 2950 psf (140 kPa) at the full excavation depth of 49 feet (15 m).

The lateral pressure interpreted from the instrumentation is comparable to that predicted using the strength data interpreted from dilatometer testing and Peck's recommended lateral pressure diagram. An undrained shear strength value of 480 psf (230 kPa) for the A1 clays would yield a maximum lateral pressure of 2950 psf (140 kPa).

Slope inclinometers were installed adjacent to the excavation as a means of monitoring the deflection of the shoring redesign during the excavation process. The measured deflections are shown in Figure 5 for various stages of the excavation. The results indicate lateral movement of the soil on the order of 10 inches (25 cm) toward the excavation. The magnitude of vertical movement outside the excavation was on the order of one foot.

Although the system functioned without failure, the magnitude of movement is of some concern. The tendency still appears to be toward base failure. A more appropriate shoring design may have been to use sheet piling and thus control the base failure mechanism.

The concept of slump failure and flow around the piling can be confirmed using an adaptation of a circular arc failure analysis that accounts for the resistance of the piling. An approximation of the ultimate static lateral resistance P_u can be given as:

$$P_u = N_p S_u D$$

where N_p = resistance coefficient
          S_u = undrained soil strength (from DMT)
          D = pile diameter

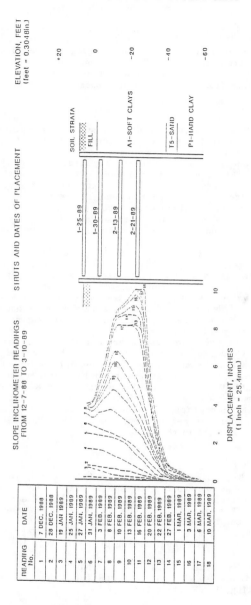

Figure 5. Slope Inclinometer Readings During Reconstruction

The value of N_p would vary with depth below the excavation level (Matlock, 1985). Circular arc failure surfaces were analyzed assuming full support from the lagging and only lateral bearing support from the flange of the piling below the lagging. A factor of safety of one was obtained using the undrained strength determined from the dilatometer probes.

CONCLUSIONS

This case study illustrates the value of subsurface exploration techniques that obtain large amounts of subsurface data for site characterization studies. Variations in strength, OCR and deformation properties are extremely important for projects involving excavations over large areas. Often strength and deformation parameters are assumed or extrapolated from the results of a limited number of conventional laboratory tests. In the present study, the dilatometer results clearly indicated trends with depth and location in the subsurface stratigraphy that would generally go unnoticed using only a laboratory testing program.

Design engineers should rely heavily on field data when parameters are marginal. Occasional low laboratory strength values are often regarded as questionable; possibly affected by sampling procedures or disturbance in the laboratory. A large number of test values, such as those available with dilatometer testing, offer the designer a more detailed view of subsurface conditions and therefore a better basis for accepting conventional trends or requesting additional definition.

The dilatometer probe proved to be a useful tool for the delineation of subsurface parameters related to shoring design in soft clay encountered on this site. The absolute value of the undrained strength based solely on the dilatometer values is subject to question. Lacasse and Lunne report on a four-year research program to evaluate correlations for clays and sands, and give recommended correlations which would actually predict slightly lower shear strengths based on K_d than Marchetti's original correlation used in this study.

REFERENCES

Peck, Ralph B., "Deep Excavations and Tunneling in Soft Ground", State of-the-Art Report, 7th ICSMFE Mexico, 1969.

Matlock H., "Correlations for Design of Laterally Loaded Piles in Soft Clay", Proceedings of the II Offshore Technical Conference, Houston, TX. Vol. 1, 1970.

Marchetti, S., "In-Situ Tests by Flat Dilatometer", ASCE Journal of the Geotechnical Engineering Division, Vol. 106, GT3, 1980.

Lutenegger, A. J., "Current Status of the Marchetti Dilatometer Test", Proceedings of the 1st International Symposium on Penetration Testing, ISOPT, FL., 1988.

Lacasse, S. and Lunne T., "Calibration of Dilatometer Correlations", Proceedings of the 1st International Symposium on Penetration Testing, ISOPT, Fl., 1988.

THE BEHAVIOUR OF EMBEDDED RETAINING WALLS

Ian F Symons[1]

ABSTRACT: Increasing use is being made of embedded retaining walls
for the construction of roads below ground level in urban areas of
the United Kingdom. This paper briefly reviews some of the results
obtained from an ongoing programme of research being carried out by
the Transport and Road Research Laboratory (TRRL) into the behaviour
of these types of structure. The work includes field measurements
on full scale walls, both during construction and in service
together with centrifuge model tests and analytical studies.

INTRODUCTION

Improvements to the existing road network form a major component
of the programme of urban renewal and development which is now
taking place in the United Kingdom. The location of roads below
existing ground level in retained cuts or in cut and cover tunnels
can offer considerable benefit in terms of reducing the required
land take and lessening the impact on the urban environment. As a
consequence increased use is being made of embedded retaining walls
of the bored pile or diaphragm types for permanent works and of
driven sheet pile walls for temporary works.

A coordinated programme of research is being carried out on these
types of structure by the TRRL on behalf of the Department of
Transport. The aim of this research is to improve understanding of
the behaviour of embedded walls which will lead to greater economy
in their design. The work carried out to date includes centrifuge
model tests, analytical studies and field investigations of the
behaviour of full scale structures, both during construction and in
permanent service.

Many urban areas in southern England are founded on heavily
overconsolidated clays of high plasticity which are particularly
susceptible to swelling and softening following release of stress
caused by excavation. The research is therefore concentrating on
the behaviour of embedded walls in these deposits. This paper
briefly reviews some of the results from the studies carried out to
date and in particular focusses on the influence of the construction
process on movements and stress changes in the ground.

1 -   Ground Engineering Division, TRRL Old Wokingham Road,
      Crowthorne Berks England RG11 6AU

562

FIELD STUDIES DURING CONSTRUCTION

An improvement scheme to the A1(M) was carried out during 1985/86 at Hatfield in Hertfordshire, England. A major part of the work comprised the construction of an 1150m long cut and cover tunnel, with the approaches to the tunnel in retained cuts. The soil conditions over the depth of the tunnel consisted of glacial sands and gravel underlain by a layer of stiff Boulder Clay. The water table was located about 8m below ground level. Over 2km of temporary anchored sheet pile wall was used for lateral support during construction of the permanent reinforced concrete retaining walls and tunnel structure. Upon completion, the space between the permanent and temporary walls was backfilled, the ground anchor cables cut and the sheet piles extracted.

A 20m long section of sheet pile wall was extensively instrumented to measure the behaviour during construction and over a ten month period when it was in service. A local road located within about 6m of the wall on the retained side carried heavy traffic throughout most of the construction period. The field observations are described by Symons et al (1987) and the ground movements discussed by Symons et al (1988).

Fig 1 shows the ground and wall movements which occurred during each of the main stages of construction. The sheet pile wall was formed by driving interlocking pairs of Larssen 4/20 sheet piles sequentially to full depth of 13m in one operation. The driving was carried out percussively using a Hushrig and large ground movements were measured during this operation. The pattern of movement suggests that the vibration and shear forces induced by driving, caused densification of the granular materials close to the piles. Only small ground and pile movements occurred during the initial excavation to a depth of 3.2m to permit access for the installation of the ground anchors. The anchors were inclined at 25° to the horizontal and placed at 3.05m centres. Each anchor comprised four prestressing tendons pressure grouted over a fixed anchor length of 8m. Tensioning of the anchors to a working load of 560kN caused an inward movement of the sheet piles and retained ground which extended to about 5m from the wall at the surface. An average lateral movement of the sheet piles of about 7mm was measured midway between the anchors during tensioning and this compares with an inward movement of about 12mm recorded on the walings at the anchor positions.

After anchor stressing, the excavation was extended to a depth of 7.1m in front of the wall, well points installed at 2m centres and dewatering carried out to a depth of about 10 metres below original ground level. Excavation was then carried out over an 8m wide strip to a final depth of 9.3m in front of the sheet piles. During this period an outward deflection of the piles of up to 12mm was measured at approximately midway between anchor and excavation level. Negligible vertical movement of the piles was measured and the pattern of ground movement during excavation suggests appreciable differential movement between the piles and adjoining ground. Only small increases in ground movement and pile deflection occurred during the five month period of construction of the permanent retaining wall. After backfilling between the permanent and temporary retaining walls to waling level, the anchor cables were released and large ground movements occurred as the sheet piles deflected by up to 25mm towards the excavation. In spite of the large ground movements caused by installation of the sheet pile wall and release of the ground anchors, the total measured settlements fell within the limits of the Zone 1 given by Peck (1969).

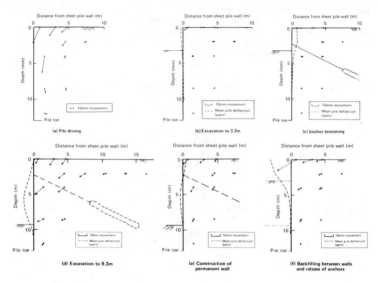

Fig. 1 Ground movements measured during each stage of construction on the A1(M) at Hatfield (after Symons et al 1988)

The design of the sheet pile wall was based on the assumptions of unfactored limiting active and passive pressures acting over the minimum height required for limiting equilibrium, and gave a maximum bending moment of 420 kN m/m. The maximum bending moments deduced from strain measurements made on two pairs of piles were 75 kN m/m on completion of excavation and 110 kN m/m prior to backfilling. A reappraisal of the maximum bending moments was carried out using Rowe's method of design (Barden 1974) which enables account to be taken of the relative stiffness of the wall through the application of moment reduction factors for fixity above anchor level and below excavation level. This gave much closer agreement with the measurement, indicating a maximum bending moment of about 140 kN m/m for the Larssen 4/20 sheet piles based on the 'as constructed' geometry and revised soil profile at the instrumented section.

Another study where field measurements illustrate the effect of

wall installation on the behaviour of the adjoining ground was
carried out in 1982/83 during construction of the Bell Common
tunnel. This 470m long cut and cover tunnel carries the M25 London
Orbital Motorway through the northern edge of Epping Forest in
Essex, England. The tunnel is formed from two secant pile walls
propped apart by the roofing slab. Fig 2 shows a half section
across the eastbound carriageway.

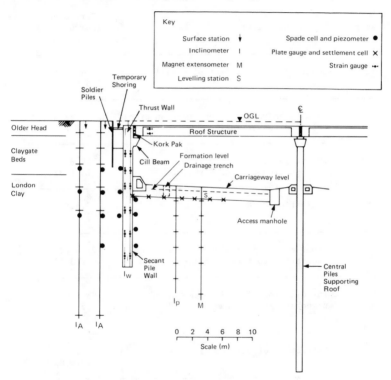

Fig. 2 Section showing instrumentation layout
(after Symons & Tedd, 1989)

The ground conditions at the site comprise London Clay overlain
by Claygate beds and Older Head. During tunnel design there was
considerable uncertainty concerning the magnitude of the earth
pressures and resulting bending moments and deflections of the
embedded walls (Hubbard et al 1984). Extensive instrumentation was
therefore installed to monitor the behaviour of a section of the
retaining wall and adjoining ground. The instrumentation layout is
also shown in Fig 2 and the principal stages of construction

together with the dates when the work was carried out are listed in Table 1. The behaviour during Stage I to VI of the construction is described in Tedd et al (1984) and over the subsequent four year period by Symons and Tedd (1989).

TABLE 1   MAIN STAGES OF TUNNEL CONSTRUCTION

Stage	Description	Period at Instrumented Section
I	Construction of secant pile wall	26.4.82 to 6.5.82
II	Excavation to 3.5m, construction of cill beam and thrust wall	24.5.82 to 11.6.82
IIIa IIIb	Excavation to 5m Construction of Roof Structure	1.7.82 to 11.8.82
IV	Excavation to 8m	23.8.82 to 31.8.82
V	Minor excavations within tunnel and excavation to formation level	21.9.82 to 2.3.83
VI	Installation of drain, removal of temporary shoring and backfilling behind wall	28.4.83 to 5.5.83
	Extraction of soldier piles behind wall	20.5.83 to 10.6.83
	Construction of road pavement	25.8.83 to 10.9.83
	Landscaping over tunnel roof 0.75m of fill	7.9.83 to 19.9.83
	Tunnel opened	25.1.84

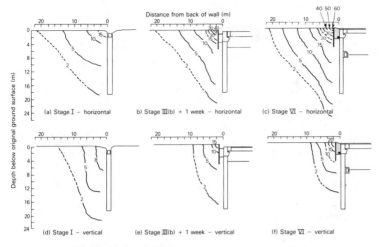

Fig. 3 Magnitude and distribution of subsurface movements (mm)
(after Tedd et al 1984)

The field measurements drew attention to the substantial ground movements which can occur during installation of the wall (Stage I) as illustrated in Fig 3. These movements extended to a considerable depth and distance from the wall and amounted to about 30 per cent of the total movements which took place throughout the construction period. Earth pressure measurements at distances of up to 6m from the wall also showed that significant reductions in the horizontal total stress took place during installation of the wall. The magnitude of these ground movements and stress changes are likely to be governed by the particular method used to form the walls. In this case the pile bores through the London Clay at the instrumented section were generally left unsupported for several hours between withdrawal of the casing and pouring of the concrete.

The deflected shapes of the wall measured at a number of stages during construction and four years later are shown in Fig 4. Excavation to a depth of 5 metres and placing of the precast roofing beams in Stage III caused the wall to bend towards the excavation over its upper half. A 75 mm thick layer of compressible material was installed between these beams and the thrust walls (Fig 2). On top of and between the roof beams a 150mm thick continuous reinforced concrete decking slab was then constructed. The roof structure applied an eccentric loading on the walls via the cill beams and the maximum bending moments during the construction period were measured in the wall at about this time.

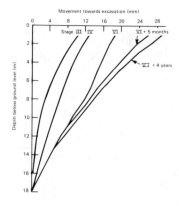

Fig. 4 Movement of Secant Pile wall during
and after construction

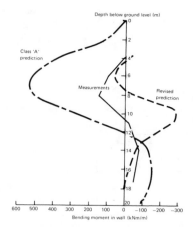

Fig. 5 Measured and predicted bending moment
profiles for Stage VI (after Higgins et al 1989)

Excavation to 8m depth in front of the wall during Stage IV caused an inward rotation of the wall but with a fairly uniform increase in lateral movement over the upper half. The reduction in curvature in this region was reflected by a decrease in the measured bending moments. This mode of deformation continued with most of the increase in wall movement taking place in the first two months after excavation to a depth of 8m.

Drain installation and backfilling behind the wall in Stage VI caused a temporary reversal in curvature to give bending moments of opposite sign over the upper half of the wall. The outward movement which then occurred as landscaping took place is probably a result of the increase in vertical loading from the roof. Over the four year period since Stage VI of the construction further small increases in movement have taken place at the top of the wall. Superimposed on the longer term movement is an annual variation which results in an increase in concave curvature towards the tunnel during the winter and a corresponding reduction in summer. These seasonal fluctuations in deflected shape, which may result from thermal expansion and contraction of the tunnel roof, are confirmed by the changes in bending moments determined from the strains in the wall.

The presence of the compressible packing at the roofing slab-wall interface is believed to have had a significant influence on the observed behaviour of the wall, since the stress-strain properties of the Kork Pak were found to be markedly time dependent (Tedd and Charles 1984).

ANALYTICAL STUDIES

Finite element analyses are being used increasingly for the design of embedded retaining walls to assess ground movements, wall deformations and bending moments under working conditions. Elastic finite element analyses were used for the design of the Bell Common tunnel and these were extended to provide a Class 'A' prediction (Lambe 1973) of the behaviour of the wall at the instrumented section. These latter analyses which modelled the soils as linear elastic perfectly plastic materials are fully described by Potts and Burland (1983). The soil stiffness profile used was derived from previous back analyses of similar structures constructed in the London Clay and the initial in-situ stresses in the ground were deduced from laboratory tests and in-situ measurements.

Comparisons between the measured and predicted behaviour during construction (Symons et al 1985) indicated that differences were due to a number of causes. First the process of wall installation had not been modelled in the analyses, so that the ground movements and reductions in horizontal total stress measured during this stage of the construction were not predicted. The measurements also indicated that the combined stiffness of the roofing slab and compressible packing were less than had been assumed for the Class A prediction. There were also differences between the actual and assumed sequence of excavation at the instrumented section.

Further finite element analyses have recently been carried out which examine the effect on the predictions of modelling wall installation, of more closely following the actual construction sequence and of allowing a softer response from the roofing slab (Higgins et al 1989). Fig 5 shows the bending moments on completion of Stage VI of the construction from one computer run in which the soil properties and initial stresses in the ground were the same as used in the Class A analyses. Also shown are the corresponding Class A prediction and measurements.

FIELD STUDIES OF WALLS IN SERVICE

The critical design condition for embedded walls often occurs in the long term, many years after completion of construction, when full porewater pressure equilibrium has been achieved under the new stress regime caused by the construction. Field studies are therefore being carried out at a number of sites where embedded walls have been in service for a number of years. The objectives are to measure the actual pressures acting on the walls and to carry out back analyses of wall stability so that the factors of safety and bending moments under working conditions can be assessed and compared with current design recommendations (Padfield and Mair 1984).

One such site is located on the A329(M) near Reading in Berkshire England, where an unpropped diaphragm wall was constructed in 1972 to retain London Clay overlain by about 3m of gravel (Fig 6).

Neg. no. R236/84/6

Fig. 6 Diaphragm retaining wall on the A329(M)

Results from this study are fully described by Carder and Symons (1989). Information from a second site on the A3 at Malden in Surrey, England, where a contiguous bored pile wall, propped below carriageway level, was constructed through London Clay in 1975 is given by Symons and Carder (1989). Push-in pressure cells and piezometers have been installed both behind and in front of the walls, and self boring pressuremeter and dilatometer tests were carried out in the retained ground to establish the pressures acting on the walls. At the A329(M) site similar measurements have also been made at positions remote from the wall in an attempt to establish the pre-construction in situ horizontal stresses in the London Clay.

The measured distribution of total lateral pressures are compared with limiting active and passive pressures used in the reassessement of the stability of the A329(M) wall in Fig 7. On the retained side, the pressures measured at 1.4m from the wall are substantially greater than the maximum active values. Garrett and Barnes (1984) also report that working stress levels were much higher than the active design values for an unpropped wall in heavily overconsolidated Gault Clay on the M26 in Kent, England. On the carriageway side the results from spade cells installed at 2.25m from the A329(M) wall are close to the upper limit of the calculated passive pressures. This is consistent with predictions from finite element analyses of the behaviour of this wall, of the wall on the M26 (Clarke and Wroth 1984) and of the propped walls at Bell Common tunnel.

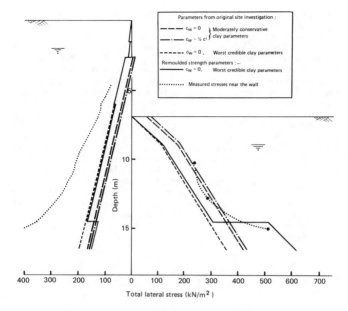

Fig. 7 Comparison of measured stresses with those used for the reassessment
of wall stability (after Carder and Symons, 1989)

CENTRIFUGE MODELLING

Centrifuge model tests of embedded retaining walls in stiff clay
have been carried out for the Laboratory by Cambridge University and
the results are described by Bolton and Powrie (1987, 1988) and
Powrie (1986). The principle involved is that if a model of a
prototype structure with a scale of 1:N is tested in a centrifuge
at an acceleration of N times earths gravity, then the self weight
stresses in the model will be identical to those in the full scale
structure. In addition the time required to achieve a given degree
of excess porewater pressure dissipation is reduced in the model by
a factor of N^{-2}. This meant that the effect on wall behaviour of
long term equilibration of porewater pressures, which would take
many years to achieve in the field, could be examined in the model
tests over a reasonably short period of time. A further advantage
of the centrifuge work was that it enabled conditions leading to,
and at collapse of the model walls to be studied and compared with
conventional theory.

Model tests were carried out mainly to simulate the behaviour of
rigid walls retaining 10m of overconsolidated clay. The principal

variables studied were the depth of penetration of the wall, the propping condition and the position of the water table. The tests showed that one of the severest conditions of stability for an unpropped wall in the short term can result from ingress of water into a tension crack which may form as the top of the wall rotates away from the retained ground. If a supply of water is available which can maintain the water level in the crack at a high level, then the crack can deepen causing further wall movement. Stability under these conditions might be viewed as a minimum requirement for any wall in stiff clay.

The failure of the model walls provided valuable insight into the movement zones and rupture pattern in the adjacent ground for comparison with conventional assumptions. The pattern of rupture lines observed in a test modelling an unpropped prototype wall with an initial penetration of 15m below excavation level is illustrated in Fig 8. From the study, simplified admissible strain fields were derived which idealized the soil behaviour in terms of uniformly deforming triangular blocks. The use of these in conjunction with relevant shear strain-stress data provides a potential method for assessing the movements of rigid walls in service.

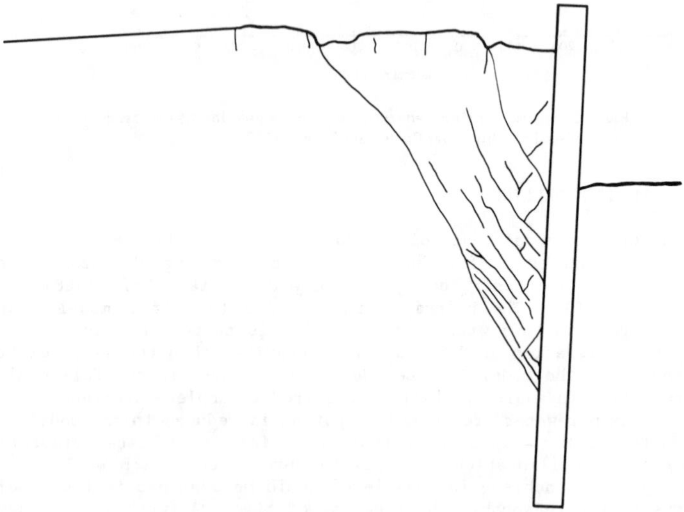

Fig. 8 Rupture pattern developed in the longer term
(after Powrie 1986)

SUMMARY AND CONCLUSIONS

This paper has briefly reviewed some of the results from a research programme being carried out by TRRL into the behaviour of embedded retaining walls.

The field studies during construction have shown that substantial ground movements can be caused by the process used to form the walls. At Hatfield these were a result of vibration and downdrag during driving of the sheet piles through the glacial sands and gravels. The bending moments measured on completion of excavation at this site also showed better agreement with values from a design method which takes account of interaction between flexible walls and the soil.

At the Bell Common tunnel these ground movements were a consequence of the reductions in in-situ lateral stress in the heavily overconsolidated clay during formation of the secant pile walls. Measurements at this site suggest that the inclusion of a compressible packing between the roof and wall had a significant effect on the subsequent behaviour. Results from finite element studies of the Bell Common tunnel indicate that improved predictions of wall behaviour can be obtained from numerical analyses which model the wall installation and closely follow the actual construction sequence.

Centrifuge modelling of stiff embedded walls in overconsolidated clay has enabled conditions leading to and at collapse, in both the short and long term, to be examined and compared with conventional theories. Analysis of the test data has suggested an approach for assessing the movement in service of unpropped walls and walls propped near the top, based on the use of simplified 'geostructural mechanisms' in conjunction with relevant stress-strain data.

Measurements of the pressures in London Clay close to an existing retaining wall near Reading together with the results from finite element analyses suggest that embedded walls may reach long term equilibrium with pressures in excess of the active values on the retained side and with full passive pressures acting, at least over the upper part, on the excavated side. This is a likely consequence of the reduction in vertical stress caused by excavation in soils with high initial in situ horizontal stresses.

Further studies of the type outlined in this paper are required particularly concerning the effect in stiff clays of different construction techniques used to form embedded walls and the long term pressures acting on such structures in service.

ACKNOWLEDGEMENTS

The work described in this paper forms part of the research programme of the Ground Engineering Division of the Structures Group of TRRL and is published by permission of the Director. The field measurements at the Bell Common tunnel were carried out in conjunction with the Building Research Establishment. The study on the A1(M) was a collaborative investigation carried out with Tarmac

Construction Ltd, Cementation Research Ltd, the Hatfield Polytechnic with funding from the Science and Engineering Research Council.

Crown Copyright 1990. The views expressed in this paper are not necessarily those of the Department of Transport. Extracts from the text may be reproduced, except for commercial purposes, provided the source is acknowledged.

REFERENCES

1. L Barden.   Sheet pile wall design based on Rowe's method. Technical Note 54 London: Construction Industry Research and Information Association 1984.
2. M D Bolton & W Powrie.   The collapse of diaphragm walls retaining clay. Geotechnique 37(3) 1987 335-353.
3. M D Bolton & W Powrie.   Behaviour of diaphragm walls in clay prior to collapse. Geotechnique 38(2) 1988 167-189.
4. D R Carder & I F Symons.   Long-term performance of an embedded cantilever retaining wall in stiff clay. Geotechnique 39(1) 1989 55-75.
5. B G Clarke & C P Wroth.   Analysis of Dunton Green retaining wall based on results of pressuremeter tests. Geotechnique 34(4) 1984 549-561.
6. C Garrett & S J Barnes.   The design and performance of the Dunton Green retaining wall. Geotechnique 34(4) 1984 533-548.
7. K G Higgins, D M Potts & I F Symons.   Comparison of predicted and measured performance of the retaining walls of the Bell Common Tunnel CR 124 Crowthorne: Transport and Road Research Laboratory 1989, 25p.
8. H W Hubbard, D M Potts, D Miller & J B Burland. Design of the retaining walls for the M25 cut and cover tunnel at Bell Common. Geotechnique 34(4) 1984 495-512.
9. T W Lambe. Predictions in Soil Engineering Thirteenth Rankine Lecture, Geotechnique 23(2) 1973 149-202.
10. C J Padfield & R J Mair.   Design of retaining walls embedded in stiff clay. Report 104 London: Construction Industry Research and Information Association 1984.
11. R B Peck.   Deep excavations and tunnelling in soft ground: state of the art report.   Proc 7th Int. Conf. Soil Mech. Foundn. Engng. Mexico State of the art volume 1969 225-290.
12. D M Potts & J B Burland.   A numerical investigation of the retaining walls of the Bell Common Tunnel SR 783 Crowthorne: Transport and Road Research Laboratory 1983 40p.
13. W Powrie.   The behaviour of diaphragm walls in clay.   PhD thesis Cambridge University 1986.
14. I F Symons, D M Potts & J A Charles.   Predicted and Measured behaviour of a propped embedded retaining wall in stiff clay Proc. 11th Int. Conf. Soil Mech. Foundn. Engng (4).   San Francisco 1985 2265-2268.

15. I F Symons, J A Little, T A McNulty, D R Carder & S
     G O Williams. Behaviour of a temporary anchored sheet pile
     wall on Al(M) at Hatfield RR99 Crowthorne: Transport and Road
     Research Laboratory 1987 40p.
16. I F Symons, J A Little & D R Carder. Ground movements and
     deflection of an anchored sheet pile wall in granular soil.
     Engineering Geology of Underground Movements, Geological
     Society Engineering Geology Special Publication 5 1988 117-127.
17. I F Symons & D R Carder. Long term behaviour of embedded
     retaining walls in over-consolidated clay.   Proc. Conf.
     Instrumentation in Geotechnical Engineering London. Thomas
     Telford Ltd. 1989, 155-173.
18. I F Symons & P Tedd. The behaviour of a propped embedded
     retaining wall at Bell Common Tunnel in the longer term.
     Geotechnique 39(4) 1989. 701-710.
19. P Tedd & J A Charles. Interaction of a propped retaining wall
     with the stiff clay in which it is embedded. Soil Structure
     Interaction Symposium ICE, I Struct. E. London 1984 47-49.
20. P Tedd, B M Chard, J A Charles & I F Symons. Behaviour of a
     propped embedded retaining wall in stiff clay at Bell Common
     Tunnel. Geotechnique 34 (4) 1984 513-532.

DIAPHRAGM WALLS - UPDATE ON DESIGN AND PERFORMANCE

William C. Kerr [1], M.ASCE and George J. Tamaro [2], F.ASCE

ABSTRACT: Following a brief summary of methods currently used to analyze diaphragm (slurry) walls, a parametric study is presented in which the effects of wall stiffness and subgrade modulus on bending moment and strut loads is shown. The special problems of cylindrical diaphragm walls and long walls forming narrow trenches are discussed.

BACKGROUND

The authors will share some of their experience in the design and construction of diaphragm walls for temporary and permanent use on a variety of projects located throughout the world.

The authors have been involved in projects of varying size and scope; some structures have had less than a dozen panels and form deep shafts of small area. Other have consisted of several dozen panels enclosing large plan areas. A few of these structures are circular in plan and the special problems will be discussed.

On these projects, the authors have employed methods which range from the ultimate in simplicity (continuous beam) to the reasonably complex (beam on elastic foundation) and have reviewed results developed by others and derived by advanced methods (FEM analysis). This paper discusses and compares the methods provided and draws conclusions from these comparisons. Recommendations and guidelines for a systematic design process for diaphragm walls are offered.

INTRODUCTION

The design/analysis process which leads to the successful construction of a diaphragm wall either for temporary or permanent construction must satisfy many special interests and concerns. For example:

The Structural Engineers want to provide structural elements which have thickness and reinforcing which satisfies code, and has sufficient reserve capacity to cover construction error.

1 - Associate, Mueser Rutledge Consulting Engineers, 708 Third Avenue, New York, NY 10017
2 - Partner, Mueser Rutledge Consulting Engineers, 708 Third Avenue, New York, NY 10017

They desire relatively high accuracy on computed reaction and bending moments and prefer loadings which are provided as "givens" by someone else.

The Soil Engineers want to control movements of the structure in order to minimize settlement and other damage-producing effects. They make conservative recommendations of soil strength parameters and desire watertight walls and joints in order to prevent groundwater drawdown.

The Contractor wants details to be simple and easily installed in the field. He desires to work as fast as possible and depart the site as early as possible. He sees large wall movements portending stop-work orders, extra costs and possible claims.

The Owners want no action taken by any of the above which will involve them in a cost or damage claim. They want all of the above to keep costs to a minimum.

In order to address these legitimate concerns, several decisions must be made during the design process and a clear thought out model must be established to form the basis for analysis. Some decisions are critical and of primary importance to the performance of the wall, other decisions are much less crucial than we might think and great effort in refinements in those areas may not be warranted.

DECISION TREE

Before proceeding with the design of the wall, a number of basic functional decisions must be made; these are:

1. Is the wall intended for temporary or permanent use, can the wall be designed to sustain temporary construction stresses which exceed levels permitted for permanent construction? Will residual stresses in the wall remain from construction, and will these redistribute with time due to creep and stress relaxation phenomena?

2. Are the structures behind the wall sensitive to movement and will they require particularly demanding control of support placement and excavation staging?

3. What are the groundwater conditions behind the wall and what particular problems with subgrade stability must be addressed?

4. What interferences between temporary strutting and permanent construction have to be addressed? Should top-down methods be used to eliminate the need for temporary supports?

Once the staging is worked out and the strut levels are established, the analysis should proceed stage-by-stage. Strut reactions, bending moments, shears and possibly deflections should be computed for each stage.

In order to proceed with the analysis, several technical decisions are made; these are:

1. Wall thickness is determined, always matching width of standard grabs. Associated with this is the question of what EI is to be used for wall analysis, gross section? cracked section? and in the case of a soldier beam reinforced wall, how is the moment shared between the steel WF member and the reinforced concrete?

2. Profiles of lateral pressure acting on both sides of the wall are developed; these should include active and at-rest pressures on the active side and at-rest and passive pressures on the passive side. The passive side pressures will of course change from stage to stage. Most diaphragm wall applications are in weak soils with a high water table. The distinction therefore between at rest and active pressure may not be of major importance. The authors generally take the loading to be active since some "at rest pressure" is lost during panel excavation and there is a tendency to place the first strut level well below top of wall.

    A number of organizations favor the use of "apparent pressure" distributions which are either trapezoidal or uniform pressures whose ordinate is proportional to the depth of excavation. Many designers analyze sheeting walls only for the final stage, assuming a reaction at or slightly below subgrade. Regardless of the pressure diagram used, this will surely underpredict the lower brace loads and the negative moment at the supports. In general, the authors feel that analysis of diaphragm walls by the so-called apparent pressure diagram is ill-advised. The use of these diagrams for design of rigid structures like diaphragm walls is open to question except for the simplest structures.

3. Toe penetration below final subgrade must be determined: This may be governed by a whole host of factors such as water cutoff considerations; bottom heave and basal stability; bearing capacity at the bottom of the wall to resist vertical loading or passive toe kickout. Each of these possibilities must be checked thoroughly before the final decision on toe penetration is made.

4. Strut (tieback) reactions must be modeled. For short cross-lot braces or rakers, the assumption of a rigid pinned support is usually adequate. Wales spanning large distances between struts may be modeled as an elastic spring where the spring constant is proportional to the Modulus of Elasticity and Moment of Inertia and inversely proportional to the length for the wale. The authors have also represented the permanent concrete slab in top-down schemes as a spring and shrinkage gap. Tiebacks should always be represented as elastic springs.

5. The passive subgrade must be modeled. A traditional method of analysis is to adopt the point of zero net pressure (level

where pressure on active side equals passive pressure) as a
point of zero bending moment and, therefore, an equivalent
simple beam support.   More sophisticated approaches dubbed
"soil-structure interaction methods" require the soil to be
represented elastically (with a passive yield limit) by
springs. This of course gives need for determination of
spring constants, a decision which may appear to be crucial!
The value of the spring varies with soil type, depth of
penetration, water levels, stress level, width of excavation
and cannot be determined in the laboratory.   A number of
techniques exist for determination of subgrade springs. The
pressuremeter test is currently in favor and has been used
successfully for piles subject to lateral load. Baguellin (1)
recommends against use of pressuremeter for diaphragm walls.
Rarely does the budget and schedule for a job permit full
scale measurement of subgrade modulus but an interesting test
was described by McClelland (4).   The senior author has had
some success in back figuring spring constants from inclinometer data and this will be reported on in a later paper.

The authors have sought to develop a systematic study of the
passive subgrade as an elastic continuum, perhaps using the
Finite Element Method (FEM), and determining influence coefficients which could replace the Winkler foundation concept.
This work is in progress and not reported on at this time.

6. Finally as the analysis proceeds, a reliable method for
   accounting for the wall movements at the support prior to
   support installation must be a part of the numerical procedure.

The authors will discuss each of these decisions in a later
section of the paper.

METHODOLOGY OF DESIGN

Over the years, several methodologies have evolved for designing temporary earth retention structures.   The earliest of these
was developed for open-sheeting systems through which groundwater
would be allowed to percolate and which were relatively flexible.
Further deflections were controlled by relatively stiff strutting
near the top of the excavations and sensitive structures behind
the walls were underpinned. Of these, the Terzaghi & Peck (7)
apparent pressure diagram method is the most well-known; it was
developed for situations with open lagging and no control of
seepage. It is used when strut spacing is regular and lagging
does not extend below subgrade.   Characteristically the toe
penetrates into stiff competent stratum.   These methods were
suitable for hand calculation (slide-rule) and the soldier beam
design and strut loads could be based on them.   They had the
further advantage that they could be substantiated and verified
with field measurements which had been obtained notably in
connection with subways being built in Chicago and Berlin.

Later, more advanced concepts were developed which recognized the concept of "soil-structure-interaction." The beam-on-elastic-foundation problem was formulated and solutions to many problems of practical interest are given by Hetenyi (3). Haliburton (2) developed a useful program which allows the Winkler foundation to be non-linear, and most importantly, to have a passive limit. These techniques were adopted in response to the demand for more watertight and more rigid systems such as diaphragm walls. These systems usually require a staged analysis particularly with the advent of top-down-construction or when the strut levels may not be regularly spaced. This staged analysis often indicates that the temporary construction stages are more critical to the wall than the permanent loading.

These methods introduced the soil modulus as a parameter in the analysis and allowed the structure to be modeled as a continuous beam. Although the model is possibly more rigorous and satisfying to the purist, it does not possess the ease of application offered by the earlier methods and the uncertainty in the selection of elastic properties for the soil is always with us.

At the present time, a number of very powerful methods and programs exist which make it possible to simulate the behavior not only of the wall but the complex mass of soil with which it interacts. These finite-element models needless to say are exclusively computer oriented and not limited to linear elastic response of the soil. In order to derive reasonable results from these programs, it is necessary to input parameters which represent the full stress-strain properties of the soil for each layers and represent its non-linear behavior including plastic yield. The inherent danger in the use of these methods is that the essential parameters are not really known, therefore the results of analysis are the result of initial "Guesses" and can be incorrect by orders of magnitude.

The authors describe briefly the four approaches to design; these are termed:

a. Equivalent Beam on Rigid Supports Methods (RIGID)
b. Beam on Elastic Foundation Method (WINKLER)
c. Finite Element Method (FEM)
d. Limit Analysis (LIMIT)

In all of these methods, the wall is presumed to be elastic and continuous, usually with a constant rigidity (EI). After a brief discussion of these methods, the authors will advance the notion of applying limit design principles (such as those used for super-structures) to design diaphragm walls. Plastic or ultimate design is well recognized and the correct analytical procedure for these methods requires "Limit Design" analysis not elastic analysis.

RIGID ANALYSIS

The salient features of the so-called rigid analysis are listed below:

* Fictitious support below subgrade at zero net pressure.
* Wall/beam below this point is neglected in analysis.
* Supports are presumed to be rigid and movement prior to installing strut is neglected.
* Loading can be given by Rankine pressure distribution or can represent an apparent pressure distribution. The load, however derived, is a following load and no effort is made to redistribute the pressure as a result of wall movement.
* Method is simple and requires knowledge only of the geometry and of the limiting pressure profiles on each side of the wall.
* Uses conventional structural analysis and widely accepted programs that are readily available.
* The model is reduced to an elastic beam which is continuous over rigid supports.

The method gives the structural engineer answers which he needs to design the wall but the soils engineer is never completely satisfied because there is no prediction of the ground movements occurring behind the wall. Contractor likes this approach because its conservatism of design tends toward heavy reinforcing and braces and these are profit-making items. The owner is not a part of the decision making process.

WINKLER ANALYSIS

The features of the WINKLER analysis are summarized below; they are:

* Assumes WINKLER springs acting as passive foundation resistance.
* Braces need not be assumed rigid, can be given a K value.
* Total depth of wall is included in analysis.
* Method requires an estimate of "elastic" properties of soil.
* Limits can be placed on maximum WINKLER spring value.
* Method will probably give more realistic strut reactions at the lower levels.
* Method gives more insight into magnitude and character of inward wall movement and resulting surface settlement.
* The method gives the structural engineer answers he needs to design the wall but the soils engineer is never satisfied with spring selection. The contractor sees delay is receiving bid documents because of this controversy.

Several approaches have been adopted to obtain values for the Winkler Subgrade. These are:

1. Full scale measurement by load-testing of prototype pile or caisson, as described by McClelland (4). However, the application of results obtained by this test may be misleading when applied to diaphragm walls.

2. Pressuremeter tests as described by Baguellin (1) are used to determine subgrade modulus at a single depth. The author discourages application of results to diaphragm walls without giving any alternative procedure.

3. Cross-trench measurement by jacking laterally against the sides, basically a plate loading test.

4. Backfigure subgrade module from measurement, using in-clinometer data gathered during the excavation of a site (other than the site in question). A technique is presently under development in which the beam equation is solved directly for the spring constant at any number of points along the wall.

Each of these approaches suffers from some limitation: full-scale measurements are notoriously expensive; pressuremeter tests may be inconclusive; cross-trench measurements are impractical at large depths, particularly below the water table and good in-clinometer data from nearby sites may not be available. In the absence of actual test values, a widely referenced source of information on the subject is given by Terzaghi (8).

As the authors will show, precise values may not be as critical to the design as one might think and having only order-of-magnitude values may quite adequate.

The authors favor use of the WINKLER method provided there can be some rational method of selection of the soil modulus values to be used. A particularly interesting application of this method is in the case of diaphragm walls which form a circular enclosure, to be described in a later section of this paper.

Before proceeding further, the authors suggest that an analysis is available which is intermediate in difficulty between the RIGID and WINKLER analyses. This analysis is basically the RIGID analysis with the deflection computed at the brace points prior to installing the braces. These deflections are held constant thereafter. The results should be more realistic than they would be with these deflections neglected.

FEM ANALYSIS

Finite element analyses have received wide attention and are acclaimed by some to be the state-of-the-art analysis for dia-phragm walls. Certainly they are the most advanced techniques available for predicting soils movement in the vicinity of the wall. They are not yet a design tool but can be useful in para-metric studies and as an aid in selection of soil parameters.

Some observations which can be made are the following:

* Generates too much data to be useful as a design tool.
* May seriously underpredict bending moment in the wall and underpredicts brace loads. (Implication of this is that it makes passive soil do too much work in resisting total thrust.)

* Model requires input of parameters for which there may be no widely accepted test methods (or budget and time to run such tests, if they are available.)

The soils engineer is generally attracted to this method because he may think he is getting "precise" estimates of settlement and other movements behind the wall but the structural engineer is not happy because the results often run contrary to past experience.

The contractor is not happy because the analysis indicates the need to instrument and monitor surrounding structures adjacent to the site and this makes the owner uneasy about the fact that his neighbors are now being made aware of the potential for damage to their properties from his construction activity.

LIMIT ANALYSIS

A totally novel approach to diaphragm wall design is the application of limit principles. Limit design is well known to superstructure designers and is well accepted when applied to building frames. The authors are not aware of any previous use of limit principles in foundation work, but have often wondered if significant economies could be achieved by their applications.

The method is simple, requiring no more information than is required for RIGID analysis.

* Allows balancing of steel on both sides of wall.
* Simple analysis (paper and pencil??)
* Structural engineer likes it, he is familiar with it.
* Soils engineer not happy because it does not give him information about ground movements behind wall.
* Contractor likes this approach because it can be turned around quickly.

Ironically the LIMIT method brings us almost full circle in which the bending moment at the struts is assumed (to be the yield moment) and a statically determinate system is analyzed.

COMPARISON BETWEEN COMPUTED RESULTS

A common design requires the analysis of a watertight wall installed in a generally granular deposit with the water table near the existing grade level. Frequently this ground can be assumed to be homogeneous in density and PHI-angle and to have a constant coefficient of subgrade modulus, Nh. It is instructive to examine the effects of various assumptions which are made prior to doing the analysis. The analytical methods chosen for comparison are the RIGID method and the WINKLER method. Also included in the table are results of an apparent pressure diaphragm analysis for the final stage only, (based upon an apparent effective pressure of 65% of active at subgrade). This is one of those formulations attributed to Terzaghi and Peck (7) and is denoted as such. A point of support was assumed at subgrade. The parametric choices which are chosen for comparison are soil

loading, as characterized by density, and PHI-angle; brace stiffness, wall stiffness (EI), and subgrade modulus (Nh).

For simplicity the groundwater table is taken to be at existing
grade elevation on the active side and at subgrade elevation on
the inside of the excavation as shown in Figure 1.

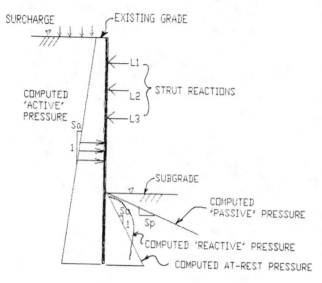

Figure 1: Section Through Diaphragm Wall
With Loading and Reactions

The following factors must be considered:

Soil Loading.   Table 1 shows loading and passive resistance
parameters for granular soils ranging from very dense (PHI=36) to
very loose (PHI=28).   The subgrade modulus parameters Nh are
suggested as values which result in reasonable deflections for the
soil considered. The parameters Sa and So are rates of total
pressure (effective plus hydrostatic) increasing with depth.   It
is seen that these rates are essentially the same for all of the
soils included in the table.   Thus we can immediately conclude
that uncertainty in the choice of loading parameters is of little
concern and the only question that needs to be resolved is whether
the wall will experience enough movement to mobilize an active
state or not.   The difference between at-rest and active is about
15%. Of course, if the water table were lower the difference
would be greater since that loading above the water table is
directly proportional to the lateral pressure coefficient.

Table 1: Soil Loading Parameters

File	Wt (pcf)	PHI (o)	Ka	Ko	*Kp	Sa (pcf)	So (pcf)	Sp (pcf)	Nh (kcf)
DOO1	135	36	0.24	0.41	7.5	80	92	607	34
DOO2	130	34	0.28	0.44	6.5	81	92	502	32
DOO3	125	32	0.31	0.47	5.5	82	92	407	30
DOO4	120	30	0.33	0.50	4.9	81	91	345	28
DOO5	115	28	0.36	0.53	4.4	81	90	294	26

1 pcf = 164 N/m^3
Sa= Ww + Ka Wb   *WALL FRICTION ANGLE = PHI/2
So= Ww + Ko Wb   Ww= 62 pcf (density of water)
Sp= Ww + Kp Wb   Wb= Wt - Ww

Subgrade Resistance.   The uncertainties in choosing parameters
that represent passive resistance are greater both in choice of
PHI and assumptions of wall friction angle. The parameter Sp is
the rate of increase of total passive resistance and this increases markedly with increased soil friction angle. The values
given here are for the assumption of wall friction angle equal to
PHI/2.   There is concern among some practitioners that this value
is too high considering a bentonite film between the wall and the
soil but it can be reasoned on the other hand that it is really a
bentonite/cement film on a rough wall which intrudes into the soil
and provides a positive shear connection with the soil, which
would provide an even greater wall friction angle, possibly equal
to PHI!

Brace Stiffness.   The practical problem with incorporating
brace (tieback) elastic properties into the analysis is that the
contractor often designs the brace system based upon loads given
to him by the designer of the wall. This means that the designer
has had to "guess" stiffness values based upon a preliminary
design or what is commonly used in the field.   Small differences,
even a factor of 2 or 3 should not be a problem since brace
systems are stiff relative to the soil.   A factor of 10 or more
can be significant.

Another interesting issue arises in top-down schemes when
permanent concrete slabs are used as struts.   It is well-known
that shrinkage and creep as well as temperature change will occur
and reduce the effectiveness of the slab as a strut as far as wall
movement is concerned.   In one case a FEM analyses run by others,
for a deep excavation where slab shrinkage was expected to have a
major effect on wall movements.   The results were found to be
sensitive to the way in which shrinkage was being modeled.

Wall Stiffness. Knowing that a concrete diaphragm wall may
crack as the bending moment increases, it seems quite natural to
include the effect of reduced EI in the analysis. As will be
shown, however, this is not as important as was thought. A

rigorous analysis from stage to stage with EI constantly reflecting the variations of moment is simply not a practical design
office effort. It is highly iterative and not as straight forward
as one might expect. For example, there is the question of what
EI is appropriate following cracking and then moment reversal?

Soil Stiffness. The WINKLER elastic foundation concept
provides a way for structural engineers to obtain reasonable
answers even though it has no fundamental basis in theory and
should not be expected to provide exact answers.

The results of parametric studies are presented in the following tables. Each table shows the results of a staged analysis of
a 50 ft. deep excavation in homogeneous granular submerged soil.

Table 2 below shows the results of five staged analyses and the
T&P analysis for final stage made for a 50 ft. excavation in dense
soil with a PHI=36. The table includes the maximum strut reaction
at each level, the maximum positive and negative bending moments,
maximum deflection and the volume of displacement for a running

Table 2:   Summary Of   Results (PHI=36)

	DOO1	EOO1	FOO1	GOO1	HOO1	T & P
Reaction L1 (kip/ft.)	12.2	13.5	13.5	12.9	14.2	15.0
Reaction L2 (kip/ft.)	49.1	32.9	32.6	31.2	35.2	15.6
Reaction L3 (kip/ft.)	47.2	44.8	43.9	44.1	44.3	21.1
Reaction L4 (kip/ft.)	74.8	64.3	64.8	63.8	65.3	39.8
M(-) (k-ft/ft)	151	108	110	105	110	44.4
M(+) (k-ft/ft)	129	158	153	152	160	40.6
Wmax (ft)	---	0.033	0.049	0.027	0.062	----
Volume Behind Elastic Curve (cu. ft./ft.)	---	1.5	1.9	1.1	2.7	----

DOO1:   RIGID analysis (deflections only between struts)
EOO1:   WINKLER analysis; Nh=34kcf, EI=gross value
FOO1:   WINKLER analysis; Nh=34kcf, EI=(1/2)gross value
GOO1:   WINKLER analysis; Nh=68kcf, EI=gross value
HOO1:   WINKLER analysis; Nh=17kcf, EI=(1/2)gross value
1 ft.=.3048 m; 1 kcf=164 kN/m^3; 1 kip/ft.= 4.448 kN/m;
1 kip-ft/ft=4.448 kN-m/m

foot of wall, computed from the staged analyses. The reactions
and moments computed by the T&P analysis are also shown for
comparison.

The tabulation shows that in general, the RIGID method gives
the most conservative estimate of strut reactions. The rigid
method also tends to overpredict the negative bending moment with
the result that the steel provided on the outside face is
probably excessively conservative. Examining the other results in
this table it can be seen that the WINKLER analysis yields smaller
strut reactions and thus assumes more of the load is supported by
the soil below subgrade. This causes a larger positive moment in
the wall near the subgrade level.

When one considers the wide range represented by an uncracked
wall and stiff soil (doubling Nh) at the high end and a fully
cracked wall with loose soil (1/2x Nh) at the low end; the results
are essentially the same. Of course estimates of maximum deflections and volume of displacement are going to be considerably
influenced by these parametric variations.

Tables 3 and 4 are presented for the same depth of excavation
in moderately dense soil (PHI=32) and loose soil (PHI=28), respectively. Similar trends are noted as for the more dense soil and
the conclusions that are to be drawn are similar. There are three
conclusions to be drawn from these comparisons; they are:

1. A standard continuous beam/rigid support analysis may result
    in a wall and strut system that is needlessly overdesigned and
    have excessive thickness and reinforcement. Certainly this
    produces a safe design for the intended purpose and has excess
    capacity in the event a problem occurs during construction
    (When using this method the authors routinely use 20%
    increase in allowable stress for temporary stages).

2. A WINKLER analysis can give reasonable predictions of struc-
    tural quantities of interest to the structural designer of the
    wall and large errors in modeling soil stiffness have little
    effect on the results. The authors recommend that no increase
    in allowable stress be used in this analysis.

3. The T&P final stage analysis far underpredicts the strut loads
    and bending moments are unbelievably low. At issue here is not
    so much what pressure diagram to use but rather the fact that
    each excavation stage must be investigated. It is the ex-
    perience of the authors and others that a strut experiences its
    highest load at the end of the excavation stage following
    installation.

Table 3:   Summary Of   Results (PHI=32)

	DOO3	EOO3	FOO3	GOO3	HOO3	T & P
Reaction L1 (kip/ft.)	12.1	12.0	12.0	11.7	12.4	15.2
Reaction L2 (kip/ft.)	26.1	29.7	29.4	28.5	31.0	9.0
Reaction L3 (kip/ft.)	44.5	38.2	37.8	37.7	38.2	19.0
Reaction L4 (kip/ft.)	62.8	50.1	51.1	49.9	51.3	17.8
Reaction L5 (kip/ft.)	113.9	68.4	69.5	67.7	70.4	34.6
M(-) (k-ft/ft)	280	124	125	125	130	31.9
M(+) (k-ft/ft)	192	180	175	175	180	28.8
Wmax (ft)	---	0.044	0.070	0.038	0.080	----
Volume Behind Elastic Curve (cu. ft./ft.)	---	2.1	2.8	1.7	3.6	----

DOO3:   RIGID analysis (deflections only between struts)
EOO3:   WINKLER analysis; Nh=30kcf, EI=gross value
FOO3:   WINKLER analysis; Nh=30kcf, EI=(1/2)gross value
GOO3:   WINKLER analysis; Nh=60kcf, EI=gross value
HOO3:   WINKLER analysis; Nh=15kcf, EI=(1/2)gross value
1 ft.=.3048 m; 1 kcf=164 kN/m^3; 1 kip/ft.= 4.448 kN/m;
1 kip-ft/ft=4.448 kN-m/m

Table 4:   Summary Of   Results (PHI=28)

	DOO5	EOO5	FOO5	GOO5	HOO5	T & P
Reaction L1 (kip/ft.)	4.5	6.0	6.0	5.6	6.6	4.2
Reaction L2 (kip/ft.)	20.5	20.9	21.8	20.5	22.8	10.5
Reaction L3 (kip/ft.)	39.1	33.6	33.7	33.4	33.6	13.7
Reaction L4 (kip/ft.)	51.0	48.4	49.3	48.2	49.4	18.2
Reaction L5 (kip/ft.)	86.5	65.3	66.8	65.2	67.0	20.8
Reaction L6 (kip/ft.)	147.5	62.5	63.4	61.9	64.2	29.8
M(-) (k-ft/ft)	416	140	140	140	140	22.5
M(+) (k-ft/ft)	258	238	233	236	235	17.8
Wmax (ft)	----	0.076	0.130	0.072	0.140	----
Volume Behind Elastic Curve (cu. ft./ft.)	----	3.0	4.7	2.7	5.3	----

DOO5:   RIGID analysis (deflections only between struts)
EOO5:   WINKLER analysis; Nh=26kcf, EI=gross value
FOO5:   WINKLER analysis; Nh=26kcf, EI=(1/2)gross value
GOO5:   WINKLER analysis; Nh=52kcf, EI=gross value
HOO5:   WINKLER analysis; Nh=13kcf, EI=(1/2)gross value
1 ft.=.3048 m; 1 kcf=164 kN/m^3; 1 kip/ft.= 4.448 kN/m;
1 kip-ft/ft=4.448 kN-m/m
==

CYLINDRICAL DIAPHRAGM WALLS

A case of particular interest occurs when a diaphragm wall cofferdam encloses a circular area. The individual panels are practically never given a curvature but are cast either as straight segmental panels, or in some cases, a panel consisting of three monolithic planes. Figure 2 shows a 30" thick diaphragm wall constructed in Boston, Massachusetts, Tamaro (6) consisting of twelve panels enclosing a circle of 75 foot diameter. The bottom of the panels were toed into firm Argillite rock. The wall is 30" thick with end joints formed by 30" wide-flange beams.

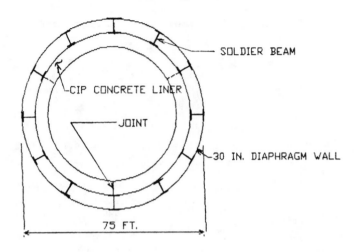

Figure 2: East Boston Pump Station -- Plan

The design required that the diaphragm wall would serve only as a temporary cofferdam and that the pump station be built therein with its own cast-in-place (CIP) concrete wall in direct contact with the diaphragm wall as shown in Figure 3. The rings were to be cast in sequence from ground level downward with one "ring" of excavation immediately followed with the casting of the ring, repeating the process until the base slab is cast. It was realized that the permanent liner rings could serve to limit radial movement of the panels due to circumferential shrinkage as well as functioning as wales to limit vertical bending during construction of the pump station. The designers could foresee two extreme possibilities which would have profound effect on the behavior of the wall: (1) the construction could be of exceptionally good quality resulting in tight joints with contact between the panels, (2) the construction could be of such a poor quality that all of the joints were open and the panels would not transmit hoop stress.

In the "best case" scenario, the diaphragm wall itself would be the principal structural system, acting as a short cylindrical shell under radial load. The inner CIP wall would pick up very little if any of this load. Conversely, in the "poor case" scenario the diaphragm wall panels would act as a series of unconnected barrel staves which would act structurally as beams supported at or below subgrade and at the inner wall rings. These rings would then have to be designed as a wale support for the diaphragm wall.

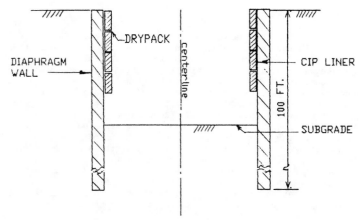

Figure 3ı East Boston Pump Station -- Cross Section

Construction Stage

For this case the designers selected the Winkler approach, modeling the soil, ring beams and slurry wall as a beam on elastic foundation. A cylindrical shell under symmetric radial load can be reduced to an equivalent beam on elastic foundation, as shown by Timoshenko (9), where the subgrade modulus is given by E, (Young's modulus of the shell), multiplied by H: (thickness) and divided by R: (radius). Subgrade moduli were estimated from pressuremeter tests for the Boston Blue Clay and the Argillite that underlie the site. These turn out to be rather incidental terms compared to the stiffness of the shell itself. When the "best case" analysis was run for each construction stage, the results confirmed the notion that the diaphragm wall could function without benefit of interior ring beams, and that the stresses in the wall were nominal and well within the allowable limits of the ACI code. When the "worst case" analysis was run for the same stages the results showed that substantial bending moments could be expected in the diaphragm wall panels requiring reinforcement by high capacity soldier beams.

Since the designers, had no real control over what the construction practices and quality would be, they recommended that the worst case analysis be the basis for the design shown on the contract drawings. This job represents a dramatic illustration of the case where the performance of the wall, the bending moments and the deflection were totally influenced by the quality of the contractor's work in the field and by the selection of construction procedures. It shows once again that seemingly large uncertainty in selecting some of the constants of subgrade modulus, are really over-shadowed by the practices adopted by the contractor.

NARROW COFFERDAMS

   The problem of the narrow cofferdam has not received sufficient attention in recent literature. In a narrow cofferdam the passive resistance and movement of the diaphragm wall is influenced by the opposite wall, usually also a diaphragm wall. It is reasonable to expect that in a trench having a width which is substantially less than the toe penetration of the diaphragm walls the passive resistance and the subgrade reaction will be affected by the opposite wall. In many designs, the passive resistance and the subgrade soil stiffness have been underestimated resulting in requirements for an excessive toe penetration of the diaphragm wall. It is unreasonable to consider that the subgrade modulus does not reflect the aspect ratio of penetration depth to width, and the confinement offered by the opposite wall. Field measurements reported by Miyoshi (5) tend to support this point of view. Further theoretical studies are needed to better define the effect of the opposite wall and to evaluate the effect of the aspect ratio.
   The authors recently developed a procedure for estimating the effect of confined subgrade while preparing criteria for the design of the Taipei MRT. Figure 4 shows the empirical approach which was adopted at that time. Three zones are defined; they are:
Zone 1: Passive pressure a function of subgrade level and completely unaffected by opposite wall.
Zone 2: Transition zone between zone 1 and zone 3.
Zone 3: Passive pressure totally influenced by opposite wall and not a function of subgrade level.

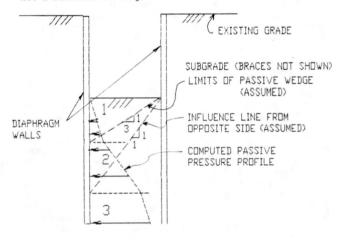

Figure 4 Narrow Trench Cross-Section
              Construction Stage

A major difficulty with this approach is the fact that the subgrade modulus is not really a fundamental property but is a computational device developed by structural engineers to analyze structures in contact with soil   The continuum nature of the subgrade makes the real problem much more complicated.   A number of FEM analyses of passive subgrade of varying rigidity width and depth would provide a better understanding of the subgrade behavior.   In these studies now being carried out in our office, we impose a deflection which is measured from actual diaphragm walls on the vertical face between the wall and the subgrade soil in order to back-figure the spring constant.

CONCLUSIONS

Exhaustive parametric studies for a deep excavation in granular soils show that RIGID analyses generally are more conservative than WINKLER analyses; furthermore when temporary overstresses are allowed, the RIGID results provide reasonable structural design of the wall.   The studies also show that the results are relatively insensitive to large variations in the parameters which represent subgrade stiffness, strut stiffness and wall flexural rigidity. Lastly, the studies illustrate the importance of performing the analysis of each stage of construction and not just the final stage.

The authors have discussed two important technical problems; the problem of ring supported cylindrical diaphragm walls and the problem of determining the subgrade behavior of a narrow cofferdam and described their approaches to solving those problems.

Finally it may be concluded that watchful control over the contractor and careful attention to construction details and practices are in the long run far more important to the performance of a diaphragm wall than are some of the more esoteric details of analysis.

REFERENCES

(1)   F. Baguellin, et al., The Pressuremeter and Foundation Engineering, Trans Tech. Publ.,Clausthal, Germany, 1978, 617 pp.

(2)   T. Haliburton, Soil Structure Interaction, T.P. 14, Okla. State Univ., Stillwater OK, 1979, 181 pp.

(3)   M. Hetenyi, Beams on Elastic Foundation, Univ. of Mich. Press, Ann Arbor, 1946, 255 pp.

(4)   B. McClelland & J. Focht, Soil Modulus for Laterally Loaded Piles, Proc. ASCE, Vol 82, sm-4, 1956, Paper no. 1081, pp 1-22

(5)   M. Miyoshi, Mechanical Behavior of Temporary Braced Walls, Intl. Conf. of SM&FE (1), Tokyo, 1977, pp. 655-658.

(7)   K.   Terzaghi   &   R.B.   Peck,   Soil   Mechanics   in Engineering
      Practice, 2nd Ed., Wiley, New York, 1967, 566 pp.

(8)   K. Terzaghi, Evaluation of Coefficients of Subgrade Reaction,
      GEOTECHNIQUE, Vol. 5, 1955, pp 297-326

(9)   S. Timoshenko, Theory of Plates and Shells, 2nd Ed., McGraw-
      Hill, New York, 1959, 580 pp.

GEOTECHNICAL DESIGN AND PERFORMANCE OF A PRECAST CANTILEVERED RETAINING WALL

William H. Hover[1], M.ASCE, Franklin M. Grynkewicz[2],
Rupert Hon[3], M.ASCE and Laleh Daraie[4]

ABSTRACT: A proprietary precast, braced retaining wall up to 20 feet (6.1 m) high and 350 feet (106.7 m) long was used for the first time in the United States in North Attleborough, Massachusetts. Precast concrete panels were erected on unreinforced concrete footings, which bear in weathered bedrock. An inclined tension member connects the panel to a reinforced concrete base slab.

Inclinometers indicated a lateral movement of 0.3 inch (7.6 mm) at the top of the wall after backfilling, but no lateral movement below grade at the toe. Sufficient rotation occurred to develop active loading conditions, and earth loads were transferred to the tension member, as indicated by concrete embedment strain gage data.

The average axial load in the tension member after backfilling, interpreted from strain gage data, was approximately 15 tons (13.35 kN). This load is 60 to 70 percent of design axial loads based on triangular Rankine and Coulomb active lateral earth pressure distributions, respectively. The data suggests that arching active condition and mobilization of wall friction occurred, since the tension member did not experience the theoretical active load. The strain gage data indicated that internal lateral forces at the base of the vertical panel were very low, thus the triangular Rankine active design pressure diagram is conservative for this case.

INTRODUCTION

Proprietary pre-cast, braced, cantilevered retaining wall systems have been used in Europe for more than twenty years to support highway cuts and fills, as bridge abutments and wing walls, and for other earth retaining applications (1). This paper presents a case history of the geotechnical design and short-term performance of a 20 foot (6.1 m) high proprietary internally braced, precast, cantilevered wall system. This installation is the first in the United States.

1,2,3,4 - Associate, Sr. Project Manager, Technical Specialist, Geotechnical Engineer, Goldberg-Zoino & Associates, Inc., Newton Upper Falls, Massachusetts, 02164

BACKGROUND

Project Description. The project comprised of 350 lineal feet
(106.7 m) of precast retaining wall installed to support
differential earth cuts of up to 20 feet (6.1 m) between upper and
mid-level parking facilities at a new regional mall facility.
Figure 1 shows the plan configuration of the U-shaped wall, and
original and final site grading. The structural components of the
wall were installed between November, 1988 and January, 1989. The
wall was backfilled between January and April, 1989.

Wall System. The braced wall system has three primary structural
components, described in order of their assembly on-site. Figure 2
presents a plan and typical wall section. A continuous unreinforced
concrete foundation block or footing was constructed by excavating a
trench to design depth (typically 4 to 6 feet (1.2 to 1.8 m)), and
maintaining stable side-wall slopes. The trench was backfilled with
unreinforced structural concrete of 3,000 psi (20.7 MN/m^2) minimum
compressive strength. The top of the foundation block was about 3
inches (76 mm) above grade.

Tee-shaped panels of reinforced concrete, each 4 feet (1.2 m)
wide and up to 20 feet (6.1 m) high, were pre-cast at a plant under
quality control conditions and delivered to the site. They were
placed in a vertical position on the foundation block using a small
crane, and were aligned vertically and laterally using wedges and
temporary timber shoring.

A curvilinear precast tension member unfolds from the panel to an
inclined position when the vertical panel is lifted. The tension
member is connected to the stem of the vertical panel using a steel
pipe as a hinge.   The pipe was inserted into space formed by
reinforcing steel protruding from both the panel stem and tension
member, such that the steel loops about the pipe.   The pipe and
reinforcing steel were surrounded by forms which were filled with
non-shrink waterproofing. Bituminous waterproofing was then applied to the
joint.   After panel erection the subgrade was cleaned, and a
reinforcing cage was tied to reinforcing steel protruding from the
base of the vertical panel and from the bottom of the inclined
tension member.   A structural concrete base slab was then cast,
followed by backfilling operations.

Subsurface Conditions. The site is located in a glaciated
region. At the retaining wall the site was originally underlain by
15 to 25   feet (4.6 to 7.6m) of dense to very dense glacial till,
above severely weathered, highly fractured conglomerate bedrock that
could generally be excavated with large backhoes. The glacial till
is comprised of fine to medium or fine to coarse sand, with 20 to 30
per cent silt, 20 to 25 per cent gravel and cobbles, and less than 5
per cent clay. Pre-construction groundwater levels were within 5
feet (1.5 m) of original grades.

Cuts of 5 to 25 feet (1.5 to 7.6 m) resulted in complete removal
of the glacial till on the toe side of the wall. The foundation
block and base slab bear on weathered bedrock, and the cut face
behind the heel is comprised of very dense glacial till. Figure 2
illustrates typical subsurface conditions at the wall.

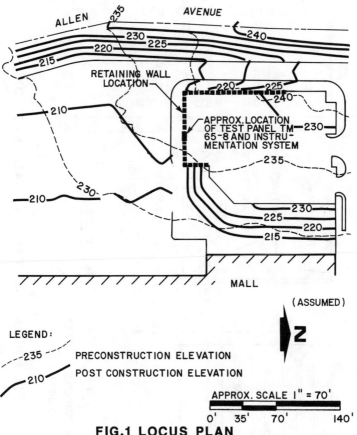

FIG.1 LOCUS PLAN

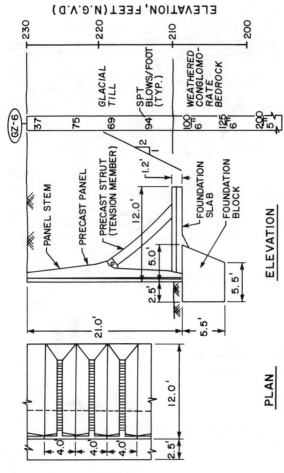

PLAN

NOTE: DIMENSIONS IN FEET.

I FOOT = 0.3048 METER

FIG.2 PLAN AND TYPICAL WALL SECTION

GEOTECHNICAL ASPECTS OF WALL SYSTEM DESIGN

   Properties of Foundation and Backfill Materials.   The foundation
stratum is very dense weathered conglomerate bedrock with a design
net allowable bearing capacity of 10.0 tsf (957.6 kN/m^2) based on
standard penetration tests and judgement.   Wall backfill consists
primarily of glacial till from on-site excavations. A 2-foot (0.61
m) wide sand-gravel drainage zone and non-woven fabric filter extend
the full height of the wall panels. A 1.5-foot (0.46 m) thick sand-
gravel drainage blanket and non-woven geosynthetic filter were
placed above the base slab.
   Prior to general backfilling, compaction tests were performed in
accordance with ASTM D-1557 Method C (2) on representative samples
of glacial till fill materials. The tests determined a maximum dry
density of 138.5 pcf (21.76 kN/m^3) at an optimum moisture content of
6.5 percent.
   During backfilling, consolidated isotropically drained (CID)
triaxial tests (3) were performed on two samples of glacial till
backfill reconstituted to a dry density of 133 pcf (20.89 kN/m^3).
This corresponds to 96 per cent of maximum dry density at a moisture
content of 7 percent, or a total unit weight of 140 pcf, which was
selected based on in-situ density tests after compaction.   The CID
tests determined a drained friction angle of 37 degrees.
   Wall Stability Analyses.   Estimated factors of safety against
overturning are 2.2 and 1.5, for the cases of static earth pressure
plus compaction and traffic surcharge, and static earth pressure
plus seismic loads, respectively.   Estimated factors of safety
against sliding are 1.6 and 1.1, likewise. These factors are based
on the earth pressures shown on Figure 3 and are within the range of
normally accepted practice.
   Design Earth Pressures. Design earth pressures specified for the
instrumented panel portion of the wall system are shown on Figure 3.
The design pressure diagrams include Rankine active earth pressures
(4) for drained conditions, K_a=0.25, friction angle = 35 degrees,
angle of friction between panel and backfill = 20 degrees, unit
weight = 135 pcf (21.2 kN/m^3), and surcharges from compaction
equipment, vehicular traffic, and seismic loads in accordance with
the Massachusetts State Building Code (5). The compaction equipment
loads shown on Figure 3 assume a vibratory plate compactor weighing
no more than 0.6 tons (1.33 kN) within 6 feet (1.8 m) of the rear
face of the wall panels and a vibratory drum roller with a static
weight not exceeding 2.5 tons (22.2 kN) with edge of closest drum at
least 6 feet (1.8 m) from the rear face of the wall.
   Theoretical Tension Member Loads.   After backfilling, a
theoretical load of 25.35 tons (225.5 kN) was calculated for the
tension member based on a Rankine active loading condition, a unit
weight of 140 pcf (22 kN/m^3) and a drained friction angle of 37
degrees. Wall friction was considered to have a negligible effect
on the active earth pressure coefficient (4). A theoretical active
load of 22.0 tons (195.7 kN) was calculated using a Coulomb
analysis, with an assumed friction angle between smooth concrete and
sand-gravel of 17 degrees (6).   These loading cases simulate the
field condition during construction, assume backfilling completed,

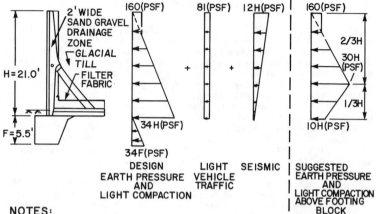

NOTES:

1) WALL BACKFILL CONSISTS OF ON-SITE GLACIAL TILL.

2) DESIGN PARAMETERS FOR GLACIAL TILL BACKFILL:

   • TOTAL UNIT WEIGHT, ι_T = 135 PSF

   • INTERNAL FRICTION ANGLE, ϕ = 35°

3) LATERAL EARTH PRESSURE COEFFICIENT FOR UNBRACED CONDITION.

4) LIGHT COMPACTOR WITHIN 6 FEET OF BACK FACE OF WALL.

   I FOOT = 0.3048 METER
   1000 PSF = 47.88 KN/M²

5) SHAPE OF SUGGESTED EARTH PRESSURE DIAGRAM IS BASED ON STRAIN GAGE READINGS.

6) MAGNITUDE OF SUGGESTED EARTH PRESSURE DIAGRAM IS VERIFIED BY STRAIN GAGE READINGS IN TENSION MEMBER, ASSUMING SIMPLE SUPPORT.

FIG.3 DESIGN EARTH PRESSURES

and no surcharge loads beyond those induced by light compaction equipment.

INSTRUMENTATION PROGRAM

Monitoring Program Description.   An instrumentation program was undertaken to evaluate a section of the wall system during backfilling to compare with anticipated design performance.   One braced wall section at the plan location shown on Figure 1 was instrumented.

The instrumentation program was designed to allow measurement of structural and geotechnical parameters and to provide some data to evaluate the design assumptions.   The parameters monitored include:

1.   Internal strains within a wall panel (tee and stem), the reinforced base slab and the tension member using vibrating wire strain gages embedded in the concrete.

2.   Lateral and rotational movements of the wall using inclinometers attached to the rear of the panel and tiltplates attached to the exterior face of the panel.

3.   Trends in earth pressures below the toe and heel of the base slab, using earth pressure cells.

Inclinometers.   Four inclinometer casings were installed to monitor lateral movements.   Two casings were attached to the back face of the wall panel, and two were installed about 1-foot (0.30 m) outside the heel of the base slabs, extending to the bottom of the foundation block.   Refer to Figure 4 for instrument locations.

Tilt Plates.   Four tilt plates were installed on the front face of the test panel to monitor the angular rotation of the test panel. The tilt plates were epoxied directly to the face of the wall with a temperature stable epoxy to reduce thermal effects.   The tilt plates are a redundant measurement system for the inclinometers.   Tilt plate locations are shown on Figure 4.

Strain Gages.   Forty concrete embedment vibratory wire strain gages were installed at the locations shown on Figure 4.   The gages are used to monitor strains and to estimate loads in the structural components for comparison to theoretical loads.   The strain gages within the wall panel and tension member were installed at the precasting facility by instrumentation specialists.

Earth Pressure Cells.   Two earth pressure cells were installed to measure vertical pressures below the base slab of the test panel. One of the earth pressure cells was located near the heel of the wall and one was located near the intersection of the foundation slab and foundation block.   Each cell was installed on a levelling pad of 1 inch (25.4 mm) of fine concrete sand.

POST CONSTRUCTION ANALYSES OF WALL PERFORMANCE

Wall Movement.   As shown on Figure 5, after 5 feet (1.5 m) of fill had been placed, inclinometer W2 attached to a panel indicated about 0.1 inch (2.5 mm) of lateral movement near the top of the wall towards the backfill side (inward).   At completion of backfilling, about 0.3 inches (7.6 mm) of outward movement had taken place at the top of wall.   Most of the deflections were within the upper 10 feet

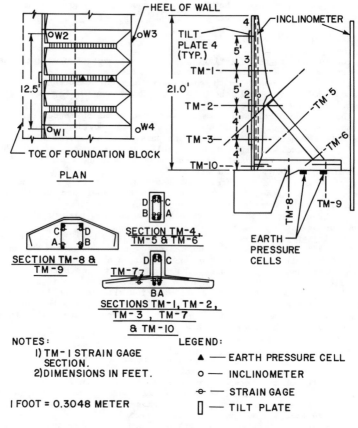

PLAN

SECTION TM-8 & TM-9

SECTION TM-4, TM-5 & TM-6

SECTIONS TM-1, TM-2, TM-3, TM-7 & TM-10

NOTES:
1) TM-1 STRAIN GAGE SECTION.
2) DIMENSIONS IN FEET.

1 FOOT = 0.3048 METER

LEGEND:
▲ — EARTH PRESSURE CELL
o — INCLINOMETER
⊕ — STRAIN GAGE
▯ — TILT PLATE

FIG.4 INSTRUMENTATION LOCATION PLAN AND SECTION

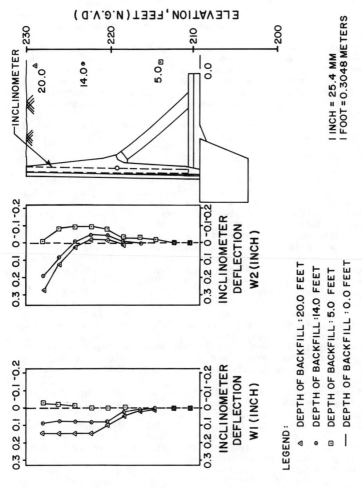

FIG.5 LATERAL PANEL MOVEMENT vs DEPTH

I INCH = 25.4 MM
I FOOT = 0.3048 METERS

LEGEND:

△ DEPTH OF BACKFILL : 20.0 FEET

⊙ DEPTH OF BACKFILL : 14.0 FEET

□ DEPTH OF BACKFILL : 5.0 FEET

— DEPTH OF BACKFILL : 0.0 FEET

(3.0 m) of the wall, with essentially zero movement in the lower zones of the wall below the intersection of the tension member and panel.

Inclinometer W1 installed near the edge of the test panel, showed maximum lateral outward movements at top of panel of about 0.15 inches (3.8 mm), at completion of backfilling. The data indicate most of the movement took place in the upper 8 feet (2.4 m) of the wall at this location. Inclinometers W3 and W4 were installed for post-construction monitoring, which is beyond the scope of this paper.

The inward movement near the upper portion of the panel shown in inclinometer W2 is due to the restraint provided by the tension member, and resulting panel rotation about the pin. The influence of the tension member is not prominent at inclinometer W1, which is located near the edge of the test panel.

After completion of backfilling, the net rotation of the test panel ranged from 0.03 degrees at the lowest tilt plate, to 0.05 degrees at the uppermost tilt plate. Rotation at the tilt plate located at mid-depth of the panel was -0.04 degree, indicating movement toward the backfill side of the wall. These data are consistent with inclinometer W2 and strain gage data. Tilt meter data collected three weeks after completion of filling indicated a rotational movement of 0.05 degree at the upper tilt plate, and 0.03 degree at the lowest plate, suggesting minor readjustments in deflections within three weeks of completion of backfilling.

Tension Member Loads.   The inclined tension member is the critical structural component of the wall system. Four internal strain gages were installed at each of three cross-sections located near the upper, middle and lower end of the tension member. Gage locations are designated TM-4, 5 and 6 on Figure 4. At each section, gages TM-A and B are located near the lower face of the tension member, and gages TM-C and D are located near the upper face. Stress versus backfill elevation for gages TM-4, 5, and 6 are shown on Figure 6.

At the completion of backfilling, data from each gage at section TM-4 (near the upper end of the member) indicated tension, with an average calculated tensile stress of 400 psi (2.76 MN/m^2). A combined modulus of elasticity of 4,000 ksi (27596 MN/m^2) was used for reinforced concrete. This is based on an elastic modulus of concrete of 3,900 ksi (26906 MN/m^2) from cylinder tests and a modulus of steel of 29,000 ksi (185,624 MN/m^2) (7).

Data from gages at location TM-5 (middle of member) indicated a measured decrease in tensile strain near the upper face of the member during backfilling, and a continuous increase in tensile strain near the lower face of the member, inferring development of bending stresses. Bending likely results due to transfer of some loads from backfill above the member, and possibly due to inability to achieve full compaction of backfill in a limited zone immediately below the member.

At location TM-6, data from the upper gages indicated compression and data from the lower gages indicated tension of sufficient magnitude to develop the cracked section (8). Therefore, the composite modulus cannot be used to estimate stresses at this

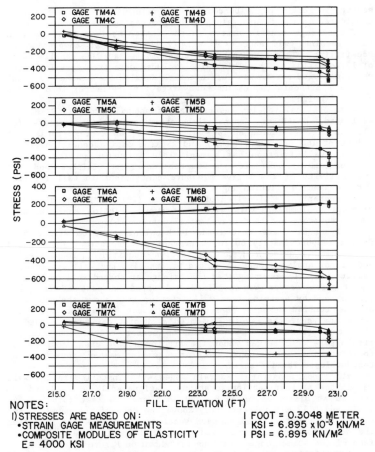

NOTES:
1) STRESSES ARE BASED ON:
   • STRAIN GAGE MEASUREMENTS
   • COMPOSITE MODULES OF ELASTICITY
   E = 4000 KSI

1 FOOT = 0.3048 METER
1 KSI = 6.895 x 10^{-3} KN/M^2
1 PSI = 6.895 KN/M^2

FIG. 6 STRESS vs. BACKFILL ELEVATION

section.   Strain gages bonded to the reinforcing steel should be
considered in the future, along with embedment gages.

The average tensile stress in the member estimated from strain
gage readings was about 430 psi (2.97 MN/m^2) at completion of
backfilling.   This corresponds to an estimated average tensile force
of 15 tons (133.5 kN). This value excluded data from gage locations
where measured strain was sufficient to develop the cracked section.
Figure 7 presents the estimated average tension member load versus
backfill height.

A theoretical tension member load of about 19 tons (169 kN) was
estimated, based on drained laboratory strength parameters for
glacial till (friction angle - 37 degrees), an average total unit
weight of 140 pcf (22.0 kN/m^3) and Rankine active conditions (K_a =
0.25).   Wall friction has a negligible influence on the Rankine
active   loading,   due   to   the   smooth   concrete   finishes   and   the
insensitivity of the active pressure coefficient to wall friction
for the assumed friction angle (4).

A theoretical active load of 22.0 tons (197.5 kN) was calculated
using a Coulomb analysis with an assumed friction angle between
smooth concrete and sand-gravel of 17 degrees (6).

Analyses indicate that loads back-calculated from strain gage
data are approximately 60 to 70 percent of theoretical loads based
on assumed Rankine and Coulomb active conditions, respectively.
Additional analyses were conducted by treating the wall system as a
gravity wall, by assuming that the triangular shaped soil mass below
the tension member was confined and acted as part of the wall
system, and by assuming that wall friction acts above the upper end
of the tension member.   Based on the above assumption and using
wedge analyses, the result indicated a theoretical tensile load of
12.4 tons (111.3 kN) in the tension member.   This analysis and
strain gage data suggest that the arching active case occurred due
to rotation about the upper end of the tension member, thus reducing
the developed tensile stresses.

Wall Panel Behavior.   Strain gages were installed in the wall
panel stem at four section locations, designated TM-1, 2, 3 and 10,
located at distances of 11.8, 8.0, 4.5 and 0.4 feet (3.60, 2.44,
1.37, and 0.12 m), respectively, above the base of the panel.   Four
strain gages were placed at each section, gages TM-A and TM-B near
the outer face of the stem, and gages TM-C and TM-D near the inner
face of the stem.   Gage locations are shown on Figure 4.

As backfilling occurred, strain gage data at location TM-1 showed
an increase in compressive stress at the outer face and tensile
stress at the inner face of the panel stem above the connection of
the tension member to the panel.   These data suggest anticipated
normal bending behavior of a structural panel as a cantilever
subject to active case lateral earth pressures.

Locations TM-2, TM-3 and TM-10 are between the base of the panel
and the intersection of the tension member with the panel.   Data
from gages near the outer face of the panel show stresses changing
from tension to compression, and data from gages near the inner face
show a continuous increase in compression, with increasing backfill
elevation.   This indicates that the lower portion of the panel
(below the connection between the tension member and the panel)

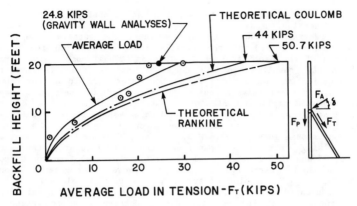

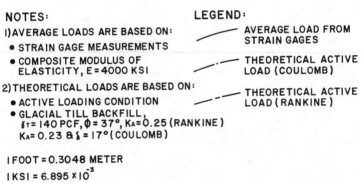

FIG. 7 TENSION MEMBER LOAD vs FILL HEIGHT

behaves as a simply supported member. As backfill proceeded above
the connection, a tensile force developed in the tension member
which restrained the wall panel.   Thus, the vertical component
(compressive force) of the tensile force exerted on the lower
portion of the panel increased. This additional compressive force
ultimately resulted in a change in stress from tension to
compression at the outer face of the panel, and in a continuous
increase in compressive stress at the inner face of the panel. Gage
10 indicated negligible strain and lateral load in the base of the
panel stem. An illustration of the deformed shape of the structure,
based on instrumentation data, is shown on Figure 8.

Four strain gages (TM7A, TM7B, TM7C and TM7D) were also installed
in the wing portion of the panel near the inner face at distances of
3, 6, 13 and 16.5 feet (0.91, 1.83, 3.96 and 5.03 m) from the bottom
of the panel, respectively. Refer to Figure 4 for gage locations
and Figure 6 for a plot of stress versus backfill elevation. The
data indicated moderate increases in tensile stress and strain near
the inner (back) face of the panel, corresponding to increased
backfill elevation. This suggests normally anticipated behavior of
the structural panel being subject to horizontal bending.

At the completion of backfilling, the interpretation of gage data
indicated an approximately linear increase in strains (and stresses)
downward, from the top of the panel to the tension member (TM7D to
TM7B), and a decrease at the location of TM7A. The gage data
obtained from TM7A, 7B, 7C and 7D indicated the bending stresses
near the inner face of the test panel corresponded to theoretical
lateral soil pressures at corresponding elevations. The lateral
soil pressures acting across the panel created a horizontal moment
which resulted in a tensile bending stress on the inner face of the
panel. Using an estimated panel section modulus, back-calculated
moments and resulting stresses, a theoretical lateral soil pressure
was estimated at various elevations. These pressures indicate a
trend of changes in lateral soil pressures suggesting a trapezoidal-
shaped lateral pressure distribution, with maximum earth pressures
approximately at the one-third point of the wall above the base.
Figure 3 presents a hypothetical back-calculated trapezoidal earth
pressure distribution near the edge of the panel. Further
instrumentation of test panels are required to assess the validity
of this hypothetical distribution, and to assess the effect of the
stem of the panel tee on design earth pressures distribution.

Base Slab Behavior. Strain gages were installed at two sections
(TM8 and TM9), as shown on Figure 4, near the upper and lower
surfaces of the base slab. TM8 and TM9 are located at distances of
7.6 feet (2.32 m) and 2.8 feet (0.85 m), respectively, from the heel
of the slab.

The gages at TM8 (near the heel of the foundation block)
indicated the upper surfaces of the base slab to be in tension, and
the bottom surface to be in compression. The estimated tensile and
compressive stresses were 68 psi (469 kN/m^2) and 76 psi (524 kN/m^2),
respectively. This indicates that the base slab was subject to
negative bending when slight rotation around the toe of the base
slab occurred. This behavior was confirmed by the rotational
movement measured by the lowest tilt plate (No. 1).

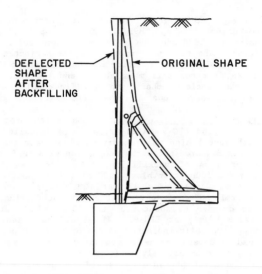

DEFLECTED SHAPE AFTER BACKFILLING

ORIGINAL SHAPE

NOTE :

THIS FIGURE PRESENTS BEHAVIORAL
DEFLECTED SHAPE OF THE WALL AT
COMPLETION OF BACKFILLING.

FIG. 8 DEFLECTED STRUCTURE SHAPE

Data from gages at TM9 (near the bottom connection of the tension member) showed that TM9A and TM9D (bottom and top in diagonal) were in compression and TM9B and TM9C were in tension (refer to Figure 4 for locations). The base slab apparently was subjected to bending stresses as a result of the rotation of the wall about the toe, the earth pressure from the soil mass above, and the influence of the stress from the tension member. No specific conclusions could be drawn from this data, due to the complexity of the loading situation.

Earth Pressure Below Base Slab. Two earth pressure cells, designated EPC-8 and EPC-9 on Figure 4, were installed underneath the base slab just behind the foundation block near the heel. The earth pressures measured below the foundation slab indicated a trend related to progress of backfilling. The total earth pressures recorded near the foundation block (EPC-8) and the heel (EPC-9) were 0.47 tsf (44.8 kN/m^2) and 0.31 tsf (29.7 kN/m^2), respectively. However, several weeks after completion of backfilling, the earth pressure had dropped to about 0.36 tsf (34.3 kN/m^2) at EPC-8 and 0.25 tsf (23.8 kN/m^2) at EPC-9. It should be noted that earth pressure cells typically indicated on the order of 40 percent of the vertical load calculated assuming geostatic conditions. This value is lower than would be normally anticipated.

Although the earth pressure cells did not determine the correct magnitude of load, the trends in earth pressures measured are consistent with the anticipated behavior of the wall. That is, the earth pressures gradually increased as backfill was placed. At the full backfill condition, the measured earth pressures decreased in response to the rotation of the wall about the toe, as confirmed by the data from the inclinometers.

CONCLUSIONS

1. The maximum measured lateral movement was 0.3 inches (7.6 mm) at the top of the wall upon completion of backfilling. Most movements occurred in the upper 10 feet (3.0 m) of the panel, above its connection to the tension member.
2. The inclinometer and tilt plate data indicate that the test panel pivoted about the base of the wall panel, at its intersection with the foundation block, during backfilling. As earth load increased, rotation occurred that was sufficient to develop the active condition, and some loads were transferred to the tension member. The foundation block did not move measurably, due to the rigid weathered bedrock subgrade.
3. Earth pressure cells installed below the base slab at the test panel indicated trends consistent with placement of backfill and anticipated wall behavior. Earth pressures under the base slab first increased as fill was placed, and then decreased as the tension member began to pick up load and rotation occurred about the base of the wall panel.
4. The average calculated axial load in the tension member at the completion of backfilling was approximately 15 tons (133.5 kN) based on strain gage data. This load is of the order of 60 to 70 percent of the theoretical axial load of 25.35 tons (225.5 kN) calculated

using Rankine and 22.0 tons (195.7 kN) using Coulomb active loading conditions, respectively, laboratory strength parameters and in-situ density parameters. Wedge analyses conducted by treating the system as a gravity wall, assuming that the soil mass below the tension member was confined and acted as part of the wall system, indicated a theoretical tension member load of 12.4 tons (111.3 kN), considering wall friction. The analyses and data suggests that the arching active cause occurred due to rotation about the upper end of the tension member, reducing the developed tensile stresses.

5. A trapezoidal earth pressure diagram should be considered for future designs similar to this case. This hypothetical lateral pressure would increase linearly from zero at the top of wall, to a maximum at the one-third point above the top of the base slab, and then decrease linearly to one-third of the maximum pressure at the bottom of the wall. The shape of the diagram was inferred from strain gage data. The trapezoidal diagram should yield the full magnitude of an active earth loading, but should be redistributed in the shape of a trapezoid. In addition to the current instrumentation, a future wall panel and footing block should be instrumented with calibrated earth pressure cells, to assess their interaction with the system, the effect of the stem of the panel tee on the distribution, and the validity of a trapezoidal earth pressure distribution for use in engineering practice.

6. Bonded strain gages should be used in conjunction with embedment strain gages in the tension member and base slab, where strains are expected to exceed those required to develop the cracked section.

REFERENCES

1.   Personal communications, Herbert Fletcher and Peter Mazza, The Mazza Consulting Group.

2.   American Society for Testing and Materials, 1989 Annual Book of ASTM Standards, Volume 4.08, Soil and Rock; Building Stones; Geotextiles, 1989.

3.   U.S. Army Corps of Engineers, Engineer Manual EM-1110-2-1906, Laboratory Soils Testing, Appendix X, 1970.

4.   Department of the Navy, Naval Facilities Engineering Command, Design Manual - Soil Mechanics, Foundations and Earth Structures, NAVFAC DM-7, March, 1971, Sec. 7-10, p.5.

5.   Massachusetts State Building Code, Building Code Commission, Sixth Ed., Rev. 7.0, July 2, 1988, Article 7, Sec. 716.6.10.

6.   JE Bowles, Foundation Analysis and Design, third ed, McGraw-Hill, 1982, p 389.

7.   American Institute of Steel Construction, Manual of Steel Construction, 1980, p. ix.

8.   KL Leet, Reinforced Concrete Design, McGraw-Hill, 1984, p. 44.

THE BEARING BEHAVIOUR OF NAILED RETAINING STRUCTURES

Manfred F. Stocker [1], Georg Riedinger [2]

ABSTRACT: Soil nailed walls have been used over a period of
more than 15 years with very positive results. The bearing
mechanisms are fairly well known and the analysis and design
methods render suitable and safe results. As an example, long-time
performance of a 15 m high wall in a cohesive soil is discussed,
based on a very extensive measurement program which has been
carried out over a period of 10 years. The development and
improvement of the design methods are discussed, and the results
of analyses are compared with measurements.

INTRODUCTION

The development of retaining structures received a new impetus
in 1958 through the introduction of ground anchors. High and
relatively slender walls as pile, diaphragm and sheetpile walls,
could now be constructed prior to excavation and tied back with
ground anchors during excavation. A new type of retaining
structure - the element wall - was developed about 10 years later.
Pre-cast   or   in-situ-cast   concrete   elements   were   placed
checkerboard-like onto the excavated soil surface and tied back
with anchors. The construction took place simultaneously with the
excavation (Figure 1).

A new idea was born at the end of the sixties. Gravity walls
constructed with artificially placed soils and strengthened with
steel reinforcement could replace anchored structures. This
method, known as reinforced earth, became very economical, since
soil is used for the main part of the structure. The disadvantage
of this method is that the retaining wall has to be built from
bottom to top, which means that the full excavation has to be
completed in advance of the construction of the wall.

1 - Technical Director, Bauer Spezialtiefbau GmbH,
    8898 Schrobenhausen, Germany
2 - Head of Design Department, Bauer Spezialtiefbau GmbH,
    8898 Schrobenhausen, Germany

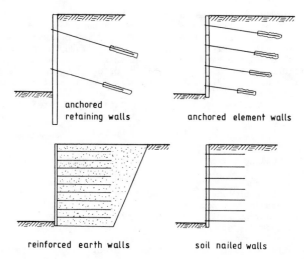

anchored
retaining walls

anchored element walls

reinforced earth walls

soil nailed walls

Figure 1. Development of Retaining Wall Systems.

The consequent criticism of this idea led to the method of soil nailing in the beginning of the seventies. Instead of constructing the wall from bottom to top the opposite way was taken. The natural insitu soil was used for the gravity wall. Together with the proceeding excavation, which was carried out in steps of 1 to 1.5 m, the soil was reinforced with steel bars, called nails, which were drilled and grouted into the ground. The excavated wall face was protected with a relatively thin shotcrete lining. Such nailed retaining structures may even be combined with anchors in special cases.

Today the technique of soil nailing is far spread and advanced in Germany, France, Great Britain, Japan and the United States.

SCIENTIFIC RESEARCH

The scientific studies began in Germany in 1975 with several series of model tests and seven large-scale tests. Bearing and failure mechanisms were observed, earth pressures were measured, the influence of nail length and spacing was examined and the internal and external stability of the "gravity wall" were studied both in non-cohesive and cohesive soils. Based on these observations and measurement results, calculation and design methods have been developed.

The most important results were:
1) The nailed soil structure behaves like a gravity wall.
2) The required nail length for the general case of a vertical wall face and a horizontal ground surface lies in the range of 0.5 to 0.8 times the height of the wall.

3) The spacing of the nails should be less than 1.5 m, i.e. the reinforcement ratio should be at least 1 nail per 2.25 m^2.
4) The earth pressure onto the wall face may be assumed with a uniform rectangular distribution. Its magnitude is on the order of 0.4 to 0.7 times the active Coulomb's earth pressure.
5) No negative effects on the stability or deformation of the wall due to dynamic loading, for example from road traffic on top of the wall, were found.

The above results are discussed in the literature listed at the end of this paper.

VALUE OF SOIL NAILING

Soil nailing cannot replace all other methods of retaining structures, neither technically nor economically, but is has several advantages:
1) Only small equipment is required.
2) The method is very economical, if
   a) it is not possible to use large machines
   b) the geometry of the wall is complex
   c) there is little space for the construction.
3) The nails consist of low-strength steel. Thus the problem of corrosion protection is extremely reduced compared to the use of permanent anchors.
4) The bottom of the wall is equal to the depth of the excavation. This saves a lot of material.
5) The failure mode is good-natured, i.e. the retaining structure does not collapse suddenly and without large deformation.
6) The construction may be carried out with little environmental disturbance, which means little noise and hardly any vibration.

Soil nailing also has disadvantages compared to other methods:
1) The horizontal deformations of the wall may reach the order of 0.2 to 0.4 % of the wall height and are usually larger than those of anchored structures.
2) Without additional measures soil nailing cannot be used for underpinning of large buildings.
3) The aesthetic form of the wall face with plain shotcrete is not satisfying. Additional measures have to be taken, e.g. covering with pre-cast elements or greening with plants.

EXPERIENCE WITH SOIL NAILING

A large number of soil nailed walls have been constructed during the past 13 years, both for temporary and permanent structures. The experience with this type of retaining structure has been very positive. One project which has been studied and extensively monitored shall be described in detail as an example to show the bearing behaviour of large structures.

A nearly 15 m deep excavation had to be designed in Stuttgart (Germany) in a sloping terrain in 1979. One side of the excavation bordered directly upon a city street, another side upon private

houses. No anchors were allowed to be driven under the neighbouring buildings. Soil nailing proved to be the only economical
solution (Figure 2).

Figure 2. View of the Site.

An extensive instrumentation program was installed to monitor
the deformations of the wall, the ground movements and the distribution of the nail forces. The measurements have been continued
over a period of 10 years. Four control sections M 1 to M 4 were

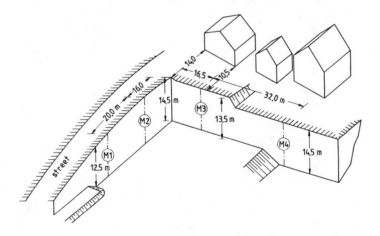

Figure 3. View of the Site and Layout of Control Sections.

installed along the wall with slope indicators, load cells for
the nail heads, extensometers and strain-gage-equipped nails
(Figure 3).

The soil consisted of a 0.8 to 1.8 m thick upper layer of fill
material (silt, sand and cinders), underlain by a layer of
"Wanderschutt" of stiff to medium consistency (clayey, sandy silt
with small sandstones and gravel). Below these layers Keuper Marl
was found (layers of siltstone and claystone).

The following soil parameters were obtained in the laboratory:

	Density (kN/m^3)	Friction $(°)$	Cohesion (kN/m^2)	Modulus of Compressibility (MN/m^2)
Fill	19	30	-	-
Clayey, sandy silt	20	27,5	5 - 10	15 - 25
Siltstone and claystone	21	23	> 50	40 - 80

A typical cross section is shown in Figure 4.

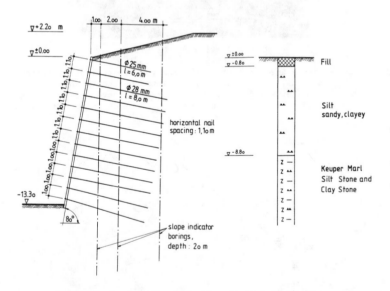

Figure 4. Typical Cross-Section through the Wall.

The excavation was carried out in steps of 1.1 or 1.0 m over the
whole length of the wall. The nails consisted of deformed steel
bars with diameters of 25 and 28 mm, a yield strength of 420 N/mm[2]
and a failure strength of 500 N/mm[2].

The corrosion protection consisted of corrugated PVC-sleeves with a wall thickness of 1 mm, drawn over the whole length of the nails. The spaces between the steel nails and the PVC-sleeve and the PVC-sleeve and the soil were grouted with cement mortar.

The wall face was sprayed with shotcrete, 250 mm thick and reinforced with two layers of wire mesh.

Nail forces.   After having run suitability tests with nails in all three typical soil layers, approximately 5 % of all nails installed were performance tested according to Figure 5. The design working load was kept constant for 15 minutes in order to observe the creep behaviour of the nail.

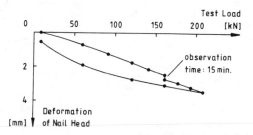

Figure 5. Characteristic Performance Test on Nails.

In section M 3 four nails were equipped with strain gages applied to the steel at intervals of 0.5 to 1.5 m. Thus, stresses could be measured along the axis of the nail during the excavation procedure. The distribution of the nail force and its increase with the depth of the excavation are shown in Figure 6. The nail force does not increase significantly below the excavation level H. This means that nail 1 does not substantially contribute to the retaining force of the wall system below this level. This agrees very well with the theoretical idea of the bearing behaviour. The force distributions of the other nails at the final excavation

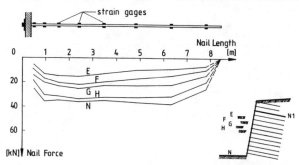

Figure 6. Distribution and Change of the Nail Force at Control
Section M 3 with Proceeding Excavation.

depth are plotted in Figure 7a. It is interesting to note that the
stress distributions in the nails 1 and 4 are rather uniform,
whereas the nails 2 and 3 have got distinct maxima. This indicates
locations of higher shear forces within the soil and thus possible
failure planes if the structure would be loaded to failure (Figure
7b).

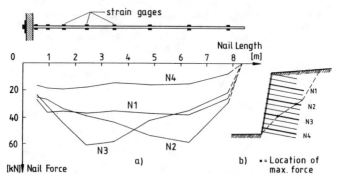

Figure 7. Distribution of the Nail Forces at Control Section M 3
           for the Final Excavation Stage.

The stresses in the nails have been monitored over a period of
10 years after finishing the wall. The development of the maximum
nail forces are plotted in Figure 8. During the last years the
forces remained fairly constant. Of interest is the rapid increase
of the nail force 1 m behind the wall face in January of the first
year. The wall at that time was still completely exposed to the
climatic conditions. Obviously ice pressure or frozen soil behind
the wall led to a temporary increase of the nail forces (Figure
9). The temperature plotted in the diagram was the daily minimum
value measured 50 mm above ground surface. By the end of February
the nail forces had decreased to almost the same levels measured
at the end of November.

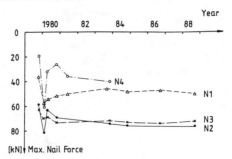

Figure 8. Change of the Nail Forces with Time.

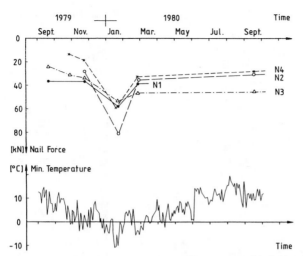

Figure 9. Influence of Frost on the Nail Forces Behind the Wall.

Horizontal deformations.   The horizontal deformations are
caused by shear and bending of the nailed "gravity wall" and by a
horizontal deformation of the soil below the excavation due to the
unilateral earth pressure (Figure 10). The summation of such
deformations results in a deformation pattern measured in all four
control sections of the site. An example is shown in Figure 11 for
section M 2.

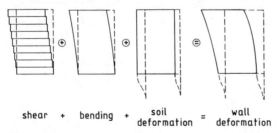

Figure 10. Various Factors Contributing to the Wall Deformation.

The following observations may be summarized:
1) The main deformation occurs at the top of the wall.
2) The deformation pattern remains similar at each further excavation step.
3) The depth of the horizontal deformation of the soil below the
   excavation level varied between 20 to 60 % of the momentary
   excavation depth. This is a function of the strength of the
   soil below the excavation level.

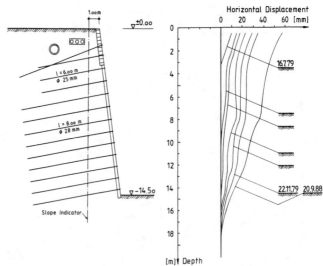

Figure 11. Horizontal Wall Deformations of Control Section M 2.

4) The ratio between the maximum horizontal deformation of the wall and the momentary excavation depth, measured in three control sections and at various excavation depths, was on the order of 0.1 to 0.36 %, with an average of about 0.25 % (Figure 12).

5) The increase of the horizontal displacement of the wall over a period of 10 years, shown for control sections M 1, M 2 and M 3, was on the order of another 0.06 to 0.15 % (Figure 13). The slope indicator sleeve of control section M 4 had been

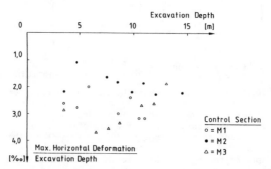

Figure 12. Relationship between Maximum Horizontal Deformation and Excavation Depth.

destroyed during construction. The main additional deformation took place during the first 3 years. This was certainly influenced by the construction and the additional load of the new building. No significant change in the deformation could be observed during the last 5 years.

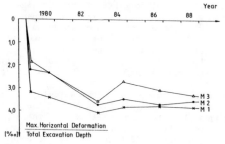

Figure 13. Change of the Horizontal Wall Deformation with Time.

6) The horizontal deformation of the soil within the nailed soil body decreased with increasing distance from the wall face (Figure 14). It is interesting that even at a distance of more than 7 m from the wall face, the maximum horizontal displacement was still 13 mm.

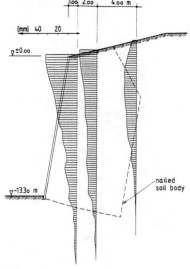

Figure 14. Soil Deformations within the Soil Nailed Wall at Control Section M 3.

It may be noticed that several horizontal deformations were measured with extensometers. The readings compare very well with the values of slope indicators. Extensometers have the disadvantage that displacements can be measured only below the level of their installation, which means not from the beginning of the excavation. Thus important information is not obtained.

Earth Pressure. Attempts were made to measure the earth pressure directly with the aid of load cells during the construction of the wall. The results, however, turned out to be very erratic. It is difficult to place load cells on a nearly vertical wall face and still obtain a uniform stress distribution between the shotcrete and the soil at and around the location of the load cells.

Very good results, however, were obtained using the sum of the nail forces measured approximately 0.2 m behind the shotcrete wall with strain gages glued to the steel nails. The apparent earth pressure distribution thus obtained, and its change over the years, is plotted in Figure 15. The nearly rectangular shape of the earth pressure distribution is very interesting. This type of stress distribution has also been found at other projects, where the soil behind the wall face was rather homogeneous over the whole depth.

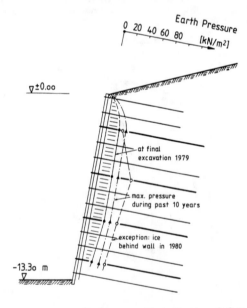

Figure 15. Apparent Earth Pressure Distribution 0.2 m Behind Shotcrete Wall at Control Section M 3 Calculated from Nail Forces.

DESIGN PROCEDURES

The failure mechanisms of nailed gravity walls may be divided into two categories:
1) External failure mechanisms are a result of insufficient stability of the complete gravity wall block. They are independent of the structure of the wall itself. Therefore the usual design methods may be applied for the analysis of sliding, overturning and overall failure. This design will produce the required minimum geometric dimensions of the wall block.
2) Internal failure mechnanisms take place if the tensile and shear strength of the nailed soil block is insufficient. In this case the block itself disintegrates. The design for the internal safety must, therefore, produce an optimized scheme of the strength, length, bond capacity, and horizontal and vertical spacing of the nails, taking into account, of course, the soil parameters and external loading conditions.

The design is based on the observation that internal failure occurs on inclined planes starting at the lowest point of the wall. The potential failing soil wedge must be held in place by retaining forces exerted upon the wedge by those ends of the nails which cross beyond the failure plane into stable soil. The action of the soil nails may be described as micro-piles anchoring the downward pull of the failing soil wedge.

The internal safety factor may be defined as the ratio of the sum of the possible counterforces of the nails beyond the failure plane over the sum of those counterforces of the nails required for equilibrium.

$$\text{s.f.} = \frac{\text{pos. N}}{\text{req. N}}$$

Only tensile forces of the nails are regarded. Dowel or shear forces of the nails are not taken into account because of safety reasons.

Lump Safety Factor Procedure. The design for internal safety involves the following steps:
1) Estimation of a suitable nail pattern.
2) Determination of the above safety factor for the most critical failure plane. This can be done only by repeating the calculations for several possible failure planes. Such calculations have to be carried out for each individual excavation level.

It is evident that such a design requires a great amount of calculation effort. Computer programs are desireable, especially for optimizing the results.

In Figure 16 the design procedure for the project described above is shown. At the time of construction (1979) the design was based on a very simplified assumption that the failure planes were straight. The assumptions included an average angle of friction of $\Phi = 32.5°$ with the cohesion set to 0. By investigating possible failure planes with inclination steps of 5°, the lowest safety factor was determined for a plane inclined at $\theta = 47°$. The calculated minimum safety factor was 1.75 for intermediate excavation levels and 2.15 for the final excavation.

The chosen safety standard required a safety factor of 1.5 for all short-time construction phases, i.e. for excavation levels with the nails installed but not yet effective, and a safety factor of 2.0 for final excavation levels, i.e. all nails installed and effective.

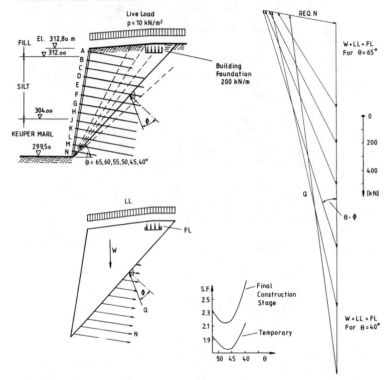

Figure 16. Failure Mechanism, Polygon of Forces and
Safety Factors for Straight Failure Planes.

The design procedure described above is rather simplified. It does not take into consideration the observation made in many model tests and large-scale field tests that failure occurs most frequently in a two-body mechanism, at least in homogeneous soils.

Observing the law of classical soil mechanics, the failure plane in the undisturbed soil behind the nailed wall should have an angle of inclination of 45° + Φ/2. This means that a Coulomb soil wedge forms behind the nailed block and exerts an active

earth pressure onto the rear of the block. The shear or failure
plane within the nailed block may have all possible inclinations
ranging from 0° to 45 + $\Phi/2$° with respect to the horizontal plane.

An analysis of the above project on the basis of a polygonal
failure plane is presented in Figure 17. For reasons of comparison
this analysis was also carried out assuming an average angle of
friction of 32,5° and no cohesion. The minimum safety factors
calculated decreased to 1.65 and 2.09 for intermediate and final
excavation levels, respectively. The critical failure plane
through the nailed block had an inclination of $\theta = 43$°.

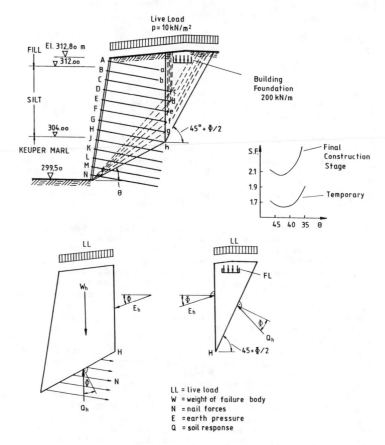

Figure 17. Failure Bodies, Forces and Safety Factors
for Polygonal Failure Planes.

Partial Safety Factor Procedure.   Major progress with the
design method was achieved   with the introduction of a system of
partial safety factors. Dead loads, live loads, soil friction,
cohesion and nail friction were introduced into the design with
individual safety factors both   for temporary and permanent
construction stages. The following table shows the safety factors
now in use:

Excavation	Partial Safety Factor				
Stage	Dead Load	Live Load	Soil Friction	Cohesion	Nail Friction
Temporary	1.20	1.10	1/1.15	1/1.55	1/1.25
Permanent	1.30	1.30	1/1.20	1/1.60	1/1.30

The values   pertain to a failure probability of 1/100 000 for
temporary use and 1/1 000 000 for permanent purposes. The final
excavation stage is regarded as permanent even if the soil nailed
wall is used for temporary purposes only, for instance for a use
of one half year.
     The computer program is set up to find an equilibrium of
driving and holding forces on the inclined failure plane. The
input requires the geometry of the terrain and the wall face, the
external loads, the spacing of the nails, the levels of the
different soil strata and the soil parameters. The program
multiplies the different parameters with the above safety factors
and iterates the calculation for the optimum nail lengths until
equilibrium is reached.
     Using the described calculation program for the project, and
using again an average friction angle of 32.5° and no cohesion as
input,   the required length of nails per square meter of wall
face was calculated to be 6.2 m. The comparable required length
obtained using the lump safety factor method was 7.1 m.
     Using the method of partial safety factors and the real soil
parameters as listed on page 5 (in this case with cohesions of 5
and 50 kN/m^2, respectively), the required length of nails per
square meter of wall face is reduced to 5.4 m. In practice it
would have to be questioned if the cohesion of 50 kN/m^2 could be
fully utilized.
     The big advantage of this method is the fact that, depending on
the scope and carefulness of the project and the soil
investigation prior to construction, the safety factors may be
weighted. In addition, the failure mechanism of this program
automatically takes care of sliding, overturning and, with a few
exceptions, overall stability. One disadvantage has to be
mentioned, too. If the slope above the wall is steeper than the
angle of friction multiplied with the partial safety factor, the
program does not work. In this case the slope above the wall has
to be taken into account as surcharge load.

Earth Pressure Acting on Wall Facing. The wall face is designed as a reinforced concrete slab supported by single supports, i.e. nailcaps. The magnitude of the earth pressure acting on the slab is difficult to calculate, since the active earth pressure is reduced due to a dowel effect. It has been standard procedure to design the wall slab for an earth pressure of 85 % of the active earth pressure at the back of the wall. Several measurements have shown that the use of even less pressure is sufficient. An additional reduction is derived from an arching effect between the nails.

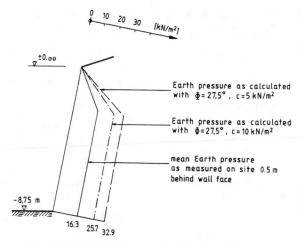

Figure 18. Earth Pressure Acting on Wall Face
at Control Section M 3.

The project in Stuttgart proved to be a very suitable opportunity to compare actual earth pressures with calculated values. As it is not known exactly how much cohesion was activated in the Keuper Marl, the comparison makes sense only for the upper 9 m of the wall face. Results are shown in Figure 18. The effective earth pressure 0.5 m behind the wall face was measured as 16.3 kN/m^2. The calculated 100 % value for a rectangular pressure distribution is 25.7 kN/m^2. This means that the actual earth pressure is only 63 % of the calculated value. In this calculation the cohesion was assumed to be 10 kN/m^2. If the cohesion is assumed to be 5 kN/m^2, the reduction is equal to 16.3/32.9 = 50 %.

CONCLUSION

Soil nailing is a relatively new system of building retaining walls. Though the horizontal displacements mostly are larger than those of anchored structures, the values are well within the limits of suitability and serviceability. The important advantages of the method are that the natural soil deposits are used as main wall material and that the wall is built from top to bottom. Thus, hardly any relaxation or loosening of the soil takes place during the excavation. The failure mechanism is good-natured. Experience and measurements have shown that the long-time stability both of the whole structure and the wall face is ensured. The methods of analysis are well advanced and have rendered enough safety margin in practice.

REFERENCES

1. M Stocker, Bodenvernagelung, Vorträge der Baugrundtagung, Nürnberg, 1976, Deutsche Gesellschaft für Erd- und Grundbau e. V., Essen, 1977
2. G Gäßler, Large Scale Dynamic Test of Insitu Reinforced Earth, Proc. Dynamical Methods in Soil and Rock Mechanics, Vol. 2, p. 333, AA Balkema, Rotterdam, 1978
3. M Stocker & G Gäßler, Ergebnisse von Großversuchen über eine neuartige Baugrubenwand-Vernagelung, Tiefbau, Ingenieurbau, Straßenbau, Bertelsmann Verlag, Sept. 1979
4. M Stocker & G Körber & G Gäßler & G Gudehus, Soil Nailing, Proc. Coll. Int. Reinforcement des Sols, pp. 469-474, Paris, 1979
5. G Gäßler & G Gudehus, Soil Nailing - some Soil Mechanical Aspects of Insitu Reinforced Earth, ICSMFE X Proc. Vol. 3, pp. 665-670, Stockholm, 1981
6. G Gäßler & G Gudehus, Soil Nailing - Statistical Design, Proc. 8th Eur. Conf. on Soil Mech. and Found. Eng., pp. 491-495, May 1983

DESIGN, CONSTRUCTION AND PERFORMANCE OF A
SOIL NAILED WALL IN SEATTLE, WASHINGTON

Steven R. Thompson[1], A.M.ASCE and Ian R. Miller[2]

ABSTRACT: Soil nailing has recently been introduced in the Seattle area as a viable alternative to soldier piles and tiebacks for temporary and permanent support of excavations. This paper describes the design, construction and performance of one of Seattle's first soil nailed walls. Instrumentation data, including wall deflections and nail forces are discussed in detail. The results of finite element modeling of the wall system are also presented. The performance predicted by the original design methodology and by the finite element analysis are compared critically to the measured performance.

INTRODUCTION

The introduction of soil nailing to the Seattle area has been quite successful. Soil conditions generally consist of heavily overconsolidated glacial soils, many of which are ideal for soil nailing. The use of tieback shoring systems has been extensive over the past 25 years, and much of that experience has transferred directly to the design and construction of soil nailed walls. The first use of the system in the Seattle area was for temporary support of a building excavation in 1987.

PROJECT DESCRIPTION

The project described in this paper represents the first soil nailing project in Seattle to be designed and constructed by local firms. It was part of a temporary shoring system for a building excavation just east of the downtown core. The project consisted of two nailed walls, with heights of 35 feet (10.7 m) and 55 feet (16.8 m) adjacent to city streets, and two soldier pile and tieback walls adjacent to existing buildings. Soil nailing was not used to support the existing buildings because of the lack of experience with the system in local soil conditions, and a concern with the potential for excessive movements associated with an unstressed shoring system.

1 - Senior Engineer, Golder Associates Inc., 4104 148th Avenue N.E., Redmond, WA 98052

2 - Associate, Golder Associates Inc., 4104 148th Avenue N.E., Redmond, WA 98052

The soil conditions consisted of fill to a depth of 8 feet (2.4 m), underlain by very dense glacial outwash sand and gravel and very dense lacustrine fine sand and silt. The contact between the outwash and lacustrine deposits was encountered at the base of the excavation on the high wall, and at about mid-height on the low wall. Groundwater was below the base of the excavation. Soil properties used in the design analysis are shown on Figure 1.

Nails were generally installed on 6 foot (1.8 m) centers horizontally and vertically, to a maximum length of 35 feet (10.7 m). Holes were drilled with an 8 inch (203 mm) diameter, continuous flight, hollow stem auger. Nail bars consisted of Grade 150 Dywidag bars ranging from 1 inch (25 mm) to 1 1/4 inch (32 mm) in diameter. Nails were typically installed at an inclination of 15 degrees, although the first row on the high wall was installed at 20 degrees to avoid utilities. A typical section of the high wall is shown on Figure 1.

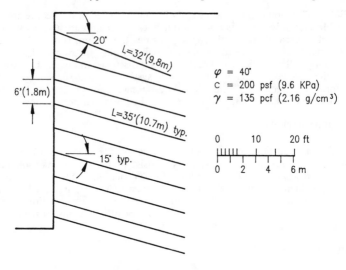

Figure 1. High Wall Section

DESIGN METHOD

A force limit equilibrium analysis was chosen for the design of this wall. It is based on the "Davis method" (10), but has been extensively modified to include more versatile modeling of a variety of wall and backslope geometries, variable nail lengths and spacings, and a treatment of anchor capacities more consistent with local tieback design practices.

The limit equilibrium method has a number of limitations as a design tool. Perhaps the most significant is that limit equilibrium analyses can only provide an indication of the total nail force required to maintain a given factor of safety. The distribution of the total nail force among the individual nails cannot be solved for intrinsically, but must be assumed. The major shortcomings with existing limit equilibrium formulations for soil nailing design, including the Davis method, have been poor assumptions with regard to the nail force distribution.

In addition to difficulties with nail load distribution, limit equilibrium models do not address displacements of the reinforced soil mass. For applications such as excavation shoring systems in urban environments, the displacements of the soil nailed wall are critical to its successful performance.

INSTRUMENTATION PROGRAM

Instrumentation was installed to provide a better understanding of the aspects of soil nailing performance not well predicted by the limit equilibrium analysis, and to provide baseline information regarding the performance of soil nailing in specific local soils.

Vibrating wire strain gages were installed on five nails, in a vertical section on the high wall (Figure 2). Four to six gages were spot welded to the steel along the length of each nail. Economic constraints forced the use of a single gage at each location. Thus, bending moments could not be measured, and the effect of bending on the measured strains could not be explicitly determined. All gages were located at the 3 o'clock position on the bar (Figure 2) to minimize the potential for bending moment interference. All strain gages were read daily during construction of the shoring wall, and monthly thereafter until the permanent basement wall was completed.

Two inclinometers were installed at a distance of three feet behind the face of the wall. One was installed on the high wall, ten feet away from the strain gaged section, and one was installed on the low wall. Inclinometers were read weekly during construction of the shoring wall and monthly thereafter until the permanent basement wall was completed.

DATA INTERPRETATION

Inclinometer Data. Deflections measured on the high wall are shown on Figure 3. The total deflection at the top of the wall after the last lift of soil was excavated (6/14/88) was 0.59 inches (15 mm). One week later (6/21/88), after the nails and shotcrete facing had been installed on the last lift, the deflection at the top of the wall had increased to 0.70 inches (18 mm). Monitoring of the inclinometer casings continued until the end of August. No additional movements, within the accuracy of the instrument, were measured during that time.

The deflections were slightly higher than what might have been expected for a typical soldier pile and tieback wall in similar soil conditions. This is not surprising, since the soil nailed system is initially unstressed and requires some deformation to mobilize the nail forces. The maximum movement at the top of the wall is close to 0.1 percent of the height of the wall, or about what would be required to mobilize active earth pressure in dense granular soil (8).

Strain Gage Data. The strain gage data provided considerable insight into the behavior of the nailed wall, including the short-term development of load

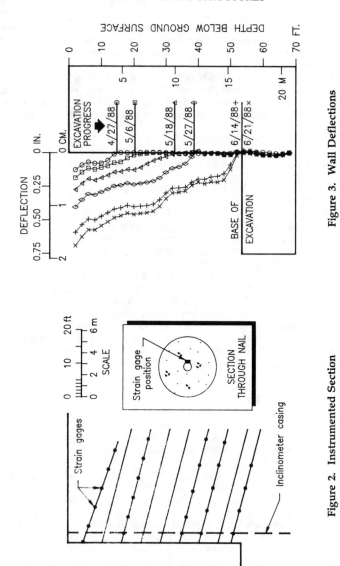

Figure 3. Wall Deflections

Figure 2. Instrumented Section

on the nails, the change in nail loads with time, and indications of the combined influence of the steel and concrete portions of the nail section.

Figure 4 shows early strain measurements from nail level 6, plotted as a function of time. The excavation of each successive lift below nail level 6 can be seen as an initial rapid increase in strain related to the actual excavation of the next lift, followed by a slow increase in strain with time as the nails and shotcrete facing were installed. The construction of each of the three lifts below nail level 6 can be seen clearly in the strain data. This type of behavior was seen on all nail levels. The influence of the advancing excavation was most significant when the bottom of the excavation was within about three lifts (20 ft (6.1 m)) of the nail. The excavation effect decreased progressively after that but was still noticeable during every lift.

Figure 5 shows the complete record of strain measurements from nail level 6. The strain increases rapidly at first, as active excavation takes place. Upon completion of the excavation (6/20/88), the rate of strain increase slows down in a pattern consistent with long-term creep of the nails. At the time monitoring was completed (3/28/89), strains had increased by a factor of more than two over their values at the end of construction. This behavior is consistent with other cases where long-term measurements have been obtained (6).

The long-term increase in strain has been attributed to creep in the soil mass which results in increased load on the nails (6). However, one of the most highly stressed regions in the soil, and thus most susceptible to creep, is the soil immediately around the circumference of the nail. Soil creep in this area would result in a reduction, not an increase, in nail load.

Creep of the concrete, on the other hand, can cause an increase in measured steel strain without requiring an increase in nail load. Creep deformations of cured (28-day) concrete under a constant load can be two to three times larger than the instantaneous deformations (1). Concrete loaded within the first one to three days, which is not usual in a soil nailing application, may show creep deformations two to three times higher than cured concrete (2). As the concrete creeps under the tensile forces, load is shed from the concrete onto the steel, resulting in increased strain in the steel.

Load is also shed onto the steel when the concrete grout cracks under tension. The strain data presented on Figure 5 provide additional insight into this behavior. In two cases, strains show a significant and uncharacteristic "jump" between two successive readings. Similar jumps were observed at one or more gage locations on every bar except nail level 9. These jumps appear to be related to cracking of the concrete under tension, which is associated with a virtually instantaneous increase in steel strain.

Creep and cracking of the concrete reduce the axial stiffness of the nail with time. If nail loads increase with time or even remain constant, this should result in increased wall deflections with time. However, long term movements were negligible on this project (Figure 3) as well as on others reported in the literature (3, 6). In the absence of the increasing wall deformations which would be caused by increasing nail loads, it appears more likely that the increase in steel strain with time (Figure 5) is related to a redistribution of the existing total nail load from the concrete to the steel.

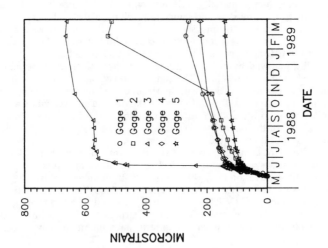

Figure 5. Long-term Strain Nail Level 6

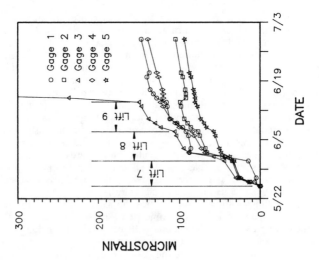

Figure 4. Short-term Strain Nail Level 6

Clearly, the determination of nail loads from the strain data is not a simple matter. It requires an assessment of the time-dependent effect of the concrete grout on the overall stiffness of the nails. Concrete has a low modulus in comparison to steel, but the area of concrete in the composite nail section is greater than the area of steel by a factor of 50. Thus the stiffness of the nail is strongly affected by the modulus of the concrete as long as the concrete remains effective in tension. When the concrete cracks, the local stiffness of the nail decreases substantially.

Two approaches were taken to calculating loads from the strain data. The first approach was to calculate a composite stiffness for the steel/concrete section based on reasonable assumptions regarding the section geometry and concrete modulus. A time-dependent concrete modulus was used, based on 3-day, 7-day, and 28-day strengths from 17 cylinders of the nail grout. The resulting maximum nail loads at each level are shown on Figure 6, as indicated by the curve titled "Indirect Method."

The disadvantage to this approach is that the area and modulus of the concrete are based on assumptions that may not reflect actual conditions. Furthermore, it is assumed that the load is shared in direct proportion to the stiffness ratio of the concrete and steel, with no allowance made for creep in the concrete. As indicated previously, creep may account for half or more of the deformation, particularly early on when the strength of the concrete is low and the nail loads are increasing most rapidly. This will result in less of the load being carried by the concrete than the stiffness ratio suggests, and actual nail loads will be lower than predicted. Thus, loads determined by this method are considered to be limiting upper bound values.

The second approach was to assume that the jumps in strain caused by cracking of the concrete represented the full release of tensile stress in the concrete. Thus, the strain on the low side of the jump was related to the stiffness of the composite section, while the strain on the high side of the jump was related to the stiffness of the steel alone. By assuming that the total load remained constant over the strain jump, the stiffness of the uncracked composite section could be back-calculated. The resulting maximum nail loads are shown on Figure 6, as indicated by the curve titled "Direct Method."

The advantages to this approach are that no assumptions need be made regarding the variation of concrete modulus with time, and any transfer of load to the steel which may have occurred by creep in the concrete is accounted for directly. The disadvantage is in the assumption that the strain jump results from the full release of the tensile stress previously carried by the concrete. This is only true if the gage is located at the crack. If the strain gage were located some distance away from the crack, the increase in strain would reflect only a portion of the tensile stress released by the concrete. The jumps in strain do provide a lower bound indication of the tensile stress in the concrete, and thus a lower bound indication of the total nail load.

Loads determined by the indirect method are substantially in excess of those required under active conditions. However, the magnitude of the wall deflections suggests that active conditions have been established. Consequently, loads determined by the indirect method may not represent the best

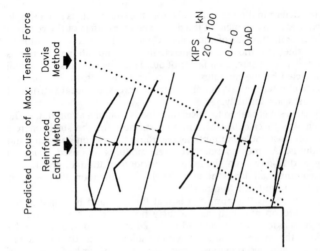

Figure 7. Load Distribution on Nails

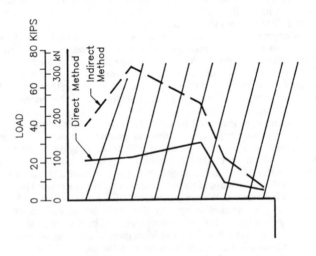

Figure 6. Measured Load vs. Depth

estimate of actual nail loads. Nail loads calculated by the direct method are consistent with active conditions, and thus may be more realistic.

The location of the maximum bar force is not affected by the assumptions made in converting from strain to force. The distribution of load along the length of each bar is shown on Figure 7, using the loads calculated by the direct method. Two maxima seen on some bars. The peak at the front of the bar may be related to bending forces caused by the weight of the shotcrete facing hanging from the nails during excavation of underlying lifts. The peaks farther away from the face map the locus of maximum strain in the soil mass. This should, in theory, coincide with the critical failure surface predicted by the limit equilibrium analysis used to design the wall. The surface predicted by the analysis is shown on Figure 7. The maximum tensile force line typically used in the design of reinforced earth walls (4) is also shown and is actually in better agreement with the measurements than the limit equilibrium prediction.

FINITE ELEMENT ANALYSIS

A finite element model was assembled to assist in the understanding of the mechanics of the wall behavior and in the interpretation of the field measurements. A variety of parameters were adjusted until the model was able to produce a reasonable match of the measured deflections. The changes in the stress field and the development of nail forces could then be analyzed as the staged excavation was simulated.

Program Features. The wall was modeled using FES2D, a two-dimensional, non-linear finite element program originally written by the secondary author. The program can analyze plane strain, plane stress, or mixed two-dimensional stress problems.

A number of elements are available to simulate the various components of a soil/structure system. Those pertinent to the analyses performed in this study are a continuum element with peak and residual strength characteristics; a bar element which can be defined with or without bending resistance; and a joint or interface element with shear and normal stiffness, a limiting friction angle, and the ability to simulate dilatant behavior.

Model Calibration. The measured wall deflections were used to calibrate the finite element model. A number of parameters control the horizontal and vertical deformations of the model. These include the horizontal and vertical soil moduli, and the coefficient of lateral earth pressure. These parameters are interrelated and similar behavior of the model might be obtained by several non-unique combinations of these parameters.

The coefficient of at-rest earth pressure was fixed at unity. This is generally consistent with the glacial stress history of Seattle soils, which tend to have somewhat elevated values of K_o. Even higher values of K_o could be defended on the basis of in situ stress measurements in the Seattle area. However, the specific value of this parameter is not as important as the overall deformation behavior determined by the combination of factors described above.

The horizontal and vertical soil moduli were treated as the main variables, and were adjusted to create reasonable agreement between the model deformations and the measured wall deflections. Initially, an isotropic soil modulus was used, but resulted in overestimates of the vertical heave in response to excavation. Anisotropy was then introduced by doubling the vertical modulus. This is also consistent with the glacial stress history of the Seattle soils, and produced much improved deformation patterns.

Analytical Approach. A series of three models was developed using a progressively more realistic representation of the nails and the nail/soil interaction. The purpose of this progression was to determine whether simplified approaches could provide accurate predictions of behavior.

The first model of the wall was assembled using a fairly coarse mesh and only continuum elements (i.e., the nails were not modeled explicitly). The excavation was simulated by progressively removing layers of elements in front of the wall. The nails were simulated by increasing the horizontal stiffness of the continuum elements in the layer immediately above the base of the excavation. Nail forces were calculated by using a line of nodes at the bar location as strain gages. Knowing the relative displacement of each of the nodes and the stiffness of the nail, equivalent nail forces could be calculated. This model produced deformations which agreed quite well with measured wall deflections. However, the resulting nail forces were considerably higher than those measured. Numerical problems related to high stress concentrations were also apparent around the progressing toe of the excavation.

A second model was assembled with a finer mesh in the immediate vicinity of the wall to accommodate the stress concentrations. Actual bar elements were included, with cross-sectional properties based on the actual nail materials and geometry. The bar elements were fixed to the soil at the common nodes. Bar forces were then a direct output of the program. This model was also able to reproduce actual deflections quite well, but nail forces were still overestimated.

A third model was assembled using joint elements between the nail elements and the soil elements. The shear stiffness of these joint elements was back-calculated from the results of over 40 production load tests conducted during construction of the wall. Wall elements were created using the properties of the shotcrete facing, and the front end of each nail was fixed to the wall element. The joint elements proved to be a critical detail in the wall model. With the joint elements in place, both deformations and nail forces were in reasonable agreement with field measurements.

Finite Element Results. A comparison of the wall deflections calculated by the model and actual measurements at the end of construction is shown on Figure 8. The good agreement is not entirely surprising as the model was adjusted specifically to produce good agreement between calculated and measured deflections.

On the other hand, the nail loads calculated by the finite element analysis are strictly a function of the ability of the model to simulate the interaction between the nail and the surrounding soil. As discussed above, discrete nail elements and joint elements, with properties back-calculated from field measurements, were required to achieve satisfactory agreement between

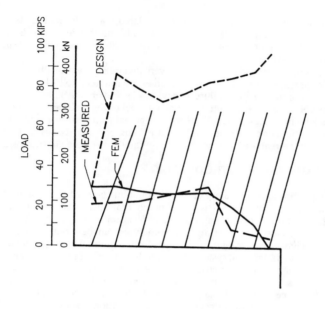

Figure 9. Maximum Nail Load vs. Depth

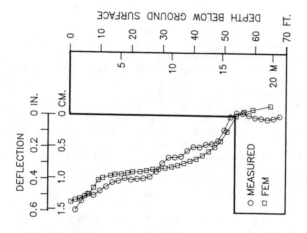

Figure 8. Wall Displacements

calculated and measured nail loads. The calculated maximum nail loads at each nail level are compared to the measured loads and the design loads on Figure 9. The locus of maximum tensile forces is shown on Figure 10, in comparison to measured values.

All continuum, bar, and joint elements were given linear elastic properties with no limiting strength values. Future studies are intended to include non-linear as well as failure/post-failure behavior. However, the linear elastic model was able to provide a quite reasonable model of the wall performance. While this does not necessarily indicate that the actual wall performed entirely within the elastic range of the soil and structural components, it does suggest that stresses in the real wall may not be substantially outside the elastic range.

CRITIQUE OF DESIGN METHODOLOGY

Limit equilibrium analyses are currently common tools for the design of soil nailed walls in the U.S. It is well known that such analyses do not provide good estimates of the magnitude and location of the maximum nail forces (5, 7, 9). However, working stress analyses can be cumbersome as design tools. Hence, limit equilibrium analyses are commonly used to estimate nail forces, and to design wall systems accordingly. Experience with the Davis method indicates that the predicted nail forces are conservative.

Figure 9 presents a compilation of the maximum nail force at each nail level as determined by the original design methodology (Davis method), the field measurements (only the direct method interpretation is shown), and the finite element analysis. Clearly, the nail loads determined using the Davis method are excessive in comparison with loads obtained from the field measurements and the working stress analysis.

A number of factors contribute to the overestimation of nail loads. In the first place, the design loads calculated using the Davis method incorporate a factor of safety of 1.5 on soil shear, resulting in more load being carried by the nails. The other loads do not include factors of safety. However, the factor of safety alone does not explain the difference in nail loads.

The primary shortcoming of the Davis method is in the assumptions made regarding the distribution of the total nail force among the individual nails. The method assumes that the force carried by each nail is proportional to the length of the nail behind the critical failure surface. This bears little resemblance to the distribution of nail forces predicted by working stress analyses or measured in case histories.

During the early stages of the excavation, when the nail length may be considerably greater than the height of the excavation, this assumption leads to excessively high nail forces, and high overall factors of safety. As the excavation is lowered, the failure surface extends deeper within the soil mass and the effective anchor length of the nails decreases. Often, this results in predicted nail loads which decrease with increasing depth of excavation. The load distribution also results in higher nail forces being carried by progressively lower nails.

This is not a good model of soil nailed wall behavior. Field data (6) indicate that the soil strength is mobilized fairly quickly. In other words, the

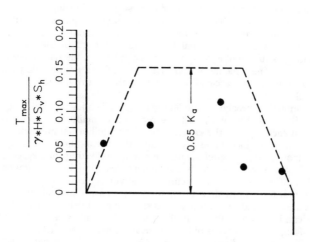

Figure 11.   Normalized Nail Force vs.
Braced Cut Pressure Distribution

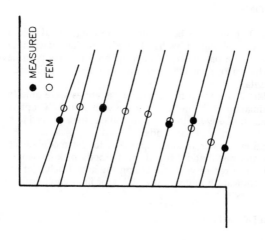

Figure 10.   Locus of Maximum Tensile Forces

factor of safety with respect to soil shear is close to one during the entire excavation process. The forces mobilized in the nails are just sufficient to maintain the system at a factor of safety of one. As conditions change, eg., the excavation is lowered, additional forces are mobilized in the nails as needed. As long as sufficient excess capacity is available in the nails the wall will remain stable. The force mobilized in a nail is related predominantly to successive lowering of the excavation. Thus, loads in the lower nails are typically small.

A more rational approach to modeling the mobilization of nail forces would be to compute the total nail force required to maintain a factor of safety of one at each stage of the excavation. Limit equilibrium methods are capable of this analysis since it is essentially an ultimate condition. For design purposes, suitable factors of safety could be applied reflecting the degree of uncertainty in the various strength parameters, resulting in a somewhat higher total nail force than for the factor of safety of one condition.

This total nail force must then be divided between the individual nails. Limit equilibrium methods are not able to distribute nail loads explicitly, but assumptions can be made on the basis of empirical data as has been done for years in the design of other types of braced excavations. In fact, field data (6) indicate that the pressure diagram for a conventional braced excavation (11) may provide a reasonable approximation of the nail load distribution. As shown on Figure 11, our data also fit this model. By accepting the limitations of the limit equilibrium method and applying an empirical understanding of load distributions, more reasonable designs may be achieved.

CONCLUSIONS

Considerable insight has been gained into the detailed mechanics of the performance of soil nailed walls through the instrumentation program described above. In particular, a better appreciation has been developed of the creep behavior of concrete grout and its effect on nail loads calculated from strain data.

Appropriately calibrated finite element analyses were able to provide a close approximation of the behavior of the wall system. Properly character-ized joint elements between the soil and nail elements proved to be critical to the accurate calculation of nail loads.

Incorporating a more rational approach to the modeling of nail force mobilization and distribution, based on the body of knowledge which has been developed in recent years regarding the actual behavior of soil nailed walls, may significantly improve the ability of limit equilibrium analyses to predict nail forces.

ACKNOWLEDGEMENTS

The authors wish to extend special thanks to DBM Contractors, Inc., the shoring wall contractor for this project. They provided research funds to assist in the work reported here, as well as a great deal of cooperation and forbearance during construction. John Byrne and David Cotton of Golder

Associates participated in many lengthy and animated discussions, and reviewed the final manuscript. Their contributions were significant and greatly appreciated.

REFERENCES

1. ZP Bazant, and L Panula, Practical Prediction of Time-Dependent Deformations of Concrete, Materials and Structures, RILEM, Vol. 11, No. 65, Sept.-Oct. 1978, pp. 307-328.

2. J Byfors, Plain Concrete at Early Ages, Swedish Cement and Concrete Research Institute, Stockholm, 1981, 464 p.

3. CL Ho, HP Ludwig, RJ Fragaszy, and KR Chapman, Field Performance of a Soil Nail System in Loess, Proc. ASCE Foundation Engineering Congress, Vol. 2, Evanston, Illinois, 1989, pp. 1281-1292.

4. I Juran, Reinforced Soil Systems - Application in Retaining Structures, Geotechnical Engineering, Vol. 16, 1985, pp. 39 - 82.

5. I Juran, Nailed-Soil Retaining Structures: Design and Practice, Trans. Research Record 1119, Trans. Research Board, Washington, D.C., 1987, pp. 139-150.

6. I Juran and V Elias, Soil Nailed Retaining Structures: Analysis of Case Histories, Soil Improvement - A Ten Year Update (GSP 12), Ed JP Welsh, ASCE, New York, 1987, pp. 232-244.

7. JK Mitchell and WCB Villet, Reinforcement of Earth Slopes and Embankments, NCHRP Rpt. No. 290, Trans. Research Board, Washington, D.C., 1987, 323 p.

8. RB Peck, WE Hanson, and TH Thornburn, Foundation Engineering, 2nd Edition, Wiley, New York, 1974, 514 p.

9. F Schlosser, Behavior and Design of Soil Nailing, Symposium on Recent Developments in Ground Improvement Techniques, Bangkok, 1982, pp. 399-413.

10. CK Shen, LR Herrmann, KM Romstad, S Bang, YS Kim, and JS DeNatale, An In Situ Earth Reinforcement Lateral Support System, Report No. 81-03, Dept. of Civil Engrg., U.C. Davis, March 1981, 187 p.

11. K Terzaghi and RB Peck, Soil Mechanics in Engineering Practice, 2nd Edition, Wiley, New York, 1967, 729 p.

DESIGN OF SOIL NAILED RETAINING STRUCTURES

Ilan Juran[1], M.ASCE, George Baudrand[2], S.M.ASCE,
Khalid Farrag[2], S.M.ASCE, and Victor Elias[3], M.ASCE

ABSTRACT: Soil nailing is an in-situ reinforcement technique
which has been used during the last two decades to retain excavations or stabilize slopes. The design methods that have been most
commonly used in Europe (the French and German methods) and the
United States (the Davis method) consider only a global stability
analysis of the structure and the retained ground. They involve
different assumptions with regard to the shape of the failure
surface, the mode of soil-reinforcement interaction and type of
resisting forces generated in the nails.

More recently, a kinematical limit analysis design method has
been developed to provide an estimate of maximum tension and shear
forces in the nails and assess the location of the potential
internal failure surface. This method permits an evaluation of
the local stability at each reinforcement level which can be
more critical than global structural stability.

This paper is focused on the evaluation of the available design
methods (i.e., the French, Davis and kinematical design methods)
through comparative analyses of design schemes established for
typical soil nailed retaining structures. It outlines a recommended design procedure and provides useful charts for preliminary design.

INTRODUCTION

The fundamental concept of soil nailing consists of placing in
the ground passive inclusions, closely spaced, to restrain displacements and limit decompression during and after excavation.
The design of soil nailed retaining structures is currently based
on limit equilibrium approaches. Slope stability analysis procedures have been developed to evaluate the global stability of

---

1 - Associate Prof., Dept. of Civil Engineering, Louisiana State
    Univ., Baton Rouge, LA 70803.
2 - Graduate Student, Dept. of Civil Engineering, Louisiana State
    Univ., Baton Rouge, LA 70803.
3 - V. Elias and Assoc., P.A., Bethesda, MD.

the soil nailed mass and/or the surrounding ground, taking into account the shearing, tension or pull-out resistance of the inclusions crossing the potential failure surface. The available design procedures involve different definitions of safety factors and different assumptions with regard to the shape of the failure surface, the type of soil-reinforcement interaction and the resisting forces in the inclusions.

Limit force equilibrium methods were developed by Stocker and coworkers (1979) (the "German" method) assuming a bilinear sliding surface, and by Shen, et al. (1981) (the "Davis" method) considering a parabolic sliding surface. Both methods take into account only the tension resistance and pull-out capacity of the inclusions.

A more general solution, integrating the two fundamental mechanisms of soil-inclusion interaction (i.e., lateral friction and passive normal soil reaction) was developed by Schlosser (1983) (the "French" method). This solution considers both the tension and the shearing resistance of the inclusions as well as the effect of their bending stiffness. A multicriteria analysis procedure is conducted using a modified slope stability slices method (e.g., Bishop's modified method or Fellinius' method) to evaluate the factors of safety with respect to:

1. Shear resistance of the soil
2. Pull out resistance of the reinforcement
3. Passive soil resistance to nail deflection at both sides of the potential failure surface
4. Strength of the inclusion that has to withstand both tension and shear forces as well as bending moments

A detailed discussion of the design assumptions and analysis procedures used in the "French", "German" and "Davis" methods has been outlined in several references (Shen, et al., 1981; Schlosser, 1983; Villet, et al., 1987) and will therefore not be repeated here. However, for the sake of clarity, the main features of these global stability design procedures (i.e., French, German, and Davis methods) are outlined in Table 1.

More recently a kinematical limit analysis design method has been developed (Juran, et al., 1989) to provide an estimate of the maximum tension and shear forces mobilized in the nails and assess the location of the potential internal failure surface. A local stability analysis is conducted at each reinforcement level to verify the required safety factors with respect to the pull-out, tension and shear resistances of the selected reinforcement. The basic assumptions of this method are summarized in Table 1. This method permits an evaluation of the effect of the main design parameters (i.e., structure geometry, inclination, spacing and bending stiffness of the nails) on the resisting forces generated in the nails during construction. Its applicability has been verified through the analysis of failure mechanisms observed on model walls and comparison of predicted and measured nail forces in full scale structures.

This paper is focused on the evaluation of the available design methods (i.e., the French, Davis and kinematical design methods)

Table 1. Basic Assumptions of the Different Design Approaches

Features	French Method (Schlosser, 1983)	German Method (Stocker, et al., 1979)	Davis Method (Shen, et al., 1981)	"Modified" Davis (Elias and Juran, 1988)	Kinematical Method (Juran, et al., 1989)
Analysis	Limit moment Equilibrium Global stability	Limit force Equilibrium Global stability	Limit force Equilibrium Global stability	Limit force Equilibrium Global stability	Working stress Analysis Local stability
Input Material Properties	Soil parameters (c,ϕ') limit nail forces Bending stiffness	Soil parameters (c,ϕ) lateral friction	Soil parameters (c,ϕ') limit nail forces Lateral friction	Soil parameters (c,ϕ') limit nail forces Lateral friction	Soil parameters $(c/(\gamma H),\phi')$ Non-dimensional bending stiffness parameter (N)
Nail Forces	Tension, shear, moments	Tension	Tension	Tension	Tension, shear, moments
Failure Surface	Circular, any input shape	Bi-linear	Parabolic	Parabolic	Log-spiral
Failure Mechanisms	Mixed[1]	Pull-out	Mixed	Mixed	Non-applicable
Safety Factors[2]					
Soil strength F_c, F_ϕ	1.5	1 (residual shear strength)	1.5	1	1
Pull-out resistance[2] F_p	1.5	1.5 to 2	1.5	2	2
Tension[3] Bending	Yield stress Plastic moment	Yield stress	Yield stress	Yield stress	Yield stress Plastic moment

Table 1. (continued)

Features	French Method (Schlosser, 1983)	German Method (Stocker, et al., 1979)	Davis Method (Shen, et al., 1981)	"Modified" Davis (Elias and Juran, 1988)	Kinematical Method (Juran, et al., 1989)
Design Output	[4]GSF [5]CFS	GSF CFS	GSF CFS	GSF CFS	Mobilized nail forces CFS
Ground Water	Yes	No	No	No	Yes
[6]Soil Stratification	Yes	No	No	No	Yes
[6]Loading	Slope, any surcharge	Slope surcharge	Uniform surcharge	Slope, uniform surcharge	Slope
[6]Structure Geometry	Any input geometry	Inclined facing Vertical facing	Vertical facing	Inclined facing Vertical facing	Inclined facing Vertical facing

[1]Mixed failure mechanisms: limit-tension force in each nail is governed by either its pull-out resistance factored by the safety factor or the nail yield stress, whichever is smaller.
Pull-out failure mechanism: limit tension forces in all the nails are governed by their pull-out resistance factored by the safety factor.

[2]Definitions of safety factors used in this analysis:
For soil strength, $F_c = c/c_m$, $F_\phi = (\tan\phi)/(\tan\phi_m)$; where c and ϕ are the soil cohesion and friction angle, respectively, while c_m and ϕ_m are the soil cohesion and friction angle mobilized along the potential sliding surface.
For nail pull-out resistance, $F_p = f_l/f_m$, f_l and f_m are the limit interface shear stress and the mobilized interface shear stress, respectively.

[3]Recommended limit nail force.
[4]GSF: Global safety factor
[5]CFS: Critical failure surface
[6]Present Design Capabilities

through comparative analysis of design schemes established for
typical soil nailed retaining structures.

The German method is mainly based on limited number of model
tests where failure was caused by substantial surcharge loading.
However, both full scale and model tests have illustrated that the
assumed bi-linear failure surface is not consistent with the
observed behavior of soil nailed retaining structures under self-
weight loading. Gassler (1988) has shown through comparative
stability analyses that the bi-linear failure mechanism appears to
be applicable only in cohesionless soils under high local
surcharges, while the assumption of a circular sliding surface,
considered by Gassler (1988), seems to be more consistent with
experimental observations and for most cases, yields more critical
engineering solutions (i.e., lower factors of safety). Therefore,
the German method assuming a bi-linear failure surface appears to
be applicable to limited cases of high surcharges and will not be
further considered in this study.

The main variables considered in the analysis of the available
design methods are the structure geometry, the pull-out interface
parameters, the bending stiffness of the nails, and ground water
level in the retained soil. As global stability analysis procedures can rigorously be verified only at failure (i.e., global
safety factor equals one), their assessment requires comparisons
of predicted and observed failure heights of reduced scale model
walls and full scale experiments conducted to failure. The paper
outlines a recommended design procedure and an attempt is done to
evaluate the applicability of non-dimensional design parameters in
the development of useful charts for preliminary design.

COMPARATIVE ANALYSIS

  Structure Geometry. Figure 1 illustrates a comparison between
design schemes obtained for three typical geometries of soil
nailed retaining structures using the different design methods
outlined in Table 1. The structure, 12 m high with nail spacing
of $S_v = S_H = 1.0$ m, is constructed in a typical silty sand with a
friction angle of $\phi = 35°$ and an apparent cohesion of $c = 10.8$ kPa
corresponding to $c/(\gamma \cdot H) = 0.05$ (where γ is the unit weight of the
soil and H is the total structure height). A lateral interface
shear stress of 120 KN/m^2 is assumed.

The comparative analysis indicates that assuming the nails to
be perfectly flexible:
1. The kinematical method and the Davis method, considering
    $F = F_\phi = 1$, yield similar structure geometry defined by
    the L/H ratio (where L is the total reinforcement length
    calculated for $F = 2$). However, the geometry of the
    potential failure surface predicted by the kinematical
    method is characterized by an S/H ratio (where S is the
    maximum width of the active zone), which is significantly
    smaller than that predicted by the Davis method.

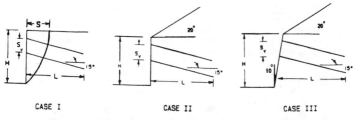

	CASE I		CASE II		CASE III	
CASE I						
CASE II						
CASE III						

METHOD	CASE I		CASE II		CASE III	
PERFECTLY FLEXIBLE	L/H	S/H	L/H	S/H	L/H	S/H
KINEMATICAL	0.47	0.39	0.63	0.47	0.52	0.39
FRENCH METHOD	0.43	0.72	0.49	0.88	0.45	1.00
DAVIS (Fc=Fφ=Fp=1.5)	0.58	0.61	0.70	0.92	0.63	0.84
DAVIS (Fc=Fφ=1; Fp=2)	0.47	0.48	0.61	0.77	0.54	0.72
ACTUAL STIFFNESS	L/H	S/H	L/H	S/H	L/H	S/H
KINEMATICAL	0.43	0.37	0.70	0.53	0.48	0.36
FRENCH METHOD	0.40	0.70	0.47	0.96	0.43	1.06

Figure 1. Design Schemes Obtained for Typical Geometries.

2. The kinematical method yields L/H design values which are within the range bounded by the global stability analysis procedures commonly used (i.e., the Davis and the French methods with $F_c = F_\phi = F_p = 1.5$).

3. The Davis method (with $F_c = F_\phi = F_p = 1.5$) yields the most conservative design scheme.

4. In order to evaluate the effect of nail bending stiffness on the design, the kinematical and the French methods were used considering the actual bending stiffness of #8 rebars. In spite of the significant difference in the S/H values, the two methods yielded for cases I and III similar design results indicating a relatively small stiffness effect.

5. All methods indicate an effect of slope surcharge and facing inclination on the design which is particularly significant with the kinematical method.

Pull-out Interface Parameters. The pull-out resistance of the reinforcement is governed by the soil-reinforcement lateral friction (i.e., ultimate interface shear stress, f_ℓ) and the grouted diameter, D_g. It is useful, in the development of design charts, to use a non-dimensional pull-out interface parameter:

$$\mu = \frac{f_\ell \cdot D_g}{\gamma \cdot S_v \cdot S_H}$$

The application of this interaction parameter implies that a unique L/H value can be related to the specified design μ value. In order to evaluate the applicability of this design parameter, design schemes for a vertical wall (12.0 m high with nail spacings of $S_v = S_H = 1.5$ m) have been established considering a range of design μ values. The μ values were obtained by varying (1) the f_ℓ values, and (2) the $S_v \cdot S_H$ product, keeping all other parameters consistent. Figure 2 illustrates that both the Davis and the French methods yield consistent results, specifically for the practical range of $\mu = 0.1$ to 0.7.

The kinematical design procedure consists of estimating the working nail forces, TN, (where $TN = T_{max}/(\gamma \cdot H \cdot S_v \cdot S_H)$ is the normalized tension force in the reinforcement) and using equation 1 directly to establish the structure geometry with the appropriate input design value of μ. At any reinforcement level, the design criteria with respect to pull-out can then be expressed as:

$$\frac{L}{H} = \frac{S}{H} + \frac{TN}{\pi} \cdot \frac{F_p}{\mu} \tag{1}$$

Figures 3a and 4a illustrate comparative design charts obtained using the different methods and considering the range of μ values used in practice, for: (1) vertical walls (Case I) in a silty soil, $\phi = 35°$ and $c/(\gamma \cdot H) = 0.05$, and (2) for inclined walls (Case III) with a slope surcharge in a finer silty soil of $\phi = 25°$ and $c/(\gamma \cdot H) = 0.10$. Nails are assumed to be perfectly flexible and lateral interface shear stress of 120 KN/m² is assumed. Figures 3b and 4b present the corresponding S/H values characterizing the predicted geometries of the potential failure surfaces. These comparisons indicate that:

1. For the vertical walls, with no surcharge, the French method (with $F_c = F_\phi = F_p = 1.5$), the Davis method (with $F_c = F_\phi = 1$ and $F_p = 2$) and the kinematical method (with $F_c = F_\phi = 1$ and $F_p = 2$) yield significantly different S/H values but similar design geometries.

2. The Davis method, with $F_c = F_\phi = F_p = 1.5$, yields the highest reinforcement length.

3. For inclined walls with a slope surcharge, the kinematical local stability analysis yields design geometries which are significantly more conservative than those established with the global stability analysis procedures using the French

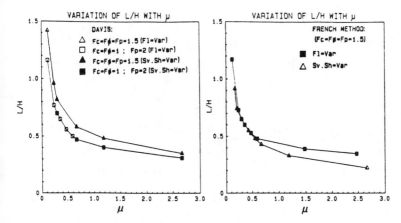

Figure 2. Evaluation of the Applicability of the
Pull-out Interface Parameter.

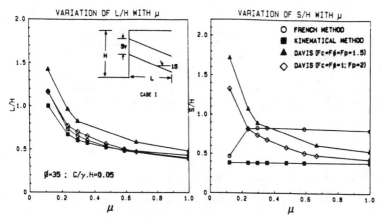

Figure 3. Variation of (L/H) nad (S/H) with μ for
a Vertical Wall with no Surcharge.

method or the modified Davis method (with $F_c = F_\phi = 1$ and $F_p = 2$).

4. The study of the S/H values for the latter case illustrates that the global stability analysis leads to an external stability failure mode with a potential failure surface passing partially or entirely behind the reinforced zone,

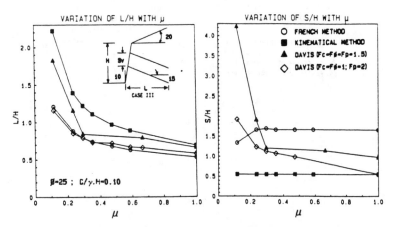

Figure 4.   Variation of (L/H) and (S/H) with μ for
an Inclined Wall with Surcharge.

while the kinematical method considers only an internal
stability failure mode.
It should be noted that differences between the design geometries L/H established with the different methods are highly dependent upon the surcharge, cut slope geometry and soil characteristics. It is therefore difficult to draw a general conclusion.
For the purpose of rational and safe design, it is recommended to
conduct both a local stability analysis using the kinematical
method and a global stability analysis using the French method
(with $F_c = F_\phi = F_p = 1.5$) or the Davis method (with $F_c = F_\phi = 1$,
$F_p = 2$).

EVALUATION OF GROUND WATER EFFECT

The two methods which currently permit an evaluation of ground
water effect on the structure stability are the French global
stability analysis procedure and the kinematical method.
Figure 5a presents the L/H ratios obtained for a vertical wall
($\phi = 30°$, $c/(\gamma \cdot H) = 0.05$) with various ground water levels (Z_w/H),
using these two design methods. Figure 5b shows the corresponding
S/H values. In spite of the substantial difference in S/H values,
indicating an external failure mode for the global stability
analysis as compared with the internal failure mode considered in
the kinematical analysis, the two methods yield similar structure
geometries.
It is of interest to note that the French method yields a continuous decrease in the required structure geometry with the
increase of the Z_w/H values, while the kinematical method yields a
minimum L/H value for $Z_w/H = 0.5$. A detailed analysis of the

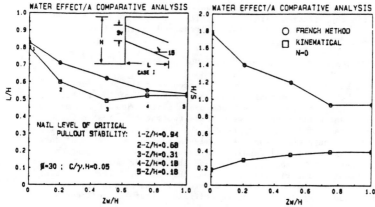

Figure 5. Variation of (L/H) and (S/H) for a Vertical
Wall with Various Ground Water Levels.

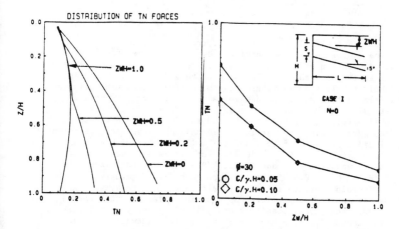

Figure 6. Distribution and Variation of TN for Different
Water Levels Predicted by Kinematical Method.

ground water effect on the location and distribution of maximum
tension forces in the nails (shown in Figure 6a), conducted with
the kinematical method, illustrates that as the ground water depth
increases, the level (Z_c/H) of the most critical local nail pull-
out stability decreases. As the Z_w/H value exceeds 0.50, the nail
level of critical local stability lies above the ground water
level and any further increase of the Z_w/H value will have almost

no effect on the the required structure geometry. Figure 6b illustrates the effect of ground water level on the maximum tension
forces mobilized in the nails, as predicted by the kinematical
method.

EVALUATION OF THE EFFECT OF NAIL BENDING STIFFNESS

The kinematical and the French methods are the only design
approaches which presently permit an evaluation of the effect of
nail bending stiffness on structure stability. One of the most
significant differences between those two methods lies in the
input data of material properties (i.e., soil and reinforcement).
In the French global stability analysis procedure, the input data
pertaining to the nail bending stiffness consists of dimensional
parameters, including the product of the elastic modulus (E) times
the moment of inertia (I) and the product of the lateral reaction
modulus (K_s) times the nail diameter (D). The kinematical method
uses a non-dimensional relative stiffness parameter defined as:

$$N = \frac{K_s \cdot D}{\gamma \cdot H} \cdot \frac{\ell_o^2}{S_H \cdot S_v}$$

where $\ell_o = [4EI/K_s D]^{1/4}$ is the nail transfer length.
The use of this non-dimensional design parameter facilitates
the preparation of design charts. Therefore, an attempt has been
made to evaluate the applicability of this design parameter in
global stability analysis and determine if a unique L/H ratio can
be related to a specified design value of N. For a typical
vertical wall (H = 12.0 m, ϕ = 35°, c/($\gamma \cdot$H) = 0.05), values of the
global safety factor were calculated for increasing N values,
which were obtained by varying: (1) nail spacings with a constant
nail diameter (D = 2.5 cm), and (2) nail diameter with constant
nail spacings ($S_v = S_H = 1.50$ m). For all cases, the structure
geometry was kept constant (i.e., L/H = 0.59 for the kinematical
method and L/H = 0.65 for the French method). Figure 7 shows the
relationships between the safety factors and the relative stiffness parameter N as predicted by the kinematical and the French
methods. The global stability analysis yields only a global
safety factor (i.e., $F_c = F_\phi = F_p = F$). The kinematical local
stability analysis yields for each reinforcement level two
independent safety factors corresponding to: (1) failure by nail
pull-out, F_p (equation 1), and (2) failure by nail breakage, F_b,
defined as:

$$F_b = \frac{K_{lim}}{K_{eq}}$$

where: $K_{eq} = [4TS^2 + TN^2]^{1/2}$; $TS = T_c/(\gamma \cdot H \cdot S_H \cdot S_v)$, is the normalized maximum shear force; $TN = T_{max}/(\gamma \cdot H \cdot S_H \cdot S_v)$, is the normalized maximum tension force; $K_{lim} = [4R_c^2 + R_n^2]^{1/2}$, R_c and R_n

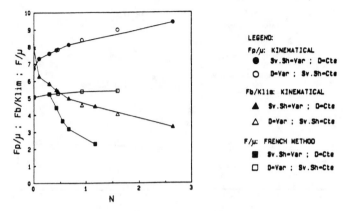

Figure 7. Variation of the Safety Factor with
N as Predicted by the Kinematical
and the French Methods.

are, respectively, the normalized shear and tension resistances of
the nail.

The kinematical analysis yields unique relationships between
the normalized safety factors (i.e., F_p/μ and F_b/K_{lim}) and the
relative stiffness parameter N. It indicates an increase of the
(F_p/μ) value and a decrease of the (F_b/K_{lim}) value, with
increasing N values. With the French method, the relationship
between the safety factor and the design N value is not unique.
For a selected nail diameter, increasing N values by decreasing
nail spacings results in a decrease of the normalized safety
factor (F/μ), while increasing nail diameter with constant nail
spacings results in a slight increase of (F/μ). Therefore, the N
value cannot be used to evaluate the effect of nail bending
stiffness on the global structure stability.

Figure 8a and 8b illustrate the variations of TN and S/H values
with N calculated for both the French and the kinematical methods.
The TN values obtained with the French method were calculated for
the lowest nails and correspond to their pull-out resistance. It
can be shown that for the French method, an increase of the
relative bending stiffness, N, due to an increase of the nail
diameter, D, has only a limited effect on the predicted TN and S/H
values, while an increase of the N value due to a decrease of the
nail spacings (i.e., at a constant diameter) will result in a
substantial increase of both the S/H and TN values. As nail
spacing decreases, the potential sliding surface rotates inwards
thereby decreasing the effective adherence length of the nails.
However, the decrease in the $S_v \cdot S_H$ product will lead to a
significant increase of the normalized maximum tension force, TN.

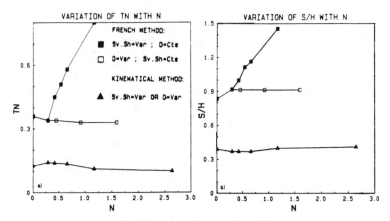

Figure 8.   Variation of TN and S/H with N for the
French and Kinematical Methods.

This analysis further demonstrates that the N parameter is not
applicable for global stability analysis. The kinemtical method
yields unique TN = f(N) and S/H = g(N) relationships indicating
that the TN value decreases while the S/H value increases as the
relative nail bending stiffness increases.

In order to further evaluate method predictions with regard to
the effect of nail bending stiffness on the structure stability,
the French and the kinematical methods have been used to analyze
the failure mechanism observed on reduced scale laboratory model
of vertical soil nailed walls (Juran, et al., 1984). Figure 9
illustrates the comparison between predicted and measured failure
heights of the model walls, which were constructed with three
different types of reinforcements, namely: flexible aluminum
strips, relatively flexible polysterene strips, and relatively
rigid polysterene strips. It should be noted that all the reinforcements have an equivalent tensile strength (i.e., tension
resistance times cross-sectional area) but substantially different
bending stiffnesses. The experimental results show that the nail
bending stiffness has a significant effect on the failure
mechanism. The larger the bending stiffness, the smaller is the
failure height. Post failure observations on the model walls
showed that the flexible aluminum and polysterene strips failed by
tension breakage, while the rigid polysterene strips failed by
excessive bending.

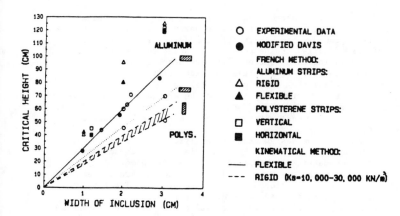

Figure 9.   Effect of Nail Rigidity on the
Critical Height of Model Walls.

The comparative analysis based on model test results indicates that:

1. The kinematical method permits an appropriate evaluation of the nail bending stiffness on the failure mechanism. The predicted failure mode (i.e., tension failure vs. excessive bending), location of the failure surface and model failure height (calculated for K_s = 10,000 to 30,000 kN/m^3) are consistent with the experimental results.

2. The French method shows no apparent bending stiffness effect on the structure stability and therefore results in a significant overestimate of the failure heights of model walls with relatively rigid nails.

3. The kinematical and the Davis methods predict fairly well the experimental results obtained on model walls with flexible reinforcements, while the French method, yields slightly overestimated failure heights.

CONCLUSIONS

The main practical conclusions that can be drawn from this comparative study is that a rational design procedure for soil nailed retaining structures requires three basic steps:

1. Estimate of working nail forces and the location of a potential internal failure surface.

2. Evaluation of the local stability of each nail with respect to failure by breakage or pull-out.

3. Evaluation of the global stability of the soil nailed structure and the surrounding ground with respect to general sliding along potential failure surfaces that may pass within or outside the soil nailed mass.

The kinematical method enables the engineer to adequately assess the effect of the main design parameters (i.e., structure geometry, ground water, inclination, spacings and bending stiffness of nails) on the tension and shear forces generated in the nails during construction. It provides the means to evaluate the local stability at the level of each reinforcements which can be more critical than the global stability.

The global structure stability can be evaluated with the design methods currently used (i.e., the French and the Davis methods), which provide a global safety factor with respect to general sliding.

For most of the design examples analyzed under this study, the Davis method with $F_c = F_\phi = F_p = 1.5$ yields more conservative structure geometries as compared with the French method. The Davis method with $F_c = F_\phi = 1$ and $F_p = 2$ and the French method yield similar design schemes. In geotechnical engineering, the concept of the safety factor is directly related to the incertitude pertaining to the material properties used in design. In practice, for most soil nailing applications, the incertitude related to the pull-out resistance of the nails is significantly greater than that associated with the soil parameters. It is therefore recommended that a larger safety factor be associated with the design pull-out resistance of the nails.

The kinematical and the French methods are the only design approaches which currently permit an evaluation of the effect of ground water and nail bending stiffness on the structure stability. They predict similar ground water effect but substantially different bending stiffness effect. The comparison between model test results and predicted failure heights suggest that the kinematical method can be adequately used to analyze the effect of nail bending stiffness while predictions based on the French method are not consistent with experimental results. However, due to significant difficulties encountered in any rational attempt to formulate and verify similitude requirements between the laboratory models and the prototype, further observations of full scale structures are required for more rigorous assessment of these methods.

REFERENCES

1. F Blondeau, M Christiansen, A Guilloux & F Schlosser, "TALREN, Methode de calcul des ouvrages en terre reforcee," Proc. Intl. Conf. on Soil and Rock Reinforcement, Paris, France, 1984, 219-224.
2. V Elias & I Juran, "Manual of Practice for Soil Nailing," Federal Highway Administration, Report FHWA/RD-89/198, 1989.
3. G Gassler, "Soil-Nailing-Theoretical Basis and Practical Design," Proceedings of the International Geotechnical Symposium, Theory and Practice of Earth Reinforcement, Kyushu University, Japan, 1988, 283.

4. I Juran, G Baudrand, K Farrag & V Elias, "Kinematical Limit Analysis for Soil Nailed Structures," J. Geot. Eng. (ASCE), Vol. 116, No. 1, pp. 54-73.
5. I Juran, J Beech & E DeLaure, "Experimental Study of the Behavior of Nailed Soil Retaining Structures on Reduced Scale Models," Int. Symp. In-Situ Soil and Rock Reinforcement, Paris, France, 1984.
6. I Juran & V Elias, "Soil Nailed Retaining Structures: Analysis of Case Histories," Soil Improvement, ASCE Geot. Special Publication No. 12, 1987, 232-244.
7. JK Mitchell, et al., "Reinforcement of Earth Slopes and Embankments," National Cooperative Highway Research Program Report 290, Transportation Research Board, 1987.
8. P Pfister, G Evers, M Guillaud & R Davidson, "Permanent Ground Anchors Soletanche Design Criteria," FHWA Report No. RD-81/150, 1982.
9. F Schlosser, "Analogies et differences dans le comportement et le calcul des ouvrages de soutennement en terre armee et par clouage du sol," Annales de l'Institut Technique de Batiment et des Travaux Publiques, No. 148, 1983, 26-38.
10. CK Shen, S Bang & LR Herrmann, "Ground Movement Analysis of an Earth Support System," J. Geot. Div. (ASCE) 107(GT12), 1981.
11. MF Stocker, GW Korber, G Gassler & G Gudehus, "Soil Nailing," Intl. Conf. on Soil Reinforcement, Paris, France, 2, 1979, 469-474.

FRENCH NATIONAL RESEARCH PROJECT ON
SOIL NAILING : CLOUTERRE

BY C. PLUMELLE[1], F. SCHLOSSER[2], P. DELAGE[3], and
G. KNOCHENMUS[4]

ABSTRACT: the focus of this paper is to introduce a research project, initiated by the French Government and to present the results of a full-scale soil nailed wall, taken to failure by progressively saturating the reinforced soil mass. The nails, soil and wall facing were instrumented to permit measurements of the displacements, forces and stresses generated in both the soil and reinforcements. To determine the soil-nail lateral friction, pull-out tests, under two different modes, were performed at the site. Analysis of the full-scale experiment was performed using both finite element calculations and the TALREN method.

INTRODUCTION

Soil nailing is a reinforcement technique which has descended from the ancient method of rock bolting, and more recently, from the method of Reinforced Earth. The first soil nailed wall was constructed in France, in 1973 (4).

A soil nailed wall is an in-situ retaining structure, constructed by successive excavation phases, while placing passive bars in the soil, as one progresses along the excavation. The bars are either driven, or placed in boreholes and grouted with liquid cement. The nails are placed in subhorizontal layers, in which the spacing is a function of the construction method utilized. The facing, either vertical or inclined, is generally constructed of shotcrete, whose role is to locally retain the soil between the nails.

There are two principal advantages in utilizing the technique of soil nailing. The first is the normally low construction cost, with respect to the other types of retaining walls, and the second is the relatively rapid

[1]Centre Expérimental du Bâtiment et des Travaux Publics, St Remy les Chevreuse, France.
[2]Professor, Ecole Nationale des Ponts et Chaussées, Paris, France.
[3]CERMES (Soil Mechanics Research Center) Ecole Nationale des Ponts et Chaussées, Paris, France.
[4]Engineer, Terrasol, Puteaux, France.

speed of construction. However, there are also limitations in employing this method. In a cohesionless sand, thefrictional resistance of the soil is not sufficient to provide stability to the unreinforced section of the wall, immediately following an excavation phase.

Up to the present, soil nailing in France, despite its possibilities, has been limited to temporary works. This is the result of the problem of the durability of the bars, related to the eventual corrosion of the steel, and also to the lack of specifications.

In 1985, under an initiative from the French Ministry of Equipment, a national research project, "CLOUTERRE", was undertaken, relating to soil nailing. The objectives of the CLOUTERRE project are as follows : better knowledge of the behavior of soil nailed walls, determination of the limitations of the method, the improvement of the methods for dimensioning the works, and an elaboration in the design recommendations. The total budget of the project is 3.5 million U.S. dollars, and is being performed in cooperation with the FHWA (U.S.A.), in regards to the specifications, and more recently with the Ministry of Quebec (Canada).

The research is composed of performing simultaneously, laboratory studies, centrifugal tests, instrumentation of actual structures, and full-scale experiments. The full-scale experimentation is comprised of :

1. A nailed wall, 7 m high, built of Fontainebleau sand and loaded until failure by saturation of the soil, from the top. The wall failed by breakage of the inclusions.

2. A nailed wall, 7 m high, built of Fontainebleau sand and conducted until failure by decreasing the adherence length of the bars.

3. A full-scale study of the effect of the excavation height, in which a soil nailed wall was failed by progressively increasing the height of the excavation phase.

The aim of this research is to study the three principal failure modes of a soil nailed wall, specifically: breakage of the inclusions (metal bar plus grout), slippage of the inclusions, or the development of an excessive excavation height, during the construction phase. In addition, the results will be used to adjust the actual calculation methods at failure, in which the TALREN method is standardly used in France (5, 6).

To study the mechanism of a soil nailed structure, it is necessary to determine the interface lateral friction mobilized along the reinforcement. Therefore, pull-out tests under both displacement rate and load controlled modes were performed as part of the CLOUTERRE project. Furthermore, two-dimensional finite element analysis and stability analysis based on the TALREN method were performed to check the experimental results obtained from the full-scale experiments.

Because of space limitations, this paper will only deal with the first full-scale experiment, which has been conducted by the Centre Expérimental du Bâtiment et des Travaux publics (C.E.B.T.P.) in St Rémy, near Paris. The objectives of this full-scale experiment are as follows: construction of a soil nailed wall under the same conditions as those occuring at an actual job site, with precise control of the construction parameters, specifically the homogeneity of the soil mass, and correctly defining the geometry of the wall and the nails; complete instrumentation of the soil, the facing and the nails; and the study of the wall displacements during construction and at failure.

FULL-SCALE EXPERIMENT ON A SOIL NAILED WALL

Site Preparation. The full-scale experimentation was performed at the C.E.B.T.P., at Saint-Rémy Les Chevreuse (2). Before construction of the nailed wall, a slightly cohesive sand embankment, 7m high, was constructed on a Fontainebleau sand foundation. The sand utilized was Fontainebleau sand with a small amount of fine grained material, and was placed and compacted to obtain a medium dense sand at a relative density $Dr = 0.6$. At this density, the principal mechanical properties of the sand were as follows:
· the Standard Penetration Test (SPT) values (Fig. 1) vary from 8 blows at 1 m depth to 15 blows at 6m depth;

· the limit pressure (p_1) and pressuremetric modulus (E_M), measured with a MENARD type pressuremeter, are also presented in Fig. 1;

· soil shear strength characteristics, determined in the laboratory, of $\phi = 38°$ and $c = 3kPa$.

To ensure plane deformations during the test, a double polymeric layer was placed on each lateral face of the future wall.

Construction of the Wall. The spacing of the nails in the horizontal and in the vertical directions was 1.15 m and 1.00 m respectively. The wall was constructed by stepped excavation, with an excavation increment of 1 m, as shown in Fig. 2. The choice of utilizing aluminium tubes to nail the sand mass permitted the inclusions to resist a normal stress and bending moment simultaneously.

To ensure failure of the wall by breakage of the reinforcements, the grouted length was dimensioned to avoid any risk of pull-out. On the other hand, the nails were dimensioned to produce a safety factor of 1.1, with respect to the failure of the aluminium tube by breakage. The tubes were grouted in the soil, with liquid concrete, injected under low pressure. Each tube was fixed to the shotcrete facing (80 mm thick) by utilizing a plate , which was then

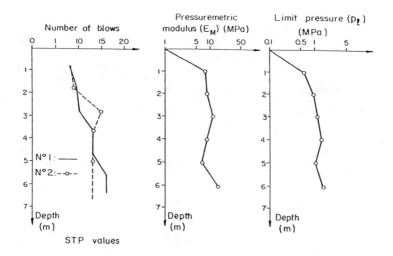

FIG. 1. Results from SPT an Pressuremeter Tests Performed
         at the C.E.B.T.P. Test Site
         (Plumelle, 1986)

bolted without pre-load onto a steel Dywidag bar, grouted to
the tube. A cross-sectional view of the grouted nails,
including the location of the strain measurements, is
presented in Fig. 3.

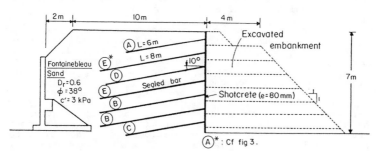

FIG. 2. Schematic of the Full-Scale Experiment on Soil
         Nailing

Instrumentation of the Nailed Mass. During the
construction of the sand embankment, horizontal bands of
black sand were placed along the axis of the wall. This
allowed visual determination of the failure zones after
testing. During the construction of the shotcrete facing,

three targets were connected to the facing at each 1 meter
depth of excavation. The objective of these targets was to
follow the wall displacements by microtriangulation. The
displacements of the nailed sand mass were measured using
vertical inclinometers placed in the tubes at a spacing of
2, 4 and 8 meters from the facing. The tensile forces in
the nails were measured using 8-10 deformation gauges per
nail, which were placed along the entire length of the nail
at an average spacing of 0.5 m.

Observations on the wall during construction. Fig. 4
indicates the horizontal displacements of the facing and the
soil, during construction and at failure. One can note a
rotation of the wall, with respect to the toe, with a final
displacement at the top on the order of 0.3% of the wall
height (21 mm). However, it must also be noted that there was

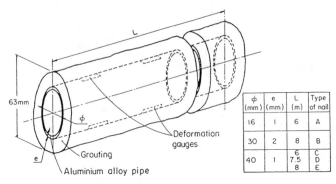

FIG. 3. Sectional View of the Grouted Bars Used in the
C.E.B.T.P. Full-Scale Experiment.

an acceleration of the displacements, with the height of
the wall, for each subsequent phase of excavation. From the
horizontal displacements of the nailed mass, it can be
observed that there is a zone extending between two and
four meters from the facing, which is moving as a rigid
block. This zone corresponds to the active zone of the soil
nailed wall. Beyond this zone, only negligible displacements
occured in the nailed mass. The small horizontal
displacement occuring at the back of the wall relative to
the frontal displacement demonstrates that the embeddment
length was sufficient to limit displacements practically to
those resulting from elastic deformations. The vertical
displacements were of the same order of magnitude as those
obtained for the horizontal displacements. The maximum
horizontal ($\delta_h$max) and vertical displacements ($\delta_v$max)
occured at the top of the wall, and the ratio
($\delta_v$max/$\delta_h$max) was approximately one for the final
excavation phases, and greater than 2 for the early phases.

The transformation from deformations to forces, in the
grouted nails is difficult because under a given load, one

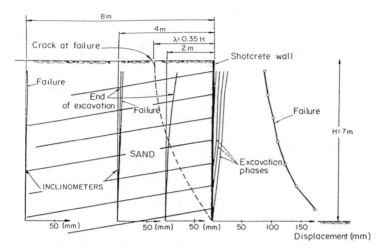

FIG. 4. Horizontal Displacements of the C.E.B.T.P. Soil Nailed Wall

can distinguish up to three different stress zones (Fig. 5). The first zone corresponds to where the aluminium is in the linear elastic phase and the concrete grout is intact. In the second case, the aluminium is still in the linear elastic phase, but the grout is fractured. For the third case, the aluminium is in a non-linear elastic phase, followed by a plastic phase. To be able to directly obtain the forces from the measured deformations, a nail with a 40 mm tube diameter (63 mm grouted diameter), under the same conditions as those at the job site, was tested and used as a reference. The force-deformation behavior for this test nail is shown in Fig. 5.

Fig. 6 presents the distribution of the tensile forces along the nails, at the end of construction. It can be noted that no tensile force was generated in the bottom row of nails. It can also be observed that the maximum tensile force is not located at the front of the wall, but rather at some distance away from the facing. As in reinforced soil structures, the soil mass is divided into two zones, the active zone and the resistant zone. In the active zone, the shear stresses acting on the nails are directed toward the wall facing while in the resistant zone, the shear stresses are directed away from the front of the wall. The maximum tensile force line was found to intersect the top of the wall at less than half the distance predicted by the Mohr-Coulomb failure wedge, and the ratio of the distance from the facing (λ) to the height of the wall (H) is equal to 0.35.

During construction, there was a large tensile force generated in the upper nail, which resulted in breakage of

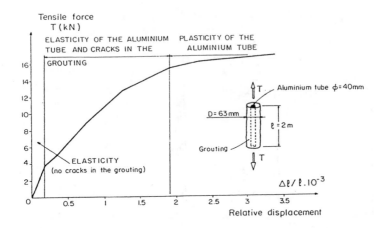

FIG. 5.   Tensile Force-Displacement Curve of a Test Nail

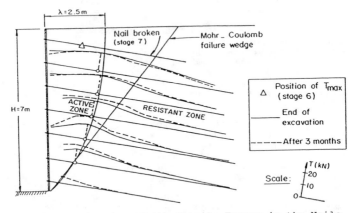

FIG. 6.   Distribution of the Tensile Forces in the Nails

the inclusion during the final construction phase. On the other hand, there was no tensile force generated during construction, in the lower nail. From the measurement of the tensile forces generated at the front of the inclusion and the maximum tensile force developed along the nail, it was observed that in general, the load in the middle and lower sections of the wall is generated primarily during the two excavation phases immediately following the grouting of the nail (normally 65-70% of the tensile force measured at the end of construction).

Fig. 7 shows the variations in the maximum mobilized tensile forces in the bars at the end of construction, as a function of the depth. It can be noted that the nails in the upper part of the wall correspond to a state of stresses in the sand greater than $Ko=1-sin\phi$, while the nails in the lower section of the wall develop tensile forces relating to

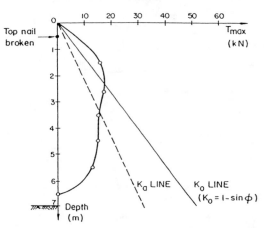

FIG. 7. Variation of the Measured Maximum Tensile Forces in the Nails with Depth (End of Construction)

a lateral pressure below the active (Ka) state of stress. In the three months following the end of construction, the tensile forces in the nails increased, generally by 15% as a result of creep within the soil, and there was a tensile force generated in the lower nail. During these three months, the creep displacement initially increased and then stabilized and remained approximately constant.

Failing the Nailed Mass. Three months after the end of construction, the nailed sand mass was progresively saturated, by a constant flow water basin located at the top of the wall, to increase its volumetric weight and decrease its cohesion. This method of failure provides the advantage of not disturbing the displacement and the stress fields, which can occur when a surcharge is applied to the top of the wall. The saturation of the sand permitted failure of the soil nailed wall without completely destroying the structure. At failure, the facing subsided by a total of 0.27 m and became embedded in the underlying sand. The wall had also advanced horizontally by 0.09 m at the top and 0.17 m at the toe, as shown in Fig. 8. Unfortunately, it was impossible to make measurements on the soil nailed wall during failure.

At failure, a fissure was observed at the top of the wall, 2.5 m behind the wall facing. This corresponds to a distance of 0.35H, which is approximately equal to the theoretical

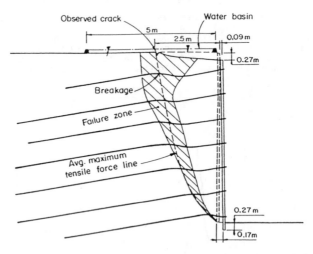

FIG. 8. Failure Zone Developed in the C.E.B.T.P. Soil
Nailed Wall

value of 0.3H, used in Reinforced Earth structures to locate
the intersection of the maximum tensile force line with the
top of the soil nailed wall. Flexure developed within the
nails was measured prior to failure of the wall. It was
determined that the flexure in the nails began after 36
hours of saturation, and accelerated during the six hours
prior to failure. Failure occurred after approximately 54
hours.

**Observations of the Nailed Mass after Failure and
Excavation.** After failure, the soil nailed wall was
excavated, and it was possible to observe the flexure within
the nails. In particular, it can be noted that the nails had
suffered from large bending deformations and consequently
some of the nails were broken just prior to failure,
particularly at the bottom of the wall.

A complete account of the bands of black sand and the
deformation of the nails, determined during excavation,
permitted the reconstruction of the failure zone within the
nailed mass. The failure zone did not follow a unique
failure surface, but rather followed a zone of variable
width between the top and the toe of the nailed mass. It was
observed that both the maximum tensile force line and the
maximum bending line lie within the failure zone, and that
they practically coincide (Fig. 8). It can also be observed
that the location of the observed crack coincides with the
point where the maximum tensile force line intersects the
top of the wall.

Behavior of the Soil Nailed Wall before and at Failure.
During construction, the first resisting force mobilized, as

a result of the excavation process, is the tensile force in the nail. This force is the primary component in the resistance of the inclusion and can be mobilized even under small displacements. The tensile force is mobilized primarily in the two excavation phases following the placement of the nail. After construction, and with time, the tensile force can continue to increase as a result of creep displacements within the soil. In the C.E.B.T.P. Wall, large creep displacements were observed as a result of the low safety factor (F.S. = 1.1.), with respect to similar walls designed with a higher safety factor. However, it must be noted that in general, creep has the ability to occur at any excavation phase, if the construction is stopped for a sufficient amount of time. Shortly prior to failure and under large deformations, the bending stiffness of the nails is mobilized, which prevented the soil nailed wall from completely failing.

DETERMINATION OF THE SOIL-NAIL LATERAL FRICTION

In studying and dimensioning the nails for the full-scale experiment, it is first necessary to determine the limit value of the lateral friction mobilized along the length of the nail. To determine this essential parameter, pull-out tests on several types of nails were performed by C.E.B.T.P. at St Rémy, under both constant displacement rate (1mm/min) and load controlled (10 loading levels, each sustained for a maximum of 1 hour) conditions (3). The four types of nails tested were : nails grouted under low and high injection pressure, and driven nails, with or without grout.

The pull-out behavior for a driven L-shaped nail, under both displacement controlled and load controlled conditions are shown in Fig. 9. In the load controlled test, the slope of the lines of stabilization (α) is defined as the slope of the displacement versus log time curve for each loading level. It can be seen that the maximum pull-out force developed in the constant displacement rate test was equal to 59 kN, and is equal to the creep limit determined from the load controlled test. However, it must be noted that this behavior corresponds to this specific soil and nail, and should therefore not be taken as a general rule for all soil and nail types.

This study resulted in two principal conclusions. First, the placement procedure significantly affects the pull-out capacity of the nails. The bored nails grouted under high pressure gave the highest pull-out resistance, although grouted driven nails gave lower but comparable results. Driven nails, on the other hand, produced a capacity equal to or greater than bored nails grouted under low pressure. However, it must be noted that the pull-out capacity for grouted nails also depends on the shape of the borehole. If the borehole is cylindrical, with smooth walls, the normal stress acting on the inclusion remains approximately zero, corresponding to the initial state resulting from the boring process, and consequently the pull-out capacity is low. On the other hand, if the boring is irregular in shape, the

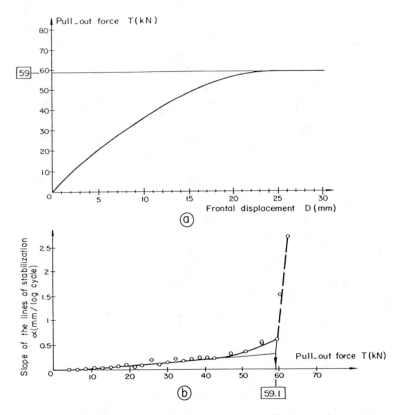

FIG. 9. Pull-Out   Behavior   of   a   Driven   Nail   under
Displacement (a) and Load Controlled (b) Modes
(Plumelle and Bel Hadj Amor, 1988)

surface texture of the grouted nail produces a rib effect,
which in turn can mobilize the effect of restrained
dilatancy (7). This results in an increase of the normal
stress on the inclusion, thereby increasing its pull-out
capacity. Furthermore, the measured normal stress in the
surrounding soil was small (10-15 %) with respect to the
pump injection pressure, even under high pressures. The low
measured normal stress is caused by stress relaxation in the
soil, under controlled deformation.
Secondly, it was also concluded from the pull-out tests
that for the Fontainebleau sand, the pull-out capacity can
be accurately determined by a constant displacement rate
test, which is more economical to perform in-situ.

DESIGN METHOD FOR SOIL NAILED STRUCTURES

The C.E.B.T.P. soil nailed wall was designed using the TALREN program, which is currently the standard procedure used in France to design soil nailed walls. This method is an at limit equilibrium slices analysis and is based on a multicriteria method proposed by Schlosser (5, 6). The four failure criteria taken into account relate to the different failure modes of a soil nailed wall and to the different soil nail interactions. The method considers the following resistances: the tensile and bending strength of the inclusion, the shear strength of the soil, the maximum soil-inclusion lateral friction, and the maximum lateral earth pressure on the reinforcement.

The TALREN design method does not calculate the tensile forces or bending moments mobilized in each row of nails, but rather the design is generally made by considering the resistance of the nails in the stability of all construction phases and the constructed wall. It is a general program, which is capable of taking into account all kinds of inclusions, and is not only restricted to retaining structures, built of reinforced soil.

The TALREN design method has been checked on a large number of structures at failure, in addition to slope stabilization and Reinforced Earth walls designed with classical methods, and generally there is a good agreement. As previously discussed the C.E.B.T.P. Soil Nailed Wall was designed with a safety factor of 1.1, with respect to breakage of the reinforcements. The wall was progressively saturated until failure occured at a saturation height of approximately 5 to 6 meters. When the factor of safety was calculated as a function of the saturation depth using the TALREN program, with $c' = 3$ kPa, it was predicted that failure (F.S. = 1) would occur when the saturation depth reached 6.2 m, as shown in Fig. 10. This demonstrates that the TALREN method is capable of accurately predicting the field behavior.

FINITE ELEMENT ANALYSIS OF THE C.E.B.T.P. WALL

To check the experimental results obtained from the C.E.B.T.P. full-scale wall, two dimensional finite element analysis was performed, considering perfectly elastic-plastic soil behavior (Mohr-Coulomb) and nails, 7.5 m long, oriented horizontally (1). Observations of the princpal stresses directions demonstrated a rotation as a result of the excavation. When the nail and facing element were placed, the principal stresses returned to their initial positions. This cyclic behavior of the principal stresses may induce non-negligible plastic shear strains (8).

The calculated stresses developed in the lower nails were found to have good agreement with the experimental values, while the calculated values tended to underpredict the stresses in the upper rows of nails. However, it is important to note that at phase 7, the upper nail, which was shorter and less resistant than the other nails, was broken,

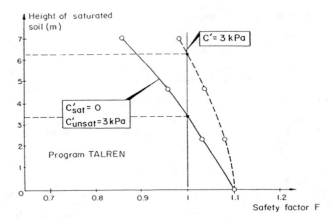

FIG. 10. Calculated Variation of the Safety Factor with the
         Saturation Height (C.E.B.T.P. Soil Nailed Wall)

resulting in a load transfer to the nearby rows of nails.
Consequently, the tensile force in the upper nails should be
higher than predicted from calculations considering the
upper nail. The distribution of tensile stresses, for the
seventh excavation phase, is presented in Fig. 11. The
accurate prediction of the stresses in the lower nails was

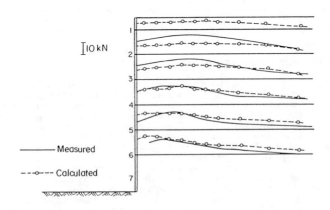

FIG. 11. Measured and Calculated Tensile Forces Generated in
         the Nails

observed at each stage of excavation. In addition, the location of the maximum tensile forces also showed a good agreement with the experimental values.

On the other hand, the calculated horizontal displacements did not show as good a correlation to the experimental values. Fig. 12 presents the predicted horizontal displacements calculated at a distance of 2 m behind the wall, during the fourth and sixth excavation phases, in addition to the measured inclinometer readings at these phases. From this figure, it can be observed that the calculated horizontal displacements are somewhat underpredicted. The poor quality of the predicted wall deformations is most likely the result of the inability of the classical elastic-plastic model to accurately represent the complex soil behavior under cyclic shear stresses, and the difficulty in correctly determining the elastic modulus (E).

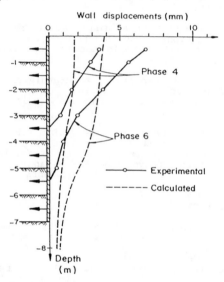

FIG. 12. Variation of the Experimental and Calculated Horizontal Displacements with Depth (measured 2 m behind Wall Facing).

CONCLUSIONS

Based on the experimentation performed by C.E.B.T.P. in the framework of the CLOUTERRE project, on a full-scale soil nailed wall in Fontainebleau sand, designed with a safety factor of 1.1, in addition to analysis of the wall using finite element analysis and the TALREN method, the following conclusion were obtained:

- the line of maximum tensile forces and maximum bending generated in the nails coincide with the actual failure zone observed in the wall;

- the failure surface intersected the top of the wall at a distance of approximately 2.5 m (0.35 H) which corresponds relatively to the theoretical value for reinforced soil structures of 0.3 H;

- the tensile force is the first mechanism mobilized, and it is developed progressively during excavation. The tensile forces can increase with time as a result of creep, especially if the safety factor is low. Close to failure and under large deformations, the bending stiffness is mobilized, giving an additional safety factor. For the wall, the lateral deformations are approximately equal to the vertical deformations and are of the order of 0.3% of the wall height.

- pull-out tests under both displacement rate and load controlled conditions produced equal pull-out capacities, although this behavior should not be generalized to all soil and nail types;

- the TALREN method is capable of accurately predicting the actual behavior of a soil nailed structure, taken to failure, which is characterized by large wall deformations and breakage of the nails;

- finite element analysis of the C.E.B.T.P. wall, was generally able to accurately predict the stresses and maximum tensile forces in the wall, especially for the lower rows of nails. Conversely, F.E.M. analysis tended to underperdict the horizontal displacements developed in the soil mass.

APPENDIX. REFERENCES

1. Delage P., Chiguer M. and Nanda A. (1988). Modélisation numérique du mur de St Rémy, Internal Report CLOUTERRE Research Project, Ecole Nationale des Ponts et Chaussées - CERMES, Paris, 53 p.
2. Plumelle, C. (1986). Compte rendu de l'Expérimentation en Vraie Grandeur de la Paroi Clouée du C.E.B.T.P., Internal Report CLOUTERRE Research Project, C.E.B.T.P., Paris, 111 p.
3. Plumelle, C. and Bel Hadj Amor, W. (1988). Essais de Tractions en vraie Grandeur de différents Types de Clous dans du Sable de Fontainebleau, Internal Report CLOUTERRE Research Project, C.E.B.T.P., Paris, 49 p.
4. Rabejac, S. and Toudic, P. (1974). Construction d'un mur de soutènement entre Versailles-Chantiers et Versailles- Matelots, Revue Générale des Chemins de Fer, 93rd year, 232-237.

5. Schlosser, F. (1980). Internal Report to l'Ecole
   Nationale des Ponts et Chaussées on the TALREN Design
   Method for Soiling Nailing, 20 p.
6. Schlosser, F. (1982). Behavior and Design of Soil
   Nailing, Proc. Symposium on Recent Developments in
   Ground Improvement Techniques, Bangkok, 399-413.
7. Schlosser, F., Jacobsen, H.M, and Juran, I. (1983). Soil
   Reinforcement, General Report-Speciality Session 5,
   Proc. 8th Intl. Conf. Soil Mechanics and Foundation
   Engineering (3), Helsinki, 1159-1180.
8. Symes, M.J.P.R., Gens, A. and Hight, D.W. (1984).
   Undrained Anisotropy and Principal Stresses Rotation in
   Saturated Sands, Géotechnique, 34(1), 11-27.

STABILITY ANALYSES FOR SOIL NAILED WALLS

J. H. Long[1], W. F. Sieczkowski Jr.[2], E. Chow[3], and E. J. Cording[4]

ABSTRACT: Analyses of the global stability of soil nailed walls using classical slope stability (limit equilibrium) methods are studied to show the importance of variables such as shape of the assumed failure surface, wall inclination, wall height, soil strength, nail capacity, nail inclination, and nail length. Results of several analyses employing a 3-part linear failure surface requiring force-equilibrium are summarized graphically using normalized parameters for soil strength, nail capacity, and nail length.

INTRODUCTION

Soil nailing is the term for a technique of reinforcing the earth in-situ to provide stability for excavations and slopes. Soil nailing employs inclusions, called nails, which provide tensile reinforcement to the soil mass. The nails are installed by driving, drilling and grouting, or by driving and grouting. Tensile capacity of the reinforcements is provided by steel in the shape of bars or angles. Soil between the nails is retained by a structural facing.

Construction of a soil nailed wall progresses incrementally in a top down fashion by repeating three stages of construction. The first construction stage begins with the natural soil being excavated to a depth, typically 1m to 2m, at which the exposed face is temporarily stable. Stage two consists of spraying shotcrete on the newly exposed face of the excavation to prevent soil losses from local instability. Finally (stage 3), the nails are installed and connected to the face of the wall. Stages 1 through 3 are repeated until the full height of the wall is attained. Stages 2 and 3 may be reversed.

Soil nailing is becoming more common in the United States and

---

1 - Asst. Prof. of C.E., and Shell Faculty Fellow, Univ. of Illinois, Urbana, IL 61801
2 - Asst. Proj. Engnr., STS Consultants, Northbrook, IL, formerly ACTC Fellow, Univ. of Illinois
3 - Grad. Res. Asst., Univ. of Illinois, Urbana, IL 61801
4 - Prof. of C.E., Univ. of Illinois, Urbana, IL 61801

abroad because the method offers economy in time, materials, and
labor for specific soil conditions. The stability of a soil nailed
wall depends upon the geometry of the wall, strength and geometry of
the nails, and strength properties of the soil. Since the same
requirements for equilibrium that govern stability of slopes and
other reinforced soil also govern the overall stability of soil
nailed walls, stability analyses commonly used and accepted for earth
masses should also be applicable to soil nailed walls. Results of
parametric studies are reported to illustrate the influence of
important parameters on the stability of soil nailed walls.

DEFINITIONS

Three important components of a soil nailed wall are, 1) the
structural face of the wall, 2) the stable composite mass of soil and
reinforcements, and 3) the unreinforced soil behind the composite
mass. The function of the face is to retain the soil between nails
while the role of the reinforced material is to resist loads
subjected on the wall by the unstable soil.

The geometry of a soil nailed wall can vary significantly depending on requirements for safety, soil properties, nail properties,
etc. Components that characterize important features of a soil
nailed wall are shown in Fig. 1 and are defined as follows: 1) β,
angle the wall face makes with a horizontal line, 2) N_{nails}, the number
of rows of soil nails, 3) L, length of the soil nails, 4) H, vertical
height of the wall, 5) L/H, length to height ratio, 6) ϕ_{mob}, the
mobilized friction angle for a cohesionless soil ($\phi_{mob} = \tan^{-1}[(\tan \phi_{soil})/FS]$, where FS is the factor of safety), and 7) F*, a non-
dimensional parameter representing the mobilized nail force in the
soil nailed wall.

The non-dimensional parameter, F*, identifies the ratio between
the maximum pressure that can be applied by the soil nail (maximum
nail capacity divided by the area it supports) and the vertical
stress, σ_v, at the elevation of the nail. The magnitude of F* is
assumed to be constant for all rows of nails and is expressed
mathematically as follows:

$$F* = \frac{T_{nail}}{\sigma_v \cdot s_v \cdot s_h} \tag{1}$$

where s_v and s_h are vertical and horizontal spacings between nails.

Results of full-scale tests on some soil nailed walls have shown
that the nail capacity, T_{nail}, is approximately constant with depth
(Cartier and Gigan, 1983). If nail capacity and nail spacing (s_v and
s_h) remain constant, F* decreases with depth according to Eqn. 1.
However, the real variation of F* with depth for a full scale soil
nailed wall depends upon the method of construction, the geometry of
the nail, the soil profile and the nail spacings (s_v, s_h) and is
inevitably variable. The value of F* can be controlled during
construction by varying s_v and s_h in response to measured nail
capacity and vertical stress.

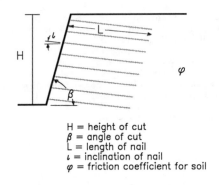

H = height of cut
β = angle of cut
L = length of nail
ι = inclination of nail
φ = friction coefficient for soil

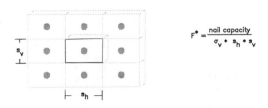

$$F^* = \frac{\text{nail capacity}}{\sigma_v * s_h * s_v}$$

Figure 1. Parameters Influencing Stability of Soil Nailed Wall

Limit equilibrium analyses conducted on soil nailed walls exhibit potential failure surfaces that extend from near the base of the excavation to the ground surface. Lengths of soil nails behind the failure surface are greatest for the lower rows of nails and least for the upper rows. Portions of nails extending behind the failure surface act to provide resistance to failure of the soil nailed wall. Therefore, the lower rows of nails contribute significantly to the stability of the soil nailed wall. If F* is assumed to be constant, it should be based upon the T_{nail}, σ_v, s_v and s_h for the lower rows of nails. Assuming F* constant represents a conservative approach (as compared with assuming T_{nail}, s_v and s_h constant), but allows a more convenient procedure for normalizing results of analyses.

USE OF LIMIT EQUILIBRIUM METHODS FOR SOIL NAILED WALLS

There are numerous limit equilibrium procedures available for determining stability of earth masses and are distinguished by the requirements for equilibrium, interslice force assumptions, and shape of the failure surface. Three common methods are 1) Spencer's procedure (Spencer, 1967), 2) The Corps of Engineers' or the Modified Swedish procedure (Corps of Engineers, 1970), and 3) Lowe and Karafiath's procedure (Lowe and Karafiath, 1960).

The procedure for determining stability of a soil nailed wall involves selecting a potential failure surface, subdividing the soil mass into a finite number of vertical slices and iteratively solving the equations of equilibrium to calculate the factor of safety. The potentially unstable soil mass is bounded below by an assumed failure surface and above by the ground surface. Numerous trial failure surfaces are analyzed to determine the failure surface which yields the minimum resistance to failure. The factor of safety, FS, for soil nailed walls is based on a frictional soil strength with no dependable long term cohesion and is defined herein as:

$$FS = \tan\phi / \tan\phi_{mob} \tag{2}$$

where ϕ is the friction coefficient of the soil, and ϕ_{mob} is the friction in the soil required for stability.

Prediction for the stability of a soil nailed wall depends upon the shape assumed for the failure surface. Many failure surface shapes are discussed in the literature with the simplest failure surface being a single wedge, such as would form in the active zone of a homogeneous cohesionless soil. Simple surfaces that have been used for analyses of soil nailed walls include a bi-linear failure surface such as the one proposed by Stocker, et al. (1979), a circular failure surface used by Schlosser (1983), a log-spiral surface (Juran, et al., 1990), and a parabola (Shen, et al., 1981). Parabolic and log-spiral failure surfaces have shown to be reasonable approximations to failure surfaces observed in some model and full-scale tests (Shen, et al., 1981; Juran, et al., 1990).

The "true" failure surface for a soil-nailed wall has no predetermined shape and will depend upon the slope geometry, soil strength, nail spacing, nail capacity, nail inclination, etc. A 3-part linear failure surface is used for the analyses conducted in this study because it provides the flexibility to approximate different shapes. For instance, the 3-part linear failure surface can approximate the shape of a parabola or log spiral for conditions similar to the model- and full-scale tests discussed by Shen, et al. (1981), and it may also model the bi-linear wedge observed and used by Stocker, et al. (1979).

Soil nails provide resistance to wall failure through the development of tensile stresses, shear stresses, and bending moments. The contribution to stability of a soil-nailed wall by the shear and moment developed in the nails is a subject of considerable interest and is discussed by Juran, et al. (1990). However, including the proper magnitude of resistance due to effects of shear and bending of the nail to the stability of a soil nailed wall requires knowledge of the deformations that will occur in the soil mass, the bending stiffness of the nail, the moment capacity of the nail, and the lateral soil stiffness provided to the nail. For this study, analytical efforts to model stability of the wall include only the tensile contribution of the reinforcement since significantly less deformation of the soil mass is required to mobilize the tension resistance in a soil nail than the lateral resistance (Schlosser, 1982; and Juran, 1985), and contribution of only nail tension simplifies the analyses.

The resistance to global failure provided by the nail is calculated as the tensile force in the nail at the intersection with the failure surface. Tensile force at the intersection between failure surface and soil nail, T_x, is calculated as:

$$T_x \quad = \quad T_{nail} \cdot x \, / \, L \tag{3}$$

where T_{nail} is the pullout capacity of the nail, x is the embedded length of nail behind the failure surface, and L is the total length of the nail. Although spatial distribution of soil nails is three-dimensional, the soil nailed wall is modelled in two-dimensions for simplicity. The force provided by each level of soil nails is approximated by an equivalent force per unit width and determined by dividing the pullout capacity of the nail by the horizontal spacing.

EFFECT OF ANALYSIS METHOD, SHAPE OF FAILURE SURFACE, AND N_{nails}

Results of analytical studies are presented to illustrate the effect of the method of the analysis and the geometry of the wall on predicted factors of safety. The analytical studies are conducted using geometry and properties similar to a wall built in Fredricksburg, Virginia. Although both soil nails and tiebacks were used in the full-scale wall, only soil nails are considered for these analyses. The soil nailed wall is taken to be a vertical and 9.75 m in height. Eight rows of nails are assumed to be 4 m in length and horizontally inclined. The soil friction angle is taken to be 35° with a total unit weight of 17.2 kN/m³. The key parameters studied herein are methods of analysis, interslice force inclination, shape of failure surface, nail capacity, and number of nails.

Method of Analysis. Three methods of analysis (Spencer, Lowe and Karafiath, and Corps of Engineers) were investigated to determine how assumptions used by each method were applicable for determining the stability of soil nailed walls. The three methods of analysis can be used to analyze circular as well as non-circular failure surfaces. Differences between methods are compared for five different values of F* using a bi-linear failure surface (Table 1). The critical failure surface is determined by fixing point 1 and incrementally moving point 2 vertically and point 3 horizontally until the failure surface with the minimum factor of safety is found. The results of this study are shown in Table 1.

Although Spencer's method satisfies force and moment equilibrium, the interslice forces required for equilibrium were inclined at angles steeper than the mobilized friction angle of the sand. The error associated with the use of high interslice forces appears minor, probably due to requirements of both force and moment equilibrium. The factor of safety and the angle of the failure plane, θ, are directly affected by changes in interslice force inclination. For all values of F*, the interslice force inclination for Spencer's method is greater than Lowe and Karafiath's interslice force inclination. The interslice force inclination for the Corps of Engineers' method is the lowest because it was selected to be zero

Table 1. Comparison of Methods of Determining Factor of Safety

F*	Method	Factor of Safety	Interslice Force Inclination	Angle of Failure Plane θ
0.60	Spencer	0.899	49.8	56.5
	L&K	0.893	31.6/39.5	58.8
	COE	0.878	0.0	64.0
0.80	Spencer	1.014	48.5	55.5
	L&K	0.997	24.8/37.8	57.2
	COE	0.963	0.0	62.6
0.99	Spencer	1.113	46.3	52.7
	L&K	1.097	24.7/35.2	54.6
	COE	1.026	0.0	62.4
1.34	Spencer	1.230	41.9	50.1
	L&K	1.187	17.1/33.4	52.8
	COE	1.102	0.0	61.5
2.01	Spencer	1.299	36.4	47.8
	L&K	1.220	8.6/32.8	52.2
	COE	1.094	0.0	61.3

degrees. As the interslice force inclination increases, the factor of safety increases. Conversely, the angle of the failure plane behind the wall, θ, decreases as the interslice force inclination increases.

Interslice Force Inclination. The Corps of Engineers' procedure was used to determine the sensitivity of factor of safety to interslice force inclination by varying the interslice force inclination for the bi-linear failure surface with F* of 1.22. The factor of safety is shown to increase as the interslice force inclination increases as illustrated in Table 2 below.

Table 2. Effect of Interslice Force Inclination on Factor of Safety

Interslice Force Inclination (degrees)	Factor of Safety	ϕ_{mob} (degrees)
0	1.10	32.6
5	1.12	32.0
10	1.14	31.5
15	1.16	31.2
20	1.17	30.9
25	1.18	30.6
30	1.20	30.3
35	1.21	30.1

The factor of safety increases approximately 10 percent as the interslice force angle approaches ϕ_{mob}. The factors of safety determined using an interslice force angle equal to ϕ_{mob} agreed reasonably with factors of safety computed using Spencer's method.

Shape of Failure Surface and Nail Capacity. The effect and interaction between shape of the failure surface and the non-dimensional nail capacity factor, F*, were analyzed and compared for a circular, bi-linear, or 3-part linear failure surfaces. The study was conducted on a soil nailed wall with a soil friction angle of 30 degrees, 8 rows of nails with a length equal to 70 percent of the height of the vertical face, and nails inclined at 10 degrees.

Critical failure surfaces were determined using different shapes (circular, bi-linear, or 3-part linear) until a minimum factor of safety was computed. Regardless of the shape assumed, all failure surfaces exhibiting the minimum factor of safety exited at the face of the wall directly above the bottom row of nails. Therefore, the analyses are appropriate for determining stability of the wall during the last excavation sequence before installing the last level of nails, or by neglecting the strength of the facing. Both Spencer's method and the Corp's of Engineers method were used to compute factor of safety.

Factors of safety determined with circular, bi-linear, and 3-part linear failure surfaces are shown in Table 3. As the nail capacity factor (F*) increases, the factor of safety increases and the failure surfaces moves away from the face of the wall toward the unreinforced soil mass and exhibits greater curvature. Regardless of F*, the factor of safety is highest for the circular failure surface because it represents the least flexible failure surface for avoiding the influence of the soil nails. Factors of safety for all three failure surface assumptions agree reasonably well at low values of F* (F* = 0.25) because the influence of the nails are minimal and the shape of the failure surface is almost linear.

As the nail capacity factor increases to 1.25, the influence of the soil nails becomes more important and the failure surface becomes more curved. The increased curvature is a result of the failure surface trying to avoid the stabilizing influence of the soil nails. The curvature causes differences to increase between computed factors of safety with the circular and non-circular failure surfaces. Factors of safety using a circular failure surface were about 5 percent greater than determined with the non-circular failure surface at F* of 1.25, and would be greater than 5 percent for values of F* greater than 1.25. In general, differences between factors of safety and required shapes for the failure surface are most pronounced for larger values of F*.

Minimum factors of safety compared within 1 percent for the bi-linear and 3-part linear failure surface for all values of F*. The position of the critical bi-linear failure surface is determined by prescribing the position of 3 points. The first point is unmovable and located on the face of the wall, just above the bottom row of nails. The second point is located just below the next lowest row of nails, and a search routine allows the point to move parallel with the inclination of the nails. The third point is located on the

ground surface behind the wall and is allowed to move along the
ground surface.   Several trial surfaces are tried with different
positions for points 2 and 3 until a minimum factor of safety is
adequately assured.   Methods used to search for the shape of the 3-
part linear failure surface are described below.

Table 3.   Values of FS for Different Shapes of Failure Surfaces

	Spencer's Method			Corps of Engineers Method		
	Circular	Bi-linear	3-part	Circular	Bi-linear	3-part
F*	Surface	Surface	Surface	Surface	Surface	Surface
0.25	0.63	-	-	0.55	0.54	0.55
0.50	0.87	0.80	0.81	0.86	0.84	0.84
0.75	1.17	1.09	1.09	1.15	1.12	1.12
1.00	1.46	1.36	1.36	1.42	1.37	1.36
1.25	1.71	1.61	1.60	1.63	1.57	1.55

Method to Search for Minimum Factor of Safety.   Considerable
attention is required to determine an overall minimum value for the
factor of safety because there may exist several local minimums.   The
existence of local minimums are illustrated by conducting stability
analyses varying the angle, α, of the lowest portion of a bi-linear
wedge, while keeping the upper wedge inclined at $\theta = 45 + \phi/2$ to the
horizontal.   The individual failure surfaces are shown in Fig. 2a
with the resulting factors of safety shown in Fig. 2b.

The presence of many local minimums hinders the successful
application of any elegant mathematical solution for determining the
global minimum factor of safety.   Procedures that have been used
successfully for adjusting the shape of non-circular failure surfaces
for earth slopes (Celestino and Duncan, 1981) require considerable
attention for soil nailed walls because the presence of reinforcements cause many local minimums.

The spatial position of the points describing the 2- and 3-part
linear failure surfaces affected the factor of safety significantly
and many different schemes for selecting trial surfaces were
investigated.   The most successful procedure for determining the 3-
part failure surface was to place 1) a fixed point at the face of the
soil nailed wall, just above the bottom row of nails, 2) a moveable
point just below the next lowest row of nails, 2) a moveable point
just below the third lowest level of nails, and 3) a moveable point
at the ground surface.   Movement of points 2 and 3 were prescribed to
be parallel with the inclination of nails, whereas movement for point
4 was prescribed to be level to the ground surface.

A procedure termed the method of collapsing envelopes was
developed to determine the 3-part linear failure surface that results
in minimum stability.   The method uses 4 points to define the failure
surface and boundaries limiting the spatial movement of the points.
Directions and magnitudes for incremental shifts in the positions of
each point are prescribed.   Factors of safety are calculated for all
failure surfaces possible with the initial grid.   Within these

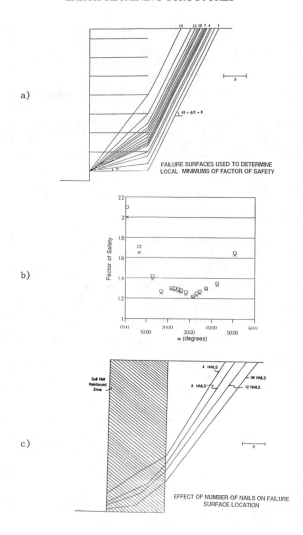

Figure 2.   Influence of Number of Nails on Shape of Failure Surface
and Factor of Safety

prescribed boundaries, factors of safety are calculated for many trial failure surfaces. New boundaries are defined by the failure surfaces of the lowest 100 factors of safety. With the new boundaries, the search procedure is repeated 2-4 more times until convergence is assured. Usually, three searches are adequate to determine the minimum factor of safety. The method of collapsing envelopes appears to converge on the global minimum for geometries common to soil nailed walls.

Effect of Number of Rows of Nails. The number of rows of nails can influence significantly the shape of the failure surface, its location, and the factor of safety. A 3-part wedge is used to analyze the stability of a soil nailed wall constructed with 4 rows of nails, 8 rows, 12 rows, and 36 rows. The normalized nail capacity, F*, is identical for each wall. The final failure surfaces for each number of nails are shown in Fig. 2c. The minimum factor of safety for each number of nails is shown in Table 4. The study shows that as the number of nail rows increase, the factor of safety increases because the failure surface can no longer "avoid" intersecting nails. Additionally, the failure surface becomes more "curved" and is forced further back into the soil mass as the number of rows increase.

Table 4. Effect of Number of Nails on Factor of Safety

Number of nails	Vertical Spacing (feet)	Factor of safety
4	8.00	0.89
8	4.00	1.13
12	2.67	1.21
36	0.89	1.36

Effect of β and F*. Results illustrating the effect and interaction of β and F* on the stability of a soil nailed wall is shown in Fig. 3. The soil is assumed to have a friction angle of 40°, with 8 rows of nails inclined at 10° and a non-dimensional length, L/H of 0.3. Small values of F* have minimal effect on the position and shape of the failure surface which passes through many rows of the soil nails (Fig. 3). Generally, for any wall inclined at an angle, β, the stability increases as F* increases. At small values of F*, stability increases as the wall angle decreases from 90° to 60°.

Large values of F* affect the shape and location of the failure surface significantly because of the contribution to stability provided by the nails. Large values of F* force the failure surface deeper into the wall, behind the nails (Fig. 3), resulting in increased stability. The soil nailed wall begins to behave similarly to an intact gravity structure as F* increases in value above 1.1; therefore, resistance to sliding increases as β increases which results in greater stability of the wall (Fig. 3).

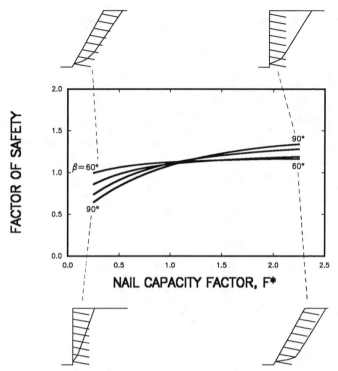

Figure 3. Effect of Nail Capacity and Wall Angle on the Factor of
Safety and Position of Failure Surface (N_{nails} = 8, ϕ = 40°,
and Nail Inclination, i = 10°).

PARAMETRIC STUDY WITH LIMIT EQUILIBRIUM ANALYSIS

Designing a soil nailed wall may require many iterations to
determine suitable geometry and nail characteristics for a specific
level of safety; therefore, a procedure for graphically determining
the dimensions and strengths required for adequate global stability
would be useful. Non-dimensional graphs are very versatile and can
be used to demonstrate effects of specific parameters on stability
and to assess important variables, dimensions, and parameters
necessary for a successful design. Although the use of non-dimensional charts can be helpful in the preliminary design of soil nailed
walls, the stability of the final design should be checked by more
detailed analyses.
    Schmertmann, et. al. (1987), Bonaparte, et. al. (1987), and
Jewell, et. al. (1985), have developed charts for determining the
stability of reinforced earth embankments (steel and geosynthetic).

However, reinforcing elements in reinforced earth are horizontal, whereas in soil nailing they are inclined.

The non-dimensional stability charts presented herein are based on limit equilibrium analyses for 1080 idealized soil nailed walls with various geometries and nail properties. The non-dimensional graphs enable the assessment of variables important to the stability of soil nailed walls. Six main variables influence soil nailed wall slope stability: L/H, β, ι, ϕ, F*, and N_{nails} (the number of rows of nails in a vertical cross-section) as shown in Fig. 1.

The parameter, F* (F* = nail capacity/$\{\sigma_v \cdot s_h \cdot s_v\}$), is assumed to be constant. Use of F* allows the non-dimensional solution for the factor of safety to be independent of γ_{soil} and height of the wall. The force provided by the soil nail intersecting a potential failure plane is assumed to be calculated according to Eqn. 3. Furthermore, the yield strength of the soil nail is assumed to be greater than the pullout capacity. All nails are assumed to have a constant vertical spacing with the top row of nails at $s_v/2$ below the crest and the bottom row of nails at $s_v/2$ above the base.

Results of additional analyses are shown in Fig. 4 to illustrate the importance of the parameters L/H, β, ι, ϕ, F*, and N_{nails}. Fig. 4a and 4b illustrate effects of β, L/H, F*, on ϕ_{mob}. Each of these parameters can have a great effect on the required value of ϕ_{mob}. While Fig. 4a shows both F* and L/H affect stability significantly, Fig. 4b demonstrates that the effect of both F* and β can be small for values of F* greater than approximately 1.

Inclination of the nails has a small effect on stability (Fig. 4c) for F* less than 0.8, while there is progressively less soil strength required as the number of rows of nails increases from 4 to 12 (Fig. 4d). In both cases, the effect of inclining the nails and increasing the number of nails forces the failure surface deeper into the wall which results in increased stability.

The results of 1080 stability analyses are summarized graphically in Fig. 5 showing the effect of the parameters L/H, β, F*, on ϕ_{mob} for a soil nailed wall with 8 rows of nails inclined at 10°. The chart allows the determination of ϕ_{mob} required for a prescribed F* and geometry of a soil nailed wall. Additionally, the figure can be used to determine requirements for F*, β, and L/H to attain a suitable factor of safety.

Given the soil strength and specific requirements for wall height, the figure may be used to determine the other parameters such as F*, β, and L/H. Additionally, the figure can be used to determine what effect changing the nail capacity factor, F*, nail length, wall angle will have on the overall safety of the soil nailed wall.

EXAMPLE

A vertical soil nailed wall is to be constructed using a factor of safety of 1.5 for the soil resistance (ϕ_{soil} = 35°), 8 rows of nails inclined at 10°, and a L/H of 0.75. The value of non-dimensional nail capacity, F*, required for stability must be determined.

The value of ϕ_{mob} ($\tan^{-1}[\{\tan 35\}/FS]$) is determined to be 25°. The remaining steps are iterative and repeated until F* assumed

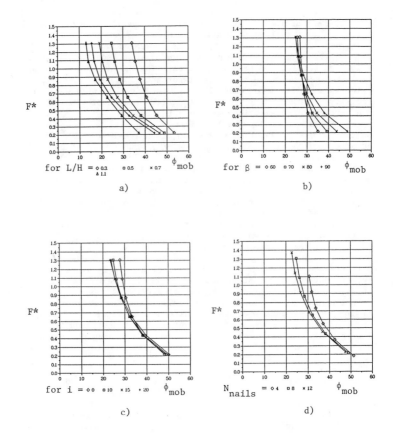

Figure 4.   Effects of Soil Nailing Parameters on Relationship between
            Mobilized Friction Angle and Normalized Nail Capacity (for
            L/H = 0.5, N_{nails} = 8, β = 90°, Nail inclination = 10°

equals F* obtained.
     F* is assumed to be 0.5, and a line is drawn vertically upward
from the diagonal in quadrant 3 to quadrant 2 (Fig. 5), until the
drawn line intersects with the line for L/H = 0.75. Another line is
drawn horizontally from this intersection from quadrant 2 to quadrant
1 until it intersects with the ϕ_{mob} = 25° line in quadrant 1.   Now
draw a line vertically downward from the point of intersection in
quadrant 1 until it intersects with the line β = 90° in quadrant 4.

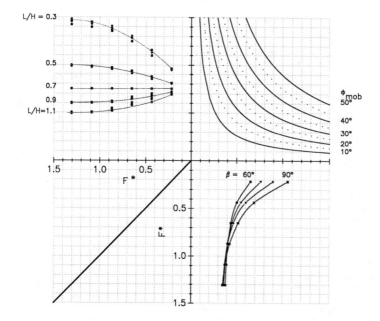

Figure 5. Non-dimensional for Determining Stability of a Soil Nailed Wall with N_{nails} = 8, Nail Inclination = 10°

The line is then drawn horizontally to the left until it intersects with the diagonal F* axis in quadrant 3 and this is F* obtained. If F* obtained is different from F* assumed, the procedure is repeated with F* obtained as the new value of F* assumed.

Using the above outlined procedure, a value of F* required for stability is determined to be 0.72. Required values of nail force and nail spacing can be determined from F* using reduction factors to account for uncertainties in the structural and pullout capacities of the nails.

Fig. 5 can also be used to determine required values of L/H for specified values of slope angle β and ϕ_{mob} as well as numerous other contributions.

CONCLUSIONS

Predictions of the stability of soil nailed walls depend on the slope of the face, the capacity of the nails, the length of the nails, the number of nails and their inclination to the horizontal, the strength characteristics of the soil, and the methods adopted to analyze the soil nailed wall. Several limit equilibrium studies have been conducted on walls with various geometries and nail characteristics to determine their effect on stability. Four of the most important variables are the strength of the soil, ϕ, the non-dimensional resistance provided by the soil nails, F*, the length/-height ratio, L/H, and the slope of the wall, β.

Other variables such as the number of nails and their inclination can also influence the stability of the wall. The number of rows of nails in a wall have a great effect on the failure surface location and the factor of safety. The more rows of nails in a vertical section of wall, the more difficult it is for the failure surface to avoid the nails. Additionally, inclined nails make it more difficult for the failure surface to avoid intersecting rows of nails.

The method of analysis affects factors of safety, as does the interslice force inclination: higher interslice force inclinations result in higher factors of safety. Stability analyses using the Corps of Engineers procedure (force equilibrium) with an interslice force inclination equal to ϕ_{mob} agrees well with stability analyses using Spencer's method (force and moment equilibrium). The Corps of Engineers' method is adopted as the analysis method for these parametric studies because predictions of stability are similar to the more sophisticated method (Spencer's method), yet is more computationally efficient.

Factors of safety for many failure surfaces were compared, showing there are numerous local minimums encountered in the search for the global minimum factor of safety. Circular, bi-linear, and 3-part wedge failure surfaces were compared with the 3-part wedge being the least constrained failure surface; however, for low values of F*, all three failure surfaces predicted similar stability. The bi-linear failure surface and the 3-part linear surface predicted similar factors of safety for all values of F* from 0.25 to 1.25.

ACKNOWLEDGMENTS

This research was supported by the U. S. Army Research Office through grants DAAL 03-87-K-0006, DAAL 03-86-G-0186, and DAAL 03-86-G-0188 to the University of Illinois Advanced Construction Technology Center. Their support is gratefully acknowledged.

REFERENCES

R Bonaparte, RD Holtz, and JP Giroud, Soil Reinforcement Design Using Geotextiles and Geogrids, Geotextile Testing and the Design Engineer, ASTM STP 952, Philadelphia, 1987, 69-116.

G Cartier, and J-P Gigan, Experiments and Observations on Soil Nailing Structures, C. R. 8th Congr. Europ. Mec. Sols. Trav. Fond., Vol. 2, Helsinki, 1983, 473-476.

TB Celestino, and JM Duncan, Simplified Search for Non-circular Slip Surfaces, Proceedings of the Tenth International Conference on Soil Mechanics and Foundation Engineering, Vol. 3, Stockholm, June 15-19, 1981, 391-394.

Corps of Engineers, Engineering and Design - Stability of Earth and Rock-Fill Dams, Engineer Manual EM 1110-2-1902, Dept. of the Army, Corps of Eng., Off. of the Chief of Eng., 1970.

I Juran, Reinforced Soil Systems - Application in Retaining Struc-
    tures, Geotechnical Engineering, Vol. 16, 1985, 39-83.
I Juran, G Baudrand, K Farrag, and V Elias, Kinematical Limit
    Analysis for Design of Soil-Nailed Structures, ASCE Jrnl. of
    Geot. Eng. Div., Vol. 116 (1), Jan. 1990, 54-72.
RA Jewell, N Paine, and RI Woods, Design Methods for Steep Reinforced
    Embankments, Polymer Grid Reinforcement, 1985, 70-81.
J Lowe III and L Karafiath, Stability of Earth Dams Upon Drawdown,
    Proc. of 1st PanAm. Conf. on Soil Mech. and Fnd. Eng., Vol. 2,
    Mexico City, 1960, 537-552.
F Schlosser, Behavior and Design of Soil Nailing, Symp. on Recent
    Dev. in Ground Imp. Tech., Bangkok, Nov. 1982, 399-413.
F Schlosser, Analogies et differences dans le Comportement et le
    Calcul des Ouvrages de Soutenement en Terre Armee et par
    Clouage du Sol, Annales de L'Institut Technique du Batiment
    et des Travaux Publics, No. 418, 1983.
GR Schmertmann, VE Chouery-Curtis, RD Johnson, and R Bonaparte,
    Design Charts for Geogrid-Reinforced Soil Slopes, Geosynthetics
    '87 Conf. Proc., Vol. 1, New Orleans, 1987, 108-120.
CK Shen, S Bang, and LR Herrmann, Ground Movement Analysis of Earth
    Support System, ASCE Jrnl. of the Geot. Eng. Div., Vol. 107
    (12), Dec. 1981, 1609-1624.
WF Sieczkowski, Soil Nailing- Insitu Soil Reinforcement, report
    submitted to the Advanced Construction Technology Center,
    University of Illinois at Urbana/Champaign, 1989, 81p.
E Spencer, A Method of Analysis of the Stability of Embankments
    Assuming Parallel Inter-slice Forces, Geotechnique, Vol. 17,
    1967, 11-26.
MF Stocker, GW Korber, G Gassler, and G Gudehus, Soil Nailing, Intl.
    Conf. on Soil Reinf., Vol. 2, Paris, 1979, 469-474.

GROUND ANCHORAGE PRACTICE

Stuart Littlejohn[1]

ABSTRACT: Guidelines on current practice are presented in relation to nomenclature, responsibilities of designer and contractor, ground investigation, design, corrosion performance and corrosion protection, construction, quality controls, acceptance testing, service monitoring and future research. The topics illustrate the scope of a modern standard code of practice.

INTRODUCTION

As the use of ground anchorages has expanded for both temporary and permanent applications, various countries and organisations(1-11) have established standards of practice in order to ensure reliable performance and fitness of purpose.

To ensure safe, cost effective anchorage solutions there is a need for a detailed knowledge of the ground and a proper design related to static and dynamic loads, location of anchorages, load transfer lengths and overall stability. Given the trend towards limit state design, the loads and accompanying deformations under service conditions should be considered as well as the failure mechanisms. For both temporary and permanent works, the need for corrosion protection should be assessed.

With regard to anchorage construction, the importance of skilled operatives cannot be over-emphasised, since quality of workmanship greatly influences subsequent performance. Quality controls and record keeping are therefore strongly recommended and further, each anchorage, once installed, should be subjected to on-site acceptance tests.

1 - Prof, Dept of Civil Engineering, University of Bradford, England

NOMENCLATURE

In accordance with international recommendations (ISSMFE; ISRM; FIP) a ground anchorage is an installation that is capable of transmitting an applied tensile load to a load bearing stratum. The installation consists basically of an anchor head, free anchor length and fixed anchor (Fig. 1). The term 'anchor' is used exclusively to denote a component of the anchorage, e.g. anchor head. A temporary anchorage is often used during the construction phase of a project to withstand forces for a known short period of time, usually less than two years.

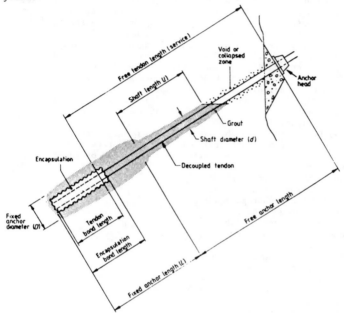

Figure 1. Ground Anchorage Nomenclature

RESPONSIBILITIES

Since ground anchorages form only one part of the design process for an anchored structure, it is prudent to define the responsibilies of the designer and anchorage contractor in the contract documents. An appropriate separation of duties is shown in Table 1 which assumes that only the designer is in a position to judge how the ground anchorage will interact with the ground/structure system. The designer is also the most appropriate person for judging the degree of risk in the use of ground anchorages and is the person who obtains agreement from the client to employ ground anchorages. In the case of design and construct contracts the responsibilities are normally combined.

Table 1. Recommended Design and Construction Duties

Designer*	Contractor
(1) Site investigation data for ground anchorages (borings near fixed anchor locations and outside the site working area if necessary). (2) Decision to use ground anchorages, required trials and testing and provision of a specification (Risk assessment). (3) Overall design of anchored structure, calculations of anchorage force required. Definition of safety factors to be employed. (4) Definition of anchorage life (permanent/temporary) and requirement for corrosion protection. (5) Anchorage spacing and orientation. Free anchor length and anchorage load. (6) Anchorage behaviour monitoring system, (structure/anchorages) and interpretation of results. (7) Supervision of the works (inspection by testing and sampling). (8) Maintenance specification for anchorages. (9) Instruction of all contracting parties of key items within design philosophy to which special attention should be directed.	(1) Anchorage components and details (2) Determination of fixed anchor dimensions (3) Anchorage spacing/orientation free anchor length and anchorage loads (instead of by the designer) (4) Detailing of the corrosion protection system for anchorage (5) Supply and installation of anchorage monitoring system (6) Quality control of works (7) Anchorage maintenance as directed by the designer

*The designer may be employed by the client, the main contractor, a specialist subcontractor or by a consultant. Contractual arrangements between the parties will vary with each contract and be specified in the contract document.

GROUND INVESTIGATION

General. The ground is one structural component of the ground anchorage system, and the importance of a good quality site investigation cannot be over-stated. Lack of adequate information on the ground remains the most common cause of individual anchorage failures at the acceptance testing stage.

The geometry of a ground anchorage and its mode of operation require, in particular, a detailed knowledge of ground conditions local to the fixed anchor. For example, the presence of thin partings of silt or sand within a clay can have a marked effect on the behaviour of the soil and on the softening action of drilling water (Fig. 2). Whilst in rocks, discontinuity frequency and orientation data together with fracture continuity and roughness can be vital in determining the size and shape of the rock mass mobilised in any overall stability analysis.

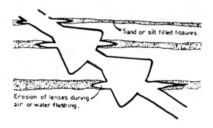

Figure 2. Influence of Sandfilled Fissures on Underream Configuration

Classification of the ground with particular reference to shear strength, compressibility, density, ground water conditions and chemical analysis should be carried out in the normal way, but guidelines are required on the extent and intensity of the ground investigation for ground anchorages.

In essence, the aim of the investigation should be to determine, by the most economic means, the nature of the block of ground that is influenced by, or influences, the installation and behaviour of ground anchorages. Since inclined anchorages are installed as commonly as vertical anchorages, lateral variations in ground properties should be investigated as thoroughly as the more easily investigated vertical variations (Fig. 3). Key boreholes should be located at site extremities, so that the strata profile over the relevant zone can be initially interpolated between the boreholes rather than extrapolated outside the area investigated. The depth of these boreholes should be adequate to ensure that either:

(a) a known geological formation is proved; or
(b) no underlying stratum will affect design.

The number of locations investigated will depend upon the
confidence that can be placed on the uniformity of the strata,
but for soil anchorages it is recommended that the maximum
centres at which the ground is investigated in detail should
not exceed 20 m. The locations should be sited both along the
line of the structure and, where practicable, along the line of
the probable fixed anchor zone.

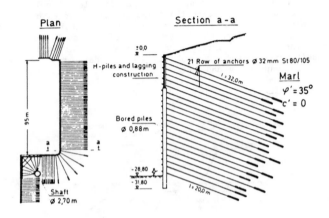

Figure 3. Excavation Allianz Stuttgart

Sampling. Emphasis should be placed on obtaining samples
that can identify the fabric or structure of the stratum in
which the fixed anchor may be installed.

In soils, samples should be taken from each stratum and at
maximum intervals of 1.5 m in thick strata.   Intermediate
disturbed samples, suitable for simple classification tests,
should also be obtained, so providing a specimen of the ground
at a maximum of 0.75 m intervals of depth.   In variable strata
continuous undisturbed samples may be necessary in the probable
vicinity of the fixed anchor zone.

In rocks, emphasis should be placed on obtaining maximum
continuous core recovery, which generally implies cores of not
less than Nx size (55 mm) and, in weaker rocks, of larger
diameter.

Ground Water. Determination of ground water conditions is
essential, not only for the overall project, e.g. where
excavations are proposed, but also for the design and
construction of anchorage systems. This area of investigation,
particularly the recording of long term ground water
conditions, is too often given scant attention during routine
site investigation works.

All observations of water conditions during boring should

be carefully recorded, as these often permit an initial assessment of true ground water conditions. The long term ground water conditions can only be measured satisfactorily by the installation of standpipes or piezometers.

Field Testing. The static cone penetrometer which is used from the surface can provide an accurate indication of the in situ properties of granular soils, and is useful in determining the presence of minor anomalies in apparently uniform soils, e.g. thin clay seams in a granular deposit or sand partings in a stratum of clay.

In boreholes, standard penetration tests should be made at regular intervals of depth in all granular strata to obtain their in situ density classification. Within the probable fixed anchor zone tests at 1 m intervals of depth are appropriate.

The in situ strength of cohesive soils can also be assessed by the test and it is particularly applicable where the soil possesses a structure or fabric that may preclude representative undisturbed sampling, e.g. clays containing partings of water-bearing sand.

The value of any empirical in situ test depends critically on its being performed and then interpreted in a standard manner, and care must be paid to maintaining stability of the bottom of the borehole by attention to size of bore, boring methods and hydrostatic balance.

The test is also of value in obtaining a relative measure of the in situ quality of weaker rocks, although experience is required in the interpretation of such tests. For example, the recording of penetration of the probe per 50 blows, say, allows the disturbance created by previous chiselling or boring operations to be identified.

Although not yet common in routine ground investigations, the radial stress/strain characteristics of the ground mass can be obtained in granular and cohesive soils, as well as in weak rocks, by pressuremeter test. For anchorages where the tendon bond length is protected by an encapsulation, bursting stresses are developed as the tendon is tensioned and the load transfer mechanism in the fixed anchor is influenced strongly by the degree of lateral constraint provided by the surrounding ground.

Where the ground investigation proves strata which may lead to potential grout loss, then permeability tests supplemented by fabric description may be required to quantify the problem. In fractured ground masses, a knowledge of the size and frequency of fractures is more important than mass permeability, since the former usually dictates the need for pregrouting(12).

Laboratory Testing. It is recommended that the grading of granular soils and the liquid and plastic limits of cohesive soils be determined for every stratum encountered in the investigation. Grading can give an empirical guide to permeability which in turn influences the radius of grout travel by permeation. The determination of low plasticity indices in a cohesive soil, even in very localised zones, can influence both the type of anchorage and the method of drilling, e.g. underreaming may be precluded or cased drilling techniques may be required, as opposed to uncased.

To determine the in situ shear strength of granular soils, in situ penetration tests combined with grading and particle shape assessments from laboratory classification tests are required. Alternatively, laboratory shear tests are acceptable where the samples are tested at the density and stress level approximating that in situ. Direct shear tests may be appropriate for cemented soils.

For cohesive soils, the shear strengths should be obtained by triaxial compression tests, the type of test depending on design method, mass permeability of the ground and the probable rate of stressing for the anchorage. Where stressing of the anchorage is rapid, the undrained shear parameters should be used in the necessary total stress analysis. For a slow rate of stressing or where the long term anchorage behaviour is required, the shear strength in terms of effective stresses should be obtained from (a) a consolidated drained test or (b) a consolidated undrained test with pore pressure measurement.

Where a cohesive soil possesses a relatively high mass permeability, e.g. clayey silts, chalk and marl, it may be prudent to determine both undrained and effective shear parameters.

In all tests, the stress/strain characteristics of the soil during shear should be recorded. In soils exhibiting a marked fabric, e.g. laminated clays, the shear strength of the soil in the plane of shear induced around the fixed anchor length, should be studied, if possible.

For anchorage systems that will apply a high average stress to a clay lying between the fixed anchor length and the structure, the compressibility characteristics (Cv and Mv) should be determined. Such data, although rarely exploited in current anchorage practice, may influence the overall design and will in the future provide guidance as to probable loss of prestress through case history comparisons.

The shear strength of a rock mass, used in the assessment of overall stability, is commonly estimated from a detailed study of fracture geometry and roughness together with a knowledge of material characteristics. The design of an individual anchorage is, however, usually based on borehole information and is often dependent upon a confining system local to the fixed anchor. The results of laboratory tests on intact rock cores can therefore be useful, if carefully interpreted. Unconfined compressive or tensile strengths are commonly used to estimate the bond or skin friction in the fixed anchor zone. Deformability of the rock material can be determined by measuring stress/strain relationship in the unconfined compression test, but the characteristics of the rock mass can only be measured in situ, e.g. by pressuremeter.

The slake durability test apparatus assesses the susceptibility of weak rock to softening in the presence of water, and therefore gives an indication of the potential loss of strength of the rock around the borehole and the period which a hole may be left open prior to tendon installation. In weak mudstones and shales, the results of this test may have a considerable influence on the design of the fixed anchor length. In this regard, a microscopic examination of minerals can also be helpful.

Chemical Testing. The purpose of any chemical analysis of the ground or ground water should be to determine its aggressivity to the constituents of the anchorage system and in particular the tendon and its grout surround.

Sulphate and chloride contents are established as a routine and dictate choice of cement, but since the overall corrosion hazard is seldom quantified, Table 2 is included for guidance.

Table 2. Aggressivity of Ground Water with respect to Cement

Ground water environment	Remarks on aggressivity
Very pure water (CaO<300mg/litre)	Such waters dissolve the free lime and hydrolyse the silicates and aluminates in the cement
pH<6,5	Acid waters attack the lime in the cement, but pH values of 9-12 are passivating
Selenious water $(SO_3)>0,5g/litre$ (stagnant) $>0,2g/litre$ (flowing) Magnesium water $(SO_3)>0,25g/litre$ (stagnant) $>0,1g/litre$ (flowing)	These sulphates react with the tricalcium aluminate to form salts which disarrange the cement by swelling

With reference to aggressivity towards metals, redox potential and resistivity are useful indicators (Table 3).

Table 3. Corrosiveness of Soils related to Values of
Resistivity and Redox Potential (after King (1977))

Corrosiveness	Resistivity	Redox potential (corrected to pH = 7) Normal hydrogen electrode
	Ω.cm	mV
Very corrosive	< 700	< 100
Corrosive	700 to 2000	100 to 200
Moderately corrosive	2000 to 5000	200 to 400
Mildly corrosive or		
non-corrosive	> 5000	> 430 if clay soil

NOTE. In the absence of the above tests, ground and ground
water samples should be taken for detailed chemical analysis
e.g. chloride and sulphate ions, in order to judge
aggressivity.

Redox potential provides guidance on the risk of
microbiological corrosion, frequently characterised by pitting
attack and most commonly encountered in heavy clay soils.

If the site is adjacent to an electrical installation, the
ground should be checked for stray electrical currents.

Investigation During Construction. Where the initial
investigation shows that ground conditions are liable to random
variations, e.g. glacial drift, then all ground data obtained
during anchorage drilling should be recorded and subjected to
daily analysis. Such a system can act as an early warning
device, should variation in strata levels or ground type
require changes in design or installation method.

It should be emphasised, however, that production drilling
associated with anchorage installation is not geared to
investigate the ground in detail, and cannot be expected to
highlight, for example, silt or sand lenses in clay or
variations in density in gravelly sand.

Any adjacent activities on the site that may influence
anchorage behaviour should be monitored, recorded and their
possible influence assessed at an early stage in the work.
Such activities include local excavation, ground water
lowering, piling, blasting, freezing and mining subsidence.

DESIGN

General. Design calculations are required in order to
judge in advance the technical and economic feasibility of a
proposed anchorage solution. In retaining wall tie-backs, for
example, anchorage dimensions can be varied in the calculations
to optimize such factors as anchorage load and spacing in
relation to wall design and cost considerations. Design rules

also permit assessment of the sensitivity of the load holding capacity to variations in anchorage dimensions and ground properties, the results of which may indicate working loads and choice of safety factors.

Anchorage construction technique and quality of workmanship greatly influence the pull-out capacity and the latter, in particular, limits the designer's ability to predict accurately solely on the basis of empirical rules. As a consequence, the calculated figures should not be used too dogmatically. In anchorage technology, practical knowledge is just as essential to a good design as the ability to make calculations.

Design rule predictions of ultimate load holding capacity are invariably created by assuming that the ground has failed along a shear surface, postulating a failure mechanism and then examining the relevant forces in a stability analysis. Fixed anchors should be designed to fail in local shear for minimal disturbance and interference, and to provide sufficient constraint from the surrounding ground, the depth:diameter ratio of the top of the fixed anchor usually exceeds 15. In practice fixed anchors are founded typically at depths in excess of 5 m, given the small diameters involved. Since design rules have been developed, primarily through systematic full scale testing of individual anchorages, designers ensure that the fixed anchor spacing is not less than four times the fixed anchor diameter, which usually means a minimum spacing of 1.5 m. In this way interference effects are avoided.

For the interested practitioner, several anchorage design reviews are available for both soils and rocks(13-19).

Anchorage Types. Since anchorage pull-out capacity for a given ground condition is dictated not only by anchorage geometry, but also by construction technique, e.g. grouting procedure, it is important to recognise the main anchorage types used in current practice (Fig. 4).

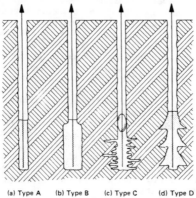

(a) Type A   (b) Type B   (c) Type C   (d) Type D

Figure 4.   Main Types of Cement Grout Injection Anchorage

Type A anchorages consist of tremie (gravity displacement), or packer grouted straight shaft boreholes, which may be temporarily lined or unlined depending on hole stability. This type is most commonly employed in rock and very stiff to hard cohesive deposits. Resistance to withdrawal is dependent on side shear at the ground/grout interface.

Type B anchorages consist of low pressure (typically grout injection pressure $p_1 <$ 1000 kN/m^2) grouted boreholes, via a lining tube or in situ packer, where the diameter of the fixed anchor is increased with minimal disturbance as the grout permeates through the pores or natural fractures of the ground. This type is most commonly employed in weak fissured rocks and coarse granular alluvium, but the method is also popular in fine grained cohesionless soils. Here cement based grouts cannot permeate the small pores but under pressure the grout compacts the soil locally after boring to increase the effective diameter and enhance the shearing resistance. Resistance to withdrawal is dependent primarily on side shear in practice, but an end bearing component may be included when calculating the ultimate capacity.

Type C anchorages consist of boreholes grouted to high pressure (typically $p_1 >$ 2000 kN/m^2), via a lining tube or in situ packer. The fixed anchor diameter is enlarged by hydrofracturing of the ground mass to give a grout root or fissure system beyond the core diameter of the borehole. Often pressure is applied during a secondary injection after initial stiffening of primary grout placed as for Type B anchorages. Secondary injections are usually made via either a tube à manchette system or miniature grout tubes incorporated within the fixed anchor length; the former is advantageous if several injections are envisaged. A relatively small quantity of secondary grout is needed. Continuous flow or a sudden drop in initial injection pressure might indicate hydrofracture after which only relatively limited pressures can be attained. Whilst this anchorage type is commonly applied in fine cohesionless soils, success has also been achieved in stiff cohesive deposits. Design is based on the assumption of uniform shear along the fixed anchor.

Type D anchorages consist of tremie grouted boreholes in which a series of enlargements (bells or underreams) have previously been formed. This type is employed most commonly in firm to hard cohesive deposits. Resistance to withdrawal is dependent on side shear and end bearing, although, for single or widely spaced underreams, the ground restraint may be mobilized primarily by end bearing. Although not common this type can be used in cohesionless soils in conjunction with some form of side wall stabilization over the enlargement length. Typically this may be by pre-injection of cement or chemical grout in the ground around the fixed anchor or by pumping polymer drilling fluid into the borehole during the drilling and underreaming operation.

Rock/Grout Interface. For Type A anchorages, designs are commonly based on the assumption of uniform bond distribution, although this is unlikely to be the case where E grout/E rock is less than 10. Few failures are encountered however at the rock/grout interface and new designs are often based on the successful completion of former projects, i.e.former working bond values are re-employed or slightly modified depending on the judgement of the designer.

For strong rocks (UCS > 20 N/mm² the ultimate bond may be taken conservatively as 10% UCS of the rock up to a maximum bond of 4 N/mm², assuming that the grout crushing strength ≥40 N/mm². Increasing grout strength beyond 40 N/mm² will not lead to a significant increase in rock/grout bond.

For weak rocks (UCS < 7 N/mm²) the ultimate bond should not exceed the minimum shear strength, based on shear tests on representative samples. The degree of weathering of the rock is a major factor that affects not only the ultimate bond but also the load-displacement characteristics. In chalk and clay filled fissured rock, the bond may be affected by the presence of smear.

Degree of weathering is seldom quantified but for design in weak weathered rocks standard penetration test data are exploited occasionally to predict ultimate bond, for example ultimate bond (kN/m²) = 10 N (where N = blows/0.3m)   (1) for stiff to hard chalk(14).

Care is required in interpretation since N values can be subject to considerable scatter and the empirical design relationships are also dependent on grout injection pressure(20). As a consequence, proving tests (proof tests in U.S.) are recommended to verify design assumptions.

With the exception of rock bolts, the fixed anchor length should not be less than 3 m (2 m in rock if Tw < 200 kN, where Tw = safe working load).

Under certain conditions, it is recognised that much shorter lengths than 3 m would suffice, even after the application of a generous factor of safety. However, for a very short fixed anchor, any sudden drop in rock quality along the fixed anchor length can induce a serious decrease in ultimate load holding capacity.

In general, there is a scarcity of empirical design rules for the various categories of rocks and too often bond values are quoted without provision of strength data or a proper classification of the rock and cement grout. In spite of this situation rock anchorages have been installed to safely withstand working loads of 20,000 kN, although values of 500 to 1000 kN are more common.

Cohesionless Soil/Grout Interface. Type A anchorages are not used generally in cohesionless soils except where they are cemented. The most common forms of anchorages are Types B and C, the latter growing in popularity in the weaker soils, where high pressure primary grouting or post-grouting techniques are used to improve local soil properties in order to create safe working loads of 500 to 600 kN.

For low pressure type B grouted anchorages, design equations relate ultimate load holding capacity to anchorage dimensions and soil properties. An accurate assessment of the fixed anchor diameter (D) is not generally possible but approximate estimates can be made from grout takes in conjunction with ground porosity. For borehole diameters (d) of 0.1 to 0.15 m, D values of 0.4 to 0.5 m can be attained in coarse sands and gravels ($k_w > 1 \times 10^{-3}$m/sec), say 3 to 4 d. Where grout permeation is not possible ($k_w < 5 \times 10^{-4}$m/sec) and only local compaction is achieved, D values for the above borehole diameters and an injection pressure of 1000 kN/m² may range from 0.2 to 0.25 m in medium dense sand (1.5 to 2 d, say), and D values of 0.18 to 0.2 m have been attained in very dense sand (1.2 to 1.5 d, say).

The ratio of contact pressure at the soil/fixed anchor interface to the average effective overburden pressure, depends to a large extent on construction technique but values of 1.4 to 2.3 have been recorded in dense alluvium, where tremie grouting has been employed.

The rules have been generally established in normally consolidated materials, and can therefore underestimate the pull-out capacity if applied to dense overconsolidated allluvium. In this regard, the overconsolidation ratio (OCR) should be quantified in ground investigation reports to permit more field studies into the effect of OCR and relative density on pull-out capacity.

In current practice, working loads up to 1000 kN can be mobilised safely in coarse sands and gravels, although values of 400 to 600 kN are more common in retaining wall tie backs.

It is a feature of Type C anchorages where the ground is hydrofractured that calculations are based on design curves created from field experience, since the magnitude and extent of fracturing are virtually impossible to quantify. Designs are normally based on the assumption of uniform skin friction and values can be as high as 500 kN/m² and 1000 kN/m² for sand and sandy gravel, respectively. Contact pressures at the soil/fixed anchor interface equivalent to 2 to 10 times the effective overburden pressure have been recorded but it should be noted that the results are influenced strongly by relative density, particle size distribution and grouting pressure. As a consequence, proving tests are recommended to verify the design assumptions for any site.

Where load is transferred by bond, fixed anchor lengths lie within the range 3 to 10 m for Type B and C anchorages. Beyond 10 m, even in weak soils, there is no significant gain in load-holding capacity unless a distributed stress transfer method is adopted, e.g. decoupled strands of different lengths each coupled only at its distal end within the fixed anchor length.

Cohesive Soil/Grout Interface. For tremie grouted straight shaft anchorages of Type A, design rules are similar to those developed for bored piles and are based on the use of undrained shear strengths.

The actions of drilling and grouting cause stress changes within the ground which cannot be accurately modelled by either an effective stress or total stress analysis. An effective stress analysis will indicate a higher calculated ultimate load holding capacity, but a total stress analysis yields results more closely resembling actual ultimate capacities. Therefore, bearing in mind the short duration of anchorage testing and the fact that effective stress analysis implies deformation that is accompanied by loss of prestress, total stress analysis is considered more appropriate.

Since relatively low adhesion factors ($\alpha = 0.35 - 0.45$) are invariably recorded for Type A anchorages, other types are now preferred which can mobilise a greater proportion of the inherent undrained shear strength.

Where high grout pressures can be safely permitted Type C anchorages, with or without post-grouting, are growing in popularity. Based on full scale tests(21), theoretical skin frictions[+] for borehole diameters of 0.08 to 0.16 m are known to increase with increasing consistency* and decreasing plasticity. In stiff clays ($l_c = 0.8 - 1.0$) with medium to high plasticity, skin frictions may be as low as 30 to 80 kN/m^2, while the highest values (> 400 kN/m^2) are obtained in sandy silts of medium plasticity and very stiff to hard consistency ($l_c = 1.25$). The technique of post-grouting is also known to generally increase the skin friction of very stiff clays by some $25 - 50\%$. Type C anchorage designs are based on the assumption of uniform skin friction and safe working loads of 300 to 500 kN are common.

For multi-underreamed anchorages of Type D, design rules are similar to those used for underreamed piles, but in the absence of proving tests in the field empirical multiplier reduction coefficients ranging from 0.75 to 0.95 are sometimes applied to the undrained shear strength to allow for factors such as underreaming technique and underream geometry.

Of vital importance in cohesive deposits is the time during which drilling, underreaming and grouting take place. This should be kept to a minimum in view of the softening effect of water on the clay. The consequences of delays of only a few hours include reduced load holding capacity and significant short term losses of prestress. In the case of sand filled fissures, for example, where water flushing is employed, a period of only 3 to 4 h may be sufficient to reduce the C_u strength to near the fully softened value. In such ground it is prudent to apply a reduction coefficient of 0.5 at the design stage(14).

+ The theoretical skin friction is calculated using the ultimate load holding capacity, the borehole diameter and designed length of the fixed anchor.

* Consistency Index, $l_c = \dfrac{L_L - m}{L_L - P_L}$

where L_L is the liquid limit (%), P_L is the plastic limit (%) and m is the natural moisture content (%).

Underreaming is ideally suited to clays of C_u greater than 90 kN/m², whilst occasional difficulties in the form of local collapse or breakdown of the neck portion between the underreams should be expected for C_u values of 60 to 70 kN/m². Underreaming is virtually impracticable below a C_u of 50 kN/m², and is also difficult in soils of low plasticity, e.g. plasticity index < 20.

The success of multi-underreamed anchorages over straight shafts can perhaps be illustrated best by reference to Figure 5. Based on the same augered hole diameter of 0.15 m, the straight shaft Type A anchorage with a fixed length of 10.7 m failed at 1000 kN, whereas the underreamed anchorage with a fixed length of only 3 m withstood a load of 1500 kN without any sign of failure. In current practice working loads of up to 1000 kN can be mobilised safely using Type D anchorages.

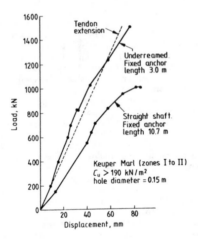

Figure 5. Comparison of Load Displacement Responses of an Underreamed Anchorage and a Straight Shaft Anchorage

As for cohesionless soils, the fixed anchor length for all anchorage types in cohesive soils should not normally be less than 3 m, nor more than 10 m.

Grout/Tendon Interface. Recommendations pertaining to grout/tendon bond values commonly take no account of the length and type of tendon, tendon geometry or grout strength, and for these reasons it is still advisable to measure experimentally the required tendon bond length for known field conditions.

As a guide, bearing in mind the compressive strength of

30 N/mm² often required for cement grouts prior to stressing, the ultimate bond stress assumed to be uniform over the tendon bond length should not exceed:

(a)   1.0 N/mm² for clean plain wire or plain bar;
(b)   1.5 N/mm² for clean crimped wire;
(c)   2.0 N/mm² for clean strand or deformed bar;
(d)   3.0 N/mm² for locally noded strands

For cement grouted anchorages, the bond length should not be less than 3 m and 2 m for tendons bonded in situ, and under factory controlled conditions, respectively, unless full scale tests confirm that shorter bond lengths are acceptable.

Bond strength can be significantly affected by the surface condition of the tendon and clearly loose or lubricant materials should be removed. A film of rust is not necessarily harmful and may improve bond but tendons showing signs of pitting should not be used.

Since tendon density (area of steel tendon/area of borehole) can affect debonding, it is normally limited to 20% for single unit tendons and noded multi-strand tendons, and 15% for parallel multi-unit tendons.

Cementitious Grouts. For ground anchorage construction, Portland cements are generally acceptable provided that the total sulphate content does not exceed 4% (by mass) SO_3 of cement in the grout. For cement used in tendon bonding, the total chloride content of the grout derived from all sources should not exceed 0.1% (by mass) of cement. Sometimes blast furnace cement is rejected in order to minimize sulphide content. The use of high alumina cement should be restricted to temporary anchorages with a service life up to 6 months and to anchorages used for testing ground holding capacity.

Water fit for drinking is acceptable as mixing water, provided that it does not contain more than 500 mg of chloride ions per litre.

Admixtures should only be used if tests have shown that their use improves the properties of the grout, e.g. by improving workability or durability, reducing bleed or shrinkage or increasing rate of strength development. No admixture that contains in total more than 0.1% (by mass) of chlorides, sulphides or nitrates should be used.

Before grouting the following properties should be established, namely (1) type of cement, (2) water/cement ratio, (3) type and concentration of admixture and/or filler, (4) flow reading, (5) unconfined compressive strength development, e.g. at 3, 7, 14 and 28 days, and (6) notes on amount of free expansion or shrinkage, bleed and setting time.

For borehole anchorage grout, the mix should attain a crushing strength of 40 N/mm² at 28 days in order to ensure good bond and shear strength. In pregrouting operations strength requirements are insignificant.

The bleeding of tendon bonding grout should generally not exceed 2% of the volume 3 h after mixing and have a maximum of 4%. Higher values may be permitted in the case of permeable ground where the bleed water is filtered from the grout during injection under pressure.

The water/cement ratio of tendon bonding grouts generally lies in the range 0.35 to 0.60. In certain circumstances, higher ratios may be used to facilitate tendon homing, provided that the dilute mix is displaced from the fixed anchor zone using a grout of suitable strength and consistency.

Tendon. Tendons usually consist of steel bar, strand or wire either singly or in groups. In current practice, popular sizes are non-alloy steel wire (7 mm dia.), non-alloy 7-wire strand (12.7 to 18 mm dia.), and low alloy steel bar (25 to 40 mm dia.). Working loads should not exceed 62.5% and 50% of the characteristic strength of the tendon for temporary and permanent works, respectively, under normal circumstances.

To distribute load to the ground more uniformly strands of differing lengths are sometimes used within the fixed anchor zone. When these are stressed simultaneously, displacements at the anchor head are the same for all strands, and thus the strains and hence stresses differ in individual strands. In such cases, the stress in the shortest strand should limit the acceptable working load. If the design requires uniform stresses within the tendon, mono-strand stressing is essential.

The maximum prestress loss allowed for in design should be taken as the 1000 h relaxation loss, for a force equal to the initial residual load, times the appropriate multiplier coefficient given in Table 4. Since relaxation is temperature sensitive it may be necessary in tropical climates to further increase the estimate of relaxation loss, e.g. the 1000 h relaxation loss of low relaxation strand at 20°C is doubled at 40°C.

Table 4. Recommended Multiplier Coefficients for Estimating Maximum Relaxation Loss in Design

Anchorage category	Recommended multiplier		
	Normal relaxation (class 1)	Bar	Low relaxation (class 2)
Permanent	2.0	1.75	1.5
Temporary	1.75	1.50	1.25

For all multi-unit tendons spacers should be incorporated to ensure separation between individual components of the tendon and thus the effective penetration of grout to provide adequate bond. A clear spacing of 5 mm is usually required for parallel multi-unit tendons. The spacer should not be compressible nor cause decoupling of the fixed anchor grout. In other words, the spacer should not be bulky and it may be advantageous to keep the spacer arms of small dimension so as not to inhibit grout flow.

Centralization should be provided on all tendons to ensure that the tendon is centred in the grout column, with a minimum grout cover of 5 to 10 mm. Centralizer design in soft cohesive ground, e.g. clay, requires special attention with respect to component dimensions, since thin sections may penetrate the clay. In current practice, tendon centralization receives insufficient attention and grout/tendon bond failures due to contamination have been recorded at the acceptance testing stage.

Centralizers and spacers should be made of materials having no deleterious effect on the tendon itself, and the use of metals dissimilar to the tendon should be avoided.

Anchor Head. The anchor head normally consists of a stressing head in which the tendon is anchored and a bearing plate by which the tendon force is transferred to the structure or excavation. Secondary distribution systems in the form of concrete blocks or steel wallings then transfer the force to the main structure. For earth retaining structures, the secondary distribution systems should be considered as part of the wall design.

The stressing head should be designed to permit the tendon to be stressed and anchored at any force up to 80% of the characteristic tendon force and should permit force adjustment up and down during the initial stressing phase. The stressing head should also be designed to anchor the tendon without damaging it and also permit an angular deviation of ±5° from the axial position of the tendon. Excess deviation reduces load transfer efficiency, and creates difficulties in wedge pull-in. In this regard, no problems should be anticipated provided wedges are homed within a 5 mm depth band.

Safety Factors. In current practice, the load safety factor of an anchorage is the ratio of ultimate load holding capacity to the working load. The proof load factor which provides a measured margin of safety in the field is the ratio of proof load to the working load. For guidance, Table 5 lists load safety factors considered appropriate for the design of individual anchorages. Since the minimum safety factor is applied to those anchorage components known with the greatest degree of accuracy, the minimum values listed invariably apply to the characteristic strength of the tendon or anchor head.

CORROSION PERFORMANCE

35 case histories of failure by tendon corrosion have been published recently(22), 24 related to permanent anchorages (protected and unprotected), and 11 related to temporary anchorages with no designed protection other than cement grout cover for the fixed length and on occasion a decoupling sheath over the free length.

Analysis of the results shows that the corrosion is invariably localised and appears to be independent of tendon type in that 9 incidents involved bar, 19 involved wire and 8

Table 5. Minimum Safety Factors Recommended for Design of Individual Anchorages

Anchorage Category	Minimum Safety Factor			Proof Load Factor
	Tendon	Ground/Grout Interface	Grout/Tendon or Grout/Encapsulation Interface	
Temporary anchorages where a service life is less than six months and failure would have no serious consequences and would not endanger public safety, e.g. short term pile test loading using anchorages as a reaction system.	1.40	2.0	2.0	1.10
Temporary anchorages with service life of say up to two years where, although the consequences of failure are quite serious, there is no danger to public safety without adequate warning e.g. retaining wall tie-back.	1.60	2.5*	2.5*	1.25
Permanent anchorages and temporary anchorages where corrosion risk is high and/or the consequences of failure are serious, e.g. main cables of a suspension bridge or as a reaction for lifting heavy structural members.	2.00	3.0+	3.0*	1.50

*Minimum value of 2.0 may be used if full scale field tests are available.

+May need to be raised to 4.0 to limit ground creep.

NOTE 1. In current practice the safety factor of an anchorage is the ratio of the ultimate load to design load. Table 5 above defines minimum safety factors at all the major component interfaces of an anchorage system. NOTE 2. Minimum safety factors for the ground/grout interface generally lie between 2.5 and 4.0. However, it is permissible to vary these, should full scale field tests (trial anchorage tests) provide sufficient additional information to permit a reduction. NOTE 3. The safety factors applied to the ground/grout interface are invariably higher compared with the tendon values, the additional magnitude representing a margin of uncertainty.

involved strand, the period of service before failure ranging
from a few weeks to many years for each tendon type. Short
term failures (after a few weeks) have been due to stress
corrosion cracking or hydrogen embrittlement.

In terms of duration of service, 9 failures occurred within
six months, 10 in the period six months to two years, and the
remaining 18 beyond two years and up to thirty one years.

With regard to failure location 19 incidents occurred at,
or within 1 m of the anchor head, 21 incidents in the free
length and 2 incidents in the fixed length. Both fixed anchor
problems were caused by inadequate grouting of the tendon bond
length which exposed the tendon to an aggressive environment.
Failures in the free length were recorded under a variety of
individual and combined circumstances such as

  (a) tendon overstressing caused by ground movement
      leading to tendon cracking, sometimes augmented by
      pitting corrosion or corrosion fatigue,
  (b) inadequate or no cement grout cover in the presence
      of chlorides, e.g. industrial waste fills or organic
      materials,
  (c) breakdown of bitumen cover due to lack of durability.
  (d) inappropriate choice of protective material, e.g.
      chemical grout containing nitrate ions and
      hygroscopic mastic, and
  (e) use of tendon stored on site for a long period in an
      unprotected state.

Failures at, or adjacent to the anchor head were due to
causes ranging from absence of protection (even for only a few
weeks in aggressive environments) to inadequate cover due to
incomplete filling initially or slumping of the protective
filler during service. From all the case histories reviewed,
it is apparent that corrosion incidents are somewhat random in
terms of cause, with the possible exception of choice of steel.
Quenched and tempered plain carbon steels and high strength
alloy steels are more susceptible to hydrogen embrittlement
than other varieties. Accordingly, those named steels should
be used with extreme caution where environmental conditions are
aggressive.

The fact that 19 failures occurred within two years of
installation confirms that where the environment is aggressive,
temporary anchorages should be given appropriate protection.
However, there is no evidence to suggest that the current limit
of two years for the service period of temporary anchorages,
should be reduced or extended.

While the mechanisms of corrosion are understood, the
aggressivity of the ground and general environment are seldom
quantified at the site investigation stage. In the absence of
quantified agressivity data it is unlikely that case histories
involving tendon corrosion will provide reliable information
for the prediction of corrosion rates in service.

CORROSION PROTECTION

General. Since there is no certain way of predicting
localised corrosion rates, where aggressivity is recognised,
albeit qualitatively, some degree of protection should be
provided by the designer. For corrosion resistance, the
anchorage should be protected overall as partial protection of
the tendon may only induce more severe corrosion of the
unprotected part. Junctions between the fixed length, free
length and anchor head are particularly vulnerable, as are
joints and couplers.

Choice of class of protection (Table 6) is the
responsibility of the designer. The choice depends on such
factors as consequence of failure, aggressivity of the
environment and cost of protection. By definition single
protection implies that one physical barrier against corrosion
is provided for the tendon prior to installation. Double
protection implies the supply of two barriers where the purpose
of the outer second barrier is to protect the inner barrier
against the possibility of damage during tendon handling and
placement.

Grout injected in situ to bond the tendon to the ground
does not constitute a part of a protection system because the
grout quality and integrity cannot be assured. Furthermore,
when smooth bar, wire or strand tendons in cement fixed anchor
grouts are stressed, cracks tend to occur at about 50 to 100 mm
apart and of widths up to 1 or 2 mm.

Non-hardening fluid materials such as greases also have
limitations such as ease of removal, leakage and susceptibility
to oxidation. Thus, although greases fulfill an essential role
in corrosion protection systems, in that they act as a filler
to exclude the atmosphere from the surface of a steel tendon,
create the correct electrochemical environment and reduce
friction in the free length, they should not be regarded as a
permanent physical barrier to corrosion. An exception is the
use of grease for protection in a restressable anchor head,
since the grease can be replaced or replenished.

Table 6. Proposed Classes of Protection for Ground Anchorages

Anchorage category	Class of protection
Temporary	Temporary without protection Temporary with single protection Temporary with double protection
Permanent	Permanent with single protection Permanent with double protection

Use of thicker metal sections for the tendon, with

sacrificial area in lieu of physical barriers, gives no effective protection, as corrosion is rarely uniform and extends most rapidly and preferentially at localized pits or surface irregularities. Sacrificial metallic coatings for high strength steel (>1040 N/mm^2) should not be used when such coatings can cause part of the steel tendon to act as a cathode in an uncontrolled manner in a galvanitic process.

Tendon Free Length. Continuous diffusion impermeable polypropylene or polyethylene sheaths (wall thickness ≮ 0.8 mm) applied in factory conditions are acceptable and commonly used for both temporary and permanent anchorages. Tendons should be greased before application of sheaths and the fit should not result in significant friction between sheath and tendon during stressing. Where in a double protection system a sheath acts as an outer sacrificial coating it is not necessary to fill the annulus between the inner and outer sheaths.

Greases should not contain any substance that could provoke corrosion, e.g. unsaturated fatty acids and water, and individual contents of sulphides, nitrates and chlorides should not exceed 5 x 10^{-4}% by mass. Greases should be stable against water and oxygen, and hydrophobic greases are preferable. Other important factors include bacterial and microbiological degradation resistance, low moisture vapour transmission and high electrical resistivity.

Tendon Bond Length. In addition to providing the same degree of protection as the free length, the protective elements have all to be capable of transmitting high tendon stresses to the ground. Requirements of no creep and no cracking are in conflict and few materials are available which can provide perfect compliance. A common solution in practice is to encase the tendon bond length inside a plastic corrugated duct using a polyester resin or cement grout (Fig. 6). To ensure effective load transfer the pitch of corrugations should be typically within six and twelve times the duct wall thickness, and the amplitude of corrugation should not be less than three times the wall thickness. The minimum wall thickness is 0.8 mm, but consideration of material type, method of installation and service required, may demand a greater thickness.

Where protection has not been specified, and the conditions are known to be benign, the cement grout cover over the fixed length may be deemed appropriate for temporary anchorage proposals, on the basis that nothing more stringent has been required.

Anchor Head. Unlike fixed anchors, anchor heads cannot be wholly prefabricated, since friction grips for wire or strand and locking nuts on bars cannot fix the tendon until extension has been achieved following tensioning. Existing locking arrangements leave two sections of the tendon, above and below the bearing plate, which require separate protective measures in addition to the protection of the bearing plate itself. If the environment is aggressive, early protection of the anchor head is recommended for both temporary and permanent

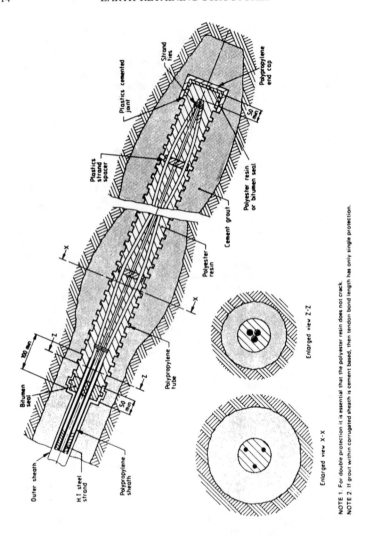

NOTE 1. For double protection it is essential that the polyester resin does not crack.
NOTE 2. If grout within corrugated sheath is cement based, then tendon bond length has only single protection.

Figure 6.   Typical Double Protection of Bond Length of Strand Tendon Using a Single Corrugated Sheath and Polyester Resin

anchorages.

The essence of inner head protection is to provide an effective overlap with the free length protection, and to protect and isolate the short exposed length of tendon within and below the plate. The protective measures usually have to allow free movement and grease-based corrosion protection compounds may be required. They may be preplaced or injected and should be fully contained within a surrounding duct and retained by an end seal (Fig. 7).

Outer head protection of the bare tendon and the friction grips or the locking nuts above the bearing plate generally comprises a plastic or galvanised steel cap, which is filled with grease or a setting sealant, the former being applicable to restressable heads. If the anchor head is to be totally enclosed by the structure, the outer head components may be encased in dense concrete or epoxy mortar, where the cover is usually 50-60 mm.

The bearing plate should be painted with bitumastic or other protective materials, care being taken to ensure that any coatings applied are compatible with the materials selected for both inner head and outer head protection.

CONSTRUCTION

General. During ground anchorage construction the method of drilling, with or without flushing, the tendon installation, the grouting system and the time period of these operations may influence the capacity of the anchorage. Anchorage construction should be carried out in a manner whereby the validity of design assumptions is maintained and a method statement detailing all operations, including drilling and grouting plant information, should be prepared prior to site anchorage work. Anchorage work is specialised and should always be carried out under the supervision of experienced personnel.

Drilling. Any drilling procedure may be employed that can supply a stable hole that is within the permitted tolerances and free of obstructions in order to accommodate easily the tendon. Drilling necessarily disturbs the ground and the method should be chosen relative to the ground conditions to cause either the minimum of disturbance or the disturbance most beneficial to the anchorage capacity.

Care should be taken not to use high pressures with any flushing media, in order to minimise the risk of hydrofracture of the surrounding ground, particularly in built-up areas. In this connection, an open return within the borehole is desirable to limit pressures and it also permits the driller to monitor major changes in ground type from the drill cuttings or flush.

Unless otherwise specified, the drill hole entry point should be positioned within a tolerance of ± 75mm. The drilled hole should have a diameter not less than the specified diameter, and allowances for swelling may be necessary if the

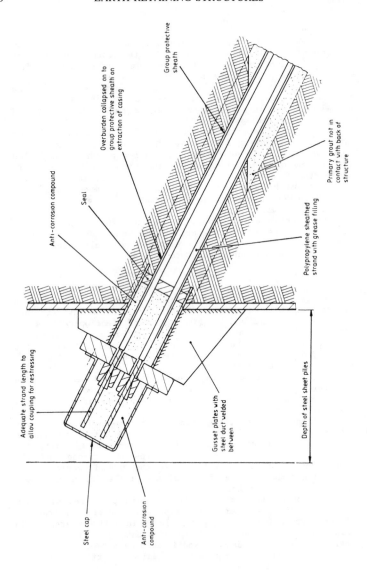

Figure 7.   Typical Restressable Anchor Head Detail for Double
Protection of Strand Tendon

hole is open for several hours in, for example, overconsolidated clays and marls. For a specified alignment at entry point, the hole should be drilled to an angle tolerance of ± 2.5°, unless, for closely spaced anchorages, such a tolerance, could lead to interference of fixed anchor zones in which case the inclination of alternate anchorages should be staggered. Ground anchorages should have a minimum inclination of approximately 10° to the horizontal to facilitate grouting.

Assuming an acceptable initial alignment, overall drill hole deviations of 1 in 30 should be anticipated. On occasions, ground conditions may dictate a relaxation of this tolerance and for downward and upward inclined holes, it is probable that the vertical deviations will be higher than lateral deviations.

After each hole has been drilled its full length and thoroughly flushed out to remove any loose material, the hole should be probed to ascertain whether collapse of material has occurred and whether it will prevent the tendon being installed completely. For downward inclined holes up to 1 m of overdrill may be added to cater for detritus that cannot be removed.

Tendon installation and grouting should be carried out on the same day as drilling of the fixed anchor length, since a delay between completion of drilling and grouting can have serious consequences due to ground deterioration, particularly in overconsolidated, fissured clays and weak rocks.

During the drilling operations, all changes in ground type should be recorded together with notes on water levels encountered, drilling rates, flushing losses or gains and stoppages.

Tendon. Ideally, tendon steel in the bare condition should be stored indoors in clean dry conditions, but if left outdoors such steel should be stacked off the ground and be completely covered by a waterproof tarpaulin that is supported and fastened clear of the stack so as to permit circulation of air and avoid condensation.

Bare or coated tendons should not be dragged across abrasive surfaces or through surface soil, and only fibre rope or webbing slings should be used for lifting coated tendons. In the event of damage, tendon which is kinked or sharply bent should be rejected because load-extension characteristics may be adversely affected.

Over the bond length, bar tendon, multi-unit tendons and encapsulations should be centralized in the borehole to ensure a minimum grout cover to the tendon or encapsulation of 5 mm between centralized locations and 10 mm at centralized locations (Fig. 8).

For multi-unit tendons where the applied tensile load is transferred by bond, spacers should ensure a minimum clear spacing of 5 mm. Given tendons with local or general nodes that provide mechanical interlock, occasional contact between tendon units is permissible.

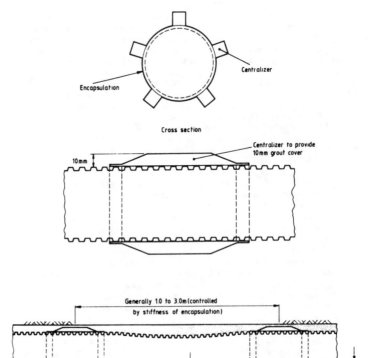

Figure 8. Centralizers

A minimum of three spacers should be provided in each fixed anchor length, and both centralizers and spacers should be provided at centres according to the angle of inclination of the ground anchorage and the possible sag between points of support, in order to provide the minimum clear spacing or cover.

At the bottom of the tendon, use of a sleeve or nose cone will minimize the risk of tendon or borehole damage during homing.

Immediately prior to installation the tendon should be carefully inspected for damage to components and corrosion, after which the tendon should be lowered at a steady controlled rate. For heavy tendons weighing in excess of 200 kg, approximately, mechanical handling equipment should be employed, as manual operations can be difficult and hazardous. The use of a funnelled entry pipe at the top of a cased hole is also recommended to avoid tendon damage as it is installed past the sharp edge of the top of the casing.

On occasion, particularly at the start of a contract, the tendon may be withdrawn after the installation operation, in order to judge the efficiency of the centralizer and spacer units and also to observe damage, distortion or the presence of smear, e.g. in chalk or clay. Where significant distortion or smear is observed, improvements in relation to the fixing or design of the centralizers, or the borehole flushing method may be necessary.

Grouting. Grouting performs one or more of the following functions:

(a) to form the fixed anchor in order that the applied load may be transferred from the tendon to the surrounding ground;

(b) to augment the protection of the tendon against corrosion;

(c) to strengthen the ground immediately adjacent to the fixed anchor in order to enhance anchorage capacity;

(d) to seal the ground immediately adjacent to the fixed anchor in order to limit loss of grout.

In general, if the grout volume exceeds three times the borehole volume for injection pressures less than total overburden pressure, then general void filling is indicated which is beyond routine anchorage construction. This extra grout merits additional payment.

The need for functions (c) and/or (d) should be highlighted by the ground investigation and/or as a result of pregrouting or water testing. To check that the loss of grout over the fixed anchor length is insignificant during injection for anchorage Types B and C, it is normally adequate to observe a controlled grout flow rate coupled with a back pressure. The efficiency of fixed anchor grouting can be finally checked by monitoring the response of the ground to further injection when the back pressure should be quickly restored.

Where pressure grouting is not carried out as part of routine anchorage construction, a falling head grout test can

be used where the borehole is prefilled with grout typically
having w/c = 0.4 - 0.6, and the grout level observed until it
becomes steady. If the level continues to fall it should be
topped up and after sufficient stiffening of the grout (but
prior to hardening), the borehole should be redrilled and
retested. The test may be applied to the entire borehole or
restricted to the fixed anchor length by packer or casing over
the free anchor length.

The likelihood of cement grout loss in rock can also be
assessed from an analysis of a water injection test(12),
although the test is particularly rigorous and interpretation
of the results demands care, since the leakage may be due to a
single fracture of consequence, or many micro fractures which
might not accept cement. Routinely, a falling head test is
applied to the borehole or the fixed anchor length, and
pregrouting is not required if the water loss is less than 5
litres/min at an excess head of 0.1 MPa (one atmosphere) over a
period of 10 min. Where there is a measured water gain under
artesian conditions, care should be taken to counteract this
flow by the application of a back pressure prior to grouting.
If the flow cannot be stabilised in this way, pregrouting is
required irrespective of the magnitude of the water gain. The
acceptable water loss in current US recommendations(8) is 0.49
millilitres/mm diameter of borehole/m of depth at an excess
head of 0.034 MPa, and again care is required in
interpretation, since for rock with occasional but significant
fractures the borehole depth and diameter are, strictly
speaking, not relevant to the estimation of potential grout
loss.

For the preparation of cement grout, batching of the dry
materials should be by mass, and mixing should be carried out
mechanically for at least 2 minutes in order to obtain a
homogeneous mix. Thereafter, the grout should be kept in
continuous movement, e.g. slow agitation in a storage tank. As
soon as practicable after mixing, the grout should be pumped to
its final position, and it is undesirable to use the grout
after a period equivalent to its initial setting time (Fig. 9).

High speed colloidal mixers (1000r/min minimum) and paddle
mixers (150r/min minimum) are permissible for mixing neat
cement grouts, although the former mixer is preferred in water
bearing ground conditions since dilution is minimized.

Pumps should be of the positive displacement type, capable
of exerting discharge pressures of at least 1000 kN/m^2, and
rotary screw (constant pressure) or reciprocating ram and
piston (fluctuating pressure) pumps are acceptable in
practice.

Before grouting, all air in the pump and line should be
expelled, and the suction circuit of the pump should be
airtight. During grouting, the level of grout in the supply
tank should not be drawn down below the crown of the exit pipe,
as otherwise air will be injected.

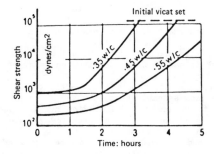

Figure 9. Setting Times for Type 1 Grouts (18°C)

An injection pressure of 20 kN/m² per metre depth of ground is common in practice. Where high pressures that could hydrofracture the ground are permitted, careful monitoring of grout pressure and quantity over the fixed anchor length is recommended. If, on completion of grouting, the fluid grout remains adjacent to the anchored structure then the shaft grout should be flushed back 1 to 2 m to avoid a strut effect during stressing.

In regard to quality controls, emphasis should be placed on those tests that permit grout to be assessed prior to injection. As a routine, initial fluidity by flow cone or flow trough, density by mud balance and bleed by 1000 mL graduated cylinder (75 mm diameter) should be measured daily along with 100 mm cube samples for later crushing at 7 and 28 days, say. These quality controls relate to grout batching and mixing and the tests do not attempt to simulate the properties of the grout in situ. For example, water loss from grout when injected under pressure into sand creates an in situ strength greater than the cube strength for similar curing conditions.

Records relating to each grouting operation should be compiled, e.g. age of constituents, air temperature, grouting pressure, quantity of grout injected and details of samples and tests, as appropriate.

Anchor Head. The stressing head and bearing plate should be assembled concentrically with the tendon with an accuracy of ± 10 mm and should be positioned not more than 5° from the tendon axis.

After final grouting or satisfactory testing, cutting of the tendon should be done without heat, e.g. by a disc-cutter, in which case the cut should not be closer than one tendon unit diameter from the face of the holding wedge or nut.

Projecting tendons, whether stressed or not, should be protected against accidental damage. This protection is not common in practice and if individual tendon components are mechanically damaged, e.g. kinking of strand, then these components should be considered redundant, when assessing a safe anchorage capacity, unless tests confirm adequacy.

Stressing. Stressing is required to fulfill two functions,
namely:
    (a) to tension the tendon and to anchor it at its secure
        load; and
    (b) to ascertain and record the behaviour of the anchorage
        so that it can be compared with the behaviour of
        control anchorages.
    A stressing operation means an activity involving the
fitting of the jack assembly on to the anchor head, the loading
or unloading of the anchorage including cyclic loading where
specified, followed by the complete removal of the jack
assembly from the anchor head.
    Stressing and recording should be carried out by
experienced personnel under the control of a suitably qualified
supervisor, since any signficant variation in procedure can
invalidate comparison with control anchorages.
    At the present time equipment calibration is not carried
out regularly and discrepancies between jack and load cell
readings are not uncommon on site. Jacks should be calibrated
at least every year, using properly designed test equipment
with an absolute accuracy not exceeding 0.5%. The calibration
should cover the load rising and load falling modes over the
full working range of the jack, so that the friction hysterisis
can be known when repeated loading cycles are being carried out
on the tendon. Load cells should be calibrated after every 200
stressings or after every 60 days in use, whichever is the more
frequent, unless complementary pressure gauges used
simultaneously indicate no significant variation, in which case
the interval between calibrations may be extended up to a
maximum of one year. Pressure gauges should also be calibrated
regularly, e.g. after every 100 stressings or after every 30
days, whichever is the more frequent, against properly
maintained master gauges, or whenever the field gauges have
been subjected to shock. If a group of three gauges is
employed this frequency does not apply.
    On every contract the method of tensioning to be used and
the sequence of stressing should be specified at the planning
stage. In general, no tendon should be stressed at any time
beyond either 80% of the characteristic strength (equivalent to
80% GUTS in U.S.) or 95% of the characteristic 0.1% proof
strength. In addition, for cement grouted fixed anchors,
stressing should not commence until the grout has attained a
crushing strength of at least $30N/mm^2$ ($24N/mm^2$ in USA).
However, in sensitive ground which may be weakened by water
softening or disturbance during anchorage construction, it may
be necessary to stipulate a minimum number of days before
stressing.
    Details of all forces, displacements, seating and other
losses observed during stressing and the times at which the
data were monitored should be recorded for every anchorage.
    Finally, it is worth noting that when a stressing operation
is the start point for future time-related load measurements,
stressing should be concluded with a check-lift load

measurement.

During stressing safety precautions are essential and operatives and observers should stand to one side of the tensioning equipment and never pass behind when it is under load. Notices should also be displayed stating 'DANGER — Tensioning in Progress' or similar wording.

TESTING

General. There are three classes of tests for all anchorages as follows:

(a) proving tests (proof tests in U.S.);

(b) on-site suitability tests (performance tests in U.S.);

(c) on-site acceptance tests (performance tests in U.S.).

Proving tests may be required to demonstrate or investigate in advance of the installation of working anchorages, the quality and adequacy of the design in relation to ground conditions and materials used and the levels of safety that the design provides. The tests may be more rigorous than on-site suitability tests and the results, therefore, cannot always be directly compared, e.g. where short fixed anchors of different lengths are installed and tested, ideally to failure.

On-site suitability tests are carried out on anchorages constructed under identical conditions as the working anchorages and loaded in the same way to the same level. These may be carried out in advance of the main contract or on selected working anchorages during the course of construction. The period of monitoring should be sufficient to ensure that prestress or creep fluctuations stabilize within tolerable limits. These tests indicate the results that should be obtained from the working anchorages.

On-site acceptance tests are carried out on all anchorages and demonstrate the short term ability of the anchorage to support a load that is greater than the design working load and the efficiency of load transmission to the fixed anchor zone. A proper comparison of the short term service results with those of the on-site suitability tests provides a guide to longer term behaviour.

On-Site Acceptance Tests. Every anchorage used on a contract should be subjected to an acceptance test. As a principle, acceptance testing should comprise standard procedures and acceptance criteria which are independent of ground type, and be of short duration. In this regard the maximum proof loads are dictated by Table 5, but acceptable load increments and minimum periods of observation have gradually been reduced over the years to save time and money (Table 7). At each stage of loading, the displacement should be recorded at the beginning and end of each period, and for proof loads the minimum period of 1 min is extended to at least 15 min with an intermediate displacement at 5 min, so that any tendency to creep can be monitored.

Table 7.   Recommended Load Increments and Minimum Periods of
           Observation for On-site Acceptance Tests

Temporary anchorages		Permanent anchorages		Minimum period of observation
load increment (T_w)		load increment (T_w)		
1st load cycle*	2nd load cycle	1st load cycle*	2nd load cycle	
%	%	%	%	min
10	10	10	10	1
50	50	50	50	1
100	100	100	100	1
125	125	150	150	15
100	100	100	100	1
50	50	50	50	1
10	10	10	10	1

*For this load cycle, there is no pause other than that
necessary for the recording of displacement data.

In order to establish the seat of load transfer within the
anchorage, the apparent free length of the tendon may be
calculated from the load-elastic displacement curve over the
range of 10% Tw to 125% Tw (temporary anchorages) or 10% Tw to
150% Tw (permanent anchorages), using the manufacturer's value
of elastic modulus and allowing for such effects as temperature
and bedding of the anchor head.   The analysis should be based
on the results obtained during the destressing stage of the
second or any subsequent unloading cycle, as shown generally by
Fig 10.   For simplicity in practice the following equation is
employed

$$\text{Apparent free tendon length} = \frac{A_t\ E_s\ \Delta X_e}{T} \qquad (2)$$

where A_t is the cross section of the tendon, E_s is the
manufacturer's elastic modulus for the tendon unit, ΔX_e is the
elastic displacement of the tendon (ΔX_e is equated to the
displacement monitored at proof load minus the displacement at
datum load, i.e. 10% Tw say, after allowing for structural
movement) and T is the proof load minus datum load.

On completion of the second cycle, the anchorage is
reloaded in one operation to 110% Tw say, and locked-off, after
which the load is reread to establish the initial residual
load.   This moment represents zero time for monitoring
load/displacement-time behaviour during service. Where loss of
load is monitored accurately using load cells with a relative
accuracy of 0.5%, readings can be attempted within the first

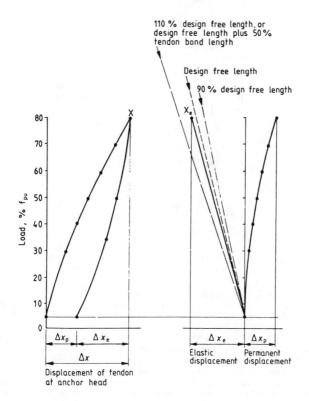

Figure 10.   Acceptance Criteria for Displacement of Tendon at
Anchor Head

50 min. Where monitoring involves a stressing operation, e.g.
lift-off check without load cell, an accuracy of less than 5%
is unlikely and longer observation periods of 1 day and beyond
are required.   Where displacement-time data are required, a
dial gauge/tripod system (Fig. 11) is suitable for short
duration testing, given that the tripod base should be surveyed
accurately for movement.

     For the testing procedures outlined above,   acceptance
criteria based on proof load-time data, apparent free tendon
length, and short term service behaviour, are proposed for
temporary and permanent anchorages.   These criteria are

discussed in the following paragraphs.

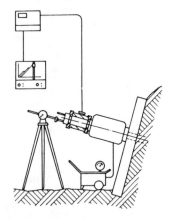

Figure 11.   Typical Method of Measuring Tendon Displacement
Using a Dial Gauge

Proof Load-time Data.   If the proof load has not reduced
during the 15 min observation period by more than 5% after
allowing for any temperature changes and movements of the
anchored structure, the anchorage may be deemed satisfactory.
If a greater loss of prestress is recorded the anchorage should
be subject to two further proof load cycles and the behaviour
recorded.   If the 5% criterion is not exceeded on both
occasions the anchorage may be deemed satisfactory.   If the 5%
criterion is exceeded on either cycle the proof load should be
reduced to a value at which compliance with the 5% criterion
can be achieved.   Thereafter, the anchorage may be accepted at
a derated proof load, if appropriate.

As an alternative to these recommendations, the proof load
can be maintained by jacking and the anchor head monitored
after 15 min, in which case the creep criterion is 5% ΔXe.

For anchorages that have failed a proof load criterion,
tendon unit stressing may help to ascertain location of
failure, e.g. for a temporary anchorage, pull-out of individual
tendon units may indicate debonding at the grout/tendon
interface, whereas, if all tendon units hold their individual
proof loads, attention is directed towards failure of the fixed
anchor at the ground/grout interface.

Apparent Free Tendon Length.   The apparent free tendon
length should be not less than 90% of the free length intended
in the design, nor more than the intended free length plus 50%
of tendon bond length or 110% of the intended free tendon
length (Fig. 10).   The latter upper limit takes account of
relatively short encapsulated tendon bond lengths and fully
decoupled tendons with an end plate or nut.

Where the observed free tendon length falls outside the limits a further two load cycles up to proof load should be carried out in order to gauge reproducibility of the load-displacement data. If the anchorage behaves consistently in an elastic manner, the anchorage need not be abandoned, provided the reason can be diagnosed and accepted. In this regard it is noteworthy that the E value of a long multi strand tendon may be less than the manufacturer's E value for a single strand, which has been measured over a short gauge length between rigid platens. A reduction in the manufacturer's E value of up to 10% should be allowed in any field diagnosis.

Short Term Service Behaviour. Using accurate load cell and logging equipment, the residual load may be monitored at 5, 15, and 50 min. If the rate of load loss reduces to 1% or less per time intervals for these specific observation periods after allowing for temperature, structural movements and relaxation of the tendon, the anchorage may be deemed satisfactory. If the rate of load loss exceeds 1%, further readings may be taken at observation periods up to 10 days (Table 8).

Table 8. Acceptance Criteria for Service Behaviour at
Residual Load

Period of observation	Permissible loss of load (% initial residual load)	Permissible displacement (% of elastic extension Δe of tendon at initial residual load)
min	%	%
5	1	1
15	2	2
50	3	3
150	4	4
500	5	5
1500 (≈ 1 day)	6	6
5000 (≈ 3 days)	7	7
15000 (≈10 days)	8	8

If, after 10 days, the anchorage fails to hold its load as given in Table 8, the anchorage is not satisfactory and following an investigation as to the cause of failure, the anchorage should be (1) abandoned and replaced, (2) reduced in capacity or (3) subjected to a remedial stressing programme.

Where prestress gains are recorded, monitoring should continue to ensure stabilization of prestress within a load increment of 10% Tw. Should the gain exceed 10% Tw, a careful analysis is required and it will be prudent to monitor the overall structure/ground/anchorage system. If, for example, overloading progressively increases due to insufficient anchorage capacity in design or failure of a slope, then additional support is required to stabilize the overall

anchorage system.   Destressing to working loads should be
carried out as prestress values approach proof loads, accepting
that movement may continue until additional support is
provided.

As an alternative to load monitoring, displacement–time
data at the residual load may be obtained at the specific
observation periods in Table 8, in which case the rate of
displacement should reduce to 1% Δe or less per time interval.
This value is the displacement equivalent to the amount of
tendon shortening caused by a prestress loss of 1% initial
residual load, i.e.:

Δe = <u>initial residual load x apparent free tendon length</u>
          area of tendon x elastic modulus of tendon   (3)

If the anchorages are to be used in the work and, on
completion of the on–site acceptance test, the cumulative
relaxation or creep has exceeded 5% initial residual load or 5%
Δe, respectively, the anchorage should be restressed and
locked–off at 110% Tw, say.

<u>Monitoring Service Behaviour</u>.   As for buildings, bridges
and dams, monitoring of structure/ground/anchorage systems will
be appropriate on occasions.   In general, monitoring is
recommended for important structures where the following
circumstances apply:

(a) wherever the behaviour of anchorages can be ascertained
    safely by monitoring the behaviour of the structure as
    a whole, e.g. by precise surveying of movements;

(b) wherever the malfunctioning of anchorages could
    endanger the structure and cause it to become a hazard
    to life or property, and where problems would not be
    detected before the structure became unserviceable
    other than by monitoring;

(c) due to the nature of the ground and/or the protective
    system, tendons cannot be bonded to the walls of their
    holes, so that breakage of a tendon at any point
    renders it ineffective throughout its length;

(d) where anchorages are of a pattern that has not been
    proved adequately in advance, either by rigorous
    laboratory tests or by site performance under
    similar circumstances;

(e) where anchorages are in ground liable to creep.

Two methods of monitoring are in common use, namely,
measurement of loads on individual anchorages or measurement of
the performance of structures or excavations as a whole, the
latter being preferable.

When monitoring individual anchorages, the maximum loss or
gain of prestress that can be tolerated during service should
be indicated, taking into account the design of the works.
Variations up to 10% of working load do not generally cause
concern.   Prestress losses greater than 10% should be
investigated to ascertain cause and consequence, and for
prestress gain, remedial action, which may involve partial

destressing or additional anchorages, is recommended when the increases exceed 20% Tw and 40% Tw for temporary and permanent anchorages, respectively.

In general, monitoring should initially be at short intervals of 3 to 6 months, with later tests at longer intervals depending on results. The number of anchorages to be monitored should be indicated by the designer of the works;   5 to 10% of the total is typical in current practice.

RECOMMENDATIONS FOR FUTURE RESEARCH

General. Appraisal of failures is an essential component of the growth of knowledge and provides the seeds for future development. As a consequence, routine acceptance tests should be augmented whenever possible by proving tests taken to failure.

Systematic full scale testing and monitoring remain the finest source of information on the behaviour of anchorages, but field research can often be site specific and should therefore be supported by physical and mathematical modelling in an effort to produce generally applicable solutions for the prediction of anchorage system behaviour.

Overall Stability. Full scale field tests on vertical and inclined anchorages are required in a variety of ground types to investigate the critical depth of fixed anchor to avoid general shear failure of the ground. Group and interaction effects should also be studied. The results could lead to more cost-effective designs, e.g. vertical anchorages holding down basements subjected to hydrostatic uplift.

For anchored retaining walls, the shape of the 'sliding block' at failure is estimated using the results of laboratory tests in cohesionless soils. The validity of such model tests for use in field conditions needs to be checked, and the work should be extended to study cohesive soils.

Resistance to Withdrawal of Fixed Anchor. Full scale pull-out tests are required in a variety of ground conditions to check current empirical design rules and to extend our knowledge into weaker materials such as soft shales and mudstones, which may be sensitive to construction technique and result in low skin frictions. For all such tests a detailed engineering classification of the ground should be provided.

Pressuremeter tests are recommended in compressible weak rocks and loose or soft soils, since a lack of lateral constraint in the ground can lead to a bursting mechanism at failure (Fig. 12).

For pressure grouted anchorages the main distinction between Types B and C relates to magnitude of grout injection pressure, and more guidance is required on the injection pressure limits that determine if the ground is to be permeated, compacted or hydrofractured. In this regard, the beneficial effects of post grouting should be quantified for different ground types.

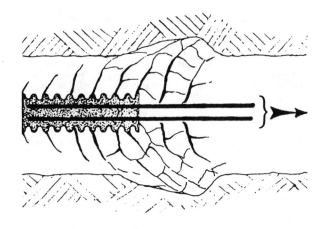

Figure 12. Bursting Mechanism of Compression Type Fixed Anchor

At the grout/tendon interface of fixed anchors, debonding requires study, particularly in high capacity anchorages (> 2000 kN). The influence of tendon density (area of steel/area of borehole), centralisers and spacers on load transfer and microcracking should be investigated.

Effects of Dynamic Loading. Where anchorages have been subjected to acceptance testing and locked-off at 110% Tw, the influence of cyclic loading should be measured at different frequencies and amplitudes, e.g ± 10% Tw, ± 20% Tw and ± 30% Tw. The information should indicate the sensitivity of anchorages to dynamic loading, including seismic effects.

Corrosion. A means of monitoring tendon corrosion should be developed, e.g. by electrical resistivity, ultrasonic or acoustic measuring techniques.

Durability. Fundamental research and field monitoring are needed in relation to durability of cement based grouts in aggressive ground and ground water conditions.

Tendon. To improve the accuracy of estimation of the apparent free tendon length in practice, a comparison of load-extension graphs and elastic modulus values is required between the standard 610 mm test length used by tendon manufacturers and the longer lengths (e.g. 10 m, 20 m, 30 m) used in the field. Tests should cover single and multi-unit tendons, where load distribution between tendon units is a further variable.

Service Behaviour. Field monitoring of anchorage loads and the movements of the overall structure/ground/anchorage system should be organised to study service behaviour and, in particular, the effect of prestress on deformations. Data on

loss of prestress or creep displacement with time will
determine optimum overload allowances for acceptance testing.
The effect of individual anchorage detensioning (simulating
failure) on adjacent anchorages or the structure also merits
study.

FINAL REMARKS

Experience indicates that higher quality and more detailed
ground investigations are required at the planning stage of
many anchorage projects to permit their economic design and
construction. In addition, there is a need to define clearly
in contract documents the design responsibilities of the
designer and specialist anchorage contractor in order to
minimize contractual problems.

In the field of permanent anchorages, corrosion protection
ranges from double protection (implying two physical barriers)
to simple cement grout cover. The latter solution is not
considered acceptable when the safety of people and property in
the event of anchorage failure is balanced against the cost of
providing protection. The required degree of protection should
be specified at the time of bid, and single protection should
represent the minimum standard for permanent anchorages.

Given the specialised nature of ground anchorage work and
the wide variety of anchorage types and construction
procedures, coupled with the variability of ground, more
reliance should be placed on performance specifications related
to choice of materials and acceptance testing of all
anchorages, compared with control of construction. Such
testing should involve proof loading to show a margin of
safety, load-displacement analysis to confirm that the
resistance to withdrawal is mobilised correctly in the fixed
anchor zone, and short term monitoring of the service behaviour
to ensure reliable performance in the long term.

If reliable performances are to be maintained, a technical
appraisal of anchorage systems is required by the practising
engineer, in addition to routine comparisons on the basis of
cost and duration of contract. Modern codes facilitate these
technical appraisals but more importantly, the adoption of code
recommendations should ensure both the safety and satisfactory
performance of the anchorage system.

Millions of anchorages have been installed successfully for
temporary and permanent works throughout the world, the
development in anchoring techniques having been dramatic over
the past forty years. With an absence of serious failures,
there is a strong base upon which anchorage specialists can
build and expand their market with confidence. There is no
room for complacency however; engineers must rigorously apply
high standards and much field development remains to be
tackled.

ACKNOWLEDGEMENT

Extracts from BS 8081:1989 are reproduced with the permission of BS1. Complete copies can be obtained through national standards bodies.

REFERENCES

1. Austrian Standards Institute, Prestressed Anchors for Soil Rock, Onorm B4455, Osterreichisches Normungsinstitut, Wien, 1976
2. British Standards Institution, Ground Anchorages, BS 8081, British Standards Institution, London, 1989.
3. Bureau Securitas, Recommendations Regarding the Design, Calculation, Installation and Inspection of Ground Anchorages, TA7, Editions Eyrolles, Paris, 1977.
4. Czechoslovak Standard, Prestressed anchors in ground, ON73 1008, Úřad pro normalizaci a měření, Praha, 1980.
5. Deutsche Industrie Norm, Soil and Rock Anchors, Bonded Anchors for Temporary Use, DIN 4125 Part I, 1974; Permanent Anchors, Part 2, 1976, Fachnormen–ausschus Bauwesen, Berlin, 1976.
6. Fédération Internationale de la Précontrainte, Recommendations for the design and construction of prestressed ground anchorages, FIP 2/7, Cement and Concrete Association, Slough, England, 1982 (under review).
7. International Society for Rock Mechanics, Suggested Method for Rock Anchorage Testing, J. Rock Mech. Min. Sci. Geomech. Abst., 22(2), 71–83, 1985.
8. Prestressed Concrete Institute, Recommendations for Prestressed Rock and Soil Anchors, Post–Tensioning Institute, Pheonix, Arizona, 1986.
9. Schweizer Norm, Ground and Rock Anchors,, SN533 191, Schweizerischer Ingenierung Architekten–Verein, Zurich, 1977.
10. South African Institution of Civil Engineers, Lateral Support in Surface Excavations, SAICE, Johannesburg, 1972. (under review)
11. Standards Association of Australia, Prestressed Concrete, CA35 Section 5 – Ground Anchorages, Sydney, 1973.
12. G S Littlejohn, Acceptable Water Flows for Rock Anchor Grouting, Ground Engineering, 8(2), 46–48, 1975.
13. G S Littlejohn & D A Bruce, Rock Anchors: State-of-the-Art, Foundation Publications Ltd, England, 1977.
14. G S Littlejohn, Design Estimation of the Ultimate Load Holding Capacity of Ground Anchors, Symp. on Prestressed Ground Anchors, Concrete Society of South Africa, Johannesburg, 1979
15. T H Hanna, Foundations in Tension – Ground Anchors, Trans. Tech. Publications, Germany 1982
16. L Hobst & J Zajic, Anchoring in Rock and Soil, Elsevier Scientific Publishing Co., Amsterdam 1983

17. H Schnabel Jr., Tie Backs in Foundation Engineering and Construction, McGraw-Hill, New York 1982
18. D E Weatherby, Tiebacks, US Dept of Transport, Federal Highway Admin. Report FHWA/RD-82/047, 1982.
19. R S Cheney, Permanent Ground Anchors, US Dept of Transport, Federal Highway Admin. Report FHWA-DP-68-1R, 1984
20. A D Barley, 10,000 Ground Anchorages in Rock, Ground Engineering, 21(6), 20-21, 23, 25-29, (7), 24-25, 27-35, (8) 35-37 and 39, 1988.
21. H Ostermayer, Construction, Carrying Behaviour and Creep Characteristics of Ground Anchors, Proc. I.C.E. Conf. on Diaphragm Walls and Anchorages, London, 1974.
22. Fédération Internationale de la Précontrainte, Corrosion and Corrosion Protection of Prestressed Ground Anchorages, Thomas Telford Ltd., London, 1986.
23. R A King, A Review of Soil Corrosiveness with Particular Reference to Reinforced Earth, TRRL Supplementary Report 316, Transport and Road Research Laboratory, Crowthorne, England, 1977.

DESIGN AND CONSTRUCTION OF A DEEP BASEMENT
IN SOFT RESIDUAL SOILS

Peter Day[1], A.M.ASCE

ABSTRACT: The northern area of central Johannesburg is situated
on a graben containing deeply weathered andesite. The residual
soil is a firm silt which extends to a depth of 50m or more.
This paper describes the design, construction and monitoring of
an 18m deep basement for a 13 storey office block immediately
adjacent to an existing 10 storey building. The value of simple
methods of analysis based on the results of standard field and
laboratory tests is emphasized.

INTRODUCTION

In 1979, a public utility company decided to develop a 13 storey
office complex in central Johannesburg. The development was to occupy
the eastern half of a city block adjacent to an existing 10 storey
structure owned by the company. The andesitic rock underlying the
site has weathered to a considerable depth, resulting in a highly
compressible residual soil profile extending well beyond the reach of
conventional piling equipment. As a result, it was decided to provide
a deep basement thereby floating the structure in the residual soil
mass.

This decision gave rise to a number of challenging problems
including:
- the support of the soft soil excavation,
- dewatering of the site,
- underpinning of the existing structure and
- control of ground movement.

The methods used to overcome these problems are discussed in this
paper.

PROPOSED DEVELOPMENT

The site of the proposed development measures $1500m^2$ (47,2 x 31,4m)
with street frontages on three sides as shown in Figure 1. The
proposed structure includes five parking basements, a ground floor and
first floor serving as a public post office, eleven floors of computer
and office accommodation and a thirteenth floor recreation area and
restaurant. The building is clad with local granite and is topped by
a clock tower 56m above pavement level.

1 - Director, Jones and Wagener Inc, Consulting Civil Engineers.
    P.O. Box 1434, Rivonia, 2128, South Africa.

734

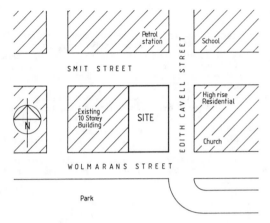

Figure 1. Location of Site.

The brief received from the architect required full coverage of the site from boundary to boundary. The maximum distance from the boundary to the inside face of the basement wall was restricted to 450mm adjacent to the existing building and 350mm elsewhere. This effectively precluded the use of a diaphragm wall.

The adjacent 10 storey building on the western side was built in the 1960's and is a concrete framed structure with brick infill panels. This building is situated hard against the boundary of the site and is founded on a forest of driven cast in situ piles extending to a depth of 10m - 12m. A section through the proposed basement is shown in Figure 2.

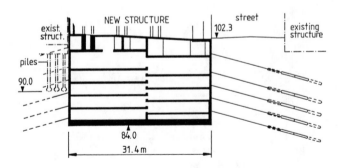

Figure 2. Section through Proposed Basement.

SOIL CONDITIONS

   Soil Profile and Water Table.   The Johannesburg graben is a down-
faulted   block   of   andesite   from   the   Ventersdorp   Supergroup
approximately 1km wide striking east-west though the northern part of
the city centre. Due to the topography and protection of the flanking
quartzites, the depth of decomposition is far greater than elsewhere
on the andesites.
   The typical soil profile recorded in a 750mm diameter auger hole is
shown in Figure 3. Diamond drilling indicated the decomposed andesite
extends to a depth of 55m beyond which depth there is a rapid
transition to highly weathered, soft rock andesite.
   The water table was recorded in 1979 at a depth of approximately
23m. By 1983, the water table had risen to 17m. An average water
table depth of 13m-14m was recorded during construction in 1987, some
4m-5m above the bottom of the excavation. The reason for the rise in
water table is not known.

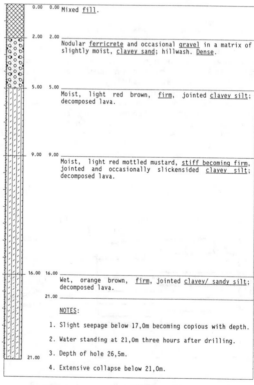

Figure 3. Typical Soil Profile.

Soil Properties. Some of the more important soil properties are summarized in Figure 4. These include Atterberg limits (liquid limit and plastic limit), moisture content, bulk and dry densities, SPT 'N' values, over-consolidation ratios based on consolidometer tests and elastic moduli based on plate load tests.

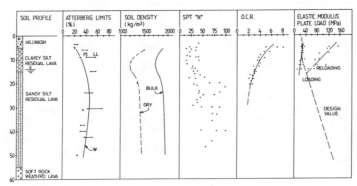

Figure 4: Summary of Soil Properties.

Shear Strength. The effective shear strength parameters of the residual lava were determined by means of slow, undrained triaxial tests with pore water pressure measurements and drained shear box tests. A total of 25 triaxial tests and 9 shear box tests were carried out over the full depth of the residual soil profile. These tests were carried out on Shelby tube samples from the diamond holes and on hand-cut block samples from the auger holes. The results of the triaxial tests are summarized in Figure 5.

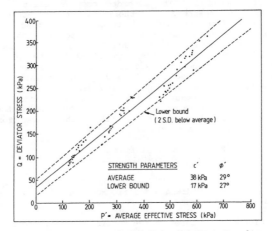

Figure 5. Summary of SCU Triaxial Test Results.

Despite the wide depth range over which samples were taken, the scatter of results was surprisingly small. The results of the tests are summarized in Table 1.

Table 1.   Strength Parameters for Residual Andesite.

TEST METHOD	AVERAGE SHEAR STRENGTH		LOWER BOUND	
	ø'	c'	ø'	c'
SCU Triaxial	29 deg	38 kPa	27 deg	17 kPa
Drained Shear Box	25 deg	39 kPa	-	-

The shear strength tests were generally conducted on intact specimens whereas the soil mass contains numerous relict joints. As these joints are clean and free of clayey infilling, their effect is to reduce the cohesion while leaving the angle of friction unaltered. The cohesive strength assumed for design was therefore reduced to approximately a quarter of the recorded average value. The following strength parameters were assumed for lateral support design:

Table 2.   Design Strength Parameters

MATERIAL	DEPTH	DESIGN SHEAR STRENGTH	
		ø'	c'
Hillwash	0-5m	35 deg	0 kPa
Andesite	5-20m	27 deg	10 kPa

Compressibility.   Consolidometer tests were conducted on Shelby tube and block samples.   The recorded compressibility showed an unacceptably wide scatter.   Furthermore, the pre-consolidation pressure inferred from tests on "undisturbed" Shelby tube samples was consistently higher than that from tests on block samples.

A number of horizontal, in situ plate load tests were conducted on the andesite above the water table using 200mm diameter plates.   The results of these tests showed considerably less scatter as can be seen in Figure 4.   The drained elastic modulus under initial loading averaged 20 MPa decreasing slightly with depth in response to the increase in moisture content. The elastic modulus under reloading was significantly higher and showed a more marked decrease with depth.

The variation in elastic modulus with depth assumed for design purposes, shown in Figure 4, was as follows:

$$E' = 2,6 \ z \quad \text{(Minimum 25 MPa)} \qquad \text{(1a)}$$
$$E_u = 3,2 \ z \quad \text{(Minimum 30 MPa)} \qquad \text{(1b)}$$

where E' = Drained Elastic Modulus   (MPa)
          E_u = Undrained Elastic Modulus   (MPa)
and z = Depth   (m)

This relationship was based on the results of the plate load tests, the analysis of SPT N values using the recommendations of Stroud

(1974) and back-analyses of two major structures founded on similar soils by Jaros (1981).

DESIGN OF LATERAL SUPPORT AND UNDERPINNING

Lateral Support. In view of the sensitivity of the adjacent buildings, roads and services to settlement, the use of continuous soldier piles was preferable to hand-over-hand installation of short lengths of soldier.

Steel soldier piles were employed on the northern, eastern and southern sides with driven replacement cast in situ piles forming part of the underpinning system below the adjacent building on the western side.

The steel soldiers comprised two 356x171x51kg/m universal beam sections encased in a weak grout. These sections were designed using the beam on elastic foundation method. In retrospect, this method of design is conservative and the weight per metre of soldier pile required has been reduced by approximately 40% in subsequent designs with no adverse effects on the performance of the wall. This reduction was achieved using ultimate limit state design methods in which the assumed failure mode requires bearing failure of the soil behind the soldier and of the formation of sufficient plastic hinges in the soldier to permit a mechanism to develop. Using this method, the required soldier pile section is usually governed by the cantilever moment or the moment required for stability of the wall after only one row of anchors has been installed.

Anchor forces were evaluated using a simple wedge analysis as shown in Figure 6. On the street faces a surcharge of 10 kPa was assumed. On the face adjacent to the neighbouring structure, the effective density of the soil over the upper 10m was increased to 35 kPa to allow for the load shed by the friction piles in this zone (See Figure 6).

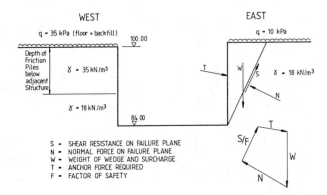

Figure 6. Wedge Failure Method of Anchor Load Determination.

The anchor loads were distributed uniformly over the height of the wall with a reduced load in the top row of anchors to prevent passive failure behind the top of the wall.

An attempt was made to apply the method of Littlejohn et al (1971) for the estimation of required anchor forces. This method takes account of the embedded length of soldiers by analyzing the multi-anchored wall using the free earth support method, regarding the resultant force of all anchors already installed as a single anchor. However, the soldiers were found to be too weak in bending and shear to justify the assumptions made in this method.

A summary of the anchor force provided on each face is shown in Table 3. These forces are based on a factor of safety of 1,5 for the design shear strength parameters given in Table 2. In addition a factor of safety of at least 1,3 was required in the case where the effective cohesion was assumed to be zero. The vertical component of anchor load was assumed to be taken down the soldiers which were embedded 6m below the base of the excavation. The statistical reliability of the support system has subsequently been checked using the point estimate method (Harr, 1987) and was found to be 95% which is regarded as acceptable for temporary works (Kirsten, 1983).

Table 3.   Summary of Anchor Force Provided.

Face	Average Depth	Total Anchor Force	Soldier Spacing	Anchors / Soldier
North	18,1m	1550kN/m	1,36m	1x300kN 4x450kN
East	17,1m	1400kN/m	1,50m	1x300kN 4x450kN
South	16,3m	1400kN/m	1,50m	1x300kN 4x450kN
West	17,1m	2700kN/m	0,90m varies	2x300kN 3x450kN 2x600kN

The overall stability of the excavation was checked using the Janbu method of slope stability analysis (Janbu, 1954). The fixed lengths of the lowest two rows of anchors were positioned beyond the critical failure surface to ensure a factor of safety of 1,25 against overall instability.

Underpinning. The ten storey building on the western side of the excavation is built hard against the site boundary. The structure is founded on friction piles and any relaxation of the ground around the piles would have led to a reduction in pile capacity. The piles extend to a depth of only 12m, 6m above the bottom of the excavation. (See Figure 2)

The obvious solution involved the use of the soldier piles to underpin the structure. However, the architect's restriction of 450mm to the inside face of the finished basement wall was too tight to accommodate conventional augered piles. It was therefore decided to excavate 3m - 4m below the existing pile caps and to install 430mm diameter driven replacement cast in situ piles using small tripod rigs operating below the existing pile caps. This pile type has the advantage of temporary casing (installed in lengths of 1,2m and screwed together) which prevents collapse of the pile hole.

These underpinning piles were designed to take the full load of the first row of columns on the east side of the building plus the vertical component of the anchor load, assuming no contribution to pile capacity from soils above the excavation level. The piles were installed at 900mm centres and socketed 5m below the bottom of the excavation. A bulbous base was provided and the lower 5m of the pile shaft was constructed using hammer compacted, low slump concrete. In this way a pile capacity of 500kN was achieved despite the close spacing of the piles.

On completion of the pile installation, a 3m deep capping beam was cast with corbels extending below the existing pile caps. This beam formed part of the new basement wall. Sleeves were cast into the beam through which two rows of anchors were installed at a later stage. Reinforced concrete walers were used for the lower rows of anchors on this face.

CONSTRUCTION TECHNIQUES

Soldier Pile Installation. The method of pile installation employed on the west face is described above. The piles along the remaining faces comprised steel sections encased in a weak sand cement mix (2-5MPa) cast in a 750mm diameter auger hole. These piles extended to a depth of 24m, about 10m below the water table. The stand up time of the auger holes at this depth was observed during the site investigation to be 15 to 30 minutes due to the copious ground water inflow.

To overcome the problem of collapse, the installation of these piles was streamlined. The steel sections, which were supplied in two 12m lengths, were joined in a vertical position and suspended from a crane. The holes were then augered to the level of the water table using a Hughes LLDH 120 auger rig. Augering to final level commenced only once the ready mixed concrete trucks had arrived on site. Augering the remaining 10m took less then 15 minutes after which the auger was slewed clear of the hole. A jig was positioned over the hole and the steel soldier was lowered into place and rested on the bottom of the hole, the method of suspension from the crane hook ensuring that the 24m long soldier was hanging plumb. Thereafter the lower 6m of the auger hole was filled with 20MPa concrete to enhance the bending strength of the soldier and provide a load bearing socket. The remainder of the hole was filled using a sand cement mix with a 2 - 5 MPa cube strength. The entire operation below the water table was completed within 30 minutes and no pile holes were lost due to collapse.

Anchor Installation. Prior to commencement of the contract, six re-injectable ground anchors were installed on the site each with a fixed length of 6m. These anchors were installed from surface through a concrete pad. Two anchors were installed vertically and the remainder were raked at 45°. Half the anchors were installed above the water table and half below.   The anchors were provided with sufficient strand to enable stressing to 1050kN.

Each anchor was stressed incrementally to 750kN and then rebounded. After two loading/unloading cycles the anchors were locked off at 660kN.   Lift off tests after 28 days showed no untoward loss of load. Two of the anchors were then loaded to 1050kN without causing failure. On the basis of these tests, the use of 600kN anchors in the lateral support design was acceptable.

The lateral support contract was let as part of the main building contract.   The successful contractor elected to appoint the same subcontractor who had undertaken the anchor tests.   Hence the working anchors were installed using the same techniques and equipment and the clause in the specification requiring site suitability tests could be waived.

All anchors installed on the site were re-injectable anchors with a 6m fixed length.   Anchors cables were made up of 15,2mm stands (with a working load capacity of 150kN per strand) placed around a central high pressure grout injection tube.   Four manchettes were provided in the tube over the fixed anchor length through which high pressure grout could be injected.   Spacers were provided to spread the strands over the fixed length.   The free length of the tendons was sheathed with plastic tube sealed at the lower end to prevent grout ingress. As the life of the anchors was less than 18 months and the environment is not aggressive, no corrosion protection was specified.

Anchors were installed by drilling a 75mm diameter hole inclined at 5° to the horizontal on the west face (to limit the vertical load component on the piles) and 15° elsewhere.   On the west face the anchors were positioned between the existing piles as far as was possible.   The anchor lengths were staggered 3m either side of the average length to reduce the group effect around the fixed anchors. The anchor holes were filled with a low viscosity cement grout and the anchors were homed into this grout.   The small diameter anchor holes proved surprisingly stable even below the water table.

After setting of the homing grout, a high pressure grout with a water cement ratio of 0,4 was injected through the central grouting tube at a pressure of between 1000kPa and 2000kPa.   Using a rapid hardening cement, the anchors could be stressed 4 days after high pressure grouting.

The stressing specification required that the anchors be stressed incrementally to 125% working load over two cycles, rebounding to 25% of working load after each cycle. Thereafter the anchors were locked off at 110% of working load and the lock-off load confirmed by a lift off test.   Lift off tests were carried out after 1 day, 7 days and 28 days.   An anchor was regarded as acceptable if the apparent free length of the anchor was neither less than 90% of the intended free length nor greater than the intended free length plus 50% of the fixed length.   In addition, the lift off load was not to be less than 105% of working load.   Unsatisfactory anchors were either regrouted or restressed.

In view of the satisfactory performance of the anchors during the early stages of the contract, the specification was relaxed to allow stressing and locking off in a single load cycle.   This speeded up operations considerably as the locking collets could be installed prior to positioning the stressing head, avoiding the need for removing the stressing head and jack after the initial load cycles.

Dewatering. During the initial investigation in 1979 the water table was below the bottom of the proposed excavation. However, by 1987 the water table had risen to a depth of 13m. Isolated seepage zones were encountered as shallow as 9m below pavement level during construction. No special provision was made for dewatering in the contract documents, the intention being that any seepage water would be led to a sump and pumped to surface. However, the rate of excavation gave rise to unacceptably high pore pressure gradients at the base of the excavation and resulted in significant softening of the underlying material. Eventually the floor of the excavation became a quagmire.

Tests conducted on a nearby site in the graben indicated the residual andesite to be an isotropic aquifer with an average permeability of $1,3 \times 10^7$ m/s, a transmissivity of $0,4$ m^2/d and a storativity of 3×10^{-3} (Terblanche, 1984). These results were confirmed by limited pumping tests from boreholes in the basement excavation. Such conditions were found to be conducive to relatively rapid dewatering using a small number of deep dewatering wells. Furthermore, the possibility of settlement due to dewatering was judged to be small as the water table had only recently risen to its present level, after erection of the adjacent buildings. Drawing down of the water table would thus not result in conditions worse than those which existed prior to the rise in the water table.

Thirteen deep dewatering wells were installed near to perimeter of the excavation extending to a depth of 45m below pavement level. Each well was lined with a 125mm steel casing surrounded by a graded sand filter with a 3m long well screen at the bottom. Each well was equipped with a 1500 l/hr submersible pump controlled by limit switches.

Within 10 days of completion of the installation, the water table at the centre of the site had been drawn by 5m to below the bottom of the excavation. Pumping continued until the permanent structure had been completed to above the water table. No significant settlement of the surrounding buildings was recorded.

MOVEMENT PREDICTIONS AND MONITORING

Movement Predictions. The success of the entire project hinged on the control of movements of the adjacent development. The prediction of movements therefore formed part of the feasibility study. As the adjacent building to the west was the most vulnerable, the analysis of ground movements concentrated on this side of the excavation. This analysis included the prediction of vertical and horizontal movements of the existing structure both during excavation and during subsequent construction of the proposed building.

In view of the uncertainty surrounding the assumed elastic modulus profile and initial horizontal stress field, the use of sophisticated techniques of analysis was not warranted. Instead, an axisymmetric elastic finite element analysis was undertaken in which the excavation was modeled as a circular hole. The soil was assumed to be isotropic and boundary loads were applied to the sides and bottom of the excavation equal to the expected stress change.

Determination of these boundary loads involved the calculation of the horizontal stresses in the ground prior to commencement of work on site, requiring an estimate of K_o, the coefficient of earth pressure

at rest.   This estimate was based on testing carried out on a nearby
site the results of which were published by Terblanche (1984):

$$K_0 = 1,8 - 0,04z \qquad (2)$$

where z = depth (m)

The elastic modulus profile was assumed to be as given in Equation
1 with a Poisson's ratio of 0,3 in the drained state and 0,5 in the
undrained state.   The use of the elastic modulus under initial loading
represented by Equation 1 was recognized as being conservative as much
of the soil around the excavation would experience unloading during
the excavation stage.

The predicted deformations on completion of the excavation are
shown in Figure 7a.   Despite the conservatism of the assumptions, the
predicted deformations below the adjacent building were regarded as
acceptable.

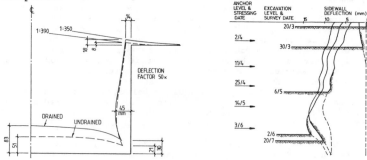

A. PREDICTED DEFORMATIONS USING   B. RECORDED DEFORMATIONS OF INCLINOMETER TUBE
   INITIAL LOADING MODULUS   ON EAST WALL

Figure 7.   Predicted and Observed Deformations

As a check on the base heave of the excavation, a further estimate
was made using Butler's charts (Butler, 1974).   The long term heave of
the base of the excavation was calculated as 80mm at the centre of the
hole and as 40mm at the middle of each side.   This agreed well with
the results of the finite element analysis.

During plate load testing, the elastic modulus under reloading was
found to be between two and four times the initial loading modulus.
The base heave of the excavation and subsequent settlement of the
proposed structure were thus expected to be between one quarter and
one half of the above predictions.   This was regarded as acceptable.

Monitoring Surveys.   Prior to commencement of construction, a
comprehensive damage survey was conducted of all buildings within 50m
of the excavation.   Reports were compiled containing photographic
records of existing cracks and a description of the general state of
each building.   Levelling points were installed on the street faces of
all adjacent structures and a precise level survey was conducted prior
to construction.

To avoid damage to the multitude of services around the site, hand dug trenches were excavated at each end of the three street faces. The level, position and condition of all services was recorded.

Ground movement surveys included line and level surveys of the top of the soldiers and along the curb line. Level traverses were also carried out at right angles to the excavation face as shown in Figure 8. Four inclinometers were installed alongside the steel soldiers on the three accessible sides of the excavation. An attempt was made to monitor the base heave of the excavation but this proved unsuccessful.

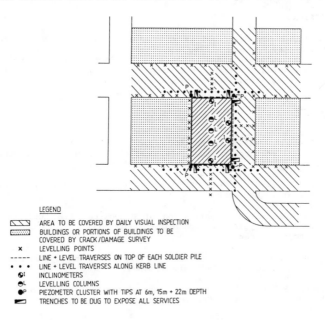

LEGEND

▨	AREA TO BE COVERED BY DAILY VISUAL INSPECTION
▨	BUILDINGS OR PORTIONS OF BUILDINGS TO BE COVERED BY CRACK / DAMAGE SURVEY
x	LEVELLING POINTS
-----	LINE + LEVEL TRAVERSES ON TOP OF EACH SOLDIER PILE
• • •	LINE + LEVEL TRAVERSES ALONG KERB LINE
◉I	INCLINOMETERS
◉L	LEVELLING COLUMNS
◉P	PIEZOMETER CLUSTER WITH TIPS AT 6m, 15m + 22m DEPTH
▬	TRENCHES TO BE DUG TO EXPOSE ALL SERVICES

Figure 8. Details of Monitoring Points.

Recorded Deformations. No untoward movements were observed during excavation of the basement structure even after dewatering. The maximum settlement of the adjacent structure was 5mm accompanied by a movement of 3mm towards the excavation. No damage other than hairline cracking of the cable cellar below the existing building has been noted.

The maximum inward movement of the top of the soldiers was 18mm on the eastern face and 8mm on the shorter northern and southern faces. The ratio of crest movement to depth of excavation thus ranges from 0,03 to 0,10 which is fairly typical for a well supported excavation. Settlements behind the wall ranged from 5 to 22mm giving a ratio of settlement to depth of 0,03 to 012.

The deformed shape of the eastern face recorded by means of an inclinometer is shown in Figure 7b. It is interesting to note the effect of anchors on the deflected shape, often giving rise to a reverse in the direction of curvature on stressing of the anchor. All significant movement of the excavation ceased approximately 1 week after stressing of the lowest row of anchors. This situation has also been observed in other well supported basement excavations.

The new building was completed early in 1990. The measured settlement of this structure is less than 20mm.

CONCLUSIONS

The case history described is an example of a large scale lateral support project designed using simple analysis methods and the results of standard laboratory and field tests. Deformations were predicted using relatively conservative assumptions. Deformations measured on site have been within the predicted limits.

REFERENCES

1 Butler F.G. (1974), Review Paper Session III: Heavily Over-consolidated Cohesive Materials, Conf on Settlement of Structures, British Geotech Soc, Cambridge. Pentech Press.

2 Harr M.E. (1987), Reliability Based Design in Civil Engineering, McGraw Hill, New York.

3 Janbu N. (1954), Application of Composite Slip Surfaces for Stability Analysis, European Conf on Stability of Earth Slopes, Stockholm, Sweden.

4 Joros M.B. (1981), The Settlement of Two Multi-storey Buildings on Residual Ventersdorp Lava, Unpublished.

5 Kirsten H.A.D. (1983), Significance of the Probability of Failure in Slope Engineering, The Civil Engineer in South Africa, January 1983.

6 Littlejohn G.S., Jack B.J. and Sliwinski Z.J. (1971), Anchored Diaphragm Walls in Sand - Some Design and Construction Considerations, J of Ins of Highway Engineers, April 1971.

7 Stroud M.A. (1974), The Standard Penetration Test in Insensitive Clays and Soft Rocks, European Conf on Penetration Testing, Swedish Geotech Soc, Stockholm.

8 Terblanche E.H (1984), Instrumentation of a Deep Basement Structure, Symp on Monitoring for Safety in Geotechnical Engineering, Int Soc for Rock Mech, South African National Group.

A FIELD STUDY OF A TIEBACK EXCAVATION WITH A FINITE ELEMENT ANALYSIS

Joseph A. Caliendo[1], M.ASCE, Loren R. Anderson[2], M.ASCE
and William J. Gordon[3], M.ASCE

ABSTRACT: This paper documents the performance of a soldier pile, tieback retention system for a four story below ground parking structure in downtown Salt Lake City, Utah. Of significance is the relatively soft lacustrine clay profile in which most of the tiebacks were anchored.

The field measurements consisted of: (1) slope inclinometer measurements on several of the soldier piles, (2) strain gage measurements at various points along the lengths of a number of the bar tendons.

The field results are discussed and analyzed. Displacements of the wall as measured with the slope inclinometer are presented. The maximum displacement thus measured was on the order of one inch. Load distribution curves from tieback strain gages are shown. The load distribution appeared to be fairly linear along the anchor portion of the tieback. The strain curves for two of the tiebacks were integrated in order to establish movement at the back of the anchor.

Each of the 300+ tiebacks was tested in the field before being put into use. Three types of tests were employed, performance test, creep test, and proof test. The test results are discussed along with the acceptance criteria.

A finite element analysis using the program SOILSTRUCT was performed. The results are presented and compared with the field measurements and good correlation was obtained with the field displacements.

INTRODUCTION

In January of 1984, excavation began for the 14 Story Block 257 Office Tower and Parking Facility located on the northwest corner of 200 South Street and 300 East Street in downtown Salt Lake City, Utah. The project included four stories of below grade excavation and it is the deepest supported excavation to date in the Salt Lake City area. The excavation was made in medium stiff clay and lateral support for the excavation walls was provided by tieback anchors.

The site is adjacent to the existing multistory Mountain Bell building on the west and a three story office building, parking deck, and the Salt Lake Athletic Club on the east side. A plan view of the excavation is shown on figure 1. Also shown are the design horizontal tieback loads. Complete documentation is provided by Caliendo (1986).

CONSTRUCTION DETAILS

Soldier Piles. Over 150 soldier piles were installed along the entire perimeter of the proposed excavation at 5 to 6 ft (1.524 to 1.829 m) spacings. These consisted of steel H

---

1 . State Geotechnical Engineer, Florida Department of Transportation, 605 Suwannee Street, Mail Station 33, Tallahassee, FL 32301.

2 . Professor of Civil Engineering and Associate Dean of Engineering, College of Engineering, Utah State University, Logan, UT 84322-4100.

3 . Vice President, Sergent, Hauskins & Beckwith, 4030 South 500 West Suite 90, Salt Lake City, UT 84123.

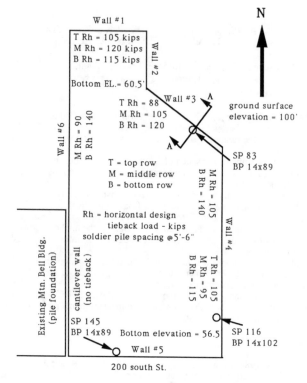

Figure 1. Plan View of Project

piles which were placed in 24 inch (.6096 m) diameter shaft excavations. The water level
was maintained near the top of the excavation to prevent caving. The piles were 60 ft
(18.28 m) long BP 14x102 and BP 14x89 sections. Wales were generally not used; the
tieback preloads were transferred directly to the soldier piles. Three inch wood lagging
was placed behind the front flanges of adjacent soldier beams. Because of the high vertical
component of the tieback preload, it was necessary to insure that the soldier piles were
founded on a suitable stratum to avoid exceeding the vertical load capacity of the piles.
Therefore, the 24-inch (.6096 m) diameter shafts were predrilled to a gravel load bearing
stratum. The depth to this stratum varied from approximately 64 ft to 100 ft (19.51 m to
30.48 m) and it was therefore necessary to hang the 60 ft (18.29 m) H piles in the shaft
until the concrete was placed by means of a double tremie (i.e., one pipe on either side of
the beam's web). Two different concrete mixes were used for the soldier piles. A "hard
rock" mix was used for that portion of the shaft below the bottom of the building
excavation. A leaner "soft rock" mix was used above the building excavation to facilitate
removal of the concrete from the front flange of the soldier pile.

Tiebacks. Up to three horizontal rows of tiebacks were used for this project. The number of rows are indicated on the plan view (figure 1) and illustrated on section A-A (figure 2). Over 300 tiebacks were installed and tested. The tendons used were, for the most part, 1-1/4 and 1-3/8 inch 60-ft long (31.75 and 34.93 mm, 18.28 m) Dywidag threadbars. Couplings were used to extend the tendons to 70 to 80 ft (21.34 to 24.38 m). At several locations, the test loads for the bottom row of tiebacks exceeded the allowable capacity of the 1-3/8 inch (34.93 mm) bars. At those locations, multi-strand cable tendons were used.

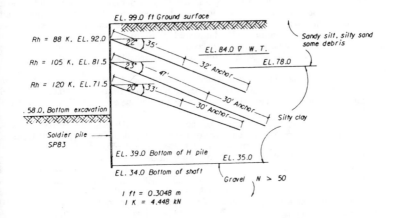

Figure 2. Section A-A

The tendons were installed by means of a continuous flight hollow stem auger at angles ranging from 8 to 30 degrees from horizontal. Grout was pumped through the annular space inside the auger at pressures ranging between 100 and 150 psi (1035 kPa) as measured at the front of the auger. After 25 to 35 ft (7.620 to 10.67 m) of anchor had been pumped, and always before the auger tip was withdrawn past the assumed failure plane, the grout pressure was reduced to zero and the auger completely withdrawn. PVC pipe was used as a bond breaker for both types of tendons. After curing, the tiebacks were each tested and locked off at 80 to 90 percent of the design load. The lockoff loads were transferred directly to the soldier piles by means of an anchorage plate and nut.

SOIL AND GROUNDWATER

Soil Profile. The soil profile at the site was developed from borings made by Dames and Moore as part of their design recommendations for the project, and from exploration borings previously made for the Mountain Bell Building, located immediately west of the site (Caliendo, 1986). All borings typically extended to depths exceeding 100 ft (30.48m). A generalized soil profile is shown on figure 2. The bearing stratum for the soldier piles and drilled shafts is the gravel layer (GM). The anchor portion of the tiebacks were formed within the overlying silty clay (CL) stratum.

Measurements made one week after completion of drilling showed the groundwater to be approximately 15 ft (4.57 m) below ground level. At the time of construction (several years later) the depth to groundwater was unchanged.

Soil Strength. In order to determine the undrained shear strength of the fine grained soils encountered during the initial investigation, nine laboratory vane shear tests were performed. The maximum undrained shear strengths of the clay soils above the final excavation level ranged from 600 to 1000 psf (28.73 kPa to 47.88 kPa). Field vane measurements made below the bottom of the excavation during drilled shaft excavation ranged from 1000 to 1900 psf (47.88 to 90.97 kPa) and generally increased with depth.

TIEBACK TESTING

Tieback tests are performed to verify that the tieback will carry the design load without excessive movements and that the capacity is not developed in soil that may fail as a result of soil movement (Schnabel, 1974). Tiebacks are one of the few structural systems in which each member can normally be tested prior to being placed into service. Recommendations for tieback testing have been established by the Post Tensioning Institute (1980). Three types of tests were performed on the tiebacks, performance tests, proof tests, and creep tests.

The tests are constant load tests in which a hydraulic jack is used to apply and maintain axial tensile loads on the tiebacks. The tieback movement is measured by means of a dial gage calibrated to 0.001 inches (.0254 mm). The jack and gage are usually calibrated as a unit.

Performance Tests. The purpose of the performance test is to: verify capacity, establish the load deformation behavior for the tieback, identify the different types of movement, and to check that an unbonded length has been established.

Performance testing involves load/elongation measurements in a sequence of loading and unloading increments. The load increments include an alignment load (AL) followed by loads of 25%, 50%, 75%, 100%, 120%, and 133% of the design load. The AL is usually about 4 kips (17.79 kN) which takes the initial "slack" out of the system. The maximum load is held for 10 minutes, if the movement that occurs between 1 and 10 minutes exceeds 0.04 inches (1.016 mm), the load is maintained for 60 minutes.

The load and deformation for each increment are recorded, the load is released back to the AL and the dial reading recorded. When the load is reduced to the AL, that portion of the load associated with elastic movement is recovered, the remainder is residual movement. With the elastic movement established, the length of the tendon associated with that movement (unbonded length) can be calculated by conventional means.

Creep Tests. These tests are performed to establish long term behavior in clay soils. Creep tests are usually run as part of the performance test. The creep curve is a semi-logarithmic plot of total movement versus time. If a creep rate per decade of time (log cycle) can be established, then the anchor load at any future time can be estimated. Weatherby (1982) recommend an acceptable creep rate in cohesive soils as 0.08 inches (2.032 mm) per log cycle at a load of 133% of the design load.

Proof Tests. Each tieback that is not otherwise tested is proof tested, this includes most tiebacks on a typical project. The proof test verifies the capacity of a tieback and at the same time preloads the tendon. The increments are the same as the performance test except that the maximum load is usually 120% of design load. The maximum load is held for a minimum of 5 minutes and the creep recorded.

Acceptance Criteria. Three acceptance criteria are in common usage and were adapted for the Block 257 project. These are:

1. The total elastic movement of the tieback must be greater than 80% of the theoretical elongation of the unbonded length.

2. The total elongation must be less than the theoretical elongation of the tendon length measured from the jack to the midpoint of the bonded zone. Larger values may indicate that the bond zone is moving excessively relative to the soil and pullout may occur.

3. The acceptable amount of elongation under constant load was taken to be 0.08 inches (2.032 mm) as measured between 5 and 50 minute readings during a performance test and between the 0.5 and 5 minute readings during a proof test.

After the testing was completed, the load was reduced and the tieback locked off at the preload value. The lockoff loads (preloads) were typically 80% for the top row of tiebacks and 90% of design load for the lower rows. Schnabel (1982) points out that if the maximum load that a tieback is subjected to is 133% of design load, then all testing can be done on production tiebacks. Otherwise, special tiebacks would need to be fabricated for higher test loads. This is because the tendon manufacturers specify a design stress of 60% of yield stress and a maximum test stress of 80% of yield stress.

TIEBACK TEST RESULTS

Performance Test Results. Performance tests were run on 24 of the approximately 300 tiebacks installed for the Block 257 project. Figure 3 shows typical performance test results in which the top graph shows total movement versus load and the bottom graph is residual movement versus load. Elastic movement is taken as the difference between the total and residual movement.

The performance test results showed that in general, the minimum elastic movements were achieved. However, the anchor lengths were longer than those estimated in the field by the contractor. The contractor reported the anchor length as the distance the auger was withdrawn before the grout pump was switched off. Elastic movements associated with several of the tiebacks were not enough to satisfy minimum movement. Since the contractor always provided PVC pipe as a bond breaker along the front portion of the tendon, at least as far as the assumed failure plane, the performance tests for these several tiebacks suggest that not only did the grout extend into the assumed failure wedge, but that the PVC did not function adequately as a bond breaker.

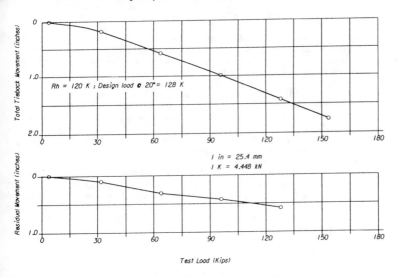

Figure 3. Typical Performance Test Results

Creep Test Results. Because most of the anchor length was cast within the firm gray clay stratum, the creep behavior of the tiebacks is of significant interest. The loss of tieback load with time according to Weatherby (1982), may be caused by anchor creep, debonding of the tendon, tendon relaxation, or structural deformation.

Figure 4 shows the results of a typical creep test. In order to establish a convenient log cycle of time, the loads were typically held for a minimum of 100 minutes at 133% of design load. The results all showed a decrease in the rate of movement with an increase in time. The final slope of the creep curve was near zero for several of the tiebacks tested. Several of the tests may indicate debonding at the grout/tendon interface as evidenced by jumps in the curve with the slope of the curve being approximately equal on both sides of the jump. This is apparent on figure 4 at a time of about 40 minutes and again at 70 minutes.

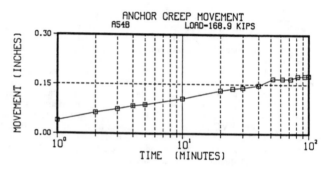

Figure 4. Creep Test Results for Anchor A54B

Proof Test Results. Proof load tests were performed on 293 tiebacks. The proof load was 120% of the design load and was held for 10 to 20 minutes until all discernible movement had ceased. The walls exhibited little movement after the tiebacks had been locked off and this may be attributed to the relatively long "holds". Typical proof test results are shown on figure 5. The curves nearly always showed a linear displacement versus load relationship. A small amount of permanent anchor movement is normally imposed as evidenced by the slightly flatter slope of the rebound portion of the curve.

The total movement was limited to the theoretical elongation of the tendon length measured from the jack to the midpoint of the anchor. Since the free lengths as reported by the contractor, were usually longer than the actual free lengths as verified by performance testing, it was conservative to compare the measured elongations with maximum allowable elongations based on the contractor's data. If the total movements were less than the allowable, the proof test was successful. Using the familiar relationship for uniaxial loading, the maximum allowable elongation was calculated and compared with the measured value. The test data indicated that none of the total movements exceeded the maximum allowable. On the other hand, it was important that enough movement be observed in order to insure that a proper free length had been established, as was verified by performance testing.

FIELD INSTRUMENTATION

The field instrumentation consisted of slope inclinometers for measuring deformation of the soldier piles and strain gages measuring load and load distribution on the tieback tendons.

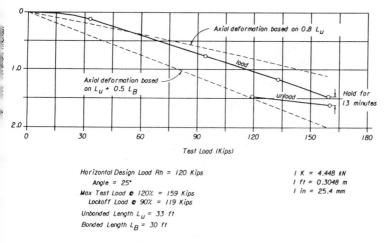

Figure 5. Typical Proof Test Results

Slope Inclinometer Measurements. Lateral deflections of several soldier piles designated as SP83, SP116, and SP145 were made by means of the Dames and Moore Earth Deformation Recorder" (EDR), the locations of which are shown on figure 1. The results of SP116 are discussed herein.

The instrumentation consisted of flexible casing mounted on the back flanges of the soldier piles and a slope recording probe. The field recording unit yields a continuous profile of soldier pile orientation versus depth. Since the recording commences after the probe reaches the casing bottom and subsequently withdrawn, all measurements are relative to the bottom of the casing (i.e., bottom of the pile).

The EDR results are shown on figure 6a as plots of deflection versus depth from June 1984 through 15 March 1985. During this time, the excavation progressed from its initial state to full depth and all below grade parking decks were installed. The data points represent the deflections as established from the field profiles. A positive deflection indicates movement towards the excavation. The lines represent a "smoothing" of the data and were obtained by taking a central moving average of five points along the pile. Backward and forward values were used to establish average values at the ends. Two iterations were made along the entire length which then yielded the smooth continuous curves shown on figure 6b.

Soldier Pile Movements. In general, the displacement vs depth curves show a gradual movement toward the excavation as the depth increases. The maximum deflection for the last curve is approximately 0.95 inches (24.13 mm) at a depth of 30 to 35 ft (9.144 to 10.67 m). The top of the pile moved approximately 0.4 inches (10.16 mm) away from the excavation as shown on figure 6a. In order to provide access into the excavation, the general contractor had to provide a temporary earth ramp along the south wall and adjacent to SP116. At the location of SP116, the ramp was 11.5 ft (3.505 m) below the ground surface. The ramp remained in place from June 1984 through mid December 1984. The effect of removing the ramp is indicated on the 27 December and 15 March profiles, with the wall moving a maximum of approximately 0.95 inches (24.13 mm) towards the excavation. The third tieback row was never installed in this vicinity because the contractor opted to provide temporary bracing until the underground structure was completed.

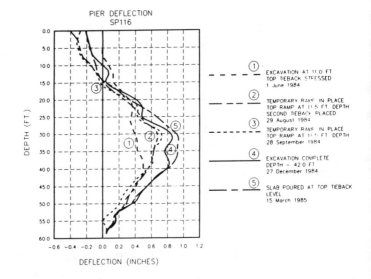

Figure 6a. Pier Deflection SP116

EDR Moment Approximations. After the final deflected shapes had been developed
the bending moments in the soldier piles were calculated using the finite difference
approximation of the moment equation:

$$(E\ I\ )/(d^2y/dx^2) = M \qquad\qquad (1)$$

A fourth order central difference approximation was used whenever possible. The
deflection values used were those derived from the smoothed curves described above.
Even so, the moments varied significantly along short increments of beam length. In order
to arrive at a better estimate of the bending moments in the final deflected beam shape, the
final beam profiles were smoothed by means of a french curve. The resulting curves were
digitized and a regression analysis performed. The analysis yielded polynomial equations
with deflection and depth as the dependent and independent variables respectively. Direct
application of equation (1) provided bending moment values.

The degree of polynomial used to describe the deflected shape is of interest. Typical
trapezoidal lateral stress distributions normally assumed for preliminary design purposes
would result in a fifth order polynomial. The regression curve as well as the sixth degree
polynomial used to describe the curve are shown on figure 7. Realizing that the trapezoidal
pressure distribution is an approximation based on "apparent" pressure diagrams, a sixth
degree polynomial seems reasonable. This would correspond with a parabolic pressure
distribution.

Strain Gage Measurements. Bonded foil type strain gages were mounted on the steel
Dywidag tendons, these were mounted in pairs on opposite sides of the bar and connected

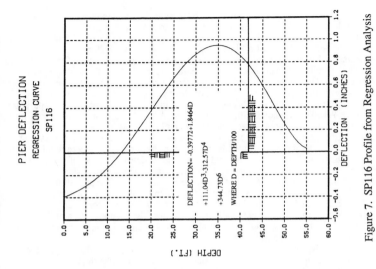

Figure 7. SP116 Profile from Regression Analysis

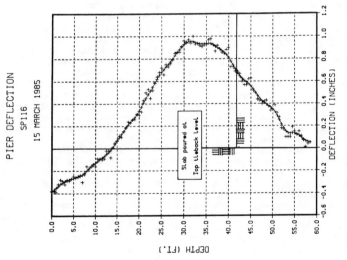

Figure 6b. Pier Deflection SP116, 15 March 1985

in series so that an average axial strain could be measured. The purposes of the gages were to monitor the tieback preload with time as well as to measure the load distribution along the tieback. The gages were mounted at distances from the wall face of 15 ft (4.572 m), 30 ft   (9.144 m) and at 5 ft (1.524 m) intervals thereafter to 60 ft (18.29 m).

Instrumented tieback tendons were installed at the three soldier piles for which EDR data was being collected. Seven tiebacks were initially instrumented with strain gages. As anticipated and despite the caution exercised by the contractor, many of the gages were destroyed during the installation. All gages on three of the tiebacks were destroyed. Of the remaining four only two of the tiebacks retained enough gages for interpretation purposes. These were A146T (Anchor #146, top row) and A116M (Anchor #116 middle row). Locations of the instrumented tiebacks are shown on Figure 1.

Load Distribution. Figures 8 and 9 are load distribution curves for tiebacks A116M and A146T respectively as calculated from the strain gage data. The gage readings were taken after the tieback was locked off and at subsequent dates. The lockoff load is shown on Figures 8 and 9 and was measured at the load test jack. The lockoff loads exceeded the measured load for both tiebacks. The loads were measured before and after the lockoff load had been transferred to the soldier piles. Evidently, slippage occurred between the lockoff nut and an anchor plate as indicated by the 18 June readings for A116M. As can be seen, the loads are rapidly attenuated along the length of the anchor portion of the tieback. By definition, the free length is the length along which there is no load transfer. The free lengths are obvious for both tiebacks.

Calculating displacement by integration of the strain curves is valid as long as the back of the anchor does not move. In order to establish any movement at the end of the anchor, the displacement calculated from the strain gage data was compared with the total movement measured at the front of the tieback during tieback testing. Since the extrapolated strain distribution curves indicated zero strain at some point in front of the anchor end, there should have been no movement beyond that point. The total movement measured by the dial gage should then equal the area under the strain curve.

Integration of the strain distribution curve for A146T indicated 0.878 inches (22.30 mm) of movement. A performance test was done on that particular tieback and a total movement at the lockoff load of 80.4 kips (357.6kN) was 0.9 inches (22.86 mm).

Long Term Behavior. Strain gage readings were taken at the time of significant construction activities in the vicinity of a particular tieback. Readings were also taken over a period of several months after the excavation had been completed. The time dependent movement can be calculated by integration of the strain curves corresponding to a given time interval. The difference in area is the movement that occurred within that time period. Integration of the strain curves for A146T shows that 0.0255 inches (.6477 mm) of movement occurred from 1 August through 26 September. This corresponds to a time period of significant construction activity and excavation. At that point, the excavation was completed and there was no further construction activity in that vicinity. From 26 September through 21 December, the corresponding movement was calculated to be 0.0906 inches (2.301 mm).

These results imply that the time dependent movements are small and that they result from anchor creep as opposed to construction or excavation operations. The small creep values are consistent with creep tests performed during the performance testing. During construction and excavation, the load tended to increase toward the back of the anchor with time. This is similar to results reported by Shields et al. (1978).

Tieback A116M was installed at the portion of the wall that was temporarily buried in order to provide a ramp for access to the bottom of the excavation, as discussed above. A comparison of the strain curves for the periods 31 August through 27 December (ramp in place) and 27 December through 15 January (ramp removed and excavation to full depth) show that very little movement occurred during either time interval and that the excavation had little effect on the small movement.

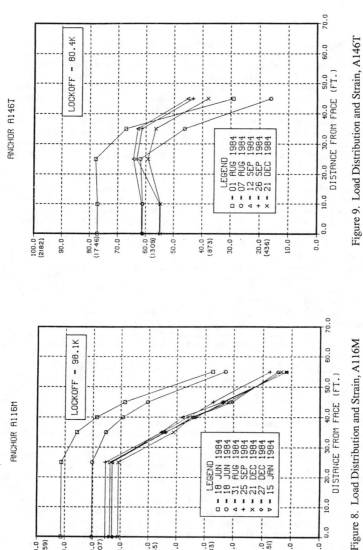

Figure 9. Load Distribution and Strain, A146T

Figure 8. Load Distribution and Strain, A116M

FINITE ELEMENT ANALYSIS MODELING

The program SOILSTRUCT was used for this finite element analysis (FEA) study. This program was first developed by Clough (1969) and has subsequently been modified by a number of researchers including Tsui (1974) and Hansen (1980). The program has been widely used in both academic and engineering practice to simulate braced and tieback retention systems. The non-linear stress strain properties of soil are modeled in SOILSTRUCT by the hyperbolic relationships suggested by Duncan et al. (1980).

Quadrilateral Element. The two dimensional quadrilateral element was used to model both soil and wall behavior. Tsui (1974) indicates that the quadrilateral element is superior to a beam element for finite element analyses because the beam element has no thickness and no mesh space is therefore allocated for the actual wall location. Furthermore, since the same type of element is used to model both wall and soil, there is no problem with incompatibility between adjacent soil elements.

Bar Element. Bar elements (springs) without soil-structure interaction were used to model the behavior of the tiebacks. If instead, the tieback would have been modeled with a bar placed in the mesh, several problems would have occurred: (1) the mesh would have been very complicated; and, (2) the two dimensional plane strain analysis would have modeled the tieback rods as laterally continuous plates that would not simulate the true soil-structure interaction between the tie-back rods and the soil.

In cases where the anchor is set into a rigid material or unmoving soil mass, the problem reduces to that of a wall connected by a simple axial member to an unmoving point. In this instance, the tieback may be modeled as a spring support located on the excavation side of the support wall. The tieback performance test results may be used to approximate the input stiffness since, in performing these tests, elastic movements are discerned from the total movements.

Site Specific Modeling. In order to verify modeling the Block 257 retention system with finite elements, a portion of the wall was chosen for analysis. The wall and soil adjacent to soldier pile SP116 was selected because this section provided relatively good instrumentation results, reliable proof and performance test results, and the soil profile was well documented.

Wall Modeling. Excavation walls were included in the finite element mesh as two adjacent columns of quadrilateral elements. The wall was assumed to behave elastically. The rigidity (EI) of the actual wall divided by the tieback spacing to simulate plane strain conditions was first established. As mentioned, two adjacent columns of quadrilateral elements were used to model the wall (the model wall thickness was established by two elements each being one foot thick). The model moment of inertia (per foot of wall) was calculated and the rigidity (EI) of the model set equal to the rigidity of the actual wall. The model modulus was then calculated.

Lateral Support Modeling. The lateral support at the section chosen for modeling consisted of top and middle rows of tiebacks and a bottom row of struts. All were modeled as strut (spring) supports. The stiffness (AE/L) per foot of wall was estimated from the performance test results obtained in the vicinity of soldier pile SP116. The stiffness was calculated from the elastic deformation, the free length of the tieback, and the jack load.

Point loads equal to the horizontal component of the tieback lock off loads, divided by the spacing, were applied to the wall at the elevation of the tiebacks. As previously mentioned, the contractor elected to use a large diameter steel pile strut in lieu of a bottom row of tiebacks. This was modeled as a spring with a very large stiffness and with a nominal lock off load.

Soil Modeling. In order to correctly model the soil response, SOILSTRUCT requires that conventional as well as hyperbolic soil parameters be included in the input. These procedures are documented in detail by Duncan et al. (1980). The hyperbolic values were determined from undrained triaxial tests performed on samples collected during the course of the excavation. In the absence of laboratory testing, it was necessary to estimate the hyperbolic values for both the dense underlying gravel stratum as well as the near surface soil which consisted of a variety of soil and non-soil types. Estimating the hyperbolic parameters for this stratum was difficult. It was decided that since most of the wall and nearly all of the anchorage portions of the tiebacks were founded in the gray silty clay, for which reasonably good laboratory results had been established, those same hyperbolic parameters would be used for the top portion of the profile as well.

Input parameters. The input parameters for the finite element analysis are shown in table 1; materials 1-3 are for the three soil types which comprise the idealized soil profile; material 4 is the steel soldier pile; and material 5 is air. Where applicable, the parameters are expressed in feet and pound units.

Table 1. Material Property Input for SOILSTRUCT

Parameter	Fill(1)	Silty Clay(2)	Gravel(3)	Steel(4)	Air(5)
Unit weight	110	113	130	490	0
Poisson's ratio before failure	0.30	0.30	0.30	0.30	0.01
Poisson's ratio at failure	0.48	0.48	0.48	0.48	0.01
At rest coefficient	0.5	0.6	0.4	0	0
Friction angle	10	0	40	0	0
Cohesion	600	900	0	0	0
Modulus number	150	150	600	0	1
Modulus number load-reload	250	250	820	0	1
Modulus exponent	0.86	0.86	0.40	0	0
Failure ratio	0.7	0.7	0.7	0	0

FINITE ELEMENT ANALYSIS RESULTS

A mesh consisting of 440 nodes and 399 elements was used for the analysis. The mesh along with boundary constraints and material types is shown on figure 10. The following load steps were applied sequentially in order to model the actual construction.

1. Excavate to 8.0 ft depth (2.438 m).
2. Install and stress tieback at 8.0 ft depth (2.438 m).
3. Excavate to 18.0 ft depth (5.486 m).
4. Install and stress tieback at 18.0 ft depth (5.486 m).

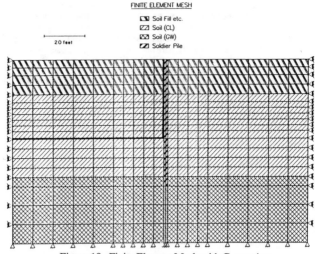

Figure 10. Finite Element Mesh with Constraints

5a. Excavate to 28.0 ft depth (8.534 m).
5b. Install strut (nominal stress) at 28.0 ft (8.534 m).
6. Excavate to 41.0 ft depth (12.50 m).

Comparison with EDR Results. In order to verify the FEA results, the displacement of SP116 as measured by the EDR on completion of the excavation was compared with the FEA results after completing load step 6 above. To make a proper comparison, the fact that the EDR results were relative to the bottom of the soldier pile had to be taken into consideration. Since the FEA results were based on the bottom of the wall being free to move, that movement was subtracted from each nodal movement along the wall including the bottom node.

The FEA displacement profile and the EDR profile (field curve) as shown on figure 11 are both relative to the bottom of the wall and correspond to the final load step. A positive value indicates movement into the excavation. As the figure shows, the magnitude of wall movement predicted by FEA is nearly identical to that measured by the EDR. However, the location of maximum deflection is shifted approximately 5.0 ft (1.524 m).

Results of Complete Analysis. The sequential wall deflections resulting from the various load steps are shown on figure 12. The deflections corresponding to load steps 5a and 5b are essentially the same since only a nominal load was applied through the strut. These are both shown on the figure as load step 5. Positive movements occur as a result of excavation, the tieback preload then forces the wall back into the soil. This cycle is evident during the first 5 load steps. The rigid strut at depth 28.0 ft (8.534 m) was successfully modeled as evidenced by the fact that the node at that location did not move even after an additional 13.0 ft (3.962 m) excavation transpired after strut installation (i.e., no change from load step 5 to load step 6). A maximum computed deflection into the excavation of nearly 1.4 inches (35.56 mm) occurred at load step 6. The base of the wall moves into the excavation approximately 0.4 inches (10.16 mm) while the top, after an initial positive movement, is shoved back to its original position after the loading is completed.

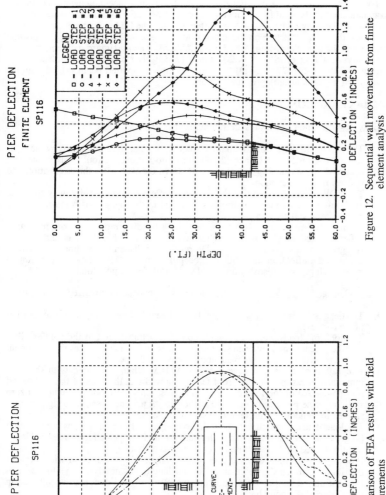

Figure 12. Sequential wall movements from finite element analysis

Figure 11. Comparison of FEA results with field measurements

Total vertical stresses are calculated at the midpoints of adjacent elements and may be extrapolated out to the wall face. The total stresses from the FEA analysis were resolved into axial and bending components for load step 6. Since the FEA output is for plane strain conditions, the moment acting on the actual soldier pile is the product of the plane strain moment and the tieback spacing, i.e., 5.5 ft (1.676 m). The maximum moment, from the FEA analysis, acting on SP116 is 202 kip-ft (273.9 kN-m) compared with a design moment of 205 kip-ft (277.9 kN-m) based on a simply supported beam loaded with the conventional trapezoidal earth pressure diagram.

SUMMARY AND CONCLUSIONS

EDR Results. The EDR results are summarized as follows:

- The maximum deflection for all three soldier piles that were monitored by the EDR was on the order of 1.0 inch (25.40 mm) into the excavation. This movement occurred approximately 2/3 of the way down the wall after the excavation had reached its final elevation.
- The EDR results showed that the pile tops deflected into the soil (away from the excavation). This may possibly be attributed to the design loads exceeding the actual lateral soil pressure.
- A finite difference approach for establishing bending moments from measured deflections did not yield reasonable results despite "smoothing" of the field curves.
- A regression analysis done on the EDR results established the relationship between depth and deflection as a 6th degree polynomial. The commonly assumed trapezoidal pressure distribution would yield a 5th degree polynomial.

Strain Gage and Tieback Test Results. These results may be summarized as follows:

- The tieback loads are attenuated rapidly in a linear fashion along the anchor portion of the tieback.
- The lock off loads were not always completely transferred to the soldier piles. Despite precautions to prevent such occurrences, in several instances the anchor nut moved relative to the anchor plate upon removal of the jack load.
- Integration of the strain gage data for A116M when compared with the total movement measured during proof testing showed that the back of the anchor did not move. All of the load distribution curves, when extrapolated indicated zero strain at some point in front of the anchor tip. It may be possible to significantly shorten the tendon lengths on future similar projects.
- The load distribution curves show the loads generally decreasing as construction progressed. Anchor creep, although relatively slight, appeared to be reasonable for the diminished loads as opposed to construction activity.
- Proof and performance test results when analyzed in conjunction with the load distribution curves showed that the anchor length was nearly always longer than that estimated in the field and that the PVC pipe did not always function satisfactorily as a bond breaker.

Finite Element Results. The FEA results provided good correlation with the measured field results and may be summarized as follows:

- After adjusting the wall deflection output to be relative to the bottom of the wall (to be consistent with the EDR results) the FEA displacements are nearly identical to the EDR measurements.

- The FEA displacement results mirror the load steps, with the wall alternatively moving into and away from the excavation as the excavation and tieback preloading occur respectively.
- The maximum bending moment from the FEA analysis was in good agreement with a design moment based on a simply supported beam loaded with a conventional trapezoidal distribution.
- The absolute movement from the FEA indicates a maximum deflection of nearly 1.4 inches (35.56 mm) into the excavation. The base of the wall is displaced approximately 0.4 inches (10.16 mm) into the excavation.

REFERENCES

1. J. A. Caliendo, A Field Study of a Tieback Excavation with a Finite Element Model Analysis, PhD Dissertation, Utah State Univ., Logan, UT, 1986, 321 p.
2. G. W. Clough & J. M. Duncan, Finite Element Analysis of Port Allen and Old River Locks, Rpt S-69-3, U.S. Army Eng Wtrwys Exp Sta, Corps of Engineers, Vicksburg MS, 1969.
3. J. M. Duncan, P. Byrne, K. Wong, & P. Mabry, Strength, Stress-Strain and Bulk Modulus Parameters for Finite Element Analysis of Stresses and Movements in Soil Masses, Rpt VCB/GT/80-81, Univ of Calif, Berkeley, CA, 1980, 70 p.
4. L. A. Hansen, Prediction of the Behavior of Braced Excavation in Anisotropoic Clay, PhD Dissertation, Stanford Univ., Stanford, CA, 1980, 439 p.
5. Post Tensioning Institute, Recommendations for Prestressed Rock and Soil Anchors, Phoenix, AZ, 1980.
6. H. Schnabel Jr., Procedures for Testing Earth Tiebacks, Meeting preprint 2278, ASCE National Structures Engineering Meeting, Cincinnati, OH, April 1974, pp. 1-28.
7. H. Schnabel Jr., Tiebacks in Foundation Engineering and Construction, McGraw-Hill, New York, 1982, 170 p.
8. D. R. Shields, H. Schnabel Jr., & D. Weatherby, Load Transfer in Pressure Injected Anchors, ASCE Geotechnical Journal (GT9), pp. 1183-1196.
9. Y. Tsui, A Fundamental Study of Tied-Back Wall Behavior, PhD Dissertation, Duke University, Durham NC, 1974, 258 p.
10. D. E. Weatherby, Tiebacks, FHWA/RD-82/047, Federal Highway Administration, Washington, D.C., 1982.

PACIFIC FIRST CENTER
PERFORMANCE OF THE TIEBACK SHORING WALL

David G. Winter[1], A.M. ASCE

ABSTRACT: An excavation seventy-five feet (23 meters) deep in downtown Seattle was monitored using load cells on all tieback anchors of two soldier piles, optical survey points on each soldier pile and on adjacent city streets, and slope inclinometers on four soldier piles. Data from the monitoring program are reported and conclusions are drawn. Recommendations for design modifications are made and areas of additional study are suggested.

THE PACIFIC FIRST CENTER DEVELOPMENT

In May 1989 the 44-story Pacific First Center was opened approximately one year after the bottom-of-hole was reached for one of the largest excavations ever in Seattle, Washington. The seven levels of underground parking required excavation of more than 150,000 cubic yards (115,000 cubic meters) of soil. The building site is surrounded by Seattle city streets except in the southwest corner where an existing 10-story building (the Logan Building) juts into the site. Traffic flow continued uninterrupted throughout the excavation and construction process.

Excavation depths varied from 66 to 77 feet (20 to 24 m) below Fifth Avenue and Pike Street on the west and north, respectively, to 77 to 82 feet (24 to 25 m) below Sixth Avenue and Union Street on the east and south, respectively. Because of space constraints and building surcharge loads, the shoring conditions adjacent to the Logan Building were significantly different from other areas of the site. That portion of the excavation shoring is not addressed in this paper.

SOIL CONDITIONS

Eight deep hollow-stem auger borings were drilled to explore the site, at the locations shown on Figure 1.

The subsurface conditions at the site were typical of the general conditions across downtown Seattle and much of the Puget

---

[1]Associate, Hart Crowser, Inc., 1910 Fairview Avenue East, Seattle, WA, 98102.

Sound area. Primary support soils are glacial in origin and at
this site consisted of a lacustrine clay. The clay is heavily
overconsolidated with Standard Penetration Test blow counts at
this site in the upper 20's to 40's, and a compressive strength
of about 12,000 to 15,000 psf (575 to 718 kPa) as measured by
both unconfined compression and unconsolidated undrained
triaxial compression. The clay is moderately to highly plastic,
with liquid limits ranging from 46 to 72, and plasticity index
from 23 to 43.

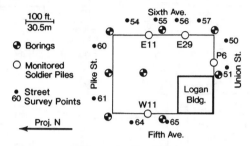

Figure 1. Site Plan

The clay was overlain by clean to silty sand ranging from
about 10 to 30 feet (3 to 9 m) thick. Average SPT blowcounts in
the sand were 45, with an estimated angle of internal friction
of 38 degrees. Loose fill with scattered debris was present
near the ground surface to a depth 10 to 15 feet (3 to 4.5 m).

Typically, groundwater in downtown Seattle is encountered as
perched water above the clay unit, and in water-bearing zones
below the clay. Perched water is usually depleted rapidly when
encountered in excavations. At this site, however, substantial
and persistent groundwater seepage occurred along the Sixth
Avenue wall. Whether the water was naturally occurring or was
being artificially recharged was never conclusively determined.
Whatever the source, the water, coupled with the great thickness
of sand on the east side of the excavation, resulted in running
ground conditions during installation of the shoring system
components and the excavation itself. Ultimately a temporary
dewatering system was installed through the shoring wall. The
system remained until the permanent wall drainage system was
activated.

Subsurface conditions are illustrated on Figures 2 through
5. Given elevations are relative to zero (0) at the corner of
Fifth and Pike. The profiles shown are based on the logs of the
soldier piles. This construction mapping confirmed the
conditions suggested by the design borings.

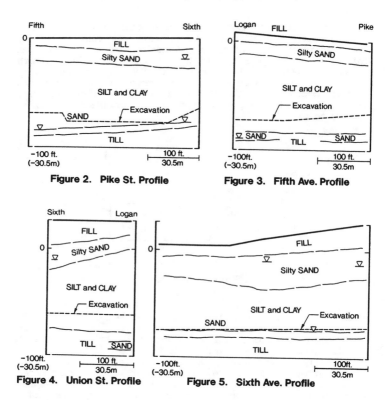

Figure 2. Pike St. Profile

Figure 3. Fifth Ave. Profile

Figure 4. Union St. Profile

Figure 5. Sixth Ave. Profile

SHORING SYSTEM: COMPONENTS, DESIGN, MONITORING

Structural Components. Adjacent to the city streets, 101 soldier piles were installed in 3-1/2 to 4 feet (1 to 1.2 m) diameter predrilled holes on a center-to-center spacing of about 10 feet (3 m). Each pile consisted of dual channel or wide-flange sections extending about 25 feet (7.5 m) below the base of the excavation. This deep penetration was required to develop sufficient resistance to the vertical load component of the tiebacks. Each soldier pile hole was backfilled with structural concrete up to the excavation level and with lean mix concrete to the ground surface. Wide flange sections ranged in size from W14x38 to W14x74. Channel sections were C15x50.

Tieback anchors were drilled at the center of the soldier piles, between the dual steel members, at a vertical spacing of about 5 feet (1.5 m). The tight vertical spacing resulted primarily from high loads and street and adjacent property right-of-way restrictions. The anchors were typically 16 inches (406 mm) diameter and were inclined 20 degrees below horizontal. Each anchor consisted of multiple strands with spacers, and structural grout bonding to the soil. The area of tieback hole between anchor and excavation, the no-load zone, was filled with non-bonding slurry. The anchors were installed using a continuous-flight hollow-stem auger (with grouting done through the auger under a gravity head) or, in limited cases, a small diameter rotary drill using an air return system for the cuttings. The small diameter anchors were pressure grouted in two stages to pressures of about 300 psi (207 kPa). Nearly 1,300 anchors were required to support the walls surrounding Pacific First Center.

Timber lagging 6 inches (150 mm) thick was installed between each pair of soldier piles, and any spaces between the lagging and soil were backfilled with non-cohesive slurry.

Geotechnical Design. Estimations of the lateral soil pressure envelope on deep excavation shoring walls in Seattle have been developed using a combination of the theoretical relationships set forth by Terzaghi and Peck (1), Schnabel (2), and others (3,4) and local experience. Because the theoretical relationships were developed from instrumented excavations in normally consolidated soils, their direct applications may not be appropriate for overconsolidated (OC) glacial silt, clay, sand or till. Cementation of the OC granular soil and the high cohesion of the OC fine grained soil cause these materials to behave differently from normally consolidated sand, silt, and clay.

Ideally, each excavation retention system would be instrumented to provide strut or tieback loads, with those data available to the designers of subsequent walls. In reality, relatively few excavations are so instrumented, and the data are often unpublished. The only information available on many shoring walls is the design pressure envelope and the report that the walls performed satisfactorily. A "satisfactory" performance may, however, indicate simply that the shoring system was overdesigned. As a result of this lack of an applicable theoretical basis and non-specific performance data, design pressures on shoring walls in the Seattle OC soils have changed little in the past twenty years. During that period, more than a dozen deep excavations have been made without a single pressure-induced failure.

Expressed in terms of the excavation height, H (in feet), maximum wall pressures have typically ranged from 25H psf to 40H psf (3.94H to 6.30H kPa, with H in meters)) for Seattle excavations primarily in silt and clay, and 22H to 30H psf (3.47H to 4.73H kPa) for excavations primarily in sand and till. Variations depend on the density or consistency of the soil units, groundwater conditions, location of the excavation with respect to other buildings and utilities, construction conditions, and on the individual engineers and designers.

The design maximum soil pressure for the Pacific First Center wall was 30H psf (4.73H kPa) and was estimated based both on past shoring walls, and on the theoretical methods. A trapezoidal distribution shape was used with the pressure envelope increasing linearly from zero at the top of the wall to the maximum over the upper 20 percent of the wall, and decreasing linearly over the lower 20 percent of the wall back to zero at the base of the excavation.

Other design values for the shoring components were based on substantial local test results and included 40,000 psf (1,920 kPa) allowable unit bearing on the bottom of the soldier pile, 1,000 psf (48 kPa) allowable unit friction on the portion of the soldier pile below the excavation base, 500 psf (24 kPa) allowable unit friction for tiebacks through the upper medium dense to dense sand and 1,000 psf (48 kPa) through the overconsolidated silt and clay. The lagging was designed for a maximum of 30 percent of the wall pressure. The soldier piles, lagging, tieback locations, and required loads were designed by the structural engineer. The tieback anchor type, size, and length were determined by the shoring subcontractor.

Shoring Monitoring Program. The monitoring program was designed to provide an early warning if the shoring system was not performing as planned, and to give actual data on system deflections and loads. All components except the load cells and slope inclinometers are routinely required by the City of Seattle.

Each soldier pile and tieback installation was continuously logged by the geotechnical engineer's representative. Each tieback anchor was tested to 130 percent of the design load. Nine anchors were tested to 200 percent of the design unit friction. Acceptance criteria for all anchors included load resistance, total movement, and creep resistance.

Hydraulic load cells were attached to 13 tiebacks on soldier pile W11 (along Fifth Avenue) and to 14 tiebacks on pile E11 (along Sixth Avenue). The cells had a doughnut shaped sensor which fit around the tieback strands behind the lockdown nut. Steel plates on each side distributed the load evenly. Pressure was recorded on a dial gauge positioned so that it could be read from the base of the excavation through binoculars. The load cells were accurate to about plus or minus 2,000 pounds (907 kg).

To test the hypothesis that the actual wall pressures would result in loads smaller than the design values, each instrumented tieback was prestressed and locked off to about 50 percent of the design load compared to the typical lock-off load of 100 percent of design. To create a test section of wall all tiebacks on soldier piles W9, W10, W12, and W13, and E9, E10, E12, and E13 were also locked off to 50 percent of design. The results were two 50-foot (15.2 m) long sections of wall prestressed to 50 percent of the design pressure. The intent of the lower lock-off was to create a condition whereby the tiebacks would be required to resist (and the load cells record) actual loads, not just the prestressed loads. If the actual wall loads ever exceeded the 50 percent lock-off, the load cells would record the increase. Without the lower lock-off, the

actual loads would have to be higher than the design values to
register on the load cells.

Soldier piles W11 and E11, E29 (Sixth Avenue) and two other
piles, and P6 (Union Street) were fitted with slope inclinometer
casing. The casings were attached to the backs of the soldier
piles using U-bolts and were therefore grouted securely into
place. Both the load cell readings and the inclinometer data
were collected by the geotechnical engineer.

An independent surveyor established monitoring points on the
top of each soldier pile, on the surface of surrounding streets,
and on buildings across the streets from the excavation. The
surveyor recorded both vertical and horizontal deformation of
each monitored point.

RESULTS FROM INSTRUMENTATION AND MONITORING

Load Cells. Figures 6 and 7 illustrate the progressive
changes in load cell readings from the lock-off load (dashed
line) as the excavation was deepened. From the time of lock-off
until the excavation was about 70 percent of the total depth,
the cell loads generally remained constant at the lock-off load
or decreased slightly. (The actual lock-off load averaged about
45 percent of the design load.) After that the loads increased
steadily until the excavation was complete, and generally
remained constant for the next month (end of readings). This
observation suggests that the lock-off loads were greater than
the actual wall pressures until the excavation was about 70
percent of the total depth. If it is assumed that the loads
then increased linearly with increasing excavation depth, the
final load (at the base of the excavation) would be about 65
percent of design, or about 20H psf (3.15H kPa).

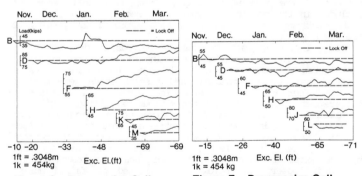

Figure 6. Progressive Cell
Loads – W11

Figure 7. Progressive Cell
Loads – E11

On E11, only anchors A and B decreased from the lock-off
value, while on W11, A, B and G decreased. However, both A
anchors were inadvertently locked off near the design load, and

the most pronounced reduction in load occurred soon after
lock-off.   After that, the anchor loads did not change
significantly. The B anchors show a similar pattern. Here it
may be concluded that the anchor loads initially decreased (by
anchor creep, slippage of the lockdown nuts, stress relaxation
in the tieback strands, soldier pile deflection), then held
steady at some equilibrium value. Because the loads did not
then subsequently increase, the data indicate the actual earth
pressures on the walls were smaller than the prestress anchor
loads. Many of the anchor loads, in fact, decreased initially,
before increasing in response to the increasing earth pressures.

A second way to view the load cell data is to plot the
maximum observed load at each tieback in terms of both actual
load and percent of the design load. Such plots are presented
on Figures 8 and 9. Note that Figure 8 presents actual recorded
and design loads, while Figure 9 presents the design pressure
envelope, and recorded loads plotted as a percentage of the
design values. The plots show that the highest recorded load is
the same as or only slightly higher than the last recorded load,
about one month after the bottom of the excavation is reached.
At W11, a pressure distribution of 19H psf (3.00H kPa), or just
under 65 percent of the design value, would envelop all observed
loads. At E11 a pressure distribution of 22H psf (3.47H kPa),
or just under 75 percent of design, would envelop all observed
loads, and 18H psf (2.84H kPa) (60 percent) would envelop all
but one observation. (That one observation was initially locked
off at about 64 percent of design.)

The load cells generally performed as intended. The readout
gauges for the cells on W11I and E11M were accidently destroyed
by construction equipment soon after installation.   Because
tiebacks W11A and E11A were inadvertently locked off near 100
percent of the design value, data from these cells were not
included in the summary analyses. It seems possible that the
high prestress load in row A may have affected the results of
the row B cells as well.

Slope Inclinometers. Progressive records of deflection for
the four soldier piles fitted with inclinometers are presented
on Figures 10 through 13.   Although the design pressure
distribution, tieback spacing, and tieback loads were similar
for all four piles, the deflection curves have significant
differences. Figure 10 shows the maximum deflection of pile W11
is about 0.7 inches (18 mm). The shape of the curve indicates
1) excessive loads have been prestressed into the upper four
anchors (as previously noted, W11A was locked off at 100 percent
of design), or 2) the pile has been bent or pushed into the soil
by some external event, or 3) the soil near the ground surface
was much looser than below and deflected or failed during
tieback stressing. Note also on W11 that the lower portion of
the soldier pile bulged toward the excavation significantly
between the 1/14/88 and 2/16/88 readings (as the excavation
progressed from the G to the L levels). By early March, when
the bottom was reached, the bulge had regressed to the shape
indicated by the 3/31/88 reading. While bulging is not unusual

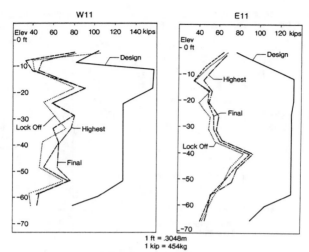

1 ft = .3048m
1 kip = 454kg

Figure 8. Recorded Cell Loads

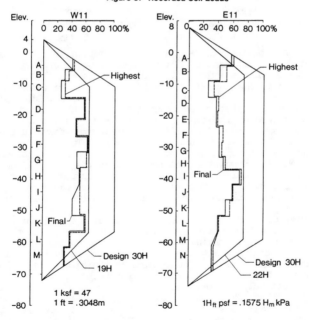

1 ksf = 47
1 ft = .3048m

$1H_{ft}$ psf = .1575 H_m kPa

Figure 9. Wall Pressure Envelopes

(since the bottom of the soldier pile is fixed) there does not seem to be a clear reason for the regression of the bulge with time.

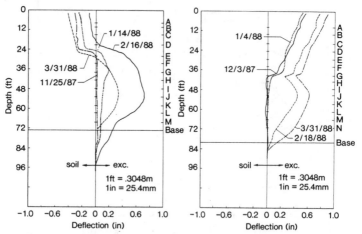

Figure 10. Soldier Pile W11 Inclinometer

Figure 11. Soldier Pile E11 Inclinometer

The shape of the E11 (Figure 11) deformation is interesting in that the predominant soil above level G was sand and below level G was silt or clay. The bulge near levels I, J, and K corresponds to the highest recorded load cell loads (see Figure 9), with the shape a partial function of the fixed soldier pile end. Note also that even though E11A was locked off to 100 percent of design, the top of the pile deflected significantly into the excavation. The difference in shape between E11 and W11 is in the upper portion of the curve. The shape of E11 suggests that the upper sand exerted more pressure on the wall than in W11 (an observation not supported by the load cell data) or that the tiebacks moved more to mobilize the necessary friction resistance. Either of these could have been the result of the groundwater and running ground condition along Sixth Avenue.

Soldier pile E29, Figure 12, deflected in a shape similar to W11. The groundwater conditions may also have been a factor at this location. If, during soldier pile installation, the soil had been excessively disturbed or if any ground had been lost into the drilled hole, the deflections of the pile under tieback loading would have increased in the negative direction.

The relatively uniform deflection of pile P6, Figure 13, indicates that both the applied pressures and tieback resistances were also uniform. In both E29 and P6, the deflections are smaller than E11 and W11, a result of the higher lock-off loads. The maximum deflection of less than 1/2 inch

(13 mm) suggests that lower tieback prestress loads (from lower
design wall pressures) are tolerable even though they may result
in correspondingly higher deflections.

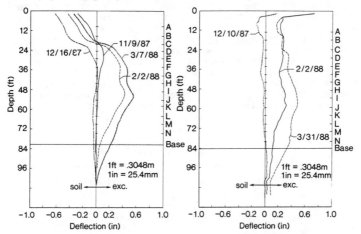

Figure 12. Soldier Pile
P6 Inclinometer

Figure 13. Soldier Pile
E29 Inclinometer

Finally it is interesting to note that the general shape of
the deflection curve is established early.   Subsequent curves
move gradually inward, but maintain the predetermined shape.
Progressive movement coupled with a cell load increase indicates
soil is moving with respect to the tieback. Movement without a
load increase indicates the wall, soil, and tieback are moving
as a unit.   In W11 the wall is apparently moving into the
excavation in early February and back into the soil in late
February and March.   The tieback loads on Figure 6 are
increasing steadily in February and more slowly in March. These
observations do not seem congruent.   On E11, however, the pile
deflects progressively into the excavation while small increases
in load are recorded by the load cells, as expected.
    Optical Survey Monitoring - Soldier Piles. Survey monitoring
of the deflection of soldier piles at the ground surface is
interesting but not usually conclusive.   In the ideal case the
maximum deflection would occur at the ground surface and could
provide an early indication of excessive deflection caused by
tieback or soldier pile failure.   In reality, the near surface
deflections are often the result of near surface occurrences
such as overdesigned tiebacks, loosely compacted fill behind the
wall, or surcharges from equipment or materials.
    The observed pattern of progressive deflection was for the
top to move into the soil as the upper tiebacks were drilled and
stressed, then move into the excavation as the lower ties were
stressed. Typical deflections were less than 1/4 to 1/2 inch (6

to 13 mm). Shoring system deflections as great as 1 inch (25 mm) or more are not uncommon, and do not usually cause concern about the stability or adequacy of the design. It may be concluded from the observed deflections that the actual pressures applied to the shoring wall were much smaller than the design values.

Optical Survey Monitoring - Streets. Four monitoring points were set on Sixth Avenue, about 29 feet (8.8 m) from the shoring wall. Two were set on each of the other streets: 40 feet (12 m) from the wall on Pike Street, 16 feet (4.9 m) from the wall on Union Street, and 26 feet (7.9 m) from the wall on Fifth Avenue. The locations of the street monitoring points are shown on Figure 1. The observations are summarized in the table below.

Table 1. Summary of Street Deflections

Location	Point	Horizontal Deflection (feet) (minus sign indicates movement toward excavation)			Vertical Settlement (feet) (minus sign indicates settlement)			
		11/87	2/88	5/88	11/87	12/87	2/88	5/88
Pike E	60	.005	-.012	-.035	.005	.005	.004	-.014
W	61	-.005	-.018	-.037	-.020	-.010	-.024	-.039
Sixth N	54	-.036	-.042	-.058	-.027	-.011	-.012	-.031
NC	55	-.046	-.043	-.069	-.013	-.106	-.123	-.149
SC	56	-.041	-.045	-.073	-.027	-.141	-.188	-.204
S	57	-.041	-.043	-.045	-.035	-.127	-.169	-.191
Union E	50	0	-.005	-.008	-.008	-.004	-.022	-.030
EC	51	-.010	-.019	-.023	-.040	.014	-.006	-.016
Fifth N	64	.005	-.003	-.013	.002	.010	.010	-.003
NC	65	.010	-.005	-.031	-.005	.003	-.004	-.015

Except for Sixth Avenue, the final horizontal deflections of the street points were about the same as the closest soldier pile deflection, possibly indicating the soil mass was tied together by the shoring system for some distance behind the wall. There was no apparent relationship between the street settlement and the wall deflections. For example, on Pike Street, point 61 settled nearly three times as much as point 60, even though the shoring wall deflected more in front of point 60 than in front of point 61. Since the observations were made from autumn to nearly summer, it is also possible that the observed street movements were more a result of normal seasonal variation than a function of the shoring wall movements.

Conditions and observed deflections were clearly different along Sixth Avenue. Horizontal deflection began to occur in

October, before any tiebacks were installed. Nearly two-thirds of the total horizontal deflection occurred before the first of November. Settlement, on the other hand, began in early November, with two-thirds of the total complete by early December. Soldier pile installation in this area occurred between September 9 and October 19. The first tiebacks in the area were installed and stressed in early to mid-November.

One might conclude that the soldier pile and tieback drilling resulted in some loss of ground, possibly leading to the observed ground surface movements. The pattern of settlement also suggests that the simple deflection of the shoring wall may not have caused the street settlement. Point 54 settled much less than points 55, 56, and 57, even though the wall deflection in front of point 54 was the most of the Sixth Avenue wall. Point 56 settled the most, even though the deflection of the wall in front was the least. Settlement of Sixth Avenue was probably tied most closely to the running ground condition in the upper half of the excavation, and the subsequent dewatering program used to stabilize the conditions.

APPLICATION OF OBSERVATIONS

Ultimately, developers look to their consultants to design a safe but cost-effective shoring wall. A "safe" design may simply duplicate a previous successful design. To be cost-effective, however, the design should be modified, based on lessons learned from observations and instrumentation.

Since 1982, three other major excavations in Seattle (5) (6) (7) have included load cells on the tiebacks of one or more soldier piles. These three excavations, together with Pacific First Center, are the four deepest in Seattle during the 1980s. Each of these excavations was through heavily overconsolidated clay, till, and sand. Combining the data from these projects with additional published data from the 1970s in Seattle (8) and Boston (9) creates a compelling case for significant reductions in the "typical" shoring wall design pressures in these overconsolidated soils.

While none of these project reports claim to have identified the actual wall pressures, the data agree that shoring walls in heavily overconsolidated glacial soils will be overdesigned based on the classical design approaches of Terzaghi and Peck and others. The problem is not in the approaches themselves, but rather in the application of empirical approaches to dissimilar soil conditions. The data from the referenced projects suggest that the actual pressures are about one-third lower than those predicted using the classical approaches.

REVIEW OF OBSERVATIONS AND CONCLUSIONS

1) The Pacific First Center excavation was through very stiff to hard, overconsolidated silt and clay overlain by medium dense to dense sand.

2) Design pressures on the shoring wall were estimated using theoretical approaches tempered by recent experience in the Seattle area.

3) Load cells recorded actual tieback loads ranging from 35 to 70 percent of the design value of 30H psf (4.73H kPa). The recorded loads typically increased slowly from the lock-off load as the excavation deepened.

4) Slope inclinometer curves were different for each pile, allowing no consistent conclusion on the deflection characteristics of the wall except that the maximum deflections were lower than expected, suggesting that the tieback prestress loads were higher than necessary to limit deflections to acceptable levels.

5) Wall deflections at the ground surface were typically about 1/4 inch (6 mm), supporting the hypothesis that the wall was designed for pressures higher than actually occurred.

6) Significant settlement of Sixth Avenue occurred during initial installation of tiebacks and continued throughout the excavation. Loss of ground during drilling, running ground conditions, and construction dewatering seem to be the likely cause. The settlement did not seem to be related to the shoring wall deflection.

7) The horizontal movement of some areas of the street in concert with the shoring wall suggest that the wall and retained soil deflected as a single mass. The tight vertical spacing of the tiebacks may have caused the wall to behave similar to a nailed (or gravity) wall.

8) A shoring wall designed based on a pressure distribution of 20H psf (3.16H kPa) would have enveloped the actual observed pressures in the test sections.

9) Data from other instrumented excavations support the findings at Pacific First Center, and should be considered in the design of subsequent shoring walls.

RECOMMENDATIONS FOR FUTURE DESIGN AND STUDY

1) Design future walls in these types of soils for 1/4 to 1/3 lower pressures than in the past. Lower pressures will result in thinner lagging, lighter soldier piles, fewer, smaller and/or shorter anchor sections, and a less costly shoring system.

2) Until more confirming data are obtained it may be practical to design tiebacks and lagging for the full recommended reduction, but soldier piles for a partial reduction (to allow easier installation of additional tiebacks, if conditions are not as expected).

3) On future instrumented sites it would be useful to isolate a much larger section of wall (50 percent lock-off) and instrument more piles. From a practical standpoint, instrumentation is expensive and often not viewed by the owner as an asset or requirement to a successful excavation.

4) Additional study should be given to the potential three-dimensional effects of wall pressures near excavation corners. A reduction in the pressures near corners would result in an additional cost savings to the shoring system.

5) The deflection of soldier piles should be studied further, considering the effect of the actual wall pressure distribution, the prestress loads, and the structural behavior of the soldier pile and tieback restraint.

ACKNOWLEDGEMENTS

The author wishes to thank the structural engineer project manager, Serge Rudchenko of Skilling, Ward, Magnusson, Barkshire of Seattle for his comments and review. Pacific First Center was developed by Prescott and designed by The Callison Partnership. We thank them for funding the instrumentation program. Significant credit for the successful performance of the shoring wall is due the shoring subcontractor, DBM of Federal Way, Washington, and the general contractor, Sellen Construction of Seattle.

REFERENCES

1. K. Terzaghi and R.B. Peck, Soil Mechanics in Engineering Practice (2nd Ed.), John Wiley and Sons, New York, 1967, 729 p.

2. H. Schnabel, Jr., Tiebacks in Foundation Engineering and Construction, McGraw-Hill, New York, 1982, 170 p.

3. G. P. Tschebotarioff, Foundations, Retaining and Earth Structures (2nd Ed.), McGraw-Hill, New York, 1973, 642 p.

4. NAVFAC, Foundations and Earth Structures, Design Manual 7.2, Dept. of the Navy, Alexandria, VA, 1982, 244 p.

5. D. Winter, A. Macnab, and J. Turner, Seattle Shored Excavation - Design, Construction, Monitoring, Proceedings, 12th Annual Member's Meeting and Seminar, Deep Foundations Institute, Hamilton, Ontario, 1987.

6. W. P. Grant, G. Yamane, and R. P. Miller, Design and Performance of Columbia Center Shoring Wall, Seattle, Washington, Proc., Intl. Conf. on Tall Bldgs., Singapore, 1984, p. 651-661.

7. Hart Crowser & Assoc., Unpublished data from First Interstate Center, 1982.

8. G. W. Clough, P. R. Weber, and J. Lamont, Jr., Design and Observation of a Tied-Back Wall, Proc., Conf. on Perf. of Earth and Earth-Supported Struc., Purdue University, 1972, (Vol. 1) p. 1367-1389.

9. T. K. Liu and J. P. Dugan, Jr., An Instrumented Tied-Back Deep Excavation, Proc. Conf. on Perf. of Earth and Earth-Supported Struc., Purdue University, 1972, (Vol. 1) p. 1321-1339.

TEMPORARY TIEBACK WALL, BONNEVILLE NAVIGATION LOCK

By Dale F. Munger,[1] A.M.ASCE, Patrick T. Jones,[2]
and Joseph Johnson[3]

ABSTRACT: During construction of the second navigation lock at
Bonneville Lock and Dam, a 50-foot-high (15.2m) temporary tieback
reinforced   concrete   diaphragm   wall   was   built   to   retain   the
foundation of the Union Pacific Railroad.   Wall deflections were a
great concern, therefore, a soil structure interaction beam-column
analysis was performed to predict wall movements.   The wall was
heavily instrumented to monitor deflections, moments, and tendon
loads. Parameter studies were conducted to calibrate the calculated
deflections and moments with the measured performance of the wall.
The   initial   predictions   of   the   deflection   and   moments   were
reasonably close to the measured values.   Increasing the subgrade
modulus brought the calculated deflections closer to the measured
deflection, but did not significantly change the magnitude of the
calculated moments.

INTRODUCTION

A   temporary   tieback   wall   was   instrumented   to   monitor   its
performance during construction of the Bonneville Second Navigation
Lock and to provide data to guide the design of future tieback walls
at the project.   Soil-structure interaction beam-column analyses
using non-linear Winkler spring supports were conducted to predict
the performance of the temporary wall.   After construction, a
parametric study of the model was conducted to match the model's
performance with the observed behavior.

1 - Geotechnical Engineer, U.S. Army Corps of Engineers, North
Pacific Division, ATTN: CENPD-EN-G, P.O. Box 2870, Portland, OR
97208-2870

2 - Geotechnical Engineer, U.S. Army Corps of Engineers, Portland
District,   ATTN: CENPP-EN-GS,   P.O.   Box   2946,   Portland,   OR
97208-2946

3 - Structural Engineer, U.S. Army Corps of Engineers, Portland
District,   ATTN: CENPP-EN-DC,   P.O.   Box   2946,   Portland,   OR
97208-2946

PROJECT DESCRIPTION AND LAYOUT

Bonneville Lock and Dam is located on the Columbia River 42 miles east of Portland, Oregon (Figure 1). The new navigation lock is being constructed between the existing lock on the north and the Union Pacific Railroad's transcontinental rail line on the south (Figure 2). Because there is no room for normal open-cut excavation slopes, 180 lineal feet (54.9m) of temporary tieback wall and 970 lineal feet (295.7m) of permanent tieback walls have been or will be constructed for the new navigation lock. The total wall areas are 3600 square feet (334.5 square meters) and 108,000 square feet (10,034 square meters), respectively.

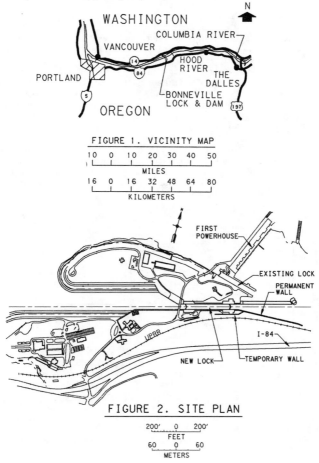

FIGURE 1. VICINITY MAP

FIGURE 2. SITE PLAN

The temporary tieback wall was completed in February 1988 as part of the lock chamber excavation contract. The lock chamber was excavated into a diabase intrusive, which extends up and downstream just enough to accommodate the main lock structure. At the upstream end of the lock chamber, the top of rock drops off rapidly, requiring installation of a temporary tieback wall to retain the soil foundation of the adjacent railroad. The maximum height of temporary tieback wall exposed during the initial lock chamber excavation contract was 48 feet (14.6m) and the exposed height of wall in future contracts will exceed 100 feet (30.48m). The railroad is only 45 to 50 (13.7 to 15.2m) feet away from the edge of the excavation.

SITE CHARACTERIZATION

Extensive exploration, sampling, and testing were performed to characterize the site. Site geology is a result of massive ancient landslides, regional volcanics, and alluvial deposits from large floods. The upper zone of material to be retained by the temporary and permanent tieback walls is Reworked Slide Debris (RSD). RSD is a heterogeneous mixture of alluvial silts, sands, gravels, cobbles and boulders, mixed with angular rock fragments of igneous and sedimentary origin from old landslide masses. Underlying the RSD is the Weigle Formation, which is composed of fine-grained, volcanic-derived, sedimentary rocks.

The Weigle formation consists of interbedded mudstones, sandstones, siltstones and claystones. In the area of the temporary wall the Weigle Formation is underlain by the Bonney Rock diabase intrusive. The diabase bedrock is a basalt-like rock with a columnar structure (Figure 3).

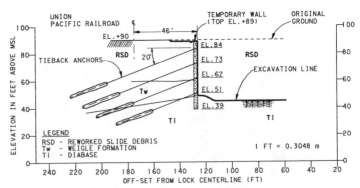

FIGURE 3. DESCRIPTIVE CROSS SECTION

The natural groundwater regime at the wall site is dominated by the Bonneville pool, with a smaller influence from hillside drainage. Dewatering was accomplished to top of rock behind the wall using wells.

MATERIAL PROPERTIES

The friction angle and cohesion values given in Table 1 are lower bound values for the RSD and Weigle Formation and are based on laboratory and in-situ testing. Samples were obtained by core and cable-tool drilling, and from test pits. Densities were obtained from core samples and block samples from the test pits. Strength data were obtained from direct and triaxial shear testing. The constant of horizontal subgrade reaction (ℓ_h) (1) was selected from values experimentally derived by Terzaghi for retaining walls embedded in soil. For the purposes of choosing ℓ_h the RSD and Weigle Formation materials were assumed to be cohesionless. The selected ℓ_h values were weighted toward the higher end of the Terzaghi values to reflect the relatively high unit weights of the materials. The RSD was assumed to behave as a medium-dense sand and the Weigle Formation was assumed to behave as a "soft" sedimentary rock. The value of ℓ_h changes with submergence for the RSD, because it is a soil, while the Weigle formation ℓ_h remains constant whether dry, moist or submerged, because it is essentially a weak sedimentary rock.

Table 1 Material Properties

Parameter	RSD	Weigle Formation
Friction angle (degrees)	30	30
Cohesion (psi)[a]	0	10
Moist Unit Weight (pcf)[b]	125[d]/118[e]	142.5[d]/130[e]
Saturated Unit Weight (pcf)	130	142.5
Constant of horizontal subgrade reaction ℓ_h		
Moist (pci)c	18	23
Submerged (pci)	13	23

[a] 1 pound (force) per square inch = 6.9 kilonewtons per square meter

[b] 1 pound per cubic foot (density) = 1.6 kilograms per cubic meter

[c] 1 pound (force) per cubic inch = 271.45 12 kilonewtons per cubic meter

[d] Unit weight used to develop earth pressure distribution for input into CFRAME or analysis

[e] Unit weight used in CBEAMC analysis, changed due to accumulation of additional in-place density data

TEST ANCHOR PROGRAM

The geometry of the geologic formations at the site dictated that most of the tieback anchors would have bonded zones in soil materials, with a few in rock (diabase). The proposed use of permanent soil anchors and the unknown strength capacity of these

formations were the reasons for conducting large scale anchor tests
at the site in the summer of 1986. The anchor tests confirmed that
high capacity soil anchors could be installed at the site using
"tremie" grouting methods. Tremie grouting is the pumping of the
grout mixture through a tube, which is installed with the tendon,
so the hole is filled with grout from the bottom up to the surface.
No additional pressure is used above that of the hydrostatic
pressure of the grout column. Tremie grouting was used for the test
anchors to obtain minimum baseline data for anchorage capabilities
of materials at this site. The allowable bond stresses obtained for
the RSD and Weigle Formations were 60 psi ($413.7 kN/m^2$) and 35 psi
($241.3 kN/m^2$) respectively. The tests showed that long-term creep
would not be a problem at the proposed design loads. The measured
creep rates were less than 0.04 inches (1.02 mm) per log cycle for
5 of the 6 test anchors. These rates were significantly less than
the allowable Post Tensioning Institute standard of 0.08 inches
(2.03 mm) per log cycle (2).

DESIGN CRITERIA

To prevent settlement of the railroad foundation and to eliminate
the possibility of initiating a landslide in this highly unstable
area, the maximum allowable horizontal wall deflection under normal
loads was set at a conservative 0.1 feet (30.5mm) riverward anywhere
along the height of the wall. A concrete diaphragm wall,
constructed by the slurry trench method, was chosen because of the
ability of these walls to resist deflections. The unbonded lengths
of the tendons were set so the bonded zones were beyond the design
failure surfaces. The design failure surfaces were determined using
slope stability limit equalibrium procedures, with a factor of
safety equal to 1.5 for static load conditions and 1.0 for
dewatering system failed and seismic load conditions. The tieback
forces were not considered in the analysis to determine the location
of the design failure surfaces. A factor of safety of 2.0 was
applied to the ultimate grout to soil bond stress to determine
required bonded lengths. The tendons were sized using ACI 318-83
with one exception to section 18.5.1(c), which was that the tendon
loads were not to exceed 75 percent, instead of the normal 80
percent, of the ultimate tensile strength during anchor testing.
Maximum performance and proof testing loads were 1.5 times the
design load. Reinforced concrete was designed using ACI 318-83.

SOIL LOAD

A combination of rectangular and triangular at-rest soil
loadings, plus a surcharge, were used for wall design (Figure 4).
A rectangular apparent earth pressure diagram for tieback (or
internally braced) walls in sand was used to account for the typical
load distribution that occurs in these walls. The at-rest
rectangular pressure distribution used was taken from Figure 29,
Chapter 3, NAVFAC DM-7.2 (3). A triangular diagram was used to
provide for the possibility that upper anchors could lose some
tension during the life of the wall, leading to greater loads at the

base of the wall.   The at-rest condition was used to limit
deformations to a minimum, because the railroad is just adjacent to
the top of the wall. The Jaky equation (4) was used to determine
a suitable value of K_o.

$$K_o = 1 - \sin \phi \tag{1}$$

where K_o = at-rest stress ratio and ϕ = friction angle.   The
at-rest triangular soil pressure was calculated using:

$$P_o = K_o \, \gamma_m \, H \tag{2}$$

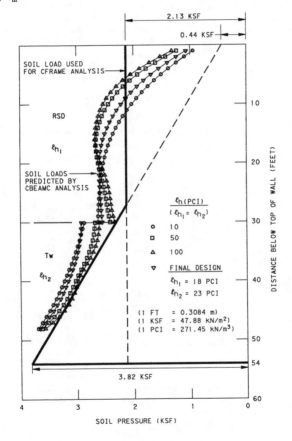

FIGURE 4. PARAMETRIC STUDY OF
SOIL PRESSURE

where γ_m = moist unit weight, H = wall height. The ordinate of the rectangular apparent earth pressure diagram was calculated using:

$$P_o = 0.5 \ K_o \ \gamma_m \ H + S \tag{3}$$

with S = surcharge load. Using elastic theory, the horizontal pressures were increased by 0.44 KSF because of surcharge loads from trains and construction equipment. The site was dewatered, thus no hydrostatic loads were considered. Panel 6 (Figure 5) was designed using a wall height of 54 feet (16.4m). Due to higher than anticipated bedrock the constructed height ended up being 48 feet (14.6m). The design earth pressure distribution was developed using the assumption that the material behind the wall consisted entirely of RSD. Later explorations and slurry trenching indicated the presence of Weigle formation beneath the RSD and above the diabase.

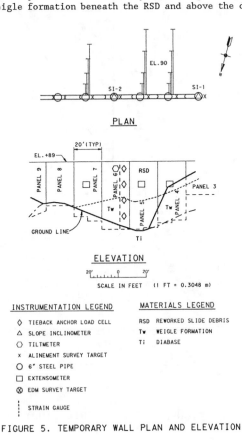

PLAN

ELEVATION

20' 0 20'
SCALE IN FEET   (1 FT = 0.3048 m)

INSTRUMENTATION LEGEND

◊ TIEBACK ANCHOR LOAD CELL
△ SLOPE INCLINOMETER
◯ TILTMETER
x ALINEMENT SURVEY TARGET
◯ 6" STEEL PIPE
☐ EXTENSOMETER
⊗ EDM SURVEY TARGET

⋮ STRAIN GAUGE

MATERIALS LEGEND

RSD REWORKED SLIDE DEBRIS
Tw   WEIGLE FORMATION
Ti   DIABASE

FIGURE 5. TEMPORARY WALL PLAN AND ELEVATION

WALL DESIGN

The wall reinforcement (Figure 6) was designed using "CFRAME" (5), to determine reactions, moments and shears. The moments were used to size the vertical reinforcement. CFRAME is a general-purpose computer program for analysis of small or medium plane frame structures. CFRAME has the capability of placing any load on the structure and can generate member shears, moments and deflections. The wall was modeled vertically as a continuous beam with supports approximately 11 feet apart at each anchor location. This beam was loaded with the earth pressure shown in Figure 4. A horizontal beam was designed to span across the anchor which was loaded with the reaction from the vertical beam. The reaction at each support was used to determine the uniform loading on assumed horizontal built-in beams. The horizontal beam was modeled as a 20-foot (6.1m)

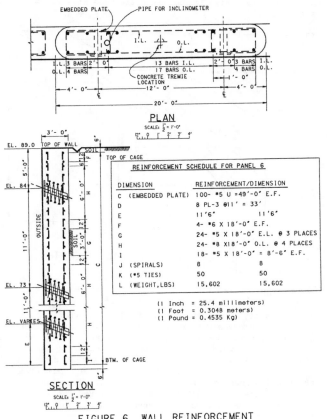

FIGURE 6. WALL REINFORCEMENT

beam with 4 foot (1.22m) overhangs over each support yielding
reactions, deflections, moments and shears. These moments were used
to size the horizontal reinforcement. The horizontal component of
the tieback design load was taken as the larger of 1) the reactions
from the horizontal beams or 2) the apparent soil pressure over the
tributary area of each anchor. There was very little difference
between the CFRAME reactions and the loads from the tributary area
method. Charts (Figure 7) were developed which show the unbonded
length of the tendon versus panel height for the various tendon
elevations on the panels. These charts were based on data derived
from the limit equilibrium slope stability analysis of three
different panel heights. Spencer's procedure as modeled in the
computer program UTEXAS2 (6) was used to perform the slope stability
analyses. The bonded zone was designed for 30 feet (9.14m) and the
unbonded lengths ranged from 27 feet (8.23m) to 83 feet (25.3m).
Tendon allowable design loads ranged from 235 kips (1,045kN) to 358
kips (1,592.5kN).

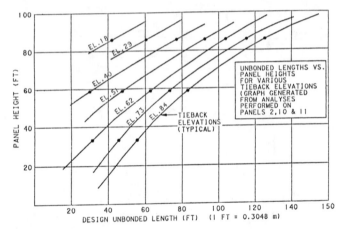

FIGURE 7. TENDON UNBONDED LENGTHS

ANCHOR DESIGN

High capacity Grade 270 KSI (1,861MN/m^2), A416 low relaxation
7-wire strand tendons were used. To increase competitiveness, the
strand diameter was not indicated in the plans. Only the design
load and criteria were specified, which allowed the contractor to
determine the number of strands using the strand diameter of his
preference. Hardware, jacking equipment and anchorages were
contractor designed to give the contractor the flexibility of using
methods and materials which he felt best suited the site. Tendons
were sized for the maximum load expected during construction which
normally was the test load of 1.5 times the design load.

INSTRUMENTATION

The wall was instrumented (Figure 5) to monitor performance during construction and throughout its service life. Instrumentation consisted of embedded inclinometers, strain gauges on the wall reinforcement steel, tiltmeters, horizontal multi-position bore hole extensometers, and tieback anchor load cells. The strain gauges were spaced every 5 feet (1.5 meters) on vertical reinforcing bars located at the inside and outside wall face.

PREDICTION OF PERFORMANCE

The performance of the temporary tieback wall was predicted using the computer program CBEAMC (7) which models soil-structure interaction using beam-column structures with one dimensional linear and non-linear spring supports, based on the Winkler hypothesis. This analysis method was selected because of the relative ease (compared to two dimensional finite element methods) with which the beam-column analysis can be made. Input soil parameters are shown in Table 1. Wall behavior was analyzed using a "one-step construction" model that did not take into consideration the construction sequence (Figure 8). Initial loads on the wall were at-rest soil loads with the tendons stressed to the design load.

TENDON ELEVATION	BONDED LENGTH (FT)	UNBONDED LENGTH (FT)	DESIGN TENDON LOAD (KIPS/LF)
84	30	74	28.1
73	30	64	28.1
62	30	53	28.1
51	30	37	35.8

STRUCTURAL PROPERTIES

CONCRETE MODULUS OF ELASTICITY = 499,590 KIPS/FT^2

CROSS SECTION MOMENT OF INERTIA = 2.25 FT^4

1 Ft   = 0.3048 m
1 KIP   = 4.448 kN
1 KIP/FT^2 = 47.88 kN /m^2

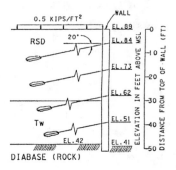

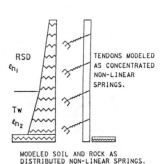

TENDONS MODELED AS CONCENTRATED NON-LINEAR SPRINGS.

MODELED SOIL AND ROCK AS DISTRIBUTED NON-LINEAR SPRINGS.

FIGURE 8. CBEAMC MODEL

Hand check procedures were used to determine moment and shear loads for the various construction loads.

Terzaghi (2) showed that deflections are sensitive to the value of ℓ_h because displacement is inversely proportional to ℓ_h and that moments were much less sensitive to ℓ_h because moments were proportional to the fourth root of ℓ_h. With this in mind, a parameteric study was done and the results showed, for even the most severe soil condition, the predicted deflection was acceptable (Figure 9).

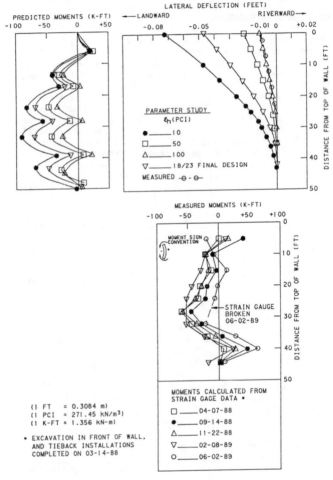

FIGURE 9. PREDICTED AND MEASURED PERFORMANCE

The stiffness of the soil springs (Figure 10a) used in the CBEAMC analysis is a function of Terzaghi's constant of horizontal subgrade reaction. Terzaghi (2) called this stiffness the coefficient of horizontal subgrade reaction (k_h) and Haliburton (8) called this soil parameter the soil modulus (E_s) shown in equation 4.

$$E_s = \ell_h \, (X/D) \tag{4}$$

where ℓ_h = constant of horizontal subgrade reaction, X = equivalent depth of material which will provide the actual effective overburden pressure acting at the elevation of interest (8), and D = contact area or effective contact dimension (8). For our case, D was taken as the average vertical distance between the tendons. E_s is the elastic portion of the non-linear soil spring and is used to establish the displacement (y) required to achieve the active (P_a) and passive (P_p) stress states for the soil spring. Equation 5 shows the relationship for finding the displacement required for the active load condition (8).

$$y_a = P_a/E_s \tag{5}$$

"Rock springs," (Figure 10b) were used for sections of the wall to be embedded in the diabase. The rock springs were modeled as a 1-foot-square (.093m²) and 2-foot-long (.609m) rock struts that only supplied passive load. The length of the rock strut was set equal to the depth of wall embedment. The maximum allowed load in the rock strut was set at 0.8 of the unconfined compressive strength of the diabase at 16,000 psi (110.3 MPa/m²) and the modulus of elasticity was 6,040,000 psi (41.641 GN/m²) per inch (.0254m).

The tendon springs (Figure 10c) were modeled with the length of the tendon equal to the unbonded length plus one half of the bonded zone. The initial load and elongation in the tendon spring at the start of the analysis was set at the design load with the corresponding elongation of the tendon at that load. Tendon failure was modeled by allowing the tendon load to go to zero after it exceeded 0.8 of the guaranteed ultimate tensile strength (GUTS) of the tendon.

CONSTRUCTION

The 3-foot thick (.9144m) concrete tieback slurry wall consists of nine 20-foot-long (6.1m) structural panels with embedded reinforcement steel, anchor plates, and instrument pipes. After all panels were completed and cured, excavation in front of the wall began. The excavation did not extend more than 5 feet (1.5 meters) below a given row of soil anchors until they were tested and locked off. A Klemm KR805 Double-head-overburden drilling unit with a down-the-hole hammer with a button bit and "lost crown" was used to drill a 5-1/4 inch (133.35mm) hole with a 5-inch (127mm) casing following immediately behind the bit.

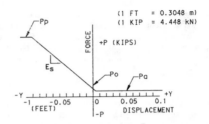

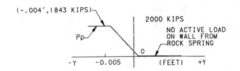

A)   P-Y CURVE FOR SOIL ON THE LEFT SIDE OF WALL.

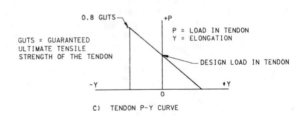

B)   P-Y CURVE FOR ROCK OF LEFT SIDE OF WALL.

C)   TENDON P-Y CURVE

FIGURE 10. TYPICAL P-Y CURVES USED
IN THE CBEAMC MODEL

The drill steel and down-the-hole hammer, minus the lost crown, were removed and tendons installed. Spacers and centralizers were installed on 10-foot centers in the bonded length to keep the strands apart, allowing grout to surround the strands.   The contractor elected to pressure grout the anchorages instead of using the specified tremie method.   A grout and flush tube was placed parallel with the strands. Grout was pumped into the bonded zone (bottom of the tendon) until it returned out of the top of the hole, then the hole was closed and pressure grouted at 90 psi until grout refused to pump. Next, water was pumped into the flush tube which was located 20 feet above the bonded zone, until water exited the top of the hole. During the performance and proof testing of the

38 installed tendons there was one pull out failure and two creep failures. All three failed tendons were successfully locked off at a reduced load. After the first tendon failed, the grout to soil interface was increased to 40 feet (12.19m) while the tendon to grout bonded zone remained 30 feet (9.144m) long by leaving the grease filled sheathing on the strands as per the original design.

MEASURED PERFORMANCE

Wall deflections were measured with inclinometers. Vibrating wire strain gauges were used to gather data to determine moments. Unfortunately the strain gauge readings taken during the construction of the wall were not taken at the appropriate stages and the data is invalid. The measured deflection at the top of the wall was 0.01 feet (3mm) landward and just slightly less in magnitude than the 0.05 feet (15mm) predicted by the CBEAMC analysis (Figure 9). The moments predicted by the CBEAMC model were approximately equal to the measured moments, but the measured trends did not indicate the reduction in the moments at the supports that were predicted by the CBEAMC analysis (Figure 9). Maximum moments for the final load condition occurred at elevation 62 (18.9m) near the point of maximum bending of the wall. The measured moments have varied with time. This variance appears to be related to changes in ambient temperature and instrument data scatter.

The load cell data (Table No. 2) show that there was a 3 to 6 percent drop in load from lock off to the initial permanent load in the tendon. The data also show the loss from the lock off load to the long-term load is between 4 to 9 percent of the lock off load. The initial permanent loads were entered into the CBEAMC model and there was no significant difference between deflections and moments from the "measured" tendon loads versus the design loads used in the original prediction.

Using the data from the upper three load cells to back-calculate soil loads on the wall shows the soil load to be in the range of 2.2 to 2.3 kips per square foot (105.3 to 110.1 kN/m^2). This compares favorably with the design load of 2.1 to 2.4 ksf (100.5 to 114.9 kN/m^2) and confirms that applied anchor loads, if of sufficient magnitude, can dictate the soil pressures applied to the wall.

TABLE 2.   LOAD CELL DATA

TENDON	LOCK OFF LOAD (KIPS)[a]	INITIAL LOAD (KIPS)	LONG TERM LOAD (KIPS)
I-84	272	265	260
I-73	292	275	265
I-62	290	-	270
I-51	356	-	-

[a] 1 KIP = 4.446kN

MODEL CALIBRATION

By increasing ℓ_h to 100 PCI (27, 145 kN/m^3), the CBEAMC one-step model was able to closely match the measured wall deflections. This was a 5 fold increase of ℓ_h from the design values and corresponds inversely to the 5 fold decrease of the measured deflection as compared to the predicted deflection (i.e., y $\propto$ 1/ℓ_h). The calculated moments were still in the range of the measured moments (Figure 8). Increasing the friction angle of the soil values used to an average value more in line with a stiffer soil had only a minor effect when compared to increases in ℓ_h (Figure 11). The soil load on the wall calculated by CBEAMC remained basically the same for all values of ℓ_h (Figure 4).

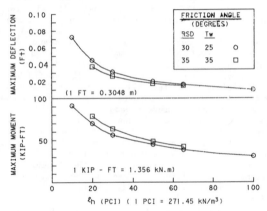

FIGURE 11. COMPARISON OF FRICTION ANGLE AND ℓ_h ON DEFLECTION AND MOMENTS

DESIGN OF REMAINING WALLS

Remaining tieback walls at the lock project were designed using ℓ_h values for RSD which were twice those used for the temporary tieback wall. The revised ℓ_h values were 36 pci (9,772.2kN/m^3) for moist RSD and 26 pci (7,057.7kN/m^3) for submerged RSD. These values were based on the comparison of the predicted deflections and moments for panel 6 with the measured deflections and moments. These data indicate that the ℓ_h values might have been tripled or quadrupled without adversely affecting wall performance. However, due to the importance of the upstream walls, the lack of similar precedents from previous projects, and the likelihood of significant variability in materials at the site, it was decided that increasing the values by more than two times would not be conservative. Both of the ℓ_h values adopted for RSD in design of the additional walls are greater than the highest ℓ_h values provided by Terzaghi (1).

With additional information obtained from the design and construction of future walls in various types of materials, it may become reasonable to suggest the use of even higher ℓ_h values for walls designed using soil-structure interaction methods.

CONCLUSIONS

The soil structure interaction beam-column method is a reasonable procedure for analysis of a stiff tieback wall that is designed for at-rest soil loads. An upper bound value of the Terzaghi constant of horizontal subgrade reaction should be used for this type of wall. Use of low bound values for the subgrade constant will overestimate movement and moments. The data reconfirm that deflections will be minimal for a tieback wall designed using the current empirical procedures to derive the at-rest soil loads.

ACKNOWLEDGEMENTS

The writers acknowledge the support and encouragement of the U.S. Corps of Engineers in writing this paper. The writers would like to thank Dr. William P. Dawkins, Oklahoma State University, Dr. J. M. Duncan, Virginia Polytechnic Institute and State University, and Reed Mosher and Virginia Knowles, U.S. Army Waterways Experiment Station for their advice during the conduct of this project and for their review of this paper. Special thanks go to Mary Theirl and Karen Savoie for typing the manuscript, to Phuong Lu for drafting the figures, and to Bill Wheeler for reducing the strain guage data to determine the moments on panel 6.

REFERENCES

1. K Terzaghi, Evaluation of Coefficients of Subgrade Reaction, Geotechnique, Vol. 5, Dec 1955, 297-326.

2. Post Tensioning Institute, Recommendations for Prestressed Rock and Soil Anchors, PTI, Phoenix, AZ, 1986, 41 p.

3. NAVFAC Design Manual 7.2, Foundations and Earth Structures, Department of the Navy, Naval Facilities Engineering Command, Alexandria, VA, May 1982, Chapter 3.

4. J Jaky, The Coefficient of Earth Pressure at Rest, Journal of the Society of Hungarian Architects and Engineers, 1944, 355-358.

5. JP Hartman and J.J. Jobst, User's Guide: Computer Program with Interactive Graphics for Analysis of Plane Frame Structures (CFRAME), Instruction Report K-83-1, U.S. Army Engineer Waterways Experiment Station, Vicksburg, MS, Jan. 1983, 62 p.

6. EV Edris and S.G. Wright, User's Guide: UTEXAS2 Slope-Stability Package, Vol. 1, User's Manual, Instruction Report GL-87-1, U.S. Army Engineers, Waterways Experiment Station, Vicksburg, MS, Aug. 1987, 220 p.

7. WP Dawkins, User's Guide: Computer Program for Analysis of BEAM - Column Structures with Nonlinear Supports (CBEAMC), Instruction Report K-82-6, U.S. Army Engineer Waterways Experiment Station, Vicksburg, MS, June 1982, 90 p.

8. TA Haliburton, Soil Structure Interaction: Numerical Analysis of Beams and Beam - Columns, Technical Publication No. 14, School of Civil Engineering, Oklahoma State University, Stillwater, OK 1971, 179 p.

DESIGN AND PERFORMANCE
OF A DEEP EXCAVATION SUPPORT
SYSTEM IN BOSTON, MASSACHUSETTS

By Robert C. Houghton[1], Member ASCE, and
Deborah L. Dietz[2], Associate Member ASCE

ABSTRACT

A deep excavation encompassing most of a Boston, MA
city block provided an excellent opportunity to monitor
and analyze the performance of a lateral earth support
system. The excavation which was approximately 60 feet
below street grade utilized three types of earth
support; soldier piles and wood lagging, bracket pile
underpinning, and a "tangent pile" system, all of which
were supported with earth tiebacks. Extensive
instrumentation was installed to monitor ground
movement and to calculate stresses in the soldier piles
above and below the excavation subgrade. A major
objective of the investigation was to analyze the moment
and load distribution along the pile and to provide
information which is useful for designing pile sections
and toe embedment lengths. The instrumentation was
plannned for research purposes, however, unexpected
soil movements on one side of the excavation caused the
instrumentation readings to become extremely valuable in
understanding the ground response during excavation.

1. Vice President of Schnabel Foundation Company,
   Bedford, NH 03102
2. Project Manager with Schnabel Foundation Company,
   Bedford, NH 03102

INTRODUCTION

The project, known as 125 High Street is a 40 story office building with 5 levels of below grade parking situated in Downtown Boston. The first phase of construction, located in the area of the Old Travelers Building, involves approximately two thirds of the project. Phase 2 of construction begins when Phase 1 is far enough along to permit relocation of the Fire Station. A row of temporary earth support between the two phases is shown as a dashed line on Figure 1. A building of historic significance (135 Oliver Street) remains on the site and is supported in place while the new building wraps around.

The 135 Oliver Street Building has a one level basement and is founded on a continuous wall footing consisting of granite blocks placed on glacial till. Excavation support for the rear or westerly portion of the 135 Oliver Street Building consists of a drilled in "tangent pile" system supported by tiebacks.

The rear portion of the existing fire station is supported by a drilled in bracket pile underpinning system supported by tiebacks, all of which is later removed along with the existing Fire Station.

Lateral support for the remainder of the site consists of a drilled in soldier pile, lagging, and tieback system. The majority of the support system is constructed off wall line (i.e., requiring 2 sided forming for the basement walls).

SOIL CONDITIONS

The geotechnical information available at bid time and during design indicated that the building site is situated in the Fort Hill area of Boston, mostly within the Fort Hill drumlin deposit, which is considered to be some of the best foundation soils in Boston. The top of the drumlin was removed in the past to provide fill material for Boston's Backbay area. A "till like" material (glacio-marine deposit), consisting of stiff to very stiff silty clay with varying amounts of sand and gravel overlying the glacial till, was encountered along Pearl Street.

The soil conditions as interpreted from the borings (location shown in Figure 2) at the site generally consist of:

o   A surface layer of fill consisting of a silty SAND with varying amounts of gravel, brick, concrete, wood, and other demolition debris. The fill varies in thickness from 2 to 13 feet.

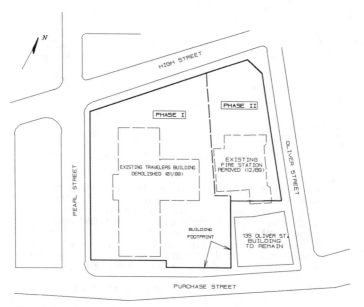

FIGURE 1: *SITE PLAN*

o   A glacio-marine deposit consisting of a stiff to very stiff, gray, silty CLAY with trace to little sand to gravel, ranging in thickness from 5 to 15 feet (encountered in two recent borings).

o   Below the glacio-marine deposit, and in most
     instances directly below the fill, lies a deposit
     of glacial till. The till is generally composed

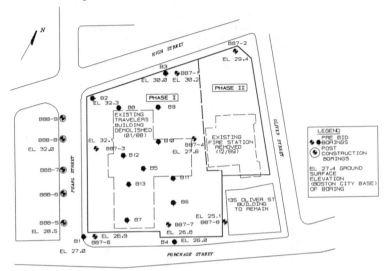

FIGURE 2: *BORING LOCATION PLAN*
of a hard, gravelly to silty CLAY with some to
trace coarse to fine sand and gravel. None of
the recent borings fully penetrated the till,
with the deepest exploration reaching approx-
mately El.-70.   Embedded in the glacial till are
lenses and layers of sand, described as a dense,
coarse to fine SAND with some to trace gravel and
silt, varying in thickness from 6 to 21 ft.

o   Based on existing information, bedrock, known as
     Cambridge Argillite, lies below the glacial till.
     (1)

The Geotechnical Engineer's report indicates that

"Water levels recorded in the observation wells
indicate a perched water condition with levels
varying from El. 8.6 to El. 21.8. The piezometric
level in the glacial till stratum varied between El. 0
to El. 15 with typical levels between El. 3 to El.
10." (1)

DESIGN CONSIDERATIONS

Due to the compact nature of the glacial till, a driven pile support system was not feasible. Soldier piles placed in drilled holes backfilled with concrete were used for all three types of support. Lean mix concrete was used as backfill for all piles except the "tangent piles" and the toe portions of the bracket piles in which 2500 psi concrete was used.

Sizing of the soldier piles and tiebacks was based on the loading diagrams shown in Figure 3. The loading depicted in Figure 3a is typical for that used in the design of the Pearl Street and High Street excavation support system. Figures 3b and 3c show loading diagrams for the bracket pile support at the Fire Station and the "tangent pile" support system at the 135 Oliver Street building. A failure plane extending from subgrade at an inclination of 3 vertical: 2 horizontal was used for selecting the tieback lengths (Figure 3a).

PILES

Soldier piles consist of paired wide flange sections (W 12 x 26 and W 12 x 30) connected with a 13 inch space between webs to allow tieback installation. The piles were designed for combined axial and bending stresses with typical toe penetrations of 5 to 6 feet. The piles were not analyzed as composite sections since the lean mix (1 sack of cement per cubic yard) is not considered strong enough to develop composite action. Wood lagging (3 inch nominal) is used to span between the soldier piles which are spaced 9 to 10 feet on center.

The bracket piles consist of HP 14 x 117 sections placed in drilled holes with structural concrete in the toe and lean mix backfill above subgrade. A steel bracket is welded to the back of the pile to support the building footing (Figure 3b).

The "tangent piles" are installed in 36 inch diameter holes drilled 48 inches apart and backfilled with 2500 psi concrete. This system was selected over a conventional pit underpinning scheme which would have been uneconomical to construct in the compact glacial till. The system stiffness is enhanced by both the close spacing of the piles and tiebacks, and the structural concrete backfill.

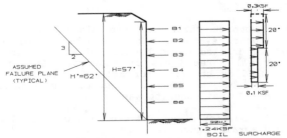

FIGURE 3a: TYPICAL LOADING DIAGRAM FOR PEARL ST. & HIGH ST.

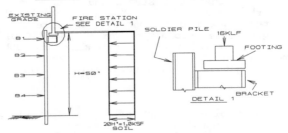

FIGURE 3b: LOADING DIAGRAM FOR BRACKET PILES

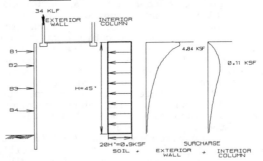

FIGURE 3c: LOADING DIAGRAM FOR TANGENT PILE SYSTEM

TIEBACKS

Tieback tendons consist of prefabricated three and four strand (0.6 inch dia., 270 ksi steel) tendons and are installed by two different methods:

A. Hollow Stem Auger (HSA)

The job was planned initially to be supported with

hollow stem auger tiebacks. HSA tiebacks are installed
by inserting the tendon inside a 12 inch diameter
continuous flight auger prior to drilling. The auger is
then drilled into the ground. When the desired depth is
reached, a grout (portland cement, sand, water, and fly
ash) is pumped through the auger as it is being
withdrawn. Because of the unexpectedly high number of
boulders and the compactness of the till the method of
installation was changed from the HSA to a smaller
diameter regroutable tieback on some portions of the
work.

B. Regroutable Tiebacks

Regroutable tiebacks are installed in 6 inch diameter
augered holes or in 6 inch diameter cased holes if
caving soils are encountered. The prefabricated
tendons and regrout tubes are inserted in tremie grouted
holes and regrouted once or twice to achieve capacity.
Tieback design loads ranged from 93 kips to 141 kips.
All tiebacks were tested to at least 125 percent of
design load and locked off at the design load.

RAKERS

Inclined rakers to concrete heel blocks were installed
at selected soldier piles along Pearl Street. The
rakers were installed and wedged in place as the
excavation approached subgrade. The purpose of the
rakers was to attempt to limit further lateral movement
of the sheeting on Pearl Street.

INSTRUMENTATION

The focus of this paper is on the contractor installed
instrumentation (Figure 4) described below. Reference
will also be made to readings obtained from various
optical survey monitoring points established on soldier
piles, buildings, and streets around the site.

1. SLOPE INCLINOMETERS
Slope inclinometers (See Figure 5) were installed on
two pairs of soldier piles (4 total) to a depth of
approximately 15 feet below the tip of each soldier
pile. The inclinometer casing within the soldier piles
was installed by:
    a. Placing a PVC sleeve within the lean mix
       concrete backfill.
    b. Drilling through the sleeve to the required
       depth.
    c. Setting the inclinometer casing.

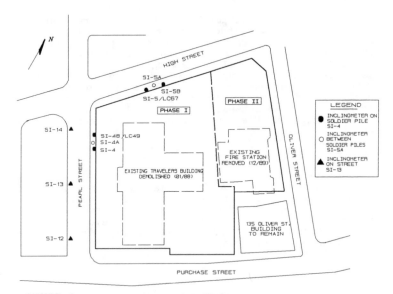

FIGURE 4: *INSTRUMENTATION LOCATIONS*

    d. Grouting the inclinometer casing in place as
    the drill casing is withdrawn.

Additionally, inclinometers were installed between the
pairs of instrumented soldier piles (48 and 49, 67 and
68) to depths of 15 feet below the tips of adjacent
piles. The inclinometer casing between soldier piles
was installed similarly except that drill casing
through the soils was used for the full depth. These
inclinometers were installed after the piles were
placed.

2. STRAIN GAGES ON SOLDIER PILES

Clusters of Geokon Model VSM-4000 vibrating wire
strain gages (total of 250±) were installed on the 4
soldier piles between tieback locations (Figure 5) and
in the soldier pile toes. The gages were installed by
welding attachment brackets to the pile and fastening
the gage to the brackets after cooling. The gages were
then protected by tack welding a protective cover before
pile installation.

3. LOAD CELLS

Thirteen Geokon Model 3000 center hole load cells
were installed on the tiebacks for soldier piles 49 and
67 to monitor change in load with time.

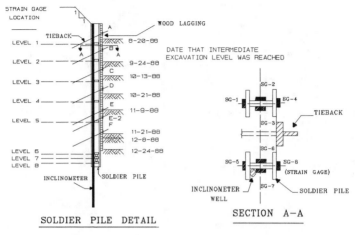

FIGURE 5: TYPICAL INSTRUMENTED PILE

4. STRAIN GAGES ON RAKERS
Two vibrating wire strain gages (one on each side of the web) were installed on each raker.

OBSERVED PERFORMANCE

Figures 6 and 7 indicate the lateral movement that occurred during excavation on Pearl Street and High Street respectively. The magnitude of ground movement on Pearl Street was several times that observed on the High Street side.

The axial load, moment and combined stress as computed from the strain gage readings are plotted for two piles in Figure 8. A family of plots representing information gathered at different times in the excavation sequence is presented for each parameter.

Optical survey readings on the streets and adjacent buildings taken by the geotechnical engineer showed greater vertical movement on the Pearl Street side than was observed on High Street. The difference in ground movement became noticeable and pronounced as the excavation approached mid depth.

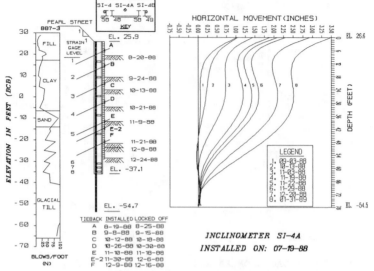

FIGURE 6a: LATERAL MOVEMENT FOR SOLDIER PILE 49

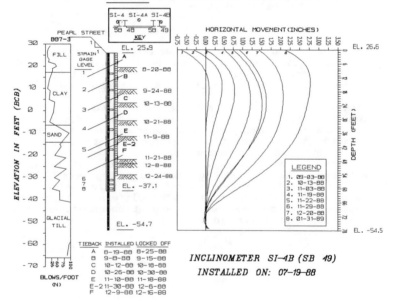

FIGURE 6b: LATERAL MOVEMENT FOR SOLDIER PILE 49

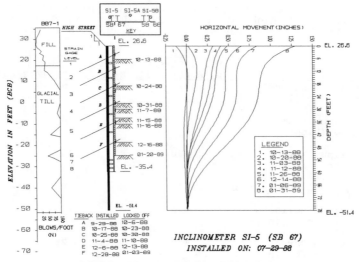

FIGURE 7a: LATERAL MOVEMENT FOR SOLDIER PILE 67

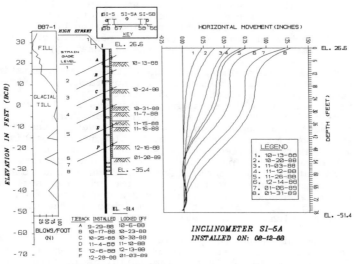

FIGURE 7b: LATERAL MOVEMENT FOR SOLDIER PILE 67

Table 1 compares performance of the piles on Pearl Street and High Street as the depth of excavation approached the midway point at the respective locations:

TABLE 1

	Pearl Street Pile 48(10/21/88)	High Street Pile 67(11/12/88)
Maximum Lateral Movement (in.)	3/4	3/8
Maximum Axial Load (kips)	75	90
Maximum Moment (kips - ft.)	80	45
Combined Stress (ksi)	15	10
Maximum Building Settlement (in.)*	0.4	0.1
Maximum Street Settlement (in.)**	1.0	0.25

 * Building reference settlement readings taken by Geotechnical Engineer. (Buildings are located approximately 45 feet and 50 feet from the Pearl Street and High Street sheeting respectively).
** Surface street settlement readings taken by Geotechnical Engineer. (Taken at points ranging from 5 feet to 35 feet from the face of sheeting).

Visual observation of the street and existing buildings revealed that little or no cracking or street subsidence was evident on High Street. Pearl Street, on the other hand, showed signs of distress in the form of non-uniform pavement settlement, cracking in the pavement, and separation between the pavement and curb. While some of the Pearl Street settlement is associated with sheeting deflection, a portion is attributable to other causes. A previously installed deep utility had apparently been backfilled carelessly and street settlements were accentuated over the trench excavation. Leaky sewer and drain lines caused a continuous recharge of water with constant seepage through the lagging boards which was not evident on the other excavated faces.

A row of three story buildings on the West side of Pearl Street began to settle as the excavation reached a depth of 15 to 20 feet. The buildings contain a one level basement below grade and are presumably supported on spread footings.

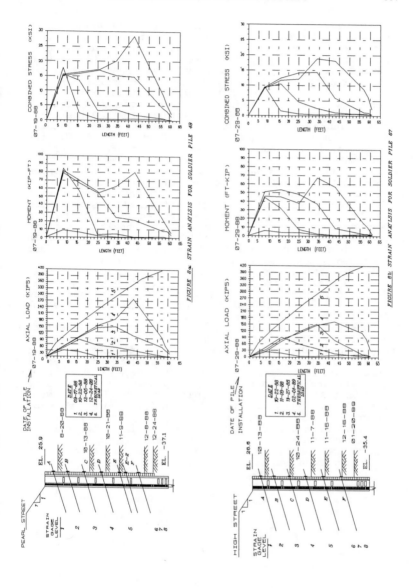

FIGURE 8a: STRAIN ANALYSIS FOR SOLDIER PILE 49

FIGURE 8b: STRAIN ANALYSIS FOR SOLDIER PILE 67

The Pearl Street inclinometers (Figure6) show successive bulges in lateral movement as each new cut was made. The magnitude of building settlement also increased as each new lift was excavated. The lateral movement peaks at approximately 2 3/4 inches. The load cell readings (Figure 9), however, generally show an initial loss of load followed by little or no load increase with time. Because of the observed difference in performance between High Street and Pearl Street, and the apparent inconsistency of increasing lateral deflection without accompanying load increase in the tiebacks, it was decided to obtain additional subsurface information.

Several additional borings were taken during construction along the West side of Pearl Street (Figure 2) closer to the area where the tieback anchors were being made. Inclinometers were installed in some of the boreholes to check for subsurface movement. The new borings revealed the "glacio-marine" deposit to be both softer (N values as low as 4) and thicker (up to 20 feet thick) than encountered in the previous borings. It appears that Pearl Street marks a dramatic change in soil conditions as the West side of the Fort Hill Drumlin falls off steeply. Discontinuous granular deposits within the layer tend to trap water and cause localized problems with soldier pile and tieback installation during construction.

Construction was still progressing as the subsurface exploration and analysis began. As the excavation approached subgrade on the southerly portion of Pearl Street, building settlements (on the West side of Pearl Street 45 feet from the sheeting face) of 1 inch or greater were measured. Inclined pre-stressed rakers to heel blocks and an additional row of tiebacks were installed. Readings from strain gages installed on Rakers 31, 33, 35, 37, and 39 in the southerly portion of Pearl Street indicate that load continued to increase after subgrade was reached. Table 2 compares the performance of the piles on Pearl Street and High Street after subgrade has been reached.

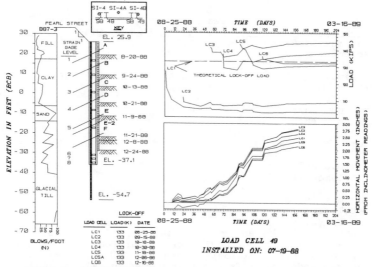

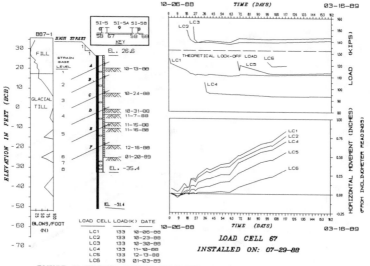

FIGURE 9a: HORIZONTAL MOVEMENT AND LOADCELL INSTRUMENTATION FOR SOLDIER PILE 49

FIGURE 9b: HORIZONTAL MOVEMENT AND LOADCELL INSTRUMENTATION FOR SOLDIER PILE 67

TABLE 2

	Pearl Street Pile 48(12/24/88)	HighStreet Pile 67(2/1/89)
Maximum Lateral Movement (in.)	2.75	0.85
Maximum Axial Load (kips)	275	160
Maximum Moment (kips - ft.)	80	68
Combined Stress (ksi)	28	18
Maximum Building Settlement (in.)	1.4	0.2
Maximum Street Settlement (in.)	4 1/4	3/4

Extra tiebacks and rakers (44, 46, 51, and 53) were installed in the northerly portion of Pearl Street (see Figure 10 for location) before the subgrade elevation was reached. Data from strain gages installed on these rakers also show that increasing load was experienced by

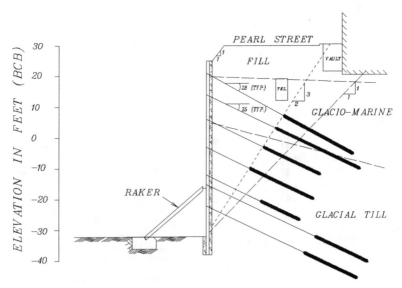

FIGURE 10: TYPICAL CROSS-SECTION OF PEARL STREET

the rakers as the excavation proceeded to subgrade. The magnitude of loads summarized in Table 3 are net loads experienced after prestressing the rakers. The benefit of the rakers is very doubtful since the pattern of settlement and deflection continued to occur after the rakers were installed.

TABLE 3

RAKERS

RAKER#	DATE OF READING/LOAD (KIPS)								
	12/8/88	12/13	12/21	12/28	12/30	1/6/89	1/10	1/13	1/16
31[1]	27	51	84	74	74	67	62	101	112
33[1]	31	64	94	88	88	67	93	120	132
35[1]		33	62	58	58	65	63	83	96
37[1]		26	45	45	45	47	44	65	71
39[1]		34	54	51	51		54	84	91
44[2]								12	27
46[2]					51	90	91	126	143
51[2]					29	47	143	180	202
53[2]						40	50	94	100

1. Rakers were installed without walers and pre-stressed on November 27, 1988. Subgrade elevation (-30) was reached approximately November 29, 1988.
2. Rakers were installed on walers and pre-stressed on December 29 - 30, 1988. Excavation level was approximately elevation -25 on December 29, 1988. Subgrade elevation (-30) was reached January 4, 1989.

ANALYSIS

The contradictory behavior of increasing deflection with decreasing tieback load build-up (Figure 9) provides an interesting dilemma. The most likely explanation is that a mass stability type movement occurred. Figure 10 shows the assumed failure plane used for design. On all streets except Pearl Street the tiebacks were anchored exclusively in the till. On Pearl Street a different set of soil conditions was encountered. The "glacio-marine" clay layer was found to be softer and thicker which suggests that the thickness of the clay layer increases toward the west as the steeply sloping drumlin side drops off. Table 2 shows that measured movements on Pearl Street were greater than on High Street.

The inclinometers (SI 12, and SI 14) installed on the West side of Pearl Street, after movement problems developed, indicate lateral deflections of approximately 1/2 inch shown in Figure 11. The extent of the deflection shown in Figure 11 indicates that once the glacio-marine deposit began to move the till beneath was also mobilized to displace laterally. Thus, the mechanism of failure involves more than the clay sliding across the top of the till. This conclusion seems to be supported by the fact that the load cells on the lower tiers behaved similarly to the upper ones, i.e. no increase in load was associated with continuing lateral deflection.

The choice of the assumed failure plane for design now appears invalid for Pearl Street when the additional thickness of the softer glacio-marine clay is considered. The upper tiebacks were not long enough to prevent a mass movement of the clay. By assuming a flatter failure plane (45°) it is apparent in Figure 10 that 15 to 30 percent of the upper four tieback anchors are located within the failure zone. Since the other side of the excavation did not experience similar magnitudes of deflection, the mass movement on Pearl Street is attributed to the thicker clay layer in the area where the ties are anchored.

The combined stress plot (Figure 8a) for soldier pile 48 shows higher than expected stress levels (approximately 30 ksi) above subgrade level. The mass movement apparently affects the piles differently at different levels as the piles adjust to the ground movement. Stresses in the lower portions of pile 48 are much less, and the magnitude of lateral movement is also less. Combined stresses in the High Street pile (67) show values peaking at less than 20 ksi (Figure 9b).

Of additional interest is the magnitude of lateral movement below subgrade, both at and between soldier piles. The inclinometers were installed to depths of 15 feet below the tips of the soldier piles. The Pearl Street side of the excavation experienced lateral movements below subgrade up to 1 1/4 inches as compared to 1/4 inch lateral movements below subgrade measured on the High Street side. The lateral movements below subgrade were uniform in each set of three inclinometers. This appears to indicate that the ground moves as a unit below subgrade and that the pile may not develop significant lateral load carrying capacity below subgrade. This also indicates that the soldier

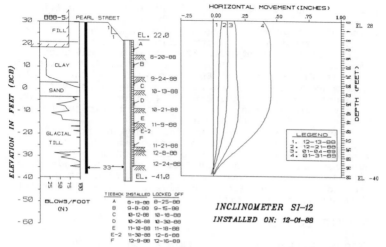

FIGURE 11a: LATERAL MOVEMENT OF PEARL ST. (33' OFFSET FROM EXCAVATION)

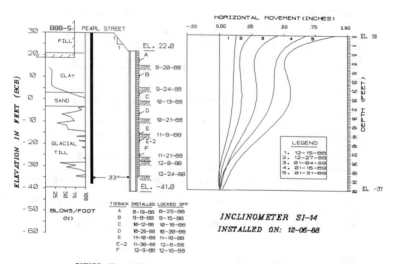

FIGURE 11b: LATERAL MOVEMENT OF PEARL ST. (33' OFFSET FROM EXCAVATION)

pile may function as a load distribution member (like a waler) which helps impart the tieback load to the soil.

One of the prime reasons for instrumenting the piles is to analyze the load distribution in the pile toes to verify design assumptions. The axial load distribution computed from the strain gages in the soldier piles is shown in Figures 8a and 8b along with the theoretical maximum axial load line derived by summing the vertical components of the tieback load with depth. The data clearly indicates that the axial load dissipates before the toe of the pile is reached.

Our experience is that soldier piles do not fail by the toe kicking out, except in cases of accidental overexcavation or soft ground at subgrade. The data confirms that load is transferred above subgrade from the pile to soil through friction and that the soldier pile does not provide significant lateral restraint below subgrade in a stiff soil.

RECOMMENDATION FOR FUTURE STUDY

The opportunity still exists to improve design techniques for tieback support systems. The following areas of research are suggested to the reader as topics for future research:

1. Instrumentation and analysis of soldier pile toe design for driven piles.
2. The presence or absence of composite action between the soldier pile and concrete backfill for drilled in piles.
3. The necessity for pile toe penetration in good soils.
4. Analysis of the design of soldier piles as load distribution members rather than moment resisting members.
5. Analysis of the effect of stiffness on the performance of drilled and driven pile excavation support systems.
6. Finite element studies of deep seated ground movements including those between piles and below subgrade which could lead to better understanding of support systems.
7. Analysis of the appropriate earth pressure diagrams for different soil conditions.
8. Stability considerations for tieback soldier pile support system.

SUMMARY AND CONCLUSION

Previous work (References 2, 3, and 4) has shown that minimum lateral displacement needed to achieve the

Active state is between 0.1% to 0.5% H depending on the soil type. The lateral support system on 125 High Street performed well within these bounds.

	Lateral Movement/Depth of Excavation
Pearl Street (clay)	0.38%
High Street	0.10%
"Tangent Pile"	0.08%
Purchase Street	0.08%

The merit of the observational approach in geotechnical engineering is proven again. What started as research on such items as soldier pile toe design and soil - pile interaction resulted in the acquisition of very timely and critical data which was needed to understand the movements that resulted.

While it can usually be said that more subsurface investigation is desirable on any given site, this project illustrates the need for proper boring coverage in areas where conditions are known to be changing. Borings are seldom taken in the area where tiebacks will achieve their anchorage. On deep excavations this oversight can be costly to the owner, engineer, and contractor. The character and thickness of the clay overlying the glacial till produced unanticipated results on the Pearl Street side. It is clear from the data presented that the assumed failure plane location for Pearl Street was incorrect for the soil conditions that existed under Pearl Street. Longer tiebacks at the upper levels would have reduced, but not eliminated, the lateral movement. The additional lower tiebacks and braces, while not adversely affecting the system, did not appreciably enhance the performance.

Clearly, design procedures which result in extremely long toe penetrations in good soils should be reviewed based on the data presented herein. Much of the axial load is lost above subgrade in friction and that load which reaches subgrade dissipates rapidly as shown. Approximately 50% of the load was dissipated on the High Street side, while about 15% of the load was taken out in friction on the Pearl Street side.

REFERENCES

1. Haley and Aldrich, Report on Subsurface
   Investigations and Laboratory Testing The
   Travelers/New England Telephone Building,
   Boston, Massachusetts,   May, 1987.

2. Sowers, George B., and Sowers, George F.,
   Introductory Soil Mechanics and Foundations,
   2nd edition, The Macmillan Co., New York,
   1961, p. 249

3. Goldberg, D.T. et. al. Lateral Support Systems and
   Underpinning Volume II - Design Fundamentals, April,
   1976, p. 68

4. Lambe, T. William, and Whitman, Robert V., Soil
   Mechanics, John Wiey & Sons, Inc., New York,
   1969, p. 184

CONVERSION TO SI UNITS

To Convert	To	Multiply by
in.	m	0.0254
ft.	m	0.3048
ton	kN	8.896
ton/sq. ft.	kPa	95.76

SEISMIC DESIGN AND BEHAVIOR OF GRAVITY RETAINING WALLS

ROBERT V. WHITMAN[1], F.ASCE

ABSTRACT: The Mononobe-Okabe equation for earth pressure is still used widely for design, although actual conditions during earthquake shaking of retaining structures are quite different from those assumed in developing the equation. This complexity is illustrated by results from shaking table tests aboard centrifuges and from numerical calculations. The proposals of Richards and Elms have stimulated use of design methods based upon allowable permanent displacement

INTRODUCTION

In 1970, at the 2nd ASCE Soil Mechanics Specialty Conference held here at Cornell, the Mononobe-Okabe equation was suggested as the standard method for evaluating dynamic lateral forces for design of retaining structures (Seed and Whitman, 1970). The behavior of gravity walls and other earth retaining structures is much more complicated than envisioned in the simple physical and mathematical model that leads to the Mononobe-Okabe equation. However, this venerable equation, when used with the proper choice of input parameters and suitable safety factors, still provides a sound basis for the design of many retaining structures. Indeed, the Mononobe-Okabe equation must rank - along with its ancestral Coulomb equation and Terzaghi's equation of consolidation - as one of the most enduring and successful theoretical contributions to geotechnical engineering.

This paper starts with a brief review of the Mononobe-Okabe equation and some of the experimental data that confirm its essential correctness. Next there is a review of studies revealing the complexity of the gravity wall-backfill system, especially the complicated situation wherein displacements result from tilting. The final main portion deals with design based upon the concept of allowable permanent displacements. While the presentation deals primarily with gravity walls - which are perhaps becoming less and less important as new types of earth retainment are developed - many of the ideas and results have much broader application, and these implications are pointed out and discussed.

[1]Prof, Dept. of Civil Engineering, Massachusetts Institute of Technology, Cambridge, MA 02139

THE MONONOBE-OKABE EQUATION

Mononobe and Matsuo (1929) and Okabe (1926) modified Coulomb's classical solution to account for inertia forces corresponding to horizontal and vertical accelerations, $k_h g$ (or $N_H g$) and $k_v g$ (or $N_V g$) respectively, acting at all points of an assumed failure wedge. They expressed the backfill's thrust against a wall as:

$$P_{AE} = (1/2)\gamma(1-k_v)H^2 K_{AE} \qquad (1)$$

where γ is the unit weight of the backfill, H the height of the wall and K_{AE}, the active stress coefficient, is a function of the friction angle of the backfill, the friction angle between backfill and wall, and of the acceleration coefficients. The full equation may be found in many references (e.g. Seed and Whitman, 1970). Fig. 1 graphs K_{AE} for the case of zero vertical acceleration; note that K_{AE}, and hence P_{AE}, includes the static as well as dynamic thrust. Seed-Whitman suggested a simple linear approximation:

$$K_{AE} = K_A + (3/4)k_h \qquad (2)$$

valid over the range of kh of practical interest, where K_A is the static coefficient of active active earth pressure. The Mononobe-Okabe equation is based upon a number of assumptions. As with the original Coulomb equation, the backfill must be deforming enough so that full shear resistance is mobilized along the failure plane in the active sense. In addition, the accelerations must be constant throughout the failing wedge.

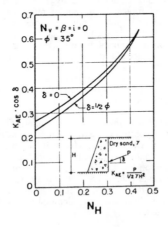

Figure 1 K_{AE} vs. friction angle of backfill according to the Mononobe-Okabe
equation, for two values of wall friction angle δ.
(From Seed and Whitman, 1970)

This derivation by itself does not indicate the distribution of lateral stress over the height of a retaining wall in contact with the failing wedge. By making additional assumptions, various authors (e.g. Prakash and Basavanna, 1969) have estimated the height of the resultant force. An upper limit for the location of the dynamic component of thrust results from the assumption that the backfill is uniform and elastic (Wood, 1973); in this case the dynamic thrust acts at 0.63H above the base. After reviewing the various available results, Seed-Whitman suggested applying the dynamic component at 0.6H.

The foregoing results are based upon the assumption of active stress conditions. If active conditions are maintained (i.e. full shear resistance along the failure surface in the backfill, directed so as to help support the weight of the wedge) while the direction of the inertia force is varied, the thrust is a maximum with the base acceleration acting toward the backfill. At this time, the inertia force acting upon the failing wedge of soil acts in the direction of the wall, thus adding to the thrust from the weight of the backfill. Conversely, the total thrust is a minimum with outward-directed base acceleration; now the inertia force on the Coulomb wedge is directed away from the wall and thus partially counteracts the effect of the weight of the backfill.

Seed-Whitman also summarized various results from model tests in which P_{AE} was measured. Fig. 2 shows more recent results by Sherif et al (1982) and Sherif and Fang (1984). In their experiments, a wall was slowly moved outward in a controlled fashion while the backfill was being shaken at its base. The measured force acting upon the wall showed a cyclic fluctuation superimposed upon a mean trend that decreased with movement and stabilized at the static active value. Once the mean value stabilized, the peak cyclic force was used to evaluate the K_{AE} plotted in the figure. Sherif and his colleagues found further that the height of the result force moved upward during shaking, reaching a height of 0.45H above the base.

The conditions existing in these experiments were just those assumed by the Mononobe-Okabe equation: essentially uniform accelerations throughout the backfill and wall movement sufficient to mobilize fully the shear resistance of the soil. Hence the results are a very good demonstration of the essential correctness of the theory. However, as discussed below, there remain questions as to the applicability of the theory in actual problems where the outward movement of the wall is not controlled as in the experiments.

A similar solution may be made for passive stress conditions. Now the maximum thrust occurs when the base acceleration is acting <u>away</u> from the backfill, and this thrust <u>decreases</u> as the base acceleration increases. The equation for K_{PE} as given in Seed-Whitman contains an error, which unfortunately has propagated through the literature. The term $\sin(\phi-\delta)$ under the radical in the denominator of the equation should be $\sin(\phi+\delta)$. The common warnings concerning the use of the Coulomb equation for static passive thrust apply to the dynamic case as well. In particular, this equation may overestimate the passive resistance in the case where there is wall friction.

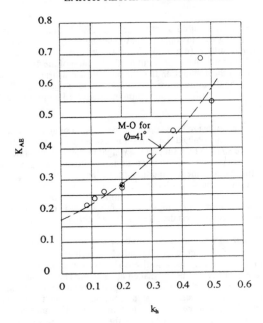

Figure 2 Prediction of Mononobe-Okabe compared to
experimental results. (Adapted from Sherif et al, 1982.)

OTHER RESULTS FOR EARTH PRESSURES

Non-Yielding Walls. If a wall moves rigidly with the underlying base, it would
be expected that earth pressures are larger than those given by the Mononobe-Okabe
theory that assumes active conditions. Using elastic theory and assuming that material
properties are constant with depth, Wood (1973) found a steady-state dynamic thrust
approximately equal to $\gamma H^2 a/g$, where a is the base acceleration. The resultant thrust
was at a height of about 0.63 above the base, corresponding more-or-less to a
parabolic variation of earth pressure with height. Finite element analyses (Nadim,
1982) in which the modulus of the soil increased with depth resulted in 5% to 15%
smaller dynamic thrust, and in a resultant height closer to 0.5H.

The most thorough experimental investigation of this case appears in Yong
(1985), who carried out shaking table tests using a wall with a height of approximately
1/2 m. He found results in good agreement with Wood's theory, while the measured
forces exceeded by a factor of 2 to 3 those predicted by the Mononobe-Okabe
equation.

Dynamic Water Pressures. For the design of new gravity retaining walls, such pressures usually are of little or no significance. Good practice calls for drainage of the backfill behind such walls. Along waterfronts, experience (Whitman and Christian, 1990) tells us that gravity walls tend to perform poorly during earthquakes, frequently because of liquefaction of loose backfills.

Almost all descriptions of the water pressures induced on a wall by an earthquake start from Westergaard's (1933) classic solution for the case of a vertical wall retaining an infinitely long reservoir of constant depth. The total dynamic force from the water is

$$P = (7/12) k_h \; \gamma_\omega \; h^2 \tag{3}$$

where γ_w is the unit weight of water and h is the total depth of water. This equation gives the effect of water standing against the exposed face of a gravity wall, and has also been used to estimate the effect of dynamic pore pressure with saturated backfill. In this connection the dynamic thrust from the mineral skeleton is given by the Mononobe-Okabe equation using the buoyant unit weight γ_b.

While this approach may be valid when the backfill is coarse sand, in general the situation will be much more complex. Steedman and Zeng (1989) have an alternative suggestion. As the permeability of the backfill decreases, it becomes impossible to evaluate effective stresses and pore pressures separately; in the limit of very small permeability, undrained conditions will prevail. Moreover, excess pore pressures may be generated cumulatively as a result of cyclic straining. In the extreme case, liquefaction may develop.

THE DYNAMIC RESPONSE OF GRAVITY WALLS

The several earth pressure theories thus far summarized assume a constant horizontal acceleration over the height of the wall, and that the movement of the wall is controlled in some manner. In actuality, neither condition may be satisfied. Depending upon the frequency of ex-citation and the height of the wall, acceleration may increase significantly upward through the backfill. The manner in which a wall moves is a complex result of the forces exerted by the backfill, the inertia of the wall and the constraints upon its movement. Moreover, the non-linear behavior of the backfill has a significant influence.

Unfortunately there are very few well-documented case studies from the field, with actual measurements of dynamic earth pressures, and the few results that do exist are very difficult to understand and interpret. Hence, it is necessary to turn to results from model tests and theoretical calculations. Several studies have indicated quite clearly the complications that can develop.

Outward Sliding of a Gravity Wall. Small-scale shaking table tests to investigate the dynamic response of gravity walls were carried out at Canterbury University in New Zealand (Lai, 1979). A wall, proportioned to slide without tilting, rested upon a layer of sand glued to the shaking table. Fig. 3 shows typical results. It may be seen that sliding occurs in steps and that the peak acceleration at the top of the wall is less than that of the shaking table. The thrust between soil and wall varied in a complex manner during the test.

Steedman (1984) also conducted model tests (on a centrifuge) of sliding walls, using a quite dense backfill. These tests primarily showed the influence upon sliding of a decrease - the result of strain-softening - of the friction angle of the backfill as testing progressed.

Movement and Distortion of the Coulomb Wedge. If a retaining wall slips outward during an earthquake, the wedge of backfill not only moves outward with it but, to satisfy kinematics, must also move downward. Thus a vertical acceleration of the wedge must accompany any rapid slip - even if there is zero vertical acceleration of the base. Zarrabi (1973) examined the implications of this kinematic requirement, and found that neglect of the vertical component of acceleration can cause a significant overestimate of the dynamic thrust.

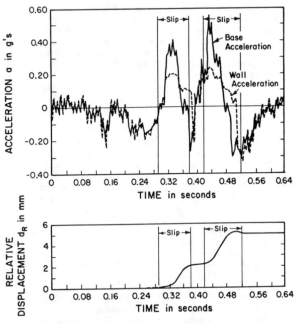

Figure 3 Accelerations and slip during shaking
table tests by Lai. (adapted from Jacobson, 1980.)

Tilting of a wall may also affect the accelerations within the Coulomb wedge. Nadim (1980) used a model assuming a series of parallel Coulomb failure planes extending from the base to the surface of the backfill, while preserving the traditional assumption of rigid-plastic behavior. Rotational acceleration of the wall must be accompanied by a linear variation with height of horizontal acceleration in the material between these planes. The main result of Nadim's work was to emphasize that the height of the resultant lateral force can vary during shaking, at times falling <u>below</u> the lower third point of the wall.

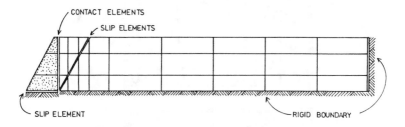

Figure 4 Finite element model with slip elements.
(After Nadim, 1982)

<u>Deformability of the Backfill</u>. As the base accelerates toward the backfill, inertia forces induced in the soil cause it to deform, a response which might be expected to force outward movement of the wall. This aspect was studied by Nadim and Whitman (1983) using the finite element grid shown in Fig. 4. Slip elements were used at the base of the wall and at the interface between backfill and wall, and also along an inclined plane representing the Coulomb failure surface. Thus the analysis incorporated both deformability of the backfill and the rigid-body failure mode of the Richards-Elms analysis to be discussed subsequently.

Fig.5 gives the computed variation of thrust from the backfill and shear at the base of the wall, using a sinusoidal variation of acceleration at the base of the backfill. When the base shear is constant at its maximum value, slip is occurring and the wedge of backfill is in the active condition. These intervals coincide with times when the base acceleration is directed toward the backfill, consistent with the Mononobe-Okabe theory, and during these intervals the thrust from the backfill follows the Mononobe-Okabe equation with the effect of the vertical component of slip included. However, the maximum thrust occurs during the intervals between slip - when the deformable backfill "rebounds" against the wall moving rigidly with the base. Meanwhile, the height of the resultant thrust varied up and down along the wall.

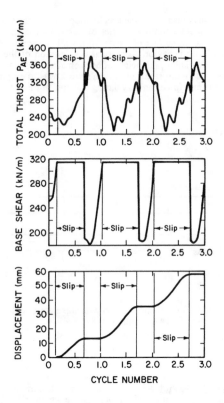

Figure 5 Computed response of wall in Fig. 4, caused by three cycles
of sinusoidal acceleration. (Nadim and Whitman, 1983)

Residual Forces. The calculations by Nadim and Whitman (1983) also revealed
a residual thrust remaining at the end of base excitation. Such a result has also been
found in several experimental programs.

In Yong's tests with unyielding walls, K_o at rest prior to shaking was 0.41, but
increased to 0.74 after repeated shaking of a loose sand and to 0.89 after shaking of
dense sand. In the centrifuge tests using tilting walls, described in the next subsection,
Andersen et al (1987) found residual forces nearly as large as the peak forces during
shaking. Steedman (1984) also observed such increases in his centrifuge model tests
upon cantilever walls. Evidence of residual increases has been observed in the field,
by (for anchored bulkheads) by Iai et al (1989).

Such increases are associated with the tendency for sands to densify, with attendant increases in minor principal stress, when shaken or vibrated. The build-up of horizontal stress during the centrifuge tests on tilting walls has been predicted well by Stamatopoulos and Whitman (1990).

Tilting Walls. Field observations suggest that, where there have been significant movements of gravity walls during earthquakes, rotation of the walls about their base has been important. The behavior of tilting walls has, until very recently, received relatively little study.

Andersen's Tilting Wall Tests. In order to understand better the dynamic response of gravity walls that experience tilting, a series of model tests was undertaken using the geotechnical centrifuge at Cambridge University in England (Andersen et al, 1987). The basic arrangement is shown schematically in Fig. 6. The wall was 0.152 m high; the dry sand backfill extended 0.483 m away from the wall and was about 0.5 m wide. The simulated, rigid wall was hinged at its heel, and was supported by springs near the toe that provided the resistance to tilting. This arrangement was selected to give a known resistance-to-tilt, so that the experiments could focus upon the interaction between wall and backfill. Three different spring stiffness were used, as given in the following table:

Test	Rotational stiffness
GA 3	5,200 lb-ft/radian
GA 6	10,500 lb-ft/radian
GA 5	31,400 lb-ft/radian

where 1 lb-ft = 1.328 m-N.

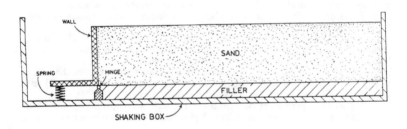

Figure 6 Schematic of test with tilting wall
(from Andersen et al, 1987)

The wall actually was made up from aluminum plates, with buttresses providing the rigidity, and looked unlike any actual wall. However, the mass of the wall and its moment of inertia about the hinge scaled to values typical for actual walls (although perhaps somewhat on the high side). Three identical walls were placed side-by-side across the width of the centrifuge test package, with all data being taken only on the central section. The sand was 25/40 Leighton-Buzzard sand, placed by pluviation to a rather dense state. The nominal centrifugal acceleration in all tests was 80g. Shaking was applied to the base of wall and backfill using Cambridge's "bumpy road", and hence consisted of 10 not-very-uniform cycles of acceleration at, in these experiments, a frequency of 125 HZ - which at prototype scale would be about 1.5 Hz.

By measuring the forces in the "foundation" springs, the reaction forces at the hinge and the accelerations of the wall, it was possible to evaluate, as a function of time during shaking, the horizontal and vertical components of the resultant force between backfill and wall and the height at which this resultant acted. Some difficulties were experienced with this instrumentation; as a result the actual magnitude of the vertical component of the thrust (i.e. the shear force on the wall) is in doubt (though the changes with time seem reasonable) and the height of the resultant could not be determined accurately. Nonetheless, I would again make the decision to measure forces and accelerations (an approach used earlier by Sherif and et al, 1982), rather than attempting to use face-mounted stress gages on the wall.

During spin-up to the 80g, the walls developed significant tilt, in large part because of the increased gravitational pull on the wall itself. Data from the end of this stage showed that active conditions had developed, with the horizontal thrust at a value consistent with active earth pressure theory and with the height of this resultant thrust being close to the lower third point of the wall.

Fig. 7 shows a typical set of results during dynamic shaking.These results emphasize the complexity of the dynamic response. The key features of these results are:

* The minimum thrust at the time of maximum outward movement of the wall (indicated by the time when the spring force is a maximum). At this time, the applied base acceleration is toward the backfill. The maximum thrust occurs when the wall has rotated back against the soil, at a time when the applied acceleration is outward away from the backfill.
* The height of the resultant earth thrust fluctuates during shaking, being highest when the wall has swung back against the soil and being lowest (typically below the lower third point!) when the wall moves outward.
* Wall friction also fluctuates during shaking, being greatest when the wall is swung outward (so that soil is moving downward with respect to the wall) and decreasing as the wall forces itself back into the soil.
* There is a permanent outward tilt of the wall, accompanied by a residual increase in the static horizontal thrust and a higher location for this resultant thrust.

Fig. 8 is a plot is maximum observed thrust vs. that predicted by Mononobe-Okabe. The striking result is that predicted and observed values are out of phase. Numerical agreement between maximum thrustsmwould not be expected, since at the time of peak actual thrust the conditions assumed in the Mononobe-Okabe theory certainly do not exist; in particular, the backfill is not in an active state. By coincidence, however, there is rough agreement.

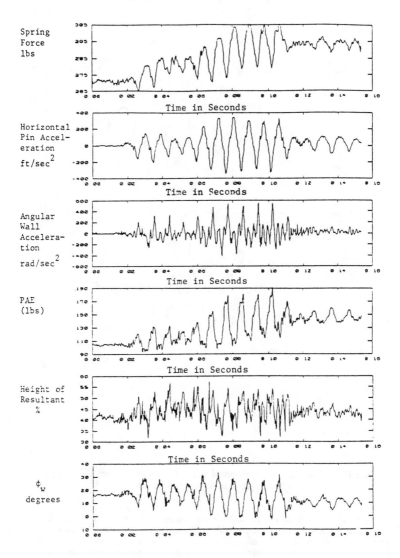

Figure 7 Results from Test GA6 (from Andersen et al, 1987).
The total time shown is 0.1 sec

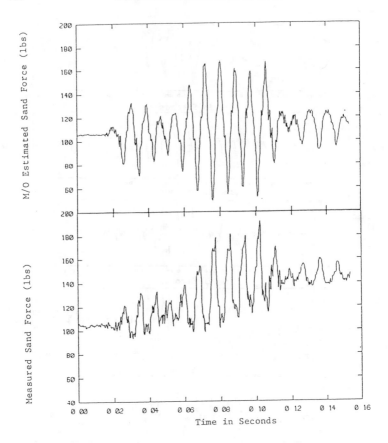

Figure 8 Thrust from backfill, as observed in Test GA6 with that predicted assuming
Mononobe-Okabe applies at all times (Andersen et al, 1987)

It might be expected that the thrust at the first minimum would be that predicted
by the Mononobe-Okabe equation; at this point the backfill is certainly in an active
condition and the base acceleration is directed toward the backfill. However, the
observed thrust is much less than the predicted value. One possible explanation is the
"Zarrabi effect", that is, the effect of vertical acceleration of the backfill accompanying
outward movement. Another explanation may be phase lag in lateral stresses behind
the upper and lower parts of the wall, as discussed by Steedman and Zeng
(1989,1990). Still a third explana-tion may lie in the complex pattern of the wall's
rotation, the base acceleration and the earth's thrust during the intervals when the wall
is near maximum outward movement.

Lessons From Simple Theoretical Models. The very simple model shown in Fig. 9 may be used to examine and understand the matter of phasing (Whitman, 1989). The mass and spring at the right represent the dynamic response of the backfill. The spring at the left provides the resistance to rotation of the wall. The spring between the masses reflects the interaction between wall and backfill. Of special interest is the magniyude of the excitation frequency relative to the natural frequencies of the system.

With values for the parameters typical for actual walls, the maximum force in interaction spring (ie. the earth thrust) occurs when the acceleration is directed away from the backfill, with a minimum when the acceleration is toward the backfill. This is just the opposite of the result reached using the Mononobe-Okabe approach, assuming active conditions at all times, and the result observed during Sherif's tests, but is just what was observed in the tests by Andersen et al. Considering the rotations, the result may be readily understood: the maximum thrust occurs when the wall has swung back against the backfill - approaching a "passive" situation - while the minimum thrust is found when the wall has its maximum rotation outward - an "active" situation. With a different relation of the exciting frequency to the apparent natural frequencies, different phasing may occur.

This model may be extended further by breaking the soil mass into two parts (Figure 10) and introducing a slider element. With this extension, the model predicts the development of residual tilt and residual earth thrust - although quantitatively such predictions are unsatisfactory.

Combined Sliding and Tilting. Nadim and Whitman (1984) undertook a very preliminary study of this case, assuming a yield point in the moment-rotation relation for the base of the wall as well as limited frictional resistance against sliding. The main result was that it seemed unlikely that sliding and tilting are both significant in any given problem.

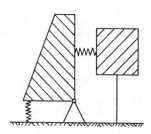

Figure 9 Simple 2-degree-of-freedom model with
essential features of tilting gravity wall

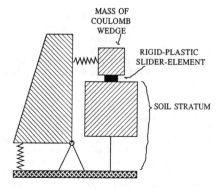

Figure 10 Extension of 2-degree-of-freedom model introducing slider element to
account for permanent tilt and residual earth thrust

A more realistic analysis is now underway by Al Homoud (1990), using the
large finite element code FLEX (Vaughan and Richardson, 1989). The wall is
modeled as resting upon a stratum of sand, as well as retaining a backfill of sand. The
more significant results thus far are: (a) the time-wise variation of pressure against the
wall differs over the upper and lower portions of the wall, with residual pressures
being largest over the upper portions; (b) phasing relations are essentially those found
in the tests on tilting walls; (c) the moment-rotation relation at the base of the wall
remains nearly linear for moments up to about 1/3 of the moment capacity of the base,
at which time the tilt is about 0.001 radians; and (d) uplift then occurs over the
innermost portion of the base of the wall, such that the moment-rotation relation at the
base is quite complex.

A somewhat similar analysis has been made by Marciano (1986). Siddharthan
and Maragakis (1989) have applied a numerical analysis to Steedman's (1984)
experiments with cantilever walls.

DESIGN FOR ALLOWABLE PERMANENT DISPLACEMENT

In what may be called the traditional approach to design, a total static plus
dynamic thrust against the wall is evaluated using the Mononobe-Okabe equation
together with an assumed horizontal acceleration coefficient. Good practice calls for
applying, in addition, an inertial force on the wall, using the same acceleration
coefficient. The wall is then proportioned to resist these horizontal forces, using a
safety factor less than that required for static forces alone. Recommended acceleration
coefficients typically range from 0.05 to 0.15, corresponding to 1/3 to 1/2 of the peak
acceleration of the design earthquake, while prescribed safety factors often are
between 1.0 and 1.2. Fig. 11 gives dimensions of a wall proportioned using this
approach, and shows the forces acting on the wall for a seismic coefficient of 0.2.
Note that the inertia force on the wall is greater than the dynamic component of the
earth thrust.

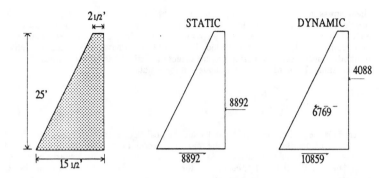

Figure 11 Wall designed by traditional approach, with friction angles of 35 degrees and seismic coefficient of 0.2. The static and dynamic forces are additive.

In general this practice has been satisfactory, as failure or excessive movement of gravity walls has occurred infrequently, even during major earthquakes - except where there has been liquefaction of backfill or foundation soils. During the recent Loma Prieta Earthquake that shook the San Francisco Bay area, retaining walls reportedly performed very well. The most likely explanation for this good performance is that very conservative assumptions - concerning strength of the backfill, wall friction, shear resistance at the base of the wall, etc. - were made in design. Nonetheless there have been instances of excessive movements (Evans, 1971; Grivas and Souflis, 1984). Moreover, it is unsatisfying to base designs upon uncertain conservatisms and arbitrarily specified safety factors and acceleration coefficients.

Choosing design accelerations. Since the traditional approach to design seemingly presumes accelerations smaller than those expected during the design event, it might be expected that some yielding and permanent deformation would occur during a major earthquake, Richards and Elms (1969) made a major advance by putting forth an explicit procedure for selecting a design acceleration coefficient, based upon the concept of allowable permanent movement. They began by drawing an analogy between the behavior of a gravity wall and that of a sliding block - see Fig. 12.

In order for a block to move with an underlying plane, a shear force must develop at the block/plane interface. The shear resistance at this interface places a limit upon the acceleration that can be transmitted upward to the block. If the accelerations of the plane exceed this limit, the block slips relative to the plane. At the end of shaking there may be a permanent relative displacement - especially if there is a static shear stress across the interface, as when the block rests on a sloping plane. This problem was first studied in detail by Newmark (1965), who developed equations and a chart for predicting permanent relative displacement. Subsequent workers (Franklin and Chang, 1977, Whitman and Liao, 1965 and others) have refined the predictive methods, considering such effects as the effect of vertical base accelerations. As part of their study, Richards and Elms suggested the equation:

$$\Delta = 0.087 \frac{V^2}{Ag} \left(\frac{N}{A} \right)^{-4}$$

(4)

where D is the permanent slip, Ag and V are the peak acceleration and peak velocity of the plane, respectively, and Ng is the maximum acceleration that can be transmitted across the block/plane interface. Eq. 4 provides a conservative (nearly an upper bound) estimate for the slip, especially for N/A < 0.4. Wong and Whitman (1982) proposed an equation for the mean slip:

$$\Delta = 37 \frac{V^2}{Ag} e^{-9.4N/A}$$

(5)

An estimate for the standard deviation was also developed.

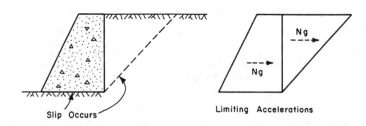

Figure 12 Gravity wall and failure wedge
treated as a sliding block

For the case of the gravity retaining wall, Richards-Elms assumed that the wall together with a failing wedge of the backfill move as a "sliding block". The maximum transmittable acceleration is reached when the shear resistance at the base of the wall just equals the inertia force on the wall plus the static-plus-dynamic thrust from the backfill. This thrust may be evaluated using the Mononobe-Okabe equation or the Seed-Whitman approximation.

On this basis, Richards-Elms proposed the following steps for design. First, an allowable permanent displacement is selected, considering the function of the wall. Several inches of permanent outward movement may well be accep-table during a major earthquake. Eq. 4 is then used, together with prescribed A and V, to find the required transmitted acceleration coefficient N. Using this N as a seismic coefficient k_h, the thrust from the backfill is evaluated. The weight of wall required to resist sliding caused by this thrust (and the corresponding inertia force on the wall) is calculated, and a suitable safety factor applied to the calculated weight. Richards-Elms originally suggested a safety factor of 1.5, but subsequently stated that a much smaller factor would suffice. They did not explicitly consider overturning, recommending that the wall be proportioned so that overturning would not be a consideration. The wall shown previously in Fig. 11, would be satisfactory by the Richards-Elms reqirements for an earthquake with A = 0.4, V = 20 in/s (0.5 m/sec), an allowable displacement of 2 in and a safety factor of 1.1.

The shaking table tests described previously (Fig. 3) demonstrate the essential correctness of the Richards-Elms method. However, there are several differences between results and predictions of the simple theory. The actual slip generally is overestimated, and the acceleration of the wall is not constant while slip is occurring.

Selecting Safety Factors. Several studies have been undertaken at MIT to refine the understanding of the behavior of a sliding wall and to investigate the safety factors appropriate in connection with the traditional and Richards-Elms methods.

The study of Zarrabi showed that neglecting this vertical acceleration led to overestimating the amount of the wall's permanent slip. By taking this effect into account, a better prediction was obtained for the data from the shaking table tests conducted at Canterbury (Jacobsen, 1980).

Another factor neglected in the Richards-Elms analysis is the deformability of the backfill. Computations by Nadim and Whitman (1983) showed that permanent wall movements calculated when deformability was included were indeed greater (Fig.13) than when only rigid-body sliding was considered, with the error increasing as the excitation frequency approached the natural frequency of the strata composing the backfill. The natural frequency is evaluated using

$$f_n = C_s/4H \tag{6}$$

where C_s is the shear wave velocity through the backfill, degraded for the expected cyclic shear strain. With a wall 30 feet high, for example, f_n during strong shaking might be about 2.5 Hz - which is within the significant range for earthquake ground motions. Because of assumptions inherent in the analysis, particularly the assumption of perfectly rigid earth below the backfill, the quantitative influence of the backfill's deformability may have been overestimated, but nonetheless it seems clear that this is an important effect.

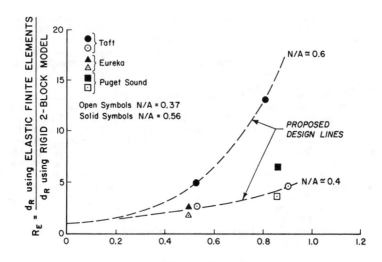

Figure 13 Increase in permanent displacement d_R as compared to d_R
by Richards-Elms analysis, from analyses using actual earthquakes.
(From Whitman and Liao, 1985)

These and other studies led to a systematic analysis, based on probabilistic concepts and methods, of the overall uncertainty in predictions of permanent slip made using the simple sliding block analysis (Whitman and Liao, 1984). Table 1 lists the potential errors and other sources of uncertainty. For each of these sources, an error in the mean and a standard deviation were assigned based upon the above mentioned studies. (At the time of the analysis, the estimate for the effect of tilting was largely an educated guess.) A log-normal distribution was assumed for the distribution of actual motions for a given problem. The conclusions were:

* The mean motion is underestimated by a factor of 3.5

* The standard deviation for the log-normal distribution ranges from 1.3 to 3, being greatest for smaller ground motions.

The largest source of error was neglect of deformability of the backfill, while the greatest contribution to the standard deviation (especially with smaller motions) came from uncertainty in the resistance parameters.

Table 1
UNCERTAINTIES AND ERRORS
IN PREDICTION OF PERMANENT WALL MOVEMENT

* ASSOCIATED WITH UNPREDICTABLE DETAILS OF GROUND
MOTION

    Frequency content, duration, directionality, vertical motions

* ASSOCIATED WITH UNCERTAINTY IN RESISTANCE PARAMETERS

    Friction angles for backfill, base of wall, backfill-wall interface

* MODEL ERRORS

    Imperfections in analytical model owing to simplications and incomplete
    knowledge

    - Vertical component of sliding of backfill wedge
    - Deformability of backfill
    - Tilting

The results of this analysis have been used in two ways. The first has been to develop a modification to the Richards-Elms method of design (Whitman and Liao, 1984). The key is an equation giving the acceleration coefficient required in design so that there will be 95% confidence that the prescribed allowable permanent displacement will not be exceeded during an earthquake with assigned k_h and V:

$$k_h = A \left[0.66 - \frac{1}{9.4} \ln \frac{\Delta A}{V^2} \right] \tag{7}$$

where $A = k_h g$. Design then proceeds in the traditional manner, but using a safety factor of unity as regards sliding.

The second usage was to study the safety factors appropriate for the traditional and Richards-Elms methods. Tables 2 and 3 give the probability that the permanent displace-ments of walls designed by the two methods will not exceed the assigned allowable displacements, should the design earthquake actually occur. While there is no general agreement as to a satisfactory probability of nonexceed-ance, a level of 95% is often used.

For the purpose of these evaluations, the traditional or "standard" method of design was defined as using an equivalent seismic coefficient corresponding to 1/2 the peak acceleration of the design earthquake. Friction angles of 30 degrees for the backfill and 20 degrees for wall friction were assumed. The calculated probabilities vary with both the assigned allowable displacement and the strength of the earthquake shaking, and for the stronger ground shaking are less than would generally be considered desirable. With the Richards-Elms method, the evaluated probability is independent of both the allowable displacement and the strength of shaking (because these factors are explicitly taken into account in design), and use of a safety factor of 1.1 or higher provides satisfactory performance. This comparison shows the value of the rational approach developed by Richards-Elms, which is now used as the basis for the design of bridge abutments in California and across the nation (ATC, 1981).

Table 2

PROBABILITIES THAT WALLS DESIGNED BY THE TRADITIONAL METHOD WILL NOT EXPERIENCE PERMANENT MOVEMENTS GREATER THAN ALLOWABLE MOVEMENTS

Safety Factor on Wall Weight	Allowable displacement inches	Probability [act.>allow.]	
		A=0.2; V=10ips	A=0.4; V=25 ips
1.1	2	97%	70%
	4	98%	84%
1.2	2	98%	77%
	4	99%	88%

Table 3

PROBABILITIES THAT WALLS DESIGNED BY RICHARD-ELMS APPROACH WILL NOT EXPERIENCE PERMANENT MOVEMENTS GREATER THAN ALLOWABLE MOVEMENTS

Safety Factor on Wall Weight	Probability [Actual>allowable]
1.0	90%
1.1	95%
1.2	>95%

Effect of Tilting. When rotation (or tilting or overturning) is considered, the height of the resultant dynamic earth pressure (that is, the distribution with height of the lateral earth pressure) becomes an important issue, and the phasing of the height of the resultant with respect to the movement of the wall becomes critical from the standpoint of design. That is to say, it may be much too conservative to assume that the time of the maximum dynamic thrust coincides with the time at which the height of the thrust is a maximum.

Given the complexity of the results discussed earlier, it is far from clear just how the force computed from the Mononobe-Okabe equation is related to the performance of a gravity wall. However, a possible approach that retains the simplicity of that equation appears when results from Andersen's tilting wall tests and Al Homoud's numerical results are examined. In these cases, a satisfactory estimate of maximum rotation is calculated using the dynamic increment of thrust, computed using the Mononobe-Okabe equation together with the actual peak acceleration within the backfill, acting at mid-height of the wall. Obviously there is need for further research to confirm or improve this rule. The stiffness of the wall's foundation against rotation may be found using (Dobry and Gazetas. 1986):

$$k_{\theta o} = \frac{\pi G B^2}{2(1 - \upsilon)} \left\{ 1 + \left[\frac{\ell n(3 - 4\upsilon)}{4} \right]^2 \right\} \tag{8}$$

Here G is shear modulus for small dynamic strains, B is the width of the base and ν is Poisson's ratio. A reduction to 40% to 80% of the small strain value is suggested to account for non-linear effects.

EXTENSIONS TO OTHER RETAINING STRUCTURES

This paper has focused almost entirely upon gravity walls, which are today used much less frequently than other types of retaining structures. Indeed, gravity walls are somewhat undesirable from the standpoint of earthquake resistance, since resisting the inertia forces upon the wall itself uses a large portion of the shear resistance at the base of the wall. However, many of the concepts developed for gravity walls may be extended to other types of retaining structures - such as the cantilever walls studied experimentally by Steedman (1984).

For example, the Mononobe-Okabe equation is widely used to evaluate earthquake-induced forces upon anchored or tied-back walls. To the extent that active conditions exist behind such walls, this is an appropriate use of the Mononobe-Okabe equation - provided that the resultant force is placed well up on the wall. In this connection, one important result of the theory is the prediction that failure surfaces may lie flatter for earthquake loading than for the static case, so that anchors must be placed farther from the wall. The equation for the inclination of the failure surface may be found in Zarrabi (1973).

Residual forces may be important in the design of such walls. An example of a wall analyzed using an approximate procedure to evaluate residual forces is sketched in Fig. 14. Here a tied-back wall supports, over its upper portion, a zone which had experienced repeated slope movements prior to construction of the wall. The soil in this zone has essentially no cohesion. Deeper strata are cohesive, and exert relatively low earth pressures against the wall. To evaluate the residual forces on the wall, and hence in the tie rods, after shaking by the design earthquake, an iterative procedure was adopted. First, slip of the slide zone was calculated as though the wall were not present. This movement, multiplied by the stiffness of the tie rods (including their anchorages) gave a first estimate for the residual forces in these rods. This increased force upon the sliding mass implies an increase in the acceleration coefficient at which the slope would begin to slip. Using this new coefficient, a reduced permanent displacement is computed. This calculation was repeated until convergence to a calculated slip consistent with the associated retaining force. The tie rods and anchors were pro-portioned to give a long-term resistance greater than this calculated force.

CONCLUDING REMARKS

On occasions in the past, I argued that - while the seismic design of gravity walls was not a major pressing national problem - this relatively simple problem would serve well for starting research into the important, broad challenge of developing methods for evaluating permanent deformations and movements of earth masses. I certainly was wrong that the behavior of gravity retaining walls is simple. Nonetheless, the studies have indeed been fruitful. They have provided experimental data against which the predictions of numerical models have been checked, and they have emphasized the important role of the residual pressures remaining after shaking. They have also given continuing support for the use of the Mononobe-Okabe equation for design of relatively simple walls with heights of 30 feet or less. For higher walls, and those with restraint against outward movement, more careful analysis should be considered.

ACKNOWLEDGEMENTS

Much of my study has been supported by the National Science Foundation as part of the National Earthquake Hazard Reduction Program. I am indebted to the many students who have worked with me in these studies.

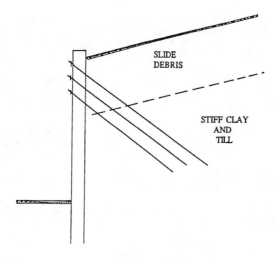

Figure 14 Schematic of tied-back wall

REFERENCES

A Al Homoud, Evaluating Tilt of Gravity RetainingWall During Earthquakes, ScD thesis (in preparation), Dept Civil Engg, Massachusetts Institute of Technology.

GR Andersen, RV Whitman and JT Germaine, Tilting Response of Centfuge-Modeled Gravity Retaining Wall to Seismic Shaking: Description of Tests and Initial Analysis of Results, Rpt. R87-14, Dept. of Civil Engrg., MIT, Cambridge, MA, 1987.

ATC, Seismic Design Guidelines for Highway Bridges, Applied Technology Council, Report No. ATC-6, Palo Alto, California, Oct. 1981.

R Dobry and G Gazetas, Dynamic Response of Arbitrarily Shaped Foundations, J. Geotechnical Eng'g (ASCE)   112, 2, 1986, 109-135.

GL Evans, The Behavior of Bridges Under Eathquakes, Proc New Zealand Roading Symp, Victoria U., Wellington, 2, 1971, 664-684.

AG Franklin and FK Chang, Earthquake Resistance of Earth and Rock-Fill Dams; Report 5, Permanent Displacement of Earth Embankments by Newmark Sliding Block Analysis, Misc. Paper S-71-17, Soils and Pavements Laboratory, U.S.Army Waterways Experiment Station, Vicksburg, Miss., Nov. 1977.

S Iai, Y Matsunaga and T Urakami, Performance of Quaywall during 1987 Chibaken Toho-Oki Earthquake, Proc. Disc. Session on Influence of Local Conditions on Seismic Response, 12th Int. Conf. on Soil Mech. and Foundation Engrg., Rio de Janeiro, 1989, 63-66.

PN Jacobsen, Behavior of Retaining Walls under Seismic Loading, ME Report 79/9, Dept Civil Engineering, Univ. of Canterbury, Christchurch, New Zealand, 1980.

CS Lai, Behavior of Retaining Walls under Seismic Loading, ME Report 79/9, Dept. of Civil Engrg., University of Canterbury, Christchurch, New Zealand, 1979.

E Marciano, Seismic Response and Damage of Retaining Structures, PhD thesis, Dept. Civil Eng'g, Purdue U.

N Mononobe and H Matsuo, On the Determination of Earth Pressures During Earthquakes, Proc. World Engrg. Congress, 9, 1929.

F Nadim, Tilting and Sliding of Gravity Retaining Walls, S.M. Thesis, Dept. of Civil Engrg., MIT, Cambridge, MA, 1980.

F Nadim, A Numerical Model for Evaluation of Seismic Behavior of Gravity Retaining Walls, Sc.D. Thesis, Res. Rpt. R82-33, Dept. of Civil Engrg, MIT, Cambridge, MA, 1982.

F Nadim and RV Whitman, Seismically Induced Movement of Retaining Walls, J. of Geotechnical Engrg. (ASCE) 109(7), July 1983, 915-931.

F Nadim and RV Whitman, Coupled Sliding and Tilting of Gravity Retaining Walls during Earthquakes, Proc., 8th World Conf. on Earthquake Engrg., San Francisco, III, 1984, 477-484.

NM Newmark, Effects of Earthquakes on Dams and Embankments, Geotechnique, 15(2), 1965, 139-160.

S Okabe, General Theory of Earth Pressures, J. Japan Soc. of Civil Engrg., 12(1), 1926.

S Prakash and BM Basavanna, Earth Pressure Distribution Behind Retaining Wall During Earthquake, Proc., 4th World Conf. on Earthquake Engineering, Santiago, Chile, 1969.

RJ Richards and D Elms, Seismic Behavior of Gravity Retaining Walls, J of the Geotechnical Engrg. Div. (ASCE) 105(GT4) April 1979, 449-464.

RF Scott, Earthquake-induced Pressures on Retaining Walls, Proc. 5th World Conf. on Earthquake Engrg., Rome, Italy, 1973.

HB Seed and RV Whitman, Design of Earth Retaining Structures for Dynamic Loads, ASCE Specialty Conf.-Lateral Stresses in the Ground and Design of Earth Retaining Structures, 1970, 103-147.

MA Sherif, I Ishibaski and CD Lee, Earth Pressure Against Rigid Retaining Walls, J. of the Geotech. Engrg. Div. (ASCE) 108(GT5), May 1982, 679-695.

MA Sherif and YS Fang, Dynamic Earth Pressures on Walls Rotating about the Top, Soils and Foundations, 24(4), 1984, 109-117.

R Siddharthan and EM Maragakis, Performance of Flexible Retaining Walls Supporting Dry Cohesionless Soils under Cyclic Loads, Int. J. Numerical and Analytical Methods in Geomechanics, 13(3), 1989, 309-326.

C Stamatopolous and RV Whitman, "Prediction of Permanent Tilt of Gravity Retaining Walls by the Residual Strain Method, Proc. 4th US National Conf. on Earthquake Engrg., Palm Springs, CA, 1990.

RS Steedman, Modelling the Behviour of Retaining Walls in Earthquakes, PhD Thesis, Engineering Dept., Cambridge Univ., UK, 1984.

RS Steedman and X Zeng, The Seismic Response of Waterfront Retaining Walls, ASCE Specialty Conf. on Design and Construction of Earth Retaining Structures, Ithaca, NY, 1990.

RS Steedman and X Zeng, The Influence of Phase on the Calculation of Pseudo-Static Earth Pressure on a Retaining Wall, Rpt CUED/D-soils TR 222, Engineering Dept., Cambridge Univ., UK, submitted to Geotechnique, 1989.

USACE, Stability of Earth and Rock-Fill Dams, EM 1110-1-1902, Dept. of the Army, US Army Corps of Engrs., Washington, DC, April 1970.

DK Vaughan and E Richardson, FLEX User's Manual, Weidlinger Assoc. Los Gatos, CA 1989.

HM Westergaard, Water Pressures on Dams during Earthquakes, Transactions, ASCE, 98, 1933, 418-472.

RV Whitman, Seismic Design of Gravity Retaining Walls, Proceedings International Conference Earthquake Resistanr Design and Construction, Berlin, June 1989, AA Balkema, Rotterdam (in press)

RV Whitman and S Liao, Seismic Design of Gravity Retaining Walls, US Army Engr. Waterways Experiment Sta., Misc.Paper GL-85-1,   1985.

RV Whitman and S Liao, Seismic Design of Gravity Retaining Walls, Proc. 8th World Conf. on Earthquake Engineering, San Francisco, III, 1984, 533-540.

RV Whitman and JT Christian, Seismic Response of Retaining Structures, Symposium Seismic Design for World Port 2020, Port of Los Angeles (in publication), 1990.

CP Wong, Seismic Analysis and Improved Seismic Design Procedure for Gravity Retaining Walls, Research Report 82-32, Dept Civil Engrg., MIT, 1982.

JH Wood, Earthquake-induced Soil Pressures on Structures, Rpt. No. EERL 73-05, CIT, Pasadena, CA, 1973.

PMF Yong, Dynamic Earth Pressures Against a Rigid Earth Retaining Wall, Central Laboratories Rpt 5-8515, Ministry of Works and Development, Lower Hutt, New Zealand, 1985.

K Zarrabi, Sliding of Gravity Retaining Wall During Earthquakes Considering Vertical Acceleration and Changing Inclination of Failure Surface, SM Thesis, Dept. of Civil Engrg., MIT, Cambridge, Mass. 1973.

SEISMIC PERFORMANCE OF TIED-BACK WALLS

Carlton L. Ho[1], M.ASCE, Gordon M. Denby[2], M.ASCE,
and Richard J. Fragaszy[3], M.ASCE

ABSTRACT: This paper presents the compilation of a survey on the response of tied-back retaining walls subjected to the Whittier, California earthquake of October 1, 1987. The data were compiled from interviews with shoring contractors, geotechnical engineers, owners, and workers in the Los Angeles region. Ten case studies are presented. Data presented include ground motion, subsurface soil conditions, excavation geometry, shoring design and installation, tieback test results, pre- and post-earthquake instrumentation, and excavation response. The tied-back walls for the ten case studies all performed well during the earthquake. The data indicate that there is a great variability in the lateral pressure used to design the shoring systems as well as the design of the anchor. Therefore, there should be a corresponding variability in the actual static and dynamic factors of safety of the shoring system.

INTRODUCTION

The purpose of this study is to research and document the response of tied-back retaining walls during the earthquake centered near Whittier, California on October 1, 1987 (Denby, et al., 1989). Shoring contractors and geotechnical engineers in the Los Angeles region were contacted to identify excavations which were retained by a tied-back shoring wall system and which were open during the earthquake. Various construction sites, owners, engineers and contractors in the Los Angeles region were visited to identify excavations and to discuss earthquake effects. Data collected on identified excavations included subsurface and excavation profiles, tieback installation and testing methods, post-earthquake testing, shoring design, and the observed and perceived earthquake effects. Anchor lift-off tests were conducted at one project site to evaluate the potential change in tieback loads due to the earthquake.

---

1 - Asst. Prof., Dept. of Civ. and Env. Engrg., Washington State University, Pullman, WA, 99164-2910.
2 - Principal, GeoEngineers, Inc., 2405 140th Ave. NE, Bellvue, WA 98005.
3 - Assoc. Prof., Dept. of Civ. and Env. Engrg., Washington State University, Pullman, WA 99164-2910.

EARTHQUAKE EVENT

The Whittier Narrows Earthquake occurred on Thursday, October 1, 1987 at 7:42 a.m. local time. The initial shock was magnitude 5.9 on the Richter scale, and was followed by a magnitude 5.3 aftershock on Sunday, October 4, 1987 (Brady, et al., 1988 and Leyendecker, et al., 1988). Significant structural and soil-related failures occurred as a result of the event (Weber, 1987).

PROJECT IDENTIFICATION AND DATA COLLECTION

Site visits. A telephone survey of shoring contractors in the Los Angeles area was used initially to identify tied-back wall excavations that were open during the earthquake. Subsequently, three visits were made to Los Angeles during April-June 1988 to meet with contractors and engineers and to visit two excavations where the shoring system was still visible. The meetings with contractors and engineers were used to identify other excavations and to interview workers about their visual perceptions of the response of the shoring systems during the earthquake. Other contacts during our work included the owners, government agencies and surveyors. Many contacts were established which assisted with the subsequent more detailed data collection for documentation of the case histories.

Only one site provided an opportunity for post earthquake testing of the tiebacks due to the construction of external walls which eliminated access to the tiebacks at other sites. Ten anchors lift-off tests were conducted at this site to evaluate load changes that occurred due to the earthquake. The tests consisted of mounting a jack on the tieback tendon and applying load until pressure was released on the lock-off nut. These loads were then compared to the original lock-off loads.

Results of project identification. Seventeen tied-back wall projects were identified where the excavations were open during the October 1, 1987 earthquake. The owners of each project were contacted to obtain permission to collect, compile and publish the project data as it pertains to the response of the excavation during the earthquake. This permission was provided subject to the condition that the owners, other project contacts and project addresses would not be identified.

Of the 17 projects, tiebacks had not been installed in three of the excavations at the time of the earthquake. Permission was denied for one project due to litigation. It was uncertain if the damage to the retaining wall was a result of the earthquake, or of inadequate construction. Permission was denied for three other projects for unspecified reasons. Data were subsequently collected and reviewed for the remaining 10 projects, which are designated as Case Histories 1 through 10 in this report.

Information provided for each case history included geotechnical reports, shoring plans, tieback testing and installation records, and displacement monitoring records.

SUMMARY OF CASE HISTORY DATA

Case history information was reviewed and compiled according to the key parameters described above (Denby, et al., 1989). Due to the length limitations of this paper, not all of the data can be presented. The complete presentation of the data can be found in the GeoEngineers, Inc. report (Denby, et al., 1989). The following paragraphs summarize data from the 10 case histories.

Ground motion. The case history sites in this study were located between 7 and 23.5 miles (11.3 to 37.8 km) from the epicenter (epicentral distance). Topography in the vicinity of these sites is generally flat. Ground accelerations measured by the U.S. Geological Survey (USGS) and the California Division of Mines and Geology (CDMG) in the vicinity of the sites covered a wide range of values: 0.09 g for Case History 3; 0.10 g to 0.19 g for Case Histories 3, 4, 5, 6, 7, 8, and 9; 0.20 g to 0.29 g for Case Histories 1 and 2; and 0.40 g for Case History 10. The Modified Mercalli intensities ranged from VI in the vicinity of Case Histories 1, 3, 4, 6, 7, 8, 9, and 10, to VII for Case Histories 2 and 5. The intensities were determined from an isoseismal study conducted by the USGS (Leyendecker, et al., 1989). Table 1 lists the epicentral distances, measured peak ground acclerations in the vicinity, and the intensities for each case history.

TABLE 1.--Ground Motion Data for Case Histories

Case History Number (1)	Epicentral Distance mi (km) (2)	Nearest Measured Peak Ground Acceleration (g)[a] (3)	Intensity (MMI) (4)
1	16.0 (25.8)	0.26	VI
2	7.0 (11.3)	0.20	VII
3	23.0 (37.0)	0.09	VI
4	10.0 (16.1)	0.18	VI
5	7.0 (11.3)	0.20	VII
6	18.5 (29.8)	0.10	VI
7	21.0 (33.8)	0.09	VI
8	23.5 (37.8)	0.09	VI
9	21.0 (33.8)	0.21	VI
10	15.0 (24.2)	0.40	VI

[a] from Brady, et al., 1989

Subsurface soil conditions. A wide range of subsurface conditions is represented by the ten sites. The subsurface materials consist of interbedded sand, silt, clay and shale bedrock. The density of the soils ranged from medium dense to dense and stiff to very stiff. Bedrock was encountered in only one of the ten excavations (Case History 4). Ground water apparently was not encountered in any of the excavations. Summaries of the soil profiles for the 10 case histories are given in Table 2.

TABLE 2.--Subsurface Soil Conditions for Case History Sites

Case History Number	Subsurface Soil Conditions		
	Depth ft[a]	USCS Classification	Soil Description[b]
(1)	(2)	(3)	(4)
1	0 - 5	SM/ML (fill)	L to MD
	5 - 8	SM/ML	L to MD
	8 - 40	SP-SM	L to MD
	40 +	SP	VD
2	0 - 5	SM/ML (fill)	D
	5 - 53	SP/SP-SM	D to VD
	53 +	SM/ML	D
3	0 - 10	SM/ML	MD to D
	10 - 20	ML	S to VS
	20 - 25	SM	D
	25 - 30	CL	VS
	30 - 45	SM	D
4	0 - 10	Variable	L to MD
	10 - 20	SM	MD
	20 -150	ML	H
5	0 - 8	SM/ML	MD-D
	8 - 38	SP	D to VD
6	0 - 4	SC/CL	L to MD
	4 - 12	ML	MS
	12 - 26	SC/CL	D
	26 - 31	SP	D
	31 - 50	SC/CL/SM	MD to D
7	0 - 10	SM/SC	MD
	10 - 25	SM/ML	MD
	25 - 50	SC/CL	D
8	0 - 6	CL	S
	6 - 25	GC	MD
	25 - 38	ML	MS
	38 - 59	GC/CL	D to VD
	59 +	GC	D to VD

9	0 -2.5	SM (fill)	MD
	2.5- 8	SM	MD
	8 - 17	SM/ML/CL	MD
	17 - 33	ML/SM	S
	33 - 35	SP	D
10	0 - 3	ML	MS
	3 - 26	SM	MD to D
	26 - 42	ML	S to VS
	42 - 49	SM	D
	49 - 62	ML	S
	62 - 88	SM	VD
	88 -103	SP	VD

^a Handled below

[a] 1 ft. = 0.305 m
[b] Soil Descriptions

| Coarse grain: | VL--very loose; L--loose; MD--medium dense; D--dense; VD--very dense. |
| Fine grain: | VS--very soft; S--soft; M--medium; S--stiff; VS--very stiff; H--hard. |

Excavation geometry. The design depth of the excavations ranged from 19 ft. (5.8 m) with one row of tiebacks to 90 ft. (27.5 m) with five rows of tiebacks. All but two of the case histories (Case Histories 4 and 10) were temporary shoring walls, and all but one (Case History 10) were rectangular excavations.

Eight of the ten excavation shoring wall systems were completed by the time of the earthquake as shown in Table 3. The systems for Case Histories 4 and 8 were about 40 percent and 70 percent completed, respectively. The maximum depth for all excavations at the time of the earthquake was 50 ft. (15.3 m). The excavation geometries are presented in Table 4.

Shoring design and installation. Shoring design and installation parameters include tieback and soldier pile spacing, anchor skin friction and soil pressure on the wall. The shoring design methodologies for the case histories are outlined in Table 5. The shoring design affects the response of the shoring system to earthquake loading--more conservatively designed systems with a higher factor of safety would be expected to perform better with lower displacement than less conservatively designed systems. The components of the shoring systems for all ten case histories were similar. Each wall was composed of tied-back soldier piles with timber lagging. The soldier piles typically consisted of steel wide flange beams set in predrilled holes, spaced 6 to 8 ft. (2.4 m), center-to-center, which were then filled with 1 to 1.5 sack cement per cubic yard "lean mix" concrete. The design skin friction value for the tiebacks was typically 600 psf (28.7 kPa), although 1000 psf (47.9 kPa) was allowed for tiebacks anchored in shale at one site.

Typically for a rectangular pressure distribution, the design lateral pressure is determined using a reduction factor on the vertical stress such that

$$p = R_f \, \gamma \, H \tag{1}$$

Table 3.--Stage of Excavation at Time of Earthquake.

Case History Number (1)	Average Depth ft[a] (2)	Tieback Rows Completed (3)	Shoring Completed (4)	Other Support (5)
1	43.5	2 to 3	100%	None
2	19.0	1	100%	None
3	33.0	2	100%	None
4	35.0[b]	3	40%	Rakers (?)
5	35.0	2 to 3	100%	None
6	24.0	1	100%	None
7	40.0	2[c]	100%	None
8	41.0($\frac{1}{2}$) 20.0($\frac{1}{2}$)	1 to 2	70%	Rakers on 20.0 ft($\frac{1}{2}$)
9	24.0	1	100%	None
10	40.0	4	100%[d]	None

[a] 1 ft. = 0.305 m
[b] along permanent wall
[c] assumed
[d] concrete lagging not in place

where p equals the design lateral pressure, R_t equals the reduction factor, γ equals the unit weight of the soil, and H equals the height of the excavation. Peck (1969) recommends that R_t is dependent on the soil type. R_t can vary from 0.15 to 0.40 (Goldberg, et al.; 1976, Peck, 1969). Typically, R_t and γ are given as a single value. The design lateral pressures ranged from 18H to 24H psf. The shape of the distribution is also a function of the soil type (Peck, 1969). The shoring designs were all based on trapezoidal pressure distribution. The design pressure distributions are outlined in Table 6.

It is interesting to note the variation in the design lateral pressures presented in Table 6. The lateral pressures range from 15H to 30H for soil conditions which are essentially similar (medium dense to dense and stiff to very stiff interbedded sand, silt and clay) at the 10 sites.

In only one of the case histories (Case History 4) was seismic earth pressure specifically considered in design of the tied-back wall. This was for a permanent tied-back wall. An inverted triangular distribution with a maximum pressure of 10H at the top of the wall was used, based on a magnitude 6.5 earthquake producing a horizontal ground acceleration of 0.26g. A seismic force for design of the basement walls was apparently used in Case History 7, but not for tied-back walls.

Tiebacks were typically installed "open hole": the tieback holes were drilled, the reinforcing steel was placed down the hole, and then grout was pumped into the anchor zone. Anchors were placed behind a no-load surface. The no-load zone between the anchor and the shoring wall was generally backfilled with a sand slurry or with lean mix concrete.

TABLE 4.--Excavation Geometry for Case Histories

Case History Number (1)	Length ft[a] (2)	Width ft (3)	Maximum Depth ft (4)	Minimum Depth ft (5)	Average Depth ft (6)
1	141.0	125.0	50.0	37.0	43.5
2	160.0	152.0	19.0	10.0	13.0
3	590.0	125.0	43.0	31.0	38.0
4	260.0	230.0	90.0[b]	40.0[c]	65.0
5	369.5	231.0	44.5	26.5	35.0
6	200.0	N/A	33.0	15.0	24.0
7	225.0	210.0	40.0	[d]	[d]
8	249.0	150.0	43.5	37.5	40.0
9	257.0	187.0	27.0	20.0	24.0
10	720.0	[d]	42.0	9.0	40.0

[a] 1 ft. = 0.305 m   [b] Permanent wall
[c] Temporary wall   [d] Data not available

TABLE 5.--Shoring Design for Case Histories

Case History Number (1)	Number of Rows (2)	Vertical Spacing (Average) ft[a] (3)	Horizontal Spacing ft (4)	Anchor Skin Friction psf[b] (5)	
1	2 or 3	12.5	8.0	600	
2	0 or 1	N/A	8.0	100 + 25h[c]	≈ 320
3	2	13.0	8.0	600	
4	5 or 6	11.0	6.0	1000	
5	2 or 3	13.5	8.0	150 + 15h[c]	≈ 400-700
6	1	N/A	8.0	600	
7	[d]	[d]	[d]	[d]	
8	2	17.0	8.0	600	
9	1	N/A	8.0	500	
10	4	12.0	6.0	[d]	

[a] 1 ft. = 0.305 m   [b] 1 psf = 0.0478 kPa
[c] h = depth to anchor in ft   [d] Data not available

TABLE 6.--Design Pressure for Case Histories

Case History Number	Design Lateral Pressure[a] psf	R_t for γ = 120 pcf
(1)	(2)	(3)
1	22 H	0.18
2	15 H	0.13
3	18 H	0.15
4	24 H[b]	0.20
4	10 H[c]	0.08
4	20 H[d]	0.17
5	20 H	0.17
6	19 H	0.16
7	30 H	0.25
8	18 H	0.15
9	19 H	0.16
10	[e]	[e]

[a] H in ft; 1 psf = 0.0479 kPa
[b] Permanent wall static design
[c] Permanent wall seismic design
[d] Temporary wall static design
[e] Data not available

Tieback test specifications.   Tieback testing typically consisted of performance testing approximately 5 percent of the tiebacks and proof testing all other tiebacks. The performance tests included loading the tieback to 200 percent of the design load and maintaining this load for periods ranging from 2 hours to 48 hours.

The proof tests included loading each tieback to 150 percent of its design load and maintaining this load between 5 and 15 minutes. Deflections were measured during loading and during the sustained maximum load.   The deflection during the sustained load is referred to as creep. A compilation of the creep test specifications are shown in Table 7.

Only six tiebacks out of approximately 2000 proof tested tiebacks in ten case histories failed, and most showed minimal creep deflections. None of the performance test anchors failed. In most cases, the load-displacement test curves were nearly linear to 150%, indicating the actual safety factors were probably higher than 2.

Pre- and post-quake instrumentation records.   The shoring walls were typically monitored during construction by measuring the vertical and horizontal offset from a predetermined reference elevation on a line. The accuracy of these measurements is generally within 0.01 ft. (3 mm). Only two sites indicated apparent movement due to the earthquake. Approximately 0.25 in. (6.4 mm) of horizontal deflection of the top of the wall was measured for Case History 1. One corner of the wall for Case History 6 moved horizontally 1.75 in. (44.5 mm), but this movement was due to a failed corner brace and not a tieback.

Table 7.-- Tieback Test Specifications: Creep Load Test

Case History Number (1)	Test Load Percentage of Design Load (2)	Hold Duration (3)	Allowable Creep Deflection in[a] (4)	Number of Failures (5)
1	200%	24 hr	0.75	0
1	200%	30 min	0.25	0
1	150%	15 min	0.10	0
2	150%	15 min	0.10	1[b]
3	150%	15 min	0.10	2[c]
4	150%	15 min	0.20	0[d]
5	150%	15 min	0.10	1[e]
6	[f]			
7	[f]			
8	150%	2 hr	0.25	0
9	150%	24 hr	0.75	0
9	150%	15 min	0.10	0
10		[g]		0

[a]   1 in. = 2.54 cm
[b]   Corner brace installed rather than replacement.
[c]   Records not available on failed anchors.
[d]   2 proof test failures (total deflection > 12 in. (305 mm)).
[e]   Replaced with anchor on opposite side of soldier pile.
[f]   Data not available.
[g]   Anchors tested to 200% of design in 25% increments; one minute with no slippage required at each increment; four minutes with no slippage required at 200%.

One site (Case History 4) had tieback load cells installed. No discernible change in the tieback loads was measured after the earthquake. This excavation was approximately 30 percent complete at the time of the earthquake and, therefore, the factor of safety would probably have been higher than the design factor of safety at full depth. Lift-off tests at the site of Case History 3 indicated a range of 5 to 32 percent decrease in lock-off load in the anchors.

Excavation response. Visual perceptions of the shoring behavior during the earthquake were recorded from witnesses for many of the sites. These reports ranged from "the wall appearing to vibrate slightly" (Case History 1) to "moving all over" (Case History 5). For the Case History 6 excavation, workmen are reported to have "run out" of the excavation. A field engineer monitoring performance testing of a tieback at the site of Case History 4 noted the occurrence of the earthquake; however, the event apparently had no effect on the progress or the results of the performance test.

DISCUSSION AND CONCLUSIONS

The tied-back walls for the case histories documented in this study performed very well and experienced little to no loss of integrity due to the earthquake. In the context of non-failure, the designs of the walls may, therefore, be considered to have performed satisfactorily for the ground motions at the sites. However, non-failure does not indicate whether the systems may be overdesigned. Before general conclusions may be made on the seismic response of tied-back walls, the data must be analyzed in more detail. Specifically, the actual safety factor for the shoring system design needs to be evaluated. If the actual safety factor is much higher than for conventional design, then there is less to be learned from the wall performance observed. In contrast, if the actual safety factor is closer to the conventional design value, then the assertion that tied-back walls are not significantly affected during earthquakes gains more support.

For instance, the tieback test results indicate only six failures in over 2000 tiebacks tested. In most cases, the creep deflections were negligible, even at 200 percent of design load. This indicates that the actual safety factor for the anchor capacity is significantly higher than the design safety factor of 2.0. A similar rationale may apply to the design earth pressures.

An interesting side benefit of this study was the opportunity to compare the recommended design lateral pressures of five different geotechnical engineering firms in the Los Angeles area. For soil conditions which may be considered to be practically similar, the design lateral pressures in psf ranged from 15H to 30H, where H equals the depth of the excavation in feet. This variation is considered to reflect the individual design philosophies of each of the firms rather than any difference in soil conditions.

ACKNOWLEDGEMENTS

This material is based upon work supported by the National Science Foundation under Grant No. CES-8804304. The authors wish to thank all of the owners of the ten projects summarized in this paper for allowing us to collect and publish the data. We also wish to acknowledge Eugene Birnbaum (Eugene D. Birnbaum, Inc.), Dan Hoffman (Converse Consultants), Marshall Lew & Perry Maljian (LeRoy Crandall, Inc.), Bob Frankian, Ken Pitcher, and Alan Wing (RT Frankian), John Meli (HCB), William Beckler (HLA),

Richard Reinhart (Kovacs-Byers), Don Young (Gerald Lehmer and Assoc.), Ed Bucher (Malcolm Drilling), Paul Meyer (Pace Eng.), Greg Hindson (Pafford Assoc.), Jon Browning (Pfeiller and Assoc.), Fred Mueller (Psomos and Assoc.), Richard Wu (Shoring Engineers), Lutz Kunz (Smith-Emery), Cheryl Zonver (Zonver Drilling), for the contribuiton of their time, effort, and data to this study.

REFERENCES

1. AG Brady, EC Etheredge, and RL Porcella, The Whittier Narrows, California Earthquake of October 1, 1987--Preliminary Assessment of Strong Ground Motion Records, *Earthquake Spectra*, 4(1), 1988, pp. 55-74.
2. GM Denby, HR Pschunder, and DE Argo, Response of Tied Back Walls During the Whittier Narrows Earthquake of October 1, 1987, Rpt. Grant No. CES-8804304, NSF, Washington, DC, 1989 29 pp.
3. DT Goldgerg, WE Joworski, and MD Gordon, Lateral Support Systems and underpinning, Rpt FHWA-RD-75-129, FHWA, Washington, DC, April 1976.
4. EV Leyendecker, LM Highland, M Hopper, AP Arnold, P Thenhaus, and P Powers, The Whittier Narrows, California Earthquake of October 1, 1987--Early Results of Isoseismal Studies and Damage Surveys, *Earthquake Spectra*, 4(1), 1988, pp 1-10.
5. RB Peck, Deep Excavations in Soft Ground, *Proc. Int. Conf. on Soil Mech. and Found. Eng.*, State-of-the-Art Vol., Mexico City, 1969, pp. 225-250.
6. Weber, F. H. (1987) Whittier Narrows Earthquakes, Los Angeles County, *California Geology*, 40(1), pp. 275-281.

SEISMIC DESIGN OF RETAINING WALLS

David G. Elms[1], M.ASCE and Rowland Richards[2], Jr., M.ASCE

ABSTRACT: A series of theoretical and experimental investigations is summarized on the seismic behavior of gravity, tied-back and reinforced-earth retaining walls. The Newmark sliding-block model is shown to be appropriate for displacement-controlled design, though modifications and limitations are discussed. Test results show that sliding block behavior only takes place after a high limiting acceleration has been passed. Where rotational failure is expected, walls will be stronger than anticipated. Elastic resonance effects are likely to be significant for full-size walls.

INTRODUCTION - - DISPLACEMENT-CONTROLLED DESIGN

In earlier papers (3, 19) a displacement-controlled approach was developed for the seismic design of gravity retaining walls. The work stemmed from a project (18) whose aim was to rationalize procedures for the earthquake design of retaining structures. For gravity walls the problem was that the very mass that gave stability in a static situation acted against the wall in an earthquake. A pseudo-static force-based design approach required that the wall should resist lateral forces due to (a) lateral earth pressures, obtained by the Mononobe-Okabe method (12, 17), and (b) the intertia effects of the wall. The lateral forces were resisted primarily by the wall base resistance against sliding or overturning, which of course depended on the wall mass. The force approach led to unacceptably high values of mass for stability.

However, it was shown (3, 19) that although in an earthquake a wall with a mass lower than some critical value would displace, the displacement was neither infinite nor indefinite. It was finite and, for a given earthquake record, calculable.

The authors made the simple assumption that outward movement of the wall would take place only when the lateral earthquake acceleration away from the wall into the soil was greater than a specific acceleration at which slipping of the wall would take place. It was also assumed that once sliding took place it would

1 - Prof., Dept. of Civil Engineering, Univ. of Canterbury, Christchurch, New Zealand.

2 - Prof., Dept. of Civil Engineering, Univ. at Buffalo, Buffalo, New York, NY 14260.

854

do so at a constant wall acceleration, and would continue at the value until the relative velocity between soil and wall reached zero. The behavior of the model is illustrated in Figure 1. The

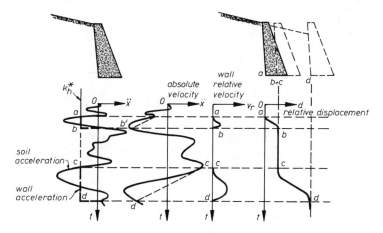

Figure 1. Incremental Displacement

figure shows that integration of the relative velocity between soil and wall gives the displacement for a particular earthquake pulse. Displacement thus takes place in an incremental manner, in a series of small outward movements.

Using this model the authors were able to calculate the displacements for specific earthquake records. They later came to realize their model was the same as the sliding-block model developed earlier by Newmark (16) for dams and embankments. A similar approach had been taken by Sarma (21). Newmark showed how the results could be standardized to compare the results for different earthquakes. The idea was taken up by Franklin and Chang (6) who produced envelope curves from the results of displacement calculations for a large number of earthquake records. The authors used Franklin and Chang's curves as the basis of a simple design formula by which displacement could be traded off against wall mass. The design approach is summarized as follows:

1. Choose an allowable design displacement d
2. Compute the design wall slipping or cut-off acceleration coefficient N from the formula

$$N = A \left[\frac{0.087V^2}{dAg} \right] 1/4 \qquad (1)$$

856   EARTH RETAINING STRUCTURES

where A is the peak acceleration coefficient for the design earthquake, V is its maximum acceleration (m/sec), d is the design displacement (m) and g is the gravitational acceleration (m/sec^2). (V, d and g could be expressed in any consistent system of units).

3. Find the active lateral earth pressure coefficient K_{AE} from the Mononobe-Okabe relation

$$K_{AE}= \frac{\cos^2(\phi-\theta-\beta)}{\cos\theta \; \cos^2\beta \; \cos(\delta+\beta+\theta) \left[1+\sqrt{\frac{\sin\,(\phi+\delta)\,\sin(\phi-\theta-i)}{\cos\,(\delta+\beta+\theta)\,\cos(i-\beta)}}\;\right]^2} \tag{2}$$

where β, i and δ are defined in Fig. 2, ϕ is the angle of friction of the soil and $\theta = \tan^{-1}k_h/(1-k_v)$, and where k_h, the horizontal acceleration coefficient, is the same as the cutoff acceleration N obtained from Eq. 1 above, and k_v is the vertical acceleration coefficient.

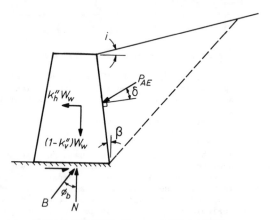

Figure 2.   Wall Forces and Notation

4.   Compute the required mass of the wall M_w from

$$M_w= \frac{0.5\rho H^2\left[\cos(\delta+\beta) \; - \; \sin(\delta+\beta)\tan\phi_b\right] K_{AE}}{\tan\phi_b-\tan\theta} \tag{3}$$

where ρ is the density of the backfill (tonnes/m^3), H is the height of the wall (m) and ϕ_b is the base friction coefficient at the bottom of the wall, assuming it fails by sliding.

The process was simple, despite the rather complex equations. Nevertheless, various questions needed to be answered. Firstly, did failing walls actually behave in such a simple manner, sliding with a constant cut-off acceleration coefficient? Probably not exactly, because the analysis was based on the simplifying Mononobe-Okabe assumptions of a uniform ϕ, a plane failure surface, uniform acceleration and no resonance. The sliding block model assumed the soil and wall were subjected to undirectional shaking and that the soil wedge moved horizontally, which is geometrically impossible. In any case, the design approach seemed only applicable to gravity walls failing by sliding and could not be applied directly to other types of walls or to alternative failure modes such as rotation about the bottom. It was necessary to explore these matters before the approach could be used with broad confidence. The following sections describe some of the explorations.

GRAVITY WALLS FAILING BY SLIDING

Model Tests. Small-scale tests were carried out (1) with the aim of confirming whether the simple sliding-block model was justified. The test set-up is shown in Fig. 3. An 810 mm wall whose mass could be varied was free to fail by sliding. It was backfilled with beach sand, placed by raining in a reasonably dense state. Excitation was by means of a large spring whose release gave a decaying sinusoidal motion. Although a shake table was available it was not used (except as a passive support) as earlier tests (10) had shown it was impossible to get a noise-free sinusoidal acceleration profile, and examination of acceleration responses to simple excitations was critical to the investigation.

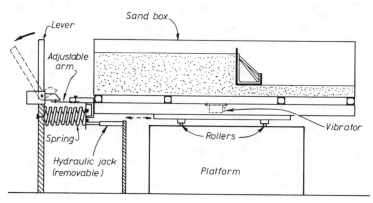

Figure 3. Experimental Set-up

Scaling effects for small-scale testing are a problem, but less for cohensionless than for cohesive soils. Essentially, the assumption had to be made that elastic effects were small, and that inelastic flow effects dominated. Even so, the sliding-block displacement levels resulting from feasible frequency levels mean that some geometric distortion must occur. For a discussion of scaling, see Schuring (22) and Kerisel (9). A further question on the use of small-scale tests is whether ϕ remains constant for low confining stresses (7).

The choice of a simple spring-driven excitation turned out to have unexpected benefits. It meant that the development of failure could be observed in detail and each stage examined at leisure, with a decision being made after each spring pulse as to what to do next.

Figures 4 and 5 show the nature of the backfill displacement during the first and second spring excitations of one test series. Long-exposure photography shows the before-and-after positions of vertical white sand markers, and also, faintly, grain movements. It can be seen in Fig. 4 that in some regions the white lines are parallel, indicating block movement, while in other places the lines are bent at an angle, indicating a zone of shear distortion rather than a distinct failure surface. In contrast, the second test, Fig. 5, shows a failure plane reaching from the toe of the wall to the backfill surface. What seems to happen is that initially, as the wall moves out, a wedge-shaped shear zone forms with its apex at the toe. With further excitation and movement, a failure surface begins to form at the toe, then grows towards the surface somewhat in the manner of a fatigue crack. In the process of failure surface formation, as will be seen later, resistance is high, and drops to a lower value when the failure surface is complete. Note that the failure surface is approximately a plane, as assumed by the Mononobe-Okabe theory.

Figure 6 shows the acceleration response to the first spring impulse of one of the tests. The wall response does not have the flat plateau presumed by the sliding block model, and neither is it constant between cycles. The only similarity is the steep final drop down to match the ground acceleration. This is in contrast with Figure 7, which gives the acceleration response for the third spring impulse on that particular test. At this stage, the predicted sliding block behavior occurs: a plateau occurs at the same height for all cycles. The response acceleration is considerably lower in Fig. 7 than the peak shown in Fig. 6. Figure 6 corresponds to Fig. 4, where a failure surface has not yet fully formed, while for the situation of Fig. 7, a complete failure surface exists. The peak acceleration values of Figs. 6 and 7 corresponded closely to theoretical values of cutoff acceleration using the peak and residual values of ϕ obtained by a direct shear test of the sand after it had been densified as much as possible.

Figure 4. Displacement During Initial Pulse

Figure 5. Displacement for Third Pulse -
Failure Plane Development

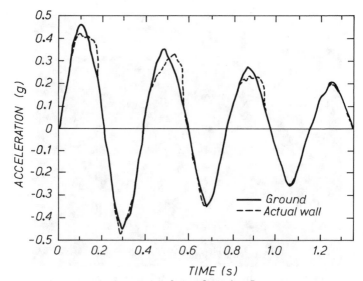

Figure 6.   Initial Acceleration Response

The tests thus lead to the following conclusions. Given the limitations of the small scale of the test, that the wall is constrained to fail by sliding, and that the test is two-dimensional, the sliding block model gives an accurate prediction of cutoff acceleration and total displacement provided the residual and not the peak value of ϕ is used. However, displacement will not commence (except for a very small movement) until a higher acceleration threshold has been crossed, at a level corresponding to a peak value of ϕ. Some initial shaking is required while sand in a failure zone densified prior to the development of a failure surface. The amount of shaking required must be the subject of future research, perhaps using an absorbed-energy approach.

Modifications to the Theory. A major modification to the simple theory was made by Zarrabi and Whitman (24, 26), who pointed out that because continuity requirements mean that the backfill wedge must drop down as the wall moves out, the accelerations and displacements of wedge and wall will not be the same. This lead to a more complex analysis in which the failure surface angle varied throughout an acceleration cycle. A modification of the Zarrabi-Whitman model used a constant wedge angle (8), based on the observation that in tests, as shown above, a fixed and stable failure surface formed.

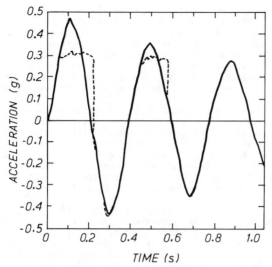

Figure 7. Acceleration Response After Failure Plane Development

However, it must be concluded from the complex behavior shown by tests and the difficulty of establishing a precise value of ϕ that despite the increased accuracy of the Zarrabi-Whitman model, its use in practice cannot be justified as it is significantly more difficult to use than the simple sliding-block model and its greater precision is swamped by the complexities of real-life behavior (8).

Whitman and Liao (25) reconsidered the sliding block model. They introduced a normalised (and therefore dimensionless) displacement

$$d' = dAg/V^2 \tag{3}$$

which has obvious advantages over the standardized displacements used earlier (16, 6, 19, 3). They also developed a curve to fit displacement data from various earthquakes which was a better fit than the Richards-Elms (19) straight line approximation implied by Eq. 1 above. Displacement was expressed by

$$d = \frac{37V^2}{Ag} \exp(-9.4 \ N/A) \tag{4}$$

However, this was a best fit relation whereas the Richards-Elms equation was an upper bound. Inverting Eq. 4 gives

$$N = 0.106 \ \ln \left[\frac{37V^2}{Agd}\right] \tag{5}$$

which would be an improvement on Eq. 1 in the design procedure out-lined above.

Franklin and Chang's results (6) showed that even standardized motions resulted in widely scattered calculated displacements. Lin and Whitman (11) attempted to narrow the scatter by incorporating ground motion characteristics into their random vibration model, as well as the physical mechanisms of displacement build-up. They investigated the distribution of displacements about an expected calculated value. They show that for N/A = 0.5, an upper bound on displacements would be about twice the mean calculated displacement.

So far, we have been assuming only one earthquake acceleration component. The effects of including vertical and transverse accelerations was considered by Sharma (23). The effect on displacement of a vertical component was negligible. There is little correlation between vertical and horizontal components, and the effects of downward and upward accelerations tend to cancel each other out. The effect of a transverse acceleration oriented along the length of the wall is, however, very significant.

The resistance of a gravity wall to sliding derives from friction at the base. If a transverse acceleration is applied at the same time as a longitudinal pulse, then unless the wall is supported by some other means against sideways movement, the base friction will have to resist a vector of longitudinal and lateral forces, so in effect decreasing the available resistance in the longitudinal direction and lowering the cutoff acceleration. In order to determine sliding-block displacements in this way, two orthogonal horizontal earthquake records are needed. They will normally be closely correlated, and because of the importance of correlation, real records must be used as suitably correlated artificial record pairs cannot easily be generated.

To find the sliding-block displacement, integration is carried out as for a single record, but with a time-varying rather than constant cut-off acceleration N. As N can only be decreased by the presence of transverse acceleration, sliding-block displacements must always increase. The amount of increase varies according to record and initial value of N/A. For the El Centro N-S and E-W records, there is an average displacement increase of 67% for N/A values between 0.5 and 0.9. Most other records examined show higher increases.

Clearly, the effect cannot occur unless a wall has no other transverse support than base friction, so that most walls would not be affected. Nevertheless, the potential increases in displacement are large enough to be a matter of concern, a concern which must also be felt for the stability of dams and embankments.

ROTATIONAL DISPLACEMENT

Rotation About the Base. This is an important failure mode. Nadim and Whitman (14) and later Elms and Richards (4) considered the problem theoretically. Some preliminary tests were carried out by Aitken (1), who found his test walls were surprisingly

hard to fail. This is explained by the fact that for failure by
outward rotation of a wall, the whole of the backfill wedge must
suffer shear distortion, for kinematic reasons. But, as shown
elsewhere in this paper, for substantial movement to take place,
the backfill must pass through a dense phase where ϕ reaches its
peak, and energy is required to do so. The requirement is
substantial for a single failure surface, and will be much more
so if a whole region has to be distorted.

<u>Rotation About the Top - Tied-Back Walls</u>. For tied-back walls
there are two main design considerations: the integrity of the
tie and anchor, and the passive resistance of the toe. For the
anchor, the most important point is that the failure surface
slope decreases with increasing lateral acceleration for both
active and passive failure, so that if the tie is too short,
interference could take place, as shown in Fig. 8.

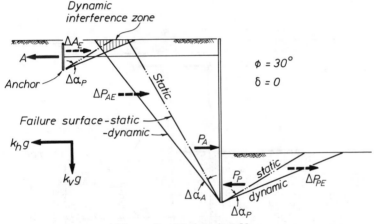

Figure 8. Tied-Back Wall

Limit analysis can be used for design and the sliding-block
analogy applied to outward displacement at the toe (20). The
active pressure is obtained from Eq. 2, while the passive
pressure at the toe is given by the Mononobe-Okabe passive
equation

$$K_{PE} = \frac{\cos^2(\phi - \theta + \beta)}{\cos\theta \, \cos^2\beta \, \cos(\delta - \beta + \theta) \left[1 - \sqrt{\dfrac{\sin(\phi + \delta) \, \sin(\phi - \theta + i)}{\cos(\delta - \beta + \theta) \, \cos(i - \beta)}} \right]^2} \qquad (6)$$

However, for calculating equilibrium, the points of action of the
active and passive forces are required. These will not be at the
third points because of the rotational rather than translational
movement of the wall: the active force will move up somewhat,
and the passive will move down.

To check that the sliding-block assumptions applied to passive
displacement, small scale tests were carried out (20) using
essentially the same test setup as that shown in Fig. 3, using
individual spring-driven pulses.   This time, however, the wall
was pin-supported at the top, and pushed into the soil at roughly
constant force using air jacks, as shown in Fig. 9. The effect
of soil friction on the wall was significant, so the actual wall
was allowed to move freely parallel to the swinging arm. Initial
pulses seemed to densify the soil, with a near-horizontal shear
plane growing from the wall toe but petering out after a short
distance.

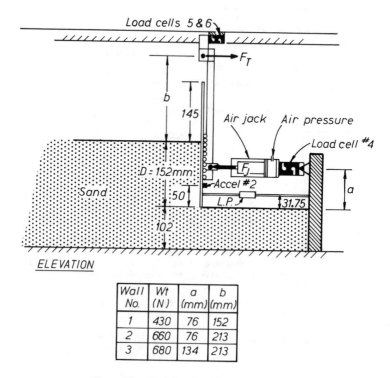

Figure 9.   Passive Pressure Test Set-up

Wall No.	Wt (N)	a (mm)	b (mm)
1	430	76	152
2	660	76	213
3	680	134	213

Pulses were kept at a low level until movement stopped, then increased to a higher level until once again movement stopped. During this phase various small shear surfaces formed radiating from the wall toe, though none broke through to the soil surface. Eventually, a complete shear surface formed, and large displacements occurred (Figure 10). The shear surfaces were either plane or convex upwards. Force, displacement and acceleration records at failure are given in Figure 11. The acceleration trace shows a plateau in the first pulse at a level corresponding to the residual ϕ, but that the subsequent cutoff acceleration drops further and movement occurs even due to the second, much lower, pulse.

Thus the sliding-block assumption applies to a passive-pressure situation if the residual value of ϕ is used. However, a higher force threshold has first to be passed, which is relatively higher than for the active, sliding wall, case. This means that the toe of a tied-back wall might have been seriously weakened even if it had survived an earthquake.

Figure 10.   Pulse 16, Test 3

REINFORCED EARTH WALLS

The displacement-controlled design approach can also in principle be applied to reinforced earth walls, though an additional and necessary design condition is that reinforcement should not fail in an earthquake. The displacement-controlled approach was first proposed by Bracegirdle (2). However, the reinforcement resistance R (Fig. 12) was undefined in his formu-

lation.   Nagel and Elms (15) assumed a 2-parameter linear
distribution for the reinforcement forces across a failure
surface for a rectangular reinforced-earth block. They went on
to investigate the kinematic behavior of small scale (300mm)
reinforced earth walls, again with a spring-loaded single-pulse
setup (Figure 13).   Face elements were full-width aluminum
strips, and the reinforcement consisted of fabric ribbon. As in
the earlier solid wall tests, a series of increasing pulses was
used, and a shear surface gradually propagated upwards towards
the backfill surface until when a complete plane had formed. At
this point large displacements occurred (Figure 14). Note that
the wall has moved out vertically, apart from rotation of the
bottom strip.

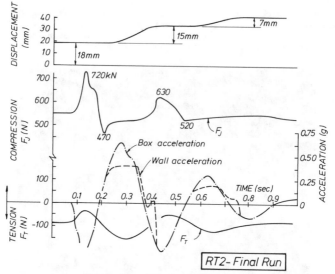

Figure 11.   Final Pulse, Test 2

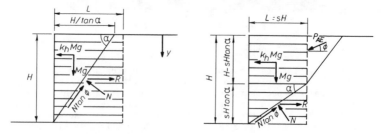

Figure 12.   Reinforced Earth Failure Types

Failure surface angles were well-predicted by theory, and where the failure surface was not contained within the reinforced volume, the bilinear shape assumed in the model could be seen (Figure 15). The acceleration time history of one of the walls at final failure is given in Fig. 16, where it can be seen a reasonable plateau response has developed, as is required for the sliding-block model to be used. Once again, the plateau acceleration corresponds to the residual ϕ.

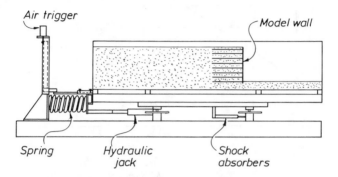

Figure 13.   Reinforced Earth Test Arrangement

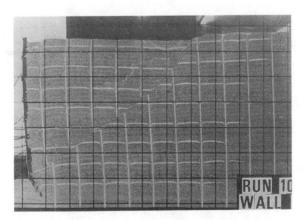

Figure 14.   Wall 6 After Run 10

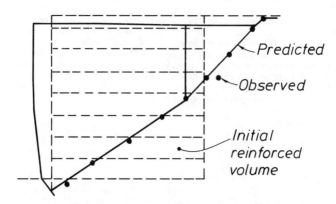

Figure 15.   Observed Behavior Failure Surface

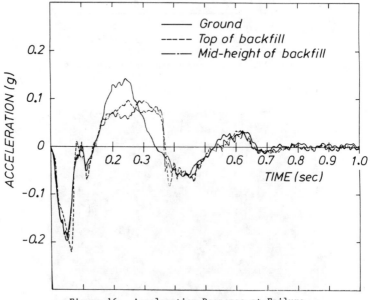

Figure 16.   Acceleration Response at Failure

The acceleration responses at the top and mid-height of the wall can be seen to be somewhat out of phase with each other in the plateau region, indicating some rocking of the wall. A series of larger scale tests was carried out (5) with 1 meter high walls, this time using a shake table rather than spring excitation. Wall rocking was more pronounced at this scale and elastic response seemed to be an important effect. The resonance effects are larger than would be expected from an elastic response using a normal value of soil modulus: they seem to occur only in conjunction with permanent outward movement of the wall. The matter needs more thorough investigation as the implications of earthquake enhancement for structures built above or near retaining walls are considerable.

A further point is that the increased acceleration levels due to elastic effects may mean that in effect a larger earthquake is applied to the wall, so increasing its displacement, as noted by Nadim and Whitman (13).

CONCLUSIONS

The implications of the results given above are that the displacement-controlled design approach can be used for the design of gravity walls, tied-back walls and reinforced earth walls, though using residual values of soil friction angle. Equation 5 should be used rather than Eq. 1. It is less conservative in that it is based on average values, it neglects amplification effects and it does not consider transverse earthquake affects. However, before a wall can slide it has first to receive sufficient excitation energy at higher acceleration levels to form a failure surface. Resonance effects during permanent displacement may be important for nearby structures.

ACKNOWLEDGEMENTS

Support for aspects of the work outlined above has been received from the NZ National Roads Board, the NZ Ministry of Works and Development and the Reinforced Earth Company (NZ) Ltd. The financial assistance received is most gratefully acknowledged.

REFERENCES

1.   GH Aitken, DG Elms & JB Berrill, Seismic Response of Retaining Walls, Res. Rept. 82-5, Dept. Civil Eng., Univ. of Canterbury, NZ, March 1982, 89p.

2.   A Bracegirdle, Seismic Stability of Reinforced Earth Retaining Walls, Bull. NZ Nat. Soc. for Earthq. Eng., 13(4), 1980, 347-354.

3.   DG Elms & R Richards, Seismic Design of Gravity Retaining Walls, Bull. NZ Nat. Soc. for Earthq. Eng., 12(2), June 1979, 114-121.

4. DG Elms & R Richards, Seismic Behaviour of Overturning Retaining Walls, Proc. 11th Australasian Conf. on Mech. of Structs. and Materials, Auckland, 1988, 335-338.

5. GJ Fairless, Seismic Performance of Reinforced Earth Walls, PhD Thesis, Dept. of Civil Eng., Univ. of Canterbury, NZ, 1989, 344p.

6. AG Franklin & FK Chang, Earthquake Resistance of Earth and Rockfill Dams: Report 5: Permanent Displacement of Earth Embankments by Newmark Sliding Block Analysis, Misc. Paper S-71-17, Soils and Pavements Lab., U.S. Army Eng. Waterways Expt. Stn., Vicksburg, Miss., Nov. 1977.

7. S Fukushima & F Tatsuoka, Strength and Deformation Characteristics of Saturated Sand at Extremely Low Pressures, Soils and Founds. (Jap. Soc. Soil Mech. & Found. Eng.), 24(4), 1984, 30-48.

8. PN Jacobson, DG Elms & JB Berrill, Translational Behaviour of Gravity Retaining Walls During Earthquakes, Res. Rept. 80-9, Dept. Civil Eng., Univ. of Canterbury, NZ, March 1980, 123p.

9. J Kerisel, Scaling Laws in Soil Dynamics, Proc. 3rd Pan-Am. Conf. on Soil Mech. and Found. Eng., (3), Caracas, Venezuela 1967, 69-92.

10. CS Lai & JB Berrill, Shaking Table Tests on a Model Retaining Wall, Bull. NZ Nat. Soc. for Earthq. Eng., 12(2), June 1979, 122-126.

11. JS Lin & RV Whitman, Earthquake-Induced Displacements of Sliding Blocks, J. Geot. Eng. (ASCE), 112 (GT1), 1986, 44-59.

12. N Mononobe, Earthquake-Proof Construction of Masonry Dams, Proc. World Engg. Conf. 1929, v.9, 275p.

13. F Nadim & RV Whitman, Seismically Induced Movement of Retaining Walls, J. Geot. Eng. (ASCE), 109(7), 1983, 915-931.

14. F Nadim & RV Whitman, Coupled Sliding and Tilting of Gravity Retaining Walls During Earthquake, Proc. 8th World Conf. on Earthq. Eng., (3), San Francisco, 1984, 477-484.

15. RB Nagel & DG Elms, Seismic Behaviour of Reinforced Earth Walls, Res. Rept. 85-4, Dept. Civil Eng., Univ. of Canterbury, NZ, March 1985, 99p.

16. NM Newmark, Effects of Earthquakes on Dams and Embankments, Geotechniques, 15(2), 1965, 139-160.

17. S Okabe, General Theory of Earth Pressure, Jour. Jap. Soc. Civ. Eng's., (Tokyo), 12(1), 1926.

18. R Richards & DG Elms, Seismic Behaviour of Retaining Walls and Bridge Abutments, Res. Rept. 77-10, Dept. Civil Eng., Univ. of Canterbury, NZ, June 1977, 39p.

19. R Richards & DG Elms, Seismic Behaviour of Gravity Retaining Walls, J. Geot. Eng. (ASCE), 105(4), Apr. 1979, 449-464.

20. R Richards & DG Elms, Seismic Behaviour of Tied-Back Walls, - - Initial Analysis and Experiments, Res. Rept., 87-8, Dept. Civil Eng., Univ. of Canterbury, NZ, 1987, 44p.

21. SK Sarma, Seismic Stability of Earth Dams and Embankments, Geotechnique, 25(4), 1975, 743-761.

22. DJ Schuring, Scale Modelling in Engineering, Pergamon Press, Oxford, 1977.

23. N Sharma, Refinement of Newmark Sliding Block Model and Application to New Zealand Conditions, ME Thesis, Dept. of Civil Eng., Univ. of Canterbury, NZ, 1989, 237p.

24. RV Whitman, Dynamic Behavior of Soils and its Application to Civil Engineering Projects, Proc. Pan-Am. Conf., Lima Peru, 1979.

25. RV Whitman & S Liao, Seismic Design of Gravity Retaining Walls, Proc. 8th World Conf. on Earthq. Eng.,(3), San Francisco, 1984, 533-540.

26. K Zarrabi, Sliding of Gravity Retaining Wall During Earthquakes Considering Vertical Acceleration and Changing Inclination of Failure Surface, M.S. Thesis, Dept of Civil Eng., MIT, Cambridge, Mass., 1979.

THE SEISMIC RESPONSE OF WATERFRONT RETAINING WALLS

R. Scott Steedman[1] and Xiangwu Zeng[2]

ABSTRACT   Advances in the understanding of the response of retaining structures to base shaking have been restricted by the use of pseudo-static calculations which have not taken into account dynamic amplification or phase effects.   By considering a finite wave speed for the propagation of elastic shear waves through the ground the conventional analysis has been developed to generate a new pseudo-dynamic analysis which throws light on many of the inadequacies of the present calculations. In particular, the pseudo-dynamic analysis shows how an initially stiff soil-wall system may deteriorate towards failure as strain softening is brought about by dynamic amplification or excess pore pressure generation. This approach is strongly supported by the experimental evidence from a series of dynamic centrifuge model tests which have also been used to illustrate the ultimate failure mechanisms for an anchored cantilever wall subject to earthquake loading.

INTRODUCTION

    Failures of retaining walls and waterfront structures during earthquakes are widely documented and common in most major events, for example in the Nihonkai-Chubu earthquake in 1983, Tohno and Shamoto (1985) or in Chile in 1964, Seed and Whitman (1970). The analysis of such structures has largely been based on pseudo-static calculations such as the Mononobe-Okabe equation which is itself an extension of Coulomb wedge theory, Seed and Whitman (1970). The wall is assumed to have moved sufficiently to generate minimum active or maximum passive pressure; a planar sliding surface is assumed to develop as a wedge slips behind the wall mobilising a maximum shear strength along its length; inertial accelerations are assumed to be constant throughout the soil body (as if it were rigid) which can then be treated as D'Alembert body forces acting in the opposite direction to the accelerations. It is this assumption that the soil can be treated as a rigid body with an infinite wave speed that reduces the problem to a pseudo-static one but which has also prevented deeper insights into the phase effects which determine the dynamic pressure distribution and the dynamic amplification that leads to failure through a degradation of strength and stiffness.

A PSEUDO-DYNAMIC ANALYSIS OF SEISMIC EARTH PRESSURE

The effect of a phase change between the base and surface

Consider a typical fixed base cantilever wall as shown in Fig. 1. The shear wave velocity $V_s = \sqrt{(G/\rho)}$ where ρ is the density of soil and G is the soil shear modulus. $T = 2\pi/\omega$ is defined as the period of lateral shaking, of angular frequency ω. It is assumed that initially G is constant with depth through the backfill and that only the phase and not the magnitude of acceleration is varying.

If the base is subject to a sinusoidal vibration, the acceleration at depth z and time t is given by

---

1 - Lecturer, Cambridge University Engineering Dept., Trumpington St. Cambridge CB2 1PZ, UK.
2 - Research Assistant, as above.

$$A(z,t) = k_h \, g \, \sin \omega \left(t - \frac{H - z}{V_s} \right) \tag{1}$$

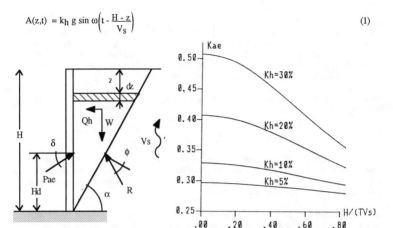

Fig. 1 Notation for a slipping wedge   Fig. 2 Variation of K_{ae} with increasing H/TV_s

The total weight of the slipping wedge is $W = \gamma H^2 / (2 \tan \alpha)$ where γ is the unit weight of the soil. Considering the mass of a horizontal element of the wedge at a depth z, the total horizontal inertia force Q_h is given by the integral

$$Q_h = \int_0^H \rho \left(\frac{H - z}{\tan \alpha} \right) A(z,t) \; dz \tag{2}$$

Substituting for $A(z,t)$ from Eqn. 1, then

$$Q_h = \frac{\lambda \gamma \, k_h}{4 \pi^2 \tan \alpha} \left[2 \pi H \cos \omega \xi + \lambda \left(\sin \omega \xi - \sin \omega t \right) \right] \tag{3}$$

where $\lambda = TV_s$ is the shear wave length and $\xi = t - H/V_s$. As $V_s \to \infty$, that is if the soil is assumed to be rigid, then

$$\left(\lim_{V_s \to \infty} Q_h \right)_{max} = \frac{\gamma H^2 \, k_h}{2 \tan \alpha} = k_h \, W \tag{4}$$

which is the definition of the Mononobe-Okabe equation. The total earth pressure on the wall can be derived by resolving forces on the wedge to give

$$P_{ae} = \frac{Q_h \cos \left(\alpha - \phi \right) + W \sin \left(\alpha - \phi \right)}{\cos \left(\delta - \alpha + \phi \right)} \tag{5}$$

and the total lateral earth pressure coefficient is then $K_{ae} = 2P_{ae}/\gamma H^2$. Substituting for W and Q_h gives an expression for K_{ae} as a function of the dimensionless groups H/TV_s, t/T and the wedge angle α. The maximum value of K_{ae} is required and optimizing K_{ae} with respect to t/T and α, it is found that K_{ae} is simply a function of H/TV_S which is the ratio of the time for a wave to travel the full height to the period of the lateral shaking. This result is plotted in Fig. 2 for a range of values of k_h. K_{ae} is seen to vary relatively slowly with H/TV_S which confirms the common view that that magnitude of dynamic earth pressure given by the Mononobe-Okabe equation is approximately correct, Seed and Whitman (1970).

Seed and Whitman separated the seismic earth pressure into a static component $P_{as}*$ and a dynamic component $\Delta P_{ae}*$, defining

$$P_{ae} = P_{as}^* + \Delta P_{ae}^* \tag{6}$$

where $P_{as}*$ is the original static force and $\Delta P_{ae}*$ the additional increment of "dynamic force". However, a dynamic 'outward' failure wedge is larger than a static wedge (ie. α is smaller) and hence $\Delta P_{ae}*$ must be a function of both the additional 'static' weight of the increased wedge size and the horizontal inertia caused by the lateral acceleration.

A more consistent definition is therefore proposed in which $P_{ae} = P_{as} + P_{ad}$ where P_{as} is the force on the wall due to the vertical weight of the wedge and P_{ad} is the force on the wall due to the horizontal inertia of the wedge. A typical example of the non-linear distribution of the lateral pressure p_{ae} is shown in Fig. 3 (for $k_h = 0.2$ and $H/TV_S = 0.3$) and given by

$$p_{ae} = \frac{\partial P_{ae}(z)}{\partial z} = \frac{\cos(\alpha-\phi) k_h \gamma z}{\cos(\delta-\alpha+\phi) \tan \alpha} \sin \omega\left(t-\frac{z}{V_S}\right) + \frac{\gamma z}{\tan \alpha} \frac{\sin(\alpha-\phi)}{\cos(\delta-\alpha+\phi)} = P_{ad} + P_{as} \tag{7}$$

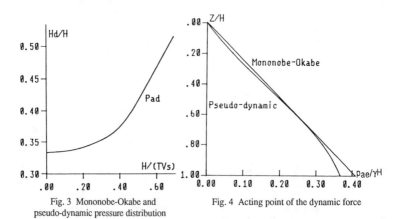

Fig. 3 Mononobe-Okabe and
pseudo-dynamic pressure distribution

Fig. 4 Acting point of the dynamic force

The acting point of the force P_{ad} is defined as to be at a height H_d above the base and is found by moment equilibrium

$$H_d = \frac{M_{d(z=H)}}{P_{ad} \cos \delta} = \frac{\int_0^H P_{ad} \cos \delta (H-z) \, dz}{P_{ad} \cos \delta}$$

$$= H - \frac{2\pi^2 H^2 \cos \omega\xi + 2\pi\lambda H \sin \omega\xi - \lambda^2(\cos \omega\xi - \cos \omega t)}{2\pi^2 H \cos \omega\xi + \pi\lambda(\sin \omega\xi - \sin \omega t)} \tag{8}$$

where $M_d(z,t)$ is the dynamic component of bending moment at depth z and time t. Fig. 4 shows the position of the acting point of the dynamic force P_{ad} at the instant when the bending moment is a maximum as a function of H/TV_s for $k_h = 0.2$. For the wall in Fig. 1 (assuming H = 10m, G = 20 MPa) the point of application of the dynamic force increment varies from 0.35 to 0.42H above the base, as the period varies from 0.5 to 0.2 seconds. The acting point of the dynamic force increment is seen to be a function of the period of shaking and the stiffness of the soil in addition to the 'pseudo-static' parameters invoked in the Mononobe-Okabe equation. The dynamic bending moment is given by

$$M_d(Z,t) = \int_0^Z p_{ad}(z,t) \cos \delta \, (Z - z) \, dz$$
$$= \frac{\cos(\phi - \alpha) \, k_h \gamma \lambda^2 \cos \delta}{4\pi^3 \cos(\delta - \alpha + \phi) \tan \alpha} \left[\lambda\left(\cos \omega\left(t - \frac{Z}{V_s}\right) - \cos \omega t\right) - \pi Z\left(\sin \omega\left(t - \frac{Z}{V_s}\right) + \sin \omega t\right) \right] \tag{9}$$

Shear modulus varying with depth

In practice, the shear modulus in the ground varies with depth, and is commonly expressed as

$$G(z) = K z^\beta \qquad \text{where } 0 \le \beta \le 1 \tag{10}$$

in which K is a constant and z is the depth. The shear wave velocity at a depth z is then

$$V_s(z) = \sqrt{\frac{K}{\rho}} \, z^{\beta/2} \tag{11}$$

The varying shear wave velocity gives a non-linear phase lag through the backfill. To investigate the effect of phase alone, the peak amplitude of the shaking was held constant and the analysis repeated except that the constant velocity V_s in the dimensionless parameter H/TV_s was replaced by the mean shear wave velocity in the backfill

$$\overline{V_s} = \frac{H^{\beta/2}}{(1 - \beta/2)} \sqrt{\frac{K}{\rho}} \tag{12}$$

For values of β between 0 and 1 the effect of the phase change on the lateral earth pressure coefficient due to different distributions of shear modulus is small, Steedman and Zeng (1989).

The amplification of ground motion

As the shear modulus reduces towards the ground surface in a cohesionless backfill there will be amplification as well as phase effects. The exact nature of such amplification is dependent on many factors, including the geometry and rigidity of adjacent structures, the damping and stiffness of the soil and so on. Again a simplifying assumption is made that the lateral acceleration varies linearly from the base of the layer to the ground surface, such that $k_{h(z=0)} = fa \, k_{h(z=H)}$ where fa is a constant. Then for a constant shear modulus in the backfill, the acceleration at depth z is given by

$$A(z,t) = \left\{1 + \frac{H-z}{H}(fa-1)\right\} k_h \, g \, \sin \omega\left(t - \frac{(H-z)}{V_s}\right) \tag{13}$$

Repeating the limit equilibrium analysis, it is found that K_{ae} is a function of H/TV_s and fa, Steedman and Zeng (1989). The result is shown in Fig. 5 for a range of values of amplification factor. The effect of an increase in the amplification factor is seen to be qualitatively similar to an increase in the lateral acceleration coefficient k_h but also the change in the point of application of the dynamic force increment H_d allowing for amplification was found to be negligible for values of fa relevant to most field problems.

Fig. 6 shows data from a fixed base cantilever wall subjected to base shaking using the Bumpy Road earthquake actuator on the Cambridge Geotechnical Centrifuge, Steedman (1984). From the phase change in the recordings of acceleration at the top of the dry sand backfill and at the base the average shear wave velocity was deduced to be 180m/s, which gave a mean shear modulus in the sand backfill of $G = 57$MPa. The accelerometers also showed an amplification of acceleration of about 100% and hence the amplification factor $fa = 2$. The non-dimensional peak dynamic moment distribution is plotted in Fig. 6 together with the results of a Mononobe-Okabe calculation and the pseudo-dynamic calculation described above. It is clear that the pseudo-dynamic approach provides good agreement with the test data, both in magnitude and distribution of bending moment.

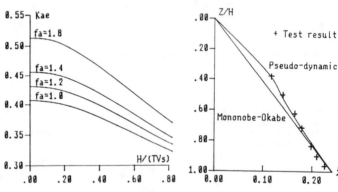

Fig. 5 The effect of amplification on the earth pressure coefficient K_{ae}

Fig. 6 Pseudo-dynamic moment distribution

HYDRODYNAMIC PRESSURE ON A WATER-FRONT STRUCTURE

Hydrodynamic pressures act on both sides of a water-front retaining wall. If the wall is rigid (i.e. the natural frequency of the wall is much higher than the base shaking frequency) then on the seaward side the hydrodynamic pressure distribution can be calculated by Westergaard's (1933) formula

$$P_s(y) = -\frac{7}{8} \gamma_w \, k_h \, \sqrt{h \, y} \tag{14}$$

where h is the depth of water in front of the wall, y the depth beneath the water surface and γ_w the unit weight of water. However, for a flexible wall the hydrodynamic pressure will be higher than that given by Westergaard because of the increased movement of the wall. The analysis proposed by Shulman (1987) for a flexible dam can be used. In the extreme the natural frequency of the wall approaches the dominant base shaking frequency and resonance may occur. Under these circumstances the hydrodynamic pressure on the wall is much higher than that given by the Westergaard formula with a phase shift of about 90°, Zeng (1989).

The calculation of the hydrodynamic pressure on the backfill side is more complex because of the existence of the soil particles. The movement of fluid is no longer free but is restricted by the soil. Amano et al (1956) assumed that the pore fluid moved together with the soil so that the water was taken into account as part of the soil. Equivalent seismic coefficients k_h' and k_v' were used instead of the original coefficients k_h and k_v where $k_h' = \gamma_{sat} k_h / \gamma'$ and $k_v' = \gamma_{sat} k_v / \gamma'$. Following this assumption, it is clear that the pressure should be linearly distributed but scaled in some way. As the analysis presumes that the water is completely restricted by soil particles it is appropriate for a soil with a very low permeability, such as a clay or silt backfill.

A drawback of this approach is that it assumes the hydrodynamic pressure on the backfill side is in phase with the earth pressure which is not the case for a flexible wall with a natural frequency near the dominant earthquake frequency. Furthermore, in soils with comparatively high permeabilities the pore fluid will be free to move under the inertial loads to some extent. Zeng (1989) proposed an alternative solution assuming that the water in the backfill is completely free. Using a simplified theoretical solution suggested by Von Karman (1933), the formula for the hydrodynamic pressure on the backfill side can be derived as $p_b = - n \, p_s$, where p_b is the hydrodynamic pressure on the backfill side, p_s is the hydrodynamic pressure on the seaward side and n is the porosity of soil (the negative sign means the two pressures are in opposite directions). This solution is for an ideal state in which the existence of soil does not affect the flow of the fluid, such as a coarse sand with a high permeability. It has the further advantage that the hydrodynamic pressure on the backfill side is in phase with that on the seaward side.

NATURAL FREQUENCY OF A FLEXIBLE CANTILEVER WALL

The response of a flexible retaining wall during earthquakes is critically dependent on its natural frequency. An energy method can be used to estimate the natural frequency of the cantilever wall by assuming a deflection mode $y = \mathcal{F}(z)$. Let the vibration of the wall be given by $u = \delta(y) \sin \omega t$ where $\delta(y)$ is the deflection of the wall under the lateral earth pressure and ω is the angular frequency of base shaking. The maximum kinetic energy is found by integrating over the whole height of the wall

$$(KE)_{max} = \int_0^{H_o} \frac{m}{2} \left(\frac{du}{dt}\right)^2_{max} dy \tag{15}$$

where m is the mass per unit height of the cantilever wall, H_0 is the total height of the wall. The maximum potential energy is given by

$$(PE)_{max} = \frac{EI}{2} \int_0^{H_o} \left(\frac{d^2\delta}{dy^2}\right)^2_{max} dy \tag{16}$$

Equating the maximum kinetic energy and potential energy, the natural frequency f_n is found to be a function of the bending stiffness EI, the mass per unit height m, the retained height of the wall H and the earth pressure coefficients K_a and K_p, Steedman and Zeng (1989). The natural frequency of a cantilever retaining a dry soil shows only a small reduction with increasing k_h.

For a free cantilever wall with saturated backfill the deflection mode is complicated by the existence of hydrodynamic pressure p_h on both sides of the wall

$$p_h = p_s + p_b = \frac{7}{8}(1 + n)\gamma_w \, k_h \, \sqrt{h \, y} \, \cos \psi \tag{17}$$

in which p_h is the total hydrodynamic pressure and ψ is the phase shift angle between the wall vibration and base shaking. As ψ is dependent on the natural frequency of the wall and the damping ratio, the solution must be found by iteration. It was found that the natural frequency of a free cantilever wall with saturated backfill reduces quite rapidly with increasing k_h, Zeng (1989).

Centrifuge test results

A series of centrifuge model tests have been carried out at 80g on the Cambridge Geotechnical Centrifuge. The free cantilever wall models were backfilled with dry or saturated LB52/100 sand and shaken at a dominant earthquake frequency of 120 Hz. The properties of the models reported here are summarised in Table 1. A section of a typical saturated model is shown in Fig. 7. The wall was fabricated by scaling the stiffness of an equivalent prototype wall and the models instrumented with strain gauges to measure bending, accelerometers, pore pressure transducers and LVDTs. Test ZENG3 was a saturated backfill free cantilever wall model. The bending moment distributions on the wall under static loading and during base shaking were calculated by the Mononobe-Okabe limit state analysis in which the soil is assumed to be mobilising a limiting strength and force and moment equilibrium are used to deduce the centre of rotation for a minimum penetration. These are plotted in Fig. 8 together with the test data recorded by bending moment transducers on the wall.

The natural frequency of the wall in Fig. 7 is predicted to be 135 Hz for very small vibrations reducing gradually to 110 Hz for a base shaking intensity of 10%. As the driving frequency of the earthquakes is 120 Hz, the vibration was dramatically amplified and a phase shift increasing from 90° to 180° between the base shaking and the wall vibration was observed during shaking. The results shown in Fig. 8 demonstrate the importance of this dynamic amplification. The model wall failed during the test, moving outwards with large uneven subsidence behind the wall.

TABLE 1 A summary of 4 centrifuge model tests

model	ZENG3	ZENG8	ZENG9	ZENG10
wall type	cantilever	anchored	anchored	anchored
retained height	90mm	90mm	90mm	90mm
penetration	90mm	30mm	30mm	30mm
wall stiffness EI	205Nm	20.4Nm	20.4Nm	20.4Nm
tie rod stiffness A_tE_t	-	4.3MNm	4.3MNm	4.3MNm
void ratio	0.740	0.705	0.647	0.647
relative density	55%	65%	80%	80%
pore fluid viscosity	80cs	80cs	dry	80cs (silicon oil)
phreatic surface	$z = 0$	$z = 0$	-	$z = H_b$
failure mechanism or deformation	large outward movements	toe rotated outwards & anchor pulled out	small outward movements	anchor pulled out

NATURAL FREQUENCY OF AN ANCHORED CANTILEVER WALL

An energy method can also be used to estimate the natural frequency of an anchored wall, provided some distribution of earth pressure and hence deflection is assumed. Fig. 9 shows a simple loading

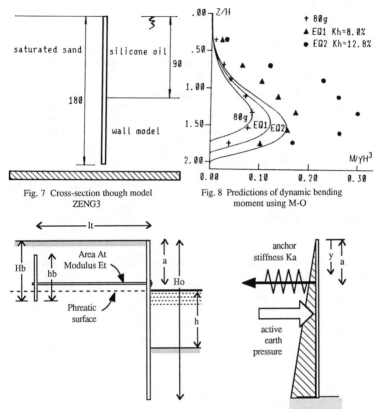

Fig. 7 Cross-section though model
ZENG3

Fig. 8 Predictions of dynamic bending
moment using M-O

Fig. 9 Cross-section through an anchored cantilever wall

condition for an anchored wall assuming that there is some degree of base fixity. Neglecting the passive pressure because it has little influence the deflection of the wall is then given by

$$\delta(y) = \frac{q_0}{EI}\left(\frac{H_o^4}{30} - \frac{H_o^3 y}{24} + \frac{y^5}{120\,H_o}\right) - \frac{F_a}{EI}\left(\frac{(H_o - a)^2}{6}(2H_o + a) - \frac{(H_o - a)^2}{2}y\right) \qquad ; y \leq a \qquad (18)$$

$$\delta(y) = \frac{q_0}{EI}\left(\frac{H_o^4}{30} - \frac{H_o^3 y}{24} + \frac{y^5}{120\,H_o}\right) - \frac{F_a}{EI}\left(\frac{(H_o - a)^2}{6}(2H_o + a) - \frac{(H_o - a)^2}{2}y + \frac{(y - a)^3}{6}\right) \qquad ; y > a$$

where EI is the bending stiffness of the wall and H_0 the total height of the wall. The value of the anchor force is determined by compatibility at the anchor position $\delta(a)=Fa/Ka$ in which Ka is the stiffness of the anchor.

$$Fa = \frac{q_0 \left(\frac{H_0^4}{30} - \frac{aH_0^3}{24} + \frac{a^5}{120\,H_0} \right)}{\frac{EI}{Ka} + \frac{(H_0-a)^3}{3}} \qquad (19)$$

Solving for Fa in Eqn. 19 shows a clear dependance on the relative stiffness of the anchor system. For the case of a rigid anchor Ka $\rightarrow \infty$ and EI/Ka $\rightarrow$ 0.

On the other hand, a failed anchor might be considered to have a greatly reduced stiffness, and letting Ka $\rightarrow$ 0 the anchor force Fa $\rightarrow$ 0 reducing the problem to a simple cantilever. These two extreme conditions provide an upper and lower bound on the anchor force and natural frequency of the structure. Assuming the displacement of the wall is given by $u = \delta(y) \sin \omega t$, equating the maximum kinetic and potential energies leads to an expression for the natural frequency f_n as a function of the bending stiffness EI, the mass per unit height m, the total height of the wall H_0, the depth to the anchor a and the stiffness of the anchor system Ka.

In the case of a rigid anchor system, the depth of the anchor a is a key factor. Figs. 10 and 11 show the variation of the natural frequency of the anchor wall model, the maximum bending moment in the wall and the tie force for different anchor positions. It is clear that the ideal depth for the anchor is around 0.4 H_0 for the anchored wall models described in Table 1, which gives the maximum reserve of natural frequency against the possibility of resonance while the bending moment and tie force are not very high.

For the three anchored wall centrifuge model tests described in Table 1, the depth of the anchor $h_a = H_0/6$ and therefore, for the rigid anchor case

$$\omega_n = \frac{21.70}{H^2} \sqrt{\frac{EI}{m}} \qquad (20)$$

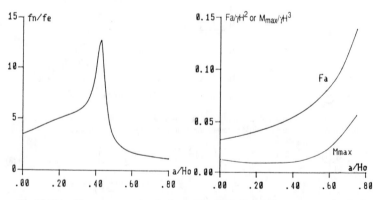

Fig. 10 Natural frequency vs anchor depth   Fig. 11 Tie force and moment vs anchor depth

Substituting the appropriate values given in Table 1 leads to f_n = 578Hz. Depending on the stiffness of the anchor system the natural frequency of the wall will be less than this, reducing to a minimum in the case of a completely failed anchor system to a cantilever wall with f_n = 97Hz.

<u>The stiffness of an anchor system</u>

Fig. 9 defined the significant parameters for an anchored wall. The displacement of the wall at the anchor position δ_A can be divided into two parts: the extension of tie rod δ_t and the movement of anchor beam δ_b and $\delta_A = \delta_t + \delta_b$. For an increase of tie force Fa, the extension of the tie rod will be

$$\delta_t = \frac{\Delta Fa}{A_t E_t} l_t \tag{21}$$

The displacement of the rigid anchor beam depends on the shear modulus of the sand, the increase in the tie force and the height of the anchor beam :

$$\delta_b = A \left(h_b\right)^{f_1} \left(\Delta Fa\right)^{f_2} \left(G\right)^{f_3} \tag{22}$$

where A is a dimensionless constant, G the mean shear modulus of sand and f_1, f_2, f_3 are constants. Assuming the sand can be considered linear elastic for a small increase in tie force then dimensional analysis gives $f_2 = 1, f_1 = 0, f_3 = -1$ and $\delta_b = A \, \Delta Fa/G$. The stiffness of the anchor system is therefore

$$Ka = \frac{\Delta Fa}{\delta_A} = \frac{1}{\dfrac{l_t}{E_t G_t} + \dfrac{A}{G}} \tag{23}$$

In this equation the only unknown is A. In practice the solution is not very sensitive to the value of A and to simplify the calculation A is assumed to be approximately equal to the factor for the settlement of a long strip footing $A = (1 - \mu) \, I\rho$ (which has a value of around 1), where μ is Poisson's ratio and $I\rho$ the influence factor.

However, the stiffness of sand ranges widely, depending on the stress and strain state. Hardin and Drnevich (1972) and Drnevich (1975) proposed an hyperbolic stress-strain equation for sand based on laboratory test data

$$\tau = \frac{G_{max} \gamma}{1 + \gamma/\gamma_r} \quad \text{where } G_{max} = 3230 \frac{(2.973 - e)^2}{(1 + e)} \sqrt{\sigma'_m} \tag{24}$$

where G_{max} is the initial tangent shear modulus based on an empirical formula, γ_r is the reference strain $\gamma_r = \tau_{max}/G_{max}$ and τ_{max} is the shear stress at failure. σ_m' is the effective mean principal stress and e is the void ratio. The tangent shear modulus G_t is then

$$G_t = \frac{d\tau}{d\gamma} = G_{max} \left(1 - \frac{\tau}{\tau_{max}}\right)^2 \tag{25}$$

For the centrifuge model ZENG9, which had a dry sand backfill, the initial stress state in front of the anchor wall, and the stress state under base shaking is shown in Fig. 12 (Fa is a function of k_h) and

$$\sigma'_{vi} = \gamma \left(H_b - \frac{h_b}{2}\right) \quad ; \quad \sigma'_{hi} = \frac{Fa}{h_b} \quad ; \quad \beta = \tan^{-1} k_h \quad ; \quad \tau_v = \sigma'_{vi} \tan \beta \quad ; \quad \sigma'_h = \frac{Fa}{h_b} \quad ;$$

$$\tau = \sqrt{\left(\frac{\sigma_h' - \sigma_v'}{2}\right)^2 + \tau_v^2} \quad ; \quad \tau_{max} = \frac{\sigma_{vi}'}{2}(K_p - 1) \tag{26}$$

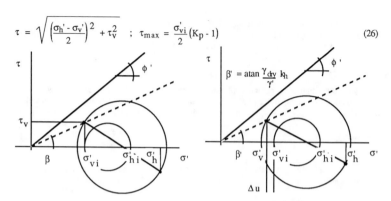

Fig. 12 Stress increment in dry backfill   Fig. 13 Stress increment in saturated backfill

Vertical and horizontal effective normal and shear stresses may be calculated in terms of the anchor force and the lateral acceleration and this leads to a solution for the anchor force as a function of the lateral acceleration coefficient k_h. Only a small variation of f_n with k_h was found for the case of the dry anchored wall, Zeng (1989). During lateral shaking, degradation of stiffness is insignificant and there is little amplification of motion. A pseudo-static analysis gives a good prediction of the bending moment distribution and anchor force, Steedman and Zeng (1989).

However, for a wall retaining saturated sand such as model ZENG8, the response of the anchor wall is complicated by the generation of pore pressure. The Mohr's circle defining the stress state in front of the anchor wall is shifted towards the origin by Δu as shown in Fig. 13. For an excess pore pressure Δu

$$\sigma_v' = \gamma'\left(H_b - \frac{h_b}{2}\right) - \Delta u \quad ; \quad \beta' = \tan^{-1}\left(\frac{\gamma_{dry}}{\gamma'} k_h\right) \quad ; \quad \tau_v = \sigma_v' \tan\beta' \tag{27}$$

Substituting the corresponding parameters for test ZENG8 from Table 1 together with the values of anchor force and pore pressure recorded during the test, the natural frequency of the model wall at different excess pore pressures predicted by the analysis is shown in Fig. 14.

For low levels of excess pore pressure the natural frequency is about 4.5 times higher than the base shaking frequency and there is only a slow reduction in the natural frequency for a further increment of pore pressure. The magnitude of shaking of the wall model should be small and this agrees with the test results.

However in earthquake 7 (Fig. 15) once the pore pressure buildup in front of the wall exceeded about 60% of the initial vertical effective stress, there a clear phase shift in the response of the wall vibration. Strong cyclic bending moments and tie forces were recorded with large outward displacements of the wall. The increase in shaking generated more excess pore pressure in the backfill, and the vibration of the wall is greatly amplified as resonance developed and the anchor wall approached failure. Cyclic bending moments during the earthquake were of the order of 15 times the peak values in previous earthquakes although the level of base shaking was only doubled. The phase of the wall vibration shifted by 180° as these large amplitude vibrations built up. Failure of the wall by gross outward movement was then inevitable. The large increase in bending moments could explain the common observation of the failure of sheet pile walls in bending such as happened in the Nihonkai-Chubu earthquake of 1983, Tohno and Shamoto (1985).

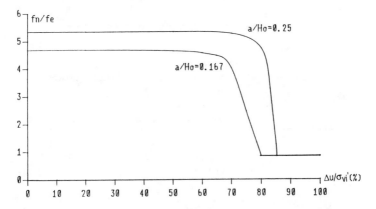

Fig. 14   The deterioration of natural frequency with build-up of excess pore pressure

This analysis shows that liquefaction of the sand backfill was not the direct cause of failure of the retaining wall.   Instead the degradation of stiffness of the soil and hence the fall in the natural frequency of the wall accelerated rapidly beyond a critical value of pore pressure generation which in this case was at around 60% of the vertical effective stress.   Vibration of the retaining structure was greatly amplified causing larger cyclic strains in the backfill and therefore even higher excess pore pressures leading to the liquefaction of the upper layers of backfill and to failure of the structure.

An important lesson for the design of an anchored wall is therefore to maintain the high natural frequency of the wall under base shaking by preventing any rapid deterioration in the stiffness of the anchor.   It was proposed above that the anchor position $a/H_O = 0.4$ would give the system an inherently high natural frequency without greatly increasing the bending moment in the wall or the anchor force in the tie rod.   Clearly a deeper anchor will also increase the threshold pore pressure and thus reduce the risk of failure of the anchor wall as shown in Fig. 14. This was demonstrated by further model tests, Zeng (1989).

FAILURE MECHANISMS FOR ANCHORED WALLS

Current practice for the design of flexible anchored walls subjected to base shaking is based on pseudo-static calculations.   Assuming that the penetration of the wall is at a minimum such that limiting active and passive lateral forces are generated in the soil, then moment and force equilibrium can be satisfied to deduce the anchor force and the position of the centre of rotation of the wall.   For an anchored flexible wall with dry sand backfill subject to base shaking, the active earth pressure coefficient K_{ae} will be increased while the passive earth pressure coefficient K_{pe} is reduced.   This leads to increased bending moments in the wall and an increase in the anchor force in the tie rod, which may provoke either local failure of the wall or of the anchor system.   Alternatively the depth of the centre of rotation of the wall may increase sufficiently to cause a loss of passive resistance and an outward movement of the toe of the wall.

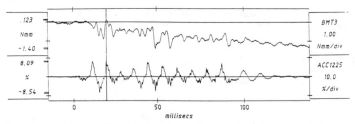

Earthquake 3, Model Test ZENG8 : Bending moment in phase with the base acceleration

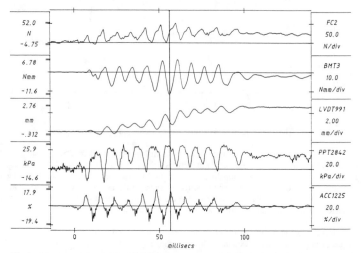

Earthquake 7, Model Test ZENG8 : Bending moment out of phase with the base acceleration

Note :   ACC1225 reference accelerometer
PPT2842 pore pressure in front of anchor   LVDT991 displacement transducer at wall top
BMT3 bending moment at mid-height ($H_0/2$)   FC2 force cell on tie rod

Fig. 15 Phase relationship of wall response with small or large excess pore pressures

Further centrifuge model tests were carried out to investigate the behavior of anchored walls as summarised in Table 1, Zeng (1989). The models included dry and saturated backfill, different depths of phreatic surface and varying anchor position.

With the exception of a global failure mechanism which would encompass a wall, anchor wall and backfill in some deep-seated failure an anchored wall can fail either by local failure of the wall itself, or by failure of the anchor wall, or by failure of the toe of the wall caused by a loss of passive resistance. In the model tests the walls had a high reserve of strength against local bending failure in order to achieve the correctly scaled bending stiffness. It has already been noted that in the field bending failures are very likely.

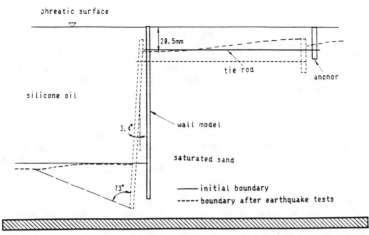

Fig. 16 Toe rotated outwards and anchor pulled out during the failure of model ZENG8

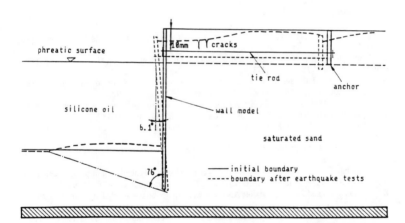

Fig. 17 Top rotated outwards and anchor pulled out during the failure of model ZENG10

Model ZENG8 failed with the wall moving seawards and the toe rotating out, Fig. 16, suggesting a loss of passive resistance. The phreatic surface in the model was at the top of the wall. Model ZENG10 had a lower phreatic surface (at the bottom of the anchor wall), and failed by rotating outwards towards the sea, dragging the anchor wall, Fig. 17, as the tie force exceeded the anchor capacity.

CONCLUSIONS

A new psuedo-dynamic calculation has been presented which allows an exploration of the parameters influencing the dynamic earth pressure on a retaining structure. In particular the pseudo-dynamic approach reveals the transition from an initially stiff soil-wall system (which may be successfully analysed pseudo-statically) to a resonant system in which degradation of strength and stiffness leads to dynamic effects of amplification, phase change and possible failure. The earth pressure distribution and the acting point of the dynamic force increment are found to be a function of the dimensionless parameter H/TV_S.

A limit state calculation may be used successfully to predict the bending moment distribution and failure mechanisms for a free cantilever wall, provided the natural frequency of the wall is much higher than the dominant earthquake frequency. However, for a flexible wall the natural frequency may be comparable with the earthquake frequency and amplification of the shaking may bring the wall to failure. Hydrodynamic pressure on a flexible wall is higher than that given by Westgaard's formula.

If the excess pore pressure in the saturated backfill behind an anchored wall exceeds a threshold value the natural frequency of the soil-wall system may drop dramatically leading to resonance and amplification of wall vibration. Liquefaction of the backfill and failure of the retaining wall may then develop simultaneously. The failure mechanism that an anchored flexible wall will adopt depends on the geometry of the structure and the position of the water table.

REFERENCES

1.   Amano R, Azuma H and Ishii Y, Aseismic design of walls in Japan, Proc. Ist World Conference on Earthquake Engineering, 1956, pp. 32-1 -32 -17.
2.   Drnevich VP, Constrained and Shear Moduli for Finite Elements, Jur. Geotech. Engng. Div., ASCE, No. GT5, 1975, pp. 459-473.
3.   Hardin BO and Drnevich VP, Shear Modulus and Damping in Soils: design equations and curves, Intr. Soil Mech. Found. Div., ACHE, No. SM7, 1972, pp. 667-692.
4.   Seed HB and Whitman RV, Design of Earth retaing structures for dynamic loads, lateral stresses in the ground and design of earth retaining structures, ASCE, 1970, pp. 103-147.
5.   Shulman SG, Seismic Pressure of water on hydraulic structures, A A Balkema, Rotterdam, 1987.
6.   Steedman RS, Modelling the behaviour of retaining walls in earthquakes, Ph.D Thesis, Cambridge University, 1984.
7.   Steedman RS and Zeng X, The influence of phase change on calculation of pseudo-static earth pressure on a retaining wall, Report CUED/D-soils TR222, Cambridge Univ.Eng.Dept., 1989.
8.   Steedman RS and Zeng X, Flexible Anchored walls subjected to base shaking, Report CUED/D-soils TR217, Cambridge Univ.Eng.Dept., 1989.
9.   Tohno I and Shamoto Y, Liquefaction damage to the ground during the 1983 Nihonkai-Chubu (Japan Sea) earthquake in Akita Prefecture, Tohoku, Japan, Proc. Natural Disaster Science 7, No. 2, 1985, pp. 67-93.
10. Von Karman, Discussion note on Westergaard's paper, Transactions of American Society of Civil Engineering, Vol. 98, 1933.
11. Westergaard HM, Water pressure on dams during earthquakes, Transactions of American Society of Civil Engineering, Vol. 98, 1933
12. Zeng X, Modelling the behaviour of quay walls in earthquakes, Thesis under preparation for the PhD Degree, Cambridge University, 1989.

DYNAMIC EARTH PRESSURE WITH VARIOUS GRAVITY WALL MOVEMENTS

Reda M. Bakeer[1], A.M.ASCE, Shobha K. Bhatia[2], A.M.ASCE
and Isao Ishibashi[3], A.M.ASCE

ABSTRACT: This paper discusses the effect of the mode of movement of a retaining wall on the dynamic earth pressure generated by horizontal ground vibrations. The discussion is based on the reported results of analytical and experimental investigations on the subject. The results of the investigations are in good agreement and they indicate that the magnitude and distribution of the dynamic earth pressure do not depend only on the characteristics of the ground motion, but also on the possible mode of movement of the retaining wall. The research shows that the distribution of the dynamic earth pressure is always non-hydrostatic and in most cases its resultant acts above the lower third-point of the wall height.

INTRODUCTION

Several investigators have reported cases of failure of bridges during earthquakes due to the rotation of their abutments and others have observed failure of retaining walls by excessive translation away from the backfill. The failure mode of an earth retaining structure is affected by the type of wall/backfill system as well as its structural configuration, such as the existence of a tieback or a bridge deck. Rotation of a bridge abutment about its top occurs when the lateral movement is restricted by the rigid superstructure and only the lower portion of the wall is relatively free to move. On the other hand, when a deep fill or a structural member is placed in front of the wall toe, it could restrain the movement of the base resulting in a probable rotation about the wall base.

A pseudo-static approach based on the Mononobe-Okabe (M.O.) formula (10, 12) is usually used for designing retaining walls subjected to dynamic conditions. It is primarily a modification of

---

1 - Asst. Prof., Dept. of Civil Eng., Tulane Univ., New Orleans, LA 70118.
2 - Assoc. Prof., Dept. of Civil Eng., Syracuse Univ., Syracuse, NY 13210.
3 - Prof., Dept. of Civil Eng., Old Dominion Univ., Norfolk, VA 23529.

Coulomb's theory of static earth pressure to account for the dynamic condition. It assumes a hydrostatic distribution for the dynamic earth pressure with its resultant acting at the lower-third point of the wall height. No consideration is made for the effect of the mode of wall movement on the resulting earth pressure. Other methods were proposed for the analysis of dynamic earth pressure but, as of know, most design codes adopt the Mononobe-Okabe approach.

Detailed studies on the effect of the mode of wall movement on the static earth pressure were conducted analytically (1, 2) and experimentally (3, 16). The results of these studies show that the mode of wall movement has a significant effect on the magnitude and distribution of the static earth pressure. Several laboratory studies were also performed to examine the dynamic earth pressure behind rigid retaining walls subjected to horizontal sinusoidal excitation (4, 5, 6, 7, 8, 9, 11, 13, 14, 15). The results of these studies indicate that the distribution of the dynamic earth pressure is always non-hydrostatic and it is affected by the mode of wall movement.

The experimental work discussed in this paper is part of the laboratory tests conducted at the University of Washington to examine the static and dynamic earth pressures on retaining walls (3, 6, 13, 14, 15, 16). The analytical results described herein are obtained from a finite element study (1, 2) undertaken to assess the effect of the different modes of deformations on the static and dynamic earth pressures. The finite element research was conducted using active and passive types of sinusoidal motions as well as actual earthquake records. However, due to space limitation, only a comparison between experimental and analytical results under sinusoidal active condition are presented in this paper.

EXPERIMENTAL TESTS

Ishibashi and Fang (6) and Sherif et al. (13, 14, 15) studied experimentally the dynamic earth pressure distribution, incremental dynamic thrust and the point of application of the earth pressure resultant during various modes of wall movement. Fig. 1 shows a sketch of the testing arrangement used in the tests conducted at the University of Washington. The apparatus consists of a rigid soil bin mounted on a shaking table (2.4 meters long, 1.8 meters wide and 1.2 meters high). The soil bin and shaking table are designed to provide one directional vibration. A movable retaining wall is attached to one end of the vibrating soil bin such that different modes of displacement conditions could be enforced. In these experiments, the wall model retained a backfill of air-dried Ottawa silica sand with average unit weight (γ) of 16.0 kN/m^3, relative density (D_r) of 53% and an angle of internal friction (ϕ) of 39.8 degrees. The average sand density was reached in all tests by shaking the soil bin filled with loosely deposited sand for 180 seconds at a frequency of 6 Hz, and a base acceleration of 0.45g. The soil bin was vibrated during the tests at a specified constant horizontal acceleration (a) and a frequency (f) of 3.5 Hz. Several

modes of wall movement were examined including translation (AT),
rotation about the wall base (ARB) and rotation about the wall top
(ART).

Several tests were conducted for each mode at different acceleration levels up to 0.5g. The wall moved away from the backfill at a constant speed when the shaking table was excited with a
steady-state vibration. The constant speed was 0.011 mm/sec for the
translation mode, 1.68×10^{-4} rad/sec for rotational movement about
the base, and 1.60×10^{-4} rad/sec during rotational movement about
the top. Two systems were used to measure the lateral earth pressure exerted on the wall model. The first system consisted of one
vertical and three horizontal load cells used to determine the total thrust, its point of application, and the wall/soil friction
angle. In addition, a second system of six soil pressure cells
(Kulite, VM-750 model) were installed along the center line of the
wall to measure the lateral earth pressure exerted on the wall at
different elevations. The data was recorded by a computer at the
rate of 200 readings per second per channel in terms of soil stress
against the retaining wall, wall displacement, and acceleration
versus time. The earth pressure distribution was studied during
different movement modes and comparisons were made in terms of total dynamic active thrust, incremental dynamic active thrust, and
the point of application of the earth pressure. For further
details of the testing apparatus and the test procedure, the reader
is referred to Refs. 6, 15 and 16.

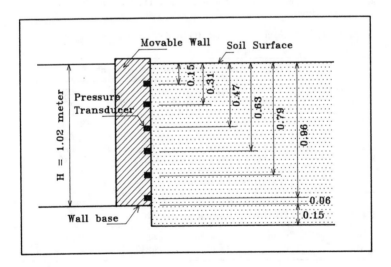

Figure 1. Experimental Retaining Wall Model

FINITE ELEMENT INVESTIGATION

The two-dimensional finite element mesh shown in Fig. 2 was designed to model a 5-meters-high gravity wall retaining a dry-cohesionless backfill. Typical dimensions of the retaining wall were selected to provide adequate safety factors against the active static earth pressure evaluated by Coulomb's theory. This assumption provides realistic dimensions that are usually used in practice. Four-noded isoparametric quadratic quadrilateral elements were used to model the soil and wall in the finite element mesh. Essentially stiff bar elements were used to connect the wall face and back to ensure that the wall displaces as a rigid body during vibration, which is the expected displacement pattern of such a massive concrete wall. Horizontal (HLI) and vertical (VLI) interface elements were used to connect the wall elements with the surrounding soil elements. The depth of the foundation soil beneath the wall was selected to be equal to the wall height (H). The backfill and the foundation soil extended a distance of 6H on both sides of the retaining wall. The infinite extension of the backfill and the foundation soil was modeled by restraining the base boundary of the mesh against vertical and horizontal movements, and the lateral boundaries of the mesh against lateral movement only. The backfill and the foundation soil were considered to be dry sand with an angle of internal friction (ϕ) of 30 degrees. The modulus of elasticity of the soil (E) was assumed to vary with the strain level in the soil elements starting with an initial value of E_t. A summary of the properties of the soil, wall, bar and interface elements are given in Table 1.

Table 1.   Properties of the different materials

(a) Properties of soil, wall and bar elements

MATERIAL		E_t kN/m^2	υ	G kN/m^2	γ kN/m^3	AREA m^2
SOIL	sand	15696	0.31	5991.0	18.64	NA
WALL	concrete	1.37E7	0.15	5.97E7	22.56	NA
BAR	NA	1E7	0.01	NA	NA	0.10

(b) Interface elements

ELEMENT	FRICTION COEFFICIENT	ANGLE	STIFFNESS (kN-m)	INITIAL STATUS
HLI	0.50	$\delta = 30$	4E13	CLOSED
VLI	0.32	$\delta = 18$	4E13	CLOSED

* NA: Not applicable

where; E_t is the initial tangent modulus of elasticity, v is Poisson's ratio, G is the shear modulus, γ is the unit weight.

A sinusoidal acceleration-time history digitized at time intervals of 0.01 second was applied at the nodes along line a-b in Fig. 2. The sinusoidal input motion had a frequency (f) of 3.5 Hz and it produced an active type of displacement in the wall away from the backfill, (from right to left in Fig. 2). Different modes of wall displacement were examined including active rocking of a free wall (ARC), active horizontal translation (AT), active rotation about the wall top (ART), and active rotation about the wall base (ARB). The required displacement mode was enforced by restraining certain nodes along the wall back and/or face. For example; the case of rotation about the wall top (ART) was achieved by restraining the node on the wall back at the ground surface level, whereas the ARB condition was enforced by restraining the node at the wall toe. Several computer runs were performed for each movement mode using vibration records with peak acceleration (a) ranging from 0.1g to 0.5g and an initial condition of a wall in the neutral state (at-rest). The plane-strain condition and a material damping of 10% of critical damping were assumed in the analysis performed using the computer program ANSYS (17).

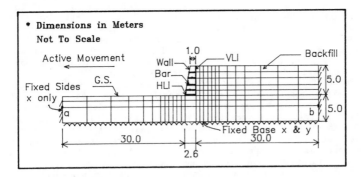

Figure 2. Analytical Retaining Wall Model

DIFFERENCES BETWEEN THE TWO MODELS

The analytical and experimental studies reported herein were performed independently with no intention to verify or duplicate the results of one another. The two models used in the studies, (laboratory wall model and finite element mesh), were designed to examine the dynamic response of a gravity wall retaining a dry cohesionless backfill. However, the two models are different in some aspects;
1. Horizontal interface elements were used in the finite element mesh to connect the wall base with the foundation soil, whereas a real hinge was used in the tests. The different boundary conditions would influence the earth pressure distribution near the wall base.

2. The finite element analysis considered a plane-strain model and an infinite soil medium while the experimental soil bin had finite dimensions.

3. The foundation soil extended on both sides of the wall in the finite element model but the laboratory model had only a backfill behind the wall.

4. A massive concrete wall was used in the finite element model while a thin, but rigid, aluminum wall was used in the laboratory experiments. The two materials interact differently with soils.

5. The boundary conditions in the finite element analysis were selected such that the wall rotates about its toe on the face side during rotation about base, and rotates about its top connection with the backfill during rotation about the top. Meanwhile, in all experimental tests the wall rotated about its center line.

Due to these differences, the response of the two models may not be identical in some cases as will be discussed in a later section.

RESULTS OF THE ANALYSIS

In the following discussion, the distribution of the dynamic earth pressure and two dimensionless parameters (K and Y/H) are examined, where K is the ratio of horizontal to vertical total earth pressures behind the wall and Y is the height of the point of application of the earth pressure resultant above the wall base as a ratio of the wall height (H). The initial values in the finite element curves at zero acceleration were obtained from a previous static analysis of the same mode of displacement (2), ie; for example the initial value for the dynamic active rotation about base (ARB) curve is obtained from a static ARB analysis.

The distribution and the order of magnitude of the dynamic active earth pressure obtained analytically at peak accelerations of 0.2g (1) are in good agreement with the distributions obtained experimentally at peak accelerations of 0.215g (6, 13, 14, 15), as shown in Figs. 3, 4 and 5. None of the analytical or experimental distributions agree with the hydrostatic distribution assumed in the Mononobe-Okabe approach. During all modes, the dynamic earth pressure is higher than the Mononobe-Okabe value in the upper third of the wall. Higher values than those of the Mononobe-Okabe are also observed along the entire height of a wall rotating about its base (ARB). Meanwhile, smaller earth pressures than the Mononobe-Okabe values develop near the wall base during translation (AT), rotation about the wall top (ART), and a rocking wall (ARC). Higher earth pressure than the Mononobe-Okabe values develop near the top third point of a wall moving in translation, as shown by the results of the finite element analysis and the experimental work of Ishii et al. (7). The distribution of the ART earth pressure obtained analytically at a rotation of 16×10^{-4} rad is similar to the distribution reported by Ishibashi and Fang (6) at a rotation angle of 20×10^{-4} rad (Figs. 3 and 5). In the finite element solution, however, the increase of the earth pressures near the wall base could be attributed to the effect of the friction interface elements, which is more likely to occur in an actual wall. The finite element distribution for ARB at a rotation of 4×10^{-4} rad also reassembles that obtained experimentally at a rotation of 2×10^{-4} rad, (Figs. 3 and 4). In active translation, the maximum dynamic earth pressure is reached experimentally and analytically (15, 1) at displacements of 0.0013H and 0.0016H, respectively.

Table 2.   Results of Analytical and Experimental Studies

Ref. No.	Wall Height (m)	Mode	Unit Weight (kN/m^3)	Frequency (Hz)	Peak Accl (g)	Earth Pressure	
						K	Y/H
$(10,12)^{1,2}$ NA	NA	NA	NA	NA	0.20	0.49	0.33
	NA	NA	NA	NA	0.50	1.02	0.33
$(10,12)^{2,3}$ NA	NA	NA	NA	NA	0.20	0.34	0.33
	NA	NA	NA	NA	0.50	0.66	0.33
$(1)^2$	5.00	ART	18.64	3.50	0.20	0.50	0.55
			18.64	3.50	0.50	0.70	0.58
		ARB	18.64	3.50	0.20	0.68	0.45
			18.64	3.50	0.50	0.90	0.42
		AT	18.64	3.50	0.20	0.45	0.40
			18.64	3.50	0.50	0.56	0.46
		ARC	18.64	3.50	0.20	0.38	0.52
			18.64	3.50	0.50	0.48	0.54
(6)	1.00	ART	16.00	3.50	0.20	0.35	0.56
			16.00	3.50	0.50	0.72	0.52
		ARB	16.00	3.50	0.20	0.42	0.32
			16.00	3.50	0.50	0.84	0.41
(15)	1.00	AT	16.29	3.50	0.20	0.37	0.45
			16.29	3.50	0.50	0.71	0.46

[1] $\phi = 30$, $\delta = 18$ degrees,   [2] Analytical,   [3] $\phi = 39$, $\delta = 18$ degrees.

The dynamic earth pressure behind the wall increases with the peak acceleration of the vibration, as shown in Figs. 6 and 7, and Table 2. In the finite element analysis, the minimum earth pressure develops behind a rocking wall, whereas the maximum earth pressure is observed during rotation about the base. Similarly, the tests by Ishibashi and Fang (6) and Sherif and Fang (14) indicate that at any peak acceleration smaller than 0.45g, the highest coefficient of dynamic earth pressure develops during rotation about the base followed by that during translation then that during rotation about the wall top, as shown in Fig. 7. The patterns obtained experimentally are similar to the Mononobe-Okabe pattern but they differ from those obtained by the finite element analyses. The experimental tests indicate higher earth pressures during all modes than the Mononobe-Okabe solution. For example in active translation, the measured values are higher by 30% than the Mononobe-Okabe values (15). On the other hand, the finite element analyses show higher

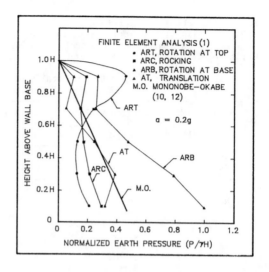

Figure 3. Analytical Dynamic Earth Pressure Distribution

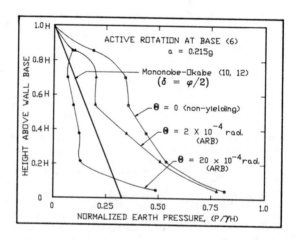

Figure 4. Experimental Dynamic Earth Pressure Distribution (ARB)

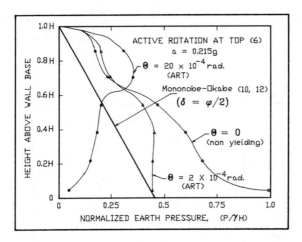

Figure 5. Experimental Dynamic Earth Pressure Distribution (ART)

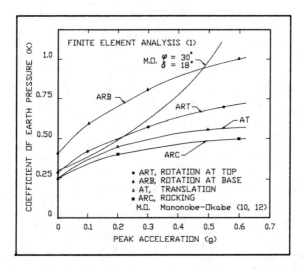

Figure 6. Analytical Coefficient of Dynamic Earth Pressure

values than the Mononobe-Okabe solution only at peak accelerations
lower than 0.5g. The results of the the finite element and the experimental tests agree well during a vibration with a peak acceleration of 0.3g.

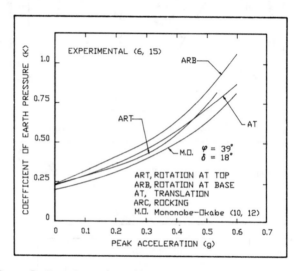

Figure 7. Experimental Coefficient of Dynamic Earth Pressure

The location of the dynamic earth pressure resultant (Y/H) is
slightly affected by the peak acceleration of vibration in addition
to the mode of wall displacement, as shown in Figs. 8 and 9, and
Table 2.   In the finite element analyses, the highest location of
the resultant occurs during a wall rotation about its top, whereas
its lowest location develops during translation at a peak acceleration lower than 0.3g or rotation about the base at higher acceleration than 0.3g. The experimental tests show similar values of Y/H
as those obtained by the finite element analyses but they indicate
different patterns of variation with the acceleration level.   Experimentally, the lowest location is reached during the case of
rotation about base, whereas the highest location is observed
during rotation about top. There is a small change in the location
of the resultant during translation with no distinctive pattern
(6), (between 0.42H and 0.47H, shown as a straight line on Fig. 9).
In most cases, the resultant of the dynamic earth pressure acts at
a location higher than the lower third point of the wall height assumed in the Mononobe-Okabe approach.   This higher location would
result in a much larger moment arm and thus produces higher overturning moment than that predicted by the pseudo-static approach.

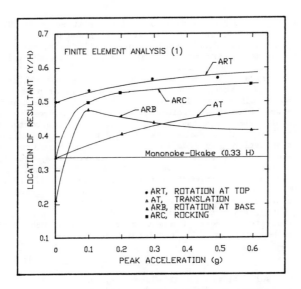

Figure 8. Analytical Location of the Resultant

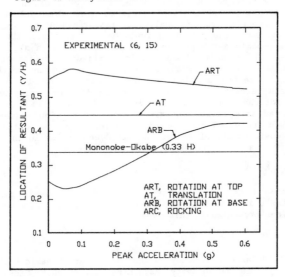

Figure 9. Experimental Location of the Resultant

The average values of the ratio Y/H for the three modes ART, AT and ARB are 0.55, 0.45 and 0.42 (analytically) and 0.55, 0.45 and 0.33 (experimentally). Consequently, for a wall restrained at its top (ART), the moment arm may be as high as 1.67 times the value calculated by the Mononobe-Okabe approach.

The finite element analysis shows that the vertical response of the system is extremely small and could be ignored as previously observed by other investigators. In the finite element analysis of a free wall with no restraints imposed (ARC), the wall moved first by rotation about its top with its top moving toward the backfill and its base away from the backfill. With time, the wall moved in pure translation away from the backfill then it returned to rotation about the top with the entire wall moving away from the backfill. Gradually, the base movement became higher than that of the top. The mode of movement of the wall, however, will be different if other boundary conditions exist such as a soil embedment at the toe (most likely rotation about toe ARB), or a bridge deck at the top (most likely rotation about top ART). Other factors may also contribute to the mode of movement in the field such as construction method, type of soil, and loading conditions.

CONCLUSIONS

The results of the finite element analysis and the experimental research show that the magnitude and distribution of the dynamic earth pressure depend on the mode of movement of the wall. The earth pressure distribution is always non-hydrostatic during all modes. Experimental and analytical research indicate that designs based on the Mononobe-Okabe formula may underestimate the magnitude of the dynamic earth pressure which may result in higher driving forces. In addition, the non-hydrostatic distribution will result in a higher location of the earth pressure resultant which in turn will increase the magnitude of the overturning moment. A design of an earth retaining structure to resist dynamic forces should account for the expected mode of movement in the field which could be predicted from field conditions and construction sequence. This research indicates the need for developing a new design approach that considers the effect of the field conditions in the design of retaining walls subject to dynamic excitations.

ACKNOWLEDGMENTS

The writers would like to acknowledge the assistance of Mr. Daniel Rau in drafting the figures and performing some of the calculations presented in this paper.

REFERENCES

1. RM Bakeer, A Study on the Static and Dynamic Earth Pressure on Gravity Retaining Walls, Ph.D. Dissertation, Syracuse University, Syracuse, NY, 1985.

2. RM Bakeer and SK Bhatia, A Finite Element Study on the Static Earth Pressure Behind Gravity Walls, Intl. J. Numerical and Analytical Methods in Geomechanics, 13(6), Nov.-Dec. 1989, 665-673.

3. YS Fang and I Ishibashi, Static Earth Pressure with Various Wall Movements, J. Geot. Eng. (ASCE), 112(3), March 1987, 317-333.

4. M Ichihara, Dynamic Earth Pressure Measured by a New Testing Device, Proc. 6th Intl. Conf. SMFE (1), Tokyo, Japan, 1965, 360-390.

5. M Ichihara and H Matsuzawa, Earth Pressure during Earthquakes, J. Soils and Foundations (JSSMFE), 13(4), 1973, 75-86.

6. I Ishibashi and YS Fang, Dynamic Earth Pressures with Different Wall Movement Mode, J. Soils and Foundations (JSSMFE), 27(4), 1987, 11-22.

7. Y Ishii, H Arai and H Tsuchida, Lateral Earth Pressure in Earthquake, Proc. 2nd WCEE (1), Tokyo, Japan, 1960, 211-230.

8. T Iwatate, T Kokusho and J Thoma, Large Scaled Model Vibration Tests of Embedded Shaft Considering the Effect of Soil Liquefaction, Tech. Report of CRIEPI (381024), 1982.

9. H Matsuo, S Ohara, Lateral Earth Pressure and Stability of Quay Walls during Earthquakes, Proc. 2nd WCEE (1), Tokyo, Japan, 1960, 165-181.

10. N Mononobe, Consideration into Earthquake Vibrations and Vibration Theories, J. Japanese Soc. Civil Eng., 10(5), 1924, 1063-1094.

11. S Niwa, An Experimental Study of Oscillating Earth Pressures Acting on Quay Wall, Proc. 2nd WCEE (1), Tokyo, Japan, 1960, 281-296.

12. S Okabe, General Theory on Earth Pressure and Seismic Stability of Retaining Wall and Dam, J. Japanese Soc. of Civil Eng., 10(5), 1924, 1277-1323.

13. M Sherif and Y Fang, Dynamic Earth Pressure on Rigid Walls Rotating about the Base, Proc. 8th WCEE (6), San Francisco, CA, 1984, 993-1000.

14. M Sherif and YS Fang, Dynamic Earth Pressure on Rigid Walls Rotating about the Top, Soils and Foundations (JSSMFE), 24(4), 1984, 109-117.

15. M Sherif, I Ishibashi and CD Lee, Dynamic Earth Pressure against Retaining Structures, Soil Engineering Research Report No. 21, Univ. of Washington, Seattle, WA, 1981.

16. M Sherif, I Ishibashi and CD Lee, Earth Pressure against Rigid Retaining Walls, J. Geot. Eng. (ASCE), 108(5), 1982, 679-695.

17. Swanson, Analysis Systems, INC., ANSYS Engineering Analysis System, Rev. 4.1, Houston, Pennsylvania, 1983.

SUBJECT INDEX
Page number refers to first page of paper.

AUTHOR INDEX
Page number refers to first page of paper.